T0269366

Book of Abstracts of the 66th Annual Meeting of the European Federation of Animal Science

EAAP - European Federation of Animal Science

The European Federation of Animal Science wishes to express its appreciation to the
Ministero delle Politiche Agricole Alimentari e Forestali (Italy) and the
Associazione Italiana Allevatori (Italy)
for their valuable support of its activities.

Book of Abstracts of the 66th Annual Meeting of the European Federation of Animal Science

Warsaw, Poland, 31 August – 4 September, 2015

EAAP Scientific Committee:

J. Conington
M. Klopčič
C. Lauridsen
G. Pollott
A. Santos
H. Sauerwein
H. Simianer
H. Spoolder
M. Tichit
G. Van Duinkerken
O. Vangen (chair)

Proceedings publication and Abstract Submission System (OASES) by

EAN: 9789086862696
e-EAN: 9789086868162
ISBN: 978-90-8686-269-6
e-ISBN: 978-90-8686-816-2
DOI: 10.3920/978-90-8686-816-2

ISSN 1382-6077

First published, 2015

© Wageningen Academic Publishers
The Netherlands, 2015

Welcome to Warsaw

Ladies and Gentlemen,

It is my pleasure to welcome you to the 66th Annual Meeting of the European Federation of Animal Science (EAAP) on the Campus of the Warsaw University of Life Sciences SGGW.

The Polish Society of Animal Production (PTZ) has had the honour to organise the EAAP Annual Meeting for the third time in its history - previously in 1975 and 1998. PTZ organised the meeting in cooperation with National Animal Breeding Centre (KCHZ) and Warsaw University of Life Sciences (SGGW), under the patronage of the Minister of Economy and the Minister of Agriculture and Rural Development.

EAAP associates, scientists and practitioners, as well as the representatives of the administration are linked to animal science from nearly all European and world countries. With the goal to promote the improvement of animal agriculture through the application of science, the Annual Meetings are the most significant forum for the exchange of information and discussion of animal production issues in Europe, and one of the most influential in the world. Just in this year alone, we have received about one thousand abstracts. Accepted abstracts will be presented during 55 plenary and two posters sessions.

The EAAP Annual Meetings give a unique opportunity to present scientific and technical achievements of animal science and its translation into practice. Special sessions will be organised on this specific issue.

On behalf of the Hosts and Organisers I cordially invite you to participate in the Congress, aimed at translating research into animal production practice.

Prof. dr. hab. Roman Niżnikowski

President of the Organizing Committee

President of the Polish Society of Animal Production

National Organisers of the 66th EAAP Annual Meeting

Polish National Organising Committee

e-mail: ko.eaap2015@kchz.agro.pl

President
- **Roman Niżnikowski**, Warsaw University of Life Sciences (SGGW)

Secretary
- **Dorota Krencik**, National Animal Breeding Centre (KCHZ)

Members
- **Dorota Lewczuk**, Institute of Genetics and Animal Breeding (PAS)
- **Robert Głogowski**, Warsaw University of Life Sciences (SGGW)
- **Marek Balcerak**, Warsaw University of Life Sciences (SGGW)
- **Joanna Makulska**, University of Agriculture in Kraków
- **Wojciech Neja**, University of Science and Technology - Bydgoszcz
- **Tomasz Strabel**, Poznań University of Life Sciences
- **Artur Oprządek**, Agricultural Property Agency (ANR)
- **Witold Rant**, Warsaw University of Life Sciences (SGGW)
- **Jolanta Oprządek**, Institute of Genetics and Animal Breeding PAS
- **Justyna Jarczak**, Institute of Genetics and Animal Breeding PAS

Honorary Patronage
- **Janusz Piechociński**, V-ce Prime Minister, Minister of Economy
- **Marek Sawicki**, Minister of Agriculture and Rural Development

Honorary Committee
- **Hanna Gronkiewicz-Waltz**, President of Warsaw
- **Adam Struzik**, Marshal of the Mazowieckie Voivodeship
- **Alojzy Szymański**, Rector of Warsaw University of Life Sciences
- **Leszek Sobolewski**, Director of the National Animal Breeding Centre
- **Leszek Świętochowski**, President of the Agricultural Property Agency
- **Radosław Szatkowski**, President of the Agricultural Market Agency
- **Andrzej Gross**, President of the Agency for Restructuring and Modernisation of Agriculture
- **Jan Jankowski**, President of the Committee of Animal Sciences PAS
- **Eugeniusz Herbut**, Director of the National Research Institute of Animal Production
- **Krzysztof Niemczuk**, Director of the National Veterinary Research Institute
- **Jarosław O. Horbańczuk**, Director of the Institute of Genetics and Animal Breeding PAS
- **Jacek Skomiał**, Director of the Kielanowski Institute of Animal Physiology and Nutrition PAS
- **Andrzej Rutkowski**, President of the Polish Branch of the World's Poultry Science Association
- **Leszek Hądzlik**, President of the Polish Federation of Cattle Breeders and Dairy Farmers
- **Bogusław Roguś**, President of the Polish Association of Beef Cattle Breeders and Producers
- **Ryszard Mołdrzyk**, President of the Polish Pig Breeders and Producers Association "POLSUS"

Polish Scientific Committee

e-mail: ssc@eaap2015.org

President
- **Tomasz Szwaczkowski**, Poznań University of Life Science

Secretary
- **Piotr Nowakowski**, Wrocław University of Environmental and Life Science

Scientific Commissions

Animal Genetics
- **Joanna Szyda**, Wrocław University of Environmental and Life Sciences
- **Maria Siwek**, University of Science and Technology - Bydgoszcz

Animal Physiology
- **Włodzimierz Nowak**, Poznań University of Life Science
- **Krystyna Koziec**, University of Agriculture in Kraków

Animal Nutrition
- **Zygmunt Kowalski**, University of Agriculture in Kraków
- **Jacek Skomiał**, The Kielanowski Institute of Animal Physiology and Nutrition PAS

Animal Health & Welfare
- **Iwona Rozempolska-Rucińska**, University of Life Sciences in Lublin
- **Anna Rząsa**, Wrocław University of Environmental and Life Science

Livestock Farming System
- **Elżbieta Martyniuk**, Warsaw University of Life Sciences (SGGW)
- **Paweł Janiszewski**, University of Warmia and Mazury in Olsztyn

Cattle Production
- **Wojciech Jagusiak**, Agricultural University in Kraków
- **Jolanta Oprządek**, Institute of Genetic and Animal Breeding PAS

Horse Production
- **Dorota Lewczuk**, Institute of Genetics and Animal Breeding PAS
- **Anna Stachurska**, University of Life Sciences in Lublin

Pig Production
- **Robert Eckert**, National Research Institute of Animal Production
- **Stanisław Kondracki**, Siedlce University of Natural Sciences and Humanities

Sheep and Goat Production:
- **Bronisław Borys**, National Research Institute of Animal Production, Experimental Station Kołuda Wielka
- **Jan Udała**, West Pomeranian University of Technology in Szczecin

Friends of EAAP

By creating the 'Friends of EAAP', EAAP offers the opportunity to industries to receive services from EAAP in change of a fixed sponsoring amount of support every year.
- The group of supporting industries are layered in three categories: 'silver', 'gold' and 'diamond' level.
- It is offered an important discount (one year free of charge) if the sponsoring industry will agree for a four years period.
- EAAP will offer the service to create a scientific network (with Research Institutes and Scientists) around Europe.
- Creation of a permanent Board of Industries within EAAP with the objective to inform, influence the scientific and organizational actions of EAAP, like proposing choices of the scientific sessions and invited speakers and to propose industry representatives for the Study Commissions.
- Organization of targeted workshops, proposed by industries.
- EAAP can represent and facilitate activities of the supporting industries toward international legislative and regulatory organizations.
- EAAP can facilitate the supporting industries to enter in consortia dealing with internationally supported research projects.

Furthermore EAAP offers, depending to the level of support (details on our website:www.eaap.org):
- Free entrances to the EAAP annual meeting and Gala dinner invitation.
- Free registration to journal *animal*.
- Inclusion of industry advertisement in the EAAP Newsletter, in the banner of the EAAP website, in the Book of Abstract and in the Programme Booklet of the EAAP annual meeting.
- Inclusion of industry leaflets in the annual meeting package.
- Presence of industry advertisements on the slides between presentations at selected standard sessions.
- Presence of industry logos and advertisements on the slides between presentations at the Plenary Sessions.
- Public Recognition by the EAAP President at the Plenary Opening Session of the annual meeting.
- Discounted stands at the EAAP annual meeting.
- Invitation to meetings (at every annual meeting) to discuss joint strategy EAAP/Industries with the EAAP President, Vice-President for Scientific affair, Secretary General and other selected members of the Council and of the Scientific Committee.

Contact and further information

If the industry you represent is interested to become 'Friend of EAAP' or want to have further information please contact jean-marc.perez0000@orange.fr or EAAP secretariat (eaap@eaap.org, phone : +39 06 44202639).

The Association

EAAP (The European Federation of Animal Science) organises every year an international meeting which attracts between 900 and 1500 people. The main aims of EAAP are to promote, by means of active co-operation between its members and other relevant international and national organisations, the advancement of scientific research, sustainable development and systems of production; experimentation, application and extension; to improve the technical and economic conditions of the livestock sector; to promote the welfare of farm animals and the conservation of the rural environment; to control and optimise the use of natural resources in general and animal genetic resources in particular; to encourage the involvement of young scientists and technicians. More information on the organisation and its activities can be found at www.eaap.org

Acknowledgements

illumına®

Accelerating agrigenomic breakthroughs.

Genetic insights for healthier crops and livestock.

Sequencing. Microarrays. Informatics.

Visit Illumina in building 23
www.illumina.com/agrigenomics

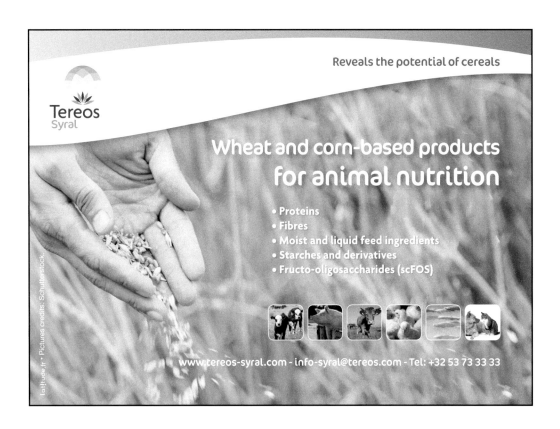

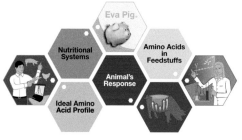

Thank you
to the 66th EAAP Annual Congress Sponsors and Friends

Gold

Silver

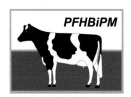 

Bronze

Friends of EAAP

European Federation of Animal Science (EAAP)

President: P. Chemineau
Secretary General: A. Rosati
Address: Via G. Tomassetti 3, A/I
I-00161 Rome, Italy
Phone: +39 06 4420 2639
Fax: +39 06 4426 6798
E-mail: eaap@eaap.org
Web: www.eaap.org

67th EAAP Annual meeting of the European Federation of Animal Science

Belfast UK, 28 August – 1 September 2016

EAAP 2016
BSAS, P.O. Box 3
Penicuik, Midlothian
EH26 0RZ
Scotland, UK

info@eaap2016.org
+44 (0)1316508784

Organised by members of the British Society of Animal Science (BSAS)

- Dr Sinclair Mayne, Agri-Food and Biosciences Institute, Belfast, EAAP 2016 Chairman
- Dr Alistair Carson, Department of Agriculture and Rural Development, Belfast
- Dr Jenny Gibbons, Agriculture and Horticulture Development Board, Stoneleigh
- Mr Jim Godfrey, Farmer, Lincolnshire
- Dr Ryan Law, Dunbia, Dungannon, Co Tyrone
- Prof Michael Lee, University of Bristol
- Dr Elizabeth Magowan, Agri-Food and Biosciences Institute, Hillsborough
- Mr Mike Steele, British Society of Animal Science
- Dr Padraig O'Kiely, Teagasc, Grange, Co Meath, Ireland
- Dr Geoff Pollott, Royal Veterinary College, University of London
- Dr Howard Simmins, British Society of Animal Science
- Dr Eileen Wall, Scotland's Rural College, Edinburgh
- Dr Stephen Whelan, Agriculture and Horticulture Development Board, Stoneleigh
- Prof Peter Williams, Independent Consultant

BSAS Presidents

- Dr Caroline Rymer, BSAS President, University of Reading
- Prof Liam Sinclair, BSAS Senior Vice President, Harper Adams Agricultural College, Newport
- Prof Richard Dewhurst, BSAS Vice President, Scotland's Rural College, Edinburgh

Conference website: www.eaap2016.org

Scientific Programme EAAP 2015

Monday 31 August 8.30 – 12.30	Monday 31 August 14.00 – 18.00	Tuesday 1 September 8.30 – 12.30	Tuesday 1 September 14.00 – 18.00
Session 1 Climate smart cattle farming and breeding – Part 1: general overviews (in cooperation with EU projects METHAGENE, RUMINOMICS, OPTIBARN) Chair: O. Vangen/M. Klopčič	**Session 8** Climate smart cattle farming and breeding – Part 2: genetic and nutritional aspects (in cooperation with EU projects METHAGENE, RUMINOMICS, OPTIBARN) Chair: Y. de Haas/C. Thomas	**8.30 – 9.30** **Welcome Ceremony**	**Session 17** Equine production, management and welfare Chair: K. Potočnik
Session 2 Strategies of national gene banks for AnGR in Europe for long term conservation purposes and to support in situ conservation of endangered breeds Chair: S.J. Hiemstra	**Session 9** How can we tip the balance between market and non-market outputs from livestock farming systems Chair: M. Zehetmeier/ C. Ligda		**Session 18** Using electronic identification (EID) and other technological advances in small ruminant farming Chair: C. Morgan-Davies
Session 3 Feasible solutions to reducing tail biting in commercial settings + rearing entire pigs Chair: A. Velarde	**Session 10** Young train session – Dairy innovative research and extension Chair: P. Aad/A. Kuipers	**9.30 – 10.30** **Session 16** Discovery plenary session part I Chair: P. Chemineau	**Session 19** Efficiency, multifunctionality and tradeoffs in livestock production Chair: R. Ripoll-Bosch
Session 4 Correctly reporting statistical genetics results in the genomic era Chair: G. Pollott	**Session 11** Integrating biological knowledge into genetic studies Chair: T. Suchocki		**Session 20** Genetics free communications Chair: E. Wall
Session 7 Nutritional and management strategies in animal disease prevention Chair: G. Savoini	**Session 12** ATF Precision livestock farming		**Session 21** Beef production, supply and quality from farm to fork Chair: J.F. Hocquette/ K. De Roest
	Session 13 Pork carcass; meat quantity or meat quality? Chair: S. Millet	**11.00 – 11.30** **Leroy presentation**	**Session 22** Genomic selection in practice Chair: I. Misztal
	Session 14 Customised nutrition taking into account the health status of farms and individual animals Chair: E. Tsiplakou		**Session 23** Non-human-edible by-products: use and added value as feed material Chair: S. De Campeneere
	Session 15 Poultry husbandry: welfare, breeding and nutrition Chair: G. Das	**11.30 – 12.30** **Session 16** Discovery plenary session part II Chair: P. Chemineau	**Session 24** Success factors for careers in the livestock industry – Industry / Young Scientists Club Chair: P. Aad/ C. Lambertz/A. Smetko
	19.00 – 21.00 **Poster Session (part I), Award Ceremony, Welcome Reception**		**Session 25** Ways of improving udder health and fertility in cattle Chair: B. Fuerst-Waltl/ M. Klopčič

Wednesday 2 September 8.30 – 11.30	Wednesday 2 September 14.00 – 18.00	Thursday 3 September 8.30 – 11.30	Thursday 3 September 14.00 – 18.00
Session 26 Innovation and research for developing the horse sector – Equine practice in Science Part 1 Chair: A.S. Santos/P. Lekeux	**Session 33** Innovation and research for developing the horse sector – Equine practice in Science Part 2 Chair: A. Stojanowska/ D. Lewczuk	**Session 41** Optimising breeding programmes in the genomic era Chair: T.H.E. Meuwissen/ A. de Vries	**Session 50** Efficient computation strategies in genomic prediction (joint session with Interbull) Chair: Z. Liu
Session 27 New sources of phenotypes in cattle production – Part 1 (with ICAR) Chair: G. Thaller/ C. Egger-Danner	**Session 34** All aspects of automatic milking including combination with grazing (in cooperation with EU project Autograssmilk) Chair: A. Van Den Pol-Van Dasselaar/C. Foley	**Session 42** In the age of genotype, the breeding objective is queen (joint session with INTERBULL) Chair: M. Coffey/R. Reents	**Session 51** Biotechnological and genomic advances in small ruminant production Chair: R. Rupp
Session 28 Integration of new technologies in livestock farming systems Chair: N. Hostiou	**Session 35** Future challenges and strategies for smallholders Chair: S. Oosting	**Session 43** Free communications animal nutrition Chair: G. Van Duinkerken	**Session 52** Data in livestock farming systems : does the science supply meet farmers' needs? Chair: S. Ingrand
Session 29 Genetics commission early career scientist competition – Part 1 Chair: C. Pfeiffer	**Session 36** Genetics commission early career scientist competition – Part 2 Chair: H. Mulder	**Session 44** Genetics and genomics in horses Chair: K. Stock	**Session 53** Cattle herd management and health Chair: B. Fuerst-Waltl/ W. Jagusiak
Session 30 New developments, techniques and research in cattle housing systems Chair: Y. Bewley/P. Galama	**Session 37** New sources of phenotypes in cattle production – Part 2 (with ICAR) Chair: G. Thaller/C. Egger-Danner	**Session 45** Health and welfare free communications Chair: M. Pearce	**Session 54** Cellular physiology of the reproductive processes Chair: C.C. Perez-Marin
Session 31 Breeding for better health and welfare Chair: L. Boyle	**Session 38** Applying physiology – improvements in animal productivity, water efficiency, welfare and behaviour Chair: C.H. Knight	**Session 47** Livestock farming systems free communications: livestock sustainability Chair: M. Tichit	**Session 55** Effects of neonate and early life conditions on robustness and resilience in later life Chair: H. Spoolder
Session 32 Towards a framework for multifunctional feeding systems Chair: J. Van Milgen	**Session 39** Role of plant bioactive compounds in animal nutrition Chair: E. Auclair/E. Apper	**Session 48** Sheep and goats free communications Chair: T. Ådnøy	
Session 57 EU project GPLUSE: Technical, economic, societal and environmental issues associated with highly productive dairy cows Chair: M. Crowe	**Session 40** Animal behaviour Chair: K. O'Driscoll	**Session 49** Nutrient sensing and metabolic signaling plus free communications – Physiology Chair: H. Sauerwein	**Friday 4 September 8.30 – 18.00**
11.30 – 12.30 **Poster Session (part II)**	**14.00 – 17.00** **Session 6** Feeding the gestating and lactating sow. Chair: G. Bee **17.00 – 18.00** **Pig Commission Business meeting**	**11.30-12.30** **Commission business meetings** **Session 46** 8.30 – 12.30 Pig genetics Chair: E. Knol	**Session 56** Local industry day Chair: Z.M. Kowalski/I. Misztal

Commission on Animal Genetics

Dr Simianer	President	University of Goettingen
	Germany	hsimian@gwdg.de
Dr Meuwissen	Vice-President	Norwegian University of Life Sciences
	Norway	theo.meuwissen@umb.no
Dr Mulder	Vice-President	Wageningen University
	Netherlands	han.mulder@wur.nl
Dr Wall	Vice-President	SRUC
	United Kingdom	eileen.wall@sruc.ac.uk
Dr Ibañez	Secretary	IRTA
	Spain	noelia.ibanez@irta.es
Dr De Vries	Industry rep.	CRV
	Netherlands	alfred.de.vries@crv4all.com

Commission on Animal Nutrition

Dr Van Duinkerken	President	Wageningen University
	Netherlands	gert.vanduinkerken@wur.nl
Dr Savoini	Vice-President	University of Milan
	Italy	giovanni.savoini@unimi.it
Dr Tsiplakou	Secretary	Agricultural University of Athens
	Greece	eltsiplakou@aua.gr
Dr De Campeneere	Secretary	ILVO
	Belgium	sam.decampeneere@ilvo.vlaanderen.be
Dr Auclair	Industry rep.	LFA Lesaffre
	France	ea@lesaffre.fr
Dr Apper	Industry rep.	Tereos Syral
	France	emmanuelle.apper@tereos.com

Commission on Health and Welfare

Dr Spoolder	President	ASG-WUR
	Netherlands	hans.spoolder@wur.nl
Dr Krieter	Vice-President	University Kiel
	Germany	jkrieter@tierzucht.uni-kiel.de
Mr Pearce	Vice-President/	
	Industry rep.	Pfizer
	United Kingdom	michael.c.pearce@pfizer.com
Dr Boyle	Secretary	Teagasc
	Ireland	laura.boyle@teagasc.ie
Dr Das	Secretary	University of Goettingen
	Germany	gdas@gwdg.de

Commission on Animal Physiology

Dr Sauerwein	President	University of Bonn
	Germany	sauerwein@uni-bonn.de
Dr Driancourt	Vice president/	
	Industry rep.	Intervet
	France	marc-antoine.driancourt@sp.intervet.com
Dr Silanikove	Vice-President	Agricultural Research Organization (ARO)
	Israel	nsilaniks@volcani.agri.gov.il
Dr Quesnel	Secretary	INRA Saint Gilles
	France	helene.quesnel@rennes.inra.fr
Dr Knight	Secretary	University of Copenhagen
	Denmark	chkn@sund.ku.dk

Commission on Livestock Farming Systems

Dr Tichit	President	INRA
	France	tichit@agroparistech.fr
Dr Ingrand	Vice-President	INRA
	France	ingrand@clermond.inra.fr
Dr Ripoll Bosch	Secretary	Wageningen University
	Netherlands	raimon.ripollbosch@wur.nl
Mrs Zehetmeier	Secretary	Bavarian State Research Center
	Germany	monika.zehetmeier@gmx.de

Commission on Cattle Production

Dr Klopčič	President	University of Ljublijana
	Slovenia	marija.klopcic@bf.uni-lj.si
Dr Thaller	Vice-President	Animal Breeding and Husbandry
	Germany	georg.thaller@tierzucht.uni-kiel.de
Dr Coffey	Vice president/	
	Industry rep.	SRUC, Scotland
	United Kingdom	mike.coffey@sruc.ac.uk
Dr Halachmi	Vice-President	Agricultural Research Organization (ARO)
	Israel	halachmi@volcani.agri.gov.il
Dr Vestergaard	Secretary	Aarhus University
	Denmark	mogens.vestergaard@agrsci.dk
Dr Fürst-Waltl	Secretary	University of Natural Resources and
		Life Sciences
	Austria	birgit.fuerst-waltl@boku.ac.at

Commission on Sheep and Goat Production

Dr Conington	President	SRUC
	United Kingdom	joanne.conington@sruc.ac.uk
Dr Ådnøy	Vice-President	Norwegian University of Life Sciences
	Norway	tormod.adnoy@nmbu.no
Dr Milerski	Secretary/	
	Industry rep.	Research Institute of Animal Science
	Czech Republic	m.milerski@seznam.cz
Dr Ligda	Secretary	Hellenic Agricoltural Organisation
	Greece	chligda@otenet.gr

Commission on Pig Production

Dr Lauridsen	President	Aarhus University
	Denmark	charlotte.lauridsen@agrsci.dk
Dr Knol	Vice President/	
	Industry rep.	TOPIGS
	Netherlands	egbert.knol@topigs.com
Dr Bee	Secretary	Agroscope Liebefeld-Posieux ALP
	Switzerland	giuseppe.bee@alp.admin.ch
Dr Velarde	Secretary	IRTA
	Spain	antonio.velarde@irta.es
Dr Millet	Secretary	ILVO
	Belgium	sam.millet@ilvo.vlaanderen.be

Commission on Horse Production

Dr Santos	President	CITAB - UTAD - EUVG
	Portugal	assantos@utad.pt
Dr Lewczuk	Vice president	IGABPAS
	Poland	d.lewczuk@ighz.pl
Dr Saastamoinen	Vice president	MTT Agrifood Research Finland
	Finland	markku.saastamoinen@mtt.fi
Dr Evans	Vice president	Norwegian University College for
		Agriculture and Rural Development
	Norway	rhys@hlb.no
Dr Holgersson	Secretary	Swedish University of Agriculture
	Sweden	anna-lena.holgersson@hipp.slu.se
Dr Hausberger	Secretary	CNRS University
	France	martine.hausberger@univ-rennes1.fr

Scientific programme

Session 01. Climate smart cattle farming and breeding – Part 1: general overviews (in cooperation with EU projects METHAGENE, RUMINOMICS, OPTIBARN)

Date: 31 August 2015; 08:30 – 12:30 hours
Chairperson: O. Vangen/M. Klopčič

Theatre Session 01

Session 02. Strategies of national gene banks for AnGR in Europe for long term conservation purposes and to support *in situ* conservation of endangered breeds

Date: 31 August 2015; 08:30 – 12:30 hours
Chairperson: S.J. Hiemstra

Theatre Session 02

Poster Session 02

Session 03. Feasible solutions to reducing tail biting in commercial settings + rearing entire pigs

Date: 31 August 2015; 08:30 – 12:30 hours
Chairperson: A. Velarde

Theatre Session 03

Poster Session 03

Session 04. Correctly reporting statistical genetics results in the genomic era

Date: 31 August 2015; 08:30 – 12:30 hours
Chairperson: G. Pollott

Theatre Session 04

Session 06. Feeding the gestating and lactating sow

Date: 2 September 2015; 14:00 – 18:00 hours
Chairperson: G. Bee

Theatre Session 06

Poster Session 06

Session 07. Nutritional and management strategies in animal disease prevention

Date: 31 August 2015; 08:30 – 12:30 hours
Chairperson: G. Savoini

Theatre Session 07

Poster Session 07

Session 08. Climate smart cattle farming and breeding – Part 2: genetic and nutritional aspects (in cooperation with EU projects METHAGENE, RUMINOMICS, OPTIBARN)

Date: 31 August 2015; 14:00 – 18:00 hours
Chairperson: Y. de Haas/C. Thomas

Theatre Session 08

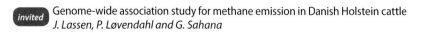

Session 09. How can we tip the balance between market and non-market outputs from livestock farming systems

Date: 31 August 2015; 14:00 – 18:00 hours
Chairperson: M. Zehetmeier/C. Ligda

Theatre Session 09

Session 10. Young train session – Dairy innovative research and extension

Date: 31 August 2015; 14:00 – 18:00 hours
Chairperson: P. Aad/A. Kuipers

Theatre Session 10

Session 11. Integrating biological knowledge into genetic studies

Date: 31 August 2015; 14:00 – 18:00 hours
Chairperson: T. Suchocki

Theatre Session 11

Session 13. Pork carcass; meat quantity or meat quality?

Date: 31 August 2015; 14:00 – 18:00 hours
Chairperson: S. Millet

Theatre Session 13

Poster Session 13

Session 14. Customised nutrition taking into account the health status of farms and individual animals

Date: 31 August 2015; 14:00 – 18:00 hours
Chairperson: E. Tsiplakou

Theatre Session 14

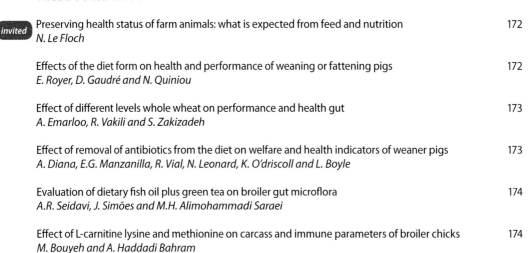

Poster Session 14

Session 15. Poultry husbandry: welfare, breeding and nutrition

Date: 31 August 2015; 14:00 – 18:00 hours
Chairperson: G. Das

Theatre Session 15

Poster Session 15

Session 16. Discovery plenary session

Date: 1 September 2015; 08:30 – 12:30 hours
Chairperson: P. Chemineau

Theatre Session 16

Leroy presentation

Session 17. Equine production, management and welfare

Date: 1 September 2015; 14:00 – 18:00 hours
Chairperson: K. Potočnik

Theatre Session 17

Poster Session 17

Session 18. Using electronic identification (EID) and other technological advances in small ruminant farming

Date: 1 September 2015; 14:00 – 18:00 hours
Chairperson: C. Morgan-Davies

Theatre Session 18

Session 19. Efficiency, multifunctionality and tradeoffs in livestock production

Date: 1 September 2015; 14:00 – 18:00 hours
Chairperson: R. Ripoll-Bosch

Theatre Session 19

Session 20. Genetics free communications

Date: 1 September 2015; 14:00 – 18:00 hours
Chairperson: E. Wall

Theatre Session 20

Poster Session 20

Session 21. Beef production, supply and quality from farm to fork

Date: 1 September 2015; 14:00 – 18:00 hours
Chairperson: J.F. Hocquette/K. De Roest

Theatre Session 21

Poster Session 21

Session 22. Genomic selection in practice

Date: 1 September 2015; 14:00 – 18:00 hours
Chairperson: I. Misztal

Theatre Session 22

Poster Session 22

Session 23. Non-human-edible by-products: use and added value as feed material

Date: 1 September 2015; 14:00 – 18:00 hours
Chairperson: S. De Campeneere

Theatre Session 23

Poster Session 23

Session 24. Success factors for careers in the livestock industry – Industry / Young Scientists Club

Date: 1 September 2015; 14:00 – 18:00 hours

Theatre Session 24

invited Important factors to consider when applying and interviewing for academic and industry jobs   275
L.J. Spicer and M.J. Pearce

Session 25. Ways of improving udder health and fertility in cattle

Date: 1 September 2015; 14:00 – 18:00 hours
Chairperson: B. Fuerst-Waltl and M. Klopčič

Theatre Session 25

invited Genetic selection of dairy cattle for improved immunity   275
B.A. Mallard

Unravelling the genetic background for endocrine fertility traits in dairy cows   276
A.M.M. Tenghe, A.C. Bouwman, B. Berglund, E. Strandberg and R.F. Veerkamp

Resumption of luteal activity in first lactation cows is mainly affected by genetic characteristics   276
N. Bedere, L. Delaby, V. Ducrocq, S. Leurent-Colette and C. Disenhaus

invited Epigenetic differences in cytokine genes of dairy cows identified with immune response biases   277
M.A. Paibomesai and B.A. Mallard

invited Breeding for improved fertility and udder health – requirements   277
M. Kargo, G.P. Aamand and J. Pedersen

invited Extension programming to address somatic cell count challenges and opportunities   278
J.M. Bewley

Detection of haplotypes responsible for prenatal death in cattle   278
X. Wu, B. Guldbrandtsen, M.S. Lund and G. Sahana

Milk and fertility performance of Holstein, Jersey and Holstein × Jersey cows on Irish farms   279
E.L. Coffey, B. Horan, R.D. Evans, K.M. Pierce and D.P. Berry

Associations of feed efficiency with reproductive development and semen quality in young bulls   279
S. Bourgon, M. Diel De Amorim, S. Lam, J. Munro, R. Foster, T. Chenier, S. Miller and Y. Montanholi

Decision support model for dairy cow management   280
J. Makulska, A.T. Gawełczyk and A. Węglarz

Poster Session 25

Economic importance of health traits in dual-purpose cattle   280
Z. Krupová, E. Krupa, M. Michaličková and L. Zavadilová

Effectiveness of teat disinfection formulations containing polymeric biguanide compounds   281
D. Gleeson and B. O'brien

Session 26. Innovation and research for developing the horse sector – Equine practice in Science Part 1

Date: 2 September 2015; 08:30 – 12:30 hours
Chairperson: A.S. Santos/P. Lekeux

Theatre Session 26

Poster Session 26

Session 27. New sources of phenotypes in cattle production – Part 1 (with ICAR)

Date: 2 September 2015; 08:30 – 11:30 hours
Chairperson: G. Thaller/C. Egger-Danner

Theatre Session 27

Session 28. Integration of new technologies in livestock farming systems

Date: 2 September 2015; 08:30 – 11:30 hours
Chairperson: N. Hostiou

Theatre Session 28

Poster Session 28

Session 29. Genetics commission early career scientist competition – Part 1

Date: 2 September 2015; 08:30 – 11:30 hours
Chairperson: C. Pfeiffer

Theatre Session 29

Poster Session 29

Session 30. New developments, techniques and research in cattle housing systems

Date: 2 September 2015; 08:30 – 11:30 hours
Chairperson: Y. Bewley/P. Galama

Theatre Session 30

Poster Session 30

Session 33. Innovation and research for developing the horse sector – Equine practice in Science Part 2

Date: 2 September 2015; 14:00 – 18:00 hours
Chairperson: A. Stojanowska/D. Lewczuk

Theatre Session 33

Session 34. All aspects of automatic milking including combination with grazing (in cooperation with EU project Autograssmilk)

Date: 2 September 2015; 14:00 – 18:00 hours
Chairperson: A. Van Den Pol-Van Dasselaar/C. Foley

Theatre Session 34

Poster Session 34

Session 35. Future challenges and strategies for smallholders

Date: 2 September 2015; 14:00 – 18:00 hours
Chairperson: S. Oosting

Theatre Session 35

Poster Session 35

Session 36. Genetics commission early career scientist competition – Part 2

Date: 2 September 2015; 14:00 – 18:00 hours
Chairperson: H. Mulder

Theatre Session 36

Session 37. New sources of phenotypes in cattle production – Part 2 (with ICAR)

Date: 2 September 2015; 14:00 – 18:00 hours
Chairperson: G. Thaller/C. Egger-Danner

Theatre Session 37

Poster Session 37

Session 38. Applying physiology – improvements in animal productivity, water efficiency, welfare and behaviour

Date: 2 September 2015; 14:00 – 18:00 hours
Chairperson: C.H. Knight

Theatre Session 38

Session 39. Role of plant bioactive compounds in animal nutrition

Date: 2 September 2015; 14:00 – 18:00 hours
Chairperson: E. Auclair/E. Apper

Theatre Session 39

Poster Session 39

Session 40. Animal behaviour

Date: 2 September 2015; 14:00 – 18:00 hours
Chairperson: K. O'Driscoll

Theatre Session 40

Poster Session 40

Session 41. Optimising breeding programmes in the genomic era

Date: 3 September 2015; 08:30 – 11:30 hours
Chairperson: T.H.E. Meuwissen/A. de Vries

Theatre Session 41

Poster Session 41

Session 42. In the age of genotype, the breeding objective is queen (joint session with INTERBULL)

Date: 3 September 2015; 08:30 – 11:30 hours
Chairperson: M. Coffey/R. Reents

Theatre Session 42

Session 43. Free communications animal nutrition

Date: 3 September 2015; 08:30 – 11:30 hours
Chairperson: G. Van Duinkerken

Theatre Session 43

Poster Session 43

Session 44. Genetics and genomics in horses

Date: 3 September 2015; 08:30 – 11:30 hours
Chairperson: K. Stock

Theatre Session 44

Session 45. Health and welfare free communications

Date: 3 September 2015; 08:30 – 11:30 hours
Chairperson: M. Pearce

Theatre Session 45

Poster Session 45

Session 46. Pig genetics

Date: 3 September 2015; 08:30 – 11:30 hours
Chairperson: E. Knol

Theatre Session 46

Session 47. Livestock farming systems free communications: livestock sustainability

Date: 3 September 2015; 08:30 – 11:30 hours
Chairperson: M. Tichit

Theatre Session 47

Poster Session 47

Session 48. Sheep and goats free communications

Date: 3 September 2015; 08:30 – 11:30 hours
Chairperson: T. Ådnøy

Theatre Session 48

Poster Session 48

Session 49. Nutrient sensing and metabolic signaling plus free communications – Physiology

Date: 3 September 2015; 08:30 – 11:30 hours
Chairperson: H. Sauerwein

Theatre Session 49

Poster Session 49

Session 50. Efficient computation strategies in genomic prediction (joint session with Interbull)

Date: 3 September 2015; 14:00 – 18:00 hours
Chairperson: Z. Liu

Theatre Session 50

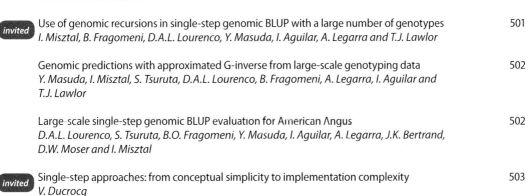

Session 51. Biotechnological and genomic advances in small ruminant production

Date: 3 September 2015; 14:00 – 18:00 hours
Chairperson: R. Rupp

Theatre Session 51

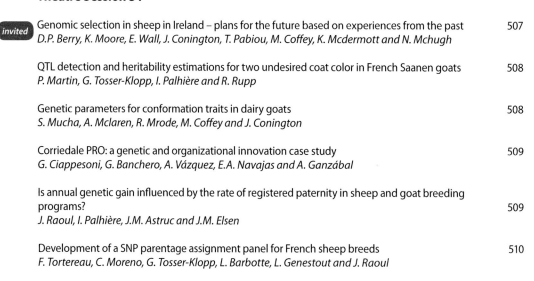

Session 53. Cattle herd management and health

Date: 3 September 2015; 14:00 – 18:00 hours
Chairperson: B. Fuerst-Waltl/W. Jagusiak

Theatre Session 53

Poster Session 53

Session 54. Cellular physiology of the reproductive processes

Date: 3 September 2015; 14:00 – 18:00 hours
Chairperson: C.C. Perez-Marin

Theatre Session 54

Poster Session 54

Session 55. Effects of neonate and early life conditions on robustness and resilience in later life

Date: 3 September 2015; 14:00 – 18:00 hours
Chairperson: H. Spoolder

Theatre Session 55

Poster Session 55

Session 56. Local industry day

Part 1. Monitoring as a tool for cattle and pig herds management
Date: 4 September 2015; 08:30 – 12:30 hours
Chairperson: Z.M. Kowalski

Theatre Session 56, part 1

Part 2. Genomic selection in practice
Date: 4 September 2015; 14:00 – 18:00 hours
Chairperson: I. Misztal

Theatre Session 56, part 2

invited Introduction to genomic selection: current knowledge and troublespots for commercial
implementations   547
I. Misztal

invited Genomic selection in practice   547
T.J. Lawlor

invited Lessons from practical implementation of genomic evaluation across species   548
D.A.L. Lourenco

invited Current status of genomic selection in Polish dairy cattle breeding   548
M. Pszczola

Session 57. EU project GPLUSE: Technical, economic, societal and environmental issues associated with highly productive dairy cows

Date: 2 September 2015; 08:30 – 12:30 hours
Chairperson: M. Crowe

Theatre Session 57

invited Herd health challenges in high yielding dairy cow systems   549
R.F. Smith

invited Phenotypic interrelationships between parameters predominately in milk – the GplusE project   549
J.K. Höglund, M.T. Sørensen and K.L. Ingvartsen

On the use of novel milk phenotypes as predictors of difficult-to-record traits in breeding programs   550
invited *C. Bastin, F. Colinet, F. Dehareng, C. Grelet, H. Hammami, H. Soyeurt, A. Vanlierde,*
M.L. Vanrobays and N. Gengler

Global perspectives for climate smart cattle farming and breeding

M.C.T. Scholten
Board of Directors, Wageningen UR, Animal Sciences Group (ASG), Co-chair GRA Livestock Research Group; President Animal Taks Force, Houtribweg 39, 8221 RA Lelystad, the Netherlands; martin.scholten@wur.nl

Accounting for an estimated 14.5% of global anthropogenic greenhouse gas emissions, livestock sector plays an important role in climate change. Ruminant products are considered to have the highest C-footprints. But for a biomass based circular economy ruminants are essential in optimizing the human edible protein production per unit of land use for agricultural biomass production. Research and innovation plays an important role for further improving the sustainability and competitiveness of the cattle production sector, with climate change adaptation and mitigation as one of the key challenges. The global perspectives for greenhouse gas mitigation in ruminant production is a more than 40% less C-footprint by: (1) genotyping low methane production for selection; (2) improving feed quality and digestibility; (3) improving animal health and husbandry conditions; (4) manure management: collection, storage and utilisation; (5) improving C sequestration soils ('4pro1000'); (6) precision livestock farming. It will be explained how cattle play a crucial role in an ecology based circular biomass economy, and thus contributing to a sustainable nutrition security: to enable to feed the world within the carrying capacity of planet earth. Special attention will be given to the importance of customized breeding for cattle that are excellent in this role by using feed alternative other biomass resources. It will also address the importance of bringing knowledge into practice, and how science can provide support to farmers in a climate smart cattle production. And finally the importance of international research cooperation at global level will be emphasized.

Achieving economic and socially sustainable climate smart farming

E. Wall
Scotland's Rural College (SRUC), Edinburgh, EH9 3JG, United Kingdom; eileen.wall@sruc.ac.uk

Half of the land in the European Union (EU) is farmed. It plays an essential role in maintaining the environment in a healthy state in terms of aesthetic appeal, social utility, sustainability for future farming and biodiversity. However, agriculture adds to greenhouse gas (GHG) emissions and so can also play a vital role in providing solutions to the EU's overall climate change challenges. The majority of the UK land area (18.6 m hec) is classed as agricultural land (including woodlands) of which 11.3 million hectares are under grass. This grass supports a ruminant animal population of 11.4 million cattle and 44.7 million sheep. Livestock account for up to 35-40% of the world methane production, a large proportion of which comes from enteric fermentation and a smaller proportion from anaerobic digestion in liquid manure. 64% of global nitrous oxide emissions are due to agriculture, chiefly due to fertilizer use. Ruminant production (cattle and sheep) needs to consider both CH_4 and N_2O, whereas monogastric production (pigs and poultry) species are mainly considered with N_2O (and NH_3). Many non-genetic farm technologies require ongoing investment of some sort to maintain the commercial benefit (e.g. dietary manipulation to improve fatty acid composition of milk). However, genetic improvement of a livestock population is effectively a permanent change and does not require additional or continuing resources, particularly given competition for feed and water resources. Further, the reduction of GHGs is only one side of the climate change challenge and breeding goals will need to consider potential adaptations required for future climate scenarios. Using economic techniques we show how GHG and climate adaptation targets can be incorporated into cattle breeding goals to help farmers future proof their production systems.

Climate smart cattle farming – management and system aspects
T. Kristensen and L. Mogensen
Aarhus University, Department of Agroecology, Blichers Allé 20, P.O. Box 50, 8830 Tjele, Denmark;
troels.kristensen@agro.au.dk

Management at farm and herd level, as well as choice of production system, has a major impact on productivity, economic return and potential emission of climate gasses (GHG) from the production process before, at the farm and later in the chain. Emission of CH_4 is the largest contributor to GHG, followed by N_2O, while use of fossil energy is only a minor part of the total GHG emission from cattle farming. This paper has focus on studies based on the LCA approach taking into account all emission in the chain from production of feed, utilization into milk and meat and manure storage and application. Dairy farming has during the last decade improved productivity, in Denmark by average 125 kg milk/cow/year, which together with improved forage quality has reduced emission of CH_4 from enteric fermentation per kg milk and together with a lower crude protein content in DMI has also reduced emission of N_2O from manure storage and application. There is no systematic effect of production systems, like organic vs conventional or confinement vs grazing, but large variation within systems indicates a large impact of management and strategic choices within the system. Emission per kg milk is reduced from farms with high herd efficiency (low herd DMI per kg milk, low replacement and longer lactation), farms with low stocking rate combined with low input of external feed and farms with large proportion of grass of total roughage both on area basis and DMI. Beef meat is basically from either dairy or beef cattle. Dairy production contributing with replacement cows and intensive fed male calves slaughtered at 8-12 month, while beef breed cattle is a combination of replacement cows, surplus heifers and male calves less intensely reared and slaughtered at 15 to 24 month of age, the oldest as steers. Type of beef production system generates a large variation in GHG per kg beef, with the lowest emission from the intensive fattening of dairy male calves and more than the double from beef breed cattle.

Dairy production and the carbon cycle: the importance of land use and land use change
C.E. Van Middelaar and I.J.M. De Boer
Wageningen University, Animal Production Systems group, De Elst 1, 6708 WD Wageningen, the Netherlands; corina.vanmiddelaar@wur.nl

The livestock sector has an important impact on the carbon cycle, and contributes to increased levels of greenhouse gases (GHGs) in the atmosphere via net emission of carbon dioxide (CO_2) and methane (CH_4). It is generally acknowledged that CO_2 emissions from animal respiration should not be included in impact assessments of livestock, because they are assumed to be balanced by CO_2 uptake by plants consumed by the animal (short term C-cycle). Carbon fluxes associated with land use and land use change (medium term C-cycle) and CO_2 emissions from the combustion of fossil fuels (long term C-cycle), however, are considered of importance. Variability in methods and data used to assess CO_2 fluxes related to land use (change), however, is large. This study aims to unravel the complexity of methodological issues related to livestock production and the carbon cycle, and evaluates the impact of methodological choices on emission calculations. We, therefore, computed a carbon footprint (CFP; cradle-to-farm gate) for the case study of a Dutch dairy farm for various time horizons. When excluding emissions from animal respiration and land use (change), the CFP per ton fat-and-protein-corrected milk (FPCM) was 1,132 kg CO_2-equivalents (CO_2-e). Including the short term C-cycle did not affect the result when correcting the global warming potential of biogenic CH_4 for inclusion of biogenic CO_2, which emphasises the importance of the new IPCC factors for CH_4 from biogenic and fossil sources. Including emissions related to land use change (e.g. deforestation), however, resulted in a CFP of 1,155 to 1,341 kg CO_2-e/ton FPCM, whereas including C-sequestration in grassland resulted in a CFP of 712 to 1,132 kg CO_2-e/ton FPCM, depending on the method used. Results show that the impact of land use (change) on the CFP of milk can be large. Carbon fluxes related to land use (change), however, are non-recurrent and only impact net emission of GHGs until new equilibrium conditions are reached.

Optimised animal specific barn climatisation facing climate change
S. Hempel[1], B. Amon[1], C. Ammon[1], G. Hoffmann[1], C. Menz[2], G. Zhang[3], I. Halachmi[4], A. Del Prado[5], F. Estelles[6], W. Berg[1], R. Brunsch[1] and T. Amon[1]
[1]Leibniz Institute for Agricultural Engineering Potsdam-Bornim, Engineering for Livestock Management, Max-Eyth-Allee 100, 14469 Potsdam, Germany, [2]Potsdam institute for climate impact research, Climate Impacts and Vulnerabilities, Telegraphenberg A31, 14473 Potsdam, Germany, [3]Aarhus University, Department of Engineering, Inge Lehmanns Gade 10, 8000 Aarhus, Denmark, [4]Agricultural Research Organization (ARO), Growing, Production and Environmental Engineering, P.O. Box 6, Bet Dagan 50250, Israel, [5]Basque Centre For Climate Change (BC3), Alameda Urquijo, 4, 4°-1%, 48008 Bilbao, Spain, [6]Universitat Politècnica de València, Institute of Animal Science and Technology, Building 7G, Camino de Vera s/n, 46022 Valencia, Spain; tamon@atb-potsdam.de

The FACCE-ERANET project OptiBarn develops region-specific, sustainable adaptation strategies for naturally ventilated dairy buildings. Appropriate construction methods and management can improve thermal control for increased productivity and welfare of animals and reduced emissions. Indicators for an optimisation of livestock buildings will be determined in an interdisciplinary approach. Field measurements, lab experiments, physical and numeric modeling are compiled to combine the pros of the different methods and evaluate the results. Companies and stakeholders are integrated from the start to ensure input from commercial farms and rapid dissemination and uptake of proposed adaptation options. First studies show spatially heterogeneous and not stationary air flow patterns. Depending on the temperature gradient and incoming wind thermo-induced effects may affect small scale turbulence or the main flow in the barn. Simultaneous measurements of temperature and gas concentrations suggest that high temperatures inside the barn result in a significant increase of emission rates.

Ruminomics – connecting the animal genome, gastrointestinal microbiomes and nutrition
C. Thomas[1] and J. Wallace[2]
[1]EAAP, Via Tomassetti, Rome 00161, Italy, [2]University of Aberdeen, Bucksburn, Aberdeen AB21 9SB, United Kingdom; cledwyn.thomas@googlemail.com

Ruminant livestock make a significant contribution to greenhouse gas (GHG) emissions. Enteric emissions in the form of methane together with losses from manures and those associated with land use changes comprise the majority of livestock related GHG. Technologies that increase rumen efficiency and lower methane emissions form a vital mitigation strategy to reduce global warming impacts. Ruminomics is an EU FP7 project. It is coordinated by the University of Aberdeen and is supported by 11 partners across Europe. The project seeks to integrate expertise and technologies to increase rumen efficiency and decrease the environmental footprint of ruminant production, significantly advancing current knowledge in this sector. Ruminomics exploits state-of-the-art –omics technologies to understand how ruminant gastrointestinal microbial ecosystems, or microbiomes, are controlled by the host animal and by the diet consumed, and how this impacts on greenhouse gas emissions, efficiency and product quality. The project relates animal genome to microbiome, feed efficiency, and methane emissions; determines host-microbe interactions in genetically identical and genetically diverse animals; relates changes in the nutrient supply of the cow with the composition and function of the ruminal microbiome, as assessed by methane and N emissions; provides tools and bioinformatics for rapid analysis of phenotypes, micro biomes and creates a public metagenomics database.

METHAGENE – towards large-scale methane measurements on individual ruminants for genetic evaluations

Y. De Haas[1] and J. Lassen[2]
[1]Wageningen UR Livestock Research, Animal Breeding and Genomics Centre, P.O. Box 338, 6700 AH Wageningen, the Netherlands, [2]Aarhus University, Department of Molecular Biology and Genetics, Faculty of Science and Technology, P.O. Box 50, 8830 Tjele, Denmark; yvette.dehaas@wur.nl

Methane is a greenhouse gas (GHG) that contributes to climate change. The livestock sector, particularly ruminants, is estimated to contribute up to 18% of total global anthropogenic GHG emissions. Preliminary data suggest that genetic selection to reduce methane emissions is possible. However, successful breeding programs require large datasets of individual animal measurements which cannot be generated by any EU country working alone. Smaller datasets of methane measurements are being generated by individual countries across the EU, which could be combined if agreement could be reached on how best to harmonise the data. Discussing harmonisation and protocols for future collection of such data is the focus of this METHAGENE network. It aims to discuss and agree on (1) protocols to harmonise large-scale methane measurements using different techniques; (2) easy to record and inexpensive proxies for methane emissions to be used for genetic evaluations; and (3) approaches for incorporating methane emissions into national breeding strategies. The network started in December 2013 and co-ordinates and strengthens EU scientific and technical research through improved cooperation and interactions, which is essential for breeding ruminants with lower environmental footprints resulting in less contribution to global warming.

Biomarking methane emissions from livestock farming using co-enzyme M: an exploratory study

H.M. Oliveira[1], M.A. Segundo[2], A.J.M. Fonseca[1] and A.R.J.B. Cabrita[1]
[1]REQUIMTE, LAQV, ICBAS, Instituto de Ciências Biomédicas Abel Salazar, Universidade do Porto, Rua Jorge Viterbo Ferreira, 228, 4050-313 Porto, Portugal, [2]REQUIMTE, UCIBIO, Dep. Química Aplicada, Faculdade de Farmácia, Universidade do Porto, Rua Jorge Viterbo Ferreira, 228, 4050-313 Porto, Portugal; hmoliveira@icbas.up.pt

Measuring methane in livestock farming is a very complex and time-consuming task because it is difficult to isolate the gas in an open environment. Thus, alternative methods based on the activity of methanogenic Archaea have been proposed. For example, molecular biology methods or archaeol (a lipid biomarker) have been used to correlate methanogens biomass and methane emissions. Nevertheless, direct and indirect methods still result in high levels of uncertainty in these correlations, which is reflected in inaccurate estimations of methane emissions, and also in drawbacks for monitoring new mitigation strategies. In this context, we will overview the current strategies and recent analytical advances for biomarking methanogenic Archaea. Furthermore, we will introduce the use of methanogenesis co-enzymes as potential methanogenic biomarkers. To this end, we will focus on the key methodological aspects for the analysis of co-enzyme M (2-mercaptoethanesulfonate) based on a current exploratory study. Therefore, the evaluation of a sample preparation strategy based on ion exchange solid-phase extraction to enhance selectivity and sensitivity of the determination, and also a chemical labeling approach based on optical properties for an easy and straightforward detection of the target compound, will be discussed.

Methane production by dairy cows measured in an automatic milking system

S. Van Engelen[1,2], H. Bovenhuis[1], P.P.J. Van Der Tol[3], J.A.M. Van Arendonk[1,2] and M.H.P.W. Visker[1]
[1]Wageningen University, Animal Breeding and Genomics Centre, Droevendaalsesteeg 1, 6708 PB Wageningen, the Netherlands, [2]Top Institute Food and Nutrition, Nieuwe Kanaal 9A, 6709 PA Wageningen, the Netherlands, [3]Lely Industries NV, Cornelis van der Lelylaan 1, 3147 PB Maassluis, the Netherlands; sabine.vanengelen@wur.nl

Dairy cows produce enteric methane, a potentially harmful greenhouse gas. Breeding could be applied to reduce methane production by dairy cattle. However, genetic selection requires observations on methane production for a large number of animals. These large number of observations can be obtained by measuring methane during milking in automatic milking systems (AMS). We measured methane with Fourier Transform infrared spectroscopy sensors installed in AMS on commercial dairy farms. Measurements on methane (CH_4) and carbon dioxide (CO_2) that were recorded in ppm twice per second were available. The aim of this study is to derive different phenotypes from these measurements and to calculate the repeatabilities of these phenotypes. The dataset consisted of measurements during 33,150 AMS visits from 538 dairy cows held on 6 farms in the Netherlands. On average, each cow was recorded for a period of 41 days. For the analysis, we will derive phenotypes that have been used in similar research by others: the mean, median, 75% and 90% quantile per AMS visit for CH_4 and for the ratio between CH_4 and CO_2. Variance components to calculate repeatabilities will be obtained from a mixed model with a fixed effect for farm-day and a random effect for cow. The results of this study will be presented.

Quantification of environmental impacts from dairy and beef production on national scale in Japan

S. Yamada, K. Oishi, H. Kumagai and H. Hirooka
Kyoto University, Graduate School of Agriculture, Kitashirakawa Oiwake-cho, Sakyo-ku, Kyoto 606-8502, Japan; syamada@kais.kyoto-u.ac.jp

Although dairy and beef cattle production is a significant component of agricultural section in Japan, the production system impacts serious environmental pollution such as nitrogen excretion from feces and urine and methane emission from enteric fermentation of cattle. Quantifying environmental impacts of cattle production systems is thus an important challenge to establish future sustainable milk and beef production in Japan. The objective of the present study was to quantify the amount of feed energy needs and environmental impacts (nitrogen and methane emissions from cattle) in major cattle categories on national scale using simulation models. Seven major cattle categories for milk and beef production were considered: (1) Holstein cows for milk production; (2) Japanese-Black breeding cows (and their calves); (3) Japanese-Black feedlot steers; (4) Holstein growing calves; (5) Holstein feedlot steers, (6) Japanese-Black × Holstein crossbred growing calves; and (7) Japanese-Black × Holstein crossbred feedlot steers for beef production. Simulation models for these cattle categories were constructed based on nutrient requirements (energy, protein and dry matter) given in Japanese feeding standards at the individual level. In the models, nitrogen excretion by cattle was predicted by subtracting retained body nitrogen from nitrogen intake (protein intake/6.25). Enteric methane emission was predicted as a function of dry matter intake. The breed and management parameters were taken from literatures. Total nitrogen excretion and enteric methane emission on national scale were then quantified by multiplying those predicted at the individual level by the present number of heads of each cattle category from Japanese Government Survey Statistics. From the simulation, total excretion of nitrogen and enteric methane emission from cattle production in Japan were estimated as 185,776 and 243,049 (ton/year), respectively.

Genetic approaches for methane mitigation strategies

J. Lassen[1], G.F. Difford[1,2], L.M. Kristensen[1], L. Zetouni[1] and P. Løvendahl[1]
[1]Aarhus University, Center for Quantitative Genetic and Genomics, Blichers Alle, 8830 Tjele, Denmark, [2]Wageningen University, Animal breeding and genomics centre, P.O. Box 338, 6700 AH Wageningen, the Netherlands; jan.lassen@mbg.au.dk

Enteric methane emission from ruminants contributes substantially to the greenhouse effect. A number of mitigation strategies are available and one of them is genetic selection. Genetic selection is cumulative and continuous so that the effect that is there today will also be there tomorrow. Few studies have focused on the genetic variation in enteric methane emission from dairy cattle. One reason for this is the limited number of methods appropriate for large scale phenotyping to generate a sufficient number of animals available to estimate additive genetic variance. The development of such methods is still ongoing and improving as well as methodologies to analyze the data. This includes defining the best trait to select for in order to decrease methane emission without jeopardizing the improvement of other traits in the breeding goal. So far several studies have indicated that genetic variance/significant sire effects exist for methane emission. This means that it is possible to select for decreased methane emission. Due to the difficulties in making accurate direct large scale measurements, a number of indicator traits are being tested. Predominantly on milk related traits such as fatty acids, metabolites or spectral profiles. Many research facilities now have methane measuring units installed, and the relation to feed intake and efficiency will be revealed in the near future. In turn the correlation to other traits of economic importance such as health and fertility deserve study. One of the limitations for all methane mitigation strategies is that there is hardly any economic incentive for the farmer to lower methane production. If greenhouse gas production from animal production is to be reduced this needs to be addressed. There is a huge international interest in collaboration on these matters amongst them the COST action METHAGENE.

Effect of body fat mobilization in early lactation on methane production from first-lactating cows

A. Bielak, M. Derno, H.M. Hammon and B. Kuhla
Leibniz Institute for Farm Animal Biology (FBN), Institute of Nutritional Physiology, Wilhelm-Stahl-Allee 2, 18196 Dummerstorf, Germany; b.kuhla@fbn-dummerstorf.de

In early lactation, energy requirements for maintenance and milk production exceed the energy supply from feed. As a consequence, high-yielding dairy cows enter a negative energy balance and mobilize their fat reserves. Mobilized long-chain fatty acids are intensively used for milk fat synthesis resulting in diminished utilization of acetate as precursor for milk fat de novo synthesis. As the synthesis of acetate in the rumen is closely linked to CH_4 production, we hypothesized that reduced acetate utilization by the host would negatively affect ruminal acetate and thus CH_4 production. To this end, eighteen heifers were monitored from week 4 prior to parturition until week 42 in lactation. Feed intake and milk yield were measured daily, whereas milk constitutes were analyzed weekly. In week -4, +5, +12 and +42 relative to parturition, animals were sampled for jugular blood plasma and ruminal fluid and transferred into respiration chambers to measure CH_4 production for 24 h. Based on plasma NEFA concentrations determined in week +5, animals were grouped into high (H; n=9; NEFA>550 μM) and low (L; n=9; NEFA<485 μM) mobilizing cows. Data were analyzed by a mixed model of SAS. Dry matter intake (DMI), milk yield and constitutes, rumen fluid pH and concentrations of ruminal acetate, propionate and butyrate did not differ between groups. However, in week +5, but not in other weeks investigated, CH_4 yield (l/kg DMI) and plasma acetate concentrations were significantly lower in H than in L cows. Also, CH_4 yield in week +5 was negatively correlated with plasma NEFA concentrations ($R^2=0.62$). Our data demonstrate that fat mobilization of the host is inversely associated with enteric CH_4 yield in early lactation and that this effect is not based on DMI. In contrast to our hypothesis, ruminal acetate concentration was not affected by fat mobilization, indicating that the mechanism of the CH_4 reducing effect remains to be investigated.

Effect of high N efficiency rations on N excretion in beef cattle
D. Biagini and C. Lazzaroni
University of Torino, Department of Agricultural, Forest and Food Sciences, largo P. Braccini 2, 10095 Grugliasco, Italy; carla.lazzaroni@unito.it

To spread among farmers the good practices in animal feeding, a study was carried out in beef cattle Italian commercial farm adopting rations with high N efficiency. Results obtained in a beef farm, located in the north-west plain of Italy (Piemonte region) and rearing Piemontese cattle, a specialised beef breed characterized by muscular hypertrophy, are reported. N balance was adopted to evaluate the reduction of N excretion using diet with low protein (LP) or ordinary protein content (HP) on two groups of 20 and 19 calves respectively. The LP diet was formulated reducing soy meal and adding urea with a low ratio release, to optimize rumen microbial protein synthesis and substrate fermentation ensuring the simultaneous availability of N and energy, with a feed gross protein reduction of 7 and 10% for the 350-450 and 450-700 kg LW feeding phases respectively. Data individually collected for the N balance were: feed gross protein (samples collected every two weeks), feed consumption, initial and final animals weight, hot dressing percentage, carcass classification (SEUROP grid). Data were analysed by GLM ANOVA procedure. No differences were found between diets in live and slaughtering performances (live weight gain 270 kg, average daily gain 1.3 kg/d, feed conversion rate 6.7, and hot dressing percentage 65%, as average), neither in carcass classification. Differences were pointed out instead for the daily N intake, gain and excretion, that was 145 vs 128 g (P<0.01), 37 vs 33 g (P<0.05), 108 vs 94 g (P<0.01) in HP and LP group respectively. Based on these results the N efficiency, calculated as N gain/N intake, was 26% for both the diets. In conclusion, the adoption of high N efficiency rations did not affect live and slaughtering performances and allowed a 13% reduction in N excretion of fattening bulls. (LIFE+ AQUA project).

High starch content in the diet lowers methane emissions from bull calves
A.L.F. Hellwing, P. Lund and M. Vestergaard
Aarhus University, Foulum, P.O. Box 50, 8830 Tjele, Denmark; mogens.vestergaard@anis.au.dk

Feed composition affects enteric methane emissions. Traditional, pelleted concentrate diets for fattening bulls are expensive. Alternative ingredients that can reduce the feed costs could influence productivity and methane emissions. The aim was to investigate the effect of alternative and cheap ingredients in the diet on methane emissions of intensively fattened bull calves. Twenty Holstein bull calves were fed identically from birth until 5½ months of age where they were allocated to 4 different diets. The methane emissions were measured by indirect calorimetry on 4 successive days when the mean age and weight of the bulls were 8.1 months and 346 kg. Feed intake was recorded daily, and the bulls were weighed every 2-4 weeks during the experiment. The 4 diets (all TMR) were CON which was made of (% DM): 78% pelleted concentrate (rye, wheat, rapeseed meal and DDGS), 10% straw, 9% barley and 3% molasses; ALT_1 which was made of: 40% maize cob silage, 17% barley, 17% sugar beet pulp, 13% rapeseed meal, 10% soya bean meal, ALT_2 which was made of: 50% rye, 25% clover-grass silage and 25% DDGS and ALT_3 which was made of: 60% clover-grass silage, 27% barley and 11% rapeseed meal. The dry matter intake (DMI) during methane measurements was 7.6 kg and was not different between diets (P=0.46). The methane emissions were 139, 161, 155 l/day for CON, ALT_1 and ALT_2, which were significantly lower than ALT_3 with 236 l/day (P<0.001). The methane emissions per kg DMI were significantly lower for CON than ALT_1 and ALT_2 which again were significantly lower than ALT_3 (P<0.001). Methane emissions were 117, 116 and 112 l/kg live weight gain for CON, ALT_1 and ALT_2, which were significantly lower than ALT_3 with 188 l/kg live weight gain (P<0.001). It was concluded that the lower methane emissions on CON, ALT_1 and ALT_2 compared to ALT_3 mainly were caused by the higher starch concentration on CON, ALT_1 and ALT_2 (326-355 g/kg DM) compared to ALT_3 (177 g/kg DM) despite the fact that the starch sources were different.

Effects of increased milk yield per cow on GHG emissions from milk and beef production

B.A. Åby[1], P. Crosson[2], L. Aass[1] and O.M. Harstad[1]
[1]Norwegian University of Life Sciences, Arboretveien 6, 1430 Ås, Norway, [2]Teagasc, Grange, Dunsasy, Ireland; bente.aby@nmbu.no

Many dairy cattle breeding schemes puts a large weight on increased milk yield. This lead to a reduction in the dairy cow population, given a constant milk demand. Thus, fewer dairy-origin calves are available for fattening, resulting in lowered beef production. To compensate beef production from the suckler cow population needs to increase to meet the demand for beef. The shift in the proportions of beef from dairy vs suckler cows may lead to increased total GHG emissions, because of higher emissions per kg of suckler beef compared to dairy beef. We investigated the effect of increased milk yield on the emissions per kg product (milk and beef) and total emissions. Several scenarios considering different domestic production targets for milk and beef were included. Within each scenario, three different increases in milk yield per year was applied (0, 1 and 2%). GHG emissions were estimated using two whole farm system models, HolosNor and BEEFGEM. Increased milk yield lead to reduced GHG emissions per kg of milk, mainly due to reduced methane emissions due to fewer dairy cows. Given a constant domestic milk demand, increased milk yield thus reduced GHG emissions from milk production. However, as emissions per kg of beef from the suckler cow population was higher than for dairy beef, GHG emissions from beef production in total increased due to the higher proportion of suckler beef. Therefore, the total emissions from the production of milk and beef increased. We conclude that when maintaining current or increasing domestic production levels of milk and beef, higher milk yields per cow leads to elevated total GHG emissions. Therefore, in the context of the environmental impact of milk and beef production and for a scenario with constant demand for milk and beef, breeding organizations may need to revise their breeding objectives, giving less emphasis on milk yield and more weight on beef production, towards a more dual-purpose cow.

Field assessment of ammonia emissions after soil application of cattle slurry amended with additives

M.Y. Owusu-Twum[1], R. Subedi[2], A.S. Santos[1], J. Coutinho[3], C. Grignani[2] and H. Trindade[1]
[1]CITAB-Centre for the Research and Technology of Agro-Environment and Biological Sciences, Department of Agronomy, University of Trás-os-Montes and Alto Douro, 5001-801, Vila Real, Portugal, [2]University of Turin, Department of Agriculture, Forest and Food Science, 10095 Grugliasco, Italy, [3]Chemistry Centre, Department of Biology and Environment, University of Trás-Os-Montes and Alto Dour0, 5001-901 Vila Real, Portugal; assantos@utad.pt

Slurry stores are a significant source of NH_3, CH_4 and N_2O. During slurry storage, organic matter is anaerobically degraded and CH_4 and CO_2 are formed as end products. Slurry storage contributes considerably to NH_3 and GHG emissions. Different slurry treatments during storage are known to influence these emissions during storage. This study intended to assess the NH_3 emmissions after soil application of different previously stored slurries. The effect of slurry treatment after field application on NH_3 emissions was evaluated using a randomised complete block design on a 1 m^2 plot with three repetitions. Seven treatments were tested: whole slurry (WS), liquid fraction (LF), composted solid fraction (SF), acidified liquid fraction (ALF), liquid fraction+JASS® (LFJ), liquid fraction+Biobuster® (LFB) and a control (C). Effluents were applied at rate of 130 kg N/ha. NH_3 volatilization was measured using the dynamic chamber technique with an acid trap (orthophosphoric acid) in washing bottles. The inlet and outlet of the dynamic chamber had controlled flow rates of 7 l/m and 3 l/m respectively. The volume of air entering and leaving the chamber was monitored using flow meters. NH_3 was measured 3 times in a day with each measurement period of at least 1 hour for three consecutive days. NH_3-N fluxes in g/ha/h were significantly higher for WS and for LFB and lower for C. When considering net fluxes, WS and LFB presented the higher net emissions whereas SF and LFA presented the lower emissions.

Identifying suitable future South African areas for milk production of Holstein cows on pastures
R. Williams[1], M.M. Scholtz[2,3], M.D. Fair[2] and F.W.C. Neser[2]
[1]*ARC Onderstepoort Veterinary Institute, Private Bag X05, 0110, Onderstepoort, South Africa,* [2]*University of the Free State, Department of Animal, Wildlife and Grassland Sciences, P.O. Box 339, 9301, Bloemfontein, South Africa,* [3]*ARC-Animal Production Institute, Private Bag X2, Irene, 0062, South Africa, 0062, Irene, South Africa; fairmd@ufs.ac.za*

A MaxEnt modelling technique was used to model the suitability of geographic areas in South Africa for optimal milk production in Holstein cows on pastures, using the locations of the top third of Holstein farms with the highest milk production as sample points. Climate variables related to temperature and humidity were identified as of significant relevance to model optimal milk producing areas, with evaporation the most important single variable in the model. The positive spatial correlation between evaporation and the temperature-humidity index, a commonly used parameter indicative of heat stress on animals, is a ratification of the validity of the model and confirms previous findings that excessive heat and humidity impacts negatively on milk production in cows. Future climate change projections (2046-2065) for South Africa were used to predict optimal milk producing areas in Holsteins on pastures, indicating progressive shrinking of currently suitable Holstein farming areas with increasing temperature and precipitation and a simultaneous geographic shift of optimal milk production areas towards the southern parts of the East Coast of South Africa.

Challenges for an integrated strategy of gene banking for farm animals in Europe
M. Tixier-Boichard
INRA, AgroParisTech, UMR GABI Animal Genetics and Integrative Biology, 78352 Jouy-en-Josas, France; michele.boichard@jouy.inra.fr

Gene banks represent the *ex situ in vitro* component of a national strategy for animal genetic resources and address several objectives which may differ according to countries. Three major challenges can be identified: the range of species to be considered; the possibility to couple genomic and reproductive material; the need to facilitate access to genetic resources. Gene banks mostly involve mammals, but other species are useful as well for agriculture and food security: birds, fishes, and even insects have to be involved. This raises questions on available reproductive technologies and their efficiency for a given species, on legal issues (sanitary, material ownership) and organization of the breeding sector. The rise in genomics has boosted research and provides major opportunities for a better characterization of old samples kept by gene banks: trace back the history of mutations, recover lost alleles, genotype founder animals, etc. A better connection between genomic information and gene bank collections is a major challenge for research, as well as for breeders, that can be met by connecting databases through the use of metadata. For new collections, coupling DNA or tissue samples with reproductive material should be a priority. Until now, the flow of samples being distributed by gene banks has remained very limited in Europe, because of the strong initial motivation for preservation. At present, a better knowledge of the genetic potential of gene bank samples together with improved reproductive technologies should result in an enhanced use. This requires clear rules regarding which information is available where and at which conditions, and who is going to benefit from the use, if any, in compliance with the Nagoya protocol for Access and Benefit Sharing. In France, the CRB-Anim infrastructure project is addressing these issues, and current initiatives such as the EUGENA network and the IMAGE H2020 research project have been undertaken to improve coordination across Europe.

Gene banks for *ex situ* conservation of animal genetic resources – a global view

B. Scherf, D. Pilling, I. Hoffmann, B. Besbes, P. Boettcher, G. Leroy and R. Baumung
Food and Agriculture Organization of the United Nations (FAO), Animal Genetic Resources Branch,
Animal Production and Health Division, Viale delle Terme di Caracalla, 00153 Rome Italy, Italy;
irene.hoffmann@fao.org

The State of the World's Animal Genetic Resources for Food and Agriculture identified significant gaps in capacity to manage animal genetic resources (AnGR), particularly in developing countries. In response, the international community developed and adopted the Global Plan of Action for Animal Genetic Resources (GPA) in September 2007. In 2014, official reports received from 128 countries, 4 regions and 15 international organizations on the progress made in the implementation of the GPA were analyzed with respect to several regional and national aspects of gene banking and cryoconservation. Among reporting countries, 54% indicated the presence of *ex situ* conservation activities for at least one species, and 45% indicated the existence of a national gene bank. However, there is great variation among regions (with coverage especially low in Africa, the Southwest Pacific and the Near and Middle East) and species. The reported proportion of world national breed populations from which material is stored in a gene bank was 6, 18, 20, 23 and 27% for chickens, pigs, goats, sheep and cattle, respectively. In total, less than 25% of reporting countries have arrangements in place for the extraction and use of conserved genetic material following loss of animal genetic resources through events such as disasters. The majority of reporting countries have identified the major barriers and obstacles to enhancing the conservation of their animal genetic resources, the most predominant being a lack of financial resources. This recent analysis shows that priorities identified in 2007 are still important to the safeguarding of genetic diversity.

Strategies for sampling and use of gene bank material

T.H.E. Meuwissen
Norwegian University of Life Sciences, Inst. Animal and Aquacultural Sciences, Box 5003, 1432 As, Norway;
theo.meuwissen@nmbu.no

Strategies for the prioritization of breeds for genetic conservation are reviewed based on molecular genetic data, phenotypic data and a combination of these. Alternative goals for such prioritisations are assessed and translated into alternative measures of diversity. Special attention will be paid to measures of diversity that are directed at the use of the genetic material after conservation. The use of gene bank material after conservation is directed at: (1) resurrecting the population; (2) aiding the *in situ* conservation of the population; (3) aiding the *in vivo* but *ex situ* conservation of the population; and (4) supplementing the genetic diversity in commercial populations. The latter will mainly be with the aim to introduce traits into the commercial populations which they are lacking or showing poor performances for. The introduction of such traits can be based on phenotypic introgression, marker assisted introgression, genomic introgression and traditional animal breeding. These alternative strategies will be compared. Also the consequences for the rates of inbreeding of the conserved and commercial populations will be addressed for the aforementioned strategies for the setting up and use of gene bank material. In addition to genetic diversity, the following motives can justify sampling for a gene bank: (1) degree of endangerment; (2) species to which the breed belongs; (3) adaptation to a specific environment; (4) possession of traits of current or future economic importance; (5) possession of unique traits that may be of scientific interest; (6) cultural and/or historical value of the breed. Any one of these motives can suffice to make a case for the conservation of a breed.

Assessing the degree of endangerment of livestock breeds: a multi-indicator approach
E. Verrier[1], G. Leroy[1] and C. Ramage[2]
[1]INRA / AgroParisTech, 16 rue Claude Bernard, 75231 Paris 05, France, [2]The French technical institutes and federations of breeders associations or breeding companies, 149 rue de Bercy, 75595 Paris 12, France; gregoire.leroy@agroparistech.fr

Preservation and development of local breeds is an issue of global importance. Risk status classification system is a key component of country-based early warning and response system, which should be established before any preservation action, such as the definition of strategies to establish collections in cryobanks or the development of *in situ* programs. Since there are many threat factors for a breed, a multi-indicator approach was developed to assess the degree of endangerment of a livestock population. A total of six indicators were used, two demographic, two genetic and two socio-economic. The six indicators were: (1) the current number of breeding females; (2) the evolution of the number of breeding females during the last 5 years or generations, according to the species; (3) the proportion of the progeny that cannot be registered as belonging to the breed (e.g. due to crossbreeding); (4) the effective population size; (5) the organization of breeders and the technical support; and (6) the socio-economic context. In order to combine these indicators of very different nature, the observed values were converted into scores on a scale of 0 (no threat) to 5 (maximum threat), the way to convert depending on the indicator. For each breed, a graphical presentation of the different scores was performed and a global score was computed by weighting the six scores. This method was applied on 178 French local breeds, from 10 different species: cattle, sheep, goats, pig, horses, donkeys, chicken, turkey, goose and Pekin duck. The results showed that a large proportion of local breeds can be considered as endangered, depending on the species: all equids and pig breeds, almost all poultry breeds, about 80% of goat and cattle breeds, and half sheep breeds. The use of such an approach for the definition of cryobanking priorities is discussed.

Strategies to conserve the potential of livestock species for adaptation
R. Wellmann[1], J. Bennewitz[1] and T.H.E. Meuwissen[2]
[1]University of Hohenheim, Institute of Animal Science, Farm Animal Genetics and Breeding, Garbenstraße 17, 70599 Stuttgart, Germany, [2]Norwegian University of Life Sciences, Department of Animal and Aquacultural Sciences, P.O. Box 5003, 1432 Ås, Norway; r.wellmann@uni-hohenheim.de

The erosion of the genetic diversity of livestock species continues, and prioritizing breeds for conservation is of high importance to halt this erosion. Current approaches usually only take neutral genetic diversity into account, but adaptation to different environments also affects the values of breeds for conservation. The objective of this study was to introduce adaptive and neutral diversity measures, and to study how the conservation value of a breed depends on the measure used. Two measures for adaptive variation were introduced. The adaptive diversity in a trait is the difference between the true variance of the genotypic values and the variance expected in the absence of selection. The adaptivity coverage of a set of breeds quantifies how well the breeds could adapt to a large range of environments within a limited time span. It is the expected total merit index that can be achieved by a breed after selecting it a number of generations towards a new randomly chosen breeding goal. Conservation values of breeds for conservation of adaptivity coverage were compared with their values for adaptive diversity and neutral diversities using simulated data. The values of breeds for conservation of adaptivity coverage showed a moderately high correlation with their values for adaptive and neutral diversities, even though the conservation values for adaptive diversity and neutral diversities were only slightly correlated. Because of the practical relevance and the desirable correlations we conclude that maintaining adaptivity coverage should be considered in conservation decisions. The results suggest that breeds selected for unique traits, breeds with high neutral genetic diversity and high additive variances, and breeds that are only slightly related with other breeds are of particular importance for conservation.

Drawbacks and consequences of including selection criteria in the design of semen banks

J. Fernández[1], P. Azor[2], M. Solé[3] and M. Valera[3]
[1]INIA, Mejora Genética Animal, Crta. A Coruña Km. 7,5, 28040 Madrid, Spain, [2]Asociación Nacional de Criadores de Caballos de Pura Raza Española, Avda del Reino Unido, n° 11 -3°-2, 41012 Sevilla, Spain, [3]ETSIA, Departamento Ciencias Agroforestales, Ctra. Utrera s/n, 41013 Sevilla, Spain; jmj@inia.es

Gene banks are supposed to be developed for the maintenance of genetic diversity. Thus, minimum coancestry criterion is commonly implemented to decide the number of doses to obtain from each potential donor. This is (more or less) clear for local breeds with a pure conservation emphasis. However, for mainstream breeds managers may be tempted to transfer some criteria used in the breeding program into the design of the semen bank. Moreover, practical or logistic considerations can also interfere with the original objective of the bank. Using data of the horse breed Pura Raza Español (PRE) we studied the consequences of three main factors: (1) selecting the candidates accounting for their breeding values for productive traits; (2) removing individuals from the pool due to their coat colour; (3) restricting the sampling to some geographical areas. Additionally, we have also compared results obtained from pedigree data with those arising from the use of molecular markers available in a subset of individuals. Results show that main factor is the inclusion of breeding values, as most 'valuable' donors will be surely related leading to a decrease in the diversity kept. The need of keeping individuals with low performance may be arguable depending on the time horizon the bank has been designed for. In any case, in parallel as it is recommended for the management of living populations, restrictions in the levels of genetic diversity (inbreeding) should be included when deciding the sampling scheme.

The impact of whole genome sequence data to prioritise animals for genetic diversity conservation

S.E. Eynard[1,2,3,4], J.J. Windig[1,4], S.J. Hiemstra[1] and M.P.L. Calus[4]
[1]Wageningen UR, Centre for Genetic Resources the Netherlands, P.O. Box 16, 6700 AA Wageningen, the Netherlands, [2]INRA, Génétique Animale et Biologie Intégrative, Domaine de Vilvert, 78350 Jouy-en-Josas, France, [3]AgroParisTech, Génétique Animale et Biologie Intégrative, 16 rue Claude Bernard, 75231 Paris 05, France, [4]Wageningen UR Livestock Research, Animal Breeding and Genomics Centre, P.O. Box 338, 6700 AH Wageningen, the Netherlands; sonia.eynard@wur.nl

Prioritisation decisions for gene bank collections can be informed by different types of data: pedigree, SNP chip or whole genome sequence (WGS) data. The availability of WGS is increasing considerably and its use allows to access all the structural genetic information of an individual and markedly the rare variants that are mostly ignored with SNP chips. Our study aims to determine what the impact of using pedigree, SNP chips or WGS data on prioritisation decisions is. For 277 bulls, from the 1000 bull genomes project, we estimated relationships for each of the data types, with for variants a minor allele frequency (MAF) above 1%. These estimated relationships were then used to select animals for conservation in a gene bank. Selection was based on minimisation of relatedness with Optimal Contribution Selection (OCS), in cases of different constraints on the number of selected individuals (20, 10 or 5). Thereafter, we calculated the loss of segregating WGS variants due to selection and especially at rare variants (MAF between 1 and 5%). When performing OCS without a constraint on the number of prioritised individuals, based on pedigree information a subset of 133 out of 277 individuals was selected. When based on SNP or WGS data all individuals were selected, because each of them is genetically unique. The three different types of data picked different subsets of individuals when selecting 20, 10 or 5 individuals. Estimated relationships from WGS performed better than from SNP at preserving rare variants genetic diversity in the group of selected individuals for gene bank collections.

Strategies of national gene banks for AnGR for long term in Slovenia

D. Kompan, D. Bojkovski, M. Simčič, M. Žan Lotrič and A. Cividini
University of Ljubljana, Biotechnical Faculty, Department of Animal Science, Jamnikarjeva 101, 1000 Ljubljana, Slovenia; drago.kompan@bf.uni-lj.si

In Slovenia, some activities relating to the animal genetic resources conservation have been systematically carried out for more than two decades. Since 2010, the biodiversity conservation in livestock has been carried out as a Public Service for Farm Animal Genetic Resources Preservation (PS-GB), financed by the Ministry of Agriculture, Forestry and Food, based on a seven-year program (2010-2016). Public service is implemented through a concession contract, where Biotechnical Faculty, University of Ljubljana (Department of Animal Science) is a contractor. One of the most important tasks are monitoring, *in situ* and ex-situ conservation, genetic and phenotypic characterization, awareness and promotion of products belonged to Slovenian autochthonous breeds, international cooperation, etc. In the case of ex-situ *in vitro* conservation is a collection of genetic material divided by two parts (locations). The first is a depository of tissues (whole blood, sire semen, hair roots, different tissues and DNA isolates) with 9.927 samples. These samples are permanently stored in a deep freezer at -80 °C. The second part is the gene bank semen collection, where semen of some species and breeds (horses, cattle, sheep, goats) are prepared and stored in containers with liquid nitrogen. The PS-GB has a contract for the storage of cattle semen with an AI centre where a core collection contains 3,248 semen straws from 36 bulls of the only indigenous breed. Likewise, there are stored 29,204 semen straws of 693 bulls of two local breeds. The semen straws of horses, sheep and goats breeds are stored in the Veterinary Faculty (contract), where semen has been experimentally taken from some breeding animals declared by PS-GB. There are 270 semen doses of nine stallions, 21 semen doses from seven bucks and 2,059 semen doses of 65 rams.

Genetic analysis to support the re-establishment of the Kempen breed

L. François[1], S. Janssens[1], F.G. Colinet[2], N. Gengler[2], I. Hulsegge[3,4], J.J. Windig[3,4] and N. Buys[1]
[1]KU Leuven, Dep of Biosystems, Kasteelpark Arenberg 30 bus 2456, 3001 Leuven, Belgium, [2]University of Liege, Gembloux Agro-Bio Tech, Passage des Déportés 2, 5030 Gembloux, Belgium, [3]Animal Breeding and Genomics Centre, Wageningen University and Research Centre, Postbus 338, 6700 AH Wageningen, the Netherlands, [4]Centre for Genetic Resource, the Netherlands (CGN), Wageningen University and Research Centre, Postbus 338, 6700 AH Wageningen, the Netherlands; liesbeth.francois@biw.kuleuven.be

The Kempen cattle breed is a local dual purpose cattle breed from the north-eastern part of Belgium. Since the 1950s, emphasis was put on milk production and Kempen cattle were crossed with Dutch MRY and Red Holstein bulls. In 1972, the herd book was renamed red-pied breed of Belgium. In 2012, due to the insistence of farmers, a herd book was re-established. However many animals are still unregistered and pedigree information is absent. Therefore the genetic diversity and population structure of the breed and the amount of genomic information shared with related breeds was evaluated. For the Kempen breed 340 samples were genotyped with Illumina 54k SNP and for 9 others breeds, 421 samples were available. PCA analysis was used to evaluate the population structure within the Kempen breed. Due to the absence of a breeding goal, there is substantial variation between the farmers. However the breeding bulls included in the analysis are positioned central in the population, ensuring a more uniform population in the next generation. Unregistered Kempen animals were identified and the effective population size was estimated to be 67. Further inbreeding values based on runs of homozygosity and F-statistics have been analysed. PCA analyses revealed that the breeds Eastern Belgium Red Pied, Improved Red Pied and Brandrood are genetically similar to the Kempen population. Furthermore the admixture events between the Kempen breed and related breeds were evaluated. These genomic analyses may help to preserve genetic diversity in the Kempen breed and ensure its uniqueness.

Inbreeding in native Norwegian poultry breeds, with partially unknown maternal pedigree

L.F. Groeneveld[1], U. Müller[2], E. Groeneveld[3], N.H. Sæther[4] and P. Berg[1]
[1]NordGen – Nordic Genetic Resource Center, P.O. Box 115, 1431 Ås, Norway, [2]Saxon State Agency for Environment, Agriculture and Geology, Am Park 3, 04886 Köllitsch, Germany, [3]FLI – Institute of Farm Animal Genetics, Hoeltystr. 10, 31535 Neustadt, Germany, [4]Norwegian Genetic Resource Center, P.O. Box 115, 1431 Ås, Norway; linn.groeneveld@nordgen.org

In 1973, a live poultry genebank for egg layers, mainly with duplicates of the Norwegian Poultry Breeding Association's (NPBA) active breeding populations was established to maintain these lines for potential future use. Until 1995, when the breeding work of the NPBA was terminated, duplicates of the active breeding lines were renewed annually in the live poultry genebank. Since 1995, all the lines and breeds have been maintained by a rotational mating scheme with approximately 23 families per line. Families consisting of 5 to 8 full-sib hens were housed with one cock. The maternal pedigree is thus partially unknown. An APIIS database, with a modified core to accommodate uncertain maternal pedigrees, was developed and populated through the LOAD system with data collected from 2006 onward. The objective of this study was to evaluate the efficiency of maintaining genetic diversity through a rotational mating scheme, by computing realized coefficients of inbreeding within lines. Inbreeding was computed with two different approaches and compared. In approach 1, animals were assumed to have multiple mothers, e.g. assuming 10% of alleles were inherited from each of 5 hens in a cage. The numerator relationship matrix was computed based on this assumption and inbreeding obtained from diagonal elements. In approach 2, a possible pedigree was constructed by randomly sampling a mother for each individual. Mean and median inbreeding was computed from estimates obtained from this repeated sampling. The two approaches resulted in similar trends in inbreeding. Rates of inbreeding are discussed relative to the risk of loss of genetic variation and possibilities for improved management routines are examined.

Supporting genetic management of live populations with gene bank material in a cattle populations

J.J. Windig[1], S.J. Hiemstra[2], I. Hulsegge[1], L. François[3] and N. Buys[3]
[1]Wageningen UR, Animal Breeding and Genomics Centre, P.O. Box 338, 6700 AH, Wageningen, the Netherlands, [2]Wageningen UR, Centre for Genetic Resources, the Netherlands, P.O. Box 338, 6700 AH Wageningen, Togo, [3]KU Leuven, Department of Biosystems, Livestock Genetics, P.O. Box 2456, 3001 Leuven, Belgium; jack.windig@wur.nl

Gene banks may not only serve as a back-up for live populations, i.e. to be used in case of calamities but gene bank collections may also be used to support current breeding programs and thereby increase the effective population size. In this paper we asses the genetic diversity in Dutch gene bank collections of two cattle populations Friesian Red and White cattle and Deep Red cattle, and explore the potential of the gene bank collections to support *in situ* breed conservation. Live populations consist of about 2,400 cows and 60 bulls each, and Dutch gene bank collections consist of material of 19 Deep Red bulls 50 Friesian Red and White bulls. We determined the average relatedness between and within the live populations and gene bank bulls. Breeding schemes were then simulated to determine to what extent the inbreeding rate decreased and the effective population size increased when semen from bulls stored in the gene bank is used in the live population. First populations were simulated without use of gene bank material. Next the populations were simulated including bulls in the gene bank, whereby the gene bank was replenished with material from offspring. We varied the number of years a bull was used and the number of cows inseminated per year, to determine their effect on the effective population size.

Conservation and sustainable development of Murboden Cattle in Austria

B. Berger[1], S. Eaglen[2], J. Sölkner[2] and B. Fürst-Waltl[2]
[1]AREC Raumberg-Gumpenstein, Austrasse 10, 4601 Thalheim, Austria, [2]University of Natural Resources and Life Sciences, Nutztierwissenschaften, Gregor-Mendel-Strasse 33, 1180 Vienna, Austria; beate.berger@raumberg-gumpenstein.at

In Austria conservation breeding started in 1982. The highly endangered Murboden cattle serve model population to analyse conservation, development and sustainable use of small populations. The breed suffered a severe genetic bottleneck from 1970 to 1979. Monitoring by the Austrian Association for Rare Endangered Breeds started in 1997. Parameters are effective population size, pedigree quality (generation equivalent) and inbreeding rate per generation. The conservation program containing planned mating as well as recording for production traits is integrated into the Austrian Agro-Environmental Program ÖPUL. With support from the Austrian Gene Bank for Farm Animals the population increased from 200 breeding females in 1982 to approx. 5,000 breeding females in 2014. 64 Murboden bulls were collected at the Gene Bank from 1997 to 2014 and the semen was used in the breeding program keeping the inbreeding rate well below 1% per generation. In 2006 a premium beef program for oxen and heifers together with one of the biggest Austrian food retailers was created. Additionally a special product – a traditional sausage – was developed to get better prices for old cows. A breeding program was developed to further develop the breed. An analysis of 25,000 calvings, 16,000 weighing data and 2,950 slaughterhouse data sets together with population data from the yearly monitoring showed weaknesses in calving ease, daily gain and carcass traits. Heritabilities were sufficiently high to estimate breeding values for heifers and cows separately and combine them with inbreeding information in a weighted index. The software allows recommendations of bulls on a much broader basis than the inbreeding coefficient of the offspring. It will be made available by integration into the national livestock database helping to further develop re-established traditional breeds.

Exclusion of admixed animals from Cika cattle *in situ* conservation program using SNP haplotypes

M. Simčič[1], J. Sölkner[2], D. Kompan[1] and I. Medugorac[3]
[1]University of Ljubljana, Biotechnical Faculty, Department of Animal Science, Jamnikarjeva 101, 1000 Ljubljana, Slovenia, [2]BOKU – University of Natural Resources and Life Sciences, Department of Sustainable Agricultural Systems, Gregor Mendel Str. 33, 1180 Vienna, Austria, [3]Faculty of Veterinary Medicine, LMU – Ludwig-Maximilians University, Veterinaerstr. 13, 80539 Munich, Germany; mojca.simcic@bf.uni-lj.si

The aim of the study was to establish an objective basis for an assessment of possibly purebred animals in an *in situ* conservation program of the indigenous admixed cattle breed Cika. For this purpose we genotyped 760 animals of 15 cattle breeds by Bovine SNP50 array. Realised genetic relationship and a degree of admixture within 106 Cika and 654 animals from 14 reference breeds were determined. The results of IBD analyses were complemented by genetic structure using the program ADMIXTURE where 15 groups (K=15) were determined. These analyses found some Cika animals significantly admixed with Pinzgauer, Fleckvieh, Red Holstein, Jersey, Limousin or Braunvieh. Paternity analyses and distribution of admixed animals suggested previously unrecognised crossing. These intentionally crossed animals should not be confused with ancient admixtures observed in all animals and all breeds. Animals identified as the most purebred were proposed for a breeding nucleus with the original genetic background. Successive inclusion of reference breeds demonstrated that power for identification of foreign haplotypes in the admixed population depends on the availability of possibly complete spectrum of haplotyped reference populations. Conservation and recovery approach should rely on the identification and removal of introgressed haplotypes what would be more powerful than the search for the original genetic background.

The strategies for animal genetic resources cryoconservation in Spain

F. Tejerina[1], M. Castellanos[1], E. Martínez[1], F.J. Cuevas[2], S. Moreno[2], J. Urquía[2] and A. Cabello[1]
[1]*Ministerio de Agricultura Alimentación y Medio Ambiente, C/Almagro 33, 28010 Madrid, Spain,* [2]*CENSYRA Colmenar Viejo, IMIDRA, Ctra. Colmenar Viejo-Guadalix de la Sierra km 1.800, 28770 Colmenar Viejo, Spain; ftejerina@magrama.es*

Spain is one of the countries with greater domestic animal biodiversity in the European region. Currently, the Official Catalogue of Livestock Breeds of Spain includes 184 breeds, being 157 of them natives, 128 of whom are in risk of extinction. The principles and objectives in relation with the AnGR cryoconservation in Spain have been laid down in the National Program for the Preservation, Improvement and Development of Livestock Breeds. The National Program binds the breeder's associations (BAs) to develop a conservation breed program, where the BAs has to set down the kind of reproductive material selected to store in their gene banks, the suitable amount of material and the number of donors. Furthermore, the National Program binds the BAs to send a 'copy' of their gene banks to the National Gene Bank. The national sanitary regulation contains some dispositions to facilitate the constitution of gene banks of native breeds. Furthermore, in 2012 the MAGRAMA developed a survey to assess the level of development of the cryoconservation activities in Spain, the main conclusion was that 78 of our breeds have any collection cryo-conserved, but only 26 have a gene bank that allows the complete recovery of a breed. Finally, subsidies from public administrations are given to BAs to support their activities, including the development of the Conservation Breed Program and building of gene banks. The MAGRAMA subsidies to BAs regulation lays down the obligations of developing a gene bank and sending a 'copy' to the National Gene Bank to get access of a specific amount of public funding. By these measures, the National Gene Bank has increased the number of breeds/varieties storage from 8 to 38; the number of donors from 77 to 506, and the number of semen doses from 12.445 to 43.258, in the last three years.

Genetic diversity of Zatorska goose population

M. Graczyk[1], K. Andres[2], E. Kapkowska[2] and T. Szwaczkowski[1]
[1]*Poznan University of Life Sciences, Department of Genetics and Animal Breeding, Wolynska 33, 60-637 Poznan, Poland,* [2]*University of Agriculture in Cracow, Department of Swine and Small Animal Breeding, Institute of Animal Science, Mickiewicz Alley 24/28, 30-059 Cracow, Poland; tszwaczkowski@gmail.com*

Over the last decades the local breeds have become important from an economic, social and cultural point of view. The Zatorska goose is one of many breeds classified in World Watch List for Domestic Animal Diversity as endangered-maintained. It is crucial to monitor the genetic diversity in endangered populations of animals. The aim of the study was to estimate the structure of the Zatorska breed in context of conservation program. In total, 5,514 individuals hatched between 1990-2013 were included into the analysis. The analysis of the population showed that 370 were founders. The average inbreeding level was low and for the whole population it was 1.46% and for inbred individuals it was 3.02%. The average individual increase in inbreeding was also low and amounted to 0.35% whereas for inbred birds 0.72%. Estimated founder genome equivalent was 130.07. This means that this number of founders with same proportion of alleles can give the same level of genetic diversity in the population analyzed. Loss of genetic diversity caused by genetic drift, bottleneck effect and unequal contribution of founder's genes was very low and amounted to 0.0097. The average effective population size was estimated from the family size variance and amounted to 67.36 individuals. The effective population size calculated form the regression of year of hatching amounted to 111.04. It can be concluded that the conservative breeding program in the Zatorska goose is sufficiently realized but continuous monitoring of the endangered population seems to be necessary.

Genetic heritage of the Eastern Belgium Red and White breed, an endangered local breed

F.G. Colinet[1], A. Bouffioux[1], P. Mayeres[2], M. Malzahn[3], L. François[4], S. Janssens[4], N. Buys[4], S.J. Hiemstra[5], J.J. Windig[5] and N. Gengler[1]
[1]University of Liege, Gembloux Agro-Bio Tech, 5030, Gembloux, Belgium, [2]Walloon Breeding Association, 5590, Ciney, Belgium, [3]Fondation rurale de Wallonie, 4950, Faymonville, Belgium, [4]KU Leuven, Department of Biosystems, Livestock Genetics, 3001, Leuven, Belgium, [5]Wageningen University and Research Centre, Animal Breeding and Genomics Centre (ABGC) | Centre for Genetic Resources (CGN), 6700 AH, Wageningen, the Netherlands; frederic.colinet@ulg.ac.be

Post-second world war, Belgian cattle were still bred in racial zones. There were two types of dual purpose red and white cattle, one in the Kempen Region and one in the East Cantons. At the beginning of the '70s, the provincial Herdbooks were merged into a single national Red and White (R&W) Herdbook. During the '80-'90s, Red Holstein sires started to become popular in order to improve type traits (size and udder shape) and the milk production. Under the holsteinization pressure and in the absence of a commitment of the R&W Herdbook to this breed type, the eastern type was considered extinct in the '90s. Nevertheless, several farmers still kept these cattle because they were perfectly adapted to their more extensive, grazing based production system on less productive medium-altitude (Ardenne-Eifel) meadows. In 2014, breeders created a commission to restart the Eastern Belgium Red and White Breed (EBRW). A Public Service of Wallonia project to support conservation and valorization of EBRW breed, involving scientists and stakeholders, was started. Genotypes (BovineSNP50 Beadchip) of 65 EBRW animals were used for comparison with reference SNP data of 12 other breeds through principal component analysis and studies on the genomic based relationship matrix, the genetic structure and the phylogeny based on distance of Nei. Results showed that the EBRW animals are clearly different from Red Holsteins, similar but still distinct, also in their breed standards, from other members (e.g. MRY, Kempen breed) of the red-pied dual purpose breed group in north-western Europe.

Main features of Lithuanian strategy for farm animal genetic resources

R. Sveistiene and V. Razmaite
Lithuanian University of Health Sciences, Institute of Animal Sciences, R. Zebenkos 12, Baisogala, 82317, Lithuania; Ruta.sveistiene@lsmuni.lt

The main purpose of the first National Programme for the conservation of farm animal genetic resources (AnGR) adopted by the Ministry of Agriculture of Republic of Lithuania in 1996 is to collect, to investigate and to conserve Lithuanian AnGR. In this programme only indigenous Lithuanian breeds were included. Later, unfavourable conditions have formed for breeds developed in 20th century and these breeds were included in a conservation programme. According to the Lithuanian conservation strategy the sequence of critical AnGR should be the formation of closed nucleus herds and maintenance of genealogical structure of breeds, evaluation of phenotypic and genetic traits, preparation of evaluation standards and search for the possibilities of their wider use. The initial priority was *in situ* conservation. Effective population size for many Lithuanian breeds is below 50 till now. There is driftless reproduction and, therefore, the survival of the population is uncertain. The inbreeding can be minimized by having a larger Ne>50 and by using special mating schemes and using sperm from the long term gene bank to maintain genealogical structure. The first decision in setting up conservation schemes is to carry forward the existing variability in the breeds. In the case of loss of a substantial number of animals in different lines and families the ex-situ conservation was initiated. This is mainly concerned with the size of available resources, which could be adjusted by choosing individuals for conservation action from different lines and by carrying out planned mating or insemination between the chosen animals.

Integrating genomic data in National Genebanks for targeted re-use of collections

E. Groeneveld[1], I. Hulsegge[2] and S.J. Hiemstra[3]
[1]Institute of Animal Genetics (FLI), Höltystr. 10, 31535 Neustadt, Germany, [2]Wageningen UR, Animal Breeding and Genomics Centre, P.O. Box 338, 6700 AH Wageningen, the Netherlands, [3]Wageningen UR, Centre for Genetic Resources, the Netherlands, P.O. Box 338, 6700 AH Wageningen, the Netherlands; eildert.groeneveld@fli.bund.de

With the advent of genomics a new class of data – SNP data from panels of increasing sizes – has become available also for samples stored in gene banks. Integration of sample and genomic data makes eminent sense. It will allow selection of genetic material from the gene bank in targeted searches for better use. CryoWEB is a gene bank management software with more than 15 installations in Europe. TheSNPpit is a new SNP management data base system to handle SNP data in highly compressed format allowing any number of panels of any size. Further it supports a generalized storage of phenotype data using the EAV model. Both are freely available under the Open Source Model. The design of the integration needs to support the two main use cases. Firstly, for a given sample from the gene bank the molecular genetic data is exported along with other sample/phenotype information. Secondly, based on genetic information in the SNP data, the sample IDs are identified. The search in the SNP data from the genebank can be done based on information derived from external annotation like SNPdb. The result of such a search will be a set of genotypes pointing to material in the genebank for possible retrieval. Data im- and export is done in the quasi standard PLINK format, while the phenotypes are available in a general spread sheet csv format for further processing. CryoWEB and TheSNPpit were integrated for the Dutch National Gene bank for farm animals. SNP data was available for cattle (50k) and pigs (80k). Furthermore, based on a full sequence the full SNP equivalent was loaded for cattle and pigs. The total number of genotypes sets available was 62 (5 cattle breeds and one pig line).

Genomic characterization of Danish farm animal genetic resources

A.A. Schönherz[1], V.H. Nielsen[2], A.C. Sørensen[1] and B. Guldbrandtsen[1]
[1]Aarhus University, Center for Quantitative Genetics and Genomics, Blichers Allé 20, 8830 Tjele, Denmark, [2]Aarhus University, Danish Centre for Food and Agriculture, Blichers Allé 20, 8830 Tjele, Denmark; vivih.nielsen@dca.au.dk

Genomic characterization is efficient for use in sustainable conservation of farm animal genetic resources. Genomic data can be used to describe genetic variation within breeds, estimate the level of inbreeding, reveal hybridization with other breeds and determine the genetic relationship among breeds and the data are valuable in selection of breeding animals and for development of breeding plans. Danish farm animal genetic resources including 3 horse breeds (The Frederiksborg Horse; The Jutland Horse; The Knabstrup Horse), 3 cattle breeds (Danish Black-Pied Cattle, 1965; Danish Red Cattle, 1970, Danish Shorthorn), 2 pig breeds (Danish Landrace Pig, 1970; Danish Black-Pied Pig), 2 sheep breeds (Danish Landrace Sheep; White Headed Marsh Sheep), 1 goat breed (Danish Landrace Goat), and 1 poultry breed (Danish Landrace Chicken) were characterized by chip typing and whole genome sequencing. Blood, semen, ear punch or feathers were sampled for DNA extraction and SNP-genotyping and whole genome sequencing. Population structure was analyzed using Principal Component Analysis and Cluster Analysis including international reference datasets. Genomic inbreeding was estimated by mapping runs of homozygosity. Results show that the Frederiksborg Horse is genetically a homogenous breed, distinct from other breeds. High diversity was observed in the Knabstrup Breed. The average chromosomal inbreeding coefficient was estimated to 0.10 in the Frederiksborg Horse and to 0.04 in the Knabstrup Horse. In cattle, the analyses show that both Danish Red Cattle, 1970 and Danish Shorthorn are distinct from other breeds but considerable inbreeding was observed in the two breeds. The average chromosomal inbreeding coefficient was estimated to 0.21 in Danish Red Cattle, 1970 and up to 0.29 in Danish Shorthorn. Genomic methods of management of co-ancestry in the animal genetic resource breeds are being developed.

Comparison of Bray Curtis and Nei's genetic distance with Mantel test for chicken diversity data

H. Onder[1], L. Mercan[2], S.H. Abaci[1] and A. Okumus[2]
[1]*Ondokuz Mayis University, Agricultural Faculty, Animal Science, Atakum, 55139, Turkey,* [2]*Ondokuz Mayis University, Agricultural Faculty, Agricultural Biotechnology, Atakum, 55139, Turkey; hasanonder@gmail.com*

In this study we aimed to examine the effects of both Nei's genetic distance and Bray Curtis distance to calculate chicken diversity data vs geographical distance with Mantel test. Many researchers used Nei's genetic distance which measure that genetic differences arise due to mutations and genetic drift but some researchers used Bray Curtis distance which used to quantify the compositional dissimilarity between two different sites. In this study, we compared both distance measures on 28 microsatellite loci taken from 45 sampling site with sample size of 364. As a result, with use of Mantel test to calculate relationship between genetic differentiations and geographic distance between populations Bray Curtis distance could be used substituted for Nei's genetic distance with great reliability on chicken diversity data.

Genetic diversity situation of sheep in Algeria

N. Adili[1], M. Melizi[1] and H. Belabbas[2]
[1]*Institute of Veterinary Sciences and Agricultural Sciences, University of El-Hadj Lakhdar, Veterinary Medicine Department, Batna, 05000, Algeria,* [2]*University of Mohamed Boudiaf, Microbiology and Biochimistry Department, M'sila, 28000, Algeria; nezar.adili@yahoo.fr*

This study aims to provide an update on the current situation of sheep breeds present in Algeria; and the various mechanisms and conditions that regulate today sheep population. The investigation focuses to collect information on different genetic resources; it also concerns the main modes of production and the difficulties restraining the development of sheep farming in Algeria. The descriptive statistical analysis of data obtained is realized by the Microsoft Office Excel 2010. The total number of ovine flock is approximately 19 millionof head with 10 million ewes; we distinguish several local breeds: the Ouled-Djellal (the most dominant with 63% of the total flock), the Rumbi (about 11.1%), the Hamra (0.31%), the Berber (estimated at 25%), the Barbarine (0.27%) the D'men (0.19%) and the Sidahou (0.13%). We have to note a strong uncontrolled progression of the crossbreeding products of Ouled-Djellal population with other types not only in Algeria, but also with sheep of neighboring countries (Tunisia and Morocco). Sheep flocks live generally in areas dominated by both extensive and semi-intensive systems; they are raised mainly for meat and wool production. Algerian sheep therefore show great diversity, which is poorly exploited and dominated by anarchical crossbreeding with reduction of number of certain breeds (like Hamra). Appropriate measures are of capital importance, and those for the preservation of these resources, which are essential components for the universal genetic heritage.

Genetic variation of the Polish cattle included into a conservation programme based on DNA markers

A. Radko, A. Szumiec and D. Rubiś
National Research Institute of Animal Production, Department of Animal Cytogenetics and Molecular Genetics, 1, Krakowska Street, 32-083 Balice, Poland; anna.radko@izoo.krakow.pl

The cattle genetic resources conservation programme in Poland covers the Polish Red (RP), Polish Red-and-White (ZR), Polish Black-and-White (ZB) and White-backed (BG) breeds. It is important to study and monitor changes taking place in the genetic structure of these breeds. Currently, the analysis of allele frequency distribution in highly polymorphic microsatellite markers (STR) is most often used to observe the genetic changes induced by breeding. The objective of the study was to identify the genetic structure of RP (383), ZR (338), ZB (210) and BG (100) cattle and to determine genetic differentiation between them based on the polymorphism of 12 STR markers (BM1818, BM1824, BM2113, ETH3, ETH10, ETH225, INRA23, SPS115, TGLA53, TGLA122, TGLA126 and TGLA227). Genetic analyses were performed in an ABI 3130xl sequencer and the results were analysed using GeneMapper Software 4.0. The identified alleles were used to estimate observed (Ho) and expected heterozygosity (H$_E$), the inbreeding coefficient (Fis), polymorphic information content (PIC) and genetic distance DN according to Nei (1987). All the markers were highly polymorphic. PIC values for each marker were high and exceeded 0.5 except for SPS115 locus in ZR (PIC=0.46) and ZB cattle (PIC=0.48), whereas Ho ranged from 51.1% to as much as 87.2%. The observed and expected heterozygosity for different loci showed similar values indicative of the lack of inbreeding in the cattle under study. The highest polymorphism, for which PIC and Ho were calculated to exceed 0.8, was observed in BM2113 and TGLA53 loci in BG cattle, in BM2113 and TGLA227 in RP cattle, and in BM2113, TGLA227, TGLA53 and TGLA122 loci in ZB and ZR cattle. The estimated coefficient of genetic distance, calculated based on 12 STR, was low and ranged from D$_N$=0.04 between ZB and BG cattle to D$_N$=0.13 between RP and ZB cattle.

Feasible solutions to producing entire pigs: an introduction

A. Velarde
IRTA, Animal Welfare Subprogram, Veïnat de Sies s/n, 17111, Monells, Spain; antonio.velarde@irta.cat

Tail-biting is a welfare problem because of the pain and suffering experienced by the bitten animal (not only due to the biting but also to secondary infections), the stress caused to the group (restlessness), and the likely frustration of the biting animal. Tail docking is carried out to prevent tail biting. However, the pain associated with tail docking procedures normally lasts a few days, but in some cases chronic pain may also result. Tail docking should not be carried out routinely, only where there is evidence that injuries to other pigs' tails have occurred, and after other measures have been taken to prevent tail-biting. As tail biting is a multi-factorial problem involving both internal and environmental risk factors (genetic background, sex, age, health status, diet, feeding management and different characteristics of the pen) the ability to control these risk factors is essential when aiming to avoid tail-docking. The main motive of castration is to prevent the boar taint which is present in the meat of some intact boars when they reach puberty. Castration is currently performed surgically and without anaesthesia during the first week of the animal's life. Although it is a rapid procedure, it induces a series of physiological and behavioural changes in the piglet which are clearly indicative of pain and stress. Post-surgical pain can last for 5 days. The scientific community is looking for alternatives to castration which eliminates the boar taint without causing the animal to suffer. As well as eliminating the pain associated with castration, breeding intact males also presents certain advantages with respect to the production of barrows. For the production of intact males to be economically viable it is necessary to have the means to control the presence of boar taint in the carcass.

Housing and management strategies to reduce tail biting

A. Valros
Univeristy of Helsinki, Department of Production Animal Medicine, Faculty of Veterinary Medicin, P.O.
Box 57, 00014 Helsinki, Finland; anna.valros@helsinki.fi

Tail docking is commonly used as a preventive strategy against tail biting, which is a painful and common behavioural problem in pigs, causing reduced health and production results of pigs. Although tail docking, reduces tail biting, it does not stop the problem. In addition, tail docking causes pain in itself. In addition, tail biting has been shown to be a sign of stress, suboptimal housing or management conditions. Thus, by docking tails, underlying welfare issues caused by the production system are partly masked, while not reduced. Much of the current focus on management of tail biting is on finding suitable manipulable materials for pigs, and often straw is mentioned as a remedy for tail biting. Even though the possibility to manipulate, root and chew is very important for pigs, and lack of rooting material is one of the main risk factors for tail biting, merely looking at enrichment will not solve the problem of tail biting. Tail biting has been suggested to have several motivational backgrounds, and identified risk factors are numerous and multifactorial. Thus, a holistic approach is needed to tackle the problem. For example, recent studies show increasing evidence that feeding and feed-related problems are important risk factors, and reduced general health status has also been indicated. Looking at tail biting prevalence in countries where tail docking is fully banned indicates that tail biting does not increase exponentially, giving promising data for a more general ban. Even though tail biting levels are slightly increased without docking, instead, no pigs have to go through the procedure of docking, and thus the vast majority of pigs do not have to endure pain. Managing and housing intact-tailed pigs needs certain improvements in basic welfare level (such as adequate space allowance, reduced competition at feeding, proper environmental management and use of manipulable materials), but tail biting can be managed to an acceptable level within modern pig production systems.

The effect of mixing after weaning on tail biting during rearing

C. Veit[1], I. Traulsen[1], K. Müller[2] and J. Krieter[1]
[1]Institute of Animal Breeding and Husbandry, Christian-Albrechts-University, Olshausenstr. 40, 24098
Kiel, Germany, [2]Futterkamp, research farm of the Chamber of Agriculture of Schleswig-Holstein, Gutshof
1, 24327 Blekendorf, Germany; cveit@tierzucht.uni-kiel.de

Tail biting in pigs is a welfare concern in intensive pig husbandry and tail docking is until now the most efficient way to avoid it. The aim of this study was to reveal the effects of housing of acquainted piglets in comparison to piglets out of mixed litters on tail biting during rearing. The study took place on a research farm from January until April 2014. The experimental groups 'litter wise' (n=240) and 'mixed litters' (n=240) were housed in five identical units with an additional daily offer of alfalfa straw. Each tail was scored regarding bite occurrence and tail losses once per week with a four-point-score (0=no damage, 3=severe damage). To investigate the behaviour of the animals in regard to a tail biting outbreak, the piglets out of four units were marked individually and observed by video. In both experimental groups, tail biting started on average 2.8 weeks after weaning and occurred in 100% of the pens. 41.8% of the piglets out of the litter wise groups and 45.7% of the piglets out of the mixed litter groups were scored on average with a score greater 0 regarding biting occurrence (P=0.096). In total 77.5% of the pens were affected by tail losses, visible on average 4.6 weeks after weaning. At the end of rearing, 53.1% of the piglets out of the mixed litter groups and 61.3% of the piglets out of the litter wise groups still had their original length of tail (P<0.05). Recorded video material is under analysis by instantaneous scan sampling and continuously observation. Preliminary results shows that pen mate directed behaviours such as 'belly-nosing' and 'ano-genital-contact' increased during five days before a tail biting outbreak. The avoidance of additional stress due to regrouping of piglets can significantly reduce the amount of tail losses during rearing.

Higher space allowance and straw rack as effective measures to reduce tail biting in fattening pigs

K. Schodl, C. Leeb, L. Picker and C. Winckler
University of Natural Resources and Life Sciences (BOKU), Division of Livestock Sciences, Gregor-Mendel-Str 33, 1180 Vienna, Austria; katharina.schodl@boku.ac.at

Tail biting in fattening pigs presents a critical welfare problem and affects productivity. Although tail docking is a painful procedure it is a widely applied measure to prevent tail biting. Several studies show that the lack of adequate manipulable material and of space are crucial factors, which provoke tail biting. In the course of a larger study the effects of increased space allowance (1 m^2 instead of 0.7 m^2/pig) and the provision of straw in racks as manipulable material on pig welfare were examined. On 3 commercial fattening farms in Austria this was implemented in half of the pens (improved pens, IP) whereas the remaining pens served as a control (CP). On Farm 1 (974 pigs/74 pens) all pigs were tail docked, tails on Farm 2 (413/47) were docked in the CP but left intact in the IP and on Farm 3 (70/10) all pigs had intact tails. In this paper the occurrence of tail biting was assessed using direct observations for each pen at the beginning (I) and in the middle (II) of the fattening period. Prevalence of tail lesions was scored at the same time and additionally a third time at the abattoir (III). Furthermore, performance data were collected. Data were analyzed for each farm individually. Differences between IP and CP and observations (I, II, III) were analyzed using a linear model (behavior), non-parametric tests (lesions) and a mixed model (production data) with P<0.05 as level for significance. Results show significantly lower tail biting behavior in IP compared to CP (2.8 vs 5.7 incidences/100 animals/10 min). Prevalence of tail lesions did not differ significantly. Animals in the IP grew significantly faster (900.5 vs 816.9 g/day) at similar lean meat content (60.7 vs 60.4%). Hence, an increase of space allowance combined with the provision of a straw rack reduced tail biting independent from whether pigs were tail docked or not and an omission of tail docking did not cause more tail lesions.

Frequent delivery of straw for slaughter pigs – effect on behaviour and welfare

A. Lind, L. Wahlund and C. Lindahl
Swedish Institute of Agricultural and Environmental Engineering, Box 7033, 75007 Uppsala, Sweden; lotten.wahlund@jti.se

The aim of the study was to investigate if frequency delivery of straw had an effect on the health, behaviour and welfare of slaughter pigs. A case-control study was carried out on a farm with conventional All-in All-out pig production with two separate sections having the same size and design. In section 1, a system for automatic straw delivery was available. Straw were given three times a day approximately 10 minutes after feeding. In section 2, the straw was delivered manually once a day. Both sections got the same amount of straw per day (1 kilo/pen). Data were collected for three weeks in each rearing period. The measurements consisted of behavioural studies, individual and in groups, and health scores. We found significant differences (P<0.0001) in time spent manipulating straw between the two sections. Furthermore, the pigs with frequent straw distribution were more active (P=0.0474), while the pigs given straw once a day spent more time lying (P=0.0107). There were no significant differences between the groups regarding social and aggressive behaviour and problem behaviours as tail biting and belly-nosing. No differences in health between the two groups were found, which could be explained by the fact that there were very few health problems on the farm in general. Since we did find an effect of frequent straw delivery on the general activity and the time spent manipulating the straw, it is possible that the welfare of the pigs improve when given straw more often. By using an automatic system for straw distribution, straw can be given more often without affecting the work time negatively. The effect of the amount of straw given and also the impact of straw length needs to be further investigated to study the effects on aggressive behaviour and welfare problems such as tail biting and belly nosing in slaughter pigs.

Curly tails: the Dutch approach

M. Kluivers-Poodt, N. Dirx, C.M.C. Van Der Peet, A. Hoofs, W.W. Ursinus, J.E. Bolhuis and G. Van Der Peet
Wageningen University and Research centre, Livestock Research, De Elst 1, 6708 WD Wageningen, the
Netherlands; marion.kluivers@wur.nl

Despite EU legislation and societal concerns, in current pig farming, most piglets are still tail docked. The pig farming sector would prefer to stop tail docking. However, without additional preventive measures, tail biting will likely increase. Several Dutch parties have designed the Declaration of Dalfsen, containing a careful road map towards curly tails. This map comprises a demonstration project, development of a toolbox and knowledge exchange, and aims at closing the gap between science and practice and relieving the anxiety and scepticism about keeping pigs with long tails in current systems. In 2014, every six weeks a batch of twelve undocked litters was included in the demonstration. Circumstances were optimized as much as possible, and additional enrichment was provided. Caretakers were coached to recognize early signs of animals at risk. Nonetheless, tail damage appeared. Mostly in individual animals, but occasionally as an outbreak at pen level (for which predictive correlates are searched). Remarkably, at three weeks of age, several piglets already showed bite marks at the tail. Attitudes of the caretakers changed during the year to a higher level of alertness and an active approach towards required management changes. A traffic light system was implemented to safeguard attention towards groups at risk. The use of some enrichment materials encountered practical problems, and labour required for adequate monitoring and providing materials was higher than expected. The toolbox is still being developed and tested, describing effective curative measures. A network of farmers keeping pigs with long tails was formed, to support exchange of knowledge and experiences. All knowledge gained will feed an educational programme for pig farmers and farm advisors to enable a responsible transition towards longer tails. The key to success of this approach is that pig farmers are at the steering wheel, with guidance from actors in the chain.

Tail biting; what we do and do not know from a genetics perspective

N. Duijvesteijn and E.F. Knol
Topigs Norsvin Research Center B.V, Schoenaker 6, 6640 AA, the Netherlands;
naomi.duijvesteijn@topigsnorsvin.com

Tail biting has economic consequences for pig farmers and animal welfare consequences for pigs. To prevent tail biting (to a certain extend) tails of pigs are docked at young age. Many different factors could cause or influence tail biting e.g. diet formulation, environmental discomfort, health problems, excessive competition for resources but also genetics. To be able to use tail biting as a trait in a breeding program, the trait should be heritable, but also easy to measure. Different feasibility studies have been undertaken. One approach which gave hopeful results was selection on indirect genetic effects (IGE) for growth. IGE are heritable effects of an individual on trait values of another. For growth this means that an individual can have a positive genetic effect on the growth of another individual. The underlying mechanism how this positive effect is transmitted was investigated and pens with a high IGE for growth showed less tail biting amongst other behavioral differences. This seems to be an approach to capture tail biting in a breeding program, but the selection on IGE for growth is not straightforward (e.g. validation on growth itself, accurate pen recording). Other approaches could result in a more direct measure on tail biting. Possibly IGE play a role in tail biting directly. Estimation of IGE on tail damage and selection on low IGE should results in less tail damage to pen mates. Tail damage can easily be measured as a categorical trait (no damage, little damage, wound) preferably at weaning or before slaughter. Pen recording is essential and also mixing of pen mates where preferably two full sib families are within one pen. Data collection has started on this approach. Other approaches such as automated detection of tail biting using video analysis, finding genetic markers, or finding phenotypic markers (f.e. general activity within the pen) are under investigation. So far, there is no straightforward measure of tail biting such that it can directly be implemented in a pig breeding program.

Relationship between sperm production and boar taint risk of purebred or crossbred entire offspring

M.J. Mercat[1], A. Prunier[2], N. Muller[3], C. Hassenfratz[1] and C. Larzul[4]
[1]IFIP, Genetics, La Motte au Vicomte BP 35104, 35651, France, [2]INRA, PEGASE, UMR1348, 35590 Saint-Gilles, France, [3]INRA, UETP, La Motte au Vicomte, 35653 Le Rheu, France, [4]INRA, GenPhySE, UMR1388, 31326 Castanet-Tolosan, France; marie-jose.mercat@ifip.asso.fr

This study focuses on 100 boars from three Pietrain varieties: 74 V1, 10 V2 and 16 V3. On average, data were recorded on 90 ejaculates collected in artificial insemination centers and boar taint measured on 15.4 entire male offspring, half purebred and half Pietrain × Large-White type. Offspring were reared at INRA UETP (Le Rheu, France) up to 110 kg. Boar taint risk was assessed by androstenone and skatole measurements in fat collected at the slaughterhouse. Sperm production data were corrected for the production site, the age and the collection frequency using a generalized linear mixed model to estimate an average sperm production count per boar and ejaculate. Boar taint odor risk was on the whole low in the tested population. Considering the three Pietrain varieties altogether, boars with the highest sperm production had a lower proportion of offspring without risk of odor compared to boars with low sperm production: the boars ranked in the top quartile of sperm production had 14% less offspring without odor (i.e. androstenone <1.0 µg/g and skatole <0.2 µg/g) compared to the boars in the bottom quartile ($P<0.05$). The sperm production of the sires of pigs with high androstenone content in fat was on average higher than that of the sires of pigs with low androstenone content: 95.3 and 86.8 billion spermatozoa per ejaculate for the sires of offspring with more than 3.0 µg/g of androstenone and non-detectable androstenone level, respectively ($P<0.05$). Differences were more pronounced on crossbred offspring than on purebred ones. With V1 sires alone, differences were smaller but still significant. As a conclusion, selection against boar taint might negatively impact semen production if reproduction traits are not included in the breeding goal.

Slaughter related factors and season and their effect on boar taint in Belgian pigs

E. Heyrman[1,2], S. Millet[2], F. Tuyttens[2], B. Ampe[2], S. Janssens[1], N. Buys[1], J. Wauters[3], L. Vanhaecke[3] and M. Aluwé[2]
[1]KU Leuven, Livestock Genetics, Department of Biosystems, Kasteelpark Arenberg 30 – bus 2456, 3001 Heverlee, Belgium, [2]Institute for agricultural and fisheries research (ILVO), Animal science unit, Scheldeweg 68, 9090 Melle, Belgium, [3]Ghent University, Department of Veterinary Public Health and Food Safety, Laboratory of Chemical Analysis, Salisburylaan 133, 9820 Merelbeke, Belgium; evert.heyrman@ilvo.vlaanderen.be

Since 2010 several Belgian farmers are keeping entire males. Measures to reduce boar taint prevalence can be a part in minimizing economic loss. In this study, a large scale screening of boar taint prevalence on farms and slaughter related factors was conducted. From 37 participating farms, boar taint prevalence was determined for a minimum of two slaughter batches, with a minimum of 50 boars per slaughter batch and with at least one month between slaughter batches. Neckfat samples from all slaughter batches were collected and each was scored for boar taint by a trained panel. Data collected: skin lesions, time of loading at farm, of arrival at slaughterhouse and of slaughter of the first boar, date of slaughter, carcass weight, lean meat percentage, meat thickness and fat thickness. For the effects on batch level separate mixed binomial models were set up for all available variables as fixed, farm as random effect and the logit (prevalence) as predicted value. The findings of this study were that the mean prevalence per farm and the range between batches can vary. In other words there are factors related to slaughter that cause variation. Slaughter batches collected in winter show lower average prevalence than fall or spring. This may be related to seasonal influence on pubertal development. We see that batches where there is more scarring or higher carcass weight there is a higher chance of boar taint. More heavily scarred boars or fatter boars have a slightly higher chance of receiving a higher score from an expert panel. Time on route and the time in the abattoir had no significant effect on boar taint prevalence.

Using genetic markers to select Canadian Duroc sires for lower boar taint levels in commercial hogs

M. Jafarikia[1,2], L. Maignel[2], F. Fortin[3], S. Wyss[2], W. Van Berkel[4], D. Cohoe[5], F. Schenkel[1], J. Squires[1] and B. Sullivan[2]
[1]*University of Guelph, Guelph, ON, Canada,* [2]*Canadian Centre for Swine Improvement, Ottawa, ON, Canada,* [3]*Centre de développement du porc du Québec, Quebec City, QC, Canada,* [4]*Western Swine Association Testing, Lacombe, AB, Canada,* [5]*Ontario Swine Improvement, Innerkip, ON, Canada; mohsen@ccsi.ca*

The feasibility of genetic selection against boar taint in sire lines to reduce levels of taint in commercial progeny was investigated. Androstenone is the main compound responsible for boar taint and in a previous study, about 50% of Duroc boars exceeded the consumer acceptance level. A total of 1,079 Duroc boars were genotyped for 97 SNP markers in candidate genes known to be involved in metabolism of androstenone and skatole, from which 60 markers had a MAF>0.05. The natural logarithm of androstenone levels in fat samples of 580 boars, weighing between 90 kg and 150 kg and less than 300 days old, were used to estimate the effects of SNPs. A two-step analysis was performed. First, the SAS GLM procedure was used to adjust phenotypes for season and the boar's age and weight at time of sampling. In a second step, residuals of the GLM procedure were used in a backward elimination in the SAS REG procedure to identify the best fitting model. The estimated effects of 17 significant SNPs were used to calculate marker-assisted estimated breeding values (MEBVs) for androstenone levels for 452 AI boars. The top and bottom 30 boars with extreme MEBVs were selected to produce commercial Duroc × Landrace/Yorkshire progeny. Average androstenone MEBVs in high and low sire groups were 1.82 and 0.94, respectively. Assuming additive effects for selected SNPs and random sampling of dams, it is expected that the difference between average levels of androstenone in commercial progeny of high and low sire groups would be 0.44 in log scale. Therefore, MEBVs for androstenone seem promising and will be validated in commercial trials.

Effectiveness of genomic prediction of boar taint components in Pietrain sired breeding populations

C. Große-Brinkhaus[1], E. Heuß[1], K. Schellander[1], C. Looft[1], J. Dodenhoff[2], K.-U. Götz[2] and E. Tholen[1]
[1]*University of Bonn, Institute of Animal Science, Group of Animal Breeding and Husbandry, Endenicher Allee 15, 53115 Bonn, Germany,* [2]*Bavarian State Research Centre for Agriculture, Institute of Animal Breeding, Prof.-Dürrwaechter-Platz 1, 85586 Poing, Germany; cgro@itw.uni-bonn.de*

As indicated by several national and European declarations it is planned to ban surgical castration of piglets without anesthesia as early as 2018. Alternatives like surgical castration with anesthesia, immune castration and fattening of entire boars are controversially discussed. Boar fattening has several advantages with respect to production. However, the risk of tainted meat has to be minimized. One possibility is to include selection for a low incidence of boar taint in breeding programs. In this study, marker effects of 35,723 SNPs on the boar taint components androstenone and skatole were estimated for 1076 Pietrain sired crossbred boars originating from five populations. These crosses reflect the heterogeneous population-structure of the Pietrain breed in Germany. Genomic BLUP was used to estimate genomic breeding values for animals in calibration and validation sets. Reliability of genomic prediction was assessed as correlation between genomic breeding value and conventional breeding values by five-fold cross validation. In a first scenario genomic prediction was performed within each population and showed reliabilities ranging from 0.25 to 0.45. In a second scenario, combining the data from the five populations resulted in slightly reduced reliabilities. Very low reliabilities (up to 0.10) were obtained when data used for calibration and for validation were from different populations. We conclude that genomic selection against boar taint using information from commercial crossbreds is promising. However, calibrations cannot be transferred between subpopulations without considerable loss of accuracy.

Fattening and carcass quality of entire male pigs as an alternative to surgical castration

I. Bahelka, P. Polák, M. Gondeková and M. Michaličková
National Agricultural and Food Centre – Research Institute for Animal Production, Institute of Animal
Husbandry Systems, Breeding and Product Quality, Hlohovecká 2, 951 41 Lužianky, Slovak Republic;
polak@vuzv.sk

Surgical castration of male piglets is a routine practice in the pig breeding aimed at elimination of boar taint. Voluntary consent of the EU countries with a stop of surgical castration from January 2018 compels the stakeholders to seek the alternatives to fattening of castrates. The aim of the study was to determine growth potencial and carcass quality of entire male pigs and to compare them with castrates and gilts. The results are the first ones obtained from testing entire males under slovak conditions. Pigs were divided to 3 groups (each of 14 animals: entire males – EM, castrates – C, gilts – G) and fed commercial diet (CP 149.81 g/kg, lysine 7.03 g/kg) *ad libitum*. They were housed in pairs (of the same sex) per pen in the control station. Test was performed from 30 to 105 kg live weight. After that, pigs were slaughtered in experimental slaughterhouse. One day after slaughter, a dissection of right half carcass of each pig was done. Entire males reached significantly higher (P<0.05) average daily gain in comparison to castrates and gilts (974 vs 890 and 854 g). It resulted in earlier achievement of slaughter weight of entire males which was significantly lower than that of castrates and gilts (159 vs 171 and 172 days). Carcass value of entire males was more better than that of other two groups. Non-castrated males had significantly lesser backfat thickness (20.56 vs 26.72 and 25.70 mm). On the other hand, entire males had higher portion of valuable meaty cuts (53.32 vs 50.90 and 52.84%) and lean meat content (59.22 vs 55.76 and 57.63%) than castrates and gilts.

Inclusion of chicory fructanes in the diet reduces fat skatole levels

M. Aluwé[1], E. Heyrman[1], S. Millet[1], S. De Campeneere[1], S. Theis[2], C. Sieland[2] and K. Thurman[3]
[1]ILVO (Institute for Agricultural and Fisheries Research), Animal Sciences Unit, Scheldeweg 68, 9090
Melle, Belgium, [2]Beneo institute, Wormser Str 1, 67283 Obrigheim, Germany, [3]Beneo animal nutrition,
Aandorenstr 1, 3300 Tienen, Belgium; marijke.aluwe@ilvo.vlaanderen.be

Main compounds responsible for boar taint, an off-odour in entire male pigs, are androstenone, skatole and indole. Chicory, inulin or fructooligosaccharides addition can reduce skatole, but results for indole are more contradictory. Hansen et al. found reduced plasma indole levels few days after chicory inclusion, but not after 6 weeks. Other studies indicate that adding chicory may stimulate the bacterial conversion of tryptophan to indole at the expense of skatole, thereby increasing indole. We therefore studied the effect of chicory fructanes on skatole and indole levels as well as the sensory evaluation of boar taint. Approximately 480 boars were allocated to 2 compartments (C1 and C2) and divided into 2 treatment groups per compartment. Until 21 days, all received the same diet. Then, depending on treatment group, animals received different finisher diets: (1) control (CONT) in C1 and C2; (2) with oligofructose (OLF: 9% Orafti® OLX) in C1; or (3) with inulin (IN: 7% Orafti® SIPX) in C2. Sensory score is based on the mean score of 4 experts by using the hot iron method (0: no to 4:strong taint). Chemical analyses of skatole and indole was performed by HS-SPME-GC/MS (ELFI, Germany). As slaughter dates varied between compartments, treatment effects were evaluated per compartment. Carcass weight did not differ between treatment groups but meat percentage increased when feeding OLF (P<0.05) or IN (P<0.1). In line with literature, OLF and IN reduced the level of skatole in fat from 180±165 to 119±185 ppb and from 265±288 to 193±262 ppb (P<0.05). In this study we found no effect on backfat indole levels. The sensory score was well correlated with indole and skatole, with r=0.51 and r=0.60 respectively. Nevertheless, mean sensory score was not significantly affected by feeding OLF or IN.

Familiarity with boar taint and previous sample affect perception by human nose methodology

E. Heyrman[1,2], S. Millet[2], F. Tuyttens[2], B. Ampe[2], S. Janssens[1], N. Buys[1], J. Wauters[3], L. Vanhaecke[3] and M. Aluwé[2]
[1]KU Leuven, Livestock genetics, Department of biosystems, Kasteelpark Arenberg 30 – bus 2456, 3001 Heverlee, Belgium, [2]Institute for agricultural and fisheries research (ILVO), Animal science unit, Scheldeweg 68, 9090 Melle, Belgium, [3]Ghent University, Department of Veterinary Public Health and Food Safety, Laboratory of Chemical Analysis, Salisburylaan 133, 9820 Merelbeke, Belgium; evert.heyrman@ilvo.vlaanderen.be

A better understanding of the sensory evaluation of boar taint by the human nose method is needed to ensure reliability when used at the slaughter line. The questions to be answered by this experiment were: (1) do training and familiarity with boar taint improve the reliability, sensitivity and specificity of assessors; and (2) is there an effect of the previous sample on the scoring of a sample? Neckfat samples were used to make three series of replicate samples in different order. One series consisted of 8 samples with boar taint and 22 without boar taint. The sample order within each series was designed to contain all possible combinations of evaluated sample and previous sample. Participants were asked to evaluate all samples using the human nose method. Participants were classified in three groups; 6 trained panelists, 6 people who have had previous contact with SKA and AND and/or boar tainted samples in general but who have had no further training, 6 people who have had no previous conscience contact with these compounds or with boar tainted samples. The unfamiliar participants gave higher average scores to both types of samples compared to trained participants, the familiar participants gave intermediate scores. Training increased reliability compared to unfamiliar participants, with intermediate reliability for the familiar group. In terms of sensitivity and specificity, comparable values can be achieved for all groups dependent on the cutoff score chosen. Even limited familiarity with boar taint seems to help in scoring tainted samples. Samples evaluated after a tainted sample resulted in a lower score.

How to present results of genomic studies in an intelligible form

H. Simianer
Animal Breeding and Genetics Group, Georg-August University, Albrecht-Thaer-Weg 3, 37075 Göttingen, Germany; hsimian@gwdg.de

Scientific presentations at conferences, both slides for oral presentations and posters, should be designed in such a way that the audience is able to understand all important aspects of the analysis and the relevant results. To pursue this goal, the following simple, but by no means always adopted, rules and guidelines can be followed: (1) Present the research question, ideally stated explicitly as a hypothesis or a set of hypotheses; (2) Give details of the used animals, materials and technologies; (3) Provide details about quality control and filtering criteria; (4) Name and describe the statistical methods you have applied. Just mentioning the used software is often insufficient because most programs allow choices among several methodological options; (5) Estimated quantities, such as effect sizes or heritabilities, should always be reported with an indication of their precision; (6) If significances are stated, report the tested hypothesis and account appropriately for multiple testing; (7) Results preferably should be presented in figures rather than tables, since quantitative differences are grasped more easily through analogue comparison; (8) Avoid the default options of graphics software. Today many packages, such as R, allow you to present results in cool and catchy graphs. Let yourself be inspired by the artwork in high ranking journals. However, 'fanciness' must not be at the cost of clarity of presentation. (9) Use clearly distinguishable colors consistently across slides. Avoid a mix of different colors, line types, and symbols in line graphs, rather restrict yourself to the minimum set of line characteristics allowing a unique identification. (10) Figures must have legends and descriptions of axes (with units) in sufficiently large fonts; (11) Finally make a clear statement about your results. If hypotheses were formulated initially, now is the time to say if they can be rejected or not.

General aspects of genome-wide association studies

P. Uimari
University of Helsinki, Koetilantie 5, 00014 University of Helsinki, Finland; pekka.uimari@helsinki.fi

Genome-wide association studies have been applied widely in livestock. Using moderate to high density SNP chips several potential chromosomal regions and genes have been linked to several traits important in animal production. Typically purebred animals with known pedigree structure are used in these studies. The dependent variable is mostly estimated breeding value with or without deregression. The choice of statistical method varies from simple t-test for single markers to the penalized regression methods for testing a multitude, possibly all, of the markers simultaneously. The most crucial aspects of the analysis are reliability of observations, adequate sample size and correction for pedigree structure. In this presentation general aspects of genome wide association study with some pitfalls are covered.

Publishing genome-wide studies in the GSA journals: GENETICS and G3: Genes|Genomes|Genetics

D.J. De Koning[1] and T. Depellegrin[2]
[1]Swedish University of Agricultural Sciences, Animal Breeding and Genetics, Box 7023, Box 7023, Sweden, [2]Genetics Society of America, 9650 Rockville Pike, Bethesda, MD 20814-3998, USA; dj.de-koning@slu.se

GENETICS and G3: Genes|Genomes|Genetics (G3), peer-edited journals with practicing scientists as editors, are published by the Genetics Society of America. While the two journals differ slightly in scope and expected impact, they both adhere to the most rigorous standards of data analysis, presentation of methods, and availability of data and tools. The level of reporting must enable readers to repeat the analysis in the same way as the authors, and to allow others to create further, more advanced, analyses. Manuscripts describing GWAS or Genomic selection require: (1) A very clear description of the quality control of your genotype data, and (if applicable) editing in your phenotype data. (2) Thoroughly describe your statistical model, including how you tested for, and handled population stratification and family relationships in the case of GWAS. (3) Describe the tool you used to analyze the data. If you used a commercial tool, best practice calls for including an example script showing how you ran the analysis. If you used home-written code to analyze the data, it's important to provide that code! (4) The same tenets apply to simulations when you present new methods for GWAS or Genomic Selection: you provide either all the simulated data files or, preferably, the code used for simulations. (5) Finally, and most crucially, include for readers all data used for analysis: genotypes, phenotypes, covariates, pedigrees, genetic maps, etc. Our data policy allows GSA journals to house a wealth of data to be further analyzed or benchmarked, which encourages new knowledge and discovery. Please help our effort to publish the most relevant genome analysis papers that are truly reproducible. DJK is Deputy Editor-in-Chief for G3 and Editor for GENETICS; TD is Executive Editor for G3 and GENETICS.

Effects of mean weight of uniform litters on sows and piglets performance

R. Charneca[1], A. Freitas[1], J.L.T. Nunes[1] and J. Le Dividich[2]
[1]*Instituto de Ciências Agrárias e Ambientais Mediterrânicas, Universidade de Évora, Apartado 94, 7002-554 Évora, Portugal, [2]INRA, 32, Avenue Kennedy, 35160 Breteil, France; rmcc@uevora.pt*

This study aimed to determine the effects of uniform litters of different mean birth weights on colostrum production of sows and piglets performance. The study involved 78 multiparous sows from a commercial lean genotype and their piglets. Simultaneous farrowings were supervised and at birth each piglet was identified, weighed (±1 g) and put in box under a heat lamp. After farrowings completion and depending on the measured weights, the piglets were then divided in experimental litters of 12 piglets each of uniform light (UL, CV=9.8%, n=27), uniform average (UA, CV=8.2%, n=23) or uniform heavy (UH, CV=8.6%, n=28) piglets and allowed to suckle. Piglets were re-weighed at 24 h and 21 d of life and deaths registered. Colostrum intake (CI) of the piglets was estimated using the Devillers et al. equation. Litter types were compared by ANOVA with batch as random factor. Mean weights of litters were all different (P<0.001), UL=1,136±23 g (SEM), UA=1,415±25 g and UH=1,649±20 g. Colostrum yield (CY) of sows was positively related to litter total weight at birth (R^2=0.30, P<0.001). The CY of UA and UH sows was similar (4.8±0.2 and 5.2±0.1 kg, respectively, P=0.32) and both higher (P<0.001) than CY of UL sows (3.9±0.1 kg). The CI of UA and UH litters was similar (400±14 and 436±12 g, respectively, P=0.31) and both higher (P<0.001) than on UL litters (335±13 g). The intra-litter CV of CI was similar between groups averaging 24%. The mortality rate of piglets until 21 d was not different between litter types averaging 9.15%. The piglets weights at 21 d were similar (P=0.11) in UA (6.4±0.2 kg) and UH (6.7±0.1 kg) litters and both higher (P=0.01) than in UL litters (5.6±0.2 kg). It was concluded that CY of sows is dependent on the total weight of the suckling litter and that the mean weight of piglets of uniform litters influences the CI and the weaning weight but not their survival.

Sow reproduction: corpora lutea characteristics as an indicator for piglet quality

S. Smits[1], N.M. Soede[1], L.J. Zak[2], M.L. Connor[3] and H. Van Den Brand[1]
[1]*Wageningen University, P.O. Box 338, 6700 AH Wageningen, the Netherlands, [2]Topigs Norsvin Research Center B.V., Schoenaker 6, 6641 SZ Beuningen, the Netherlands, [3]University of Manitoba, 201-12 Dafoe Road, R3T 2N2, Winnipeg, Canada; sascha.smits@wur.nl*

Genetic selection for female fertility in sows has focused on increasing litter size. This has led to greater variation in birth weight within a litter and lower piglet survival (i.e. low-birth weight piglets). Concomitantly, ovulation rate (OvR) also increased during this time. It is not known whether the associated increased OvR affects the quality of piglets, either because a higher OvR is associated with increased uterine crowding of foetuses or because the corpora lutea (CL) might be of lower quality. The objectives of this study were to investigate the relationships between the number and size of CL and selection for litter size, average birth weight and uniformity of piglets; and to determine genetic parameters, genomic relations and the genetic trend in OvR. The study was conducted on a genetic nucleus farm to facilitate thorough analysis of many production traits and their relations, both phenotypically and genetically. Large-White sows (≥2nd parity; n=141) of known genetics were transrectally scanned with a micro convex-array ultrasound transducer between days 22 and 35 of gestation to determine the number of CL as a measure of OvR. For each ovary the number of CL, and the diameter of the three largest CL were determined. Additionally, backfat thickness (TN-method) and body condition score were determined during gestation at time of ultrasound scanning. Preliminary results are based on 141 sows with OvR of 24.58±1.02. No differences in total number of CLs between parity number 2, 3-4 and 5-8 were found. Results indicate that litter birth weight tends to decrease with an increase in number of CL. No relationship was found between number of CL and birth weight in the sows' own birth litter. This study presents new methods and insights to study relationships between CL characteristics and production traits in the future.

Impact of increased energy and amino acids in sow lactation diets on piglet growth in large litters

A. Craig[1,2] and E. Magowan[1,2]
[1]*Queens University Belfast, IGFS, Belfast, Northern Ireland, BT7 1NN, United Kingdom,* [2]*Agrifood and Bioscience Institute, SAFSD, Hillsborough, Northern Ireland, BT266DR, United Kingdom; aleslie02@qub.ac.uk*

The EU sow herd is now highly prolific, but the wean weight of piglets has been compromised. This trial investigated the impact of tailored sow lactation diets to support a target wean weight of piglets reared in large litters. Five treatments were offered to sows (n=113) from day 108 of gestation to weaning (28 days) in a 2×2+1 factorial design. A single phase feeding regime (Diet 1: 14.4 MJ DE, 1.25% Lysine) was compared to a two phase feeding regime (Diet 1 to day 14 of lactation, followed by Diet 2 (15 MJ DE and 1.4%lysine) from day 15 to 28). A standard (0.79) or high (1.2) lysine:valine ratio was also compared. A control diet containing 13.5 MJ DE and 1.0% lysine was included. Piglets were individually weighed at birth, 14 and 28 days of age. Sow feed intake was recorded daily. Body condition scores and back fat depth at P2 was measured at farrowing and weaning. A blood sample was taken at day 21 for blood urea nitrogen (BUN). Results were analysed as a 2×2+1 using ANOVA with feed intake as a covariate. Average litter size weaned was 12.8 pigs. Average daily feed intake for sows was 7.6 kg. All experimental diets significantly (P<0.001) improved piglet daily gain from birth to weaning compared to the control. This led to an improved weaning weight of 448 g (P<0.001) per piglet, and increased (P=0.006) total litter weight weaned (110.3 kg) compared to control (102.2 kg). Two phase feeding improved piglet daily gain (P=0.009) and weaning weight (P=0.007). Lysine:valine ratio had no consistent effect. Treatment had no effect on sow backfat loss or BUN. In this study when sows had an average daily intake of 111.1 MJ DE and 31.4 g lysine they achieved an average litter weaning weight of 110.6 kg and individual piglet weaning weight of 8.6 kg. Therefore, piglets in large litters can obtain high weaning weights when the sows are offered lactation diets high in energy and lysine.

Genetic correlations between feed efficiency and longevity traits in Norwegian Landrace pigs

K.H. Martinsen[1], J. Ødegård[2], D. Olsen[3] and T.H.E. Meuwissen[1]
[1]*Norwegian University of Life Sciences, Department of Animal and Aquacultural Sciences, P.O. Box 5003, 1432 Ås, Norway,* [2]*AquaGen AS, P.O. Box 1240 Sluppen, 7462 Trondheim, Norway,* [3]*Topigs Norsvin, P.O. Box 504, 2304 Hamar, Norway; kristine.martinsen@nmbu.no*

Feed efficiency in slaughter pigs and sow longevity traits are economically important traits in pork production. The aim was to estimate genetic correlations among lean meat- and fat efficiency, stayability and body condition score in Norwegian Landrace pigs. Lean meat- and fat efficiency were calculated using an extended residual feed intake model where total feed intake in the test period was the response variable and fat (kg) and lean meat (kg) were both included as fixed and random regressions. The regression coefficients that resulted from this model represented deviations in amount of feed used pr kg lean meat and fat produced. A total of 8,161 purebred Norwegian Landrace boars from the testing station with records for feed intake, fat (kg) and lean meat (kg) were included in the analysis. In addition, 25,336 purebred Norwegian Landrace sows had records for stayability and 26,133 sows had records of body condition score. Stayability was a binary trait (1/0) and stated whether a sow was able to give birth to a second litter or not. Body condition score was registered after weaning of first ltter and was a categorical variable ranging from 1 (thin) to 5 (obese). Both stayability and body condition score were analyzed in an animal model. Significant genetic correlations was estimated between sows' body condition score after weaning of first litter and fat efficiency in slaughter pigs (0.24) and slaughter pigs overall feed intake (0.16). The results indicated that pigs with a higher body condition score were less efficient in utilizing feed for fat deposition. This may be explained by overfeeding being associated with reduced feed efficiency. Genetic correlations among lean meat efficiency, stayability and body condition score were not significantly different from zero.

Development of a new farrowing pen for individually loose-housed sows: the UMB pen

I.L. Inger Lise
Norwegian University of Lifes sciences, Animal and aquacultural sciences, P.O. Box 5005, 1432 Ås, Norway;
inger-lise.andersen@nmbu.no

The objective of the present work was to collect preliminary production data on a newly-developed farrowing pen for individually loose-housed sows and to use these results to produce a pen for commercial practice which has a high piglet survival rate and improved sow and piglet welfare. The 'UMB farrowing pen' (7.9 m^2) comprises two compartments: a 'nest area' and an activity/dunging area with a threshold in between. Forty clinically health sows (28 Australian sows and 12 Norwegian, balanced for parity), were used in the experiment. The pens in the two different countries were the same except that rubber coating was not used in the activity area in the Australian pens during summer, The pens consisted of a nest area that was separated from a dunging area with a threshold between them and hay and straw for nest-building were supplied from a hay rack. Floor heating and sloped walls were used to stimulate sows and piglets to rest in preferred locations in order to reduce the likelihood of crushing. The preliminary results show that the production results were similar to, or better than, those reported for other types of pens for individually loose-housed sows. Mortality of live born piglets was 12.1±2.9% in the Norwegian sows and 12.9±2.0% in the Australian sows whereas the number of weaned piglets in both countries was 12.1±0.4 and 9.1±0.3, respectively. Overlying and starvation were the most common causes of death and were significantly affected by parity and litter size. All sows showed a high level of communication with their piglets, and primiparous sows communicated significantly more with their newborns during the birth process than the pluriparous sows. At parturition, 33.3% of the sows were resting with the back towards the back wall whereas 41.7% rested towards the threshold. In 50% of the nursings, the sows were resting against the back wall while 30% of the nursings occurred towards the threshold. Some results from data collection in two commercial farms will be presented at the meeting.

Are the recommendations for dietary protein too high for Swiss Large White pigs?

I. Ruiz[1,2], P. Stoll[1], M. Kreuzer[2] and G. Bee[1]
[1]Agroscope, Institute for Livestock Sciences (ILS), Tioleyre 4, 1725 Posieux, Switzerland, [2]ETHZ, Universitätstrasse 2, 8092 Zurich, Switzerland; isabel.ruiz@agroscope.admin.ch

Pig production is responsible for an important portion of the nitrogen excreted by farm livestock. However, breeding efforts of the last decades improved the lean content of pig carcasses, implying that efficiency of dietary protein utilisation improved, too. The current results are part of a large study aiming to revise the nutrient composition of the whole carcass and to update the Swiss feeding recommendations for pigs. An experiment was performed with 48 entire males (EM), barrows and females, each. From 20 kg BW, they were either allocated to a control (C) grower-finisher diet according to Swiss feeding recommendations, or a low protein grower-finisher diet (R; 80% of C). Pigs had *ad libitum* access to the diets. Intake was monitored and pigs were weighed weekly. Empty body composition was determined at 40, 60, 80, 100, 120 and 140 kg BW on 4 pigs per each of the six subgroups. Pigs were stunned, exsanguinated and de-haired. Organs and empty intestinal tract from each animal were ground and homogenized together. Left carcasses were frozen until grinding. To determine the empty body composition, dry matter, protein, fat and energy content were analyzed in each blood, hair, offal and carcass samples. Energy efficiency was calculated as weight gain divided by gross energy consumed. Data were analyzed using the Anova procedure of Systat 13 considering BW, gender and diet as fixed effects. Compared to C-pigs, R-pigs were 13 and 10% less energy efficient from 20 to 60 and from 60 to 100 kg BW, respectively. These differences were even greater in EM than females and barrows. No differences were found from 100 to 140 kg BW. The empty body protein:fat ratio of R-EM was 24.6% lower (P<0.05) compared to C-EM, while in barrows and females this difference amounted to 16.6 and 14.5%, respectively. In conclusion, dietary protein could be reduced for pigs from 100 to 140 kg BW. The effect of a less severe dietary protein reduction level is currently under study.

Effect of rapeseed meal supplementation on physiological responses and reproductivity in sows
H.B. Choi, S.S. Jin, P.S. Heo, W.L. Jung, S.H. Yoo, J.S. Hong and Y.Y. Kim
*Seoul National University, Department of Agricultural Biotechnology, College of Animal Life Sciences, 1
Gwanak-ro, Gwanak-gu, 151-742, Seoul, Korea, South; nutrihong@naver.com*

The objective of this study was to evaluate the effect of rapeseed meal (RSM) levels in sow diets on reproductive performance, thyroid hormones levels of sows and performance of their progeny. A total of 55 mixed-parity (average=3.82) sows (F1, Yorkshire × Landrace) with an initial body weight (BW) of 193±1.22 kg. Sows were allotted to 1 of 5 treatments based on BW and backfat thickness (BFT) with 11 replicates in a completely randomized design. The 5 experimental treatments differed according to the dietary RSM supplementation levels (0, 3, 6, 9 or 12%) in the gestating diets. During lactation, all sows were fed a diet without RSM. The contents of total glucosinolates and erucic acid in the RSM were 47.30 μmol/g and 4.25 mg/g DM, respectively. There were no significant differences in BW, BFT, weaning to estrus interval, and growth performance of the offspring when different levels of RSM were provided. In addition, no difference among dietary treatments was observed with respect to average daily feed intake of sows. However, serum triiodothyronine (T3) concentration was affected by dietary treatments (quadratic response, $P<0.05$) and serum thyroxine (T4) concentration was markedly higher (linear response, $P<0.01$) when sows were fed the diet containing 12% RSM at 110 d of pregnancy. Consequently, sows fed gestating diets containing up to 12% RSM had affected serum T3 and T4 concentrations prior to farrowing, but there were no detrimental effects on BW, BFT, litter size and growth performance of their progeny.

Effects of dietary energy levels on physiological response and reproductivity of gestating gilts
S.S. Jin, J.C. Jang, S.W. Jung, J.S. Hong and Y.Y. Kim
*Seoul National University, Department of Agricultural Biotechnology, College of Animal Life Sciences, 1
Gwanak-ro, Gwanak-gu, 151-742, Seoul, Korea, South; nutrihong@naver.com*

This experiment was conducted to investigate the effects of increasing dietary energy levels in gestation on physiological parameters and reproductive performance of primiparous sows. A total of 52 F1 gilts (Yorkshire × Landrace) were allocated according to a completely randomized design to 1 of 4 treatments. The 4 experimental diets contained 3,100, 3,200, 3,300 or 3,400 kcal of ME/kg. As daily diet allowance during gestation was 2.0 kg, ME intake of gestating gilts amounted to 6,200, 6,400, 6,600 and 6,800 kcal of ME/kg, respectively. Other nutrients met or exceeded the requirements of NRC (1998). During the gestation period, body weight ($P<0.05$) and weight gain ($P<0.01$) of gilts increased as dietary energy level increased (linear response, $P<0.05$), but the backfat thickness was not affected by dietary treatments. Higher birth weight of progeny was observed when sows were fed high energy diets ($P<0.05$) but there was no significant difference in the total number of pigs born per litter. In the lactation period, voluntary feed intake of sows tended to be decreased when higher energy treatment diet was provided during gestation ($P<0.10$). Backfat thickness of sows was not affected by dietary treatments, but body weight (linear response, $P<0.05$) at 21 d of lactation and weight gain (linear response, $P<0.01$) linearly decreased as dietary energy level increased. These results demonstrated that sows fed high energy diets during gestation produced heavier birth weight offspring, but there were no detectable positive effects on reproductive performance of primiparous sows. Consequently, considering sow performance during lactation, 3,200 kcal of ME/kg is thought to be an adequate energy requirement of gestating gilts.

Antioxidant nutrition in dairy ruminants

A. Baldi and L. Pinotti
Department of Health, Animal Science and Food Safety, Università degli Studi di Milano, 20133 Milano, Italy; luciano.pinotti@unimi.it

Over the past decades, a number of studies investigated the importance of an adequate antioxidant status to sustain both animal health and production. In this context, the concept of oxidative stress is becoming more and more important in medical and nutritional research. Oxidative stress occurs when the generation of pro-oxidant agents, exceeds their safe detoxification by antioxidant mechanisms. Nutrition has a major influence on pro-oxidant/antioxidant balance. Several papers addressed whether antioxidants supplementation in dairy ruminants affected immune function during critical periods. The definition of antioxidants bioactivity and bioavailability is a very important issue, because it dictates the effective amounts of antioxidant requirements. Antioxidant absorption and bioavailability are influenced by several factors like structural diversity of compounds that exert antioxidant activity, the feed matrix structure and the coexistence of different nutrients. It is generally accepted that the naturally occurring form is more biologically active. However, high cost and stability in animal feeds, limit the use of natural antioxidants as feed additives. Some components of the antioxidant system are micronutrients (vitamin E, C, carotenoids); others require dietary micronutrients (Se, Zn) and dietary supply of these compounds is important in protecting tissues against free radical damage. The use of sensitive and specific oxidative stress biomarkers (e.g. superoxide dismutase, glutathione peroxidase) is essential to evaluate and monitor the redox-balance. Research on improvement of antioxidant status by nutrition in ruminants has been mainly concerned with studying the effect of one or few selected dietary antioxidants on performance and health. Some epidemiological studies also carried out retrospective analysis on the association between dietary antioxidants and disease occurrence in ruminants. These evidences seem indicate that antioxidant components work in concert and the activity of individual antioxidants should not be considered in isolation.

Effects of SOD-rich melon on inflammation, oxidative status and performance of challenged piglets

A.S.M.L. Ahasan[1], A. Agazzi[1], F. Barbe[2], G. Invernizzi[1], F. Bellagamba[1], C. Lecchi[3], G. Pastorelli[1], V. Dell'orto[1] and G. Savoini[1]
[1]Università degli Studi di Milano, Dipartimento di Scienze Veterinarie per la Salute, la Produzione Animale e la Sicurezza Alimentare, via G. Celoria 10, 20133 Milano, Italy, [2]Lallemand, SAS, 19, rue Briquetiers, Blagnac. BP 59, 31702, France, [3]Università degli Studi di Milano, Dipartimento di Patologia Animale, Igiene e Sanità Pubblica Veterinaria, via G. Celoria 10, 20133 Milano, Italy; alessandro.agazzi@unimi.it

The effects of a SOD-rich melon supplement (Melofeed®, Lallemand, France) were evaluated on inflammation, oxidative status and performance of challenged piglets. Forty-eight 24d-old piglets were individually allocated to 2 treatments for 29 days: (1) basal diet (C); (2) C plus Melofeed® (T; 30 g/ton of complete feed). From day 19 half of the animals were subjected (+) or not (-) to intramuscular injections of increasing dosage of LPS (*Escherichia coli* serotype 055:B5). Performances were evaluated weekly and blood samples collected on day 0, 19, 21, 23, 25, 27 and 29 for total antioxidant activity (TAOC), SOD, ROS, haptoglobin, IL-1β, IL-6, and TNF-α. On day 19, 25 and 29 kit radicaux libres and 8-hydroxy-2'deoxyguanosine were evaluated. Performance were analysed by a GLM procedure of SAS using pre-challenge period as covariate. Antioxidant parameters and interleukins were analysed by a MIXED procedure for repeated measurements. The piglet was the experimental unit as random effect nested with the treatment × challenge effect. LPS affected growing performance ($P<0.01$), proinflammatory cytokines ($P<0.01$) and haptoglobin ($P=0.03$), while Melofeed® increased body weight, gain, ADG ($P=0.05$), and feed intake ($P<0.01$) during challenge. TAOC was higher in T ($P<0.01$) with a tendency over – and + piglets ($P=0.08$). A significant improvement of red blood cells resistance to haemolysis was observed in T ($P\leq0.01$). These results suggest that Melofeed® in the diet of post weaning piglets might increase total antioxidant activity and growth performance during the LPS challenge period.

Slowly fermentable grains may reduce metabolic heat and ameliorate heat stress in grain-fed sheep

P.A. Gonzalez-Rivas[1], K. Digiacomo[1], V.M. Russo[2], B.J. Leury[1], J.J. Cottrell[1] and F.R. Dunshea[1]
[1]Faculty of Veterinary and Agricultural Sciences, The University of Melbourne, Parkville, Victoria, Australia, [2]Department of Economic Development, Jobs, Transport and Resources, Ellinbank, Victoria, Australia; gonzalezp@student.unimelb.edu.au

Rumen fermentation has direct relation with the metabolic heat increment in ruminants and the risk of heat stress (HS). Therefore, feeding slowly fermentable grains may reduce heat load and the impact of HS in grain-fed sheep. An *in vitro* gas production technique was used to determine the gas production kinetics of 28 wheat and maize grain samples incubated in buffered rumen fluid at 39 °C during 24 h. Then, 22 Merino × Poll Dorset wethers were housed in 2 climate-controlled rooms and were fed either maize grain plus forage (MF) or wheat grain plus forage (WF) during 3 experimental periods: P1) 7 d of thermoneutral conditions (18-21 °C/26-30% relative humidity (RH)) and restricted feed intake (1.28 times maintenance); P2) 7 d of HS (28-38 °C/40-50% RH) and restricted feed intake; and P3) 7 d of HS as P2 with unrestricted feed intake in a complete randomized block design. Rectal temperature (RT), heart rate (HR), respiration rate (RR) and left and right flank skin temperature (LST, RST) were measured four times a day. Gas production curves were fitted to a Gompertz model, gas kinetic and physiological parameters were analysed by the Restricted Maximum Likelihood. Maize grain had a slower (-15%, P<0.001) rate of gas production reaching maximum fermentation activity later than wheat (6.8 vs 8.7 h P<0.001). All physiological parameters were elevated (P<0.001) during HS especially during P3. RR, RT, LST, RST and HR were lower (P<0.05) in sheep fed MF, particularly during HS and sheep feed WF had a larger (P<0.001) difference between LST and RST. These data confirm the slower *in vitro* fermentability of maize compared to wheat and that feeding maize reduces the metabolic heat increment and ameliorates the physiological responses negatively affected by HS in grain-fed sheep.

Effects of feeding frequency on reproductive performance and stress response in gestating sows

S.W. Jung, J.S. Hong, S.S. Jin, J.C. Jang and Y.Y. Kim
Seoul National University, Department of Agricultural Biotechnology, College of Animal Life Sciences, 1 Gwanak-ro, Gwanak-gu, 151-742, Seoul, Korea, South; nutrihong@naver.com

This study was conducted to evaluate the effects of feeding frequency on reproductive performance and stress response of sows. A total of 20 F1 multiparous sows (Yorkshire×Landrace) were allotted to one of two treatments: (1) once daily feeding (OF) and (2) twice daily feeding (TF), in a completely randomized design (CRD) based on their parity, body weight (BW), backfat thickness (BF) and weaning to estrus interval (WEI). Gestating diet with 3,265 kcal of ME/kg, 12.90% of CP, and 0.75% of lysine was provided at 2.2 kg for the 2nd parity and 2.4 kg for the 3rd parity during pregnancy, while a same lactating diet with 3,265 kcal of ME/kg, 16.80% of CP and 1.08% of lysine was provided *ad libitum* regardless of treatment. In gestation, BW, BF and BF gain were not affected by feeding frequency, but OF had higher BW gain in day 35 to 90 of gestation (P<0.05). In lactation, there were no significant differences in BW gain, BF gain, ADFI and WEI. Although litter and piglet performance were not affected by feeding frequency, litter weight at birth was significantly higher in OF (P<0.05). There were no significant differences in chemical composition of colostrum and milk as well as IgG in sow colostrum and piglet serum between treatments. Average daily water consumption on overall gestating period was significantly lower in OF treatment (P<0.05). While there were no significant differences in stereotypic behaviors and salivary cortisol levels during gestation, OF treatment showed significantly lower sow activities at day 105 of gestation (P<0.05). In conclusion, OF had no negative influence on sow reproductive performance and reduced the stress-related responses of sow in gestation compared to TF. Moreover, sows showed less hunger during pregnancy when feed was provided once.

Performance of raters to assess locomotion in dairy cows

A. Schlageter-Tello[1], P.W.G. Groot Koerkamp[1], E.A.M. Bokkers[2] and K. Lokhorst[3]
[1]Wageningen Univertity, Farm Technology Group, P.O. Box 317, 6700 AH Wageningen, the Netherlands, [2]Wageningen University, Animal Production Systems Group, P.O. Box 338, 6700 AH Wageningen, the Netherlands, [3]Livestock Research Wageningen UR, P.O. Box 338, 6700 AH Wageningen, the Netherlands; peter.grootkoerkamp@wur.nl

In lameness research, five-level locomotion scores are used. The objective of this study was to determine the performance of raters to assess locomotion in dairy cows and to study relations between five traits (arched back, asymmetric gait, head bobbing, reluctance to bear weight and tracking up) and locomotion in dairy cows. Locomotion and traits were scored on a five-level ordinal scale. Ten experienced raters scored 58 video records of different cows in two scoring sessions. Intrarater and interrater reliability and agreement were calculated as weighted kappa coefficient (κw), percentage of agreement (PA) and specific agreement for individual levels of the scale (SA). The relation between locomotion score and traits was estimated by a logistic regression aiming to calculate the size of the fixed effects on the probability of scoring a cow in one of the five levels of the scale. Intrarater reliability ranged from κw=0.63-0.86 whereas intrarater agreement ranged from PA=60.3-82.8%. Interrater reliability ranged from κw=0.28-0.84 and interrater agreement ranged from PA=22.6-81.8%. The intrarater SA were 76.4% for level 1, 68.5% for level 2, 65.0% for level 3, 77.2% for level 4 and 80% for level 5. Interrater SA were 64.7% for level 1, 57.5% for level 2, 50.8% for level 3, 60.0% for level 4, and 45.2% for level 5. All traits were significantly related to locomotion, being the most important reluctance to bear weight, asymmetric gait and arched back. Interactions between rater and session, and rater and tracking up were significant. In conclusion, experienced raters showed a variable performance to assess locomotion and locomotion traits in dairy cows. Variability of raters should be considered for on-farm utilization of locomotion scoring.

A lipopolysaccharide challenge in young piglets to quantify immune competence

A. De Greeff[1], J. Allaart[2], C. De Bruijn[3], S.A. Vastenhouw[1], L. Ruuls[1], D. Schokker[4], P. Roubos[1,2], M.A. Smits[1,4] and J.M.J. Rebel[1]
[1]Central Veterinary Institute, P.O. Box 65, Lelystad, the Netherlands, [2]Nutreco Research & Development, Veerstraat 38, Boxmeer, the Netherlands, [3]Swine Research Center, Boxmeerseweg 30, St. Anthonis, the Netherlands, [4]Wageningen Livestock Research, P.O. Box 16, Wageningen, the Netherlands; astrid.degreeff@wur.nl

General health and immune status (immune competence) depend on intestinal health of animals and are determined by parameters like nutrition, genetic background, management, and housing conditions. It is difficult to quantify immune competence, and thus to measure effects of dietary interventions on health. In general, dietary interventions will not yield radical improvements, only small incremental changes can be expected. Therefore, quantifiable parameters are required to determine effects of feed interventions on immune competence. In a pilot experiment we used a lipopolysaccharide (LPS) challenge to evaluate immune competence parameters of pigs. Piglets were challenged intraperitoneally with different doses of LPS (5, 10, 25 µg) 4 days after weaning to determine the optimal dose and time-line. Blood was collected before LPS challenge and 3 and 6 hours after the challenge to determine cytokine profiles (TNF-α, IL-6 and IL-10), rectal temperature (RT)was recorded at the same time-points. Feed intake was determined 4 and 24 hours after challenge. A dose-dependent response was observed in piglets: RT increased, feed intake was reduced and a systemic proinflammatory immune response was induced. Piglets recovered within 24-48 hours after the challenge. Challenge with 25 µg LPS was considered optimal. In a follow-up experiment, the effect of early life feed interventions on the outcome of the LPS challenge were determined. Three nutritional-interventions (medium chain fatty acids, galacto-oligosaccharides and yeast beta-glucans) were included, that were either fed to sows (in lactation feed), or directly to neonatal piglets before weaning (by oral gavage). The results of this trial will be presented and discussed.

Effect of yeast supplementation on cubicle-housed dairy cows with induced bouts of acidosis
V. Ambriz-Vilchis[1], N. Jessop[2], R. Fawcett[2], D. Shaw[1] and A. Macrae[1]
[1]R(D)SVS and The Roslin Institute, The University of Edinburgh, EH25 9RG, Scotland, United Kingdom,
[2]The University of Edinburgh, KB, EH9 3FE Scotland, United Kingdom; v.ambriz-vilchis@sms.ed.ac.uk

Sub-acute rumen acidosis (SARA) is of concern to the dairy industry. Several dietary interventions have been used to counteract SARA e.g. yeast. However benefits have been inconclusive. The present study aimed to evaluate the effect of yeast supplementation on performance, rumen pH and rumination of dairy cows with induced episodes of SARA. Fourteen cows were selected and allocated to two groups of seven cows each. Cows were offered a PMR (1st cut grass silage 44.9%, wholecrop wheat silage 17.6%, 2nd cut grass silage 15.6%, dairy meal 18.5% and molasses 3.4%), with additional concentrate fed to yield. In addition, the cows were fed 2.5 kg of ground wheat per day to induce bouts of acidosis. Cows were then supplemented with yeast (4 g or 50 billion cfu/d cow day VistaCell AB Vista, UK) following a cross-over design. The trial consisted of 4 periods, each lasting four weeks: three weeks to adapt to the change in diet, and all measurements were recorded on the fourth week. A rumination collar (Qwes-HR Lely, UK) was fitted to each cow, and they received a rumen bolus (WellCowTM) to record pH. Milk yield and characteristics were determined. Acidotic bouts were determined as total minutes rumen pH$\leq$5.8. Rumination was determined as minutes per day. Data was analysed using a mixed effect model for repeated measures. Bouts of pH lower than 6 were obtained in all animals. No statistically significant differences were observed (pH: ny=543 y=482 min/d; rumination: ny=407 y=406; milk: ny=32 y=32; fat: ny 4.65 y=4.51; protein: ny=3.54 y=3.52 and lactose: ny=4.46 y=4.43). The minimal response to yeast supplementation could be attributed to stage of lactation, as positive effects of yeast have been reported on early lactation cows. An arithmetical difference was observed for rumen pH, which could be explained by the varied response observed per individual cow.

Effect of Arg and Gln supplementation of does on translocation and immune response of their litters
R. Delgado, N. Nicodemus, R. Abad, D. Menoyo, J. García and R. Carabaño
Universidad Politécnica de Madrid, Producción Agraria, Ciudad Universitaria s/n, 28040 Madrid, Spain;
rosa.carabano@upm.es

A control diet was formulated to meet nutrient recommendations of rabbit does: 18.6 CP, 1.13 arginine (Arg), 0.93 lysine, 0.33 methionine and 3.08 glutamine (Gln) %DM. Another 3 diets were formulated by adding 0.4% Arg, 0.4% Gln or 0.4% Arg + 0.4% Gln to the control diet. Six days after parturition one kit per litter (6 litters/diet) was slaughtered after milking. The mesenteric lymph nodes (MLN) were excised under sterile conditions and analyzed for cultures of aerobes, anaerobes and facultative anaerobes bacteria. In addition, the gene expression of selected cytokines (IFNg, iNOS, IL2, IL6, IL8 and IL10) were analyzed in the appendix by real time RT-PCR at 6, 25 and 35 d of age (6 rabbits/diet). At 6 d of age, translocation of aerobes, anaerobes and facultative anaerobes microorganisms were observed regardless the diet (5.73, 5.2, and 7.84 cfu/mg MLN, respectively). Kits from rabbit does supplemented with Gln tended to reduce the total number of both aerobic (2.62 vs 6.40 cfu/mg MLN; P=0.091) and facultative anaerobic bacteria (2.65 vs 6.49 cfu/mg MLN; P=0.10) than those fed unsupplemented Gln diets. Expression of IL2, IL6, and IL10 increased from 6 to 25 d of age (P<0.001). At 35 d of age, the expression decreased (IL6 and IL2) or was similar (IFNg, IL8 and IL10) to that reported at 25 d. Kits from does supplemented with 0.4% Arg, 0.4% Gln and 0.4% Arg+0.4% Gln tended to increase the expression of IL10 at 25 d and decreased it at 35 d of age compared with those from control group (P=0.061). Expression of IL2 increased in kits from does supplemented with Gln at 25 and 35 d of age compared with those fed unsupplemented Gln diets (P=0.017). These results suggest that kits from does supplemented with dietary Gln improved their gut barrier function enhancing also their immune response against bacterial translocation.

Associations of rumen structure, function, and microbiology with feed efficiency in beef cattle

S. Lam[1], J. Munro[2], J. Cant[1], L. Guan[3], M. Steele[3], S. Miller[1,4] and Y. Montanholi[1,2]
[1]University of Guelph, Guelph, ON, Canada, [2]Dalhousie University, Truro, NS, Canada, [3]University of Alberta, Edmonton, AB, Canada, [4]AgResearch, Puddle Alley, Mosgiel, New Zealand; slam02@uoguelph.ca

The rumen exhibits high metabolic expenditure and constitutes over 30% of the digestive tract mass. Our aim was to study associations of rumen fluid parameters and rumen tissue morphology with feed efficiency (residual feed intake, RFI; kg/d). Forty-eight crossbred beef cattle, from two populations of 16 steers each (708±61 kg; 640±83 kg) and one population of 16 bulls (663±72 kg), were assessed for productive performance (body weight (BW), body composition via ultrasound and individual feed intake). Using esophageal tubing, rumen fluid was sampled to compare bacterial and volatile fatty-acids (VFA) profiles. Loggers to measure ruminal pH (RpH) and temperature (RT) were inserted into the ventral-sac of the rumen 5 to 8 d prior slaughter. At slaughter, tissue samples were harvested from the ventral-sac, dorsal blind-sac, and ventral blind-sac and processed for histomorphometrics. RpH and RT were used to determine circadian patterns and RpH ranges (indicating poor to optimal pH). Empty rumen weight (RW; %BW) was also determined. Cattle were grouped into the 50% most- (HE) and 50% least-feed efficient (LE) cattle based on RFI. Mean comparisons for RFI groups were done using the GLM procedure. RpH and RT circadian measures were done using the MIXED procedure in SAS. RFI values for HE and LE were -1.82±0.3 and 1.87±0.3 (kg/d; P=0.01), and for RW were 1.57±0.1 and 1.52±0.1 (BW%; P=0.24), respectively. RpH and RT hourly circadian patterns and time (%) within RpH ranges did not differ (P>0.10) between RFI groups, suggesting that feed efficiency may be determined by post-absorptive metabolism. However, further assessment of RpH and RT in response to feeding bouts over the circadian period, rumen epithelium histology, rumen bacteria diversity and VFA profile, may provide further understanding of the variation on feed utilization.

Microbiote activity and immunity of horses fed with scFOS and subjected to an EHV1-EHV4 vaccination

E. Apper[1], L. Faivre[2], A.G. Goachet[2], F. Respondek[1] and V. Julliand[2]
[1]Tereos, Research and Development, Z.I. Portuaire, 67390 Marckolsheim, France, [2]Agrosup Dijon, 26 Boulevard Dr Petitjean, 21000 Dijon, France; emmanuelle.apper@tereos.com

Short-chain fructo-oligosaccharides (scFOS) modify intestinal microbiote activity and immune response in several species, including after a stress. No investigation has been undertaken in horses while they are very sensitive to stressful conditions. Objectives of the study were to measure simultaneously the potential changes of the intestinal microbiote activity and the immune system of horses supplemented with scFOS and subjected to an equine herpes virus 1 and 4 (EHV1 and EHV4) vaccination. Eight horses were used. They were fitted with cannulas in caecum and right ventral colon. They were fed with a basal diet supplemented either by 30 g/day maltodextrine (n=4) or scFOS (n=4) according a longitudinal study. After 22 days of supplementation, horses received an intramuscular vaccination against EHV1-EHV4. Digestive and immune parameters were evaluated from intestinal contents, blood and saliva samples collected before the dietary supplementation (D-1), the day before (D21) and 21 days after the immune challenge (D43). A methodology was adapted to quantify Immunoglobulin A (tIgA) in intestinal content, based on method used to determine blood IgA concentrations. Vaccination increased monocytes and neutrophil polynuclears numbers and salivary and serum tIgA concentrations. It also increased caecum propionate and acetate concentrations. Dietary scFOS supplementation increased serum tIgA. The present study provided a new method to assess tIgA in intestinal content. It highlighted that vaccination resulted not only in an immune response but also in a modification of intestinal microbiota activity while scFOS affected immune parameters of horses. Measuring tIgA in intestinal content allows a new approach for studying horse immunity. Understanding scFOS mechanisms need further research.

Changes in pH and microbial composition in the ruminal and reticular fluids of SARA cattle
S. Sato, R. Nagata, A. Ohkubo and K. Okada
Iwate University, Cooperative Department of Veterinary Medicine, Faculty of Agriculture, 3-18-8 Ueda,
Morioka, Iwate, 020-8550, Japan; sshigeru@iwate-u.ac.jp

Objectives; There are few reports on the relations between ruminal and reticular pH in cattle with subacute ruminal acidosis (SARA). The objective of this study was to reveal the relations between ruminal and reticular fluid, pH and microbial composition were investigated in SARA cattle. Materials and Methods; Four rumen-cannulated Holstein steers (age, 6-10 months; weight, 180-200 kg) were used. The steers were fed hay or a SARA-inducing diet (hay:concentrate, 2:8) for 7 days. The experiment was performed twice. A wireless radio-transmission pH-sensor (YCOW-S; DKK-Toa Yamagata, Japan) was placed in the ventral sac of the rumen and in the reticulum, and the pH was measured every 10 min. Ruminal and reticular fluid were collected at 20:00 of 7 days after each feeding. Microbial compositions were analyzed by 16S rRNA gene pyrosequencing and real time PCR using the bacterial DNA extracted from fluids. Results; 24-hr mean pH of the ruminal and reticular fluids were decreased by the SARA diet, and of the ruminal fluid was significantly (P<0.05) lower than that of the reticulum in both hay and SARA diets. Circadian changes in 1-hr mean pH of the ruminal and reticular fluids were similar in each feeding periods, however, the ruminal pH was lower than that of the reticulum. Microbial composition in the ruminal and reticular fluids showed same patterns at hay feeding periods, and numbers of some bacteria in the both fluids were almost same at hay and SARA diet feeding periods. Conclusions; Ruminal pH and microbial composition were similar to that of the reticulum. Therefore, the reticular pH might be used to detect SARA cattle, as opposed to using the ruminal pH. Further studies are needed to elucidate the correlations among the pH, VFA and microbial composition between the rumen and reticulum.

Effect of probiotic and prebiotic on performance, blood and intestine traits of laying hens
A. Zarei, M. Porkhalili and B. Gholamhosseini
Islamic Azad University-Karaj Branch, Animal Science, College of Agriculture,Azadi st, Eram blvd.,
Mehrshahr, Karaj, 318764511, Iran; kaa.zarei@gmail.com

In this experiment, sixty Hy-Line (W-36) laying hens were selected in 40 weeks of age. Experimental diets were consumed for 12 weeks duration by them. The experimental design was completely randomized block including four treatments and each of them with five replications and three samples in each replicate. Treatments were as follow: Basal diet, basal diet + probiotic, basal diet + prebiotic and basal diet + probiotic+ prebiotic. Performance traits were measured such as: hen production, egg weight, feed intake, feed conversion ratio, shell thickness, shell strength, shell weight, haugh unit, yolk color and yolk cholesterol. Blood parameters like; Calcium, cholesterol, triglyceride, VLDL and so morphological of intestine were determined. Results showed; shell weight was significantly greater than other treatments in probiotic treatment. Yolk weight in prebiotic treatment was significantly greater than other treatments. The ratio of height of villi to depth of crypt cells in duodenum, jejunum, ileum and cecum in probiotic treatment were significantly greater. Results from the other traits were not significant between treatments.

The use of whey in poultry industry
V. Tsiouris and E. Sossidou
Hellenic Agricultural Organization-DEMETER, Institute of Animal Research, Thermi, Thessalloniki, Greece; sossidou.arig@nagref.gr

The use in poultry diets is compromised by its high water, lactose and sodium contents, although it has been successfully used in animal feed. However, lactose, the main sugar of whey, is used as a readily available substrate for fermentation by intestinal lactic acid in broiler chicks. Lactic acid reduces the pH in the intestine and could suppress the growth of intestinal pathogens. *Campylobacter* spp. is an intestinal pathogen and is recognized as the leading cause of bacterial foodborne diarrheal disease worldwide. In EU, 9 million citizens are infected every year and the cost exceeds 2.4 billion euro. Since, 60-80% of chickens are a natural host for *Campylobacter* spp., the implementation of *Campylobacter* spp. control measures at the primary production level is of vital importance for the reduction of contamination of broiler meat and subsequently human exposure. A trial will be undertaken early May 2015 to investigate the effect of the whey on broiler chicks' performance, welfare and caecal *Campylobacter* counts. Briefly, 100 day-old broiler chicks will be randomly allocated into four treatment groups: group N, as negative control, group W, in which birds will consume whey, group P, in which birds will be experimentally challenged, and group WP, in which birds will consume whey and be challenged. Body weight, feed consumption, average daily weight gain, feed conversion ratio and European Production Efficiency Factor will be calculated and mortality will be recorded. Finally, one caecum from each bird will be used for microbiological analysis of *Campylobacter* spp. The trial is part of a research project funded under the Action 'Research & Technology Development Innovation Projects, AgroETAK', MIS 453350, in the framework of the Operational Program 'Human Resources Development'. It is co-funded by the European Social Fund through the National Strategic Reference Framework (Research Funding Program 2007-2013) coordinated by the Hellenic Agricultural Organization – DEMETER.

RNA level of selected protein or peptide genes in goat milk somatic cells supplemented with Se-yeast
E. Bagnicka[1], D. Reczyńska[1], J. Jarczak[1], M. Czopowicz[2], D. Słoniewska[1] and J. Kaba[2]
[1]Institute of Genetics and Animal Breeding, Polish Academy of Sciences in Jastrzębiec, Department of Animal Improvment, Postepu str. 36A, 05-552 Magdalenka, Poland, [2]Warsaw University of Life Sciences, Faculty of Veterinary Medicine, Division of Infectious Diseases and Epidemiology, Nowoursynowska 159c, 02-766 Warsaw, Poland; e.bagnicka@ighz.pl

The aim of the study was to evaluate the effect of inorganic (sodium selenite) vs organic selenium (Se-yeast, *Saccharomyces cerevisiae*) supplementation of Polish dairy goats on expression of selected genes. The experiment was conducted on 24 Polish dairy goats, which were equally divided into two analogous groups according to their parity. The control group was fed a sodium selenite-supplemented diet, while the experimental one was fed a diet supplemented with Se-yeast in amount of 0.6 g/day/goat. The experiment started three weeks after kidding. The expression of genes was established in milk somatic cells isolated from milk samples collected 4 times during lactation (21st, 70th, 120th, 180th day after kidding during the morning milking). The transcript levels of alphaS1 casein (CSN1S2), casein κ (CSN3), interleukin 8 (IL-8), serum amyloid A3 (SAA3), interleukin 1β (IL-1β) and bactenecin 7.5 (Bac7.5) were measured. The supplementation of Se in organic binding caused an increase of CSN1S2 and decrease of IL-8 and Bac7.5 gene expressions. The expression of the other genes did not differ between groups. Simultaneous increase of CSN1S2 and decrease of IL-8 and Bac7.5 gene expressions indicate that Se in organic binding may have positive influence on milk protein synthesis and health status of the mammary gland. Research was realized within the project 'BIOFOOD – innovative, functional products of animal origin' no. POIG.01.01.02-014-090/09 co-financed by the European Union from the European Regional Development Fund within the Innovative Economy Operational Programme 2007-2013.

Colostrum, milk yield and quality of cows supplemented with flaxseed during the dry period

N. Guzzo[1], R. Mantovani[1], L. Da Dalt[2], G. Gabai[2] and L. Bailoni[2]
[1]Dept. of Agronomy Food Natural Resources Animals and Environment, Viale dell'Universita', 16, 35020 Legnaro (PD), Italy, [2]Dept. of Comparative Biomedicine and Food Science, Viale dell'Universita', 16, 35020 Legnaro (PD), Italy; lucia.bailoni@unipd.it

This study has aimed at analyzing the effects of supplementing high yielding dairy cows with flaxseed during the dry period on the colostrum, milk yield and quality during the first month of the subsequent lactation. At the beginning of the dry period, 73 Holstein cows in the same parity, days open and production level in previous lactation, and with a close date of drying off were randomly assigned to a control (CTRL; n=35), or to an experimental flaxseed supplement dry-off diet (FLAX; n=38; 200 g/head/day). Cows were fed isoenergetic and isoproteic dry-off diets (12.3% CP, 48.8% NDF, and 0.78 milk feed units/kg on DM). After calving, a single lactation diet containing 16.7% CP, 34.0% NDF, and 0.97 milk feed units/kg on DM was given to cows. Milk yield was measured daily, and colostrum and milk samples were collected at about d 4, 15 and 30 d after calving. Milk samples were analyzed for fat, protein, casein, and MUN. All variables were analyzed using a hierarchical mixed model for repeated measures. The patter of milk yield in the first month of lactation was similar comparing CTRL and FLAX groups, particularly for the first 15 d. The overall milk yield up to 30 d was 1,043 and 1,054 kg (P=0.881) for CTRL and FLAX groups, resp. Colostrum and milk quality were not different comparing the 2 experimental groups immediately after calving or until 30 d. As expected, fat, protein, and casein levels were greater (P<0.001) in colostrum than in milk (i.e. 6.27 vs 3.97%; 4.42 vs 3.13% and 3.38 vs 2.47%; resp.). No differences in MUN were detected between CTRL and FLAX groups, but an increase from 15.5 to 20.4 mg/dl (P<0.001) was observed in colostrum and milk, resp. Flaxseed supplementation of high yielding cows during the dry off did not modify colostrum, milk yield and quality in the subsequent lactation.

Effect of prepartum pH and concentrate levels on reticuloruminal-pH levels in dairy cows

A. Steinwidder[1], M. Horn[2], R. Pfister[1], H. Rohrer[1] and J. Gasteiner[1]
[1]Agricultural Research and Education Centre, AREC Raumberg-Gumpenstein, Raumberg 38, 8952 Irdning, Austria, [2]University of Natural Resources and Life Sciences, Sustainable Agricultural Systems, Gregor-Mendel-St. 33, 1180 Wien, Austria; marco.horn@boku.ac.at

In the study we investigated: (1) the effects of two concentrate levels (Con, Low) on reticuloruminal pH values; and (2) the effects of prepartum pH values on postpartum pH values of lactating cows receiving no concentrate before parturition. An indwelling wireless data transmitting system for continuous pH measurement was given to 9 heifers and 11 cows orally 2 weeks before expected calving and continuously measured pH from between week 2 prepartum and week 6 postpartum. Prepartum all animals were fed hay and grass silage only. During the first 36 days in milk concentrate supplementation increased from 2 to 7.5 kg DM and from 1 to 3.7 kg DM for groups Con and Low, respectively. From day 36 onwards concentrate supply depended on milk yield. The dataset was analysed using a mixed model. Before parturition no significant effects of week on mean pH was found but pH values varied between the animals. During the last week before parturition the median, lower and upper quartile values of the mean pH values were 6.47, 6.41 and 6.59 for heifers and 6.29, 6.19 and 6.39 for cows, respectively. After parturition no diet effect on mean pH and max. pH values (6.35 and 6.67, respectively) was found. A significant (P<0.01) and strong correlation (r>0.8) between mean pH value before parturition and reticuloruminal pH values after parturition was found. Animals having lower pH levels before parturition continued to have lower mean pH and min. pH values during the studied period of lactation. Furthermore they had stronger pronounced short term fluctuations of H3O+-ion concentrations and a longer time span with pH values below 6.2. The results support the theory of the existence of cow-specific baselines concerning rumen environment, fermentation and metabolism and emphasise the importance of further research on this topic.

Investigation of the status of health management in dairy farms –case study: Kashmar, Iran
R. Vakili and A. Alizadeh
Islamic Azad University,Kashmar Branch, Animal Science, Seyed morteza blvd, Kashmar, Iran; rezavakili2010@yahoo.com

The purpose of this study was to assess the status of health management in dairy farms studied and compared them with existing standards (Case study Kashmar, Bardaskan, KhalilAbad) In this research to collect data from two questionnaires were used and Check Lists. The population of the study consisted of dairy cattle farms (industrial, semi-industrial and traditional) are active, which were selected by simple random sampling. Data collected at regular Excel and then to the questionnaire were scored. – The data was analyzed using SPSS statistical software and for test the hypotheses, used was the binomial test results showed that the lowest percentage of illiteracy in the dairy industry and most of the traditional dairies. The main job of 84% of industrial dairy farms and dairies traditional at only 45%. In the Check The highest and lowest percentage of participation in training courses related to industrial and semi-industrial dairy. The results of the research questions of health management compared with existing standards showed that Industrial dairy farms dairies were significantly higher compared to standard parameters. The semi-industrial dairy farms and industrial intermediates in the more traditional parameters. The traditional dairy farms also had the lowest percentage in compliance with health management.

Rumen microbiota of beef calves fed high-grain concentrate and straw, given separately or as unifeed
A. Gimeno[1], S. Schauf[1], G. De La Fuente[2], C. Castrillo[1] and M. Fondevila[1]
[1]Universidad de Zaragoza, Producción Animal y Ciencia de los Alimentos, M. Servet 177, 50013 Zaragoza, Spain, [2]Universitat de Lleida, Producció Animal, Alcalde Rovira Roure 19, 25198 Lleida, Spain; mfonde@unizar.es

Offering concentrate and straw in a mixed form might synchronize the input of starch and fibre, helping to maintain more stable rumen fermentation and microbiota and thus reducing the risk of acidosis in beef cattle. The effect of offering both ingredients separately (SEP) or as unifeed (UNI) on rumen bacteria was studied in 8 Friesian intensively reared calves, following a change-over arrangement. Rumen contents were sampled before the morning feeding and after 3, 6 and 12 h. Neither daily concentration of total bacteria (9.73 vs 9.92 $\log_{10}$ 16S rDNA gene copies/g for SEP and UNI) nor the relative abundance of *Streptococcus bovis* (0.0028 vs 0.0024%), *Selenomonas ruminantium* (0.292 vs 0.287%) and *Megasphaera elsdenii* (0.067 vs 0.057%) measured by real-time PCR were affected by feeds distribution. The microbiome profile, assayed by Ion-Torrent NGS, was scarcely affected; however, calves fed UNI showed a quantitatively ($P>0.10$) greater proportion of Bacteroidetes (51.1 vs 42.0%), and a lower abundance of Proteobacteria (29.7 vs 38.9%), as the more abundant phyla. At the genus level, feeding UNI resulted in higher proportions of Olsenella (0.131 vs 0.094%), *Eubacterium* (0.020 vs 0.014%), *Sphaerochaeta* (0.071 vs 0.034%) and *Pyramidobacter* (0.004 vs 0.002%), and in a lower proportion of *Butyrivibrio* (0.020 vs 0.025%) than SEP ($P<0.05$). Despite rumen fermentation pattern (presented elsewhere) indicated that UNI promoted more buffered rumen fermentation and a lower proportion of cases of acidosis (pH<5.6) than SEP, it seems that differences on rumen fermentation were not translated into a relevant change of the microbiome profile, which may indicate the existence of a core rumen microbiota.

Development of a *Salmonella* Typhimurium challenge model in pigs: evaluation of known interventions

J.G.M. Wientjes[1], H.M.J. Van Hees[2], A.E. Heuvelink[1], W. Swart[1], P.J. Van Der Wolf[1] and P.J. Roubos-Van Den Hil[2]
[1]GD Animal Health, Arnsbergstraat 7, 7418 EZ Deventer, the Netherlands, [2]Nutreco R&D, Veerstraat 38, 5831 JN Boxmeer, the Netherlands; a.wientjes@gddiergezondheid.nl

Our aim is to develop a *Salmonella* Typhimurium (S. Typh) challenge model in pigs suitable for evaluation of effects of dietary interventions on fecal shedding. Previously, we infected pigs with a S. Typh field strain via oral inoculation, resulting in quantifiable fecal shedding. In this study we validate whether our model is able to show differences in fecal shedding between (proven effective) dietary interventions and a positive control group. Four groups of 8 individually housed weaned piglets were used: NC (negative control; no inoculation); PC (positive control; inoculation); WI (inoculation + water intervention: 2 l/m^3 Selko pH); FI (inoculation + feed intervention: 4.6 l/ton Selko pH). Piglets of PC, WI and FI were orally inoculated with a S. Typh field strain (1 ml of 1.09×10^9 cfu/ml) for 7 days (d1-7). Rectal fecal samples were taken at d-7, -4 and 1 for detection and at d2, 3, 7, 9 and 21 for quantification of *Salmonella*. Diarrhea incidence was recorded daily. No fecal shedding of *Salmonella* was detected before inoculation, nor in the NC group during the trial. Peak fecal shedding at d3 tended to be lower for WI (4.1 log10 cfu/g) compared to PC (5.5 log10 cfu/g; P=0.06), and was intermediate in FI (5.4 log10 cfu/g). After d3, fecal shedding gradually declined to 1.4 log10 cfu/ml at d21 for all treatments. Diarrhea incidence was lower in NC (53%) compared to other treatments (67-71%; P<0.05). To conclude, we successfully infected pigs with a S. Typh field strain, resulting in significant fecal shedding. The water intervention resulted in 1.4 log10 cfu/g lower peak shedding at day 3 after inoculation, which may be relevant in practice. The model may be suitable for evaluation of effects of dietary interventions on peak fecal shedding, although further fine-tuning (e.g. more frequent fecal sampling during peak shedding) is required.

Modulation of rumen microbial fermentation by herb mixture and dietary fat in sheep

M. Wencelová, Z. Váradyová and S. Kišidayová
Institute of Animal Physiology, Slovak Academy of Sciences, Šoltésovej 4-6, 040 01 Košice, Slovak Republic; wencelova@saske.sk

In vitro and *in vivo* experiments were conducted to investigate the impacts of dry herb mixture, dietary fat and their combination on rumen fermentation characteristics, fatty acid concentration and microbial population in sheep. Treatments were: control (basal diet: meadow hay/barley grain, 400/600, w/w), HM (basal diet with 10% replacement of meadow hay by herb mixture), SO (basal diet with 3.5% DM of sunflower oil) and HMSO (basal diet with 10% replacement of meadow hay by herb mixture and 3.5% DM of sunflower oil). The selection of dry HM was based on information available about their carminative and anti-inflammatory effects. All diets were incubated *in vitro* for 24 h at 39 °C in rumen fluid inoculum from cannulated sheep. The *in vitro* dry mater digestibility of HM was the highest of all the treatments (P<0.001). Compared with the control, lower values of methane concentration were obtained for HM and HMSO (P<0.001). The concentration of polyunsaturated fatty acids (PUFA) was significantly increased by HMSO treatment. To further test the results obtained *in vitro*, four rumen-cannulated sheep were randomly assigned according to a 4×4 Latin square design experiment to be fed experimental diets consisted of control (basal diet: 700 g DM/day meadow hay and 500 g DM/day barley grain), HM (basal diet with 70 g DM/day of herb mixture), SO (basal diet with 40 g/day of sunflower oil) and HMSO (basal diet with 70 g DM/day of herb mixture and 40 g/day of sunflower oil). The concentrations of linoleic acid, α-linolenic acid, γ-linolenic acid of the control and HM were significantly higher (P<0.001) compared with SO and HMSO. We conclude that no adverse effect of HM, SO or HMSO on fermentation end products and rumen ciliate and eubacterial population were observed. Results suggest that using an herb mixture as feed additive in a high-concentrate diet could be used to improve rumen fermentation efficiency. The authors would like to acknowledge the VEGA 2/0009/14 and COST Action FA 1302.

Dynamics of fumarate metabolism in rumen environment *in vitro*

J. Pisarčíková, Z. Váradyová and S. Kišidayová
Slovak Academy of Sciences, Institute of Animal Physiology, Soltesovej 4-6, 04001 Kosice, Slovak Republic;
pisarcikova@saske.sk

In vitro experiment was conducted to investigate the effect of different doses of fumarate addition on rumen metabolism of organic acids, fermentation and methane production in presence of high-concentrate diet (HCD). The rumen fluid from ruminally cannulated sheep were mixed with McDougall´s buffer (1:1), added to fermentation vessels containing HCD substrate (meadow hay:barley grain, 400:600, w/w) and incubated for 24 h at 39 °C. Fumarate disodium salt was added to the incubation medium to achieve the fumarate concentrations of 0, 10 or 30 mmol/l (F0, F10, F30). The concentrations of fumarate, succinate, malate and lactate were recorded at 0, 4, 6, 12 and 24 h of incubation using high-performance liquid chromatography-mass spectrometry (HPLC-MS/MS). The concentrations of methane and short-chain fatty acids in the medium at 24 h were determined by gas chromatography. The incubation without fumarate addition (F0) showed a decreasing tendency of lactate and succinate concentrations until 24 h. Fumarate addition to rumen fluid showed an increasing of succinate concentration and the highest concentrations were found in 6 h (F10) and 12 h (F30) of fermentation. The rate of fumarate metabolism (F30) was much higher than metabolism of succinate. We observed slight accumulation of malate in 4 h of incubation after F30 addition. The concentration of lactate (F0, F10, F30) remained constant during 24 h because of presence high-concentrate diet in our experiment. Lower concentration of methane ($p<0.05$) of HCD substrate with F30 was accompanied by higher propionate proportion ($p<0.001$). We conclude that the higher concentration of exogenous fumarate (F30) affected *in vitro* dry matter digestibility, methane emissions and concentration of organic acids in high-concentrate diet in rumen fluid. HPLC-MS/MS method enables the analysis of the dynamics of organic acids for the first time in a rumen environment. The authors would like to acknowledge the VEGA 2/0009/14 and COST Action FA 1302.

The effect of lauric acid on inflammation response caused by cow fungal mastitis in mice

Y. Suda, N. Sasaki and Y. Masumizu
Miyagi University, School of Food, Agricultural and Environmental Sciences, Animal Breeding and Genetics, 2-2-1 Hatatate, Taihaku ku, Sendai city, 982-0215, Japan; suda@myu.ac.jp

Cow mastitis is known as the inflammation of mammary gland and is the most multiple diseases in dairy cattle. There is still no effective treatment now. Lauric acid, included richly in coco residue extracted oil has been proved to have antibacterial activity, and is contained at the about 3% level in cow milk fat. In this study, we aimed to increase lauric acid level and then moderate inflammation response by antibacterial activity. So we utilized a mouse mastitis model to study the effect of lauric acid on fungal mastitis. The commercial diet contained 5% of lauric acid was supplied *ad libitum* to 3 of mice during one week after inducing inflammation by fungus pellicle (CNA70) from contracted dairy cow in the mammary gland. 3 of induced mice as control were supplied normal diet. Blood and mammary gland tissue was collected at the day 8 from all mice, and serum, total RNA and total protein were extracted quickly under cooling. IL6, IL8, TNFα, CD4, CD8 expression in the tissues were examined, and serum CRP level as inflammation marker was evaluated. In addition, the effect of lauric acid on mastitis was researched *in vitro* by using stimulated cell line from mouse mammary gland. The results showed that serum CRP level in mice supplied lauric acid diet was low clearly, and lower IL6, CD4 and CD8 expression were confirmed in tissues compared to mice given normal diet. *In vitro* examination, the profiles of these expressions corresponded to *in vivo* results. So, to increase lauric acid level in milk fat by lauric acid intake from diet might be effective to moderate and prevent inflammation of fungal mastitis.

Genome-wide association study for methane emission in Danish Holstein cattle
J. Lassen, P. Løvendahl and G. Sahana
Aarhus University, Center for Quantitative Genetics and Genomics, Molecular Biology and Genetics,
Blichers Alle 20, 8830, Denmark; jan.lassen@mbg.au.dk

A number of methane mitigation strategies have been suggested for dairy cattle, amongst them genetic selection for low methane emitting cows. In this study we have investigated the possibility to locate QTLs affecting methane emission in dairy cattle. We recorded methane emission for 1,793 Danish Holstein cows. The cows were chosen from herds with automated milking systems. In each of the automated milking systems, a portable Fourier transform infrared (FTIR) detection based gas analyzer was installed. The air inlet was placed in front of the cow's mouth, in order to measure the breath. The instrument analyzed the CH_4 and CO_2 content of the exhaled gas. By using CO_2 as an internal marker, it is possible to calculate an estimation of the CO_2 and CH_4 excreted by the animal. Utilizing data on the animal build, it is possible to calculate the CO_2 excretion. The cows with methane records were genotyped with the Illunina Bovine SNP50 BeadChip (Illumina Inc., San Diego, CA). After quality control 52,048 SNP was used for association analysis. The sporadic missing genotypes were imputed using the software BEAGLE. The association mapping was carried out using SNP-by-SNP analysis of a linear mixed model where polygenic component was fitted in the model in addition to regression on allele doses for the SNP. The SNPs significantly associated with methane emission were selected. This work will be followed with imputation to the whole genome sequence for the targeted region followed by association analysis. The candidate gene will be prioritizing for further study to identify the causal variants affecting methane emission in dairy cattle.

Methane emission collected on Polish commercial dairy farm
M. Pszczola, M. Szalanski and T. Strabel
Poznan University of Life Sciences, Genetics and Animal Breeding, Wolynska 33, 60-37 Poznan, Poland;
marcin.pszczola@gmail.com

Methane emission from livestock is lately gaining on significance due to the environmental footprint end economic importance of this gas. Dairy cattle produce most of methane among the livestock, which is an environmental cost for digesting the feed. Methane emission is mainly driven by the feed composition and dry matter intake, however, it also has a genetic component. The estimated low to moderate heritability gives a potential for selecting low emitters and thus reducing the genetic level of methane emission. The aim of this study was to assess the usefulness of the methane emission data collected during milking for the genetic analyzes. Data was collected from the commercial farm housing 350 Polish Holstein-Frisian cows. Methane emission was recorded during the milking time at three automated milking systems. Breath samples were taken continuously in 5-seconds intervals for 24 hours at each robot for the period of two consecutive months. The infra-red based measuring equipment was employed to analyze the samples. The resulting time-point measurements of methane emission were converted to a single observation per milking as an average methane emission adjusted for the methane level at the beginning of milking. This resulted in 10,388 unique samples collected of 169 unique cows. Each cow, on average, had about 62 records with methane data. The average cow emitted 503 ppm (s.d. 107 ppm) of methane. Along with methane emission observations, production, health, fertility, nutrition and management information was collected. The obtained individual methane emission phenotypes will be further corrected for the day of the lactation, the lactation stage, nutritional information and the production level for calculating the repeatability of methane emissions. Next relation of methane to production and functional traits will be assessed.

Methane emission of dairy cows reflected by sensor measurements

M.H.P.W. Visker[1,2], P.P.J. Van Der Tol[3], H. Bovenhuis[1] and J.A.M. Van Arendonk[1,2]
[1]Wageningen University, Animal Breeding and Genomics Centre, P.O. Box 338, 6700 AH Wageningen, the Netherlands, [2]Top Institute Food and Nutrition, P.O. Box 557, 6700 AN Wageningen, the Netherlands, [3]Lely Industries NV, Cornelis van der Lelylaan 1, 3147 PB Maassluis, the Netherlands; marleen.visker@wur.nl

Reduction of methane emission by dairy cows through breeding requires large-scale collection of phenotypic data on methane emission. Such large amounts of data can be obtained by using methane sensors in automatic milking systems. Measurements recorded by these sensors can be used to generate different traits that reflect methane emission. The aim of this study was to compare these different traits with methane emission recorded in climate respiration chambers (CRC). Data was obtained from 20 dairy cows that were housed individually in one of two CRC for five days each. Inlet and exhaust air of each CRC was quantified and sampled for methane (CH_4) and carbon dioxide (CO_2) concentrations using a non-dispersive infrared method at 10-min intervals. Each CRC was equipped with a Fourier transform infrared sensor in the upper part of the feeding trough. The same sensors have also been installed in automatic milking systems for collection of methane data on commercial dairy farms. These sensors recorded CH_4 and CO_2 twice per second. Preliminary analyses suggest correlations of 0.8 between CH_4 obtained from sensor measurements and CH_4 obtained from CRC measurements, and correlations of 0.6 between CH_4/CO_2 obtained from sensor measurements and CH_4 obtained from CRC measurements. These results indicate that a substantial part of methane emission by dairy cows can be measured with sensors that can be installed in automatic milking systems. This enables large-scale collection of methane emission data. Further analysis is in progress, focussing on different traits that reflect methane emission, the influence of several fixed factors, and the impact of recording data only during a few short timeframes per day.

Assessment of breath methane measurement with an artificial reference cow in the lab

L. Wu[1], P. Groot Koerkamp[1,2] and N. Ogink[2]
[1]Wageningen UR, Farm Technology Group, P.O. Box 16, 6700 AH, Wageningen, the Netherlands, [2]Wageningen UR Livestock Research, P.O. Box 338, 6700 AH, Wageningen, the Netherlands; peter.grootkoerkamp@wur.nl

To mitigate methane emission from dairy cows, the primary issue is to evaluate the efficacy of measures to reduce methane emission. To evaluate e.g. genetic differences and feeding measures in practice, we need a technique to assess the individual methane emission from a large number of cows in the barn. First, we designed and constructed an artificial reference cow that simulated exhalation and eructations of cows, with known methane production rates and representative release patterns. The methane mass balance of the artificial cow was extensively tested and results showed strong linear relation between controlled and measured methane mass. Methane concentration release patterns with a sinusoidal curve produced by five simulated cows were compared to patterns measured from real cows. Results showed small differences between simulated and real concentrations with respect to time interval of eructations and lowest and peak concentrations. We concluded that the artificial reference cow can be used as a known reference source to develop methane measurement methods. Second, the artificial cow was used in a typical setup for measuring methane concentrations in a feed bin, where a cow eats concentrates during milking in an automatic milking system. Methane concentration was continuously analysed in the sampled breath air. The objective of this study was to validate whether variation of methane production rates (set at five levels between 200 and 400 g/d, each level repeated 5 times) between cows and within time can be assessed. This was done under stable and turbulent aerial conditions. Results under stable conditions showed that imposed daily methane production had a strong linear relation with measured average methane concentration in the feeder. But preliminary results under turbulent conditions showed that the relation was more variable, probably due to the aerial turbulence in the barn.

Proxies for methane output in dairy cattle: evaluation as indirect traits for breeding
E. Negussie[1] and F. Biscarini[2]
[1]*NRI Biometrical Genetics, Myllytie 1, 31600 Jokioinen, Finland,* [2]*PTP, Bioinformatics, Via Einstein, 26900 Lodi, Italy; filippo.biscarini@gmail.com*

Methane output in dairy cattle has become a public concern due to its contribution to climate change. Given that it is related to feed utilization in ruminants, it also turned into an emerging phenotype to be used in breeding for environmental impact as well as feed efficiency. Since accurate measurement of methane output from individual animals, especially on a large scale, has proven difficult, the inclusion of proxies for methane in the breeding goal would provide a long-term sustainable mitigation strategy. Although several techniques have been developed to measure methane output, direct accurate measurements in dairy cows is still challenging. The most accurate methods like respiration chambers are too costly, time consuming and labor intensive for routine large-scale on-farm applications. Collecting a sizable amount of accurate phenotypes is nonetheless a prerequisite for genetic selection. Under such circumstances, identifying highly correlated, simple and relatively inexpensive proxies for methane output would be the best alternative. Critical reviews of the available proxies and their suitability for breeding are still lacking. The main objective of this study was therefore to identify all available proxies for methane emissions, assess their accuracy, robustness and suitability as indirect traits for breeding in national genetic evaluations. Available literature on proxies that is published and unpublished was critically reviewed. Proxies for methane output in dairy cattle were classified into categories of lipid markers, MIR spectral analysis, fatty acids, sensors and prediction models. For each class of proxies, convenience, cost and accuracy were evaluated. Their ability to provide robust predications of methane output under different breeds, diets and production systems was also assessed. Finally, the best proxies to be used in breeding for environmental impact and feed efficiency were identified, documented and discussed.

Can chamber and SF6 CH_4 measurements be combined in a model to predict CH_4 from milk MIR spectra?
A. Vanlierde[1], M.-L. Vanrobays[2], F. Dehareng[1], E. Froidmont[1], N. Gengler[2], S. Mcparland[3], F. Grandl[4], M. Kreuzer[4], B. Gredler[5], H. Soyeurt[2] and P. Dardenne[1]
[1]*CRA-W, 5030, Gembloux, Belgium,* [2]*ULg, GxABT, 5030, Gembloux, Belgium,* [3]*Animal and Grassland Research & Innovation Centre, Moorepark, Cork, Ireland,* [4]*ETH Zürich, IAS, 8092, Zurich, Switzerland,* [5]*Qualitas AG, 6300, Zug, Switzerland; a.vanlierde@cra.wallonie.be*

Methane (CH_4) naturally produced by dairy cows during ruminal fermentation is an important greenhouse gas. An equation based on 446 reference data has been developed to predict easily individual CH_4 emissions from milk mid-infrared (MIR) spectra. This equation was based on CH_4 data measured exclusively with the SF_6 technique on 146 distinct Holstein, Jersey and Holstein×Jersey cows. As breeds, managements, diets, etc. are different from one geographical area to another, representative reference data have to be included in the calibration set before applying this equation in a location. However, the local CH_4 data needed are likely to be collected with different techniques (chambers, GreenFeed, etc.) depending on the research team and its equipment. A first study has therefore been conducted (1) to test the performance of the actual equation on data obtained in open-circuit chambers and (2) to analyse the impact of the inclusion of these data in the calibration set. A total of 60 chamber measurements of CH_4 and milk MIR spectra were obtained from 30 lactating Brown-Swiss cows. The correlation between actually measured and predicted CH_4 (C1) was 0.48. This result is in the range of expectations given the R^2c of the equation (0.75), the correlation known between SF_6 and chamber methods (~0.80), and the breed and diet differing between calibration sets. The correlation was about 0.70 after the inclusion of the chamber data (and so the inherent variability) in the calibration set (C2). As chambers are known as the gold standard method, the C1 observed confirms the relevance of using milk MIR technique. Moreover, C2 is very encouraging regarding the possibility to include data coming from chambers into the existing CH_4 equation.

Variability among dairy cows in methane, digestibility and feed efficiency

P.C. Garnsworthy[1], J. Craigon[1], E. Gregson[1], E. Homer[1], S. Potterton[1], P. Bani[2], E. Trevisi[2], P. Huhtanen[3], K. Shingfield[4] and A. Bayat[4]
[1]University of Nottingham, Sutton Bonington, Loughborough LE12 5RD, United Kingdom, [2]Università Cattolica Sacro Cuore, IT, 29122 Piacenza, Italy, [3]Swedish University Agricultural Sciences, S, 90183 Umea, Sweden, [4]MTT Agrifood Research, FI, 31600, Jokioinen, Finland; phil.garnsworthy@nottingham.ac.uk

A goal of both Methagene and RuminOmics is to identify low methane emitters through genetics. Genetic variation must be distinguished from other sources of variation. The aim of this paper is to review sources of variation in methane output by dairy cows, and to consider methane in relation to digestibility and feed efficiency. Across a wide range of values, methane output is related to dry matter intake. After allowing for intake, however, methane varies among individuals. Ultimately, methane variation is due to differences in microbial activity, but many factors are involved. Animal factors include live weight, milk yield, stage of lactation, rumination rate, passage rate, digestibility and eating behaviour. Non-animal factors include diet composition, feeding system, housing and cow management. Variation in methane can be influenced also by measurement technique. For large-scale studies, non-genetic influences on methane variation must be minimised and recorded in order to identify true genetic effects. Breeding from low methane emitters is often suggested as a way to concurrently lower environmental impact and improve efficiency. Improved efficiency is assumed to follow from less gross energy intake lost as methane, but recent evidence challenges this. For example, methane is produced during digestion of fibre, so a cow with poor forage digestion could be a low methane emitter, but would have poor feed efficiency. We conclude that methane output is variable, but understanding sources of variation and interactions with efficiency provides opportunities for improvement. Methane phenotypes and breeding goals must be defined carefully to ensure that potential benefits are complementary.

Feed intake, digestion efficiency and methane emissions of lactating cows of different age

F. Grandl[1], M. Furger[2], J.O. Zeitz[3], M. Kreuzer[1] and A. Schwarm[1]
[1]ETH Zurich, Inst. of Agricultural Sciences, Universitätstr. 2, 8092 Zurich, Switzerland, [2]Agricultural Education and Advisory Centre Plantahof, Kantonsstr. 17, 7302 Landquart, Switzerland, [3]Justus Liebig University Gießen, Inst. of Animal Nutrition and Nutritional Physiology, Heinrich-Buff-Ring 26-32, 35392 Gießen, Germany; florian.grandl@usys.ethz.ch

High milk yield is often associated with short longevity, and extensive use of concentrate is questionable from a food security perspective. Increasing longevity and reducing concentrate input can improve the sustainability of milk production, but only if older cows perform similarly with regard to feeding and digestion efficiency. Thirty cows with an age ranging from 2.4 to 10.1 yr fed a diet either with or without concentrate were studied. Individual feed intake, rumination time, mean retention time of feed (MRT) and digestibility were assessed during an 8-d period, and CH_4 emissions were measured in respiration chambers for 2 days. Diet, age, milk yield and body weight were included as effects in regression analyses. Feed and fibre intake increased ($P<0.05$) with age, though this was mainly driven by the difference between primi- and multiparous cows. Absolute rumination time was not influenced by age, but older cows seemed to process their feed more efficiently, as they tended ($P<0.1$) to spend less time chewing per unit of feed. MRT and organic matter digestibility tended to increase in older cows but the model fit was only moderate (adj. R^2 of 0.33 and 0.21). In contrast, CH_4 emissions were lower in older cows (e.g. CH_4/kg organic matter or per ingested gross energy was both 17% lower in cows older than 5.5 yr). There were differences in CH_4 emissions related to higher fibre intake between the feeding regimes, but these were less pronounced than expected, probably due to the limited use of concentrate (5 kg/d·cow). Based on these findings, increasing the average age in dairy herds has no detrimental effects on feeding and digestion efficiency and could reduce the environmental impact of milk production.

RuminOmics technologies and methane, a mixture of metagenomics and metaproteomics
T.J. Snelling[1], F. Strozzi[2] and R.J. Wallace[1]
[1]University of Aberdeen, Rowett Institute of Nutrition and Health, Bucksburn, AB21 9SB, United Kingdom,
[2]Parco Tecnologico Padano, Lodi, Lo 26900, Italy; t.j.snelling@abdn.ac.uk

The RuminOmics consortium is applying state-of-the-art technologies to understand the role of the ruminal microbiota in digestion efficiency and to decrease the environmental impact of ruminant livestock production. Methane is a potent greenhouse gas with 28 times the global warming potential of CO_2. Ruminants produce abundant methane and are significant contributors to greenhouse gas (GHG) emissions. They also excrete large amounts of N, which can lead to the formation of N_2O, an even more potent GHG than methane. Thus, measures are required to lessen the environmental impact of ruminant livestock production. Sophisticated techniques were needed to understand the mechanisms underpinning the complex functions of the rumen microbial community. Therefore, a metagenomic approach, whereby all microbial DNA is sequenced, was undertaken to describe the total genetic potential of the rumen microbiota and metaproteomics was investigated as a means of understanding how the genes were expressed. This information was then related to the genome and the metabolic characteristics and parameters of the host animal. Metagenomic pipelines are well established to describe the total genetic content of microbial communities. Conversely, metaproteomics is very much in the development phase. Separation of extracted proteins was initially carried out using 2D SDS PAGE although this was superseded by a shotgun metaproteomic approach. For each sample up to 150 unique proteins were identified. These contained examples of structural proteins and enzymes from a range of eukaryotic and prokaryotic rumen microorganisms as well as proteins from dietary plants and the host animal. From a mixture of the two technologies, an integrated approach (metaproteogenomics) can be developed as a powerful tool to describe the structure, functions and interactions of the rumen microbial community relating in particular to methane and N emissions.

Host genetic effects on the rumen microbiome as revealed by rumen content exchange in lactating cows
I. Tapio[1], A. Bonin[2], K. Shingfield[3], S. Ahvenjärvi[1] and J. Vilkki[1]
[1]Natural Resources Institute, Myllytie, 31600 Jokioinen, Finland, [2]Laboratoire d'Ecologie Alpine, 2233 rue de la Piscine, 38041 Grenoble, France, [3]Aberystwyth University, Institute of Biological, Rural and Environmental Sciences, SY23 3EE, United Kingdom; ilma.tapio@luke.fi

The intestinal microbiome is partially shaped by host genetics in rodents and humans. Few studies have considered host specificity of the rumen microbiome. This study examined the influence of host animal genetics on rumen microbial populations. Two pairs of monozygotic twins and four unrelated cows at the same stage of lactation, fitted with rumen cannulas and fed the same diet, were used in an experiment with two 42 d periods. On d 42, rumen evacuations and digesta exchange between identical twins and unrelated cows were performed. Ruminal digesta was sampled immediately before exchange and weekly thereafter. DNA was extracted from 30 mg of freeze-dried sample. Illumina technology was used for metabarcoding sequencing of amplicons from bacteria and archaea (16S rRNA), ciliate protozoa (18S rRNA) and fungi (ITS1). Sequencing data was processed using OBITools and Qiime v1.7.0. Greengenes_13_8 database was used for bacterial taxonomical identification, RIM-DB for archaea and SILVA 18S for ciliate protozoa. Fungi were assigned using in-house reference database from S. Kittelmann (AgResearch). Substantial between-animal variation in the response to rumen exchange was detected by Bray-Curtis distances comparing microbial populations at rumen exchange with later sampling times. The bacterial community appeared to be least influenced by the host. Significant differences in intra-class correlation coefficient distributions between identical twin pairs and unrelated pairs were detected for archaea and fungi, indicating a genetic influence on the occurrence of specific taxa. While the results suggest genetic control of the rumen microbiome, large between-animal variation highlights the contribution of other factors defining rumen microbial populations and function.

Enteric methane emissions from beef cattle of different genetic groups in confinement in Brazil

A. Berndt[1], L.S. Sakamoto[1], F.B. Ferrari[2], H. Borba[2], E.D.M. Mendes[1] and R.R. Tullio[1]
[1]Embrapa Southeast Livestock, Rod. Washington Luiz, km 234, PB 339, 13560970 Sao Carlos, SP, Brazil;
[2]University of Sao Paulo State, Via Paulo Donato Castellane s/n, 14884900 Jaboticabal, SP, Brazil;
alexandre.berndt@embrapa.br

At present the need to intensify meat production systems is increasing due to less area available and increasing demand for food. An alternative is to rear animals in confinement considering the environmental impact caused by methane emissions. The objective of this study was to measure the enteric methane emissions from cross-bred cattle belonging to different genetic groups, using GreenFeed. Steers offspring of Brangus, Canchim (synthetic breed 5/8 Charolais) or Bonsmara bulls and Nellore, ½ Angus + ½ Nellore or ½ Senepol + ½ Nellore cows, reared on pasture and finished in the feedlot were evaluated. The animals were confined in collective stalls equipped with GrowSafe troughs, according to weight. The diet was based on maize silage, ground maize, soybean bran and wheat bran with 51.8% of DM, 13.1% of CP, 71.0% of TDN and 3.2% of EE, provided twice daily, ensuring *ad libitum* consumption. Methane emissions were measured using GreenFeed equipment, developed by the company C-Lock, an online or real-time system to quantitatively measure CH_4 and CO_2 emissions en masse, from moment to moment, individually, while animals are attracted to the trough to receive a small amount of feed. Methane emissions were measured from two stalls with 28 animals, however it was only possible to obtain results for emissions calculations from 19 animals as a result of insufficient visits to the trough or visits of insufficient duration. Data was analyzed using the MIXED procedure of SAS and averages were compared using Tukey's test with significant differences at $P<0.05$. Statistical differences between genetic groups were not found for the enteric methane emission variable in grams per day (CH_4 g/d) or for methane yield (YM %), with average values of 166.6±30.7 gCH_4/day and 4.45±0.89% presented respectively.

Multifunctionality of the farm animal genetic resources seen through ecosystem services approach

K. Soini
Natural Resources Institute Finland, Economy and Society, Latokartanonkaari 9, 00790 Helsinki, Finland;
katriina.soini@luke.fi

It is widely agreed that the Ecosystem Services (ES) approach provides a perspective for the conservation of biodiversity. Compared with previous approaches, it has been suggested that ES expands the focus from individual resources to the full array of contributions which ecosystems make to human well-being and better recognises the interconnectedness of ecosystems across the broad temporal and spatial scales over which ecosystems and humans interact. The genetic resources of farm animals are resulted from co-evolution of nature and human and nature and they are dependent on cultural values and practices. ES approach, which integrates both ecological and cultural aspects of conservation of biodiversity, can be seen as an opportunity for promoting the multifunctionality of genetic resources in rural livelihoods and in policy making. By using examples from different empirical research, the paper will discuss the potential of ES framework to value and discuss the multifunctionality of farm animal genetic resources. The paper reveals that ES approach broadens the scope of conservation and preservation from provisioning services (genes, embryos, food, other products) and maintaining services (biodiversity of landscape) to human values that they entail. It also provides a tool to make a link to the (rural) livelihoods. The examples also reveal the challenges (for example to name and measure cultural ecosystem services) and criticism (anthropocentrism), which the ES approach has met more generally. Yet, despite of these shortcomings, it is concluded that the ES approach is a step ahead in the conservation of biodiversity, and also an opportunity for widening the scope of conservation policies of farm animal genetic resources towards multifunctionality.

Value of ecosystem services provided by livestock systems in HNV farmland: a psychographic analysis

T. Rodríguez-Ortega[1], A. Bernués[1] and F. Alfnes[2]
[1]Centro de Investigación y Tecnología Agroalimentaria (CITA), Avda. Montañana 930, 50059 Zaragoza, Spain, [2]Norwegian University or Life Sciences, School of Economics and Business, Universitetstunet 3, 1432 Ås, Norway; trodriguezo@cita-aragon.es

High nature value (HNV) farming systems, such as grazing systems in Mediterranean mountains, are highly multifunctional. These functions underpin their capacity to provide a wide range of ecosystem services (ES) of which many are non-marketable. Agricultural policy needs to evaluate the desired levels of provision of ES according to societal demands and to motivate farmers to provide them. In this study, we aimed at uncovering how different societal attitudes influence the perception and willingness to pay (WTP) for ES provided by HNV agro-ecosystems. We combined psychographics (Likert-type statements) and economic analysis (choice experiment) to: (1) segment the general population (residents in nearby regions where the study area is located) and the local population (residents in the study area) in different psychographic profiles (based on worldviews on the environment, economy, agriculture, rural environment, food consumption, quality perception and agri-environmental policy); (2) calculate the WTP for several ES of the different profiles. In both populations, we found a 'conservationist' and a 'productivist' psychographic profile. These, together with location, had a strong influence on the WTP. Productivists stated a WTP (in €/person/year) of 88€ and 141€ in general and local populations, respectively. For conservationists, WTP was 152€ and 334€, respectively. In relation to the relative preference for ES provision, all profiles were highly concerned about forest wildfires, followed by the availability of quality products for the productivists, the biodiversity maintenance for the general conservationists and a more human-intervened landscape for the local conservationists. The value of ES for different societal profiles should be considered to increase the legitimacy of the EU agri-environmental policy.

Stakeholders' expectations about services provided by livestock farming systems at territory scale

C.H. Moulin[1], C. Aubron[1], J. Lasseur[2], M. Napoléone[2] and M.O. Nozières[2]
[1]Montpellier SupAgro, UMR SELMET, 2 place Viala, 34060 Montpellier cedex 1, France, [2]INRA, UMR SELMET, 2 place Pierre Viala, 34060 Monptellier cedex 1, France; charles-henri.moulin@supagro.fr

Beyond the provision of commodities such as milk or meat, the livestock farming systems (LFS) deliver multiple non-marketed outputs. Some of those outputs are recognized and they have values for human beings enjoying them. So they may be viewed as services provided by LFS at territory scale. But the various stakeholders of a territory do not have the same perception of those outputs and it is a necessity to build the convergence of the points of view to allow the development of LFS with valuable services. To support these assertions, the aim of this presentation is to get evidences from case studies. Two contrasted territories are chosen in French Mediterranean region. The first one is in the littoral area of Provence, characterized by urbanization, touristic attractiveness, decrease of crops and livestock and large areas of forest, with the journey of mobile pastoral herds and flocks coming from the Alps. The second case study is in the hinterland of Languedoc. The altitude plateaus (Causses) and mountains (Cévennes) are used by various sheep, goat and cattle LFS, building an agropastoral landscape recognized as a patrimony by UNESCO. Three surveys have been conducted, from 2011 to 2015, with 79 stakeholders (farmers, operators of commodities chains, farmer unions' representatives, land owners, territorial and environment managers,local communities' representatives, etc.). From the interviews, we identify the valuable outputs recognized by the stakeholders, through their practices and perceptions. Those outputs are then organized in a grid of services at territory scale from livestock systems, combining two frameworks: multifunctionality of agriculture and ecosystem services. Finally, we discuss the convergences and divergences about services between stakeholders and the actions that could support the LFS with valuable services in those two territories.

The cultural value of local livestock breeds: a review

G. Gandini[1], S.J. Hiemstra[2] and I. Hoffmann[3]
[1]Università degli Studi di Milano, DIVET, via celoria 10, 20133 Milan, Italy, [2]Wageningen UR, Centre for Genetic Resource, the Netherlands (CGN), P.O. Box 338, 6700 AH Wageningen, the Netherlands, [3]FAO, Animal Production and Health Division, Viale Terme di Caracalla, 00153 Rome, Italy; gustavo.gandini@unimi.it

Local livestock breeds, within their original farming systems, can provide cultural ecosystem services. However, although much has been said about the cultural value of breeds, to date these services have been only marginally investigated. In principle, a local breed can be considered cultural property in relation to its role as an historical witness. Moreover a local breed can be, even today, a point of reference in ancient local traditions, and thus a true custodian of local rural culture. The paper reviews the literature on the cultural value of local breeds, including an unpublished survey on local goat breeds farmed on the Italian Alpine Ark. Differences observed among cultural services in various areas of the world are discussed, also in view of the need for a common methodology of analysis. The problem of identifying and developing a market for cultural ecosystem services to support the economic development of rural areas and the conservation of endangered breeds is analysed.

Ecosystem services provided by livestock species and breeds – a global view

I. Hoffmann[1], T. From[1], S.J. Hiemstra[2] and G. Gandini[3]
[1]Food and Agriculture Organization of the United Nations, AGA, Via delle Terme di Caracalla, 00153 Rome, Italy, [2]Wageningen University, Center for Genetic Resources, UR Livestock Research, P.O. Box 338, 6700 AH Wageningen, the Netherlands, [3]University of Milan, Dipartimento di scienze veterinarie e sanita' pubblica, v. celoria, 10, 20133 Milano, Italy; irene.hoffmann@fao.org

Livestock provide a range of ecosystem services which are not recognized by the market and go beyond provisioning services such as food and fibre. In order to identify the nature of all ecosystem services provided by livestock, a European and a global survey with together 120 responses were undertaken, complemented by country responses for two global agricultural biodiversity assessments undertaken by FAO, and a literature review. Among all agricultural land, grassland, shrub, herb and sparse vegetation cover more than double the area covered by cropland and slightly more than that of forests. Most of these often marginal lands are can only be used by livestock. With 33%, habitat provisioning was the most frequently mentioned supporting service provided by livestock in grazing systems, and a close link was found between nature conservation and breed conservation. This was followed by livestock's contribution to nutrient cycling and primary production. Shrub and weed control, erosion control and seed dispersal were mentioned as most relevant regulating services, followed by water and climate regulation. Contribution to landscape values, and to cultural and historic heritage together made 44% of cultural services, followed by knowledge systems and education, and recreational values. Although ecosystem service provision was recognized, a lot needs to be done to improve assessment methods, including for the valorization of ecosystem services provided by livestock, and to develop results-based incentive systems.

Enhancing actors' coordination and collective action within the supply chain at local level
F. Casabianca
INRA, SAD, LRDE Quartier Grossetti, 20250 Corte, France; fca@corte.inra.fr

For producing specialty products, local supply chain actors are mobilizing biological resources and know-how, with value-added on the market classically considered as the main issue of such systems. However, from several case studies on animal products, we assume that these dynamics are also providing several non-market values often under-estimated and an effective way to combine market and non-market issues in the livestock production systems. An emblematic situation is linking a local breed and a food quality assurance scheme. This coupling supposes to build up and share numerous points of reference along the chain. The agreements on the qualification criteria are not spontaneous, requiring new coordination and learning processes for converging on the product identity. Linking raw material characteristics with the final product in the code of practices reveals new solidarities among heterogeneous actors, breeders together with processors and retailers. The whole local society is finally required for validating the result of the dynamics. So, the perspective of market value valorizing the specialty product induces local coordination which is non-market value. Moreover, the needs for managing such resources over time produce new capacity in implementing collective action. Livestock systems are structured by requirements for the specialty product inducing new stakes for the actors within the supply chain: how to reach more frequently the targeted characteristics despite the uncertainties remaining in the production process? For achieving such objectives, local actors are led to adjust their activities and to renew the resources, remunerating their efforts thanks to the market values. We illustrate these phenomena through meats and cheeses from small ruminant production in several exemplary cases in Mediterranean area. In these situations, market and non-market values are clearly combined around some main common goods such as collective reputation whose sustainability is depending on the actors' coordination at local level.

Dairy farms in mountainous areas: synergies and conflicts between production and non-market outputs
E. Sturaro, G. Bittante and M. Ramanzin
University of Padova, DAFNAE, viale dell'Università 16, 35020 Legnaro (PD), Italy; enrico.sturaro@unipd.it

In the last decades, socio-economic drivers influenced the evolution of dairy systems in mountainous regions, with a strong reduction of farms number (-38% in Eastern Italian Alps from 1990 to 2010) and an increase of the average herd size. We aimed at discussing the synergies and conflicts between production and non-market outputs of the dairy farming systems in the Alps using case studies from the eastern Italian Alps. Typical high nature value farmland areas in mountains are extensively grazed uplands, alpine meadows and pastures. A landscape GIS approach in the Dolomites showed that only the maintenance of traditional small dairy farms was able to contrast the loss of open areas, whereas both abandonment and intensification had a negative effect. At the farm level, several indicators can be used to address the multifunctionalilty of dairy systems. An on-farm survey on 610 dairy herds of the Trento province allowed to identify four different systems, from the most traditional farms (tie stalls, local forages, summer transhumance with lactating cows) to the modern farms (free stalls, total mixed ration, transhumance only with replacement cattle). The productive gap of traditional farms was partially compensated by their strong link with Protected Denomination of Origin (PDO) and other traditional cheeses. In addition, only the traditional farms were able to use autochthonous breeds, to maintain grasslands with extensive practices, to make full use of summer pastures, and to conserve the traditional landscape. The assessment of environmental footprint of dairy farms is controversial: a partial LCA approach on a sample of 38 mountain dairy farms showed that, as respect to intensive ones, traditional farms had a higher carbon footprint per kg of FCM, but minimized external inputs and off-farm emissions. A holistic and multidisciplinary approach is recommended to evaluate the sustainability of livestock systems in mountainous areas.

Small ruminant grazing, vegetation and landscaping
P. Nowakowski, R. Bodkowski and K. Czyż
Wroclaw University of Environmental and Life Sciences, Chelmonskiego 38c, 51-630 Wroclaw, Poland;
piotr.nowakowski@up.wroc.pl

Herbivores having temporally access to grass sward are not the same as ones permanently relying on it. They are shaping landscape and sward differently. The process of learning about the landscape biodiversity from the animal's birth till the adulthood is of a paramount importance for long lasting relations between environment and animals. Due to limited livestock presence the threat exists of invasions of newcomer plants to local flora, ex. *Heracleum* sp. It is nearly 10 years' time span for one plant survival to seed. When ruminants have access to such areas *Heracleum* sp. is kept under control. Single cattle grazing (without sward cutting) leads to spreading of wild rose, hawthorn and blackthorn. The severity of invasion in the range of ca 6 thousand shrub plants per ha was limited up to 69% with the mixed group of small ruminants at stocking density of ca 1 LU/ha during 2 years. Permanent presence of herbivores weakens the most vigorous plant species and makes biodiversity richer while biomass productivity lower. Plants from Asteroideae, Fabaceae, Liliaceae, Orchidaceae families, many protected by law, require low swards height to persist and have active biological compounds which play role as pharmacy to animals. There is a positive role of diversified plant composition varying in root systems (deep or shallow) in enriching mineral supply to animals, too. Mountain pastures in the Sudetes (altitude ca 500 m) with no mineral fertilizers yielded from 4,4 t (set stocking of sheep and llamas as BW 3:1; 39 plant species, 25% grasses) to 6,9 t of DM/ha (rotational grazing of cattle, sheep and goats as BW 45:46:9; 35 plant species, ≤48% grasses) during 160 days season. In a dams + offspring grazing system animal body gain was ca 160 and 180 kg/ha, respectively. Moving herbivores between different environments to fix current problems with the landscape face problems with adaptation and the reaction of the present state of environment on animals actions.

The importance of sheep in the Carpathian region
M. Milerski
Institute of Animal Science, Přátelství 815, 10400 Prague 10 – Uhříněves, Czech Republic;
milerski.michal@vuzv.cz

Traditional sheep husbandry is deeply ingrained in Carpathian culture and environment. Shepherding plays an important role in preserving mountain communities with their rich folk culture and habits. Sheep grazing can help preserve the biodiversity of mountain meadows including endemic plants and insects and providing nutrition for birds and mammals. The sheep dungs helps to fertilise the soil and due to flocks mooving the grazed areas have enough of time to recover when the animals have gove. Tlavelling sheep help to disperse seeds. Traditional Carpathian sheep production systems demonstrate the harmony between human culture, domestic animal needs, wildlife and environment protection. These attributes can be successfully used in the marketing of traditional products and support the tourist attractiveness of the area. The importance of sheep farming for the Carpathian region is hard to measure.

Comparing cows – including dry period and lactation length in a yield measure

A. Kok, C.E. Van Middelaar, A.T.M. Van Knegsel, H. Hogeveen, B. Kemp and I.J.M. De Boer
Wageningen University, Animal Production Systems group, P.O. Box 338, 6700 AH Wageningen, the
Netherlands; akke.kok@wur.nl

To assess economic and environmental consequences of dry period (DP) length in dairy cows, we need to compare milk yields of cows that differ in DP length. Milk yield is generally defined as the sum of all milk produced during 305 days after calving (305-d yield). This measure, however, ignores additional milk yield in the previous lactation in case cows are milked longer. In addition, it does not adjust for actual lactation length, while a shorter DP may also affect calving interval. We aimed, therefore, to develop a measure of milk yield that includes DP length and actual lactation length; and to apply this measure to compare 217 cows with a conventional (49-90 d), short (20-40 d) or no DP before 2nd calving. An 'effective lactation' was set from 60 days before last calving to 60 days before the next calving and expressed in 'effective yield' in kg fat-and-protein-corrected milk per cow per day. In this way, we included possible additional milk in the previous lactation and excluded milk that depended on the choice regarding the next DP. Cows without a DP had a 22% lower 305-d yield than cows with a conventional DP, whereas the effective yield was only 9% lower. The latter resulted from accounting for additional milk yield in the previous lactation (924±352 kg) and the shorter calving interval (355 vs 408 d). The 305-d yield of cows with a short DP did not differ from cows with a conventional DP, but the effective yield was 7% higher. When 1st lactation 305-d yield was used as a covariate of genetic merit, the effective yield of cows with a short and conventional DP did not differ, while it was 2 kg/day lower (across production levels) for cows without a DP. The difference between results from 305-d and effective yield emphasizes the importance of the methodological choice when comparing milk yields of cows. The effective yield enables a sound comparison of milk yield when DP length and lactation length vary.

Genetic association between functional longevity and health traits in Austrian Fleckvieh cattle

C. Pfeiffer[1], C. Fuerst[2] and B. Fuerst-Waltl[1]
[1]University of Natural Resources and Life Sciences, Department of Sustainable Agricultural Systems, Gregor-Mendel-Strasse 33, 1180 Vienna, Austria, [2]ZuchtData EDV-Dienstleistungen GmbH, Dresdner Straße 89/19, 1200 Vienna, Austria; christina.pfeiffer@boku.ac.at

The importance of functional traits in modern dairy breeding programs is increasing worldwide. Beside a broad range of functional traits like fertility or calving traits, direct health traits gain more importance as they affect animal welfare, farm economy and consumer demands concerning food safety. In fact disease related losses are high in dairy cattle production. An increase of health disorders leads to precocious culling and decreases longevity. However, genetic correlations are not available among health traits and functional longevity because of no or limited access of direct health data in most countries and due to methodical restraints. The objective of this study was to conduct an approximate multitrait two step approach applied to yield deviations (functional longevity) and de-regressed breeding values (health traits) in order to estimate genetic correlations between functional longevity (LONG), clinical mastitis (CM), early fertility disorders (EFD), cystic ovaries (CO) and milk fever (MF). In total, 66,890 pseudo-records of Austrian Fleckvieh cattle (dual purpose Simmental) of two Austrian regions born between 2004 and 2009 were used. The pedigree included 203,430 animals. Variance components were estimated based on an animal model using ASReml 3.0. Genetic correlations between LONG and CM, EFD, CO and MF are 0.63±0.05, 0.29±0.08, 0.20±0.07 and 0.20±0.07, respectively (positive values are favourable). Among health traits significant and positive genetic correlations (ranging from 0.14 to 0.45) were observed. Concerning animal welfare, selecting for more robust disease resistant cows would imply an improvement of functional longevity.

Environmental and economic consequences of subclinical ketosis and related diseases in dairy farming

P.F. Mostert, E.A.M. Bokkers, C.E. Van Middelaar and I.J.M. De Boer
Wageningen University, Animal Production Systems group, P.O. Box 338, 6700 AH Wageningen, the Netherlands; pim.mostert@wur.nl

Subclinical ketosis (SCK) in dairy cattle is a metabolic disease that occurs around the calving period and increases the risk on other diseases. SCK and other diseases result in, e.g. milk losses, reduced pregnancy rate, culling, and therefore have environmental and economic consequences. This study aimed to estimate the environmental and economic consequences of SCK and related diseases in dairy farming. A dynamic stochastic simulation model at cow level was developed and combined with a life cycle assessment and partial budget analysis. The model was divided into four parts. In part one, cows receive a parity (1-5+) and a potential milk production. Cows subsequently have a risk on getting retained placenta or milk fever (part 2), SCK (part 3), and metritis, displaced abomasum, clinical ketosis, lameness or mastitis (part 4). The risk on diseases depends on parity and previous diseases. The model was parameterized using literature. Inputs are the number of dairy cows, prevalence of diseases and culling rate, outputs are the change in global warming potential (GWP) and profit per case of SCK. Cows with (a combination of) diseases had: a reduced daily milk yield, discarded milk if treated, an increased calving interval, and risk of culling. Monte Carlo simulation was performed to find the variation in the output. Preliminary results showed that the costs increased from €33.0 (±31.3) to €55.2 (±58.3) and GWP increased from 1.3 (±1.3) to 1.8±2.0% CO_2-e/unit milk per parity based on milk losses per case of SCK. Results differ per parity (P<0.001) due to differences in milk yield and risk on diseases. The highest contribution came from SCK (68%). Other diseases particularly had an impact on the variation of the output. Future calculations will be extended by including reproduction and culling. In conclusion, SCK has an impact on the environmental and economic performance of dairy farming.

Assessment of the mammary gland elasticity in dairy cows by using a single day once-daily milking

C. Charton[1,2], H. Larroque[1], C. Robert-Granie[1], D. Pomies[3], H. Croiseau-Leclerc[4], N. Friggens[5] and J. Guinard-Flament[2]
[1]INRA, UMR 1388 GenPhySE, 24 Chemin de Borde Rouge, 31326 Castanet-Tolosan Cedex, France, [2]INRA – Agrocampus Ouest, UMR 1348 PEGASE, 65 route de Saint Brieuc, 35000 Rennes, France, [3]INRA, UMR 1213 Herbivores, Theix, 63122 Saint Genes Champanelle, France, [4]Institut de l'élevage, – Domaine de Vilvert – Bât 211 -, 78352 Jouy en Josas Cedex, France, [5]INRA / AgroParisTech, UMR MoSAR, 16 rue Claude Bernard, 75231 Paris Cedex 05, France; clementine.charton@rennes.inra.fr

In the current uncertain and fluctuating context, identifying adaptable cows is of key importance to secure dairy systems. Adaptable dairy cows could be defined by the elasticity of their mammary gland through its ability to both tolerate disruption and to return to its original state. However, little is known about how to identify those individuals. This study aimed to determine whether a single day ODM challenge could be used to quantify mammary gland elasticity and to explore individual responses profiles, factors of influence and repeatability of these responses. The trial used 292 Holstein-Friesian cows and consisted of 3 successive periods: 1 wk control of twice-daily milking (TDM), one day of ODM, followed by 2 wk of TDM. Cows were split into 10 groups over 5 milk years. The number of observations per cows varied from 1 to 9, with no more than 3 ODM challenges per lactation. Single day ODM generated an average loss of 6.3 kg/d (-21.3%) and an average recovery of 4.8 kg/d was observed when resuming twice-daily milking (TDM). Variability in responses was high among cows (CV=62% and 98% for milk losses and recovery). Clustering lead to the identification of 4 profiles of response, corresponding to different potentials of mammary elasticity. The 4 clusters obtained were well characterized by stage of lactation and potential milk yield level. Repeatability ranged between 33% for milk yield recovery (kg/d) and 57% for milk yield loss (%), suggesting a genetic determinism of the mammary elasticity to single ODM challenge.

Environmental impact of Italian dairy industry: case of Asiago PDO cheese

A. Dalla Riva[1], J. Burek[2], D. Kim[2], G. Thoma[2], M. Cassandro[1] and M. De Marchi[1]
[1]University of Padova, Department of Agronomy, Food, Natural resources, Animals and Environment, Viale dell'Università 16, 35020 Legnaro, Padova, Italy, [2]University of Arkansas, Ralph E. Martin Department of Chemical Engineering, 3202 Bell Engineering Center, Fayetteville, AR 72701-1201, USA; alessandro.dallariva.1@studenti.unipd.it

This study presents a 'farm gate to factory gate' Life Cycle Assessment on the Asiago cheese, an Italian Protected Designation of Origin (PDO) product. Potential environmental impacts occurring from raw milk collection through ready-to-sell cheese production, at typical cheese aging, are investigated. All inputs and outputs required in one year at the dairy factory are counted. One dairy factory provided data for production of the Asiago PDO cheese, other hard cheeses and liquid whey. Milk solids of each product are used to allocate raw milk among products. Salt, and its transportation, is allocated only to cheeses. Allocation based on raw milk mass (kg) destined to each product is used to assign other inputs and outputs, such as fuels, cleaning agents and raw milk transport. The Ecoinvent v3.1 database is used for secondary data and SimaPro© 8 is the main software in the analysis. The following impact categories are investigated: Climate Change (CC), Ozone Depletion (OD), Terrestrial Acidification (TA), Freshwater Eutrophication (FE), Photochemical Oxidant Formation (POF), Human Toxicity (HT), Terrestrial Ecotoxicity (TE), Freshwater Ecotoxicity (FE) and Cumulative Fossil Energy Demand (CFED). Electricity is the first emitter, followed by heating, lubricant oil, refrigerants and wastewater treatment. In terms of CC and CFED, impacts are 0.86 kg CO_2 equivalent and 10.90 MJ per one kg of Asiago cheese, respectively. This research is one of the first to study potential environmental impacts of Italian PDO cheese production, with the ultimate goal to valorize an important resource of economic revenue and employment in Italian regions, joining to it an environmentally sustainable production.

A new threat for modern dairy farming: dirty data

K. Hermans, G. Opsomer, M. Van Eetvelde, J. De Koster, H. Bogaert, S. Moerman, E. Depreester, B. Van Ranst, J. Van De Pitte and M. Hostens
Faculty of Veterinary Medicine, Ghent University, Department of Reproduction, Obstetrics and Herd Health, Salisburylaan 133, 9820, Belgium; kristofhermans@bovinet.be

One could state that data is the biggest and most powerful asset in modern dairy farming. Data is not directly meaningful, because it put facts out of context. Only when data is put into relation with each other, information arises. Knowledge can be extracted from the collected information. Hence data only has value when transformed into information and knowledge. Computers have made it possible to collect more data than we can interpret with current analysis tools. When the value of data increases, the interest in the quality of the data increases proportional. Incorrect data has always existed, but now the effects of it are more visible and the consequences more serious. We hypothesized that the rapid evolution of soft- and hardware on dairy farms and new precision livestock farming technologies will lead to an increased interest in the quality of dairy related data. A literature review was conducted in order to find papers mentioning dirty data in dairy farming who complied with the dirty data types mentioned by Li. In total 24 articles were selected from 1900 to 2014. Four articles were published prior to the year 2000, whereas, half of the accepted articles were published between 2010 and 2014. Within the selected articles, 15 out of 24 (63%) described the quality of disease data. Other reported areas in which dirty data occur are fertility, genomics, medicine use, nutrition, sensors and mutation data. Data quality problems are present in every aspect of cattle farming.

Effect of grass silage maturity and level of intake on *in vitro* methane and gas production
F.M. Macome[1,2], W.H. Hendriks[1,2], J. Dijkstra[2], D. Warner[2], W.F. Pellikaan[2], J.W. Cone[2] and J.T. Schonewille[1]
[1]Faculty of Veterinary Medicine, Department of Farm Animal Health, Martinus G de Bruin Building, Yalelaan 7, 3584 CL Utrecht, the Netherlands, [2]Wageningen University, Animal Sciences, De Elst 1, Building nr. 122, 6808 WD Wageningen, the Netherlands; felicidade.macome@wur.nl

An experiment was conducted to evaluate the effect of quality of grass harvested at different maturities and two levels of intake on *in vitro* gas and methane (CH_4) production, using rumen fluid from adapted and non adapted animals. Grass silages were part of a total mixed ration fed to cows at different intake levels (low and high). Rumen fluid was collected from the cows in a warm insulated flask filled with CO_2, filtered through cheese cloth and mixed with an anaerobic buffer mineral solution. Each substrate was incubated *in vitro*, using the gas production technique, with each type of rumen fluid seperately. Gas and CH_4 production were recorded for 48 h. Data from each substrate were averaged before statistical analysis, using the GLM procedure. Gas production (ml/g OM) decreased with increasing maturity of the grass. It was the high digestibility of the non-soluble fraction that made the gas production from young grass higher than from old grass. The gas and CH_4 production (ml/g OM) were higher in rumen fluid from cows adapted to the low feed intake level than in rumen fluid from cows adapted to the high feed intake level. Rumen fluid from cows adapted to young ensiled grass showed a higher ability in degrading the soluble fraction, whereas rumen fluid from cows adapted to old ensiled grass showed a higher ability in degrading the non-soluble fraction of the grass silages. The synthesis of total volatile fatty acids (TVFA) was not affected by the maturity of the grass, whereas an increased level of feed intake showed a decrease in TVFA. However, the molar proportions of propionic acid, butyric acid and branched chain volatile fatty acids (BCVFA) were affected by the maturity stage of the grass and the level of intake affected the proportion of propionic acid and BCVFA.

Bacterially induced cheese blowing defects with particular attention to butyric acid clostridia
J. Brändle, K.J. Domig and W. Kneifel
Institute of Food Science, Muthgasse 18, 1190 Wien, Austria; johanna.braendle@boku.ac.at

Bacteria play a crucial role during hard cheese production and ripening. However, some organisms of the cheese microflora do not only contribute to the characteristic properties of cheeses, but are also responsible for quality defects in the final product like off flavours and irregular eye formation. In some cheese types, excessive gas formation by propionibacteria and heterofermentative lactobacilli impairs cheese quality. These variations remain largely unnoticed by consumers. Growth of clostridia, in contrast, leads to severe quality defects in hard cheeses as these bacteria are not only able to produce tremendous amounts of gas but also the malodorous and rancid tasting butyric acid. Affected cheeses are practically unsellable and the producer has to bear high economic losses. Clostridia enter the cheese via contaminated raw milk. Hence, by optimising farm management and milking practices, contaminations can be reduced but not fully prevented. Despite of many efforts, a suitable routine method for the detection and enumeration of clostridia in milk has not been found yet. To gain a general insight into the cheese microbiota, cultural and molecular methods have been combined for the analysis of Austrian hard cheese samples with blowing defects. The results showed high microbial counts for facultatively heterofermentative lactobacilli and propionibacteria but no clostridial growth. However, the low number of clostridia present in this cheese type required further optimisation of the cultural isolation method. Clostridial isolates obtained with the optimised procedure were identified by Real-Time-PCR and melting curve analysis. Moreover, rep-PCR patterns of the isolated strains were compared with reference strains. The long-term aim of this study is the development of a detection method for cheese-damaging clostridia in raw milk to prevent late-blowing. The first steps have been made by the isolation, identification and characterisation of suitable reference strains from milk and cheese.

Organic dairy production without concentrates: effects on milk yield, animal health and economics

P. Ertl[1], A. Steinwidder[2] and W. Knaus[1]
[1]BOKU-University of Natural Resources and Life Sciences, Vienna, Gregor-Mendel-Strasse 33, 1180 Vienna, Austria, [2]Agricultural Research and Education Centre Raumberg-Gumpenstein, Trautenfels 15, 8951 Trautenfels, Austria; paul.ertl@boku.ac.at

Today´s rations in dairy nutrition are often based on high concentrate rates to increase animal performance and meet animals´ nutrient requirements. However, feeding grain-based concentrates to dairy cows is inefficient in terms of net-food production and is linked to animal health concerns. The objective of the study was, therefore, to determine the effect of the abandonment of concentrates in organic dairy nutrition on milk yield, animal health and economics in comparison to different quantities of concentrates. Basic data was collected from eight organic dairy farms where no concentrates were fed (C0). This data was compared with results from 131 and 140 Austrian organic dairy farms included in a federal program (WG) for the years 2010 and 2011, respectively. These farms were divided into three groups, depending on the amount of concentrates (fresh matter basis) fed per cow annually (WG1: up to 975 kg; WG2: 976 to 1,400 kg; WG3: more than 1,400 kg). Data was analysed using PROC MIXED of SAS 9.1.3. including year and concentrate group as fixed, and the individual farm as random effect. The energy corrected milk yield increased from 5,093 kg in C0 to 6,828 kg in WG3. Calculated forage milk yield decreased by increasing concentrate supplementation from 5,093 kg (C0) to 4,412 kg (WG3). Data related to animal health did not significantly differ between groups. However, the calving interval was longer in C0 but non-return-rate and insemination index were the same. Although milk yield per cow was lowest in C0, the marginal income per cow was on the same level as in the other groups. The marginal income per kg milk decreased significantly from C0 to the other groups. In conclusion, this data showed that animal health and profitability were not negatively affected on farms without concentrate supplementation.

A dynamic mechanistic whole animal simulation model: methane predictions

V. Ambriz-Vilchis[1], R.H. Fawcett[2], J.A. Rooke[3] and N.S. Jessop[2]
[1]R(D)SVS and The Roslin Institute, EBVC, EH25 9RG Scotland, United Kingdom, [2]The University of Edinburgh, KB, EH9 3FE Scotland, United Kingdom, [3]SRUC, West Mains Road, EH93JG Scotland, United Kingdom; v.ambriz-vilchis@sms.ed.ac.uk

Livestock are the source of 33% of the protein in human diets and provide many other services. Methane (CH_4) emissions from livestock are a significant contributor to global greenhouse gas emissions. Global climate change research requires animal models that accurately predict discharges to the environment from animal feeding regimes. Our aim was: to evaluate CH_4 predictions made by a dynamic mechanistic whole animal model by comparing them with data from on-farm trials. A trial was conducted at SRUC'S Beef Research Centre, UK. Steers (39 Charolais and 39 Luing) were fed two contrasting diets: concentrate based (barley 70%, barley straw 8%, molasses 2%, wheat distillers dark grains 18%, minerals 1%) or forage based (barley 38%, wheat distillers dark grains 10%, minerals 1%, whole crop barley silage 28% and grass silage 22%). Six indirect open-circuit respiration chambers were used (No Pollution Industrial Systems Ltd., Edinburgh UK). CH_4 concentrations were measured by infrared absorption spectroscopy. Animal and feed characteristics were used as inputs for the model and daily CH_4 emissions were predicted. CH_4 outputs were compared: for individual animal: correlation coefficient and concordance correlation coefficient (CCC); and at herd level paired t-test was used to compare observed and predicted means. All statistical analyses were carried out using R Individual animal predictions were acceptable (r=0.59 CCC=0.50 P<0.01 n=69). At herd level, observed CH_4 emissions 167.2±5.0 g/day were not different to that predicted by the model 162.3±2.8 (P=0.40, n=69) Given an adequate description of the animal and the diets consumed, our model can successfully predict CH_4 emissions and be used as a tool to evaluate CH_4 discharge to the environment as CH_4 output at a herd level for different breeds and diets.

Portuguese dairy farmers' views on animal welfare

S. Silva[1], M. Magalhães-Sant'ana[2], J. Borlido-Santos[1] and A. Olsson[1]
[1]IBMC, Rua do Campo Alegre,823, 4150-180 Porto, Portugal, [2]University College Dublin, School of Veterinary Medicine, Belfield,Dublin 4, Ireland; sandra.silva@ibmc.up.pt

Farm animal welfare (AW) is important for many Europeans and increasingly a subject of public debate in Europe. Despite farmers' central role in improving AW, only recently research started to address their representations of AW. The role of the farmer may be particularly important in dairy farming given its long human-animal relationships and the absence of European legislation protecting dairy cow welfare. The aim of this study was to provide a first mapping of how Portuguese dairy farmers think about AW. We conducted semi-structured interviews with 22 dairy farmers randomly selected from 3 dairy hubs in the north-west of Portugal, addressing the farmers' personal history, how they define and assess AW, its importance within their routine and how they perceive their own role, the role of other stakeholders and different production systems in the promotion of AW. AW was clearly valued [' If they fare well, we fare well']: for economic reasons (increased milk production, decreased costs), but also in terms of relationships with animals (liking cows was the main reason for 5 interviewees to become dairy farmers).In addition to housing, feeding and management, 5 farmers mentioned engaging with the cows, by paying attention to them and respecting them, as also being important for successful dairy farming. However, tensions became evident regarding production systems [between current conditions/practices vs view of an optimal production system, and between consumers' views on AW vs their willingness to pay more]. Regarding stakeholders, most farmers considered the cooperatives to be knowledgeable and engaged. Veterinarians may play an important role as knowledge-brokers for improved AW. Farmers were generally dismissive of retailers as being merely interested in the milk price. The view of consumers was more mixed: the general impression was that they were poorly informed, but several farmers thought they were nevertheless concerned with AW.

Housing for animal welfare in cattle

L. Leso[1], M. Uberti[2], W. Morshed[1] and M. Barbari[1]
[1]University of Florence, Department of Agricultural, Food and Forestry Systems, Via San Bonaventura 13, 50145 Florence, Italy, [2]Freelance veterinarian, NN, Mantua, Italy; lorenzo.leso@unifi.it

The free stall barn has some severe shortcomings with respect to animal welfare. Cultivated pack barns (CPB), i.e. compost barns, are alternative housing believed to offer improved welfare. Aim was to study the effect of CPB housing on longevity-related parameters. The study was conducted on 30 dairy farms in Po Plain, Italy, with Holstein breed: 20 had free stall barns -10 used rubber mattresses (FSM) and 10 straw bedding (FSS) – and 10 had CPB. Herd records were obtained from the Italian Dairy Association. Each farm was visited to assess barn's characteristics. An automatic model selection procedure (R's package 'glmulti') was used to evaluate the association between housing system and the outcome variables: herd age, no. of parity and monthly herd turnover rate. Total area per cow was higher in CPB (11.0 ± 4.1 m^2/cow) than in FSM (9.0 ± 2.0 m^2/cow) and FSS (9.3 ± 4.5 m^2/cow). Number of cows in CPB (132 ± 71) was lower than in FSM (173 ± 127) and FSS (163 ± 95). ME production at 305 days ($10,500\pm1,072$ kg in FSM, $10,901\pm988$ kg in FSS, and $10,541\pm667$ kg in CPB) was similar among housing systems. The final model for herd age included housing system, DIM, calving interval and age at first calving. Cows housed in CPB (47.50 mo) were older than those in FSM (44.88 mo, $P<0.001$), and FSS (45.64 mo, $P<0.001$), with no differences between FSM and FSS. The final model for no. of parity included housing system, length of dry period, calving interval and herd turnover rate. No. of parity was higher in CPB (2.41) than in FSM (2.16, $P<0.001$) and FSS (2.23, $P<0.001$). The final model for monthly herd turnover rate included housing system, 305-d mature equivalent yield, no. of parity and days open. Monthly herd turnover rates were 2.73, 2.60 and 2.82% in CPB, FSM and FSS, respectively, showing no significant differences. Results confirm that CPB housing may improve longevity of dairy cows. Further research is needed concerning turnover rates and reasons for culling.

Milk mineral variation of Italian dairy cattle breeds predicted by mid-infrared spectroscopy

G. Visentin[1], P. Gottardo[1], C. Plitzner[2] and M. Penasa[1]
[1]University of Padova, Department of Agronomy, Food, Natural Resources, Animals and Environment (DAFNAE), Viale dell'Università 16, 35020 Legnaro (PD), Italy, [2]Vereinigung der Südtiroler Tierzuchtverbände, Galvanistraße 38, 39100 Bolzano (BZ), Italy; giulio.visentin@studenti.unipd.it

Consumer's quality perception has changed in the last years and healthy aspects are becoming more and more relevant. Cow milk is an important source of minerals, especially Ca, K, Mg, Na, and P. Large-scale monitoring of milk mineral content is difficult due to expensive and time-consuming reference analyses, and therefore phenotyping of these traits at the population level is hardly achievable. Aim of the present study was to investigate sources of variation of milk Ca, K, Mg, Na, and P, predicted by mid-infrared spectroscopy models developed after uninformative variables elimination, on a large multi-breed spectral dataset (n=123,240). Cattle breeds considered were Holstein-Friesian, Brown Swiss, Alpine Grey, and Simmental. Sources of variation were studied using mixed models, including the fixed effects of breed, month and year of sampling, days in milk, parity, and the interactions between the main effects. Random factors were herd nested within breed, cow nested within breed, and the residual. All fixed effects were highly significant (P<0.01) in explaining the variation of milk mineral concentration. Milk of Simmental cows exhibited the greatest concentration of minerals. Alpine Grey breed had greater concentration of Ca and P compared to HF. These aspects could be important for a valorization of local breeds. Variation of concentration of Ca, Mg, Na, and P across lactation was significant and it exhibited an opposite trend to that of milk yield, with the lowest values around the peak of lactation. On the contrary, K concentration resembled the trend of milk yield.

Prediction of proteins including free amino acids in bovine milk by mid-infrared spectroscopy

A. Mcdermott[1,2], G. Visentin[1,2], M. De Marchi[1], D. Berry[2], M.A. Fenelon[3], P.M. O'connor[3] and S. Mcparland[2]
[1]Department of Agronomy, Food, Natural Resources, Animals and Environment, University of Padova, Padova, Italy, [2]Teagasc Animal & Grassland Research and Innovation Centre, Moorepark, Fermoy, Co. Cork, Ireland, [3]Teagasc Food Research Centre, Moorepark, Fermoy, Co. Cork, Ireland; audrey.mcdermott@teagasc.ie

The aim of this study was to evaluate the effectiveness of mid-infrared spectroscopy (MIRS) in predicting individual milk proteins and free amino acids in bovine milk and estimating their correlations with milk processing characteristics. A total of 730 milk samples were collected from seven Irish research herds and represented cows from a range of breeds, parities and stages of lactation. Gold-standard methods were used to quantify protein fractions and free amino acids (FAA) of these samples. Separate prediction equations were developed for each trait using partial least squares regression and accuracy of prediction was assessed using both cross validation and external validation. The greatest coefficient of correlation in external validation obtained for protein fractions was for total casein (0.74), while weak to moderate prediction accuracies were observed for FAA and among these glycine had the greatest coefficient of correlation (0.75 in both cross validation and external validation). Near unity correlations existed between total casein and beta-casein irrespective of whether the traits were based on the gold-standard (0.92) or MIRS predictions (0.95). Pearson correlations between gold-standard protein fractions and gold-standard milk processing characteristics ranged from -0.48 (pH and protein) to 0.50 (total casein and curd firmness). The lactation profile of total FAA indicated that the greatest concentration of FAA in milk was during early and late lactation. Results from this study demonstrate that mid-infrared spectroscopy has the potential to predict protein fractions and some FAA in milk.

Potential of visible-near infrared spectroscopy for the characterization of butter properties

T. Troch[1], V. Baeten[2], F. Dehareng[2], P. Dardenne[2], F.G. Colinet[1], N. Gengler[1] and M. Sindic[1]
[1]University of Liege, Gembloux Agro-Bio Tech, Passage des déportés 2, 5030 Gembloux, Belgium, [2]Walloon Agricultural Research Center, Chaussée de Namur 24, 5030 Gembloux, Belgium; ttroch@ulg.ac.be

This study is aimed to investigate the potential of visible-near infrared (VIS-NIR) spectroscopy for the determination of butter characteristics. Analyzes (butter yield, texture, color, dry matter, water and fat content) were performed on 88 butters prepared in duplicate from 44 individual cow milks. In order to maximize the variability in the measurements, milk samples were collected on basis of several criteria, e.g. breeds and moment of sampling (morning or evening milking). Butter analyzes were carried out the day of the production. Butter samples were then frozen and the VIS-NIR spectra were carried out later after thawing in a refrigerator at 13 °C. Standard normal variate transformation followed by a first derivative treatment were applied on the spectra. The principal component analysis of the VIS-NIR spectra allowed to distinguish butters made from different breeds (i.e. Jersey vs Holstein). Using partial least square regression (PLS) and VIS-NIR spectra with leave-one-out cross-validation (cv), high determination coefficients (R^2_c and R^2_{cv}) were obtained for the water content and color. For the water content, R^2_c was 0.91 and R^2_{cv} was 0.88. For the color, R^2_c and R^2_{cv} were respectively 0.93 and 0.91. These results indicated the possibility of using VIS-NIR spectral analysis for the simultaneous determination of mater content. The use of VIS-NIR spectra could be a tool to determine directly characteristics of new butter samples. Additional samples should be analyzed in order to consolidate the database and to validate with an independent dataset the models constructed.

Grass silages, differing in maturity and nitrogen fertilisation, on *in vitro* gas and methane

F.M. Macome[1,2], W.H. Hendriks[1,2], J. Dijkstra[2], D. Warner[2], W.F. Pellikaan[2], J.W. Cone[2] and J.T. Schonewille[1]
[1]Faculty of Veterinary Medicine, Department of Farm Animal Health, Martinus G de Bruin Building, Yalelaan 7, 3584 CL Utrecht, the Netherlands, [2]Wageningen University, Animal Sciences, De Elst1, building nr.122, 6708 WD, the Netherlands; felicidade.macome@wur.nl

An experiment was conducted to investigate the effect of grass silages, differing in maturity and nitrogen fertilisation, on *in vitro* gas and methane (CH_4) synthesis. A further objective was to compare the observed *in vitro* CH_4 production with that determined *in vivo*. The *in vitro* experiment was conducted simultaneously to the *in vivo* trial. A total of 12 lactating dairy cows were randomly assigned to one of the six experimental treatments consisting of silages from grass receiving two different levels of N fertilisation (65 vs 150 kg N/ha) and harvested at three different maturity stages (early, mid and late maturity). Samples of rumen fluid were taken from the animals. Each grass silage was incubated with rumen inoculum of a cow receiving the same silage in their basal diet, and with rumen fluid, mixed from the six cows. Within run, data from replicate bottles of each substrate-rumen fluid combination were averaged before statistical analysis, and were analysed using the GLM procedure of SAS. *In vitro* gas production decreased with increasing maturity of the grass silages. The N-fertilisation showed lower CH_4 productions at a low fertilisation, compared to a high fertilisation, using adapted rumen fluid. Total gas production was higher for young grass than for older grass and decreased linearly with increasing maturity. Using mixed rumen fluid to incubate the grass silages, showed a similar pattern as using the adapted rumen fluid. The *in vitro* CH_4 production of the tested grass silages did not correlate well with the *in vivo* CH_4 production when expressed in g/g OM or expressed in g/kg digested OM. It is concluded that the complexity of the rumen fermentation conditions need to be considered to predict the *in vivo* CH_4 production from *in vitro* measurements.

Factors affecting coagulation properties in Sarda sheep milk

M.G. Manca[1], J. Serdino[1], A. Puledda[1], G. Gaspa[1], P. Urgeghe[1], G. Battacone[1], I. Ibba[2] and N.P.P. Macciotta[1]
[1]Univesity of Sassari, Dipartimento di Agraria, Viale Italia 39, 07100, Italy, [2]Associazione Regionale Allevatori della Sardegna, Via Cavalcanti, 8, 09128, Italy; battacon@uniss.it

Sardinia is one of the most important regions of Europe for dairy sheep industry. Around 3 millions sheep are farmed and all the milk produced is used for cheese making therefore its technological properties are of great interest. The technological ability of milk for cheese production could be assessed by milk coagulation properties (MCP) that are measured as rennet coagulation time (RCT, min), curd firming time (k20, min) and curd firmness (a30, mm). Cheese making ability could be evaluated by individual cheese micro-manufacturing experiments. This study aims to investigate the variation in MCP and individual cheese yield of milk of Sarda sheep. Individual milk samples were collected from 1018 ewes farmed in 47 flocks located in the four provinces of Sardinia between April and July 2014. MCP were measured using Formagraph, cheese yield was assessed by laboratory individual micro-manufacturing experiments (ILCY), chemical composition was also determined by MilkoScan. Data were analyzed using mixed linear model including province, parity, somatic cells count (SCC), lambing month as fixed effects, age and days in milk (DIM) as covariate and herd as a random effect. Fat and protein contents were 6.04±1.43% and 5.45±0.72%, respectively. The variability of fat was related to province, parity, SCC and lambing month, while protein content was affected by SCC and lambing month. The clotting ability of Sarda sheep milk was quite good with a RCT equal to 15.10±6.37 min, k20 of 1.33±0.53 min and a30 of 49.25±20.18 mm, while the ILCY was equal to 36.25±9.33%. Statistical analysis showed that only SCC level affects significantly ILCY, RCT and a30 ($P<0.01$). The increase in SCC results in a worsening of coagulation properties that become significantly higher in the class with SCC>1.5 million/ml. Neither parity, lambing month and province did affect milk technological properties and cheese yield in this study.

Effects of crude glycerin supplementation on intake of primiparous grazing cows

M.C.A. Santana[1,2], H.A. Santana Junior[3], M.P. Figueiredo[3], E.O. Cardoso[4], F.B.E. Mendes[3], A.P. Oliveira[3] and E.S. Cardoso[3]
[1]Faculdade Evangélica de Goianésia, Agronomia, Av. Brasil, n° 1000 – Covoa, CEP: 76.380 – 000, Brazil, [2]EMATER, Rua 227A n.331 – Setor Leste Universitário Goiânia – Goiás – CEP: 74.610-060, Brazil, [3]Universidade Estadual do Piauí, Rua Joaquina Nogueira de Oliveira, s/n, CEP: 64002-150, Brazil, [4]Universidade Estadual do Sudoeste da Bahia, Praça da Primavera, s/n, 45700000, Brazil; mcaspaz@yahoo.com.br

The objective of this study was to evaluate the effect of different levels of crude glycerin supplementation on intake of primiparous lactating cows grazing on irrigated tropical pasture. The experiment was conducted at Rancho Santana farm, located in Jequié city, Bahia, Brazil, in the period from Dec. 21, 2010, to March 16, 2011. Ten 3/4 Holstein × 1/4 Dairy Gyr lactating primiparous cows, with 109±24 d of lactation and a mean age of 30±6 mo and mean body weight of 426.2±68.29 kg were distributed into five treatments, using two simultaneous 5×5 Latin squares. Treatments consisted of inclusion levels (0, 94, 191, 289, 389 g/kg dry matter basis) of crude glycerin (CG) in the supplement. Results were statistically analyzed by variance and regression analyses at 0.05 probablity. For all intake variables no significant differences were found ($P>0.05$) between the levels of crude glycerin. The lack of effects of adding glycerol to the TDMI is consistent with other findings of researches. In this research the variables of feed consumption as studied and presented were the average of total dry matter (DM) 13,65 kg/d; forage DM intake 9 kg/d; neutral detergent fiber 6,65 kg/d; organic matter 13.10 kg/d; non-fiber carbohydrates 3,8 kg/d; ether extract 0,49 kg/d; crude protein 1,96 kg/d and total digestible nutrients 8,21 kg/d. In conclusion, this data indicated that the addition of glycerin did not cause effects on the variables that express the feed consumption, so the recommended crude glycerin level could be increased to 389 g (171 g glycerol) per kg of supplement.

Identification of technological areas for dual purpose cattle in Mexico and Ecuador

J. Rangel[1], Y. Torres[2], C. De Pablos-Heredero[3], J.A. Espinoza[1], J. Rivas[4] and A. García[5]
[1]INIFAP, KM.3 Carretera Internacional Ocozocoautla – Cintalapa, 29140, Chiapas, Mexico, [2]Universidad Tecnica Estatal Quevedo, Campus Ing. Manuel Haz Álvarez, 13300, Quevedo, Ecuador, [3]Universidad Rey Juan Carlos, Paseo de los Artilleros S/N, 28032, Madrid, Spain, [4]Univesidad Central de Venezuela, Campus UCV-Maracay, 2101, Maracay, Venezuela, [5]Universidad de Córdoba, Campus Rabanales, 14071, Córdoba, Spain; pa1gamaa@uco.es

Dual purpose cattle farms are of great economic and social importance. It is a mixed family farming system that is combined with crops. These systems are known as low-input systems and small-scale producers. The future of the system is associated with specialization based on the adoption of technologies. The aim of this research was to identify technological areas and their relationship with the variability in production. 183 technologies into six technological areas were grouped by using qualitative and participatory methods. The information was obtained from a sample of 3,603 and 130 dual purpose cattle farms from, respectively, Mexico and Ecuador. The means of technologies adopted was 123 ± 6. Technological areas showing higher degrees of implementattion were animal health (67.8%) and feeding (56%) and management (55.7%); but their adoption is not sequential or responds to independent events. This research facilitates the identification of a number of technologies that should be implemented from an organizational strategy perspective. Moreover, all technologies are seeking a dynamic balance system that allows firms migrating to more efficient processes.

Technology inventory of dual-purpose cattle farms in the Mexican and Ecuatorian tropics

Y. Torres[1], J. Rangel[2], C. De Pablos-Heredero[3], J. Rivas[4], E. Angón[2] and A. García[2]
[1]Universidad Técnica Estatal de Quevedo, Campus Ing. Manuel Haz Álvarez, 5863, Quevedo, Ecuador, [2]Universidad de Córdoba, Campus Rabanales. Madrid-Cádiz, km5, 14071, Córdoba, Spain, [3]Universidad Rey Juan Carlos, Paseo de los Artilleros S/N, 28034, Madrid, Spain, [4]Universidad Central de Venezuela, Campus UCV-Maracay, 2101, Maracay, Aragua, Venezuela; pa1gamaa@uco.es

The aim of this research was to determine the technological inventory, the level of thenological adoption and its association with the productive characteristics of the dual purpose cattle in tropical areas of Mexico and the Coast of Ecuador. The sample was of 3603 farms from Mexico and 135 from Ecuador, during the 2012-2013 period. Descriptive statistics were applied to analize the adoption of the technologies. The Spearman correlation was used to study the degree of association between the various technologies and milk production. 183 technologies were identified. The technologies of animal health and reproduction were the most adopted: 94.7% in Mexico and 91.4% in Ecuador. The technologies adoption depends ($P<0.05$) on the level of education, the expectations of continuing in the cattle activity, organization of legal aspects, land ownership and herd size. The relationship between the total number of technologies and the milk production was highly positive and significant ($r=0.78$). The dual purpose cattle farms were related to mixed and complex systems. The knowledge about the technological adoption inventory had a major impact on innovations: the level of technology adoption was associated with improved milk production.

Using of myrrh essential oil as a highly effective antimicrobial agent in processed cheese
A.G. Mohamed and T.A. Morsy
National Research Centre, 33 elbehoss st. Dokki, 12311, Egypt; ashrafare@yahoo.com

Commiphora myrrha is considered as a highly effective, natural and safe material. Antimicrobial activity tests of *C. myrrha* as essential oil against different species of pathogenic gram-positive bacteria {*Listeria monocytogenes* (EMCC 1875); *Staphylococus aureus* (ATCC13565) and *Bacillus cereus* (EMCC1080)} as well as gram-negative bacteria {*Escherichia coli* O157:H7 (ATCC51659) and *Salmonella typhimurium* (ATCC 25566)} were carried out. Minimum inhibitory concentration (MIC) was also estimated. Data revealed that all tested microorganisms were susceptible to the action of *C. myrrha*. Their MIC were (2-5 µl/ml) for all microorganisms. Processed cheese spreads (PCSs) samples were prepared by using five different ratios of *C. myrrha* essential oil to evaluate its organoleptic acceptability. Gross chemical composition, SN, TVFA as well as pH values were estimated through three months of storage at 5±1 °C. Oil separation, penetration, melting point and color parameters were also determined through the storage period. Obtained results revealed that using 0.2% (w/w) *C. myrrha*-EO for preparing PCSs gave satisfactorily sensory properties. The appearance was well shiny; gumminess and oil separation were absent. The overall preference of preferable ratio was like very much comparied to the other treatments. Color was more acceptable and it well done spread when it was compared to the control. The penetration values of satisfied treatment of myrrh were 33.5, 32.0 and 31.2 mm when fresh, and after one month and three months of storage compared to control samples: 33.0, 30.5 and 26.5 mm. On the other hand, meltability took the same trend; their values were 85.4, 81.6 and 80.0 mm for the best treatment compard with the control samples (81.6, 80.5 and 78.7 mm) during fresh, one month and 3 months. So; it could be concluded that using of Commiphora myrrha essential oil is succesful for preparing processed cheese spreads with acceptable properties and satisfactory sensory behavior.

Variation in the cattle genome – what it does and what it means
B. Guldbrandtsen, M.B. Hansen and G. Sahana
Aarhus University, Center for Quantitative Genetics and Genomics, Dept. of Molecular Biology and Genetics, Box 50, 8830 Tjele, Denmark; bernt.guldbrandtsen@mbg.au.dk

Genome sequencing has proven itself a transformative technology in animal breeding including cattle breeding. The 1000 Bull Genomes Project has provided us with comprehensive knowledge of genomic variation in most modern cattle breeds, but sequences can only be interpreted in the light of functional information. Databases provide the opportunity to interpret the biological significance of biological of sequence variation, qualitatively and quantitatively. However volumes of variation are large, so automatic tools are required. Annotations need to be presented in standardized languages, ontologies e.g. Sequences Ontology and Mammalian Phenotype Ontology to enable integration and searching. Annotations are available for elements present, functions of the elements, known polymorphisms, and known or predicted effects of variation. Sources include ENSEMBL, Online Mendelian Inheritance in Animals and Animal QTL Database. Homeology means that information from other species can be incorporated from say ENSEMBL, the International Mouse Phenotype Consortium and Online Inheritance in Man. Information about effects of variants in coding sequences is of particular interest: functional parts of genes affected, the predicted magnitude of the effect of the changes, and sites' evolutionarily conservation. However there are numerous pitfalls in the interpretation of evidence: absence of evidence is not evidence of absence, marker sets are often filtered, assemblies and alignments may be wrong, and categories of genomic elements maybe missing – or even undiscovered. Two cases where annotations on are used in a semi-quantitative fashion on a genomic scale will be presented. One will be the annotations of the top SNP for a large set of QTL mapped in dairy cattle, the other will be functional characterization of chromosomal regions highly differentiated between Danish and French cattle breeds.

Multi-breed genomic prediction from whole genome sequence data in dairy cattle

B.J. Hayes[1,2,3], I. Macleod[2,4], A. Capitan[5,6], H.D. Daetwyler[1,2,3] and P.J. Bowman[2,3]
[1]La Trobe University, Bundoora, Victoria, Australia, [2]Dairy Futures Cooperative Research Centre, Bundoora 3083, Victoria, Australia, [3]Biosciences Research Division, Department of Primary Industries Victoria, 5 Ring Road, 3086 Bundoora, Australia, [4]Faculty of Land and Food Resources, University of Melbourne, Parkville, Victoria, Australia, [5]Union Nationale des Coopératives d'Elevage et d'Insémination Animale, 75595, Paris, France, [6]INRA, UMR1313 Génétique Animale et Biologie Intégrative, 78350 Jouy-en-Josas, France; ben.hayes@depi.vic.gov.au

Discoveries from whole genome sequencing of cattle have been translated into tests that cattle breeders can use. For example a lethal recessive mutation in the SMC2 gene in Holstein-Friesian cattle that causes embryonic loss was identified as a result of sequencing several hundred bulls – this mutation can now be routinely screened on very large numbers of cattle, so that carrier matings are avoided. A more ambitious goal is to use whole genome sequence variant genotypes for prediction of quantitative traits. Genomic prediction is widely used in dairy cattle breeding programs around the world, where selection of young animals for breeding is based on genomic predictions of milk yield, fertility, mastitis resistance, longevity and other traits. While this has increased rates of genetic gain, the fact that these predictions are typically based on 50,000 SNP genotypes could limit the accuracy of predictions, the persistency of the predictions across generations, and their utility across breeds. Using whole genome sequence data might overcome these limitations, as the causative mutations underlying the variation in the complex traits should be in the data set. We describe a new method for genomic prediction that simultaneously maps causative mutations and estimates variant effects. The value of the method is demonstrated in a multi-breed dairy cattle data set of 16,000 animals with imputed whole genome sequence data, and fertility, somatic cell count and milk production phenotypes. In some cases plausible mutations affecting these traits were identified. This information can be rapidly applied in dairy cattle breeding, by including these mutations on low density SNP arrays that are now used to test hundreds of thousands of animals each year.

Gene and pathway analysis of metabolic traits in dairy cows

N.-T. Ha[1,2], J.J. Gross[2], H.A. Van Dorland[2], J. Tetens[3], G. Thaller[3], M. Schlather[4], R.M. Bruckmaier[2] and H. Simianer[1]
[1]Department of Animal Sciences, Georg-August-University Goettingen, 37075 Göttingen, Germany, [2]Vetsuisse Faculty, University of Bern, 3012 Bern, Switzerland, [3]Institute of Animal Breeding and Husbandry, Christian-Albrechts-University Kiel, 24118 Kiel, Germany, [4]Chair of Mathematical Statistics, University of Mannheim, 68161 Mannheim, Germany; ngoc-thuy.ha@vetsuisse.unibe.ch

During early lactation, dairy cows experience a severe metabolic load. Here, a failure in metabolic adaptation results in an increased susceptibility to health problems. We analyzed the genetic basis of the metabolic adaptability during the transition period. Blood samples were taken from 178 cows at three critical stages: T1 = week 3 ante-partum (no metabolic load); T2 = week 4 post-partum (lactating and high metabolic load), and T3 = week 13 post-partum (lactating and low metabolic load). Plasma concentrations of non-esterified fatty acids, beta-hydroxybutyrate and glucose, metabolites characterizing the metabolic status and adaptability, were measured at T1, T2, and T3. All cows were genotyped with the Illumina HD Bovine BeadChip. After quality control, the remaining 601,455 SNPs were annotated to genes (Ensembl) and pathways (KEGG). For each gene and phenotype, we performed a score test based on a linear regression model with all SNPs in the gene as explanatory variables, while considering also breed effects. The results were used to identify pathways enriched for significant genes using a Kolmogorov-Smirnov test. We found 99 genes significantly associated with the metabolites. For each metabolite, we found genes that are significant at T2 but not at T1/T3 or vice versa. This strongly suggests those genes to be involved in the adaptive regulation. We further identified three pathways (steroid hormone biosynthesis, ether lipid metabolism and glycerophospholipid metabolism) to jointly affect the metabolites. In conclusion, this may be regarded as evidence for the genetic basis for the adaptation performance and, at the same time, reveals its complexity.

Dissecting complex traits in pigs: metabotypes illuminate genomics for practical applications

L. Fontanesi[1], G. Schiavo[1], S. Bovo[1], G. Mazzoni[1], F. Fanelli[2], A. Ribani[1], V.J. Utzeri[1], D. Luise[1], A.B. Samorè[1], G. Galimberti[3], D.G. Calò[3], A. Manisi[3], F. Bertolini[1], M. Mezzullo[2], U. Pagotto[2], S. Dall'olio[1], P. Trevisi[1] and P. Bosi[1]
[1]University of Bologna, Department of Agricultural and Food Sciences, Division of Animal Sciences, Viale Fanin 46, 40127 Bologna, Italy, [2]University of Bologna, Department of Surgical and Medical Sciences, Endocrinology Unit, Via Massarenti 9, 40138 Bologna, Italy, [3]University of Bologna, Department of Statistical Sciences 'Paolo Fortunati', Via delle Belle Arti, 40126 Bologna, Italy; luca.fontanesi@unibo.it

Integration of omics information can produce holistic visions of biological processes underlying production traits in livestock. In this study we combined genomics and metabolomics in addition to a deep phenotyping approach in pigs to dissect complex traits of economic relevance. About 180 metabotypes were measured on plasma of 900 performance tested Italian Large White pigs that were also genotyped with the Illumina PorcineSNP60 BeadChip. All these pigs were also measured for about 40 performance, carcass and haematological traits. Metabotype information was used to obtain: (1) heritability estimation of these molecular phenotypes; (2) a Graphical Gaussian Model (GGM), that defined biological networks; (3) predictors of production traits; and (4) descriptors of sex differences. Genome wide association studies identified genetic variations in genes directly involved in the biochemical pathways of several metabolites that were also characterized by target and whole genome resequencing, opening interesting applications in pig breeding and nutrigenomics. This study demonstrated the powerful potential derived by the dissection of complex traits using a reductionist approach based on molecular phenotypes.

Towards better understanding of genetic variation in dairy cattle breeds

J. Szyda[1], K. Wojdak-Maksymiec[2], G. Minozzi[3], E. Nicolazzi[3], C. Diaz[4], A. Rossoni[5], C. Egger-Danner[6], J. Woolliams[7], L. Varona[4], R. Giannico[3], H. Schwarzenbacher[6], C. Ferrandi[3], T. Solberg[8], F. Seefried[9], D. Vicario[10] and J. Williams[3]
[1]Wroclaw University of Environmental and Life Sciences, Kozuchowska 7, 51-631 Wroclaw, Poland, [2]West Pomeranian University of Technology, Słowackiego 17, 71-434 Szczecin, Poland, [3]Parco Tecnologico Padano, vie Einstein, 26900, Lodi, Italy, [4]University of Zaragoza, Calle Miguel Servet 177, 50013 Zaragoza, Spain, [5]Associazione Nazionale Allevatori Bovini della Razza Bruna, Ferlina 204, 37012 Bussolengo, Italy, [6]ZuchtData EDV-Dienstleistungen GmbH, Dresdner Straße 89/19, 1200 Wien, Austria, [7]Roslin BioCentre, Wallace Building, Midlothian EH25 9PP, United Kingdom, [8]GENO Breeding and A.I. Association, Holsetgata 22, 2317 Hamar, Norway, [9]Qualitas AG, Chamerstrasse 56, 6300 Zug, Switzerland, [10]Associazione Nazionale Allevatori Bovini di razza Pezzata Rossa Italiana, Via Ippolito Nievo 19, 33100 Udine, Italy; joanna.szyda@up.wroc.pl

Thanks to the availability of whole genome DNA sequences it is now possible to identify most of the variation contained in individual genomes. If multiple individuals are sequenced it is also possible to test for patterns of variation in sequence using formal hypothesis testing. In our study whole genome sequences of 130 bulls and 32 cows were used for assessing patterns of inter-individual, inter-genic, and functional variation in mononucleotide polymorphisms. Bulls and cows were sequenced within the framework of the Gene2Farm project and the NADIR project respectively using the IlluminaHiSeq 2000 Next Generation Sequencing platform. Filtered reads were aligned to the reference genome using BWA-MEM and then polymorphic variants were identified using FreeBayes. The variability in the numbers of polymorphisms across individuals, breeds, but also across genes representing different functional categories, as well as separately for structural elements of genes representing exons and UTRs and different metabolic pathways was tested, revealing significant differences.

Using sequences data to identify causal variants for milk fatty acids composition in dairy cattle
A. Govignon-Gion[1,2], M.P. Sanchez[1,2], P. Croiseau[1,2], A. Barbat[1,2], S. Fritz[3], M. Boussaha[1,2], M. Brochard[4] and D. Boichard[1,2]
[1]AgroParisTech, 16 rue Claude Bernard, 75231 Paris, France, [2]INRA, UMR1313 GABI, Domaine de Vilvert, 78350 Jouy en Josas, France, [3]Allice, 149 rue de Bercy, 75012 Paris, France, [4]Idele, 149 rue de Bercy, 75012 Paris, France; didier.boichard@jouy.inra.fr

In the framework of the PhenoFinlait project, milk fatty acids were analyzed by genome-wide association study at the whole genome sequence level in three French dairy cattle breeds. Traits were estimated from Mid-Infrared spectrometry for cows in first or second parity in Montbeliarde (MO), Normande (NO) and Holstein (HO) cattle breeds. More than 8,000 cows were genotyped with the 50k Beadchip. These genotypes were subsequently imputed within breed in two steps: a first imputation step was performed from the 50k into the HD beadchip level, using a reference of 522 MO, 546 NO and 776 HO bulls genotyped on the HD chip, and followed by a second imputation step to get into the sequence level, using a multi-breed population of 1,147 bulls of the '1000 bull genomes project' as a reference panel. Individual test-day records were first adjusted for environmental effects and averaged per cow. Analyses were conducted within breed for 23 fatty acids with GCTA software, which implemented a linear mixed model with a polygenic component (based on a genomic relationship matrix), and a residual. The most significant regions were in agreement with known regions for fat components or fatty acid desaturation, e.g. on chromosomes 14 (DGAT1 gene), 19 between 51.3 Mb and 51.4 Mb (FASN) and chromosome 26 around 21 Mb (SCD1). Several additional QTLs were also found on chromosomes 5, 11 (especially in the LGB gene for unsaturated fatty acids), 20 and 27 (AGPAT6). Resolution obtained in the present study was high enough to directly pinpoint to several candidate mutations. The authors acknowledge the financial support from ANR, Apis-Gène, French Ministry of Agriculture (CASDAR), Cniel, FranceAgriMer and FGE and the contribution of the 1000 bull genomes consortium.

Whole genome mapping of genes affecting fatness in pigs using linear mixed model and Bayesian method
D.N. Do and H.N. Kadarmideen
Department of Clinical Veterinary and Animal Sciences, Faculty of Health and Medical Sciences, University of Copenhagen, Frederiksberg C, 1870, Denmark; hajak@sund.ku.dk

Backfat thickness is one of the importance traits that determine meat quality which is highly relevant for both livestock producers and consumers. Therefore, identification of single nucleotide polymorphisms (SNPs) linked to backfat thickness can provide an alternative tool to improve carcass and meat quality traits. This aim of this study was to identify genomic regions controlling backfat thickness in Yorkshire pigs. After quality control, a total of 37,192 SNPs and 596 pigs remained in the final dataset for genome wide association analyses (GWA). Deregressed EBVs were used as response variables in GWA which were performed by a linear mixed model (LMM) using DMU package and a Bayesian Variable Selection (BVS) model using Bayz package. Nine and one SNPs were detected to be genome-wide significantly associated with the traits at the threshold of $P<5e^{-05}$ and Bayes factor (BF)>10 using LMM and BVS, respectively. Among top 100 associated SNPs, only 28 SNPs were found in common between two methods. The mutation (A/G) at the locus ALGA0111801 located in the intergenic regions on chromosome 5 showed the strongest association with the trait by both methods ($P=7.77e^{-06}$ and BF=11.86). Approximately 17 kb apart from this locus, the lysophosphatidylcholine acyltransferase 3 (LPCAT3) gene involving phospholipid metabolism can be an interesting candidate gene for fatness traits. Other regions on chromosome 3, 5, 17 and 18 also harbor significant SNPs for the trait and will be further characterized. Finally, studies on pathway enrichment and systems genetics will be carried out to reveal functional importance of genes on candidate chromosomes with high numbers of significant SNPs. This study offers the list of candidate SNPs which enhances our biological knowledge of the variants controlling fatness that might be relevant not only for reducing the fatness in pigs but also for studying obesity in humans.

Molecular response to heat stress and lipopolysaccharide in chicken macrophage-like cell line

A. Slawinska[1,2], M.G. Kaiser[2] and S.J. Lamont[2]
[1]UTP University of Science and Technology, Department of Animal Biochemistry and Biotechnology, Mazowiecka 28, 85-084 Bydgoszcz, Poland, [2]Iowa State University, Department of Animal Science, 2455 Kildee Hall, 50010 Ames, USA; slawinska@utp.edu.pl

Climate change modulates the environment of the domestic animals by providing additional abiotic (e.g. heat stress) and biotic (e.g. modified activity or range of pathogens) stressors. The animals react to such stimuli through modulation of their transcriptome, including activation of stress and immune response pathways. In this study we aimed to determine the molecular response of the chicken macrophage-like cell line (HD11) to heat stress combined with immune stimulation with lipopolysaccharide (LPS) *in vitro*. The experimental treatment included heat stress (2 hours at 45 °C followed by 2, 4 and 8 h of the temperature-recovery period at 42 °C), LPS antigen stimulation (5 µg/ml) or the combination of both stressors (heat stress and LPS). Control samples were cells incubated in physiological temperature without LPS. Each experiment was performed three times followed by RNA isolation directly after heat stress and after each temperature-recovery period. For the gene expression analysis, high throughput RT-qPCR was performed with the Fluidigm BioMark HD system using 192.24 Dynamic Arrays. The gene panel included members of gene families involved in heat-stress response (e.g. heat shock proteins, HSP transcription factors), apoptosis (e.g. caspases), and different signaling pathways (e.g. MAPK, TGFB and TLR signaling pathways) as well as cytokines (e.g. IL-1B, IL-6, IFNG) and chemokines (e.g. CCL4, CCL5, IL-8). The effects of the treatments were compared using least square means method. Performing this study allowed us to analyze the regulation of the immune-related gene expression in environmental stress conditions, as well as to describe the molecular function of the heat stress proteins during immunological burden. Acknowledgements: Polish-American Fulbright Commission and USDA-NIFA-AFRI Climate Change Award #2011-67003-30228.

Data integration and network reconstruction with muscle metabolome and meat quality data in pig

J. Welzenbach, C. Große-Brinkhaus, C. Neuhoff, C. C. Looft, K. Schellander and E. Tholen
Institute of Animal Science, University of Bonn, Endenicher Allee 15, 53115 Bonn, Germany; jwel@itw.uni-bonn.de

Aim of this study was the elucidation of underlying biochemical processes and identification of potential key molecules of meat quality traits drip loss, pH1, pH24 and meat color. An untargeted metabolomics approach detected the profiles of 400 annotated and 1,597 unreported metabolites in 97 Duroc × Pietrain pigs. The levels of metabolites are helpful in order to understand the complex biological mechanisms of the underlying meat quality traits. Additionally in animal selection metabolite biomarkers might be used for prediction of economical attractive phenotypes. Four different statistical methods, namely correlation analysis, principal component analysis (PCA), random forest regression (RFR) and weighted network analysis (WNA) were applied to identify significant relationships between muscle metabolites and meat quality traits and to handle with the statistical 'large p, small n' problem. Despite obvious differences regarding the statistical approaches, parameters and command variables, the four applied statistical methods revealed similar results. These lead to the conclusion that meat quality traits pH1, pH24 and color were strongly influenced by processes of pm energy metabolism like glycolysis and pentose phosphate pathway whereas drip loss was significant associated with metabolites of lipid metabolism. In order to predict meat quality accurately and to clarify the complex molecular background of drip loss in more detail further investigations to select the most suitable statistical method are needed. Based on comprehensive understanding of drip loss selected metabolites might be better indicator than the measured phenotype itself. As a further step metabolomics data will be combined with proteomics and transcriptomics data and analyzed in an integrated GWAS.

Development of a genetic marker panel to predict reproductive longevity in Holstein cattle

K. Żukowski[1], N. Jain[1], J. Joshi[1], J.E. Koenig[1], R.G. Beiko[1] and H. Van Der Steen[2]
[1]Dalhousie University, Faculty of Computer Science, Halifax, Canada, [2]Performance Genomics Inc., Truro, Canada; zukowski.kacper@gmail.com

Reproductive longevity (RL) traits are economically relevant in both male and female cattle, where improvement of herd life and reduction of cow replacement rate are significant production factors. RL is of particular importance to the dairy cattle industry. RL is determined by a complex of traits involving ovarian function, fertility, stress resistance, robustness, lifetime production and health. The genetic basis of RL is hence very complicated; for example, the reproduction trait class in the CattleQTLdb has more than 2,300 annotated QTLs. This relatively large number of genes and implied pathways represents both a major challenge and a substantial opportunity to improve the RL trait of cattle, but large-scale sequencing and analysis are needed to identify influential genes with subtle effects. Our consortium to investigate the genetic basis of RL includes four partners: Dalhousie University as a scientific partner, CRV and Geneseek as industrial partners, and Performance Genomics Inc. (PGI) as a leader of the project. The project is founded on a unique resource, mouse lines selectively bred over a period of 30 years that reproduce for twice as long and have twice as many litters as compared to matched control lines. We are using several techniques to investigate the mouse data, including genome-wide association study/differential expression (GWAS/DE) based on microarray and sequencing (DNASeq/RNASeq) data. Candidate genes identified in this mouse study are being used to complement a detailed cattle GWAS analysis in which over 400 bulls with high and low genetic merit for longevity will be sequenced. The primary project outcome will be commercial DNA marker-based tests of RL for the Holstein cattle industry. This study is being supported by: PGI, CRV, GeneSeek, the NRC-IRAP, Canada no. 811790 and no. 832709, ACOA and the Mitacs-Accelerate Graduate Research Internship Program no. IT04818 and IT02410.

A framework to incorporate knowledge on gene interaction into genomic relationship

J.W.R. Martini[1], M. Erbe[2], V. Wimmer[3] and H. Simianer[1]
[1]Department of Animal Sciences, Georg-August-University Goettingen, 37075 Göttingen, Germany, [2]Institute of Animal Breeding, Bavarian State Research Centre for Agriculture, 85354 Freising, Germany, [3]KWS SAAT SE, P.O. Box 1463, 37574 Einbeck, Germany; hsimian@gwdg.de

The relationship matrix of different animals or plant lines is the central object in many statistical methods used in breeding science. Due to decreasing costs of genotyping, the classical pedigree has been substituted more and more by genomic relationship matrices in the last years. The current standard approach calculates a genomic relationship matrix whose structure is motivated by a model with additive marker effects only. Additive here means that the effect a marker allele has on the phenotype is independent of the configuration of alleles at other loci (no epistasis) and of the same locus on the other homologous chromosome(s) (no dominance). The overall effect is then just the sum of all allele effects across the genome. This model does not seem to be suited to capture fully the nature of the phenotype-building biological networks in which gene to construct a relationship matrix, if we assume interactions only between defined pairs of markers. This extension of the purely additive model provides the basis for building trait specific, gene-network dependent relationship matrices, which can incorporate a priori knowledge on the biology of the trait and thus may lead to products interact intensively. Here, we present a generalization of this concept by building a relationship matrix based on marker effects and pairwise interactions. Moreover, we show how more accurate predictions. As a final illustration, we demonstrate ith two real data sets that already the model with interactions between all marker pairs can increase predictive ability for quantitative traits.

Identification of genomic regions related to marbling in Nellore cattle

M.E. Carvalho[1], F. Baldi[2], M.H.A. Santana[1], R.V. Ventura[1], G.A. Oliveira Junior[1], R.S. Bueno[1], M.N. Bonin[3], F.M. Rezende[4] and J.B.S. Ferraz[1]
[1]University of São Paulo, Duque de Caxias, 225, Pirassununga, SP, 13635-900, Brazil, [2]Sao Paulo State University, Via Prof. Paulo Donato Castellani, Jaboticabal, SP, 14884-900, Brazil, [3]Embrapa Beef Cattle, Av. Radio Maia, 830, Campo Grande, MS, 79106-550, Brazil, [4]Federal University of Uberlândia, Av. Para, 1720, Uberlandia, MG, 38400-902, Brazil; jbferraz@usp.br

The aim of this study was to identify, by ssGWAS, genomic regions that potentially have association with marbling in Nellore cattle. Phenotypes were obtained according to standard USDA Quality Grade (1999), where standard levels were considered as numeric values varying between 400 and 900. Data of 1,080 Nellore bulls presented mode equal to 400 with 51.7% of the measurements. Those animals were genotyped with Illumina Bovine beadchip HD® GGPi (74k). Based on another Nellore population genotyped for Illumina beadchip BovineHD® (777k), genotypes were imputed by FImput software. Analyses were performed using a pedigree composed by 6,276 animals and, assuming contemporary group (farm and slaughter batch) as fixed effect and age at slaughter as a covariate. Single step analyses were realized by Blupf90 program considering windows of 10 markers (SNP) to estimate their effects, this procedure enables the identification of regions associated with marbling along the chromosomes. After quality control (MAF <0.05%, call rate <90%), 463,995 SNPs in autosomal chromosomes were used in the association analyses. Based on that, 19 regions in 11 different chromosomes (1, 4, 5, 7, 8, 12, 16, 21, 23, 25 and 29), that explained more than 1% of the additive variance, were explored and some genes were identified in these regions, as KIAA1797, SLC41A2, ALDH1L2, APPl2, CARD10 and OR12D3. With ssGWAS method using high density panel was possible to identify regions related to marbling in Nellore beef cattle. Posteriorly, those genes and their pathways will be investigated to evaluate their importance for meat quality traits.

Effect of pathway-based low-density SNP panels in pigs

T. Okamura[1], Y. Takahagi[2], C. Kojima-Shibata[3], H. Kadowaki[3], E. Suzuki[3], M. Nishio[1], M. Matsumoto[4], H. Uenishi[4], K. Suzuki[5] and M. Satoh[1]
[1]NARO Institute of Livestock and gGrassland Science, Tsukuba, Ibaraki, 3050901, Japan, [2]NH Foods Ltd. R&D Center, Tsukuba, Ibaraki, 3002646, Japan, [3]Miyagi Prefecture Animal Industry Experiment Station, Osaki, Miyagi, 9896445, Japan, [4]National Institute of Agrobiological Sciences, Tsukuba, Ibaraki, 3058602, Japan, [5]Tohoku University, Graduate School of Agricultural Science, Sendai, Miyagi, 9818555, Japan; okamut@affrc.go.jp

The objectives of the current study were to build low-density SNPs panels based on our previous pathway analysis for complement alternative pathway activity (CAPA) in Large White (LW) and to investigate the effects of these panels on the accuracy of genomic evaluations in LW and Landrace (LA). The phenotypes of 1,524 and 1,304 individuals, and the genotypes of 570 and 338 individuals, were studied in LW and LA populations, respectively. Three SNP panels were extracted from the 60k SNP chip. In the first panel (Full SNP), 38,591 and 42,931 SNPs with minor allele frequencies (>0.05) were extracted for LW and LA, respectively. In the second (Pathway SNP), SNPs in the neighborhood of genes belonging to the pathways were extracted, resulting in 1260 and 1539 SNPs for LW and LA, respectively. In the third (Gwas SNP), the SNPs related to CAPA in our previous GWAS in LW were extracted, resulting in equivalent number of Pathway SNP. The genomic evaluations were estimated with each panel by using single-step genomic BLUP and the accuracies were defined as the correlations between the evaluations and the adjusted phenotypes in the validation population by a 5-fold cross validation. The results show that the accuracies for the Pathway SNP in both LW and LA (0.30 and 0.11, respectively) were similar to those for the Full SNP (0.31 and 0.09, respectively), while the highest and lowest accuracies were for the Gwas SNP in LW (0.58) and LA (0.05), respectively. These results suggest that extraction of the low-density SNP panel according to the results of pathway analysis was effective even if the breed was different.

IDH3B genotypes relate with carcass traits in an F2 between Landrace and the Jeju (Korea) Black pig

S.H. Han[1,2], Y.J. Kang[3], Y.K. Kim[1,3], H.S. Oh[1] and I.C. Cho[3]
[1]*Jeju National University, Department of Science Education, Jeju National University,102 Jejudaehak-ro, Jeju-si, Jeju Special Self-Governing Province, 690-756, Korea, South,* [2]*Jeju National University, Educational Science Research Institute, Jeju National University,102 Jejudaehak-ro, Jeju-si, Jeju Special Self-Governing Province, 690-756, Korea, South,* [3]*National Institute of Animal Science, Rural Development Administration, Subtropical Livestock Research Institute, 593-50, Sanrokbukro, Jeju-si, Jeju Special Self-Governing Province, 690-150, Korea, South; hansh04@naver.com*

This study tested the association between genetic polymorphisms of iso-citrate dehydrogenase 3, beta subunit (IDH3B) gene and economic traits in an F_2 crossbred population between Landrace and Jeju (South Korea) Black pig. A 304-bp insertion/deletion mutation in promoter region was screened for determining genotypes of IDH3B gene for a total of 1,070 F_2 pigs. Total of three genotypes (AA, AB and BB) were found in the founders, F_1 and F_2 population. Association analysis results showed the significant differences with carcass weights (CW), backfat thickness in two positions of the body (BF4_5 and BF11_12) and carcass lengths (CL) ($p<0.05$), but not with meat color (MC), eye muscle area (EMA) and marbling scores (MARB) ($p>0.05$). The F_2 pigs harboring the IDH3B BB homozygote showed heavier CL (80.790±0.725 kg) and shorter CL (101.875±0.336 cm) than those of the others ($p<0.05$), respectively. In addition, the BF levels of BF measured between 4th and 5th and 11th and 12th vertebrae were thicker in the carcasses harboring the IDH3B BB genotype than those from the other genotypes AA and AB ($p<0.05$). These results indicated that dfferent IDH3B genotpes induced the different biological metabolisms in fat deposition in subcutaneous fat tissues. This study suggested that the genetic variations of IDH3B gene may assist as molecular genetic markers for improving the Jeju Black pig, Landrace and their-related crossbreeding systems.

Impact of somatic nuclear reprogramming on DNA methylation in bovine placental and foetal tissues:

M. Guillomot[1], A. Prezelin[1], H. Kiefer[1], L. Jouneau[1], F. Piumi[1], J. Tost[2], V. Renaud[1,2], C. Richard[1], D. Lebourhis[3], N. Beaujean[1], T. Aguirre-Lanvin[1], J. Salvaing[1], J.P. Renard[1] and H. Jammes[1]
[1]*INRA, UMR1198, Domaine de vilvert Batiment 230, 78352 Jouy en Josas, France,* [2]*CNG/CEA, 2, rue Gaston Crémieux, CP5706, 91057 Evry Cedex, France,* [3]*ALLICE, R&D Department, Route Departementale 73, 37380 Nouzilly, France; helene.jammes@jouy.inra.fr*

DNA methylation (5-mC) and hydroxymethylation (5-hmC), are highly dynamic during mammalian embryogenesis. In blastocyst, the first cell differentiation is associated with higher levels of 5-mC in the inner cell mass than in trophectoderm, and this profile is disrupted by somatic cell nuclear transfer (SCNT) in bovine. Whether placental pathologies observed in clones are associated with alterations of global DNA methylation remains to be determined. In this study, DNA methylation was analysed in extra-embryonic and foetal tissues obtained by artificial insemination (AI) or SCNT (1) by a global quantification using a LUminometric Methylation Assay; (2) by Methylated DNA immunoprecipitation (MeDIP)/sequencing; and (3) by immunohistochemistry with specific antibodies. Variations in global methylation levels were observed in a range of 30-90% (chorionic villi<chorion<liver<heart<brain<amnion<allantois at D60 of gestation). As observed in blastocyst, SCNT increased the global methylation level in trophoblast at D18, in chorion at D40 and in chorionic villi at D60. Only the mesenchymal part of chorionic villi is stained using 5-mC antibody, suggesting that this tissue is responsible for this increase. An unexpected 5-hmC staining was found in various placental and extra-embryonic cell types, suggesting a role for this epigenetic mark in all lineages in bovine. Genome-wide analysis by MeDIP/sequencing identified a large number of differentially methylated regions (DMR) between stages and tissues. The methylation patterns within these DMR are affected by SCNT, suggesting a deregulation of spatio-temporal methylation in clones.

Systems genetics and transcriptomics of feed efficiency in Nordic dairy cattle
S. Mohamad Salleh[1], J. Höglund[2], P. Løvendahl[2] and H.N. Kadarmideen[1]
[1]University of Copenhagen, Department of Veterinary Clinical and Animal Sciences, Grønnegårdsvej 7, 1870 Frederiksberg C, Denmark, [2]Aarhus University, Blichers Allé 20, 8830 Tjele, Denmark; hajak@sund.ku.dk

Feed is the largest variable cost in milk production, thus improving feed efficiency will result in better use of resources. The objectives of this project are to find molecular mechanisms, causal genes and biomarkers in feed efficiency of dairy cattle using transcriptomics. This study is expected to deliver potential causal genes and their networks for feed efficiency. Twenty cows (10 Jersey; 10 Holstein Friesian) were used in the experiment. These two groups of breeds were divided into two feed efficiency groups depending on their feed efficiency status which are of high or low efficiency. The efficiency of the cows were calculated based on their residual feed intake (RFI), Kleiber Ratio (KR), milk production, as well as gas emission. The cows were kept in respiration chambers for gas emission measurements for three days. Immediately after the cows left the chamber, liver biopsies and blood were sampled. mRNA were extracted from liver biopsies samples for RNA-sequencing. Blood samples were collected for genotyping. Feed efficiency, namely RFI and KR based on daily feed or dry matter intake, body weight and milk production records were also calculated. The bovine RNAseq gene expression data are being analyzed using statistical-bioinformatics and systems biology approaches to identify a list of differentially expressed genes, co-expressed genes, differentially wired networks, co-expression, transcriptional regulatory networks and biomarkers for feed efficiency. This study will provide molecular mechanisms of metabolic processes, energy balance, nutrient partitioning and deliver predictive biomarkers for feed efficiency in cattle. This study will also contribute to systems genomic prediction or selection models including the information on potential causal genes or their functional modules.

The covariance between genotypic effects in half-sib families
D. Wittenburg, F. Teuscher and N. Reinsch
Leibniz Institute for Farm Animal Biology, Institute of Genetics and Biometry, Wilhelm-Stahl-Allee 2, 18196 Dummerstorf, Germany; wittenburg@fbn-dummerstorf.de

In livestock, current statistical approaches involve extensive molecular data (e.g. SNPs) to reach an improved genetic evaluation of phenotypes. As the number of model parameters increases with a still growing number of SNPs, multicollinearity between covariates can affect the results of whole genome regression methods. The objective of this study is to additionally incorporate dependencies between SNPs due to the linkage and linkage disequilibrium on chromosome segments in appropriate methods for the estimation of SNP effects. The population structure, however, affects the extent of dependence. To consider this, the covariance among SNP genotypes is investigated for a typical livestock population consisting of half-sib families. Conditional on the SNP haplotypes of the common parent (sire), the covariance can theoretically be derived knowing the recombination fraction and haplotype frequencies in the population the individual parent (dam) comes from. The resulting covariance matrix is included in a statistical model for some trait of interest. From a likelihood perspective, such a model is similar to a first-order autoregressive process but the parameter of autocorrelation is explicitly approximated from genetic theory and depends on the genetic distance between SNPs. In a Bayesian framework, the covariance matrix can be used for the specification of prior assumptions of SNP effects. In this study, the challenges associated with model complexity and estimation of linkage disequilibrium from SNP genotypes are addressed. The approach is applied to realistically simulated data resembling a half-sib family in dairy cattle to identify genome segments with impact on performance or health traits. The results are compared with methods which do not explicitly account for any relationship among predictor variables.

LTBP2 genotypes relate with carcass traits in an F2 Population between Landrace and Jeju Black pig

I.C. Cho[1], Y.J. Kang[1], Y.K. Kim[1,2], H.S. Oh[2] and S.H. Han[2,3]
[1]National Institute of Animal Science, Rural Development Administration, Subtropical Livestock Research Institute, 593-50, Sanrokbukro, 690-150, Jeju-si, Korea, South, [2]Jeju National University, Department of Science Education, Jeju National University,102 Jejudaehak-ro, 690-756, Jeju-si, Korea, South, [3]Jeju National University, Educational Science Research Institute, Jeju National University,102 Jejudaehak-ro, 690-756, Jeju-si, Korea, South; choic4753@korea.kr

This study tested the association between genetic polymorphisms of LTBP2 gene and carcass traits in an intercross F_2 population between Landrace and Jeju Black pig (JBP) of South Korea. Genetic polymorphisms were screened by DNA sequencing and genotyping. A total of 8 nucleotide substitutions in the protein coding regions and 3 insertion/deletion polymorphisms in the intronic sequence were found in LTBP2 gene. Among those, LTBP exon 32 c.4481A>C (p.1494H>P) showed the significant associations with meat muscle area (MMA), carcass weights (CW), and carcass length (CBL) (P<0.05), but not with backfat thicknesses at three different points (BFs), meat color (MC), marbling score (MARB), and eye muscle area (EMA) (P>0.05), respectively. The F_2 animals harboring the LTBP2 c.4481 A/C heterozygote showed relatively heavier body weights for carcass weights (79.601±0.634 kg) than those of C/C homozygotes (77.590±0.917 kg) (P<0.05). The F_2 animals possessing LTBP2 c.4481 A/A genotype were shorter levels of CBL than those of A/C and C/C genotypes (P<0.05), showing approximately 2.0 cm shorter levels than those of A/C and C/C animals. In addition, the F_2 progeny carrying the LTBP2 c.4481 A allele showed significantly larger MMA levels than those of C/C homozygotes (P<0.05). These findings indicate that LTBP2 genotypes may be involved in the muscular development or body length growth, suggesting certain roles of the LTBP2 gene in mediating osteogenic differentiation, ECM network, and cell adhesion. Thus, LTBP2 genotypes can assist as molecular genetic markers for improving the JBP and Landrace-related crossbreeding systems.

Genetic parameters for indicator traits of the endoparasites resistance in Santa Ines sheep

C.C.P. Paz[1,2], E.J. Oliveira[1,2], R.L.D. Costa[2], L. El Faro[2], F.F. Simili[2] and A.E. Vercesi Filho[2]
[1]Universidade de São Paulo, Departamento de Genética, Av. Bandeirantes 3900 – Monte Alegre, 14049-900 Ribeirão Preto, SP, Brazil, [2]SAA/APTA, Instituto de Zootecnia, Centro de Bovinos de Corte, Rod. Carlos Tonani, Km 94, CP 63, 14.160-900 Sertãozinho, SP, Brazil; claudiapaz@iz.sp.gov.br

The objective of this study was to determine possible measurements that are indicative of resistance to endoparasites and to estimate genetic parameters for conjunctival staining score (CSS), fecal eggs count (FEC), packed cell volume (PCV), body condition score (BCS) and body weight (BW) in sheep Santa Ines. A total of 1,272 records from 511 animals born between 1992 and 2013 to 68 rams and 352 ewes, belonging to five herds, were used. The CSS is a technique based on the evaluation of the level of animal anemia by coloring the ocular conjunctiva, there is not necessary the use of laboratory resources. On the basis of conjunctiva color chart, the sheep were categorized in to five scales (1 to 5) as 1: red (non-anemic), 2: red-pink (non-anemic), 3: pink (mild anemic), 4: pink-white (anemic) and 5: white (severely anemic) at monthly interval. The determination of the PCV was performed by microhematocrit technique and of the FEC was performed by McMaster technique. The evaluation of the BCS was performed by visual and palpation evaluation of the lower back region, and assigned scores representing by 1: too lean animal; 2: lean animal; 3: good body condition; 4: fat animal and 5: too fat animal. The covariance components were estimated by animal model multi-trait by using Bayesian analysis approach by THRGIBBS1F90 computer package. The estimates of heritability for CSS, FEC, PCV, BCS and BW were 0.16 (0.03); 0.24 (0.09); 0.37 (0.06); 0.11 (0.03) and 0.25 (0.07), respectively, indicating that you can select animals based on the genetic value of individuals. The genetic correlations between CSS and BW, CSS and PCV, CSS and FEC, CSS and BCS and between the BW and BCS -0.40 (0.25); -0.73 (0.19); 0.60 (0.15); -0.42 (0.19); 0.71 (0.15), respectively.

A novel mutation of BMP-15 and GDF-9 gene in the Romanov sheep breed in Poland

G. Smołucha, A. Piestrzyńska-Kajtoch and B. Rejduch[†]
National Research Institute of Animal Production, Cytogenetics and Molecular Genetics of Animals,
Krakowska 1, 32-083, Poland; grzegorz.smolucha@izoo.krakow.pl

Analyses to date have indicated that mutations with a large effect on ovulation rate are not responsible for the exceptional prolificacy of Romanov sheep. The objectives of this study were to ascertain if any of 12 known mutations with large effects on ovulation rate in sheep or any other DNA sequence variants within the candidate genes BMP-15 and GDF9 are implicated in the high prolificacy of the Romanov sheep breed. Genomic DNA was extracted from whole blood samples of 71 females with a high level of fertility and 11 with low level of fertility or sterile. The PCR products were sequenced from both complementary strands. Genotyping results showed that none of 12 known mutations (FecBB, FecXB, FecXG, FecXGR, FecXH, FecXI, FecXL, FecXO, FecXR, FecGE, FecGH, or FecGT) were present in a sample of 82 Romanow sheep and, thus, do not contribute to the exceptional prolificacy of the breed. In BMP-15 gene, study revealed a three nucleotide indel CTT (leucine amino acid) in the signal peptide coding region. The deletion occurred in all animals studied except of 16 individuals which were heterozygous. This mutation seems to be a breed specific and there is no correlation with high or low fecundity level in this sheep breed. The CHI square test was used for statistical analysis. Two SNP mutations were found in the exon 2 fragment encoding the mature GDF-9 protein. First polymorphism is the transition A>G, which was named by Hanrahan et al. as the mutation G5 (GAA>GAG). The second is the transition G>A (GTT>ATT), marked as G6. Polymorphism was found in 38 animals: 5 individuals were homozygotes GG and AA while 33 sheep had heterozygous genotypes AG and GA. G5 silent mutation did not alter the amino acid sequence of the protein chain. However, these mutations can be crucial for the function of the encoded protein. The G6 mutation resulted in -substitution of valine by isoleucine (V994I) what changed the protein composition of the amino acid chain.

Screening of indigenous sheep for the prolificacy associated DNA markers

G. Smołucha, A. Piestrzyńska-Kajtoch, A. Kozubska-Sobocińska and B. Rejduch[†]
National Research Institute of Animal Production, Cytogenetics and Molecular Genetics of Animals,
Krakowska 1, 32-083, Poland; grzegorz.smolucha@izoo.krakow.pl

The aim of the study was to identify the polymorphism of genes involved (BMP-15, BMPR-1B) and probably involved (INHA, PRLR) in reproductive traits which could be a new marker to improve fertility and fecundity traits in some sheep breeds under the Polish Genetic Resources Conservation Program. The genomic DNA was extracted from whole blood samples of 120 female sheep (*Ovis aries*) of six breeds: Świniarka, Wrzosówka, Cakiel Podhalański, Olkuska, Polish Mountain Sheep color variety and the old-type Polish Merino. Gene fragments encoding the mature protein PRLR, BMP-15, INHA, BMPR-1B were amplified. The sequencing and RFLP methods were used to investigate the polymorphism. Multiple sequence alignments were performed with BioEdit Sequence Alignment Editor and ClustalW. The BMPR-1B, PRLR and INHA genes were monomorphic in all animals studied, making it unlikely that any of these genes are responsible for fecundity and fertility traits of analyzed breeds. One SNP (A>C, resulting in Asn to His amino acid change) having an impact on fertility and increased usability breeder was found in exon 2 of BMP-15 gene, only in Olkuska sheep. This mutation was located in the sequence encoding the mature BMP-15 protein. Two genotypes were found: AA and CC. The occurrence of this mutation in Olkuska homozygous ewes did not cause the infertility like it was with other mutations located so far in the BMP-15 gene. Average fertility in subjects with mutation (CC) was 327.37% compared to 279% in all animals tested. Average fertility animals without mutation was 239%. The differences in fertility between animals with and without mutations were statistically significant (ANOVA). The research has been continued.

Fourier transform infrared for breed identification and meat quality analysis of chicken meat
A. Molee[1], K. Thumanu[2], S. Okrathok[1] and S. Pitagwong[1]
[1]Suranaree University of Technology, School of Animal Production Technology, 111 University Avenue, Nakhon Ratchasima Province, 30000, Thailand, [2]Synchrotron Light Research Institute, 111 University Avenue, Nakhon Ratchasima Province, 30000, Thailand; amonrat@sut.ac.th

The objectives of this study were, applied to identify Korat meat chicken breed from another breeds by using Fourier transform infrared (FTIR) technique to measure a % relative absorbance of biochemical compound such as fat, secondary structure of protein and carbohydrate content in meat. The Six samples of Korat meat chicken (KR), Leung Hang Khaw (LK), commercial broiler (CB) from breast meat were used for this study. The samples were prepared by cryosection method at 4 microns using a microtome and placed immediately onto Kelvey IR slide. Data analysis were carried out using principal component analysis (PCA) to identify the biochemical compound using spectral ranges between 3,050 to 1000 cm^{-1} for breed identification. The results showed that this technique can be separated the samples into 3 groups, KR, LK, and CB. The highest and the lowest integral area of the Amide I protein covered between 1,656-1,648 cm^{-1} were detected with the sample of KR's, and CB's meat, respectively. While, the highest and the lowest peaks of CH stretching bands from lipid covered between 3,000-2,800 cm^{-1} were observed from the sample of CB's, and LK's meat, respectively. The integral area of Carbohydrate covered between 1,200-900 cm^{-1} was highly observed from the sample of CB's, and LK's meat. The preliminary study demonstrated that the technique can be applied for breed identification which did not identify from genome sequence, but from the end product of the sequence. Moreover, the technique is the powerful tool to enhance the ability of animal breeder which can penetrate the main obstacle of genetic improvement of meat quality since individual meat qualities data need a huge budget, and consume time.

Increasing the content of essential nutrients in pork – added value or not?
S. De Smet
Ghent University, Department of Animal Production, Laboratory for Animal Nutrition and Animal Product Quality, Proefhoevestraat 10, 9090 Melle, Belgium; stefaan.desmet@ugent.be

Over the last decades, a lot of research has been performed on optimizing the fatty acid profile and increasing the content of essential nutrients in pork. This is motivated by the growing concern of consumers for the effect of diet on their health, and the opportunity to create added value in pork production. In this overview, the potential and the limitations of strategies to increase the contents of essential nutrients in pork will be discussed, with focus on (long chain) omega-3 fatty acids and several trace elements. Feeding strategies offer the largest prospect for modifying the fatty acid profile of pork, but the outcome is largely affected by animal factors and the degree of fatness. The response in muscle to dietary supplementation with essential trace elements is strongly variable and dependent on the trace element. The contribution of current approaches aiming at improving the nutritive value of pork, to the human intake of essential nutrients will be critically evaluated. Also the implications for meat quality and the oxidative stability of meat and meat products will be discussed.

Effect of dietary lecithin and sex on finisher pig growth performance and pork texture
H. Akit[1,2], H. Frobose[3], F.T. Fahri[1], D.N. D'souza[4], B.J. Leury[1] and F.R. Dunshea[1]
[1]University of Melbourne, Faculty of Veterinary and Agricultural Sciences, Parkville, 3010 VIC, Australia, [2]Universiti Putra Malaysia, Department of Animal Science, Serdang, 43400, Selangor, Malaysia, [3]Kansas State University, Animal Sciences and Industry, Manhattan, 66506 KS, USA, [4]Australian Pork Limited, Deakin West, 2600 ACT, Australia; henny@upm.edu.my

The effect of interaction between dietary lecithin and sex on pig growth performance and pork texture was investigated. This study involved 210 pigs (in 30 pens) of three sexes (female, entire male, immunocastrated male (IC) pigs immunised against gonadotrophin-releasing hormone (Improvac®)) and two dietary treatments (0 or 8 g/kg dietary lecithin) for 4 weeks prior to slaughter. Pen weights were recorded and prior to slaughter, individual live weights were obtained. Pen feed intake was measured and feed conversion ratio subsequently calculated. The M. longissimus (loin) was removed from a pre-determined subset of 60 carcasses and stored at -20 °C prior to collagen content, Warner-Bratzler shear force and compression analyses. Dietary lecithin improved feed intake (P=0.05) and the effect was most effective in females (P=0.04). Lecithin had no effect on pork chewiness, shear force values or muscle collagen content (P>0.05, respectively) and these parameters were not influenced by sex of pigs. Immunocastrated pigs were most feed efficient and had the highest feed intake and average daily gain followed by entire males and females (P=0.001, respectively). Pork from IC pigs also had the lowest shear force values (P=0.042) and this may be associated with increased myofibrillar protein degradation during increased growth rate. The data showed that lecithin was most effective in improving feed intake in females. However, lecithin had no effect on pork texture and this lack of effect was not influenced by sex. The high growth rate in IC pigs may influence meat texture.

Acceptability of fresh pork meat with different levels of boar taint
F. Borrisser-Pairó, N. Panella-Riera, L. Domínguez-Clavería, M. Gil, M. Font-I-Furnols and M. Oliver
IRTA-Monells, Product Quality Programme, Finca Camps i Armet, 17121, Spain; francesc.borrisser@irta.cat

Meat with boar taint can affect the sensory quality of the pork and pork products, due to the off-odours and flavours that can be detected by consumers. Boar taint is mainly caused by the presence of androstenone (AND) and skatole (SKA). The aim of this study was to assess the acceptability of fresh pork with different levels of AND and SKA by consumers. The meat for the test consisted in patties made from castrated and entire male pigs. Eight types of patties from entire male pigs (N1-N8) were prepared with different combinations of levels of AND (from 0.5 to 2.0 µg/g) and SKA (from 0.10 µg/g to 0.40 µg/g). One hundred and twenty three women were selected as regular pork consumers. Patties were cooked on a cooking plate set at 170 °C and served warm to consumers. Each consumer evaluated 5 paired samples. The first serving consisted of a comparison of two patties from castrated pigs to train consumers with the procedure. The other four servings consisted of a combination of a patty from castrated pig and a patty from entire male pig with known levels of AND and SKA. Consumers evaluated 'odour and taste' rated on a Likert scale (from 1: 'dislike extremely 'to 9: 'like extremely'; without the intermediate level). Results were analyzed with GLIMMIX procedure of SAS® according to the score given to each patty. The highest score was 6.8±0.25 for N3 patties and 6.7±0.10 for 'castrated' patties (0.9 µg/g for AND; 0.15 µg/g for SKA). They both were significantly different to N6 (1.5 µg/g for AND; 0.40 µg/g for SKA) and N8 (2.0 µg/g for AND; 0.39 µg/g for SKA), scored with 5.7±0.24 and 5.7±0.26 respectively. No significant differences were found among all the other type of patties. Boar tainted meat with high levels of AND and high levels of SKA is less preferred to meat from castrated pigs. Further research is needed to find an adequate application for the meat with high levels of boar taint.

Effect of time post second injection on performances, carcass and meat quality of immunocastrates

M. Aluwé[1], I. Degezelle[2], D. Fremaut[3], L. Depuydt[2] and S. Millet[1]
[1]ILVO (Institute for Agricultural and Fisheries Research), Animal Sciences Unit, Scheldeweg 68, 9090 Melle, Belgium, [2]VIVES University College, Wilgenstraat 32, 8800 Roeselare, Belgium, [3]Ghent University, Valentin Vaerwyckweg 1, 9000 Gent, Belgium; marijke.aluwe@ilvo.vlaanderen.be

Immunocastration can be used as an alternative for surgical castration of male piglets to avoid boar taint. The vaccine is administrated twice, once around 10 weeks of age and a second time about 4 to 6 weeks before slaughter. Before this second vaccination, pigs behave and perform like entire male pigs. Afterwards, feed intake increases which may result in a lower lean meat percentage. It can be expected that the timing of the second vaccination is crucial to optimise results. In total, 180 animals (hybrid sow × Piétrain): 60 gilts, 60 early vaccinated male pigs (IM-E, 79±6 kg live weight) and 60 late vaccinated male pigs (IM-L, 87±10 kg live weight) were slaughtered at an average pen weight of 119 kg. Performances, carcass and meat quality traits were analyzed by ANOVA (R-core), with treatment as fixed factor and carcass weight as co-variable for carcass quality. Second vaccination was performed 42±5 and 31±3 days prior to slaughter, for IM-E and IM-L respectively. Daily feed intake (DFI), daily gain (DG) and feed conversion ratio (FCR) did not differ significantly between IM-E and IM-L, while gilts had a lower DFI and DG compared to both IM groups. FCR of gilts was higher compared to IM-L, but not compared to IM-E. Gilts showed a higher lean meat content compared to both IM groups. Earlier vaccination increased dressing which could partly be explained by the lower weight of the gastrointestinal tract, but not by testes weight. For meat quality traits, intramuscular fat content (IMF) tended to increase and redness and yellowness values tended to decrease with earlier vaccination. Within the studied time frame of second injection before slaughtering, earlier vaccination improved dressing percentage and tended to increase IMF, while maintaining performance and carcass characteristics.

Effect of immunocastration on carcass and meat quality in boars, barrows and gilts

A. Van Den Broeke[1], F. Leen[2], M. Aluwé[1], J. Van Meensel[2] and S. Millet[1]
[1]ILVO (Institute for Agricultural and Fisheries Research), Animal Sciences Unit, Scheldeweg 68, 9090 Melle, Belgium, [2]ILVO, Social Sciences Unit, Burg. Van Gansberghelaan 115, 9820 Merelbeke, Belgium; marijke.aluwe@ilvo.vlaanderen.be

The aim of this study was to assess the effect of sex and immunocastration and their interaction on carcass and meat quality of finishing pigs in a 3×2 study (sex × IC). Forty boars (BO), 40 barrows (BA) and 40 gilts (GI) were raised in individual pens. Each sex was divided in 2 groups of 20 animals, the control (C) and the immunocastrated (IC) group. The IC group received two injections (at 70 kg and 105 kg) of Improvac®, an anti-GnRH vaccine. The pigs had free access to water and were fed *ad libitum* with the same three phase diet. They were slaughtered at an average live weight of 135±4 kg (mean ± standard deviation). The effect of sex on carcass and meat quality was assessed in the subset of control animals with sex as fixed factor. The effect of IC was assessed separately in each gender subset (BO, BA and GI) with IC as fixed factor. The experiment confirms the known sex differences. In line with literature, dressing percentage of BO was lower compared to BA and GI. Meat percentage of BO and GI was higher and backfat thickness and IM fat content lower compared to BA. Drip loss was higher in BO compared to BA. Meat of BA was less yellow compared to meat of GI. In BO, the higher feed intake after IC led to higher backfat thickness resulting in a lower meat percentage, without affecting intramuscular fat content (IMF). The meat of IC BO was less red compared to C BO and tended to have higher cooking loss and a lower ultimate pH. In gilts, immunocastration tended to increase backfat thickness and decrease meat percentage. IC GI showed improved meat quality compared to C GI with lower shear force values and a tendency to higher IMF. In barrows, IC had only minor effects. IC BA tended to have a lower ham angle and more yellow and red meat compared to C BA.

Development of high throughput methods to predict pork quality at industrial scale

S. Schwob[1], B. Lebret[2], A. Vautier[1], B. Blanchet[3], M.J. Mercat[1], J. Faure[2], R. Castellano[2], S. Quellec[4], S. Challois[4] and A. Davenel[4]
[1]IFIP, La Motte au Vicomte, 35651 Le Rheu, France, [2]INRA, UMR 1348 PEGASE, Domaine de la Prise, 35590 Saint Gilles, France, [3]INRA, UETP, La Motte au Vicomte, 35651 Le Rheu, France, [4]IRSTEA, 17 avenue de Cucillé, 35044 Rennes, France; sandrine.schwob@ifip.asso.fr

This project aimed at providing new precocious and non-invasive predictors of pork quality usable in abattoirs, in order to orientate use of carcasses and cuts and optimize their economic value. For this purpose, the project included development, testing and validation of various methods under industrial conditions on both purebred and crossbred pigs. Usual meat quality traits (ultimate pH, colour, drip loss) were available on the Longissimus muscle (LM) from 2,679 carcasses. In addition, Magnetic Resonance Imaging (MRI) technology was used to estimate the LM intramuscular fat (IMF) content and marbling. Image analysis method was automated to improve measurement rate (70 samples scanned per hour) and ensure data traceability. This automatic image processing was validated on a large scale (2,679 samples). MRI was also used to study the representativeness of IMF content determined at the 13th rib to assess average IMF of the whole LM (38 entire loins scanned). Results showed high repeatability and good predictive ability of LM average IMF content with determination at the 13th rib level (R^2=0.88). Near Infrared Spectroscopy (NIRS) was used to predict the LM cooking yield. NIRS measurements were recorded on 157 fresh loins, of which 98 were individually processed to estimate cooking and slicing yields and structural defects. NIRS technology could predict slicing losses caused by paste-like and cohesion defects on processed loin slices. Finally, the expression level of 22 genes previously identified as quality biomarkers was quantified on LM samples taken early after slaughter on the same 157 carcasses to validate these biomarkers differentiating 3 pork quality categories: low, acceptable and extra technological and sensory quality levels.

What is the best strategy to increase the accuracy of genomic selection for meat quality traits?

E. Gjerlaug-Enger[1], J. Ødegård[2], J. Kongsro[1], Ø. Nordbø[1] and E. Grindflek[1]
[1]Topigs Norsvin, Pb 504, 2304 Hamar, Norway, [2]AquaGen, Pb 1240, 7462 Trondheim, Norway; eli.gjerlaug@norsvin.no

The weighting of meat quality represents 20% of the Norsvin Duroc breeding goal. Intramuscular fat (IMF) is measured with Near Infrared Spectroscopy (NIRS) on slaughtered relatives of the selection candidates. Prediction models have previously shown an R^2 of 0.98 to chemical methods, and NIRS thus replaced chemical determined IMF for Norsvin breeds in 2005, resulting in a higher heritability (62%). Despite this highly informative phenotype, selection for IMF is challenging due to unfavorable genetic correlations to production traits, and new strategies have therefore been evaluated. Meat quality can not be measured with the same accuracy using the *in vivo* methods Computer Tomography (CT) or Ultrasound (UL). *In vivo* predictions of IMF using CT (n=104) and UL (n=96) image analysis showed that the best prediction models gave an R^2 of 0.18 to 0.21 for CT and UL, respectively. Still, having records directly on selection candidates improves accuracy of the estimated breeding values (EBV). Using existing testing schemes, IMF of the 1,600 selection candidate boars from the testing station per year can be included using CT, and the 15,000 animals in field testing can also be *in vivo* phenotyped using UL. This will significantly increase the tested number of full- and half-sibs. In contrast, NIRS is only available for 900 Duroc's annually, and only on full- and half sibs of the candidates. Still, implementation of Genomic Selection (GS) has increased the accuracy of selective breeding for NIRS predicted IMF compared to traditional selection (from 0.36 to 0.63). The aim of the study was therefore to evaluate the added value of *in vivo* phenotypes on the candidate itself and a large number of its (partly ungenotyped) relatives under a genomic selection program. Results from a stochastic simulation study will be used to examine this issue, where different phenotypes (NIRS, CT and UL) are observed on different groups of animals (selection candidates and relatives).

Effects of the sex and halothane gene on pig carcass composition measured by computed tomography
G. Daumas and M. Monziols
IFIP-Institut du Porc, BP 35104, 35601 Le Rheu Cedex, France; gerard.daumas@ifip.asso.fr

The aim of this study was to quantify the main effects influencing both the tissue composition measured by Computed Tomography (CT) and the classification variables of slaughtered pigs. A representative sample of the French pig slaughtering was selected in 3 abattoirs and stratified according to sex (50% castrated males and 50% females). Carcasses were measured by 3 classification methods – CSB Image-Meater® (IM), CGM, ZP – and cooled. An ear sample was analysed for Halothane gene (Hal). The left sides were cut according to the EU procedure and the four main joints were CT scanned. Images were thresholded in order to determine lean meat, fat and bone weight. Among the 209 pigs, the proportions of Nn and NN alleles were respectively of 52% and 48%, leading to a well balanced design. In the analysis of variance the interaction between sex and Hal was never significant. Sex was significant on all the fat and muscle depths as well as on all the tissues proportions in the joints, except the bone % in ham and loin. Hal was significant on all the tissues proportions in the joints, except the fat % in shoulder, the bone % in belly and the lean meat percentage (LM%) predicted by IM. Hal was not significant on the IM depths taken on the splitline but was significant on the CGM lateral depths. Sex had a major effect (1 standard deviation) on the LM% in the loin and the fat % in the shoulder. The highest Hal effect (0.6 standard deviation) was on the LM% in the carcass predicted by the CGM and the bone % in the shoulder. The adjusted differences between females and castrated males were for the LM% measured by CT and predicted by CGM, ZP and IM respectively of 3.0, 2.0, 1.6 and 1.6. The adjusted differences between Nn and NN alleles were for LM% measured by CT and predicted by CGM, ZP and IM respectively of 1.5, 1.5, 0.6 and 0.4. Hal and sex have important effects on pig carcass composition, but the automatic classification by IM is less sensitive than CGM to these ones.

Effects of sex and halothane gene on the pig grading prediction equations of lean meat percentage
G. Daumas and M. Monziols
IFIP-Institut du Porc, BP 35104, 35601 Le Rheu Cedex, France; gerard.daumas@ifip.asso.fr

Automation of pig classification methods make it relevant to quantify the main effects influencing the prediction equations of the lean meat percentage (LM%). A representative sample of the French pig slaughtering was selected in 3 abattoirs and stratified according to sex (50% castrated males and 50% females). Carcasses were measured by 3 classification methods – CSB Image-Meater® (IM), CGM, ZP – and cooled. An ear sample was analysed for Halothane gene (Hal). The left sides were cut according to the EU procedure and the four main joints were CT scanned. Images were thresholded in order to determine lean meat weight. Among the 209 pigs, the proportions of Nn and NN alleles were respectively of 52% and 48%, leading to a well balanced design. The least squares means were calculated for the factors SEX and Hal in an analysis-of-covariance model per classification method including the corresponding fat and muscle depths as well as the interactions. Interactions were never significant. Hal effect was not significant for the CGM. The adjusted differences between females and castrated males were for the lean meat percentage predicted by CGM, ZP and IM respectively of 1.0, 1.7 and 1.8. The adjusted differences between Nn and NN alleles were for the lean meat percentage predicted by ZP and IM respectively of 1.0 and 1.3. Sex could be managed by a different intercept in the classification equations if it is considered of practical relevance. The knowledge of the bias size between Hal alleles is of interest for the pig chain stakeholders.

Effect of sex and sire on lean meat percentage and weight of primal cuts of pork using AutoFom data

S. Tanghe[1,2], S. Millet[2], S. Hellebuyck[3], J. Van Meensel[2], N. Buys[1], S. De Smet[3] and S. Janssens[1]
[1]KU Leuven, Livestock Genetics, Kasteelpark Arenberg 30, 3001 Leuven, Belgium, [2]Institute for Agricultural and Fisheries Research, Scheldeweg 68, 9090 Melle, Belgium, [3]Ghent University, Department of Animal Production, Proefhoevestraat 10, 9090 Melle, Belgium; sofie.tanghe@ilvo.vlaanderen.be

Manual grading probes for pig carcass classification are being replaced by the automatic ultrasonic AutoFom III, which provides the lean meat percentage and weight of primal cuts. This study aimed to examine the effect of sex and sire on the weight and lean meat percentage of primal cuts. AutoFom data from 1,735 gilts, 274 boars, 707 immunocastrates (IC) and 869 barrows were collected. Cold carcass weight, lean meat percentage and relative weight of primal cuts were analyzed for sex and sire effects and correlations were calculated. Cold carcass weight was higher for boars than for barrows and IC. Gilts had the lowest cold carcass weight. Gilts and boars had higher lean meat percentages in carcass and primal cuts compared to IC. Carcass and primal cuts of barrows had the lowest lean meat percentage, particularly in loin and belly. Barrows had heavier belly, but lighter ham, loin and shoulder compared to IC, boars and gilts. Lean meat percentages of primal cuts had a strong positive correlation. Weights of ham, loin and shoulder were negatively correlated with belly weight and positively correlated with each other, but the correlation between ham and shoulder was moderate. Cold carcass weight was negatively correlated with lean meat percentage of primal cuts for gilts, boars and IC, but not for barrows. Heavier gilt, boar and IC carcasses had relatively less ham and loin, but more belly, whereas heavier barrow carcasses had relatively less ham and belly, but more loin and shoulder. An effect of sire was observed on lean meat percentage and relative weight of primal cuts, pointing at selection possibilities. Next, heritability and genetic correlations of lean meat percentage and relative weight of primal cuts will be estimated.

Pork quality from pig genotypes of different fatness

V. Razmaitė, A. Bajorinaitė and R. Šveistienė
Institute of Animal Science of Lithuanian University of Health Sciences, Department of Animal Breeding and Genetics, R. Žebenkos 12, 82317, Lithuania; ruta@lgi.lt

In the recent years, the consumer demand for leaner pork has increased, and pig production made great strides to reduce the fat content and improve the leanness of pig carcasses. This resulted in symptoms of pork quality deterioration and decline in the pig numbers of fat local Lithuanian breeds. The objective of the study was to determine the influence of pig fatness associated with the genotype in view of its implications for meat quality characteristics and consumer health. All pigs were slaughtered at the same time in an accredited abattoir. The samples of Longissimus dorsi muscle and backfat from the left side of conventional hybrids of selected Large White and Landrace, Lithuanian indigenous wattle and old type Lithuanian White were used. The mean of backfat thickness at 10th rib of the hybrids, Lithuanian Indigenous Wattle pigs and Lithuanian White was 14.7, 30.6 and 30.9 mm, respectively. Pork composition and properties were detected according to the EU reference methods. The data were subjected to the analysis of variance in GLM procedure in SPSS 17. The differences among the means were determined using the LSD test. The lean hybrids were characterized by the lowest pH value and percentage of IMF ($P<0.010$) and the highest meat lightness ($P<0.001$) if compared with the fatty Lithuanian indigenous wattle pigs. The pigs from both Lithuanian breeds had higher contents of SFA ($P<0.01$), MUFA ($P<0.001$) and lower contents of PUFA ($P<0.001$) in the IMF and backfat lipids. However, the n-6/n-3 PUFA ratios were lower in the lipids of the fatty pigs of Lithuanian breeds ($P<0.001$). Although their atherogenic and thrombogenic indexes and hypocholesterolemic/hypercholesterolemic ratio were less favourable, the meat from Lithuanian pig breeds had lower contents of cholesterol and lower peroxidizability index ($P<0.01$). The iodine value of backfat from lean pigs showed lower backfat quality compared with pigs from Lithuanian breeds of higher fatness ($P<0.001$).

Effects of the use of *Pediococcus acidilactici* in dry feed on fattening pig performance
E. Royer[1], D. Guillou[2], L. Alibert[1] and E. Chevaux[2]
[1]Ifip-institut du porc, 34 bd de la gare, 31500 Toulouse, France, [2]Lallemand SAS, BP 59, 31702 Blagnac, France; eric.royer@ifip.asso.fr

Through colonization in liquid feed systems, *Pediococcus acidilactici* MA 18/5M (PA) can secure hygiene of liquid feeds for pigs. A study examined the effects of PA on pigs when used in dry feeds. Two diets based on cereals, wheat bran, oilseed meals were prepared without and with PA supplementation (Bactocell®, Lallemand at 10^9 cfu/kg). Diets had a medium-high fibre level (168 then 179 g NDF/kg), contained 9.5 to 9.4 MJ NE/kg and 0.9 to 0.8 g digestible lysine/MJ NE, for growing and finishing pigs, respectively, and were distributed as dry meal, *ad libitum* up to 69 kg LW. 160 (LW×Ld)×PP pigs (27.2±1.6 kg) were blocked to 16 single-sex pens of 5 pigs each, per treatment. Treatments were separated among 4 identical rooms (8 pens each), in order to measure NH_3 in extracted air, and sample manure for each treatment. PA diet increased DFI during the initial growing period from d 0 to 27 (1.76 vs 1.82 kg/d, P<0.001) resulting in a better ADG (793 vs 831 g/d, P<0.001) and a higher body weight at d 27 (48.6 vs 49.6 kg, P<0.001), whereas FCR was not modified. From d 27 to harvest, DFI, ADG and FCR were similar for both treatments (P>0.05). Within-pen heterogeneity was significantly decreased for PA pigs at d 15, 49 and 64, and tended to be lower at harvest (P=0.06). Carcass weight and yield were unaffected by treatments (P>0.05). However, PA pigs had a slightly higher fat depth (14.1 vs 14.6, P=0.07) resulting in a lower leanness (P=0.02). Analysis of faeces showed higher PA and *Lactobacillus* counts with PA feeds, and individual faecal score was slightly increased at d 85 by PA feed (2.03 vs 2.24, P=0.04). The composition of manure for DM, OM, N, P, K and pH, and excretion rates per pig were similar for both treatments. However, PA fed pigs tended to have a lower N excretion per kg of gain (P=0.10) as also illustrated by reduced NH_3 emission (NS). PA in dry feeds may influence some gut parameters, improve feed intake and daily gain and reduce heterogeneity of pigs.

Differences in volatile compounds among heated porcine muscles
M. Oe[1], I. Nakajima[1], K. Ojima[1], K. Chikuni[1], K. Sasaki[1], M. Shibata[2] and S. Muroya[1]
[1]NARO Institute of Livestock and Grassland Science, Tsukuba, Ibaraki, 305-0901, Japan, [2]NARO Western Region Agricultural Research Center, Ohda, Shimane, 694-0013, Japan; mooe@affrc.go.jp

Flavor is an important factor affecting sensory characteristics of meat. Although meat flavor varies among muscles, differences in volatile compounds among muscles is not fully elucidated. In this study, to reveal the intermuscular difference, we evaluated volatile compounds extracted from porcine five muscles: vastus intermedius, diaphragm, psoas major, longissimus thoracis and semitendinosus. Muscle samples were excised from six carcasses of three-way cross (Landrace × Large White × Duroc) female pigs at 7 days after slaughter. Muscle samples were heated at 90 °C for 30 min. Volatile compounds in heated muscles were retrieved by solid phase micro extraction (SPME) and were analyzed by gas chromatography-mass spectrometry. Approximately 100 peaks were detected, in which totally 14 compounds were identified. Fourteen compounds included 1-pentanal, 1-hexanal, 1-heptanal, 1-octanal, 1-nonanal, trans-2-heptenal, trans-2-octenal, benzaldehyde, 1-pentanol, 1-hexanol, 1-heptanol, 1-octanol, 1-octen-3-ol and 2-penthylfuran. Those compounds were detected in the five muscles. Compounds unique to each muscle were not identified. The sum of the amount of 14 compounds from longissimus thoracis and semitendinosus were higher than that from vastus intermedius (P<0.05). These results showed that the amount of volatile compounds were different among muscles. These difference in the amount of volatiles may contribute to variation in meat flavor among muscles.

Effect of SNPs in candidate genes on carcass and meat quality traits in pigs

N. Moravčíková, A. Navrátilová and A. Trakovická
Slovak University of Agriculture in Nitra, Department of Animal Genetics and Breeding Biology, Tr. A.
Hlinku 2, 94976, Slovak Republic; nina.moravcikova@uniag.sk

The aim of this study was to investigate the association between SNPs in genes encoding leptin (LEPR) and melanocortin 5 (MC5R) receptors and back fat thickness (mm), proportion of valuable meat parts (%), area of musculus longisimus thoracis (cm^2) and proportion of thigh (%). In total 180 genomic DNA samples of crossbreeds (Large White × Landrace) have been used to genotyping of selected markers by means of PCR-RFLP method. The allele frequencies were as follows: LEPR (HpaII) A 0.61 and B 0.39 (±0.03) and MC5R (BsaHI) A 0.63 and G 0.37 (±0.03). A prevalence of AA genotype (41.67%) compared to AB (38.89%) and BB genotypes (19.44%) were detected for the LEPR gene. Similarly the highest frequency for MC5R gene was observed for AA genotype (43.33%). Analysed loci had in average the medium level of polymorphic information content (0.36). Positive value of within population fixation index (0.25) was due to lower proportion of heterozygotes (H_e=0.36). One-way ANOVA was used for association analysis of SNP effects on selected production traits. The average values of analyzed traits for different LEPR and MC4R genotypes indicated positive effect of A allele or homozygote AA genotype in both loci. The statistical analysis showed significant association (P<0.001) with selected traits only for LEPR gene. However, further study about these SNPs with involving of other breeds and candidate genes could clarify its exact role in regulation of carcass and meat quality traits in pigs.

Perspectives of use of total meat percentage as a selection criterion in pig industry of Ukraine

O. Kravchenko[1] and A. Getya[2]
[1]Poltava state agrarian academy, str. 1/3 Skovorody, 36003 Poltava, Ukraine, [2]National University of Life and Environmental Sciences of Ukraine, str. Heroyiv Oborony 15, 03041 Kyiv, Ukraine; oksanakravchenko@ukr.net

The demand for lean pork is being increased in Ukraine for recent years. This tendency is taken into consideration by the pig industry which improves housing- and feeding technology. At the same time specialized genotypes, including imported animals from different countries, became more and more popular. Over the last 10 years the average daily gain in Ukrainian pig farms increased by 1.96 times. But due to an outdated payment system for the animals delivered to slaughterhouses one of the main criteria for the carcass quality assessment, namely 'total meat percentage' (TMP), is not evaluated. To analyse the carcass quality based on TMP and to prove the need of involvement of this trait into Ukrainian breeding programs 7,246 pigs from 12 farms in central Ukraine were slaughtered. The measurements of TMP were done using CGM-device. The average weight of carcasses of gilts and castrates was 88.38 kg and 89.03 kg respectively. The TMP of carcasses of gilts was 59.30% and of castrates 57.60%. Despite the fact that according to Ukrainian classification system all carcasses were related to the 2nd class, a considerable variation of total meat percentage was established. So, the average value of this trait in the worst 30% carcasses of gilts and castrates was 50.00±1.22% and 48.50±1.19% respectively while in best 30% – 64.50±0.76% and 64.00±0.91%. The difference between 30% best and 30% worst carcasses of gilts and castrates was 14.5% and 15.5% respectively (P≤0.001). Understanding the importance of trait TMP for efficiency of pig industry the implementation of European payment system is needed as well as integration of this trait into breeding programs of Ukrainian genetic companies.

Effect of increasing dietary hydrolysable tannin supply on the incidence of boar tainted carcasses
E. Seoni, S. Ampuero, P. Silacci and G. Bee
Agroscope Institute for Livestock Sciences (ILS), la Tioleyre 4, 1725, Switzerland;
eleonora.seoni@agroscope.admin.ch

The goal of this study was to determine threshold levels of supplemented dietary hydrolysable tannins (HT) on androstenone (A), skatole (S) and indole (I) tissue deposition. For the study 44 Swiss Large White entire males from 11 litters were selected at 59.4 kg BW and assigned within litter to 1 of 4 treatments: unsupplemented control finisher diet (T0); finisher diet supplemented with 1.5 (T15), 3.0 (T30) or 4.5% (T45) chestnut powder. All pigs were reared in one pen equipped with 4 automatic feeders. They had *ad libitum* access to the assigned diets. The animals were weighed weekly and individual feed intake was monitored daily. At 103.5 kg BW, pigs were slaughtered and organ weights as well as carcass characteristics were assessed. In addition, concentrations of A, S and I levels were measured in the backfat. Despite similar feed intake, T45-pigs tended ($P<0.10$) to grow slower than T15-pigs, with intermediate values for T0- and T30-pigs (0.80, 0.92, 0.88, 0.83 kg/d). T45- and T30-pigs were less (0.34 kg/kg each; $P<0.05$) feed efficient than T0- and T15-pigs (0.38 and 0.37 kg/kg). Compared to the T0-group, lean meat percentage was lower ($P<0.05$) in carcasses of boars from the T15-group (56.7 vs 58.4%), with intermediate values for the T30- and T45-group (57.3 and 57.5%). At the highest dietary HT level (T45) liver, kidney and salivary gland were the lightest ($P<0.10$). Due to large variability in the A, S and I levels in the backfat, no significant ($P>0.27$) dietary effects were observed. However, taking into account the suggested boar taint thresholds for A (<1 ppm) and S (<0.25 ppm), fewer T45- than CO-, T15- and T30-pigs were above these thresholds (2 vs 5, 6, 7). In conclusion, the current data show that despite the slight negative effect of the highest dietary HT supply on growth performance incidence of boar taint can be reduced with dietary HT.

Effects of transportation methods on pig welfare and pork carcass quality
P. Na-Lampang
Suranaree University of Technology, School of Animal Production Technology, 111 University Avenue, Nakhon Ratchasima 30000, Thailand; pongchan@sut.ac.th

The objective of this study was to compare the results of transport slaughter pigs to slaughter house by 2 methods, i.e. single confined in individual metal crate and group confined on the truck, on transport processes, wounds or injuries, behavior, physiology of stress and meat quality. It was found that: (1) Pigs transported in group had more skin bruises and scratches, from fighting, than those transported in single crates, from scratched with the crate. (2) Pigs transported in single crates had very high incidents of non-ambulatory pigs. Pigs transported in group had much lower incidents of non-ambulatory pigs than those transported in single crates ($P<0.01$). (3) Pigs transported in group showed behaviors of climbing, reverse, and slipped on the loading chute. During the journey, the majority of these pigs lied down and had a low level of agonistic behavior. The pigs transported in single crates did not have the chance to show any behavior due to being forced to be in prone position all the time. (4) Rectal temperature and respiratory rate, the physiological indicators of stress, of those transported in group and in single crates were not significantly different. However, it was found that moving on to the truck and travelling caused higher level of stress than normal in both groups of pigs. (5) Meat quality of pigs transported by 2 different methods as measured in terms of pH level (at 45 min and 48 h post mortem), color (brightness, redness and yellowness) and water holding capacity was not significantly different.

The strategy to produce hogs at the delivery heavier and more uniform

M. Sviben
Freelance consultant, Siget 22B, 10020 Zagreb, Croatia; marijan.sviben@zg.t-com.hr

Danish scientists expect that in 2020 80% of pigs in Denmark will be placed on 700-900 large farms. The typical weaner producer will have 1,500-3,000 sows. It has been analyzed how to produce heavier and more uniform hogs deliverable from suchlike pig factories. Foremost technological runts should be culled from the farrowing room. The share of such suckers, weighing up to 4.6 kg at the age of 16 days, can be 15%. They have to be moved to the recovery unit. Other weaners should be sorted in 4, equally divided, weight groups: 21.25% lightest (LT) ones, weighing 4.7-5.5 kg, 21.25% lighter (LR) ones, weighing 5.6-6.2 kg, 21.25% heavier (HR) ones, weighing 6.3-7.0 kg, and 21.25% heaviest (HT) ones, weighing 7.1 and more kilograms, at the age of 16 days. After staying in the nursery for 27 days, the piglets can achieve the mean live weight of 22.841 kg, at the age of 48 days. In one of 3 raising rooms and in one of 15 feeding rooms the LT piglets can grow up to 143.3 kg on an average. LR piglets, after staying in one of 2 raising rooms and in one of 14 feeding rooms can be 142.7 kg on an average. After staying in one raising room and in one of 13 feeding rooms, the HR hogs can achieve the mean live weight of 137.5 kg. HT piglets, fed in one of 12 feeding rooms can weigh at the end 138.5 kg on an average. The slaughterers' wanted range of the hogs' weight can be achieved delivering finishers successively during the last week of feeding.

Quicker production of hogs weighing conventionally but at the delivery being more uniform

M. Sviben
Freelance consultant, Siget 22B, 10020 Zagreb, Croatia; marijan.sviben@zg.t-com.hr

Nowadays pigs are fattened from 30 till 108 kg of their live weight on an average. Finishers are delivered in large groups, since farms are steadily enlarging. The weight dispersion of hogs delivered at equal age causes troubles at the slaughter-houses where 600-1,600 pigs can be killed to one hour. The pig producers are interested to deliver fatteners more quickly, the slaughterers ask for getting large groups of more uniform finishers. The breeding companies and the manufactories of the feed mixtures can ensure the piglets capable to grow according to the equation $Y_{C-KG} = 39.9247+6.097X_C+0.2038X_C{}^2-0.0021X_C{}^3$, calculating $X_C = (X_{day}-70)/7$. Hyotechnological analysis showed that heaviest (HT) pigs, weighing 7.1 and more kilograms at the age of 16 days, can achieve 29.524 kg at the age of 48 days and 108.700 kg at the age of 113 days. Heavier (HR) pigs, having 6.3-7.0 kg at the age of 16 days, can average 29.867 kg at the age of 54 days and 106.831 kg at the age of 123 days of age. Lighter (LR) pigs, weighing 5.6-6.2 kg at the age of 16 days, can have 31.992 at the age of 61 days and 108.302 kg at the age of 133 days. Lightest (LT) pigs, having 4.7-5.5 kg at the age of 16 days, can accomplish the mean of 33.064 kg at the age of 68 days and 107.914 kg at the age of 144 days. F.C.R. are 2.861 at LT, 2.380 at LR, 2.008 at HR and 1.623 at HT fatteners.

Hybridization *in situ* of genes associated with meat production traits on chromosomes of *Suidae*

A. Kozubska-Sobocińska[1], B. Danielak-Czech[1], M. Babicz[2], K. Kruczek[1] and A. Bąk[1]
[1]*National Institute of Animal Production, Department of Animal Cytogenetics and Molecular Genetics, 1, Krakowska Street, 32-083 Balice n. Cracow, Poland,* [2]*University of Life Sciences in Lublin, Department of Pig Breeding and Production Technology, 13, Akademicka Street, 20-950 Lublin, Poland; anna.sobocinska@izoo.krakow.pl*

Ghrelin (GHRL) and uncoupling proteins UCP2, UCP3 play a functional role in global energy metabolism in the body, growth and obesity as well as organoleptic meat quality. The studies of genes encoding these proteins, involving the genetic variation underlying economically important traits, may be useful for searching of markers associated with meat production in the domestic and wild pigs to further integrate genomics of *Suidae* species. The aim of this study was comparative cytogenetic mapping of GHRL as well as UCP2 and UCP3 genes in the Sus scrofa domestica [38,XY] and Sus scrofa scrofa [36,XY,rob(15;17)] species with the use of commercial human probes (Vysis, Q-BIOgene), specific for human chromosome regions HSA3p25-26 and HSA11q13-14 comprising loci of these genes. Chromosome localization was performed by FISH on DAPI-FISH banded metaphase plates. The following physical location was established – the GHRL was assigned to the domestic and wild pig chromosomes SSC13q31-32, and due to their proximity, UCP2 and UCP3 in both studied species were mapped to the same chromosome region – SSC9p21-24. Cross-species *in situ* hybridizations confirmed conservation of the linkage groups and high degree of homology of chromosome regions containing GHRL, UCP2 and UCP3 loci in compared species. This study was conducted as part of NRIAP statutory activity, project no. 04-006.1.

Comparative physical assignment of the heat shock protein beta-1 gene in the genomes of *Suidae*

B. Danielak-Czech[1], A. Kozubska-Sobocińska[1], M. Babicz[2], K. Kruczek[1] and M. Miszczak[1]
[1]*National Institute of Animal Production, Department of Animal Cytogenetics and Molecular Genetics, 1, Krakowska Street, 32-083 Balice n. Cracow, Poland,* [2]*University of Life Sciences in Lublin, Department of Pig Breeding and Production Technology, 13, Akademicka Street, 20-950 Lublin, Poland; anna.sobocinska@izoo.krakow.pl*

The heat shock protein beta-1 (HSPB1), from the family of small heat shock proteins (HSPB), displays a cytoprotective function involving apoptosis and inflammatory response as well as chaperone activity concerning the protein folding and aggregation control. Mutations of gene encoding this protein are responsible for desmin-related myopathies or neurodegenerative and autoimmune disorders. The aim of the study was comparative physical assignment of the HSPB1 gene in the genomes of Sus scrofa domestica [38,XY] and Sus scrofa scrofa [37,XY,rob(15;17)] species with the application of the CHORI-242 Porcine BAC Library clone as probe, selected by using information on BAC-end sequences (BES). Prior to FISH technique, carried out on DAPI-FISH banded metaphase chromosomes, the presence of this gene in the selected clone was confirmed by PCR method with gene-specific primers. In an effect of the experiments, FISH signals in the 3p15 chromosome region of both the species were obtained, which enabled to assign physical localization of the HSPB1 gene on the domestic and wild pig genome maps. The results obtained were verified by cross-species *in situ* hybridizations with the commercial (Vysis) probe specific for the human chromosome region HSA7q11.23 containing the HSPB1 gene and proved that location of this gene is in agreement with previous comparative mapping data between pigs and humans. The experiments confirmed conservative nature and homology of linkage groups and chromosome regions comprising the heat shock protein beta-1 locus in *Suidae* species and human genomes, which can be the basis of future phylogenetic and evolutionary studies. This study was conducted as part of NRIAP statutory activity, project no. 04-005.1.

Preserving health status of farm animals: what is expected from feed and nutrition
N. Le Floch
INRA, UMR 1348 PEGASE, 35590, France; nathalie.lefloch@rennes.inra.fr

Health preservation is one of the main priorities and a constant challenge for livestock production. Medication based on antibiotics have been used systematically and preventively to limit the negative consequences of poor health status. The reduction of their utilization is a challenge for the development of sustainable livestock production systems, which requires that health and performance could be maintained to limit the environmental impact and to improve the animal welfare. The role of nutrition is clearly questioned because it may be one of the solutions that contribute to limit medication through preserving health and performance. Indeed, health and nutrition are strongly connected. First, because health disturbances impact on the different components of nutrition functions, i.e. feed intake, digestion, and metabolism; and second, because nutrition determines the organism's ability to maintain or to restore homeostasis in response to health disturbances. Feeding practices like a moderate and transient feed restriction are commonly applied to limit gut disorders in young animals. Diet formulation can also be adapted to include raw materials acting at the gut level to stimulate digestive processes and select microflora. The feed has to supply nutrients adequately to support all the physiological functions, this means that nutrient deficiencies need to be identified and corrected. Additionally, specific nutrients are known to enhance body defenses. Those nutrients, when identified, could be supplied in the feed through adjustments of the feed formulation. The question is whether and how feed formulation and dietary nutrient supplies could be adapted to preserve both health and performance. A major limitation remains to determine quite early the situations during which an adaptation of feed could be relevant and efficient for preserving health and performance. Tools developed for precision livestock farming and precise feeding will be useful for the detection of health problems at an early stage making possible such adjustments of feeding practices.

Effects of the diet form on health and performance of weaning or fattening pigs
E. Royer, D. Gaudré and N. Quiniou
Ifip-institut du porc, BP 35104, 35651 Le Rheu, France; eric.royer@ifip.asso.fr

Four experiments were carried out to study the effect of dietary presentation on pigs. In a postweaning study, meal and pellets were compared in either good or bad sanitary conditions (Exp. 1, 14 pens/treatment). Over 41 days after weaning, pigs were offered *ad libitum* a phase 1 then a phase 2 diet in a dry feeder. The meal form increased DFI (+6%, P<0.01) but reduced ADG (-4%, P<0.05), then increased FCR (+10%, P<0.01). Meal improved the faecal consistency at d 20 (+8%, P<0.05), but no differences were observed between treatments at d 5, 12 and 26. No interaction between presentation and sanitary conditions was found. In the fattening periods, diet was presented either as meal or ground pellets and delivered restrictively using a liquid feeding system to gilts and barrows (Exp. 2, 16 pens/treatment) or boars (Exp. 3, 8 pens/treatment). In Exp. 2, DFI averaged 2.13 kg for both presentations, but FCR (-5%, P<0.05), carcass yield (+1%, P<0.01) and carcass leanness (+0.4%, P=0.09) were (or tended to be) improved with pellets. In Exp. 3, the tendency for a lower DFI with pellets (-2%, P=0.06) without any difference in ADG resulted in a reduction in FCR (-3%, P=0.05). With pellets, fewer pigs presented a skatole level above the minimum detectable concentration in liquid fat (P=0.01). In Exp. 4 (32 pens/treatment), meal and pellets supplied in dry feeders and meal mixed with water in the trough were given to fatteners, either under restricted or *ad libitum* feeding conditions. The DFI was higher with liquid meal than with the other forms (P<0.01). During the growing phase, the highest ADG was found with pellets and the lowest with liquid meal, whereas during the finishing period ADG was higher with liquid meal and pellets than with dry meal. The FCR was 4 and 7% lower with pellets than with dry and wet meal, respectively (P<0.01). The improved FCR observed in Exp. 1 to 4 with pellets, compared with dry or wet meal, could be attributed to the increased digestibility of nutrients induced by technology used for diet preparation.

Effect of different levels whole wheat on performance and health gut

A. Emarloo[1], R. Vakili[2] and S. Zakizadeh[1]
[1]Jihad-e-Agriculture Higher Educational Complex, Organic Animal Husbandry, TV Square, Asian Highway, Mashhad, Iran, [2]Islamic Azad University, Kashmar Branch, Animal Science, Seyed Morteza blvd, Kashmar, Iran; rezavakili2010@yahoo.com

The aim of this research was to investigate the impact of different levels whole wheat (0, 20 and 30%) on performance, gut flora and oocyst coccidia native laying hens under organic system. This experiment consisted of 135, 15 week old native, over 12 weeks with 3 treatments and 3 replicate. Evaluated traits were egg production.egg mass, feed intake, feed conversion rate, body change weight egg cholesterol, lactobacillus bacteria, coliform bacteria and OPG test. The different levels whole wheat has not significant effect of egg production.egg mass, feed intake, feed conversion rate and body change weight ($P<0.05$). whole wheat has significant effect on OPG ($P<0.05$). The lowest level of cholesterol in yolk egg observed in 0% whole wheat group ($P<0.05$). The result of lactobacilli and coliform bacteria showed lactobacilli bacteria enhance in 20% whole wheat group. The result of experiment showed whole wheat improve health gut in laying hens.

Effect of removal of antibiotics from the diet on welfare and health indicators of weaner pigs

A. Diana[1,2], E.G. Manzanilla[2], R. Vial[2], N. Leonard[1], K. O'driscoll[2] and L. Boyle[2]
[1]UCD, School of Veterinary Medicine, Dublin, Ireland, [2]Teagasc, PDD, Moorepark, Fermoy, Co. Cork, Ireland; alessia.diana@teagasc.ie

The aim was to evaluate the effects on health and welfare of removing antibiotics from the diets of 1st and 2nd stage weaner pigs. The study was conducted on a farrow-to-finish farm (300 sows). Every week for 6 wks, 70 pigs were weaned at 28 days, weighed, tagged and sorted into 2 groups of 35 pigs according to weight (10.6 ± 0.7 kg). AB were removed from the diet of one group (NO, n=6) and maintained in the other group (AB, n=6). Ten focal pigs were chosen per group. At the end of the 1st stage (4 wks and 4 d) each group was split into two pens of c. 15 pigs each (NO, n=12 and AB, n=12) for a further 4 wks. Data were recorded weekly. Skin lesions were scored on the focal animals according to severity; body (BL, 0 to 6), tail (TL, 0 to 5), ear (EL, 0 to 3) and flank (FL, 0 to 3). The number of coughs (COU) and sneezes (SN) per pen were counted over a 5 minute period and the number of health deviations (HD) was counted. Data were analysed using SAS 9.3. The BL score tended to be higher in AB than in NO pigs during the 1st (10.9 ± 0.92 vs 9.4 ± 0.92; P=0.09) and 2nd (18.8 ± 0.83 vs16.6 ± 0.83; P=0.07) weaner stages. Treatment had no significant effect on EL, FL and TL scores ($P>0.05$). There were significant differences between stages with BL ($P<0.05$), EL ($P<0.05$), and FL ($P<0.001$) scores higher in the 2nd compared to the 1st stage and TL scores higher in the 1st compared to the 2nd stage ($P<0.001$). There was a significantly higher frequency of COU in the AB treatment during the 2nd stage (0/5 min, 0-1 vs 0/5 min, 0-0; $P<0.01$) but no treatment effects were detected for SN and HD ($P>0.05$). Removing AB from the diet of weaner pigs had no effect on lesions related to welfare or on health indicators. Lower BL scores in NO pigs may have been associated with reduced growth rates in these animals but the unexpected reduction in coughing is difficult to explain.

Evaluation of dietary fish oil plus green tea on broiler gut microflora

A.R. Seidavi[1], J. Simões[2] and M.H. Alimohammadi Saraei[1]
[1]Rasht Branch, Islamic Azad University, Department of Animal Science, Pole-Taleshan, 41635, Iran,
[2]CECAV, Department of Veterinary Science, University of Trás-os-Montes e Alto Douro. Quinta de Prados,
Vila Real, 5000-811, Portugal; alirezaseidavi@iaurasht.ac.ir

In order to evaluate the effect of soybean oil substitution by fish oil (FO) plus green tea (*Camellia sinensis* L.; GT) powder at low levels on digestive tract microflora, 270 one-day old male chicks of Ross 308 strain were randomly assigned into 9 groups with 3 replicates and 10 birds each. All starter, grower, and finisher diets included a combination of 0, 1.5 or 2.0% of FO, substituting a part of soybean oil, with 0, 1.0 or 1.5% of GT powder. At day 42, the gizzard, ileal and cecal contents were independently collected, from one euthanized bird of each replicate, for microbial cultures and colony counts. The mean number of aerobic (7.396 ± 0.931; least-squares mean log10 cfu/g $\pm$ S.E.M.), lactic acid-producing (8.165 ± 0.900), coliforms (7.033 ± 0.892) or intestinal negative lactose (6.559 ± 0.823) bacteria of cecal contents from birds fed with 2% FO were similar to all other groups. No significant differences between groups were also observed from gizzard and ileal contents. These results suggest that the FO until 2.0% plus GT until 1.5% supplementation on diet don't affect negatively or even can preserve the gizzard, ileum and cecum microflora balance in apparent healthy broilers. Soybean oil may be substituted by FO until 2.0% level without apparent adverse effect on gut microflora.

Effect of L-carnitine lysine and methionine on carcass and immune parameters of broiler chicks

M. Bouyeh and A. Haddadi Bahram
Animal Science Department, Islamic Azad University, Rasht branch, Rasht. 4193963115, Iran;
mbouyeh@gmail.com

This experiment was carried out to determine the effects of dietary L-Carnitine and excess levels of lysine and methionine on carcass characteristics and some immune system parameters of broilers. 270 one-day-old chicks (Cobb 500) were used in a completely randomized design, 3×3 factorial arrangements: 3 levels of L-carnitine (0, 75 and 150 mg/kg) and 3 levels of lysine and methionine (NRC, 20 and 40% more than NRC recommendations) Data were analyzed by SPSS software and Duncan's test was used to compare the means. The results indicated that dietary carnitine + Lys and Met led to significant improve in carcass efficiency, breast muscle yield, heart and liver weight, plasma cholesterol, blood Lymphocytes and Heterophyls. A linear increase in Newcastle antibody observed parallel with increasing dietary Lys and Met ($P<0.01$). FCR, crude fat contents of breast, thigh and also plasma triglyceride was decreased by the highest levels of L-Carnitine or L-Carnitine + Lys and Met ($P< 0.05$). Results reported here support the hypothesis that it is possible to produce more healthy and economic poultry meat by adding L-Carnitine and excess Lys and Met to broiler diets.

Effects of a sequential offer of hay and TMR on feeding and rumination behaviour of dairy cows
F. Leiber, J.K. Probst and A. Spengler Neff
Research Institute of Organic Agriculture (FiBL), Ackerstrasse 113, 5070 Frick, Switzerland;
florian.leiber@fibl.org

Eighteen cows (Swiss Fleckvieh) with moderate performance (7,000 kg milk/cow/year) were investigated for their behavioural response to a change in roughage offer. They were kept in a stanchion barn with separated feeding places. In the first measurement week, cows received forages as a total mixed ration (TMR), based mainly on grass silage (0.32), maize silage (0.30) and hay (0.21), *ad libitum* 24 h/d. In the second measurement week, TMR contained a lower proportion of hay (0.06). In lieu thereof hay (2nd cut) was separately offered *ad libitum* to each cow in the morning between 6.00 am and 8.00 am Animals had been adapted to each feeding system 14 days before measurement weeks started. During both measurement weeks, cows were equipped with chewing sensors (noseband collars with pressure tubes) to record jaw movements and identify eating, rumination and idle. Roughage intake was weighed individually during the measurement weeks. Data were processed with SPSS® in a mixed linear model with measurement week as fixed and cow as random factor. Separate offer of hay in the morning led to longer intake time (min/h) between 6.00 am and 8.00 am ($P<0.01$), but also between 4.00 pm and 6.00 pm ($P<0.05$). Intake amounts (kg/d) were not affected. Rumination time (min/h) was increased between 9.00 am and 3.00 pm ($P<0.01$). Further, when hay was offered separately in the morning, the number of activity changes (between eating, ruminating or idle) per hour was decreased in every time of the day and night ($P<0.05$). The data show a clear effect of sequential roughage feeding on intake and rumination. Separately fed hay might have caused a necessity to increase intake and rumination time to maintain intake amounts and digestion. Decreased activity changes per hour indicate a calmer behaviour, which might be an advantage for animal health and welfare. The response of behaviour to feeding conditions shows the indicative potential of such data for feeding evaluation.

Effects of L-carnitine and methionine-lysine on reproductive and blood parameters of native ducks
M. Bouyeh, M. Mirzai, A. Seidavi and A. Haddadi Bahram
Animal Science department, Islamic Azad University, Rasht branch, Rasht, 4193963115, Iran;
mbouyeh@gmail.com

This study was conducted in order to investigate the effects of different levels of L-carnitine, and methionine-lysine on the reproductive performance of native parent ducks. A total 180 breeder ducks were used for seven weeks based on a completely randomized design in a factorial (3×3) arrangement (include three levels of L-carnitine × three levels of methionine-lysine) with four replicates per each treatment and five ducks per each replicate. Feed intake and egg production weekly. Fertility and hatchability were measured using transferring the collected eggs into setter during last week. To measurement of blood parameters at the end of the experiment, blood samples were taken via wing vein and analyzed in laboratory The results of this study showed that increased levels of L-carnitine and methionine-lysine had no significant effect on feed intake ($P>0.05$), but the study of interactions showed a significant difference in the period ($P<0.05$). These two factors had a significant effect on the egg production, so that increasing of L-carnitine and methionine-lysine increased the number of eggs. L-carnitine and methionine- lysine had no significant effect on feed conversion ratio ($P>0.05$). Lysine-methionine also increased egg weight significantly so that the treatment with 20% more than standard requirements showed the highest egg weight ($P<0.05$). Survey the treatments indicated the highest level in treatment 9, which was significantly more than control treatment ($P<0.05$).

Effects of kelp meal on health and productivity of mink challenged with the Aleutian disease virus

A.H. Farid[1], N.J. Smith[2] and M.B. White[2]
[1]Dalhousie University Faculty of Agriculture, Animal Science, Haley Institure, 58 River Road, Truro, Nova Scotia, B2N 5E3, Canada, [2]Perennia Food and Agriculture, Bible Hill, Nova Scotia, B4N 1J5, Canada; ah.farid@dal.ca

Aleutian mink disease virus (AMDV) is endemic in Nova Scotia (NS) and causes economic losses to the industry. More than 30 years of test-and-removal strategy failed to eradicate the virus from many ranches in this province. Since 2013, many mink ranchers in NS have started selecting their herds for tolerance to AMDV. Feed additives that can ease the negative effects of infection are of particular interest to these mink ranchers. A total of 75 AMDV-free female black mink were inoculated intranasally with a spleen homogenate containing a local strain of the AMDV in Sep. 2013. Mink were fed a commercial pellet with the Ascophylum kelp meal added at the rates of 0% (TR1, control), 0.75% (TR2) and 1.5% (TR3) of the feed. Blood was collected on days 0 (pre-inoculation), 31, 56, 99, 155 and 367 post-inoculation. Mortality rate from October 1, 2013 to November 1, 2014 were 31.8, 20.0 and 13.6% for TR1, TR2 and TR3, respectively, suggesting the beneficial effects of kelp on survival rate. No significant difference was observed among treatments for the proportion of animals positive for antibody against the virus measured by the counter-immunoelectrophoresis, viremia measure by PCR, antibody titer measured by quantitative ELISA, serum total proteins measured by a refractometer, or albumin:globulin ratio measured by the iodine agglutination test. Compared with the controls, mink on TR2 and TR3 had lower body weights at days 155 and 367 post-inoculation but not at days 0, 31, 56 or 99. A total of 62 females bred in March 2014, and those on the TR3 outperformed TR1 and TR2 for kits born alive per female exposed to males (5.18, 1.78, 1.04) and kits weaned per female exposed to males (3.54, 1.11, 1.32). It was concluded that feeding kelp could improve reproductive performance of mink infected with AMDV.

Zinc-methionine bioplex administration to pregnant and lactating sheep and selected wool parameters

K. Czyz[1], S. Kinal[2], A. Wyrostek[1], K. Roman[1], M. Janczak[1], R. Bodkowski[1] and B. Patkowska-Sokola[1]
[1]Wroclaw University of Environmental and Life Sciences, Institute of Animal Breeding, Chelmonskiego 38c, 51-630 Wroclaw, Poland, [2]Wroclaw University of Environmental and Life Sciences, Department of Animal Nutrition and Feed Management, Chelmonskiego 38c, 51-630 Wroclaw, Poland; anna.wyrostek@up.wroc.pl

The aim of the study was to examine the possibility of merino sheep supplementation with chelates preparation (Zn + methionine). The study was conducted on 22 pregnant and lactating sheep of Polish Merino breed. The sheep were divided into two equal groups. The differentiating factor was an addition of zinc-methionine preparation in amount of 0.4 g/head/d to the ewes from the experimental group. The experiment was conducted for 3.5 months of pregnancy, and additionally 2 weeks of lactation period. The wood from each sheep was cut at the left side before the beginning of the experiment, in order to compare the wool grown during the experimental period in both groups of ewes. The obtained results concerning administration of zinc and methionine preparation to pregnant and lactating sheep demonstrated that it profitably affected the rate of wool growth, which was ca. 30% higher in the experimental group, as well as its thickness (ca. 8% increase) over the examined period. Moreover, about 15% higher zinc content was noted in wool from the experimental group sheep. What interesting, also sulfur content was higher in sheep supplemented with preparation (ca. 9%), and the wool was characterized by higher strength (ca. 11%). This should be explained by methionine addition, as well as by the fact that zinc contributes in sulfur amino acids incorporation into the wool. The study conducted unequivocally pointed profitable effect of applied preparation on wool performance of pregnant and lactating sheep. However, no effect on wool histological structure was noted. Zinc-methionine bioplex administration is recommended during pregnancy and lactation in order to reduce depression in wool growth, and its strengthening in investigated periods.

Effects of dietary energy on growth performance of Korat chickens from 3 to 6 weeks of age

P. Maliwan, S. Khempaka and W. Molee
Suranaree University of Technology, Nutrition, 111 University Avenue, Muang, Nakhon Ratchasima, 30000,
Thailand; praphot382@gmail.com

The Korat chicken is Thai indigenous crossbred (50%) meat chicken, which is crossed between Thai indigenous male (Leung Hang Khoa) and SUT female breeder. So far, there is a limited information on metabolizable energy (ME) and protein (CP) requirements for the Korat chicken. Therefore, this study was conducted to evaluate the responses of Korat chickens to various dietary ME levels during 0 to 12 weeks of age by dividing into 4 phases (0 to 3, 3 to 6, 6 to 9 and 9 to 12 wk of age). Since this study still in the process, therefore, the recent report will be presented the results only the age period of 3 to 6 weeks. Four dietary ME levels were composed of 2,750, 2,900, 3,050 and 3,200 kcal of ME/kg. Seven hundred and twenty 21-day old mixed-sex Korat chicks were randomly divided into 4 dietary treatments, each containing 6 replicate pens with 30 birds per pen, and using completely randomized design (CRD). These birds were raised until 6 weeks of age. Body weight (BW), BW gain, feed intake, feed conversion ratio (FCR: feed/gain), feed cost per kilogram of BW gain, protein efficiency ratio (PER), energy efficiency ratio (EER) and blood urea nitrogen (BUN) of chickens from each pen were measured at the end of experiment. As dietary energy increased from 2,750 to 3,200 kcal of AME/kg, the feed intake was decreased ($P<0.01$), while FCR and PER were improved ($P<0.01$). The significant differences in PER among treatments was found ($P\leq0.05$). However, BW gain, feed cost per weight gain 1 kg and EER did not differ significantly ($P>0.05$) among treatments. According to the broken-line regression analysis (straight-line analysis), the ME requirements of Korat chickens for optimal FCR and PER were 3,151 and 3,194 kcal/kg, respectively. The regression equations predicted the ME requirement for optimal FCR and PER were y = 1.9150 + 0.000639 X (3,151.3 – x) [$P<0.01$, $R^2=0.85$] and y = 2.5233 – 0.00072 X (3,193.8 – x) [$P<0.01$, $R^2=0.88$], respectively.

Feather pecking and cannibalism in 107 organic laying hen flocks in 8 European countries

M. Bestman and C. Verwer
Louis Bolk Institute, Animal Husbandry, Hoofdstraat 24, 3972 LA Driebergen, the Netherlands;
m.bestman@louisbolk.nl

The aim of our study was to get insight in the frequency of feather and injurious pecking in organic laying hens, its relations with husbandry practices and give recommendations for farmers and policy makers on how to reduce feather and injurious pecking. We visited 107 organic layer farms in 8 European countries. Information was collected about management, flock, vaccinations, medical treatments, feeding, housing, range management and specific problems. At the end of lay, 50 hens per flock were assessed for plumage condition and wounds at the neck, back, belly and tail. Potential factors related to the percentages of 'hens affected' were screened by partial correlation analyses for all continuous- and categorical variables. Dichotomous variables were screened by means of linear regression. Fifteen percent of the flocks had severe feather damage, 20% had moderate and 65% had little/no feather damage. We found a higher percentage of hens with feather damage if longer pre-lay feed was fed, if more different feed phases were fed, in case of lower protein, lower methionin, lower percentage of hens in the wintergarten, lower percentage of hens in the free range area, more times dewormed, higher no of alternative treatments, no litter replacement or topping, if roughage was provided during rearing, if there was no natural light source and in case of a needle vaccination after rearing. We found a higher percentage of hens with wounds if longer pre-lay feed was fed, if more different feed phases were fed, in case of lower protein, higher degree of blood mites, no needle vaccination at placement, lower calcium and no litter topping. We recommend free-choice feeding, environmental enrichment of the wintergarten, shelter in the free range area, good litter management, provision of natural light and the prevention of red blood mites, harmonization of the regulation in the different countries.

Reducing feather pecking in commercial UK flocks – measuring success and overcoming barriers

C.J. Nicol[1], P.E. Baker[1], J. Gittins[2], S.L. Lambton[1], J. Walton[1] and C.A. Weeks[1]
[1]University of Bristol, School of Veterinary Science, Langford House, BS40 5DU, United Kingdom, [2]ADAS, Animal Health, Cefn Llan Science Park, Aberystwyth, SY23 3AH, United Kingdom; c.j.nicol@bris.ac.uk

Feather pecking (FP) causes significant welfare and economic problems on commercial laying hen farms, and is particularly hard to control in non-cage systems. In the UK, the majority of laying hens are beak trimmed which can reduce the damage caused by FP. However, beak trimming is considered a mutilation and there is growing pressure to ban this practice in EU member states. There is therefore a pressing need to consider the extent to which targeted management changes can reduce the risk of FP. There is a considerable scientific literature on the causes of FP but, until recently, this source of knowledge had not been fully utilised in commercial practice. We used this literature systematically to design a set of management strategies for commercial application. Consultation with farmers and other stakeholder groups was essential to produce a feasible package. We tested effectiveness by recruiting 53 treatment flocks and comparing their performance with 47 control flocks. A greater uptake of management strategies was associated with lower plumage damage, gentle FP, severe FP, vent pecking and mortality at 40 weeks in these (largely) beak-trimmed birds. More recently we have studied the effects of similar management strategies in commercial flocks with intact-beaks, where the risk of damage from FP is far higher. The adoption of additional management strategies reduced FP in small, experienced intact-beak flocks, and was partially effective in mitigating the risk of transition from beak-trimmed to intact-beak flocks. An overall economic evaluation of the financial costs and benefits of adopting additional management strategies will be presented and the process of developing training materials and websites to share good practice (e.g. www.featherwel.org) will be described. The barriers, challenges and remaining obstacles will also be highlighted.

Genetic analysis of feather pecking in divergent selected lines and in an F2 cross

J. Bennewitz[1], J. Kjaer[2] and W. Bessei[1]
[1]Institute of Animal Science, University Hohenheim, Germany, Garbenstrasse 17, 70599 Stuttgart, Germany; [2]Institute of Animal Welfare and Animal Husbandry at the Friedrich-Loeffler Institute, Federal Reserch Institite of Animal Health, Dörnbergstraße 25/27, 29223 Celle, Germany; joergen.kjaer@fli.bund.de

Feather pecking in laying hens is characterized by non-aggressive pecks directed towards the plumage of other hens. The underlying mechanisms are not well understood, but physiological, nutritional as well as genetic factors are known to influence this trait. The present study investigated feather pecking in two experimental populations, i.e. in divergent selected lines (high feather pecking and low feather pecking lines) as well as in a large F2- cross established from these two lines. The selection experiment was conducted for eleven generations. A screen for selection signatures in a sample of birds from the two lines from the last generation revealed 12 clusters with significant elevated Fst-values. The data from the F2 cross (966 hens with observations) were used for quantitative-genetic analysis and will be used for mapping QTL underlying feather pecking. Quantitative genetic analysis revealed a moderate heritability of the trait, depending on the length of the observation period and the statistical model applied. In a series of bi- and tri-variate models genetic correlations with a number of other traits were investigated, including fear traits, egg production, growth, feather eating, feather score and activity. These analysis were partly conducted using structural equation models to better understand the genetic correlations. The results showed a complex relationship between feather pecking and some other traits like egg production, activity, and feathers eaten.

Genetic selection on social genetic effects to reduce feather pecking in layers
E.D. Ellen[1], J. Visscher[2] and P. Bijma[1]
[1]Wageningen UR, ABGC, BO Box 338, 6700 AH Wageningen, the Netherlands, [2]ISA, P.O. Box 114, 5830 AC Boxmeer, the Netherlands; esther.ellen@wur.nl

Feather pecking is a multi-factorial problem, that depends on management and breeding factors. In this study, we used genetic selection to reduce mortality due to feather pecking in commercially housed layers. Feather pecking is a socially-affected trait, it depends on both the genotype of the individual itself (direct genetic effect) and the genotypes of its group mates (social genetic effect). Behavioural observations can be used to select against feather pecking. However, these are expensive, time consuming and difficult to apply in animal breeding. A solution can come from statistical methods that take into account both the direct genetic effect (victim effect) and the social genetic effect (actor effect). In this study, we used selection based on relatives kept in family groups to select for improved survival in layers. Using this selection method, both the victim and actor are taken into account. As founder population, a purebred White Leghorn layer line of ISA was used. For all six generations, individually housed selection candidates (SC) were selected based on survival days of relatives kept in family groups. Relatives had intact beaks and were kept in 4-bird family cages. For generation 1, 5 and 6, SC were selected to breed high (HIGH) and low (LOW) survival. Remaining SC were used to breed control (CONT). For generation 2-4, SC were selected only to breed HIGH, and there was no CONT present. The generations were kept at three different locations, therefore, it was not possible to compare HIGH across generations. In generation 1, 5, and 6, difference in survival days between HIGH and LOW ranged from 26-29 days. Difference in survival days between HIGH and CONT was 13, -12 and 19 days in generation 1, 5 and 6, respectively. These results show that selection against mortality due to feather pecking is feasible. However, results also illustrate that feather pecking is very sensitive to changes in the environment.

Breeding value predictions for survival in layers showing cannibalism
T. Brinker, E.D. Ellen and P. Bijma
Wageningen University, ABGC, P.O. Box 338, 6708 PB Wageningen, the Netherlands; tessa.brinker@wur.nl

Feather pecking is an economic and welfare problem that results in bird losses. Genetic selection for increased survival can contribute to solve this issue. This is challenging because heritability is low, censoring is high (hens still alive at the end of the laying period), and individual survival depends on social interactions. The latter means that survival time depends on an individual's own genes (direct genetic effect; DGE), and on genes of its group mates (indirect genetic effect; IGE). Until now, studies using DGE-IGE models focussed on survival time (ST). Shortcomings of this approach are that censored records are assumed known and that IGE are assumed to be continuously expressed by all individuals, irrespective of whether they are alive or dead. However, dead animals no longer express IGE so that IGE would ideally be time-dependent in the model. Neglecting censoring and timing of IGE expression may reduce accuracy of breeding value (BV) estimates. The aim was therefore to improve BV predictions for survival in layers showing cannibalism. Next to a DGE-IGE analysis of ST, we considered three alternative models that treated survival in consecutive months as a repeated binomial trait. Model 1 included direct and indirect random genetic regressions on time. Model 2 included random regressions on a function of mean survival. Model 3 was a generalized linear mixed model including a random direct and indirect genetic intercept. Accuracy of EBVs were calculated using cross validation. The models were fitted to survival data on two purebred layer lines (6,276 W1 and 6,916 WB) with 13 monthly survival records each. Data were provided by ISA. Adjusting the models for timing of IGE expression was detrimental. Compared to analysis of ST, repeatability models of survival yielded considerably higher accuracy of EBVs, in both W1 and WB. Accuracies of EBVs were improved by 10%-21% in W1, and by 10%-12% in WB compared to analysing ST. Repeatability models with DGE and IGE can be applied to select against feather pecking.

Evaluation of an additional water supply in Pekin ducks (*Anas platyrhynchos* f.d.)

L. Klambeck[1], F. Kaufmann[1], J.D. Kämmerling[1], N. Kemper[2] and R. Andersson[1]
[1]*University of Applied Sciences Osnabrück, Animal Husbandry and Poultry Sciences, Am Krümpel 31, 49090 Osnabrück, Germany,* [2]*University of Veterinary Medicine Hannover, Insitute of Animal Hygiene, Animal Welfare and Animal Ethology, Bischofsholer Damm 15, 30173 Hannover, Germany; L.Klambeck@hs-osnabrueck.de*

According to the recommendations of the standing committee of the European convention for the protection of animals kept for farming purposes, Pekin ducks without access to open/bathing water must be provided with water resources that allow them to take in water with their beak, dip their head and splash water over their bodies. The aim of the study was to evaluate a certain modified cup drinker system (BF) regarding their suitability to fulfill the EC-Recommendations. A total of 284 pekin ducks (22 d of age) were divided into 6 groups and reared for 30 days with either standard nipple drinkers or BF. Water usage, feed consumption, body weight development, feed:gain and behaviour of the birds were recorded at regular intervals. At day 30, 37 and 40, randomly selected birds (n=42) of the BF groups were placed in experimental compartments containing BF with blue-coloured water (food colour). After 90 minutes, integument of certain body regions of the birds was evaluated regarding colouring. All birds (100%) showed a blue discolouration of the head and back at each sampling date, whereas the head region showed the highest intensity (score: 5) when compared with the back region (score: 2-3). The results indicate that the BF system fulfill the EC-Recommendations as the birds have the opportunity to dip their head and splash their feathers and thus, are able to keep their eyes, nostrils and feathers clean. However, economic and hygienic aspects have to be taken into account.

Effects of physical water treatment on microbiological water quality and performance of turkeys

R. Boussarhane[1], R. Günther[1], N. Kemper[2] and J. Schulz[2]
[1]*Heidemark GmbH Veterinärlabor, Jakob-Uffrecht-Str. 20, 39340 Haldensleben, Germany,* [2]*University of Veterinary Medicine Hannover, Foundation, Institute for Animal Hygiene, Animal Welfare and Farm Animal Behaviour, Bischofsholer Damm 15, 30173 Hannover, Germany; nicole.kemper@tiho-hannover.de*

Good water quality is essential for animal health, welfare and performance. In order to reduce biofilms, with possibly adverse effects on the animals, electromagnetic fields can be used in water pipes. The aim of this study was to examine the effects of physical water treatment on water quality and performance parameters in a turkey barn. The study was conducted with two groups of 1,900 birds in two consecutive fattening periods. Drinking water from the same source was provided to each group via a separated drinking system. The water of the test group was treated with an electromagnetic device. The water of the control group remained untreated. In each group, 32 water samples were taken from the source, from the beginning and from the end of the pipe, respectively. The concentration of mesophilic bacteria and endotoxins were measured in all samples. Moreover, the performance parameters of the birds were recorded. Mesophilic bacteria and endotoxins increased significantly from the source to the end of the pipes in both drinking systems. Endotoxin concentrations varied in most cases between 1 and 100 ng/ml and showed no significant differences between treated and non-treated water. A lower bacteria concentration was observed at the end of the pipe in the test group. However, the numbers of colony forming units of the treated water (end of the pipe) were in 24 out of 32 cases higher than 1000 cfu/ml (recommended value by the German Federal Ministry of Food and Agriculture). Concerning performance parameters, no difference between the two groups was assessed. Concluding, physical treatment of drinking water was neither able to improve microbiological water quality nor production parameters in a turkey barn.

Relationships between residual feed intake and carcass traits in a population of F2 chickens

H. Emamgholi Begli[1,2], R. Vaez Torshizi[1], A.A. Masoudi[1], A. Ehsani[1] and J. Jensen[2]
[1]Tarbiat Modares University, Animal Science, Jalale-Ale-Ahmad and Chamran Highways, 14117-13116
Tehran, Iran, [2]Aarhus University, Molecular Biology and Genetics, Blichers Alle 20, P.O. Box 50, 8830
Tjele, Denmark; emamy202@yahoo.com

Feed represent about 70% of the total costs in poultry production. These costs can be reduced by improving feed efficiency through genetic selection. However, it is also important to study the impacts of selection for improved feed efficiency in other performance traits. Therefore, the aim of the present study was to evaluate the relationships between residual feed intake (RFI), and other performance traits using 118 F2 chickens. The F2 population derived from the cross between Arian broiler line and Iranian native fowl. The individual feed intake and body weight, measured every week from 2 to 10 weeks of age, were used to calculate RFI. Birds were slaughtered at 12 weeks of age. Chickens were categorized into high (n=39), medium (n=40), and low (n=39) RFI groups based on their breeding values (BV). BVs were calculated using a quadratic spline with four knots and heterogeneous residual variance. Performance traits measured included: weights of the carcass, eviscerated carcass, wing, breast muscle, heart, gizzard, spleen, liver, lung, skin, back and neck, burs and abdominal fat. No differences were detected in eviscerated carcass, wing, breast muscle, lung and burs among the high, medium and low RFI chickens. Chickens from the low RFI group had greater gizzard and lesser heart, gizzard, spleen, liver, back and neck and abdominal fat compared with the high RFI groups. Results suggest that selection of chickens for RFI did not respond appropriately on all performance traits.

Influence of non-starch polysaccharide-degrading enzymes on growth performance of broiler

A.R. Seidavi[1], S.M. Hashemi[1], F. Javandel[1] and S. Gamboa[2]
[1]Rasht Branch, Islamic Azad University, Department of Animal Science, Pole-Taleshan, 41635, Iran,
[2]Coimbra College of Agriculture, Polytechnic Institute, Department of Zootechnic Sciences, Animal Rep,
Coimbra, Animal Reproduction Laboratory, Portugal; alirezaseidavi@iaurasht.ac.ir

An experiment was conducted to evaluate the effects of dietary inclusion of commercial feed enzymes (non-starch polysaccharide-degrading enzymes, NSPases) on growth performance, carcass quality and blood parameters of broilers from 1 to 42 days of age. Three hundred male broiler chicks (Ross-308) were fed two basal diets (corn-based diet, and a wheat- and barley-based diet), two commercial enzymes (Kemin® and Rovabio®), at two levels (0.025% and 0.05%) in a 2×2×2 factorial arrangement from 1 to 42 days of age. The dietary groups were control (basal non-supplemented corn-based diet, T1, and basal non-supplemented wheat- and barley-based diet, T6), Kemin® (basal diet + 0.025%, T2 and T7; and 0.05%, T3 and T8) and Rovabio® (basal diet + 0.025%, T4 and T9; and 0.05%, T5 and T10). All the diets were isoenergetic and isonitrogenous. Feed treatment had a significant effect (P<0.05) on the weight gain, the feed conversion ratio (FCR) and on the blood serum GLU, LDL and HDL levels in birds fed wheat-barley diets. Overall, birds fed corn-based diets (T1 to T5) consumed more feed (P<0.05) over the entire experiment, experienced a higher weight gain (P<0.05) and a FCR (P<0.05) when compared with wheat- and barley-based diet (T6 to T10). Notwithstanding, the FCR was not improved in birds fed on corn-based diet plus enzymes while in birds fed on wheat- and barley-based diet weight gain and FCR were improved (P<0.05) with addition of 0.05% enzyme. The results demonstrated the bio-efficacy of feed xylanases and glucanases in poultry diets rich in soluble non-starch polysaccharides.

Is cyromazine in poultry feeds metabolized to melamine and transferred to meat and eggs?

C.W. Cruywagen, R. Basson and E. Pieterse
Stellenbosch University, Animal Sciences, Private Bag X1, Matieland, Stellenbosch 7602, South Africa;
cwc@sun.ac.za

Two studies were done to i.) determine if cyromazine (CYR) in poultry feed can be transferred to meat and eggs in the form of melamine and ii.) to measure the efficiency of dietary melamine (MEL) transfer to poultry meat and eggs. Five diets containing CYR and graded levels of MEL were formulated for broiler and layer chickens. In Experiment 1, 480 day-old broiler chickens were allocated o five treatment groups. Diets contained either 4 mg/kg of CYR, or 50, 100 or 500 mg/kg of MEL. A control (CON) diet was also included that contained no CYR or MEL. The duration of the trial was 56 days and breast muscle, kidney and liver samples were harvested on Days 11, 13, 15, 18, 22, 29 and 36, and analysed for MEL. In Experiment 2, 120 hens (24 weeks of age) were randomly allocated to five treatment groups. The treatment diets were the same as for Experiment 1. The duration of this trial was 20 days and layers received the treatment diets for the first 10 days and the CON diet for another 10 days. Data were subjected to a two-way ANOVA with treatment and days as main effects. For the CYR treatment, no MEL was detected in the meat, other tissues, or eggs. For the MEL treatment, dietary MEL was absorbed by broilers and layers and rapidly distributed to the kidneys, livers, muscles and eggs. As dietary MEL concentration increased, so did muscle, organ and egg MEL concentrations. Melamine concentration for broilers peaked at 22 days of age and decreased until day of slaughter. Kidneys contained higher MEL residue levels than muscles or liver. In layer hens, MEL plateaued in eggs between Days 1 and 4 and decreased from Days 7 to 10. The partitioning of MEL in eggs was higher to albumin than to the yolk. Upon withdrawal, MEL concentration in these tissues decreased to undetectable levels within 7 days. The distribution efficiency of MEL to meat (1.2 to 2.7%) and eggs (0.7 to 0.8%) did not appear to be dose dependent. Cyromazine is not transferred to meat and eggs in the form of melamine.

The eating quality of Mallards (*Anas platyrhynchos* L.)

P. Janiszewski[1], D. Murawska[2], A. Gugolek[1], M. Zawacka[2] and J. Folborski[1]
[1]University of Warmia and Mazury in Olsztyn, Department of Fur-Bearing Animal Breeding and Game Management, Oczapowskiego 5, 10-718 Olsztyn, Poland, [2]University of Warmia and Mazury in Olsztyn, Department of Commodity Science and Animal Improvement, Oczapowskiego 5, 10-718 Olsztyn, Poland;
janisz@uwm.edu.pl

The development of effective sales systems facilitates the purchase of seasonally harvested game birds by non-hunters. Mallard ducks (*Anas platyrhynchos* L.) are relatively common game birds, and wild duck hunting is one of the most popular types of fowl hunting in many countries. The consumption of meat, including less popular kinds of meat such as wild game meat, is affected by both availability and information on the product. The aim of this study was to determine the carcass value, the proximate chemical composition of meat and the fatty acid profile of lipids extracted from the breast muscles of mallards (10♂ and 7♀) hunter-harvested in north-eastern Poland. Despite considerable sexual dimorphism in body weight, no differences were noted between male and female mallards in the percentages of carcass tissue components. Muscle tissue, skin with subcutaneous fat and bones accounted for approximately 64, 17 and 18%, respectively, of total carcass weight in both males and females. Nearly 70% of lean meat was located in the most valuable carcass parts, i.e. the breast and legs. The concentrations of protein and fat in the breast muscles of mallards reached 23.5 and 0.81%, respectively. C12-C22:6 fatty acids were identified in the fatty acid composition of lipids extracted from the breast muscles of wild ducks. Saturated fatty acids, monounsaturated fatty acids and polyunsaturated fatty acids accounted for 39.1, 17.3 and 43.6%, respectively, of the total fatty acid pool in the breast muscle lipids of mallards. The results of our study indicate that the meat of hunter-harvested mallards can be attractive to modern consumers.

Selected growth parameters of farm-raised Mallard (*Anas platyrhynchos* L.) ducklings

P. Janiszewski[1], D. Murawska[2], V. Hanzal[3], K. Tomaszewska[2] and B.J. Bartyzel[4]
[1]University Of Warmia and Mazury in Olsztyn, Department of Fur-Bearing Animal Breeding and Game Management, Oczapowskiego 5, 10-718 Olsztyn, Poland, [2]University of Warmia and Mazury in Olsztyn, Department of Commodity Science and Animal Improvement, Oczapowskiego 5, 10-718 Olsztyn, Poland, [3]University of South Bohemia in Česke Budejovice, Department of Landscape Management, Studentska 13, Česke Budejovice, Czech Republic, [4]Warsaw Agricultural University, Department of Morphological Sciences, Nowoursynowska 159, Warsaw, Poland; janisz@uwm.edu.pl

The size of wild populations of mallards (*Anas platyrhynchos* L.) can be increased by artificial propagation. Captive-reared mallards are released from farms into natural habitats for hunting purposes. Wild ducks and geese are characterized by a fast growth rate in early life stages, which is a flight adaptation. The objective of this study was to analyze selected growth parameters of farm-raised mallard ducklings. The experimental materials comprised 70 ducklings (1:1 sex ratio) raised to six weeks of age on a farm in Ceske Budejovice, Czech Republic (Lesy a rybníky města Českých Budějovic s.r.o). Between one day and six weeks of age, the average body weight of mallards increased nearly 16.5-fold (from 39.2 g to 645.0 g), carcass weight increased 26.6-fold (from 14.5 g to 385.0 g), gizzard weight increased 34-fold (from 1.02 g to 34.7 g), liver weight increased 19.1-fold (from 0.91 g to 17.4 g), and heart weight increased 18.6-fold (from 0.29 g to 5.4 g). Among the analyzed carcass parts, the wings were characterized by the fastest growth rate (wing weight increased 87.7-fold, from 0.78 g to 67.6 g), followed by the breast (48.2-fold, from 1.72 g to 82.9 g). Slower growth rates were noted in the remaining carcass parts whose weight increased as follows: back: 23.1-fold; neck: 19.5-fold; legs: 16.7-fold. Over the analyzed period, the weights of the head, feet and gastrointestinal tract of mallards increased 6.3, 8.6 and 8.3-fold, respectively.

Use of distilled dried grain with solubles in laying hens to replace soybean and sorghum

J.M. Uriarte, J.A. Romo, R. Barajas, H.R. Guemez and J.M. Romo
Universidad Autónoma de Sinaloa, Facultad de Medicina Veterinaria y Zootecnia, GRAL. Angel Flores S/N, 80080, Culiacán, Sinaloa, Mexico; jumanul@uas.edu.mx

To determine the effect of inclusion of distiller dried grain with solubles (DDGS), in substitution of soybean meal and sorghum on laying performance and egg flavour a total of 270 hens (White Leghorns of 1.68±0.16 kg), were fed one of three diets in a totally randomized design. All three diets had approximately similar protein (16.0-16.3% CP) and energy concentrations (2.83-2.85 Mcal of ME/kg) and contained 8% of a vitamin-mineral premix. The first diet (CON) included 70% sorghum and 22% soybean meal (n=90). The second diet was composed of 66% sorghum, 19% soybean meal and 7% DDGS (DDGS7, n=90). The third diet contained 62.5% sorghum, 15.5% soybean meal and 12% DDGS (DDGS12, n=90). Hens received the experimental diets in cages (3 hens/cage) and were weighed on days 0 and 105 of the experiment. Egg production and feed intake were recorded daily. Egg weight and egg flavour were determined at weekly intervals. Body weight at day 0 (1.67, 1.69 and 1.68 kg) was similar (P>0.15) for CONT, DDGS6 and DDGS12, respectively. Feed intakes were similar (P>0.05) between dietary treatments. The egg production (70.1, 70.0 and 69.3%) was not affected (P>0.94) by the treatments. Egg weight (59.2, 59.3 and 59.6 g) was similar (P>0.30) among the treatments. Feed intake ratio/kg egg produced (2.58, 2.58 and 2.60 kg) was not affected by treatments (P=0.55). It is concluded that DDGs can be used up to 12% in substitution of soybean meal and sorghum in diets for laying hens without affecting production performance, although DDGs slightly modify the flavour of the egg, transferring a little spice to freshly baked cookies.

Taxonomic and functional caecal microbiome of hens fed with dry whey alone or with prebiotic

C. Pineda-Quiroga, A. Hurtado, A. Lanzén, R. Atxaerandio, R. Ruiz and A. García-Rodríguez
Neiker-Tecnalia, Campus Agroalimentario de Arkaute, 01080 Vitoria-Gasteiz, Spain; rruiz@neiker.net

Dietary interventions in poultry to enhance productive performance and reduce risk of infection can affect gut microbiota. Dietary addition of dry whey alone or combined with prebiotic Pediococcus acidilactici was shown to improve several productive parameters in laying hens. Here, the caecal microbiome of 67-weeks-old Isa Brown hens previously fed per 70-days with diets supplemented with 6% dry whey (W), 0.2% P. acidilactici (P), a combination of both (WP) or a control diet (C) was characterized by Illumina HiSeq shotgun sequencing of one hen per diet. Over 57 million reads pairs were generated, quality trimmed, assembled (using Ray Meta). The abundances of resulting contigs (n=1,116,363 N50=214 nt) in each of the four samples were determined by mapping individual reads to contigs, which were then submitted to MG-RAST for annotation and taxonomical and functional analyses (e-value $<1e^{-5}$, alignment length >50). Over 98% of annotated sequences were bacterial, the most predominant phyla being bacteroidetes (69%), fimicutes (25%) and proteobacteria (2.2%). archaea (0.6%) were mainly euryarchaeota of methanogenic classes (mostly methanomicrobia). The taxon distributions among diets were large for the archaeal order methanomicrobiales, which was high proportions in all supplemented diets (mainly in WP) related to the control. Functional analysis using the SEED revealed a similar functional gene content in all hen microbiomes, with amino acids and carbohydrates metabolisms being the most abundantly represented. Further studies are needed to verify that diet-associated changes in community structure observed were not affected by individual host variation.

Ancestral influences on descendant generations: a case study using the olfactory system in rodents

B.G. Dias
Department of Psychiatry and Behavioral Sciences, Yerkes National Primate Research Center, Emory University, Atlanta, GA, USA; bdias@emory.edu

Traumatic experiences impact not only the generation that directly experiences the trauma, but also descendant generations that follow. How this occurs and what effect such trauma might have on the development of neuropsychiatric disorders in descendant generations is less studied. Training F0 adult mice (ancestral generation) to associate an odor (acetophenone) with mild foot-shocks allowed us to ask how an environmental cue associated with an aversive outcome in the ancestral generation is perceived by descendant generations at the level of behavior and neuroanatomy. Fear conditioning F0 male mice to acetophenone caused subsequently conceived odor naïve F1 male offspring to display behavioral sensitivity to acetophenone (P=0.043, t(27)=2.123). Acetophenone is detected by the M71 odorant receptor and we find that the F1 generation have larger M71 glomeruli in the olfactory bulbs {Dorsal M71 glomerular area: (P<0.0001, F(2,91)=15.53). Medial M71 glomerular area: (P<0.0001, F(2,84)=31.68)}. To begin to explain these results, we queried methylation of the gene encoding the M71 receptor (Olfr151) and found that it is hypomethylated in the sperm of F0 males that had been fear conditioned with acetophenone (P=0.0323, t(16)=2.344). We conclude that ancestral olfactory experience affects olfactory neuroanatomy and consequently behavior in the descendant generation via an epigenetic mechanism. This work allows us to appreciate how ancestral trauma contributes to the development of neuropsychiatric disorders such as phobias, and post traumatic stress disorder in descendant generations.

Epigenetic mechanisms and their implications in animal breeding

K. Wimmers, N. Trakooljul, M. Oster, E. Murani and S. Ponsuksili
Leibniz Institute for Farm Animal Biology, Institute for Genome Biology, 18196 Dummerstorf, Germany;
wimmers@fbn-dummerstorf.de

`Epigenetics´ refers to mechanisms causing variation of gene activity that are not based on variation of nucleotide sequences but on modifications of DNA and chromatin such as DNA methylation and histone-acetylation and methylation. Histone modifications affect gene expression by shifting the degree of chromatin compaction, whereas (hydroxy-)methylation of cytosine influences the accessibility of regulatory sequences for transcription factors or binding proteins. Epigenetic variation explains parts of the missing heritability and genotype-environment-interactions and thus epigenetic markers can greatly improve breeding selection. Genomic imprinting mediates differential allelic expression in a parent-of-origin specific manner due to divergent epigenetic alteration of the alternative alleles during gametogenesis or early zygote development. QTL regions with parental-of-origin effects and imprinted genes have been identified in mammals. E.g. the expression of the growth associated paternal allele of IGF2 is used to assist the breeding program. In cross breeding schemes, imprinting offers a new opportunity to establish dam and sire lines with enhanced production performance and maternal skills and to produce crossbreeds according to market requirement. Epigenetic modifications driven by environmental factors (e.g. diet, stress, disease) can form an epigenetic memory programming the organismal metabolism. Our own studies suggest that gestational nutrition with varying protein or methionine content can shift gene activity of various metabolic pathways due to changes of DNA methylation pattern. Considering various diets, tissues, stages and breeds reveals that there is no common molecular path or gatekeeper mediating the epigenetic processes. The transcriptional and epigenetic responses differ depending on the genetic background/breed. Extensive knowledge of the epigenetic mechanisms and gene regulation offers a new opportunity in animal breeding based on the environment-genome-epigenome interactions.

Ancestral exposure to stress epigenetically programs preterm birth risk and health outcomes

G.A. Metz[1], Y. Yao[1,2], A.M. Shriner[1], F.C.R. Zucchi[1], O. Babenko[1,2], I. Kovalchuk[2] and D.M. Olson[3]
[1]Canadian Centre for Behavioural Neuroscience, University of Lethbridge, Lethbridge, AB T1K 3M4, Canada, [2]University of Lethbridge, Department of Biological Sciences, 4401 University Drive, Lethbridge, AB, T1K3M4, Canada, [3]University of Alberta, Departments of Obstetrics & Gynecology, Pediatrics and Physiology, 227 HMRC, Edmonton, AB, T6G2S2, Canada; gerlinde.metz@uleth.ca

In most cases of human preterm birth (PTB), the causes remain unknown in spite of 60 years of investigation into the leading cause of perinatal mortality and morbidity. Here we show that stress across generations has downstream effects on endocrine, metabolic and behavioural manifestations of PTB via miRNA regulation. Stress (restraint and forced swimming) was induced during gestational days 12-18 in pregnant Long-Evans dams and effects were examined across three generations removed from the stressor (transgenerational) or in each of three generations (multigenerational). With each generation, both stress regimens gradually reduced gestational length, maternal weight gain and behavioural activity, and increased the risk of gestational diabetes. Shorter gestational length was accompanied by delayed offspring development that was recognizable as early as postnatal day 7, with the greatest effect in the F3 offspring of transgenerationally stressed mothers. Stress in F2 mothers affected the microRNA (miRNA) expression patterns of the miR-200 family in brain and uterus which regulates pathways related to brain plasticity and parturition, respectively, and also increased placental miR-181a, a human marker of PTB. Hence the compounding effects of gestational stress propagate across three generations to influence PTB risk and outcomes via miRNA regulation. These findings indicate that a family history of stress may program central and peripheral pathways regulating gestational length and health outcomes with potentially lifelong consequences.

Epigenetic databases in cattle and the prediction of phenotypes from genotypes

H. Jammes[1], H. Kiefer[1], E. Devinoy[2] and J.P. Renard[1]
[1]INRA, UMR1198, 78350 Jouy en Josas, France, [2]INRA, UMR1313, 78350 Jouy en Josas, France; jean-paulrenard@orange.fr

We are interested in the characterization of epigenetics contribution to the construction of phenotype in cattle. For that we use sets of cloned cattle as a relevant experimental model of the phenotypic diversity that can be expressed from the reprogramming of a single initial genome. We found that the global DNA methylation level of circulating blood is more variable between adult clones than between genetically non related individuals. We also found that epigenetic profiles of key glucose metabolism genes correlate with markers of hepatic dysfunctions. We have identified sequences with a specific methylation pattern in various tissues of clones including adult muscle that are independent of the animal's health status. These methylation patterns are signatures of the epigenetic adaptation of the developing organism to the environmental constraints imposed to the genome since the onset of its reprogramming. We also found that the differential milking of each of the two compartments of the mammary gland of individual lactating cows provides a suitable situation to highlight the causal relationship between a decrease in milk production induced by herd management and both a tissue and sequence-specific methylation pattern. We are now generating a large scale tissue-specific bovine epigenomic database in cloned and non cloned animals (equivalent to 10^{11} sequenced and mapped bases) in order to identify the placenta, muscle, liver, mammary gland as well as semen and peripheral blood mononuclear cells epigenetic marks associated to the regulation of gene pathways involved in the construction of traits of economic interest including health. Providing they can be made affordable, an achievable objective today in cattle, bovine multi-tissues epigenetic databases will allow an access to the molecular information needed not only for a better prediction of adult phenotypes from genotypes in a species with both a long gestation and a long generation interval, but also to identify epigenetic modifications that may be transmitted to next generation(s).

Genomic selection: the use of genomic information in animal breeding

T. Meuwissen
Norwegian University of Life Sciences, Department of Animal and Aquacultural Sciences, Box 5001, 1432 Ås, Norway; theo.meuwissen@nmbu.no

During the past 8 years, genomic selection has to such extend entrenched animal breeding practice, that it becomes pertinent for general animal scientists and practitioners to understand the basics of this new technology. Since the early nineties, animal breeders have explored gene mapping as a means for genetic improvement. The rational was obvious: find the genes and fix them. However, the traits of interest in animal production (e.g. milk production) turned out much more complex than envisaged: there were ~100 times more genes affecting our breeding goals than expected 15 years ago. Consequently, more than 90% of the genetic variability of our breeding objectives remained undiscovered despite immense gene mapping efforts. This implied that by 2005 only few breeding companies had implement any form of marker assisted selection. Currently, almost all dairy breeding companies are using DNA information, and other species are following rapidly. This is due to three breakthroughs: (1) the genomic selection (GS) methodology; (2) the discovery of large numbers of single-nucleotide polymorphism (SNP) markers; and (3) cost effective methods to genotype an animal for many thousands of SNPs. Instead of attempting to identify single genes, GS estimates the effects of many thousands of DNA markers simultaneously. Next, when new animals are genotyped, it combines their genotypes with the known effects estimates to predict genomic breeding value estimates (GEBV), if needed, at a very young age. The benefits of GS are greatest when selection is for traits that are not themselves recorded on the selection candidates, and when the generation interval is reduced, resulting in more selection cycles per decade. In the future, genome sequence data may replace SNP genotypes as markers. The increased marker resolution holds the promise that we can make accurate predictions of GEBV across large genetic distances, possibly across breeds.

Mare's milk production with Lipizzan mares

K. Potočnik[1], M. Simčič[1] and A. Kaić[2]
[1]University of Ljubljana, Biotechnical Faculty, Department of Animal Science, Groblje 3, 1230 Domžale, Slovenia, [2]University of Zagreb Faculty of Agriculture, Department of Animal Science and Technology, Svetošimunska cesta 25, 10000 Zagreb, Croatia; klemen.potocnik@bf.uni-lj.si

The aim of this trialwas to determine mares' milk composition produced by mares of Lipizzaner breed. There is still a lack of knowledge regarding equine milk production and its composition. In most cases, published data based on milk samples taken from mares which were not included in the routinely milk production. Consequently, these samples were actually first jets of milk and could not represented the composition of the whole amount of milk from one milking. In this analysis three Lipizzaner mares in the age of seven years were included. All were foaled in April and were milked from mid-August to the end of September, three times per day. Mares were adopted on the routine machine milking before the milking trialperiod started. Milk yield (MY) were measured after the each milking on the test day (TD) once per week. Samples were taken after each milking on TD. The contents of fat (FC), protein (PC), and lactose (LC) as well as somatic cell count (SCC), total bacteria count (TBC) and freezing point (FP) were determined in 45 samples. In average mares produced 1,030 g of MY per milking where 0.42% of FC, 1.50% of PC and 6.32% of LC, 12.51 of binary logarithmic SSC, 4.74 of decimal logarithmic TBC and -0.497 of FP were. The statistical model was used considering monitored effects. Days in milk had significant negative effect on the PC, LC and TBC and significant positive effect on the MY. Successive milking had significant effect on the PC and LC. The highest PC and LC were estimated in the milk from the second and the lowest from the third milking. The highest TBC were in the milk from the first milking. Such a trial was performed for the first time and results came from the routinely machine milking Lipizzan mares which are definitely suitable for milk production what could supports the preservation of this unique breed.

The possibility of production and placement of donkey's milk in Croatia

A. Ivanković[1], J. Ramljak[1], K. Potočnik[2] and M. Baban[3]
[1]Faculty of Agriculture, Department of Animal Science and Technology, Svetošimunska cesta 25, 10000 Zagreb, Croatia, [2]Biotechnical faculty, Department of Animal Science, Groblje 3, 1230 Domžale, Slovenia, [3]Fauculty of Agriculture, Department for Animal Husbandry, Trg Sv. Trojstva 3, 31000 Osijek, Croatia; aivankovic@agr.hr

Breeding donkeys in Croatia in recent decades has suffered significant changes. In the middle of the second half of the 20th century donkeys lost their primary functions, followed by a significant decline in the size of their population. Models for their economic reaffirmation during last decade are trying to be found. Traditional use of donkey's milk in treatment of respiratory and other human chronical diseases motivate breeders and consumers for production of this functional food. The aim of this study was to determine the potential for milk production in two dominant breed, Littoral-Dinaric and Istrian donkey. The study included 30 female kept on four farms, by each jennies two control milking was conducted. The Istrian donkeys have a higher milk yield compared to Littoral-Dinaric donkey (825 vs 492 ml/milking). Milk of Istrian donkey had lower content of milk fat than milk from Littoral-Dinaric donkey (0.42 vs 0.54%), milk protein (1.38 vs 1.64%) and lactose (5.73 vs 5.92%). Favorable milk production of these breeds justifies the establishment milk production of this kind if continued placing is provided. The modest tradition of donkey milk consumption in Croatia opens the possibility for its production and market development, although certain difficulties are observed. Observations in Croatia indicate that most consumers prefer consuming fresh milk as a functional food supplement in the rehabilitation of some health disorders. A small part of the consumers have knowledge about other options of donkey milk consumption, like food products with added value (koumiss, milk powder) or cosmetics. It is possible to increase production and consumption of donkey milk in Croatia with good organization of production networks from producers to consumers.

Socioeconomic and pedigree factors affecting the Asinina de Miranda donkey breed viability

M. Quaresma[1,2], M. Nóvoa[1], A.M.F. Martins[2], J.B. Rodrigues[3], J. Colaço[2] and R. Payan-Carreira[2]
[1]AEPGA, Largo da Igreja, 5225-011 Atenor, Portugal, [2]UTAD, CECAV, Quinta de Prados, 5000-801 Vila Real, Portugal, [3]University Lusófona de Humanidades e Tecnologias, FVM, Campo Grande, 376, 1749-024 Lisboa, Portugal; miguelq@utad.pt

The present study analyzed the pedigree and herd records of the Asinina de Miranda donkey breed, identifying genealogical and human factors that may affect the breed genetic diversity in the future, predicting the progression of the breed under present management and identifying determinants for survival, by means of a population viability analysis (PVA) program. The PVA showed a high risk of extinction. The most critical factor for breed survival was the percentage of females breeding per year. Reducing female mortality and age at production of first offspring, assuring registration in the Studbook, and tracking the foals will significantly foster this donkey breed's recovery and maintenance. The breed comprised a potentially reproductive population of 589 individuals; however, just 54.1% of the adult females registered in the Studbook ever foaled, and of these 62.7% foaled just once. The estimated number of founders and ancestors contributing to the reference population was 128 and 121. The number of founder herds in the reference population was 64, with an effective number of founder herds for the reference population of 7.6. The mean age of herd owners was 65.50 ± 0.884 years, with a negative association among the herd size and owner's age (P<0.001). In contrast, the size of the herd and the ownership of a male were both positively associated (P<0.001) with the herd number of in-born foals. Both the owners' age and the herd location (Asinina de Miranda home region vs dispersal region) were negatively associated with the foaling number (P<0.001). The main identified risk factors were: low breeding rates; low number of males and their unequal contribution to the genetic pool; unequal contribution of the herds to genetic pool; and advanced age of herd owners.

Animal (at)traction in the 21st century

J.B. Rodrigues[1,2], J. Prazeres[2], M. Pequito[2], E. Bartolomé[2] and A.S. Santos[1,3]
[1]APTRAN, Portuguese Association of Animal Traction, 5210-050 Miranda do Douro, Portugal, [2]Faculty of Veterinary Medicine, University Lusófona de Humanidades e Tecnologias, 1749-024 Lisboa, Portugal, [3]CITAB, Department of Agronomy, University of Trás-os-Montes and Alto Douro, 5001-901 Vila Real, Portugal; assantos@utad.pt

According to the DAD-IS of the FAO, there are nowadays 44 million donkeys, 11 million mules and hinnies and 59 million horses worldwide, with the vast majority of these equids acting as working animals in developing countries. Working equids play a fundamental role in human livelihoods through their contribution to financial, human and social capital. Yet, the recognition of their role remains a neglected area in programs of cooperation development in sectors such as agriculture, gender equity, food security and rural development. However, a collective ecological and economical consciousness and increased awareness of public opinion on the need to reduce excessive industrialization and mechanization of agriculture and forestry has led some sectors of western society to consider the (re)use of animal traction as a valid modern source of energy. Indeed, animals are a clean source of energy, that optimally transform consumed biomass in energy and fertilizer, avoiding soil degradation and contributing to durable management of arable lands, forests and sensitive areas. The need to maintain biodiversity, reduce carbon emissions, encourage self-reliance and reduce consumption of resources also contribute to this trend. The emerging importance of animal traction as an alternative/complementary option to mechanical traction throughout Europe is highlighted by the increased use of such technology in small and medium sized farms and in urban surroundings where it has proved to be economically viable. Also, conservation of endangered local breeds comprises a sustainability value supporting the local economies and the fixation of human population in marginal areas, as well as an ecological value, allowing improvement and preservation of agrobiodiversity.

Analyzing the technical performance of trotting forms in Sweden using data envelopment analysis

K. Kataria[1], D. Gregg[2], Y. Surry[3], R. Kron[4] and H. Andersson[3]
[1]*Swedish Agency for Marine and Water Management, Gullbergs Strandgata 15, 411 04 Göteborg, Sweden,*
[2]*Central Queensland University, Centre for Environmental Management, Bruce Highway, 4702 North Rockampton QLD, Australia,* [3]*Swedish University of Agricultural Sciences, Economics, Brauners väg 3, 756 51 Uppsala, Sweden,* [4]*IVBAR AB, Hantverkargatan 8, 112 21 Stockholm, Sweden; yves.surry@slu.se*

Trainers and trotters have two main sources of income. The first is the fee charged for training horses that should cover the expenses related to training. This fee tends to be similar throughout Sweden. The second source of income is related to the performance at the racetrack where the trainer receives a share of the potential winnings. To improve their economic performance firms can either win more races and raise incomes or become more efficient in resource usage per horse and that way cut costs. The first option of more won races is not possible for all trainers at the same time. Given this, it is important to understand the efficiency level of resource usage in trotting firms and characterizes the so-called 'best practice farms' and put that in relation to race track performance to help trainers and maintain this important sector of the Swedish economy. The purpose of this paper is to measure the technical efficiency (TE) of the trotting and training firms in Sweden, using data envelopment analysis methods. This allows us to characterize the so-called 'best practice farms' by computing individual TE scores of trotting firms in Sweden. Then, we examine the determinants of the technical performance of trotting firms. A model linking TE scores to a set of explanatory variables is estimated econometrically. In this model several explanatory variables including performance on the racetrack, geographical location of the firm and other background variables of the trainer-trotter are tested for statistical significance.

Assessment of time requirements and work quality in five Swedish stables

M. Lundholm and A.-L. Holgersson
Swedish University of Agricultural Sciences (SLU), Box 7046, 75007 Uppsala, Sweden;
lundholm.marcus@gmail.com

The horse requires daily attention and care. In stables where horses are kept individually the daily work includes feeding, cleaning of boxes, keeping the stables clean, horse care, and taking the horses back and forth to the paddocks. The execution of, and time required for, daily stable work differs between horse facilities. Therefore SLU conducted a quantitative and qualitative study of time required and quality achieved in daily stable work. The aim was to develop the stable work routines in general and a training module for the Equine Studies Bachelor program in particular. The study was conducted in 5 stables of the three Nat. Equestrian Centers of Sweden. The number of horses in each stable varied from 9 to 30 horses used for school riding purposes. The quantitative part included time studies of daily stable work during a seven-day-period. The qualitative part consisted of interviews with students and teachers and aimed at answering questions on how students perceive the stable work from a pedagogical perspective, how the training module met the syllabus content, and how quality assurance of course objectives was performed. The time study results showed that total time for daily care of the horses varied between 33 and 46 minutes per horse between the different stables. When separate tasks were compared it was found that taking the horse to and from the paddock was the most time consuming activity. Distance to paddocks, stable planning, use of different stable tools and automatic feeding systems or not are some of the factors found to cause substantial time differences. The results of the qualitative part of the study showed that students requested a higher degree of integration between theory and practice, a higher extent of individual teaching and guidance, and increased participation in research projects regarding stable work. To achieve efficient stable routines and quality of teaching activities, guidelines were produced to support future developments in this area based on the results of this study.

Effect of two feeding levels on growth and plasma T3, T4 and IGF-I of Lusitano foals

M.J. Fradinho[1], R. Fernandes[2], P. Francisco[3], W. Martin-Rosset[4], L. Mateus[1], R.J.B. Bessa[1], G. Ferreira-Dias[1] and R.M. Caldeira[1]
[1]CIISA, Faculdade de Medicina Veterinária, ULisboa, Av. Universidade Técnica, 1300-477 Lisboa, Portugal, [2]FMV-ULisboa, Av. Universidade Técnica, 1300-477 Lisboa, Portugal, [3]Coud. Ferraz-da-Costa, Serpa, Serpa, Portugal, [4]INRA, Centre de Recherche, Clermont-Ferrand/Theix, 63122 Saint-Genes Champanelle, Portugal; amjoaofradinho@fmv.ulisboa.pt

The present study was performed in order to compare two models of growth and development of Lusitano foals from pregnancy until one year of age: moderate (M) vs optimized (O) growth, on improved extensive systems. Fourteen Lusitano mares from the same stud-farm were fed during the gestation and lactation periods either 100% (n=7) or 130% (n=7) of their requirements according to INRA recommendations. Until weaning, foals were kept with their dams in the pasture and mares were fed accordingly to their group. From the 4th month onward, creep feed was distributed to all foals. During weaning and until 12 months of age foals were mainly fed with preserved forages and two levels of compound feeds to maintain distinct growth rates: M vs O. Foals were periodically weighed and measured and blood samples were collected for determination of T3, T4 and IGF-I plasma concentrations. A mixed linear model allowing for repeated measures on time was used to assess the effect of feeding level on body weight (BW), withers height (WH) and blood variables. The effect of feeding level was not significant for BW, WH, T3, T4 and IGF-I. A significant effect of age was observed on T3, T4 and IGF-I plasma concentrations (P<0.0001). T3 was highly correlated with T4 (r=0.657; P<0.0001) and IGF-I was correlated with BW (r=0.443; P<0.0001) and WH (r=0.462; P<0.0001). This preliminary study suggests that, in face of forage availability, the levels of compound feeds were not sufficiently different, in order to induce a significant result on growth rate.

Effect of selenium and vitamin E supplementation on serum cortisol in horses under moderate exercise

E. Velázquez-Cantón, A. Ramírez-Pérez, L.A. Zarco-Quintero, A. Rodríguez-Cortéz and J.C. Ángeles-Hernández
Facultad de Medicina Veterinaria y Zootecnia. Universidad Nacional Autónoma de México, Av. Universidad 3000, Mexico D.F., 04510, Mexico; eliasvelcanton@hotmail.com

Physical activity increases reactive oxygen species formation, which can induce an acute inflammatory phase. Mobility and activity of inflammatory cells seem to be mediated by stress hormones such as cortisol (CORT). Twenty-four horses (450 kg, 5-15 yr), which did not exercise for a month prior to this experiment, were used in an 11-w trial to study the effect of both selenium (Se, Se-yeast) and vitamin E (E, α-tocopheryl) supplementation on serum CORT concentration in horses under moderate exercise. Horses were housed in individual stables and randomly assigned to 4 treatments in a factorial trial arrangement (2×2, Se, E levels) with repeated measures (n=6). Both, Se and E met NRC 2007 requirements, for horses under moderate or intense exercise. Study supplementation levels were low (L) and high (H). Experimental treatments were: LSeLE, HSeLE, LSeHE and HSeHE (LSe, 0.1; HSe, 0.3 mg Se/kg DM and LE, 1.6; HE, 2 IU vitamin E/kg BW). Daily ration provided was poor in Se and E (<2 μg; 14.4 IU/kg DM), therefore Se and E were fully supplemented. The experimental weeks were as follows: w0 to w3, adaptation period; w4 to w7, exercise period (3 consecutive days for 30 min (5-20-5), including: warm up-moderate gallop-cool down); w8, rest and w9, end of supplementation. W0 corresponded to the baseline of the horse CORT without oral supplements. Jugular blood samples were taken every week (w4-w7, at the end of last exercise day). CORT was quantified by ELISA. Se affected serum CORT (P=0.02; LSe, 168.9; HSe, 298.4±39.03 ng/ml) concentrations. Most important differences between weeks (P<0.001) were observed between w0-w4 (177.4±18.1 ng/ml) and w6-w8 (326.3±38.4 ng/ml); which corresponded to the exercise period. In conclusion, while exercise increases cortisol levels, higher vs lower level of Se supplementation does not decrease cortisol concentrations.

Is group-housed horses less reactive?

K. Morgan, I. Kåmark, J. Lundman, S. Sandberg and M. Eisersiö
The Swedish National Equestrian Centre, Stallbacken 6, Strömsholm, Sweden; karin.morgan@stromsholm.com

The horse's flight behaviour is of great significance for the individual and the flock. Many horse-related accidents are due to flight reactions. Therefore it is of interest to study if management can reduce flight reactions. The aim was to study if the horse's reactivity relates to the housing system. We hypothesised that an individual box housed horse is more reactive than a group-housed horse. We use 28 SWB geldings (8-19 yrs) divided into two groups. All horses were working school horses in the university Equine program. The reactivity for each horse was tested with a Novel Object test by dropping an umbrella from the ceiling. We recorded heart rate and behaviours (feeding, standing still, walk, 'trot & canter', high head and low head). The horses performed the Novel Object test twice. At the pre-trial all horses were stabled in individual boxes, ridden one hour per day and let out in a paddock four to eight hours daily. Then the test group of 14 horses was introduced to a group-housing system HIT Active Stable®. After two months the reactivity was tested again (post-trial). The data was statistically processed in a two-way RM ANOVA and where appropriate followed by a post-hoc test Holm-Sidak. There was a tendency ($P=0.052$) that the control group walked more (mean 20±20%) than the test group (mean 10±16%). Apart from that there were no significant difference for the other observed behaviours. The heart rate (HR) decreased significantly from pre-trial to post-trial for peak HR (130 vs 106 bpm; $P=0.013$), HR first minute after stimuli (120 vs 86 bpm; $P<0.001$) and HR increase (96 vs 72 bpm; $P=0.013$). The peak HR was significantly ($P=0.049$) higher for the control group (134±46 bpm) than the test group (105±36 bpm). We concluded that the horses got accustomed to Novel Object test from the pre- to post-trial based on the lower heart rates the second time. The hypothesis couldn't be accepted, since the behavioural reactivity didn't differ between groups. However, further studies are needed.

Can relaxing massage and music decrease stress level in race horses?

W. Kędzierski[1], I. Janczarek[2], A. Stachurska[2] and I. Wilk[2]
[1]University of Life Sciences in Lublin, Department of Biochemistry, Akademicka 12, 20-033 Lublin, Poland,
[2]University of Life Sciences in Lublin, Department of Horse Breeding and Use, Akademicka 13, 20-950 Lublin, Poland; witold.kedzierski@up.lublin.pl

At the beginning of training routine, young race horses are exposed to many stressful stimuli. The objective of the study was to assess the effect of relaxing massage and music featured in the stable on the horse's long-lasting stress level. In the study, 120 Purebred Arabian horses were included. The study lasted for six months. At the beginning, the horses were 28-31 months old. They were brought to a race track from their mother studs and randomly divided into three following groups: Control Group of 24 horses, Music Group listening to the relaxing music, including 48 horses, and the Massage Group of 48 horses regularly submitted to the relaxing massage. All horses were regularly trained and competed in official races. Once a month, the saliva samples were collected from each horse to determine the cortisol concentration. The obtained data were analysed with the use of multivariate analysis of variance (ANOVA GLM; SAS) considering the effect of the group (Control, Music and Massage Groups) and sex (stallions, mares). Tukey's multiple comparison test was used to identify the differences between the groups. Both relaxing methods resulted in significant decrease in cortisol release, as compared to Control Group. Because of economic and organizational conditions, particularly featuring the relaxing music in the stable may be recommended to be put into practice as a cheap and easy method improving the welfare of the race horses. The study was financially supported by the National Centre for Research and Development, Poland (N180061 grant).

Horse owners' experiences and decision making regarding equine emergencies

E. Parkinson and C.V. Brigden
Myerscough College, Equine, St Michaels Road, Bilsborrow, Preston, PR3 0RY, United Kingdom;
cbrigden@myerscough.ac.uk

Horse owners face emotional and financial trauma when confronted with equine emergencies, like fractures or colic. This study examined the decision making processes and experiences of horse owners during such events. Horse owners who had dealt with life threatening equine conditions (n=362) completed a self-administered, online questionnaire. Fifteen questions captured owners' experiences and influences during decision making stages. Chi square test of association was used to examine associations between categorical responses. Colic was the most common condition (36.74%), followed by 'other conditions' (30.39%), laminitis (21.27%) and leg fracture (11.60%). Condition was significantly associated with the initial decision to treat or euthanize (χ^2_3=27.042, P<0.001); owners dealing with leg fractures euthanized more frequently than expected, whilst laminitis was treated more often than expected. Of horses that were insured, 90% were initially treated, which was significantly higher than expected (χ^2_1=6.226, P<0.01). Entire males were limited in number (n=13), but were more likely to be euthanized than expected (χ^2_2=8.300, P<0.05). Veterinary advice was the most common influencer in the decision to euthanize (64.58%) or treat (77.40%). The next common was prognosis for euthanasia decisions (58.33%), but length of ownership or relationship (53.08%) for treatment decisions. Age and personality of horse were influential in both decisions. Decisions regarding treatment or euthanasia are influenced by many factors. Gender was surprising; possibly linked to the owners' views of a stallion's capacity to cope with rehabilitation compared to a mare or gelding. Each condition has a different treatment and recovery rate, however fractures are traditionally considered life limiting, despite veterinary advances. The cost involved, long term restraints during treatment and the horse's future use are likely to be considered. The horse: human relationship is a clear influence on the decision making of owners when facing equine emergencies.

Does releasing horses to paddocks affect emotional response to transportation to race track?

I. Janczarek[1], W. Kędzierski[2], A. Stachurska[1] and I. Wilk[1]
[1]University of Life Sciences, Department of Horse Breeding and Use, Akademicka 13 str., 20-950 Lublin, Poland, [2]University of Life Sciences, Department of Biochemistry, Akademicka 12 str., 20-033 Lublin, Poland; anna.stachurska@up.lublin.pl

The objective of the study was to assess the effect of releasing race horses to paddocks, on the emotional response related to regularly being transported to a race track. The emotional response was estimated by heart rate (HR). The study was financially supported by the National Centre for Research and Development, Poland (N180061 grant). The study included 2.5-year-old Purebred Arabian horses in their first racing season. 90 horses were divided into three equal groups. R group was trained at the race track. The other two groups were trained in off-the-race-track centers and were regularly transported to the races: T group not released to paddocks and TP group released to a paddock for an hour a day. The training-racing season lasted from March to October and since May the horses participated in 6-8 races in total. Five 3-week measuring sessions for each horse were performed throughout the whole season on the days when the horse did not participate in a race. HR was measured for 5 min at rest, 10 min at saddling and 10 min while walking under rider. The study revealed that compared to R and T groups, the level of HR in TP group was reduced. HR was heightened for first 2 months of transporting to the race track in TP group and 4 months in T group. After those months, the emotional response determined by HR was less pronounced as if the horses got accustomed to the transport. Training in off-the-race-track centers was assessed beneficial for the horses' welfare, particularly because of the possibility of releasing horses to paddocks.

Milk and blood serum concentration of selected trace elements in lactating donkeys

F. Fantuz[1], S. Ferraro[2], L. Todini[1], P. Mariani[1], R. Piloni[2], E. Malissiova[3] and E. Salimei[4]
[1]Università di Camerino, Scuola di Bioscienze e Medicina Veterinaria, Camerino, 62032, Italy, [2]Università di Camerino, Scuola di Scienze e Tecnologie (Sezione Chimica), Camerino, 62032, Italy, [3]Technological Educational Institute of Thessaly, Department of Food Technology, Karditsa, 43100, Greece, [4]Università degli Studi del Molise, Dipartimento di Agricoltura, Ambiente, Alimenti, Campobasso, 86100, Italy; francesco.fantuz@unicam.it

Donkey milk can be considered a functional food for sensitive consumers such as infants and elderly people but the concentration of minor and potentially toxic trace elements is not well documented. The aim of this study was to measure the concentrations of Ti, V, As, Mo, Cd, Cs, and Pb in donkey milk and blood serum. Sixteen clinically healthy lactating donkeys (Martina Franca derived population) were used to provide individual milk and blood samples (n=112) during a 3-months period. Milk and blood samples were collected every 2 weeks. Milk and blood serum samples were analysed for Ti, V, As, Mo, Cd, Cs, and Pb by Inductively Coupled Plasma-Mass Spectrometry. Data were processed by analysis of variance for repeated measures. More than 80% of samples were below the limit of detection for V, As, and Cd in milk and for Cd, and Pb in blood serum. The average milk concentrations ($\pm$ SD) of Ti, Mo, Cs, and Pb were 77.3$\pm$7.7, 4.5$\pm$1.6, 0.49$\pm$0.09, and 3.2$\pm$2.7 µg/l, respectively. The blood serum concentrations of Ti, V, As, Mo, and Cs averaged 12.4$\pm$3.7, 1.0$\pm$0.38, 0.49$\pm$0.12, 28.5$\pm$13.8, and 0.17$\pm$0.05 µg/l, respectively. The effect of the stage of lactation was significant for all the measured elements in milk and blood serum but only small changes or inconsistent trends were observed. The average milk concentrations of Ti and Cs was higher than those of blood serum whereas that of Mo was lower. Current results suggest that the mammary gland play a role in determining the milk concentrations of Ti, Mo, and Cs.

Lying behaviour in horses on straw and pelleted sawdust

L. Kjellberg, K. Morgan, J. Ilvonen and L. Segander
The Swedish National Equestrian Centre, Stallbacken 6, 73494 Strömsholm, Sweden;
linda.kjellberg@stromsholm.com

Domestication of the horses changed their existence from living on the steppe to be accommodated in limited areas. To ensure some aspects of the horses' welfare stabled horses should have an even surface to stand and lie down on. The horse has four different types of resting: idling, resting, drowsing, and sleeping. Adult stabled horses sleep 3-5 hours per day and the sleep can be divided in to orthodox sleep and paradoxical sleep (REM-sleep) which they only perform when lying down. Straw has been shown to be preferred as for lying down and give opportunity for ingestive behaviour. However, there are no studies done on pelleted sawdust regarding lying behaviour. The aim of the study was to investigate the horse's lying behaviour on pelleted sawdust compared to straw as control. Eight SWB-geldings (8-14 years) was used and divided into two subgroups within a cross-over design. Four boxes were used; two with straw and two with pelleted sawdust of pine and fir. The horses were filmed 18:00 to 06:00 and the movie was processed afterwards. The observed behaviours were: standing actively, standing passively, ingestive behaviour, sternal recumbency and lateral recumbency. The data was statistically processed in a one-way RM ANOVA and where appropriate followed by a post-hoc test Holm-Sidak. The horses lay down more (P=0.047) in lateral recumbency on straw (3.3%) than on pelleted sawdust (2.4%). The horses also showed more ingestive behaviour (P$\leq$0.001) on straw (40.2%) compared to pelleted sawdust (33.8%) and standing passively more (P$\leq$0.001) on pelleted sawdust (48.8%) than to straw (40.9%). No significant difference between the other behaviours. The results show that the bedding material can be an important factor for the horse's lying behaviour. The horses laid down in sternal recumbency 7 min. more on straw and thus enabled them to retrieve more REM sleep. Since there is no data on the need of the horses REM sleep pelleted sawdust can be used as bedding.

Diurnal and nocturnal concentration of essential minerals and trace elements in donkey milk

F. Fantuz[1], S. Ferraro[2], L. Todini[1], P. Mariani[1], R. Piloni[2], A. De Cosmo[1] and E. Salimei[3]
[1]Università di Camerino, Scuola di Bioscienze e Medicina Veterinaria, via Gentile III da Varano, Camerino, 62032, Italy, [2]Università di Camerino, Scuola di Scienze e Tecnologie (Sezione Chimica), via Gentile III da Varano, Camerino, 62032, Italy, [3]Università degli Studi del Molise, Dip. di Agricoltura, Ambiente, Alimenti, via De Sanctis, Campobasso, 86100, Italy; francesco.fantuz@unicam.it

Donkey milk has been proposed as a food for sensitive consumers such as infants and elderly people. Circadian rhythm are report to affect donkey milk gross composition but no information is available on essential minerals and trace elements. Aim of this trial was to study the diurnal and nocturnal Ca, P, K, Na, Mg, Zn, Fe, Cu, Mn, Co, Se, and Mo concentrations in donkey milk. Four lactating donkeys (Martina Franca derived population) were used to provide milk samples. Donkeys were machine milked during the day (at 10:00 am and 06:00 pm) and during the night (at 11:00 pm and 02:00 am). Individual milk samples were analysed for the aforementioned elements by Inductively Coupled Plasma-Mass Spectrometry. Results from samples obtained at 10:00 am and 06:00 pm were grouped as day milk whereas samples obtained at 11:00 pm and 02:00 am were grouped as night milk. Data were processed by analysis of variance for repeated measures. The average milk concentrations of investigated elements (minerals: Ca 1100, P 459, K 875, Na 172, Mg 93 mg/l; Trace elements: Zn 1097, Fe 201, Cu 70.3, Mn 5.3, Co 0.67, Se 2.5, Mo 7.8 µg/l) were similar to literature data on donkey milk except for lower Zn and Se and higher Ca and Fe. The average concentrations of the investigated variables did not differ significantly between milk obtained during the day and during the night.

Some morphological traits of Arabian and Thoroughbred horses and their future in Turkey

O. Yilmaz[1] and M. Ertugrul[2]
[1]Ardahan University, Vocational High School of Technical Sciences, Ardahan 75000, Turkey, [2]Ankara University, Faculty of Agriculture, Department of Animal Science, Diskapı, 06110 Ankara, Turkey; zileliorhan@gmail.com

The aim of this study was to define body coat colour and some morphological traits of Turkish Arabian and Thoroughbred horses in Turkey. Arabian horses are mainly used for race, but about 30 years they are also used for javelin swarm game in some regions of Turkey. In this study a total of 90 Arabian horses (87 males and 3 females), and 52 Thoroughbred horses (14 males and 38 females), was analysed. Arabian horses were investigated in four age groups (3-4, 5-6, 7-8 and 9-13 years) and descriptive statistics gave the following means: withers height 158.5±0.51 cm, height at rump 156.4±0.48 cm, body length 152.5±0.72 cm, heart girth circumferences 176.3±0.38 cm, chest depth 68.5±0.33 cm, chest width 39.5±0.26 cm, cannon circumferences 20.3±0.12 cm, head length 61.3±0.37 cm and ear length 14.9±0.15 cm. In this study the frequencies of body coat colour of the sampled horses for gray colour was 53.4%, chestnut 33.3%, and bay 13.3%. Thoroughbred horses were analysed in two age groups of 4-5 and 6-9 years respectively and descriptive statistics gave the following means: withers height 164.1±0.64 cm, height at rump 162.7±0.65 cm, body length 167.3±0.72 cm, heart girth circumferences 193.3±0.43 cm, chest depth 79±0.57 cm, and cannon circumferences 20.1±0.18 cm. In this study the frequencies of body coat colour of the sampled horses for bay colour was 34.6%, chestnut 30.8%, black 25%, and grey 9.6%. It can tentatively be concluded that Arabian Horses now have larger body size than they had in last century. The study also showed that Turkish Thoroughbred Horses were almost similar to other Thoroughbred Horses raised in other countries. According to data the population of Arabian horse decreases but population of Thoroughbred horse increases in Turkey.

Economic, social and environmental impact of Alltech FEI World Equestrian Games™ 2014 in Normandy
C. Vial[1], C. Fabien[1,2], E. Barget[2] and G. Bigot[3]
[1]IFCE, INRA, UMR 1110 MOISA, 34000 Montpellier, France, [2]CDES, OMIJ, Hôtel Burgy, 13 rue de Genève, 87065 Limoges, France, [3]Irstea, UMR Métafort, 9 avenue Blaise Pascal, CS 20085, 63178 Aubière, France; genevieve.bigot@irstea.fr

In 2014, France hosted the World Equestrian Games in Normandy. As part of a collaborative research program carried out by IFCE, INRA, and Limoges CDES, an economic social and environmental impact study of the event was conducted. A methodology has gradually been built and tested during the previous years on various small size equestrian sporting events. It combines quantitative and qualitative approaches, pushing the classical limitations of impact studies. The economic evaluation uses Base Theory and Keynesian multipliers. Social utility is addressed through the monetarization of use and non-use values of the event for spectators and inhabitants and is completed by an analysis of the reasons for these values. We study the various items that impact the environment through their specific units and the efforts made by the organizers to minimize the impact of the event. We base the results on the collection of information from the Games organizers, field observations and around 2,300 surveys conducted during the event. The Alltech FEI World Equestrian Games™ 2014 in Normandy gathered together more than 300,000 spectators and 1,060 participants. In the short term, they represent a source of economic, touristic and media benefits for the territory. Their total economic impact amount to 102 million euros for Normandy and their social utility is evaluated at more than 36 million euros. Our analysis also highlights the organizer's involvement to preserve the environment. This research will now focus on the long-term impact of the Games. This study represents both a contribution to academic research and to the horse industry, participating in the methodological and theoretical advances in impact studies and in reflections on optimizing the benefits of equestrian events for the territories that host them.

Integrating electronic identification in hill farming management
C. Morgan-Davies[1], N. Lambe[1], H. Wishart[1], F. Kenyon[2], D. Mcbean[2], A. Waterhouse[1], A. Mclaren[1], F. Borthwick[3] and D. Mccracken[1]
[1]SRUC, Hill & Mountain Research Centre, Crianlarich, FK20 8RU, United Kingdom, [2]Moredun Research Institute, Bush Loan, Penicuik, EH26 0PZ, United Kingdom, [3]SRUC, Kings Building, Edinburgh, EH9 3JG, United Kingdom; claire.morgan-davies@sruc.ac.uk

Electronic identification (EID) is based on radio frequency identification; it was investigated for livestock farming in the early 1980s to accurately monitor and track animal movements. Within Europe, legislation governing traceability requires the identification and registration of some livestock species. In the UK, EID became a mandatory requirement for sheep identification in 2010. Whilst all farmers EID-tag their animals, the potentials from using EID technology to improve animal performance are scarcely exploited by livestock farmers in more extensive conditions. This paper presents results from research exploring the possibilities of applying EID technology in extensive livestock management using a hill flock of 900 ewes under two different management systems. A fixed ear-tag reader on a weigh crate and handheld electronic devices are used to assess their potential for animal data collection. Using auto-sorting technology for handling ease, targeted management decisions have been implemented based on individual animal weight changes for winter feeding of ewes and worming of lambs (Targeted Selective Treatment). On-farm labour savings have also been quantified. Issues and potential uptake of the use of EID technology for such a management approach have been assessed through workshops and surveys. Results so far indicate that such technology: is feasible to implement on extensively managed farms without compromising productivity, can provide labour savings; has potential to simplify and improve winter feeding management; reduces reliance on anthelminthic treatments and risk of anthelminthic resistance. Whilst farmers and the wider farming industry believe that using EID for farm management could be beneficial, cost of the equipment, however, remains one of the major barriers to wider uptake.

Using RFID and other electronic devices for the management of small ruminant flocks in France

J. Holtz, J.M. Gautier, S. Duroy and G. Lagriffoul
Institut de l'Elevage, 149 rue de Bercy, 75012 Paris Cedex, France; jean-marc.gautier@idele.fr

Since 2010, the electronic identification (EID) to all sheep (adult and youth) has been generalized in France, thus going beyond the EU regulations. For the goat species, these measures do not concern the kids for slaughter. This generalization offers a priori significant perspectives for automation of tasks by linking electronic reading and the data collection or the automatic use of other individual data (weight, geographical location, measure of food, etc.) allowing a minima flock data management, sorting, counting. Hundreds of breeders commonly use a handheld reader in sheep. The association DAC (automatic concentrate feeder) + RFID reader is increasing in dairy production in order to reduce the concentrate feeding. In areas where extensive grazing dominates, goat breeders equip some animals with GPS in order to decrease the livestock guarding necessity. However these potentialities are only used by few small ruminant farmers. Several factors explain this situation: on the side of the farmers, material supply, when it is new, is considered expensive and poorly appropriate to the needs; on the side of manufacturers, the weak market outlook and the lack of knowledge of the expectations do not encourage R & D. Even so, expectations exist or appear, such as warning devices (intrusion of predators, exit of a given pasture area). Rather than precision farming per se, as is often mentioned for cattle about the use of electronic devices and sensors, the small ruminant management will be satisfied first with proper workable support tools. In this context, networking of field experiences is necessary to their development and projects begin to be implemented at regional scale or cross-border. For example, the 'Agripir' project involves the both sides of the Pyrenees. In addition to the implementation of a virtual fence test, regular meetings between manufacturers and livestock technical advisers were held to better define the expectations of R & D.

UHF electronic identification may improve efficiency and animal welfare in sheep production

G. Steinheim[1], Ø. Holand[1], B. Stevens[2], M. Mchugh[3] and R. Mobæk[1]
[1]Norwegian University of Life Sciences, Dept. Animal and Aquacultural Science, P.O. Box 5003, 1432 Ås, Norway, [2]UK Intelligent Systems Research Institute, Melton Mowbray, Leicestershire, United Kingdom, [3]Lisnashannagh, Carrickmacross, Co Monaghan, Ireland; geir.steinheim@nmbu.no

EC legislation on electronic identification (EID) for sheep can be difficult to meet with current ear tag and bolus technology using low-frequency (LF) identification. Misidentified animals mean problems for farmers, abattoirs or authorities. EID may increase efficiency and precision of farming and logistics further on in the field-to-fork chain. Use of EID could improve animal welfare by increasing precision of husbandry and reducing need for handling of individuals. Ultra-high frequency (UHF) identification systems seem superior to LF in some management situations. LF technology is reliable, but short range and slow reading makes groups of moving sheep difficult to read; UHF has longer range and faster reading, and should be a better tool in such situations. During a FP7 project we developed and tested a UHF based ear tag identification system for sheep. Adaptation to small rather than large ruminants (for which UHF is more commonly used) is through the low weight and small size of tag electronics: the tested chip was fitted inside a 24×4.5×5.5 mm tag body. Average reading range under laboratory conditions was 2.1±0.5 m. We tested the system in a 1.6 m wide race with two 30×30 cm UHF antennas, running the sheep through in a flock of one hundred. The sheep moved unhindered and mostly 2-4 abreast. We repeated the 100-sheep flock run 28 times and the 2,800 individual passes were all correctly registered. The longer read range may be problematic in some situations where specific animals must be identified, but UHF can be superior to current LF systems when the task is rapid identification of many animals. UHF should also work well for monitoring by continuous reading at fixed points such as pasture salt licks.

Influence of management and environmental factors on the movement of sheep on alpine pasture
T. Guggenberger, F. Ringdorfer, A. Blaschka, R. Huber and P. Haslgrübler
Agricultural research and education centre, Raumberg 80, 8952 Irdning, Austria;
ferdinand.ringdorfer@raumberg-gumpenstein.at

Land use conflicts in alpine pastures and the desire to revive slowly declining pastures due to encroachment of bushes need practical grazing regimes which are equitable for both animal and human. For their design, the movement behavior of sheep in different management methods and pastures in the Austrian Central and Eastern Alps was investigated. Objective: To develop a grazing regime as a compromise between the natural behavior of the animals and the necessary control measures by the shepherd. Methods: 5 years of observation, 4 GPS collars, 5 different alpine pasture areas, 4 management procedures; free grazing method: the animals are inspected once per week, but are otherwise left to themselves; guided grazing: the flock is accompanied by the shepherd. This regulates only the pasture borders on a large scale; treasured method: the flock is led by the shepherd and his dogs specifically; coupled processes: paddocks with electric fence. Results: The free-grazing sheep move at a speed of 85 meters/hour. At the beginning of dawn the activity begins and is adjourned at noon by a long rest period. Thereafter, the animals graze until dusk and then spend the night in the same few places. Temperature fluctuations, different altitudes or slope gradients have little influence. The treasured method reduces the sheep path length by 6 meters/hour, whilst the guided method extends it to 34 meters/hour. These results suggest that whenever a man leads a flock, he brings impacts from his own behavior. These include his working hours, the sensitivity in bad weather and behavior at the crossing of the slopes.

Advances in monitoring of livestock
D.W. Ross[1], C. Michie[2], C.A. Duthie[1], S. Troy[1] and I. Andonovic[2]
[1]Scotland's Rural College (SRUC), Future Farming Systems, Roslin Institute Building, Easter Bush, EH25 9RG, Midlothian, United Kingdom, [2]University of Strathclyde, Electronic and Electrical Engineering, Royal College Building, 204 George Street, G1 1XW, Glasgow, United Kingdom; dave.ross@sruc.ac.uk

Livestock monitoring technologies have continued to advance, and provide better, more accurate, and a wider range of diagnostics, based on measures extracted by sensor systems. Animal mounted sensors can now provide wider aspects of behaviour to be classified, having the capacity to extend the principles of activity monitoring (e.g. oestrus) to more detailed estimates of feeding and rumination behaviour. These can inform on both animal health and welfare status. Additional mounted sensors can provide complimentary data that informs on detection, and possibly prediction, of physiological processes (such as parturition). Body-internal sensor systems such as intra-ruminal boluses, provide relevant measures of rumen condition, such as pH, reduction-oxidation, and local temperature. These can inform on conditions such as digestive disorder status. Fixed –site animal monitoring systems are being developed to inform on individual performance, and for example, provide data that relates to metabolic health diagnosis. A number of the above described systems benefit from deployment of complimentary EID systems to identify the individual. As bovine EID is introduced, it will enhance the capacity to deploy monitoring systems, and these systems may find greater application in the small ruminant sector.

A semi-automated procedure to detect variations of mountain pastures using Landsat imagery

R. Primi, C.M. Rossi, G. Piovesan and B. Ronchi
Università degli Studi della Tuscia, DAFNE, Via S. Camillo de Lellis, snc, 01100 Viterbo, Italy;
ronchi@unitus.it

This study presents a semi-automated approach, based on the application of multi-spectral remote sensing Landsat imagery, to determine mountain pastures variation during wide periods. The sample areas was located in Central Italian Apennines (Catria-Nerone and Aurunci mountains), with traditional extensive farming system; sheep, goats, cattle and horses are the main species farmed. Using ArcGIS 10.1 (Esri, Inc.) software, 25,000 ha of hilly and mountain areas (>600 m a.s.l.) was mapped. From on-screen visual interpretation of 2010 aero-photograph images, the main land cover (2010-LC) classes was described within the 2nd-level CORINE Land Cover legend. The approach employs 1984 and 2010 free of charge Landsat imagery to determine large-scale spatiotemporal variations in 2010-LC, on a scale of 1:10,000. The semi-automated procedure is based on Normalized Difference Vegetation Index (NDVI) pixel-oriented image differencing technique. The NDVI derived from 1984 and 2010 Landsat images were stratified within the 2010-LC, highlighting the areas where changes occurred. The results were validated and rectified as a result of on-screen visual interpretation, whereby all the false-positive changes that were incorrectly mapped during the automatic procedure were identified and removed. The derived NDVI data of LC change show that about 1000 ha of mountain pastures (4% of the total area mapped) were loss during the period under review. The largest losses have occurred principally under 1,200 m a.s.l., mainly because of secondary forest succession processes (840 ha), caused by a decrease in sheep and goat farming. This semi-automated, real-time and cheap methodology can contribute to the monitoring of mountain rangelands changes and support for the pasture planning.

Plasma lactate concentration at slaughter is associated with intramuscular fat levels in lamb loin

S.M. Stewart, P. Mcgilchrist, G.E. Gardner and D.W. Pethick
Murdoch University, School of Veterinary and Life Sciences, South Street, 6150 Murdoch, Australia;
s.stewart@murdoch.edu.au

There is increasing interest in the link between plasma indicators of stress and meat quality in lamb. Selection for leaner and more muscular lamb phenotypes has been shown to reduce the level of intramuscular fat (IMF), a key driver of consumer sensory acceptability. Previous work has also shown that selection for leanness reduces the muscle response to adrenaline. Thus we hypothesise that increasing plasma lactate concentration at slaughter will be associated with a decrease in the IMF of the M. longissimus lumborum (LL). Blood was collected at exsanguination from 1652 lambs (mean age 298±57 days) from the Meat and Livestock Australia genetic resource flock. Lambs were managed on two research stations (Katanning, Western Australia and Armidale, New South Wales). Plasma lactate concentration was measured using commercial reagents. Muscle samples from the LL were collected 24 hours post mortem and IMF content was measured following freeze drying using a near infrared procedure. IMF data was analysed using linear mixed effect models. Fixed effects for flock, kill group, sex and dambreed within siretype, dam age and birth type rear type were included in the model along with the covariates for hot carcass weight and plasma lactate concentration and their interactions. There was a significant curvilinear association between IMF and lactate (P<0.01). Increasing plasma lactate from 1 to 7 mmol/l was associated with an increase in IMF of 0.23%. Conversely, increasing plasma lactate concentrations from 7 to 16 mmol/l was associated with a decrease in IMF of 0.67%. These results partly support our hypothesis in that lambs with a larger fat deposition, reflected by IMF, have a greater muscle response to stress. However, it also suggests that lambs with very high levels of lactate at slaughter may have a greater overall sensitivity to stress throughout life and as a result mobilise more or deposit less intramuscular fat.

Finishing of lambs through the autumn: carcass and meat quality using tropical grassland

B.N. Marsiglio-Sarout[1], C.H.E.C. Poli[1], N.M.F. Campos[1], J.F. Tontini[1], J.M. Castro[1], M.T. Braga[1], Z.M.S. Castilhos[2] and C. Bremm[1]
[1]Federal University of Rio Grande do Sul, Department of Animal Science, Porto Alegre, 91540-000, Brazil, [2]FEPAGRO, State Foundation of Agricultural Research, Porto Alegre, 90130-060, Brazil; bruna.sarout@sruc.ac.uk

In the south of Brazil sheep production faces limitations in the autumn, as supply struggles to meet demand due to low productivity levels. Inadequate animal nutrition may directly affect the herd performance. A better knowledge of appropriate feeding system in needed to improve productivity while using tropical grassland in the autumn. A feeding experiment was conducted with 54 castrated lambs (LW=24.9±1.47 kg). The lambs were kept in nine paddocks of 0.2 ha with tropical grass Aruana (Panicum maximum) and were assigned to three feeding systems groups: (1) continuous grazing without supplementation (CO group); (2) continuous grazing with supplementation of 1.5% Live Weight (LW) (S1 group); (3) continuous grazing with supplementation of 2.5% LW (S2 group). The supplement composition (DM basis) was 24% soybean meal, 74% corn, 1% urea and 1% minerals. A randomized block design was used. Statistical analyses were conducted using the mixed procedure of SAS. The slaughter LW was higher for S2 compared with CO and S1 groups (P<0.01). Carcass yield (P<0.05) and LW gain (P<0.01) were affected by all feeding systems, with S2 showing the highest value and CO the lowest. The muscle bone ratio was not affected by the feeding systems. Group S2 had higher Longissimus dorsi (Ld) weight than others groups (P<0.01); Ld deepness had no difference; maximum width of Ld (P=0.02) and rib eye area (P<0.01) were higher for S2 compared with CO, however did not differ from S1 group; and the Ld subcutaneous fat thickness was higher for S2 and S1 compared with CO group (P<0.02). Meat pH drop (0, 45 minutes and 24 hours after slaughter) was not affected by the feeding systems. These results suggest that the feeding system S2 had better carcass traits for finishing of lambs using tropical grassland through the autumn.

Lamb growth increases myoglobin and iron concentration in the longissimus muscle

K.R. Kelman[1,2], L. Pannier[1,2], H.B. Calnan[1,2], D.W. Pethick[1,2] and G.E. Gardner[1,2]
[1]Murdoch University, School of Veterinary and Life Sciences, Western Australia, 6150, Australia, [2]Australian Cooperative Research Centre for Sheep Industry and Innovation, Armidale, NSW, 2351, Australia; k.kelman@murdoch.edu.au

Lamb growth is a key driver of farm profitability. Growth can be crudely estimated using hot carcass weight, and this has shown a positive association with oxidative capacity and mineral content. However this precludes the ability to assess the impact of different periods of growth on these traits, hence it has not previously been investigated. Based on associations with hot carcass weight we hypothesised that increased growth would increase oxidative capacity, indicated by myoglobin concentration, and mineral content, indicated by iron concentration. Lambs were weighed at birth and throughout life until slaughter. Lamb weights at weaning (100 days) and post weaning (150 days) were estimated using random regression. Myoglobin (n=8,987) and iron (n=8,434) concentration were measured in the loin at slaughter. Data was analysed using linear mixed effects models in SAS with fixed effects for site, year of birth, sex, birth type-rear type, age of dam, sire type, dam breed within sire type and kill group within site, and the growth variable was used as a covariate. The association between myoglobin and iron concentrations and growth was similar, reducing with weight at birth and increasing with weight at 100 and 150 days. Between weaning and post weaning time points the association between myoglobin concentration and growth diminished by about 20% while for iron concentration there was no change. As hypothesised, the associations with weight at weaning and post weaning were positive. These findings indicate that the greatest potential for modulation of these meat quality traits is pre weaning. As lambs are reliant on ewes for nutrition pre weaning there may be scope to influence these traits in lambs by manipulating feed to ewes.

Zinc concentration changes in the tissues with supplementation of zinc sources in growing lambs

S. Sobhanirad[1] and E. Rabiei[2]
[1]*Islamic Azad university, Mashhad Branch, Department of Animal Science, Mashhad, 9187147578, Iran,*
[2]*Ferdowsi University of Mashhad, Azadi Square, Mashhad, 9177948974, Iran; sobhanirad@gmail.com*

Recently, organically bound Zn supplements are used in animal diets. Thus, The objectives of this experiment were to examine the effects of inorganic and organic Zn supplementation on Zn concentration in the tissues of lambs. Thirty-six male growing lambs with an average weight 27±0.3 kg were used to determine the effects of level of the zinc (Zn) supplementation in diet from inorganic and organic sources on performance of Baluchi lambs, an indigenous sheep breed in Iran. Animals were allocated to six groups, with the control group (C) receiving the basal diet (no zinc supplementation); the other five groups were offered the basal diet supplemented with 50 or 100 ppm Zn from zinc sulfate monohydrate (Zn-S) and Zinc proteinate (Zn-P) and added Zn from Zn-P+Zn-S (50+50 mg/kg, respectively). Ingredients of basal diet contained (%, DM) Alfalfa hay 55%, Barely 20%, corn 15.75%, Soybean meal 1.8%, Cottonseed meal 1.8%, Wheat bran 4.5%, Urea 0.2%, Calcium carbonate 0.45%, Salt 0.2%, and Minerals and vitamins mix 0.5%. Upon termination of experiments, lambs were killed by exsanguination and liver, heart, kidney, lung, muscle, spleen and testis were quantitatively excised. Results showed that the Zn concentration of liver, kidney, muscle, spleen and testis in lambs fed diets supplemented 100 mg/kg Zn as zinc proteinate, 100 mg/kg Zn as zinc sulfate and the mixture of inorganic and organic Zn were greater (P<0.05) than other groups. The Zn concentrations of heart and lung in lambs on treatments 100 Zn-P and ZnP+Zn-S were higher (P<0.05) than other groups. In conclusion, the results of this experiment can be concluded that the bio-availability of Zn from organic and inorganic sources did not differ.

Evaluation of the chemical composition of the lamb meat using near infrared (NIR) technology

M. Ślęzak and R. Niżnikowski
Warsaw University of Life Sciences, Department of Animal Breeding and Production, Ciszewskiego 8, 02-786 Warsaw, Poland; roman_niznikowski@sggw.pl

The aim of this study was to analyze the basic chemical composition parameters: fat, protein, collagen and moisture in lamb meat using modern spectroscopic techniques. The research was performed on the FoodScan analyzer which is based on Near Infrared Transmittance (NIR) technology and were compared with the reference methods (L). The investigations were conducted on 110 lambs of the Zelaznienska sheep (ZEL), Polish Heath sheep (PH) and F_1 crosses of both mentioned breeds with meat breed Berrichone du Cher (BC)-ZEL×BC and PHxBC. All ram lambs were slaughtered at a body weight of 40 kg (±1.5). The slaughter and carcass value and meat (mld) quality were estimated. The highest amount of protein was found in meat of PH (L-22.17%, 22.9%-NIR; P≥0.01) with the lowest fat content of L-3.51% and NIR-4.09% (P≥0.01). The lowest amount of protein was found in meat of ZEL (L-20.88%, NIR-21.57%; P≥0.01). The highest amount of collagen and fat was found in meat of PHxBC (collagen: L-1.88%, NIR-1.75%; P≥0.01; fat: L-4.79%, NIR-5.02%; P≥0.01). Both methods showed lowest amount of collagen in meat of ZEL (L-1.68%, NIR-1,55%; P≥0.01). There were no statistically significant differences between the breeds in the moisture content and dry matter. The highest level of water was found in meat of crossbreeds (PHxBC: L-73.26%, NIR-72.64%; P≥0.01; ZELxBC: L-73.40%, NIR-73.30%; P≥0,01). The highest value of correlation coefficient between methods was obtained by comparing fat content (0.964) and protein content (0.626). Other parameters also received positive correlations and they were only significantly lower (collagen-0.400, water-0.351; P≥P0.01). The results showed the suitability of B rams for crossing with ZEL and PH ewes in order to obtain slaughter F_1 lambs.

Effect of feeding system on meat quality and fatty acid profile of lambs and kids
S. Karaca and A. Kor
Yuzuncu Yıl University, Department of Animal Science, Faculty of Agriculture, 65080, Turkey;
serhatkaraca@gmail.com

Sixty Karakaş lambs and Hair goat kids were used to determine the effect of feeding system on fattening performance, meat quality and fatty acid profile of male lambs and kids. For this purpose, 30 lambs and 30 kids were randomly divided into 2 groups of 15 each as fed with concentrate lambs (CL) and kids (CK) or grazed on pasture lambs (PL) and kids (PK). The initial body weights of lambs and kids were 22.68 and 14.61, respectively. The average daily weight gains of CL were higher than PL (204.24 vs 123.08 g) ($P<0.001$). However, CK and PK kids have similar daily weight gains (109.82 vs 104.93 g). Fattening period was finished when lambs and kids attained 38 kg and 28 kg, respectively. Then the animals were slaughtered to evaluate carcass and meat quality traits. CL and CK groups have heavier and fattier carcasses than PL and PK groups ($P<0.001$). In addition to this, CL and CK have low ultimate pH and high L*, b* and H° values when compared to PL and PK groups ($P<0.001$). On the other hand, the ultimate pH of lambs were lower than kids (5.82 vs 5.95) ($P<0.001$). Drip loss and instrumental toughness of meat were found similar among groups. Fatty acid analyses were performed by samples of longissimus dorsi (LD), semimembranosus (SM), triceps brachii (TB) and subcutaneous (SC). The total saturated fatty acids (SFA) contents of LD, SM and TB of PK were higher than CK, but PL has lower SFA than CL in the same tissues ($P<0.01$). SFA contents of SC were higher in both lambs and kids pasture groups than concentrate groups ($P<0.001$). Moreover, n-6/n-3 ratios of LD, SM, TB and SC of concentrate groups were higher than pasture groups ($P<0.001$). Overall, the results indicated that many of the traits that we evaluated have significantly affected by feeding system and some of them vary between species depending on feeding system.

How to solve the problem of scales to improve the efficiency in livestock production?
P. Faverdin
INRA, Agrocampus Ouest, UMR1348 PEGASE, Domaine de la Prise, 35590 Saint-Gilles, France;
philippe.faverdin@rennes.inra.fr

It is essential to consider environmental and resource issues in livestock farming systems for the future of animal production. However, research on farming systems and their relationship to the environment are complex. Accounting for environmental and resource issues usually starts from improving the constituent elementary processes (animal, culture), applying these to the farm system and then aggregating these at the sector or territory level. Although this bottom-up approach may seem logical, the extrapolation of solutions of lower levels does not guarantee their relevance at a more aggregated level. This report illustrates the aggregation problem with examples concerning greenhouse gas mitigation in dairy production, but the problem applies to other environmental topics, such as eutrophication. In cattle, milk and meat production are inseparable issues to optimize the use of feed resources and reduce environmental impacts. Improving one sector without considering the other can lead to inefficient aggregated solutions. This lack of conservation of properties of solutions at different levels of aggregation comes from interactions among entities within the aggregation. Optimizing efficiency by sector or industry promoting the use of the best resources to improve efficiency could lead to increase competition on edible foods for humans. The proposed strategy consists to study first the optimal allocation of resources and agricultural land use by various production systems for the different territories with their agronomic potentials and environmental constraints. After defining the optimal allocation, the search for a better efficiency within each sector becomes relevant. Finally, it is necessary to ensure that improved systems fit with the objectives. The strategy of successively combining a top-down strategy with a bottom-up approach between the levels of organization is probably more consistent to identify the optimal solutions for the global food system than the up-scaling strategy.

Nutrient balance at chain level: a valuable approach to benchmark nutrient losses of milk production

W. Mu, C.E. Van Middelaar, J.M. Bloemhof, J. Oenema and I.J.M. De Boer
Wageningen university, De Elst 1, 6708 WD Wageningen, the Netherlands; wenjuan.mu@wur.nl

A nutrient balance (NB) approach is often used to quantify losses of nutrients, such as nitrogen (N) and phosphorus (P), that contribute to environmental problems such as eutrophication. An NB generally is computed at farm level, whereas nutrient losses related to off-farm processes are neglected. Using an NB at farm level to compare systems that differ in, for example, amount of purchased concentrates, however, may lead to biased conclusions. We, therefore, analysed 19 Irish grass-based dairy farms (IR) and 13 Dutch concentrate-based dairy farms (NL). For each farm, we computed N and P losses at farm and chain level, expressed per ton fat-and-protein-corrected milk (FPCM) to investigate whether an NB at chain level provides better insights than an NB at farm level. Data were analysed using an independent T-test, a Wilcoxon-Mann-Whitney test and regression analysis. Results show that on average, IR farms had higher N losses per ton FPCM than NL farms, both at farm (IR=20; NL=8 in kg N/ton FPCM) and chain level (IR=22; NL=11 in kg N/ton FPCM). P losses per ton FPCM, on the other hand, did not differ between IR and NL farms at farm (IR=0.3; NL=0.1 in kg P/ton FPCM) or chain level (IR=0.8; NL=1.0 in kg P/ton FPCM). Results of the regression analysis showed that the NB at chain level could be accurately predicted from the NB at farm level (R^2=0.992 for N; R^2=0.910 for P), whereas in case of P the slope tended to differ between IR and NL farms (P<0.10). Ranking all 32 farms based on the N farm or chain balance showed a similar pattern, whereas the ranking pattern based on the P farm balance differed from pattern based on the chain balance. We concluded that a farm level NB can be used to benchmark milk production systems if (1) the on-farm impact is relatively important compared to the off-farm impact; and (2) differences in on-farm impact between systems are large. However, a chain level balance of a sample set is required to verify these conditions.

Environmental footprint of a France – Italy integrated beef production system with a LCA approach

M. Berton, G. Cesaro, L. Gallo, M. Ramanzin and E. Sturaro
University of Padova, DAFNAE, viale dell'Università 16, 35020 Legnaro (PD), Italy; marco.berton.4@studenti.unipd.it

The beef production in Italy is conglomerated in the Po Valley, and is based on young bulls imported mainly from France and intensively fattened using total mixed rations (TMR) based on maize silage and concentrates. This study aimed to examine the environmental footprint of intensive beef sector of Veneto region through a cradle-to-farm-gate life cycle assessment (LCA). The batch (a group of animals homogenous for breed, diet, fattening period, and finishing herd) and 1 kg of Body Weight (BW) were taken as system boundary and functional unit respectively. The considered impact categories were greenhouse gases (GHG) emission (kg CO_2-eq) and eutrophication (g PO_4-eq). The study involved 198 batches (Charolaise breed) and 15 specialized beef fatteners. An on farm survey was conducted for the fattening period in Italy (225±17 d, from 392±25 to 734±21 kg BW), whereas the French cow-calf phase emissions were obtained from literature data. On-farm data for crop-to-feed production and materials (fuel, electricity, plastic, lubricant, fertilisers) were recorded during 2013; off-farm emissions were based on literature and software data (Simapro 7.3.3 – Ecoinvent database v.3.0). Mean GHG emission was 14.3±0.8 kg CO_2 eq/kg BW, while eutrophication was 59±6 g PO_4-eq/kg BW. The Italian fattening phase contributed for 27% and 35% on average to GHG emissions and eutrophication respectively. For the intensive fattening period, the factors affecting the environmental footprint were average daily gain (1.53±0.09 kg/d) and diet composition. The cow-calf phase will be further investigated in order to analyse the variability of impact categories and the ecosystems services provided by animal grazing, with the perspective to obtain a holistic evaluation of the sustainability of the integrated France – Italy beef production system.

Can the environmental impact of pig systems be reduced by utilising co-products as feed?
S.G. Mackenzie[1], I. Leinonen[1], N. Ferguson[2] and I. Kyriazakis[1]
[1]School of Agriculture, Foood and Rural Development, Newcastle University, Newcastle, NE1 7RU, United Kingdom, [2]Nutreco Canada, 150 Research Lane, Guelph, ON N1G 4T2, Canada; s.g.mackenzie@ncl.ac.uk

Life cycle assessment (LCA) studies have shown that feed production causes the majority of global warming potential (GWP) and non-renewable resource use (NRRU) resulting from pig systems. We investigated the effect of including alternative feed ingredients (co products) in Grower/Finisher (G/F) diets on the GWP, acidification potential (AP) and NRRU of pig farming systems. G/F diets with maximum inclusions of bakery meal (BM), corn DDGS (DDGS), meat meal (MM) and wheat shorts (WS) were compared individually to a control diet based on corn, soymeal, canola meal, fat blend, limestone, amino acids, minerals and additives. All diets were formulated for a four-phase G/F feeding regime and designed for optimum feed efficiency. Impacts were calculated for a functional unit of 1 kg expected carcass weight, with system boundaries of cradle to farm-gate using an LCA developed for pig systems in Canada. The system impacts for the co-product diets were compared to the control using 1000 parallel Monte-Carlo simulations, to account for shared uncertainties in the calculations. Compared to the control, the BM diet produced small reductions of <5% in the environmental impact of the system for all categories tested (P<0.001). The MM diet reduced NRRU by 9% (P<0.001), did not significantly affect GWP and increased AP by 7% (P<0.001). DDGS caused a small reduction (<1%) in AP (P=0.01), but increased GWP and NRRU by 17% and 57% respectively (P<0.001). The WS diet caused the largest reductions in GWP (11%) and NRRU (19%) (P<0.001) and did not significantly affect AP. The study showed it possible to reduce the environmental impact of pig systems through the increased inclusion of co products in G/F diets. Increased BM and WS inclusions (87 and 260 g/kg respectively) reduced the environmental impact of the system with no increase in any of the impact categories tested.

Productivity and efficiency of French suckler beef production systems: trends over the last 20 years
P. Veysset, M. Lherm, M. Roulenc, C. Troquier and D. Bébin
INRA, UMR1213 Herbivores, 63122 Saint Genès Champanelle, France; veysset@clermont.inra.fr

Over the past 23 years (1990-2012), French beef cattle farms have expanded in size and increased their labour productivity by over 60%, chiefly – though not exclusively – through capital deepening (labour-capital substitution) and simplifying herd feeding practices. The efficiency of beef-sector production systems, as measured by the ratio of volume of farm output excluding aids to volume of intermediate consumption, has fallen by nearly 20% while income per worker has held stable thanks to aids and subsidies and the labour productivity gains made. This aggregate net efficiency of beef cattle systems is positively correlated to feed sufficiency, which is in turn negatively correlated to farm size and herd size. While volume of farm output per ha of UAA has stalled, fodder feed sufficiency (added grass resource value) has lost 6 pc points. The continual increase in farm size and labour productivity has come at a cost of lower production-system efficiency – a loss of efficiency that 20 years of genetic, technical, technological and knowledge-driven progress has barely managed to rake back.

Nitrogen conversion efficiency in French livestock production from 1938 to 2010

J.P. Domingues[1], T. Bonaudo[1], B. Gabrielle[2] and M. Tichit[1]
[1]INRA, UMR SADAPT, 16 rue Claude Bernard, 75005 Paris, France, [2]AgroParisTech, UMR ECOSYS, Route de la Ferme, 78850 Thiverval-Grignon, France; johnpetter.ds@gmail.com

Environmental pollution, competition for natural resources and land are common issues in livestock production. Improved resource use efficiency in the sector is vital in order to meet sustainably the increasing demand of animal products. Therefore it is crucial to seek how the use of feed resources has evolved along the past years in order to close the efficiency gap. The objective of this study was to assess the change over time of N conversion efficiency (NCE) in the French livestock sector. The analysis was conducted at the whole-of-France scale focusing on all herbivore and monogastric species from 1938 to 2010. Collating statistical data from the French agricultural census, N content in feed resources (natural grasslands, cultivated fodder crops, cereals and by-products) and N content in output (meat, egg and milk) were calculated. An aggregated indicator of NCE was computed, as the ratio between N in animal products and N in feed resources, representing the overall efficiency of N conversion into human food of high protein value animal products. The total N in feed fed to livestock from 1938 to 2010 increased by 53.7%, reaching 1.7 Tg of N. The total N in animal products doubled, resulting in an increased NCE for the years studied, from 12.6 to 16,4% in 2010. The 2010 value of NCE corroborate with the figures found in literature, with NCE at 16% for the whole Europe. However, our analysis shows that the change of NCE was achieved at the expense of a 23% decrease for fed protein self-sufficiency. Further research is needed to downscale this analysis in order to identify if differential changes of NCE occurred according to species and regions and to better understand the drivers of such changes in NCE.

The role of human-edible concentrates for future food security, the environment and human diets

C. Schader[1], A. Muller[1] and N. El-Hage Scialabba[2]
[1]Research Institute of Organic Agriculture (FiBL), Socio-economics, Feldstrasse 118, 5070 Frick, Switzerland, [2]Food and Agriculture Organization of the United Nations (FAO), Viale delle Terme di Caracalla, 00153 Rome, Italy; christian.schader@fibl.org

Modern, intensive livestock systems are, one the one hand, highly efficient in terms of per-animal productivity of meat, milk and eggs. On the other hand, intensive livestock systems compete for arable land with food crops as a substantial fraction of energy-rich concentrates stems from crops grown on arable land. A further expansion of livestock production poses a challenge for the sustainability of future food systems, as only 33% of the global agricultural area consists of arable land. Against the background of ever increasing demand for livestock products, a strategic discussion of the role of intensive livestock systems using human-edible livestock feed is needed. This paper aims to contribute to these discussions by providing quantitative modelling results on future scenarios for food production focusing on grassland-based and reduced concentrate feed livestock systems. We used a model which builds on FAOSTAT, the food balance sheets, life cycle inventories from LCA databases and additional scientific literature. 180 plant and 35 livestock production activities are defined for 229 countries. Activity and country-specific defined inputs and outputs allow to model environmental impacts of a wide range of scenarios. We cover land occupation, energy use, greenhouse gas emissions, N and P eutrophication from a life cycle perspective as well as pesticide use potential, irrigation water use, deforestation and soil erosion potential as additional environmental indicators. Our results show that reducing the use of human-edible concentrates in livestock rations can contribute to more sustainable food systems by increasing the energy and protein availability for human nutrition and reducing environmental impacts. However, this change would limit the availability of livestock production for human nutrition.

Efficiency of Austrian dairy farms in terms of net food production and strategies for improvement

P. Ertl, W. Zollitsch and W. Knaus
BOKU-University of Natural Resources and Life Sciences, Vienna, Sustainable Agricultural Systems, Gregor-Mendel-Strasse 33, 1180 Vienna, Austria; paul.ertl@boku.ac.at

While energy-dense supplements help to meet the nutrient requirements of high producing dairy cows, they lower the animals' net contribution to the food supply, because feeding human-edible feed is inefficient in terms of net food production. The aim of the present study was to calculate the edible feed conversion ratio (eFCR) for 30 Austrian dairy farms in order to evaluate their contribution to net food production and identify strategies to improve eFCR. The eFCR was calculated at farm gate level on a gross energy and crude protein basis, and was defined as potentially human-edible output in form of animal products (milk and meat) divided by the input of potentially human-edible feedstuffs. The potentially edible fraction of all feedstuffs used on the 30 farms was estimated based on the available literature for 'low,' 'medium,' and 'high' scenarios, representing low, average, and above average extraction rates of edible nutrients from feedstuffs. The edible fraction of feedstuffs used on the selected dairy farms ranged from 0% for some fibrous feedstuffs up to 100% for some cereals in the scenario 'high.' For the scenario 'medium,' eFCR ranged from 0.50 up to 2.95 for energy and from 0.47 up to 2.15 for protein. About half of the analysed farms showed an eFCR of below 1, indicating a net loss in food supply. There was a negative correlation between eFCR and the amount of concentrates per kg milk and a positive correlation between eFCR and area of grassland required per ton of milk. In conclusion, the current contribution of dairy farms in Austria in terms of net food production is only marginal. Strategies to improve eFCR for dairy farms include reducing concentrate supplementation, substitute common concentrates with more or less human inedible by-products from the food processing industry, or focus on forage-based dairy production.

Will beef and dairy have a place on our plate in the future?

M. Patel[1], J. Spångberg[2], G. Carlsson[3] and E. Röös[2]
[1]Swedish University of Agricultural Sciences (SLU), Dept of Animal Nutrition and Management, Box 7024, 75007 Uppsala, Sweden, [2]SLU, Dept of Energy and Technology, Box 7032, 75007 Uppsala, Sweden, [3]SLU, Dept of Biosystems and Techology, Box 103, 230 53 Alnarp, Sweden; mikaela.patel@slu.se

There is need for change in dietary patterns to reach sustainable food systems, and although studies have shown environmental benefits with vegan diets, such diets may not be the best option since grazing could be important for biodiversity conservation. In Sweden, the area of semi natural pastures is decreasing and reforested due to fewer animals and rationalization. Many of the red-listed species are found here; hence preservation of pasture is one of Sweden's most important environmental goals. The optimum degree of national self-sufficiency is also discussed; hence our objective was to calculate the amounts of beef and milk that can be produced if the basic criterion is that all semi natural pastures should be grazed. Two different cattle scenarios were created; intensive dairy and extensive dairy. Since cattle's nutritional requirements vary with production level, they can utilize semi natural pastures to different extents which restricted the number of animals in each scenario. The annual production of milk and meat was calculated in relation to the number of inhabitants and thus the number of servings was obtained. Current consumption is 3.5 weekly servings of beef, 5 daily slices of cheese and 2.9 dl of milk. In the intensive- and extensive dairy scenarios the corresponding figures were 1.5 and 0.7 servings of beef, 3 and 1 daily slices of cheese and 7.9 and 1.5 dl of milk. If cattle production would be justified mainly for preservation of the Swedish agricultural landscape and to sustain biodiversity, beef consumption would need to be reduced to 20-40% of today's consumption, while there would be surplus milk if an intensive dairy production system is practiced. It can be discussed if the surplus milk should be consumed within the country or exported to countries with few available protein sources.

Crop-livestock integration and diversification to enhance farming systems sustainability
I. Sneessens[1,2,3], P. Veysset[2], M. Benoit[2], A. Lamadon[2] and G. Brunschwig[2,4]
[1]French Environment and Energy Management Agency, 20 avenue du Grésillé, BP 90406, 49004
Angers Cedex, France, [2]INRA, UMR1213, Route de Theix, 63122 Saint-Genès-Champanelle, France,
[3]Institut Polytechnique LaSalle Beauvais, PICAR-T, 19 rue Pierre Waguet, BP 30313, 60026 Beauvais
Cedex, France, [4]Clermont University, VetAgro Sup, 89, avenue de l'Europe, 63370 Lempdes, France;
ines.sneessens@lasalle-beauvais.fr

In the past half-century, global intensification and specialization of agriculture is observed, mainly in developed region, and widely criticized both from an environmental point of view and in case of input price volatility. Diversification and integration of crop and livestock productions appear to be a valuable way to enhance farming system sustainability. However, the socioeconomic and environmental performances of those mixed systems aren't well understood at farm scale. In this work, we posit that they must be conditions of diversification and integration that permit mixed crop-livestock systems to be more sustainable than specialized systems. To test this hypothesis, we focused on the specific case of French sheep-for-meat production and we designed a whole-farm model – Sheep'n'Crop – that permits simulating contrasted mixed crop-livestock systems and evaluating their sustainability through economic, productive and environmental (MJ, GHG, N balance) indicators. Results permit confirming that crop-livestock integration has potential to be more favorable than specialization regarding economic performances (+8.7%), productive performances (Livestock: +4.4%; crop: +2.9%), and environmental performances as nitrogen balance (-36%), GHG emissions (Livestock: -7%; crop: -15%) and MJ consumption (Livestock: -3%; crop: -10%). However, even if integration improves global performances, a high percentage of crops influences the performances in contrasted direction, highlighting the existence of trade-off between sustainability objectives.

Applying agroecological principles to redesign and to assess dairy sheep farming systems
V. Thénard[1], J.P. Choisis[2] and M.A. Magne[3]
[1]INRA, UMR AGIR, 31326 Castanet Tolosan, France, [2]INRA, UMR DYNAFOR, 31326 Castanet Tolosan,
France, [3]ENFA, UMR AGIR, 31326 Castanet Tolosan, France; vincent.thenard@toulouse.inra.fr

For a few decades the ruminant feeding systems have increased some negative environmental impacts. Farms' efficiency is directly depending on input prices and climate variability. In southern France, a group of dairy-sheep farmers are sharing experiences to test innovative practices to become less vulnerable to global change and more sustainable. Recently agroecology provided a framework to improve adaptive capacities to global change using resources diversity, and to ensure the sustainability of the farming systems. While agroecological principles were formulated, they must become operational to develop new livestock systems. A research project was developed between dairy sheep farmers, advisors and researchers to seek sustainable practices improving feed self-sufficiency. The aim was to analyse the livestock farming systems in an agroecological viewpoint and to assess the their practices with trade-offs among farms' multi-performances. The study was based on 20 farmer's interviews about practices, farming management and collection of animals' performances and economic results. The analysis revealed 4 types of farming management according to the milking duration and feed self-sufficiency, the diversity of forage resources, and the breeding selection. Practices are related to agroecological principles and farmers implement them to reach a compromise between three targets: being productive, economical and self-sufficient. Four various self-sufficiency managements, with different trade-offs, were identified. Their environmental impacts were assessed with a set of 23 indicators concerning main environmental stakes (soil preservation, no renewable resources use, wild biodiversity and cultivated biodiversity). Our findings contribute to better insight agroecologically management of dairy sheep and should be incorporated in a tool to support farmers' transition towards agroecological farming systems.

Towards an integrated index for sustainability in multifunctional dairy farms: a case study

S. Salvador[1], A. Zuliani[1], M. Corazzin[1], E. Piasentier[1], A. Romanzin[2] and S. Bovolenta[1]
[1]*University of Udine, DISA, Via Sondrio 2, 33100 Udine, Italy, [2]CRITA, Via Pozzuolo 324, 33100 Udine, Italy; zuliani.anna.2@spes.uniud.it*

Traditional and organic mountain farms appear unsustainable from an environmental point of view when emissions are allocated only to the quantity of milk produced. However, mountain farms deliver to the community also co-products and ecosystem services (ES) that need to be taken into account. Animal welfare can be considered an additional service provided by mountain farming systems to urban consumers through more ethical foods. 8 organic (ORG) and 8 conventional (CON) dairy farms were assessed in the eastern Italian Alps. Different functional units and allocations were considered in a Life Cycle Assessment (LCA). On-farm emissions were computed using 3 allocation methods: a conventional allocation to milk, a physical allocation to milk and beef (in accordance with the IDF) and an economic allocation to milk, beef and ES (on the basis of regional agri-environmental payments). Welfare Quality® (WQ) protocol and Animal Needs Index (ANI 35L) were used to score dairy cow welfare. Comparing ORG and CON farm for kg of Fat and Protein Corrected Milk, emissions were always higher for the ORG (warming potential: 1.03 vs 0.92 kg CO_2 eq.) When the environmetal burden is either physically or economically allocated, GHG emissions attributed to milk are reduced on avarage by 15 or 34% respectively. All farms resulted in 2 welfare categories (acceptable and enhanced) according to the WQ. Similarly, according to the ANI 35L score, medium-high values were recorded (25.8 vs 26.9). The findings of the survey show that welfare scores are flatten to medium values even when different management practices are taken into consideration. The capability of LCA and welfare assessment to capture the beneficial effects of multifunctional mountain farms is limited. Agreed and integrated methodologies are needed in order to properly assess environmental sustainability of mountain farms.

Dairy farm productivity according to herd size, milk yield, and number of cows per worker

L. Krpálková[1], V.E. Cabrera[2], J. Kvapilík[1] and J. Burdych[1]
[1]*Institute of Animal Science, Department of Cattle Breeding, Přátelství 815, 10400 Prague 10 – Uhříněves, Czech Republic, [2]University of Wisconsin, Department of Dairy Science, 1675 Observatory Dr., Madison, WI 53716, USA; lenka.krpalkova@centrum.cz*

This study evaluates associations of farm herd size (HERD), milk yield (MY, kg/cow per year), and cows per worker (CW) with production, reproduction, and economic traits in 60 commercial dairy herds (34,633 cows) in the Czech Republic (year 2012). Data were analyzed using a PROC MIXED model in SAS 9.2 on the independent variables HERD, MY, and CW. Each parameter was split into 3 groups. The largest herds (≥750 cows) had the highest mean MY (8,255 kg), conception rate at first service (43.07%), and CW (57 cows). Herds with lowest MY (≤7,499 kg) had the longest mean days open (127 d), longest calving interval (411 d), lowest overall conception rate (39.3%), highest death rate of calves (8.2), highest total loss of calves (14.8%), and oldest age at first calving (820 d). Herds with lowest CW (≤39) had the most movement disorders (19.9%) and highest death rate of calves (7.8%). Herds with lowest CW vs highest CW (≥60) also differed in the mean number of lactations (2.4 vs 2.7) and total loss of calves (13.9 vs 11.2%). The largest herds had the highest profitability of costs without subsidies (-3.8±4.3%) but the highest costs for cereal grains and concentrates (2.4 CZK/l milk). The largest and smallest herds also differed in labor costs (1.11 vs 1.56 CZK/l milk, respectively). Herds with lowest MY had the lowest net profit without subsidies (-1.58±0.37 CZK/l milk). Herds with highest MY and herds with lowest MY also differed in costs of cereal grains and concentrates (2.39 vs 2.00 CZK/l milk, respectively). Herds with highest CW had lowest costs for veterinary services (0.2 CZK/l milk). Herds with lowest CW had the highest mean labor costs (1.51 CZK/l milk) and highest costs for breeding operations (0.22 CZK/l milk). Increasing herd size tended to be – accompanied by higher milk yields and overall efficiency.

Sustainability evaluation of beef cattle production systems in the Pampa Biome
C.S. Nicoloso, V.C.P. Silveira, F.L.F. Quadros and R.C. Coelho Filho
University Federal of Santa Maria, Av. Roraima, n° 01, 97105900, Brazil; carolinanicoloso@hotmail.com

The Pampa biome covers approximately 2% of the Brazil territory, 63% of the Rio Grande do Sul State (RS), part of Argentina and whole Uruguay. The introduction and expansion of cultures and exotic forage (mainly soybeans) are leaving the rapid degradation of the natural Pampa grasslands, as well as social and cultural changes. The aim of this study is to evaluate the sustainability of family beef cattle production systems in biome Pampa, in RS. The evaluation of systems management of natural resources by incorporating indicators sustainability (MESMIS) was used to study the three beef cattle production systems: GPN-family farmer, with production systems in natural pastures and no commercial crops; GPNC-family farmer with production systems in natural pastures and the presence of commercial crops (except soybeans); GPNS-family farmer with production system on natural pasture and soybean cultivation. The analysis of variance was used with SPSS (19.0). There were no significant differences between the three production systems, to the attributes of MESMIS (productivity, stability, adaptability, equity and self- sufficiency). We found significant differences in the indicators 'Crop presence' (productivity) and 'Source of income' (stability).The three systems presents income from Other Activities, but GPNC and GPNS are high values reflecting the appreciation of the grains in the market. The more sustainable system was the GPN (60% of income originating in Animal production). The Crop presence indicator (productivity) considers the team and percentage of crops in the production system. The GNP had better results by having a lower percentage of crops in the system. Although there is no statistical differences in MESMIS sustainability attributes, significant differences in the indicators are representative and point to a greater or lesser sustainability for the systems, in different ways.

Sustainability assessment of Spanish sheep and Welsh beef mixed crop-livestock farming systems
A.M. Olaizola[1], C. Nicoloso[1,2], O. Barrantes[1], I. Blasco[1], R. Reiné[1], S. Moakes[3] and P.K. Nicholas[3]
[1]Zaragoza University, Ciencias Agrarias y del Medio Natural, Miguel Servet 177, 50013 Zaragoza, Spain, [2]Universidade Federal de Santa María, Avenida Roraima 1000, 97105-900 Santa Maria, Brazil, [3]Aberystwyth University, Gogerddan Campus, Aberystwyth SY23 3EE, United Kingdom; carolinanicoloso@hotmail.com

There is widespread agreement regarding the relevance of sustainability in agricultural production. The numerous definitions of sustainability emphasise the need for taking into account the ecological, economic and social consequences of development choices for present and future generations. The sustainability of Spanish sheep and Welsh beef mixed crop-livestock farming systems was evaluated using MESMIS framework. MESMIS is organised around sustainability attributes (productivity, stability, adaptability, equity and self-reliance), although these attributes are related to the three sustainability pillars (economic, social and environmental). The information was collect by means of a direct survey carried out on 40 sheep farms located in Aragón (Spain) and 38 beef farms located in Central and South Wales (UK). A hierarchical Cluster Analysis on five sustainability attributes was performed in order to classify the farms. Four types or 'groups' of farm sustainability were obtained in each mixed farming system studied. 45% of mixed sheep farms were included in Group 1 'More self-sufficiency and stability' and only 15% of farms were 'Less sustainable'. 56% of mixed beef farms (Group 1 and Group 2) were more sustainable than 44% of farms (Group 3 and Group 4). Mixed sheep farms presented higher levels of self-sufficiency and equity than productivity. In terms of sustainability pillars, the social and economic sustainability was lower than environmental sustainability. Mixed beef farming systems obtained higher scores in self-sufficiency and productivity than in the other sustainability attributes. These farms scored lower for social than economic and environmental sustainability.

Milking practices of Breton dairy farmers with reduced milking intervals

V. Brocard[1], G. Trou[2], J. Flament[3], F. Gervais[3], N. Muller[3] and K. Quéméneur[3]
[1]Institut de l'élevage, BP 85225, 35652 Le Rheu Cedex, France, [2]Chambres d'Agriculture de Bretagne, Pôle Herbivores, Rue Maurice Le Lannou CS 74223, 35042 Rennes Cedex, France, [3]INRA- Agrocampus Ouest, UPSPA, UMR 1348 Pegase, 65 route de St Brieuc, 35000 Rennes, France; valerie.brocard@idele.fr

Milking represents 50% of the compulsory working time of dairy farmers and refrains many people from settling in dairy production. Moreover, the increase in herd size leads to a greater resort to salaried workers who do not wish to experience long 'working days' on farms; that is why experiments were carried out by INRA researchers to test reduced milking intervals (MI). B. Rémond showed lowt effect of reduced MI on dairy production as long as they remained over 5 h 30 min. But do the farmers apply such strategies on farms? To better understand practices, a study was done on the database of the milking times registered on the 4,600 farms at milk performance recording in Brittany, France, on the official control days. The milking times and intervals appeared to be very 'traditionnal' with only 0,1% of farms below 8 h 30 of milking intervals. Half of the farmers start milking between 7 am and 8 am, and between 5 pm and 6 pm. The average MI is 10 h 25 min. An enquiry was made by the 68 (0.45%) farms below 9 h of milking interval to better understand their motivations for ' shorter' intervals. This study shows the reluctance of farmers to decrease their MI and change their practices. The motivations for more flexible milking times (shorter intervals, but also once a day milking, no milking on Sunday evenings, etc.) according to their relation to milking were described: more time for family, salaried worker milking for instance. The main restraint for changing appears to be related to fears around animal welfare ou udder health. It underlines the need for further experiments and communication around milking intervals. This study was carried out thanks to BCLEO and EYLIPS organisations.

Incidence of process management program on the technological areas and the viability of mixed system

J. Rivas[1], A. García[2], C. De Pablos-Heredero[3], M. Morantes[1], C. Barba[2] and R. Arias[4]
[1]Universidad Central de Venezuela, Campus UCV-Maracay, 2101, Maracay, Venezuela, [2]Universidad de Córdoba, Campus Rabanales, 14071, Córdoba, Spain, [3]Universidad Rey Juan Carlos, Paseo de los Artilleros S/N, 28032, Madrid, Spain, [4]CERSYRA, Avenida del Vino 10, 13300, Valdepeñas, Spain; pa1gamaa@uco.es

The Manchega breeding program systematizes production levels with record procedures, reproduction and quality of milk, across a process management program (PGP). The aim of the study was to evaluate the impact of PGP on the level of adoption technological areas and the viability in the mixed system cereal-sheep. Six technological areas were identified: TA1 Management, TA2 Feeding, TA3 Biosecurity, TA4 Land use, TA5. Equipments and TA6 Reproduction-genetic. Information from 157 dairy sheep farms in the Castilla-La Mancha was used. The effect of the implementation of PGP was evaluated by t test for quantitative variables, and χ^2 test for qualitative variables. Logistic regression was used to assess the impact of PGP on the economic viability of the farm. Results showed PGP was predominantly used by farmers with higher levels of education. Their farms are larger (flock and surface) and tend to improve economic viability (47.8%) three times the average profit per ewe (30.3 vs 98.1 €/ewe) and improved gross margin by 77% (107.0 vs 190.1 €/ewe). The PGP promoted technology adoption (65.8%), with significant differences mainly in the TA6 Reproduction-genetics, TA1 Management, TA5 Equipment and TA3 Biosecurity. According to the initial hypothesis, the logistic regression model showed that the probability of being economically viable is 5,059 times greater when it has implemented a PGP. This study confirms implementing a PMG improves production and economic performance of mixed cereal-sheep farms, due to the better integration of processes, resulting in a more efficient management.

Dairy farm size in relation to performance and economic efficiency measures
I. Pavlík, J. Huba, D. Peškovičová, M. Záhradník and S. Mihina
NPPC, Research Institute for Animal Production Nitra, Hlohovská 2, 951 41 Lužianky, Slovak Republic;
ivan.pavlik@vuzv.sk

The aim was to study relations among the size of dairy farms and selected performance and economic efficiency measures in Slovakia. The information was obtained from complex survey study (280 Holstein farms with 84,537 cows in total represented 91% of Holstein population in 2014). The farm size in the study varied between 20 an 3,324 with average number of cows 302. Significant relations were found between size of farm and milk yield (r=0.4034***) as well as between size of farm and purchase price of milk in €/kg (r=0.4001***). The highest milk yield was found in farms with more than 800 cows (9,342 kg) while the small farms with less than 50 cows achieved 5,562 kg/cow/year. The same trend was observed in purchase prices (0.3404 €/kg for large farms and 0.3082 €/kg for small farms with less than 50 cows). Positive correlations were found between size of farm and daily gains of heifers (r=0.2362) and size of farm and culling of cows (r=0.1854**). Negative correlations were observed between size of farm and age at first calving (r=-0.2162***) and between size of farm and calving interval (r=-0.1248*). Smaller farms (100-200 cows) achieved the highest average total costs (0.4083 €/kg); the lowest costs were observed in large farms (0.3778 €/kg). Except the quantitative parameters subjective perception of the future perspective was part of the survey as well. Over 40% of farmers with less than 50 cows and more than 800 cows plan to increase the size of farm after the abolition of milk quota in 2015. Relations of declared performance and economic parameters and subjective perception of the farmers were analyzed more in detail.

Comparative analysis of livestock sector policy in European Union and Turkey
Y. Tascioglu[1], C. Sayin[1], M.G. Akpinar[1] and M. Gul[2]
[1]Akdeniz University, Faculty of Agriculture, Department of Agricultural Economics, 07058 Antalya, Turkey,
[2]Suleyman Demirel University, Agricultural Economics, Faculty of Agriculture, Department of Agricultural Economics, 32260 Isparta, Turkey; ytascioglu@akdeniz.edu.tr

Agriculture is a sector that should be given special attention because it provides employment and the necessary food for the maintenance of human life. Agricultural support policies are implemented to regulate agricultural production, to help increase producers income located in the sector, agriculture has an important role play in governments programs in many developed and developing countries. The purpose of the implementation of agricultural support policies are to prevent the loss of producers and consumers' income, to ensure that small family farm are able to compete with the big farms. It is observed that there are significant differences between sub-sectors and countries as their type of agricultural subsidies. Livestock activities are important for a balanced diet and protein needs by people. Livestock policies are implemented in different ways in the various activities and support in Turkey and the European Union (EU). Turkey has introduced new regulations in all areas of agriculture in order to adapt to the EU Common Agricultural Policy (CAP). Livestock production activities are also one of the most important compliance issues. However, at this point, to meet the demand for animal protein in Turkey in the foreground, while the EU is subject to excess. Turkey is implementing policies constantly encouraging the sector to close the gap. The EU is implementing the trouble-shooters excess production policies. On the other hand, there are differences between the EU and Turkey such as the organization of livestock enterprises, support policies, marketing structure, food safety and structural issues. In this case, this paper is a comparative analysis of implemented livestock policies with EU and Turkey.

Economic efficiency of breeding suckler cows and fattening bulls in the Czech Republic

J. Syrůček[1], J. Kvapilík[1], L. Bartoň[1], M. Vacek[2], L. Stádník[2] and L. Krpálková[1]
[1]Institute of Animal Science, Department of Cattle Breeding, Přátelství 815, 10400 Prague 10 – Uhříněves, Czech Republic, [2]Czech University of Life Sciences, Department of Animal Husbandry, Kamýcká 129, 16521 Praha 6 – Suchdol, Czech Republic; lenka.krpalkova@centrum.cz

The study deals with the evaluation of the economic efficiency of breeding suckler cows and fattening bulls in the Czech Republic on data from farms from different regions of the country. Data were obtained by questionnaire method in 2013 from 20 farms, with breeding suckler cows and from 19 companies' deals with fattening bulls' beef cattle and combined yields breed. The data included the production and economic indicators from 2,164 suckler cows and from 3,670 bulls included in fattening. Model calculation is determined by the level of profitability including and without taking subsidies received. The breakeven points define production and economic parameters to ensure, other things equal conditions, zero profit. The sensitivity analysis reveals how changes of main factors affect to overall economic efficiency of the system. The results indicate a low level of profitability breeding cow suckler (2.15%) and loss rates in disregarding of subsidies. On the basis of model calculations for zero profit would have been sufficient sales price of calves 54.63 CZK or sell 81 weaned calves from 100 cows per year. Fattening bulls in 2013 was based on the monitoring results achieved by loss of 1,289 CZK per fattened bull. Breakeven points that balance revenues and expenditures, for fattening bulls at an average selling price for the year 2013 in the Czech Republic were found for daily weight gain 1.16 kg, length fattening 412 days a selling price of CZK 87.11 per kilogram carcass weight. Sales prices of calves, number of weaned calves, subsidies and calving interval in breeding suckler cow according to sensitivity analysis have a great Influence on the economy. In addition to output prices, the most fattening bulls affect the length of fattening, weight gain and weight of the animals when entering fattening.

Haplotype construction methods to enhance genomic evaluation

D. Jónás[1,2,3], V. Ducrocq[3] and P. Croiseau[3]
[1]ALLICE, 149 rue de Bercy, 75012 Paris, France, [2]AGROPARISTECH, 16 rue Claude Bernard, 75231 Paris, France, [3]INRA, UMR1313 Génétique Animale et Biologie Intégrative, Domaine de Vilvert, 78350 Jouy-en-Josas, France; david.jonas@jouy.inra.fr

Haplotypes have been used in practical genomic selection in France since 2008, improving the selection efficiency compared to other, SNP-based methods, as a result of higher linkage disequilibrium (LD) between haplotypes and QTL. The aim of our study was to build haplotypes around QTL-SNPs (i.e. SNP that are in strongest LD with a QTL within a region) which could outperform in genomic selection the haplotypes built using flanking markers. A criterion was developed to select the haplotype with the largest number of 'predictable' alleles within a window of N SNP around the QTL-SNP, where allele frequencies were used as indicators of good predictability. The selected haplotypes were then used in a modified Bayes-Cπ model, suitable to use haplotype markers as independent variables and their overall effect on the correlation between observed daughter yield average (DYD) and genomic breeding values (GEBV) was measured in a Montbeliarde population of 2,235 bulls for 5 production traits in a regular validation study. Haplotypes of 4-, and windows of n=10 SNP were found to be optimal. The haplotype selection method increased the number of alleles with 'acceptable' allele frequency by 15%, while their average frequency decreased by 16%, compared with haplotypes built from flanking markers. Merging the selected SNP into haplotypes led to a gain of 2% in correlations compared to an analysis where the same selected SNP were used as SNP markers in genomic selection. The selected haplotypes outperformed both an analysis using only the QTL-SNP as SNP markers, increasing the average correlation between DYD and GEBV by 3.1% (absolute value) and an analysis using the flanking-marker haplotypes by 0.5%, ranging from -0.2 till 1.1%. The haplotype selection method efficiently increased the number of haplotype alleles for which the allele frequency is sufficiently high for a proper effect estimation.

Predicting mutation carriers from unbalanced data

F. Biscarini[1], H. Schwarzenbacher[2], H. Pausch[3] and S. Biffani[1]
[1]PTP/IBBA-CNR, Via Einstein, 26900 Lodi, Italy, [2]ZuchtData, Dresdner Str 89, 1200 Vienna, Austria,
[3]TMU, Liesel-Beckmann Str 1, 85354 Freising, Germany; filippo.biscarini@tecnoparco.org

Binary classification problems often deal with unbalanced data: e.g. carriers of a recessive mutation are rare compared to non-carriers. This leads to unbalanced training datasets, which violates the assumption of evenly distributed observations among classes. In such contexts, identifying rare instances is difficult and the false-negative rate tends to be high. In classification tasks it is often more important to identify instances belonging to the minority class: it is more critical to misclassify carriers than non-carriers, since this would spread the defect in the population. We analysed data from 3,116 Fleckvieh and 175 Brown Swiss cattle for the mutation underlying the BH2 haplotype on BTA19, associated with perinatal mortality. The proportion of carriers in the dataset was 0.04 in Fleckvieh and 0.72 in Brown Swiss; the two datasets were therefore unbalanced in opposite directions. A ridge regression logistic model was used to predict the carrier status from SNP genotypes on BTA19 (1,318 SNPs after editing). First, 20% of the data were randomly selected out and used for validation. The remaining 80% of the data were used in 5-fold cross-validation to fine-tune the regularization parameter λ and train the predictive model. The predicting equations were then applied to the validation set to estimate the prediction error. The process was repeated 20 times. The total predictive ability (TPA), true negative (TNR: non-carriers correctly identified) and true positive (TPR: carriers correctly identified) rates were measured. TPA was 0.99 ($\pm$0.01) and 0.97 ($\pm$0.03) in Fleckvieh and Brown Swiss. However, the classification of carriers and non-carriers differed in the two breeds: TPR was 0.99 in Fleckvieh, and 0.89 in Brown Swiss; TNR was 0.95 in Fleckvieh and 1.00 in Brown Swiss. This highlights the relative difficulty of correctly classifying instances in the minority class, where the prediction accuracy tends to be lower and more variable.

Accuracy of imputation from SNP array data to sequence level in chicken

G. Ni[1], H. Pausch[2], T.M. Strom[3], R. Preisinger[4], M. Erbe[1] and H. Simianer[1]
[1]Georg-August-University Goettingen, Animal Breeding and Genetics Group, Goettingen, Germany,
[2]Technische Universität München, Chair of Animal Breeding, Freising, Germany, [3]Helmholtz Zentrum München, Institute of Human Genetics, Neuherberg, Germany, [4]Lohmann Tierzucht GmbH, Am Seedeich 9-11, Cuxhaven, Germany; gyni.ni@agr.uni-goettingen.de

It is now possible to generate whole-genome re-sequencing data for different species. As sequencing is still expensive, usually only a few individuals of a population are sequenced completely and imputation is used to obtain genotypes on all sequence-based single nucleotide polymorphism (SNP) loci for other individuals, which have already been genotyped with SNP arrays. This study focuses on assessing the quality of imputation from SNP array data up to sequence level in chickens. Sequence data (Illumina HiSeq2000, coverage ~ 8×) of 25 key ancestors from a commercial brown layer line and SNP array data (Affymetrix Axiom® Chicken Genotyping Array, 580k) of 1051 individuals from the same line was available. Imputation programs used were Minimac, FImpute and IMPUTE2. Quality of imputation was checked for three chromosomes in different scenarios. With all imputation programs, correlation between original and imputed genotypes was very high (>0.95 in average with randomly masked 1000 SNPs from the SNP array and >0.85 for a leave-one-out cross-validation within sequenced individuals). FImpute performed slightly worse than Minimac and IMPUTE2 in terms of genotype correlation, especially for SNPs with low minor allele frequency, while it had lowest numbers in Mendelian conflicts studied in available father-son pairs. Genotype correlation of real and imputed genotypes stayed constantly high even if individuals to be imputed were several generations away from the sequenced key ancestors. The results of this study show that imputing up to sequence level is possible with high accuracy, even along several generations.

Sequence-based maps reveal similar recombination patterns on macro- and microchromosomes in chicken

S. Qanbari[1], M. Seidel[2], G. Mészáros[3], T.M. Strom[4], R. Preisinger[5] and H. Simianer[1]
[1]Animal Breeding and Genetics Group, Georg-August University, 37075 Göttingen, Germany, [2]Institute of Plant Genome and Systems Biology, Helmholtz Zentrum München, Neuherberg, Germany, [3]Division of Livestock Sciences, University of Natural Resources and Life Sciences, Vienna, Austria, [4]Institute of Human Genetics, Helmholtz Zentrum München, Munich, Germany, [5]Lohmann Tierzucht GmbH, 27472, Cuxhaven, Germany; sqanbar@gwdg.de

The available pedigree based genetic maps of the chicken genome have poor resolution in terms of quantifying frequency and distribution of recombination hotspots. We constructed the first fine scale recombination map of the chicken genome by re-sequencing of 50 hens from two populations of commercial brown (BL) and white (WL) layers. The population scaled recombination rate (ρ) was averaged as 3.30 ± 0.01 kb^{-1} in WL compared to 3.62 ± 0.01 kb^{-1} in BL that corresponds to the ~3.5 cM/Mbp of the chicken genome. The length of the autosomic map was estimated as 2,560 cM in WL compared to 2,732 cM in BL which concords the available linkage maps. The hotspot analysis revealed a comparable frequency between populations with an average density of one every ~100 kb across the autosomic genome. Recombination rates vary between populations (r=0.31) and across the genome. We estimated LD for SNPs in <25 kb interval as an indicator of recombination frequency and found a subtle, but irregularly lower LD for micro ($r^2=0.13$, SD=0.03) compared to macro-chromosomes ($r^2=0.18$, SD=0.06). A consistent trend also emerged in comparison of ρ in micro (2.88 cM/Mbp, SD=3.15) vs macro-chromosomes (2.64 cM/Mbp, SD=0.97). Therefore, through two lines of evidence derived by examining LD and measuring population based recombination frequencies we show insignificant differences between macro vs microchromosomes (P<0.05). This observation contrasts with the common belief that microchromosomes show higher rate of recombination than macrochromosomes and therefore warrants further investigation.

Imputation accuracy from high density genotypes to whole genome sequence in cattle

D.C. Purfield[1], J.F. Kearney[2] and D.P. Berry[1]
[1]Teagasc, Animal & Grassland Research and Innovation Centre, Moorepark, Fermoy, Co. Cork, Ireland, [2]Irish cattle breeding federation, Bandon, Co.Cork, Ireland; deirdre.purfield@teagasc.ie

Whole genome sequencing (WGS) provides the most comprehensive information of an individual's variome. The high cost associated with sequencing has resulted in genome-wide association studies and genomic predictions relying on single nucleotide polymorphism (SNP) marker arrays, which may be subject to ascertainment bias. WGS data, however, does not suffer from ascertainment bias and contains information on rare variants which may affect phenotypic variation. The aim of the study was to quantify the accuracy of imputation from the Illumina bovine high density (HD) genotype panel to WGS in a Holstein-Friesian population. WGS from Run 4 of the 1000 bull genomes project was available on 1,147 sequenced animals from 27 breeds. In total, 289 Holstein-Friesian sequences were available, of which 73 had HD genotypes available. The accuracy of imputation was determined on BTA 18 and 29. Imputation was undertaken with FImpute2 exploiting both family-based and population-based information. The 73 test animals were removed from the sequence population and using their HD genotypes, were imputed to sequence density. Imputation accuracy was evaluated using: (1) the 215 Holstein-Friesian sequences in the reference population; and (2) using the 1,074 multi-breed sequenced individuals as the reference population. Accuracy of imputation was determined as the concordance between the actual and imputed genotypes in the 73 animals. The mean concordance between the raw sequence genotype and HD genotype was 99.09%;the genotype discrepancies were mainly heterozygous genotypes called as homozygotes or vice-versa (95.78%). Slightly greater imputation accuracy (96.79%) was obtained when using the multi-breed reference population in comparison to the single breed reference population (96.68%).Imputation accuracy for rare alleles was low; though this has improved from previous studies due to the expansion of reference population.

Can optimal contribution selection with pre-selection realise its full potential for long-term ΔG?

B. Ask[1], T. Ostersen[1], P. Beerg[2] and M. Henryon[1]
[1]SEGES, Pig Research Centre, Breeding and Genetics, Axeltorv 3, 1609 Copenhagen V, Denmark, [2]NordGen, Nordic Genetic Resource Center, P.O. Box 115, 1431 Ås, Norway; bas@seges.dk

We tested the hypothesis that optimal contribution selection (OCS) with pre-selection realises most of the long-term genetic gain achieved by OCS without pre-selection. We tested this hypothesis by comparing a reference selection scheme (full-OCS) with three alternative schemes with pre-selection carried out before OCS using stochastic simulation. The three pre-selection strategies were truncation selection (trunc), truncation selection with full-sib family restrictions (trunc$_{FS}$), and OCS (pre$_{OCS}$). Generations overlapped and selection was for a single trait (h^2=0.3), which was observed for all selection candidates prior to (pre-) selection. We expect that OCS with pre-selected candidates can realise most of the long-term genetic gain realised by full-OCS regardless of pre-selection strategy, but only until pre-selection intensity becomes too strict. When pre-selection becomes too strict, we expect that pre$_{OCS}$ can still realise most of the long-term ΔG achieved by full-OCS, whereas trunc and trunc$_{FS}$ cannot. If our expectations are confirmed then the potential benefits of OCS in practical breeding schemes with pre-selection may not be realized fully, if OCS is only implemented in the final selection stage(s) and pre-selection is too strict. This is important because many practical breeding schemes have at least two selection stages, and the proportion of pre-selected candidates can be as low as 1% or even less.

Identification of causal variants for milk protein composition using sequence data in dairy cattle

M.P. Sanchez[1], A. Govignon-Gion[1], P. Croiseau[1], A. Barbat[1], M. Gelé[2], S. Fritz[3], G. Miranda[1], P. Martin[1], M. Boussaha[1], M. Brochard[2] and D. Boichard[1]
[1]INRA, AgroParisTech, UMR 1313 GABI, 78350 Jouy en Josas, France, [2]Idele, 149 rue de bercy, 75012 Paris, France, [3]Allice, 149 rue de bercy, 75012 Paris, France; marie-pierre.sanchez@jouy.inra.fr

A genome wide association study (GWAS) was performed at the sequence level to identify candidate mutations for six major milk proteins in Montbéliarde (MO), Normande (NO) and Holstein (HO) dairy cattle breeds. The concentration in the six major proteins was estimated from Mid-Infrared (MIR) spectrometry on almost 600,000 test-day milk samples from 116,495 cows in the first three lactations (PhénoFinlait project). Out of these, 8,080 (2,967 MO, 2,737 NO and 2,306 HO) were genotyped with the 50k Beadchip. For each breed, genotypes were first imputed at the HD level using HD genotypes of 522 MO, 546 NO and 776 HO bulls. The resulted HD genotypes were subsequently imputed at the sequence level using 27 millions of highly confident sequence variants selected from the latest run of the 1000 bull genomes project (1,147 bulls). Phenotypes were average test-day measurements adjusted for environmental effects. Within breed association studies were performed for each sequence variant using a mixed model including also a mean and a random polygenic effect. Numerous QTL were identified in each breed. Three QTL, located on chromosomes 6, 11 and 20, had very significant effects and were shared by the three breeds. Other significant effects, partially overlapping across breeds, were found on almost all autosomes. Potential causal variants were identified in several QTL regions, including mutations previously described in LGB, DGAT1 or GHR genes. These results show that GWAS applied to whole sequence genotypes is a promising approach to the identification of QTL with a better resolution than lower density genotypes. The authors acknowledge the financial support from ANR and Apis-Gène, and the contribution of the 1000 bull genomes consortium.

Expression levels of 14 genes found as RNA in cells extracted from milk

G.E. Pollott, S. Mirczuk, R.C. Fowkes, C. Lawson and Z. Thorp
RVC, PPH, Royal College Street, London NW1 0TU, United Kingdom; gpollott@rvc.ac.uk

Milk production throughout lactation follows a characteristic curve. It has been postulated that three basic processes determine the level of milk production on any given day of lactation; the number of mammary cells produced up to that day, the number dying off through apoptosis and the secretion rate of the cell. Attempts to monitor these processes have been made in a commercial herd by measuring extracellular vesicles in milk as an indicator of apoptosis. This paper reports work to investigate cell activity by monitoring the expression levels of key genes, as found from RNA levels in mammary cells extracted from milk. Previous reports of GWAS and mammary studies were used to select the 14 genes most likely to be linked to mammary cell activity throughout lactation. Twenty fully-recorded Holstein dairy cows in early to mid-lactation were used to provide milk samples. RNA was extracted from the milk using the trizol method and checked for concentration and quality using a NanoDrop Spectrophotometer. RT-PCR was carried out on the 14 milk production genes and 2 housekeeping genes and electropherograms were generated. These were used to provide data on expression levels per unit volume of milk for the genes ARNTL2, CYSLTR2, DGAT1, FOXH1, JAK2, SCD, UGDH, UTRN, LALBA, B4GALT1, UGP2, ACACA, FASN and CSN1. Expression levels of the 14 genes were related to the production and milk composition data recorded on the 20 cows. There was considerable variation in the relationship between gene expression and daily milk production (mean 26.3, SD 8.2 kg/d). Correlations ranged from 0.62 (DGAT1) to 0 (UGP2). Relationships with fat% (3.88±0.77), protein% (3.12±0.29) and SCC (128±298,000 cells/l) showed a similar range. The relationships between the 14 genes and 'days in milk' were also examined. As a preliminary trial this work looks promising as a means of monitoring mammary cell activity throughout lactation but further work is required to investigate the factors which affect gene expression and its relationship with milk production.

Season affects expression and heritability of in-line recorded estrus traits in Danish Holstein

A. Ismael[1,2], E. Strandberg[1], B. Berglund[1] and P. Løvendahl[2]
[1]Swedish University of Agricultural Sciences, Department of Animal Breeding and Genetics, P.O. Box 7023, 750 07 Uppsala, Sweden, [2] Aarhus University, Center for Quantitative Genetics and Genomics Dept. of Molecular Biology and Genetics, P.O. Box 50, 8830 Tjele, Denmark; ahmed.ismae@mbg.au.dk

The objective of this study was to investigate the seasonality of return to cyclicity after calving expressed as the interval from calving to first high activity (CFHA) with respect to the month of calving. For duration of first estrus (DFE) and strength of first estrus (SFE) the seasonality was investigated regarding to the high activity episode month. Estrus traits records were derived from physical activity measurements obtained by electronic activity tags for 10,404 Holstein cows in 102 commercial Holstein herds. Data were analyzed with a linear mixed animal model. Fixed effects in the model included herd, parity, year month of high activity episode, and month of calving in case for CFHA, or month of episode in case for DFE and SFE. The effect of month of calving on CFHA was found to be significant (P<0.0001). The highest least square mean for CFHA of 57 days was recorded for December calvings, while the lowest mean of 32 days was recorded for August calvings. Episode month was found to have a significant effect on both DFE and SFE (P<0.0001). The longest duration mean of 9.0 hours was found for November while the shortest duration of 8.2 hours was found in July. The weakest estrus expression of 1.02 ln-units was found in July and the strongest estrus expression of 1.11 ln-units was found in December. The highest heritability for CFHA was found for autumn calvings (0.08), while the lowest (0.01) was found in spring calvings. Heritability for DFE in winter was higher than the heritability in summer, although both estimates were low, 0.05 and 0.03. Heritability for SFE in winter was twice the heritability estimate found in summer (0.08 vs 0.04). The obtained results show the existence of G × E effects for estrus related traits in dairy cattle with respect to season.

pH24 of lamb drives fresh meat redness more than myoglobin or iron content

H.B. Calnan[1,2], R.H. Jacob[2], D.W. Pethick[1,2] and G.E. Gardner[1,2]
[1]*Murdoch University, Veterinary and Life Sciences, Murdoch, WA 6150, Australia,* [2]*Cooperative Research Centre for Sheep Industry Innovation, University of New England, Armidale, NSW 2351, Australia; h.calnan@murdoch.edu.au*

Fresh meat colour was measured in 8,165 lambs produced over 5 years (2007-2011) at 8 sites across Australia in the Sheep Cooperative Research Centre's information nucleus flock experiment. The lambs were raised on extensive pastures before slaughter at a carcass weight of ~22 kg. Isocitrate dehydrogenase (ICDH) activity was measured in the m. longissimus within 4 hours of slaughter. At 24 hours post slaughter the loin was measured at the level of the 12th rib for pH (pH$_{24}$) and meat lightness (L*), redness (a*) and yellowness (b*) using a Minolta colorimeter after the fresh-cut loin surface was allowed to oxygenate for 30 minutes. The loin was then sampled for measurement of myoglobin and iron. L*, a* and b* were analysed simultaneously in a multivariate analysis to define a final model which was then applied to each of the fresh colour measures separately in linear mixed effects models in SAS. The base model included fixed effects for site, year of birth, slaughter group, sire type, sex and dam breed within sire type. In subsequent analyses measures of pH$_{24}$, myoglobin, iron and ICDH were included in the model as covariates. Increasing pH$_{24}$ from 5.4 to 6 was associated with a reduction (P<0.01) in meat L*, a* and b* by 1.9, 2.5 and 2.1 units. Increasing myoglobin and iron had a greater impact on meat lightness; reducing L* by 2.8 and 2.6 units, but a lesser impact on meat redness; increasing a* by 1.3 and 0.85 units, and yellowness; reducing b* by 0.2 and 0.3 units. Higher muscle ICDH activity also increased a* while reducing L* and b*, though to only half the magnitude of impact as myoglobin. These results suggest that while myoglobin and iron content are important determinants of fresh lamb meat colour, particularly lightness, controlling pH$_{24}$ is the most important factor for ensuring premium colour and thus consumer acceptance of fresh lamb meat.

Genetic analysis of patellar luxation in dogs

K. Nilsson[1], S. Zanders[1] and S. Malm[2]
[1]*Swedish University of Agricultural Sciences, Department of Animal Breeding and Genetics, P.O. Box 7023, 75007 Uppsala, Sweden,* [2]*The Swedish Kennel Club, Rinkebysvängen 70, 16385 Spånga, Sweden; katja.nilsson@slu.se*

Patellar luxation (PL) is a condition where the knee cap is dislocated, mainly during locomotion. PL is one of the most common orthopaedic diseases found in dogs today. The condition is also frequently found in other species such as cattle and horses. PL is included in the central registration of disorders by the Swedish Kennel Club (SKK) since 2002. Several breeds in Sweden are part of a genetic health program where both parents of a litter must be examined for patellar luxation before a litter can be registered. These data have so far not been genetically analysed, and the aim of this study was to estimate genetic parameters for PL. Data was provided by the SKK for two breeds frequently affected by the condition: the Chihuahua and the Bichon Frise. PL was scored on a scale from 0 to 3, where 0 represents an unaffected dog and 3 a severely affected dog. Because the very severe cases were quite uncommon, scores of 2 and 3 were grouped together in the analyses. The edited data included 7,225 records for the Chihuahua and 1,136 records for the Bichon Frise. PL was analysed using a mixed linear animal model, including fixed effects of sex, birth month, birth year, age at examination and a random effects of examining veterinarian, genetic effect of the individual and residual. The prevalence of PL was 23% for the Chihuahua and 12% for the Bichon Frise. In the absolute majority of cases the patella luxated towards the inside of the leg (medial luxation) for both breeds. Female dogs were more frequently affected compared with male dogs. Heritabilities were estimated at 0.25 for the Chihuahua and at 0.21 for the Bichon Frise. We conclude that there is genetic variation in patellar luxation, and genetic progress should be possible to improve leg health status of the studied breeds.

Genetic heterogeneity of residual variance in GIFT Nile tilapia

J. Marjanovic[1,2], H.A. Mulder[2], H.L. Khaw[3] and P. Bijma[2]
[1]Swedish University of Agricultural Sciences, Department of Animal Breeding and Genetics, Box 7023, 75007 Uppsala, Sweden, [2]Wageningen University, Animal Breeding and Genomic Group, P.O. Box 338, 6700 AH Wageningen, the Netherlands, [3]WorldFish, Bayan Lepas, 11960, Penang, Malaysia; jovana.marjanovic@wur.nl

In animal breeding, ample success has been achieved in improving mean levels of trait values by means of genetic selection. However, the same has not been realized for variability of trait values among individuals. Selection for more uniform individuals is possible only in the presence of genetic variation in trait variability. This phenomenon is known as genetic heterogeneity of environmental (residual) variance. The aim of this study was to investigate the potential for genetic improvement in variability of harvest weigh (HW) in the GIFT strain of Nile tilapia, by applying double hierarchical generalized linear models. This approach offers simultaneous estimation of genetic parameters for the mean and the residual variance, which is modelled on the exponential scale. Both untransformed and Box-Cox transformed HW were analysed to study the impact of non-normality on estimates of variance components. Analysis involved 6,090 fish with records on HW. We found substantial genetic variation in the residual variance of 0.41, which decreased to 0.27 after the transformation. The given estimates are on a log scale. The genetic coefficients of variation for residual variance were 64% and 52% on untransformed and transformed scale respectfully. These values describe the potential decrease in residual variance when achieving one genetic standard deviation by selection. The genetic correlation between mean and variance was 0.67 for untransformed HW, which points to a strong relationship between both traits. The genetic correlation for Box-Cox transformed HW was 0.12. The results obtained in this study indicate a good opportunity to improve variability of HW. The high genetic correlation of 0.67 between HW and residual variance suggests necessity for index selection in order to increase HW and reduce its variation at the same time.

Population genomics and signatures of selection in resequenced honey bee drones

D. Wragg[1], B. Basso[2], J.P. Bidanel[3], Y. Leconte[4] and A. Vignal[1]
[1]INRA, UMR1388 GenPhyse, 31326 Castanet-Tolosan, France, [2]ITSAP, UMT PrADE, 84914 Avignon, France, [3]INRA, UMR1313 GABI, 78352 Jouy-en-Josas, France, [4]INRA, UMR406 Abeilles et environnement, 84914 Avignon, France; jean-pierre.bidanel@jouy.inra.fr

The haplodiploid nature of the honey bee *Apis mellifera*, and its small genome size, permits affordable and extensive studies of population genomics by sequencing drones. Sequencing of drones facilitates more reliable variant detection compared to diploids, and consequently can be achieved at lower depths of coverage. Here we demonstrate the utility of this approach by re-sequencing drones sampled from two populations selected for different traits of commercial interest: honey production and royal jelly. Analyses of single nucleotide polymorphisms (SNPs) reveals signatures of positive selection in both populations. Specifically, the results identified two odorant receptors (OR151, OR152) in bees selected for honey production, and orthologs to the coenzyme Q biosynthesis protein 2 (Coq2) and king-tubby in bees selected for royal jelly. Creating diploid drones in silico enabled an assessment of population structure with reference to a diploid dataset comprising representatives of each of the four main evolutionary lineages. Phylogenetic analysis grouped the diploid drones with the C lineage, the dominant genetic background throughout the American and European bee keeping industries. Down-sampling of sequence data for two *A. mellifera* bees sequenced at high depth of coverage (15-18 X), allowed variant detection to be compared in the two haploids and an equivalent diploid reconstructed genome at varying depths of coverage. In all pair-wise comparisons from 3X to 10X in haploids vs 6X to 20X in the diploid drone, the haploids outperformed the diploid drone. This study presents the results of preliminary analyses arising from the SeqApiPop project, and highlights the benefits of sequencing drones compared to workers/queens in *A. mellifera*.

Litter size and survival rate at weaning in mice divergently selected for birth weight variability

N. Formoso-Rafferty[1], I. Cervantes[1], N. Ibáñez-Escriche[2] and J.P. Gutiérrez[1]
[1]Universidad Complutense de Madrid. Facultad de Veterinaria, Producción Animal, Avda. Puerta de Hierro s/n, 28040 Madrid, Spain, [2]Centre IRTA_Lleida, Genètica i Millora Animal, Lleida, Spain; n.formosorafferty@ucm.es

There is an increasing interest in the homogeneity of animal production. It would decrease the cost of handling and production increasing the profitability of the farm. In addition, homogeneity of weight could lead to reducing the mortality and increasing the animal welfare. Less sensitivity with respect to environmental effects, as indicated by a low variation, has benefits in the animal welfare and it is one of the main targets of selection. The relationship between homogeneity and robustness traits, such as survival or litter size, has been reported. The aim of this work was to study the changes that selecting for birth weight environmental variability can bring in other interesting traits in livestock such as litter size (LS) and survival rate at weaning (SW). Data from eight generations of a divergent selection experiment for birth weight environmental variability were analysed to estimate genetic parameters and to explore signs of selection response. A total of 1,265 LS records from 720 litters and 10,587 SW records from 722 litters together with 11,393 pedigree records were used. Each record was assigned to the mother of the pup. The model included generation, parity number and litter size (only for SW) as systematic effects and the litter and additive genetic effect as random effects besides the residual. Heritabilities were 0.21 and 0.06 respectively for LS and SW. Both phenotypic and genetic trends showed clear superiority of the low line over the high line, with roughly 2 pups more of litter size and 4 points higher of survival rate at weaning. The greater robustness and welfare of the homogeneous line was confirmed.

Birth and weaning weight in a divergently selected mice population for birth weight variability

N. Formoso-Rafferty[1], I. Cervantes[1], O. Lizarraga[1], M.A. Pérez-Cabal[1], N. Ibáñez-Escriche[2] and J.P. Gutiérrez[1]
[1]Universidad Complutense de Madrid. Facultad de Veterinaria, Producción Animal, Avda. Puerta de Hierro s/n, 28040 Madrid, Spain, [2]Centre IRTA_Lleida, Genètica i Millora Animal, Lleida, Spain; n.formosorafferty@ucm.es

The success obtained when selecting the environmental variability of birth weight (BW) in an experimental mice population could have been affected other important productive traits such as weaning weight (WW) and its environmental variability. The aim of the present study was to analyse the correlated genetic trends on BW, WW and its standard deviation (SDWW) from a divergent selection experiment for BW environmental variability. Individual BW, WW and SDWW of eight generations of selection were analyzed as a maternal trait but BW and WW was also analyzed under a direct-maternal animal model. A total of 12,051 records of BW from 1,266 litters and 10,587 records of WW from 1,232 litters together with a pedigree file containing 11,393 and 12,637 records were respectively used for maternal and individual models. The model included generation, litter size, sex and parity number as systematic effects. Genetic correlation between the mean of the trait and its environmental variability was 0.31 for BW and -0.32 for WW. A divergent phenotypic trend of WW was observed between lines showing that animals from the high line weighed 2.64 g more than those from the low line in the last generation. Despite the negative genetic correlation obtained, SDWW was also superior in the high line. Furthermore, although animals were heavier in the high line, total litter weight at birth and at weaning was greater in the low line. Therefore, it seems that selection for environmental BW affects positively from a productive perspective. However, before a direct application in farm animals, implications on other traits regarding robustness, longevity and welfare would be worthwhile to confirm.

Influence of maternal genetic effect on the genetic evaluation of peripheral blood immunity in mice

K. Suzuki, K. Yoshimura, K. Miyauchi, D. Ito, Y. Miyazaki, E. Sakai, K. Yamamoto and M. Takemoto
Graduate School of Agricultural Science, Tohoku Univerisity, 1-1 Tsutsumidori amemiya machi, Sendai, Japan, 981-8555, Japan; k1suzuki@bios.tohoku.ac.jp

We have done the selection experiment for peripheral blood immunity in mice for the breeding of disease resistance. Mice were selected for phagocyte activity (PA) at 5 weeks of age as natural immunity, sheep red blood cell specific IgG (IgG) at 9 weeks of age as acquired immunity and both traits over 26 generations and non-selected control line, and named as N line, A line, NA line and C line, respectively. To examine the correlated response, these two traits were measured in all mice. The objective of the present study was to clarify the direct genetic and maternal genetic effect on the selection for peripheral blood natural and acquired immunity. Genetic parameter and breeding value were estimated by two traits animal model and maternal genetic effect model using VCE6 program. The heritability of direct genetic effect for PA ad IgG were estimated to be 0.33 and 0.22 in the N line, 0.64 and 0.34 in the A line, 0.22 and 0.34 in the NA line, and 0.47 and 0.19 in the C line. On the other hand, by the model including maternal genetic effect, the heritability of direct genetic effect for PA and IgG were 0.14 and 0.20 in the N line, 0.72 and 0.32 in the A line, 0.14 and 0.21 in the NA line, 0.30 and 0.08 in the C line. In addition, though the heritability o the maternal genetic effect of PA in A, NA and C line was less than 0.10, those of PA in N line and IgG in all lines were more than 0.1. The selection response of additive genetic effect for PA in the N and NA line, IgG in the A and NA line were higher than those in the C line. However, in the maternal genetic effect model, the selection response of the direct genetic effect of PA in the N and NA line, and IgG in NA line was relatively small. These results suggest that the direct genetic effect concerning peripheral blood immunity is necessary to estimate by the model including the maternal genetic effect.

Effect of divergent selection for intramuscular fat on lipid metabolism traits in rabbits

S. Agha[1,2], M. Martínez-Álvaro[1], V. Juste[1], A. Blasco[1] and P. Hernández[1]
[1]Universitat Politécnica de Valencia, Institute for Animal Science and Technology, 46013 Valencia, Spain,
[2]Ain Shams University, Faculty of Agriculture, Animal Production Department, Cairo, 11241, Egypt;
safsoof23@hotmail.com

A program of divergent selection for intramuscular fat (IMF) in the longissimus muscle (LM) in rabbits was carried out. The difference in IMF between High (H) and Low (L) lines after 5 generations was a 27.7% of the mean at 9 weeks of age. Enzymes activities involving lipid metabolism of two muscles (semimembranosus propius (SMP) and LM) and in the perirenal fat (PF) were studied in H and L lines at 9 and 13 weeks of age. Bayesian inference was used. SMP showed higher G6PDH and FAS activities than LM while LM showed a greater glycolytic activity (LDH). Lipogenic enzyme activities increased with age in all studied tissues. Differences between H and L lines were detected for lipogenic and catabolic activities at 13 weeks but not at 9 weeks. Greater lipogenic activities in SMP, LM and PF were observed in H line compared to L line. Our results indicate that lipogenic and metabolic activities are influenced by the metabolism of each muscle. Differences in lipogenic and oxidation activities between H and L lines can be attributed to the selection for IMF.

Genotype by feeding interaction and genetic parameters for growth rate and carcass traits in rabbits

M. Ragab[1,2], J.P. Sanchez[1], J. Ramon[1], O. Rafel[1] and M. Piles[1]
[1]*Institute of Research and Agro-food Technology, Animal Breeding and Genetic, Torre Marimon, IRTA, 08140 Caldes de Montbui, Barcelona, Spain,* [2]*Kafr El Sheikh University, Poultry Prod, El Geish Street, 33516, Kafr El Sheikh, Egypt; juanpablo.sanchez@irta.cat*

The interaction between the genotype and feeding regimen (G×E) for carcass traits was estimated from data corresponding to 4,981 animals under *ad libitum* (AD; 2,557 individuals) or restricted (RF; 2,424) feeding. Body weight at slaughter (BWS), carcass weight (CW) and dressing out percentage (DoP) were analyzed by using Bayesian multivariate Linear Animal Models in which records obtained under different feeding regimes were treated as different traits. Animals belonged to a paternal line (Caldes line) selected for average daily gain (ADG) under *ad libitum* feeding. Therefore, the complete information of the selection process involving data from 134,419 animals was included in the analysis. Kits from the same litter were equally distributed into both treatments ensuring that animals in a cage had a similar initial weight. Model for ADG included the effects of year-season, parity order, number of born alive, and the additive genetic and common litter effects. For carcass traits, the model included parity order, number of kits born alive, batch, size of the animal at weaning, number of animals per cage, age at slaughter, and the additive genetic, common litter and cage effects. Marginal posterior mean±sd for the heritabilities were 0.16±0.01 for ADG, and 0.34±0.07, 0.28±0.05 and 0.16±0.03 for BWS, CW and DoP under AD feeding, the respective figures under RF were 0.18±0.03, 0.15±0.03 and 0.32±0.06. Genetic correlations between ADG and carcass traits were positive and moderate except for DoP which was low and negative. The estimated genetic correlation between BWS, CW and DoP under different feeding regimen were different from 1: 0.90±0.06, 0.78±0.09 and 0.93±0.02, respectively; which, jointly with the differences in genetic variances, indicates the existence of relevant G×E interaction for these traits.

Genotype by feeding regimen interaction for common diseases in rabbits

M. Ragab[1,2], J.P. Sanchez[1], J. Ramon[1], O. Rafel[1] and M. Piles[1]
[1]*Institute of Research and Agro-food Technology, Animal Breeding and Genetic, Torre Marimon, IRTA, 08140, Caldes de Montbui, Barcelona, Spain,* [2] *Kafr El Sheikh University, Poultry Prod, El Geish Street, 33516, Kafr El Sheikh, Egypt; juanpablo.sanchez@irta.cat*

The study aimed at assessing the possible existence of an interaction between genotype and feeding regimen (G×E) for presence/absence of enteropathy (EN) respiratory symptoms (RS), poor body condition (PBC) and mortality (MO) due to unspecific causes during the fattening period. Animals were submitted to two feeding regimens: *ad libitum* AD (4024 individuals) or restricted (RF; 3,840). They belonged to Caldes line, selected since 1985 for average daily gain (ADG) under AD feeding. The kits of the same litter were distributed into both treatments ensuring that animals in a cage had a similar initial weight. For each disease, records obtained under different feeding regimes were treated as different traits. Bayesian linear-threshold animal models were used for the analyses including all the information of the selection criteria (ADG; 134,419 animals). The model for ADG included the effects of year-season, parity order, number of born alive, and the random additive genetic and common litter effects. For every disease the model included parity order, number of born alive, batch, and size of the animal at weaning as fixed effects, and the additive genetic, common litter and cage random effects. Animals under RF had lower MO and fewer incidences of EN and RS than under AD. The estimated heritabilities on the liability scale were converted to the observed scale. Estimates were between 0.02 and 0.08, and for all the diseases and they were equal under both treatments. The genetic correlations between ADG and all of the diseases under both feeding regimens were moderate to low, ranging from -0.18 (0.08) to -0.49(0.13). The genetic correlations between the two feeding regimens were 0.72(0.09) for MO, 0.55 (0.13) for RS, 0.24 (0.2) for PBC and 0.03 (0.13) for PBC, which indicates an important magnitude of the G×E interactions for those traits.

Using of genomic data to assess genetic differentiation in cervids

R. Kasarda, N. Moravčíková, A. Trakovická, O. Kadlečík and R. Kirchner
Slovak University of Agriculture in Nitra, Department of Animal Genetics and Breeding Biology, Tr. a. Hlinku 2, 94976 Nitra, Slovak Republic; nina.moravcikova@uniag.sk

The aim of this study was to assess the suitability of bovine genotyping array for cross species applying and analysis of genetic differentiation between species from family *Cervidae*. The SNP data were obtained using Illumina BovineSNP50 BeadChip in total from 67 individuals of 8 cervid species. The call rate was 60.91%, which was higher than expected. Within 12,802 successfully genotyped autosomal SNPs in at least 90% of individuals 9.83% of loci were polymorphic. The polymorphic SNPs distribution across and within cervid species was not uniform and autosomes consisted of significantly (P<0.05) different number of loci. Overall the average value of MAF was 0.19±0.12. Most of the genetic variability was distributed among populations (F_{ST}=96.92). The genetic differences between individuals explained only 3% of variation. The lower level of average observed heterozygosity (0.05±0.01) suggested also the value of F_{IT} (0.98) which indicated excess of homozygotes. The genetic differentiation between cervids was estimated also based on identity by descent and principal component analysis. Both of these methods clearly showed division of cervids into the separate clusters, unique per species. The use of commercially development assay for livestock in analyses of evolutionary related, non-model species provides sufficient information for diversity studies. *In situ*ations when the genome sequences for non-model species are unavailable use of commercially available SNP is one of the successful ways for looking inside the genetic variation of species from family *Cervidae*.

Comparison of microarray-based detected runs of homozygosity with next-generation sequencing result

T. Szmatoła, A. Gurgul, T. Ząbek, I. Jasielczuk and M. Bugno-Poniewierska
National Research Institute of Animal Production, Laboratory of Genomics, Krakowska 1, 32-083, Poland; tomasz.szmatola@izoo.krakow.pl

Runs of homozygosity (ROH) are defined as contiguous homozygous regions of the genome where the two haplotypes inherited from the parents are identical. It has been shown that the length and frequency of ROH may describe the history of the individual's population and may contribute to the level of inbreeding within the population, recent population bottlenecks or signatures of positive selection. Currently the most commonly used method for ROH detection in cattle are genotyping microarrays which allow to genotype about 54 thousand SNPs evenly distributed across the genome. However, it is largely unknown if this genotyping platform is able to precisely identify ROHs, since distance between neighboring SNPs commonly exceeds 50 kb and may not precisely reflect variability present across the genome. To dissipate these doubts, in this study we compared ROH deduced from Illumina Bovine SNP50 genotyping platform with sequencing data obtained from whole-genome sequencing of the Polish Red bull. A 4,008,696 high-quality SNPs were used for this comparisons. We used two most common sliding window-based approaches implemented in Plink software for ROH detection from microarrays: 50 SNP which is more restrictive, allows for 1 heterozygote inside ROH and treats the region as ROH when it has a minimum of 50 SNP in 1 Mb of length; 15 SNP which is less restrictive, allows for 0 heterozygotes inside ROH and requires a minimum of 15 SNP in 1 Mb of length. We found that detection of ROH based on both calculation approaches on microarray data provides reliable results in the case of long ROHs (over 4 Mb) representing recent inbreeding. Shorter ROHs, however, can be significantly over (15 SNP approach)- or underestimated (50 SNP approach) depending on the ROH calculation method. The obtained results will be further used to establish the most reliable array-based ROH calculation approach.

Multibreed association studies at the sequence level: comparison between meta and joint analyses
M. Tessier, M.P. Sanchez, E. Venot, R. Letaief and P. Croiseau
INRA, AgroParisTech, UMR1313 GABI, Domaine de Vilvert, 78350 Jouy en Josas, France;
pascal.croiseau@jouy.inra.fr

With new sequence data availability in cattle, linkage disequilibrium can now be studied at population level: this could help identifying causal mutations shared by several breeds and more particularly, the ones found non significant at the within breed level. Thanks to the 1000 bull genomes project, imputation at the sequence level of 2,199 Normande (NO), 2,462 Montbéliarde (MO), 6,375 Holstein (HO) bulls was performed. These data were used to carry out within breed and multi-breed genome wide association studies (GWAS) on five production traits (milk quantity, fat yield, fat content, protein yield and protein content). Within breed GWAS results were also combined in a meta-analysis. The purpose of this study was to compare multi-breed GWAS and meta-analysis ability to detect QTL shared by the three breeds (a QTL is defined as the SNP with the highest –log(p-value) (with a minimal threshold) in a given window size on the chromosome (different combinations of these two parameters were tested). Compared to within breed GWAS, both multi-breed GWAS and meta-analysis led to a higher number of significant tests in MO and NO (the two breeds with smallest reference populations). When focusing on specific regions, several cases of non significant tests in within breed GWAS but significant tests in both meta-analysis and multi-breed GWAS were found. Moreover, when a QTL was identified in within breed GWAS for the three breeds, both meta-analysis and multi-breed GWAS allow reducing the confidence interval around the QTL. Therefore, at the sequence level, meta-analysis and joint GWAS are promising tools to improve QTL detection power, more particularly when the reference population size is limited. The authors acknowledge the contribution of the 1000 bull genomes consortium.

Development of method for the isolation and sequencing of the mitochondrial genome
I. Jasielczuk, A. Gurgul, M. Bugno-Poniewierska, K. Pawlina, E. Semik and T. Szmatoła
National Research Institute of Animal Production, Laboratory of Genomics, Krakowska 1, 32-083, Poland;
igor.jasielczuk@izoo.krakow.pl

Mitochondrial DNA (mtDNA) encodes a number of genes responsible for the synthesis of enzymes of the respiratory chain and is a necessary element for the functioning of every living cell. MtDNA analysis is used in population genetics, evolutionary and forensic medicine and mutations in mitochondrial genes can cause serious genetic diseases. Existing methods for the analysis of mitochondrial DNA are primarily based on amplification of selected fragments of mtDNA and sequencing with Sanger method, which does not guarantee high performance of analyzes. We have attempted to develop a non-species specific enrichment method for the analysis of mtDNA fraction conducted on the basis of next generation sequencing. Isolation of mtDNA from blood of five cows was conducted using the method proposed by Lindberg'a et al. Isolation procedure is based on the lysis of cell membranes except the nuclear envelope in the presence of a nonionic detergent (Triton X-100) and separating the cell nuclei from mitochondria by differential centrifugation. From the purified mitochondria, DNA was extracted using commercial DNA isolation kit (Sherlock AX). Quality control of isolates showed the presence of high-molecular-weight DNA, congruent with the size of the mtDNA molecule (approx. 16,000 bp). Obtained DNA was used to prepare the libraries for the next generation sequencing. Libraries were sequenced in a single 50 bp read on Illumina system. More than 93% of reads were mapped against the reference genome. In relation to the control genomic DNA we achieved enrichment of mtDNA in the range of 2,651-7,179 times. Coverage analysis showed that mtDNA in a test samples was covered from 140 to 466 times, whereas in the case of control samples mtDNA had a coverage of 0.061. Collected data allowed for the identification of 42 unique sequence variants within the mtDNA, of which 12 can be considered as new.

Estimation of genetic parameters for reproductive traits in alpacas

A. Cruz[1], I. Cervantes[2], A. Burgos[1], R. Morante[1] and J.P. Gutiérrez[2]
[1]Fundo Pacomarca – INCA TOPS S.A., Miguel Forga 348, Arequipa, Spain, [2]Dpto. de Producción Animal, Universidad Complutense de Madrid, 28040 Madrid, Spain; gutgar@vet.ucm.es

One of the main deficiencies affecting animal breeding programs in Peruvian alpacas is the low reproductive performance leading to low number of animals available to select strongly decreasing selection intensity. Leaving aside management issues, some reproductive traits could in turn be improved by artificial selection, but very few information about genetic parameters exist for these traits. The aim of this study was to estimate genetic parameters for six reproductive traits in alpacas both in Suri and Huacaya ecotypes, as well as their genetic relationship with fiber and morphological traits. Dataset belonging to Pacomarca experimental farm collected between 2000 and 2014 were used. Number of records for age at first service (AFS), age at first calving (AFC), intercourse time (IT), pregnancy diagnosis (PD), gestation length (GL), and calving interval (CI) were respectively 1,704, 854, 19,770, 5,874, 4,290 and 934. Pedigree consisted of 7,742 animals. Estimated heritabilities, respectively for HU an SU were 0.19 and 0.09 for AFS, 0.45 and 0.59 for AFC, 0.04 and 0.05 for TI, 0.07 and 0.05 for PD, 0.12 and 0.20 for GL, and 0.14 and 0.09 for CI. Genetic correlations between them ranged from -0.96 to 0.70. No important genetic correlations were found between reproductive traits and fiber or morphological traits, in HU being the highest 0.24 between AFS and DE. However several favorable genetic correlations higher than 0.34 in absolute value were found in SU, being AFS and IT the trait more related to productive traits and AFC, IT and GL that more related to Morphological traits. Reproductive performance would be indirectly selected in SU but some reproductive traits could be included as selection objectives in HU.

Inbreeding in Swiss dairy cattle populations

A. Burren[1], A. Kreis[1], J. Kneubuehler[2], E. Barras[3], A. Bigler[2], U. Schnyder[4], M. Rust[5], U. Witschi[2], H. Joerg[1] and F. Schmitz-Hsu[2]
[1]Bern University of Applied Sciences BFH, School of Agricultural, Forest and Food Sciences HAFL, Länggasse 85, 3052 Zollikofen, Switzerland, [2]Swissgenetics, Meielenfeldweg 12, 3052 Zollikofen, Switzerland, [3]Holstein Association of Switzerland, Route de Grangeneuve 27, 1725 Posieux, Switzerland, [4]Swissherdbook, Schützenstrasse 10, 3052 Zollikofen, Switzerland, [5]Braunvieh Schweiz, Chamerstrasse 56, 6300 Zug, Switzerland; alexander.burren@bfh.ch

In recent years, many studies have shown that increasing inbreeding results in declining performance, fertility and health of dairy cattle, resulting in inbreeding depression. In this context, the inbreeding development of seven dairy cattle populations was studied: Brown Swiss (BS), Original Braunvieh (OB), Holstein (HO), Red Holstein (RH), Red-Factor-Carrier (RFC), Swiss Fleckvieh (SF=RH × SI) and Simmental (SI). All herd book animals with birth years from 1990 to 2012 and their ancestors were included. The mean pedigree completeness, considering 1 to 6 generations, varied between 83% to 99% (OB=99%, SI=95%, BS=92%, SF=91%, RH=89%, RF=85%, HO=83%). In the HO, RH and RFC animals, the mean pedigree completeness is lower than for the other populations, which is why the inbreeding coefficients are less meaningful. This finding is not surprising as sires from other No-Swiss populations with low pedigree information is still common today. For all seven breeds average inbreeding coefficients showed a varying increasing trend, and were found at 4.7% (BS), 4.2% (HO), 3.7% (SI), 3.5% (OB), 3.1% (RH), 2.6% (RFC) and 2.4% (SF) for the year 2012. The increase in the mean inbreeding coefficient (ΔF) was found in the period 1990-2012 in the following populations: SI (2.8%), HO (2.7%), BS (2.5%), OB (2.3%), RH (1.8%), RFC (1.5%) and SF (1.3%). For all seven breeds no specific management measures are required but a regular monitoring of inbreeding coefficients is proposed.

Genetic analysis of hyperthelia in Brown Swiss cattle
A. Butty[1], J. Moll[2], S. Neuenschwander[1], C. Baes[1,2] and F. Seefried[2]
[1]ETH, Tannenstrasse 1, 8006 Zurich, Switzerland, [2]Qualitas AG, Chamerstrasse 56, 6300 Zug, Switzerland; adrien.butty@gmail.com

Hyperthelia (supernumerary teats, SNT) is an abnormal morphological phenotype describing phenotypes other than the regular number of 4 functional teats on a cow's udder. SNT vary in developmental stage, from rudimentary epithelial abnormalities to well-formed functional teats with additional mammary gland. Therefore SNT can cause mastitis, negatively affect milking ability and have economic relevance. Previous studies in different cattle populations have shown that SNT are a heritable, poly-/oligogenic congenital abnormality. Phenotypic SNT data has been routinely collected in Switzerland since 1995 during conformation scoring of cows. Animals without SNT are coded with score 9, whereas the presence of SNT, depending on severity, is scored from 1 to 8. Three main categories of records were observed in the data: clear udder, non-functional SNT and complete SNT with prevalence of 80.1%, 14.1% and 3%, respectively. The scoring system permits different trait definitions of SNT. The objective of this study was to analyse the genetic architecture of hyperthelia in Swiss Brown Swiss cattle using dense genomic information. Firstly, routinely collected phenotypes were used together with pedigree information for estimation of genetic variance. A univariate genetic model was evaluated including expert by year and farm by year as random effects and the dam's lactation by coding first lactation vs combining all following lactations together as fixed effect. Similar to previous studies on other populations, heritability estimates from 0.18 to 0.33 with a standard error below 0.01 were observed. In a second step, genome-wide association studies were applied with dense genotype information from approximately 1,700 Brown Swiss dairy cows on low input and organic farms in Switzerland. SNP data was generated in the context of the project LowInputBreed (EU Grant agreement no. 222623) and combined with data from a national genotyping project.

Gene flows and dog breeding practices in France, Sweden and UK
S. Wang[1], G. Leroy[2], S. Malm[3], S. Lewis[4], E. Strandberg[1] and W. Fikse[1]
[1]Swedish University of Agricultural Sciences, Animal Breeding and Genetics, Uppsala Sweden, [2]AgroParisTech, Paris, France, [3]Swedish Kennel Club, Spånga, Sweden, Unknown, Sweden, [4]The Kennel Club, London, United Kingdom; shizhi.wang@outlook.com

Dog breeding is characterized by various breeding practices across countries, as well as by substantial exchange of breeding animals with potential impact on genetic variability and health of national populations. Pooling data from different countries may improve the efficiency of genetic evaluation of complex disorders, such as hip dysplasia, and result in a larger pool of animals to select from, which can be exploited to increase gain or control inbreeding. The aim of this study was to explore the gene flows and mating practices for three dog breeds and three countries: France, Sweden and UK. Société Centrale Canine (France), Swedish Kennel Club, and the Kennel Club (UK) provided pedigree files for the three dog breeds: the Bernese Mountain dog (pedigree file including 59,446 individuals in France, 22,569 in Sweden and in 25,164 UK); Bullmastiff (2,532 individuals in France, 4,364 in Sweden and in 48,119 UK); and English Setter (181,644 individuals in France, 12,806 in Sweden and in 33,799 UK). For each breed, the pedigree databases from all three countries are merged into one combined pedigree database, by identifying and replacing duplicated dog IDs. We use the combined pedigree databases to quantify the gene flows and investigate inbreeding evolution and practice (mating between close relatives) over time and countries.

Genetic markers with non-additive effects on milk and fertility in Holstein and Jersey cows

H. Aliloo[1,2,3], J.E. Pryce[1,2,3], O. González-Recio[1,3], B.G. Cocks[1,2,3] and B.J. Hayes[1,2,3]
[1]Dairy Futures Cooperative Research Centre (CRC), AgriBio, 5 Ring Road, Bundoora, VIC. 3083, Australia,
[2]School of Applied Systems Biology, La Trobe University, Bundoora, VIC. 3083, Australia, [3]Biosciences
Research Division, Department of Economic Development, Jobs, Transport and Resources, AgriBio, 5 Ring
Road, Bundoora, VIC. 3083, Australia; hassan.aliloo@ecodev.vic.gov.au

It has been suggested that traits associated with fitness, such as fertility, may have proportionately more genetic variation arising from non-additive effects than traits with higher heritability, such as milk yield. Here, we performed a large genome scan with 408,255 single nucleotide polymorphism (SNP) markers to identify chromosomal regions associated with dominance and epistatic (pairwise additive × additive) variability in milk yield fertility (measured by calving interval), using 7,055 genotyped and phenotyped Holstein cows. The results were subsequently validated in an independent set of 3,795 Jersey cows. We identified genome regions with validated dominance effects for milk yield on *Bos taurus* autosomes (BTA) 2, 3, 5, 26 and 27 whereas SNPs with validated dominance effects for fertility were found on BTA 1, 2, 3, 7, 23, 25 and 28. A number of significant epistatic effects for milk yield on BTA 14 validated across breeds. However, these were likely to be associated with the mutation in the diacylglycerol O-acyltransferase 1 (DGAT1) gene, given that the associations were no longer significant when the full additive effect of the DGAT1 mutation was included in the epistatic model. The results of our study suggest that individual non-additive effects make a small contribution to the genetic variation of milk yield and fertility.

Genetic parameters for reproductive traits of Africanized drones *Apis mellifera* L. (*Apidae*)

F.M.C. Maia[1], E.N. Martins[1], E. Migliorini[1], T.E. Stivanin[1], M. Souza[1], M.C. Rodrigues[1], D.A.L. Lourenco[2],
E.L.R. Silva[1], M. Potrich[1], F.T. Diniz[1], R.L. Fantin[1], L. Bonetti[1], N.V. Santina[1], G.D. Maia[1] and F. Raulino[1]
[1]University Technological Federal of Paraná, Animal Science, E. Boa Esperança, km 4, C. São Cristóvão,
85660000 Dois Vizinhos, PR, Brazil, [2]University of Georgia, Animal Science, 425 River Rd., Athens, GA
30602, USA; dandilino@gmail.com

Genetic parameters were estimated for drone emergence weight (EW), weight (WM), area (AM), and volume (VM) of the mucous gland, and weight (WS), area (AS), and volume (VS) of the seminal vesicule. The drones were emerged from nine colonies under laboratory conditions; the emergence process was monitored over 24 hours. A total of 1,108 drones had their weight recorded after birth and 463 were dissected for measuring the traits related to the reproductive system. Also, the mucous gland and seminal vesicule were pictured in real scale. The seminal vesicule volume was calculated as formula volume cylinder; the area was obtained through the Image Tool Software (v.3.0). Heritabilities and genetic correlations were estimated under the Bayesian approach using single-trait and two-trait models with EW. In single-trait models, heritabilities were 0.70 for EW, 0.68 for WM, 0.69 for AM, 0.63 for VM, 0.67 for WS, 0.68 for AS, and 0.70 for VS. The genetic correlation between EW and VS was 0.44, which indicates that selection for havier drones at birth would also select for drones with higher capacity of sperm storage. The genetic correlation between EW and the other reproductive traits ranged from 0 to 0.26. The genetic parameters obtained in this study indicate that there is a potential for improving the reproductive traits in drones. This is especially important when instrumental insemination is applied for honey bee breeding. Reproductive traits in drones are difficult to measure and further studies on genetic correlations with easy-to-measure traits may be helpfull for improving the reproductive potential of drones. Financial Support: CNPq (483146/2013-7), Fundação Araucária.

morphometric analysis of the two different honeybee (*Apis mellifera* L.) populations
F. Gurel
Akdeniz University, Animal Science, Akdeniz University Agricultural Faculty Antalya, 07058, Turkey;
fgurel@akdeniz.edu.tr

A total of thirteen morphological characters were determined for two geographic Turkish honeybee populations Caucasian (*Apis mellifera caucasica*), and Aegean ecotype of Anatolian honeybee (*Apis mellifera anatoliaca*) used intensively for commercial queen rearing. The multivariate discriminant analysis was performed on 350 worker bee samples from 14 colonies representing two populations. The multivariate discriminant analysis correctly identified 76.6% of the samples to their actual populations. The results of the discriminant analysis showed that the length of cubital vein b and forewing width were the most significant morphological variables to discriminate these populations.

Genetic parameters for behaviour and type traits in Charolais cows
A. Vallée, J.A.M. Van Arendonk and H. Bovenhuis
Wageningen University, Animal Breeding and Genomics Centre, P.O. Box 338, 6700 AH Wageningen, the
Netherlands; amelie.vallee@wur.nl

In the last decades, breeding strategies in beef cattle mainly included production and reproduction traits. Nowadays, there is a growing interest to include behaviour because it is associated with human safety and workability, and type traits because they might be indicator for longevity. The objective was to estimate heritabilities and genetic correlations among behaviour and type traits in Charolais. For behaviour traits we used a simple on-farm recording system to enable large scale of phenotypes collected by farmers. Aggressiveness at parturition and during gestation period were scored on a scale from 1 (docile) to 7 (aggressive). Maternal care was scored from 1 (rejecting) to 7 (attentiveness). Type traits were scored by ten trained classifiers. Udder traits (n=3), teat traits (3), and feet and leg traits (5) were scored from 1 to 7. Locomotion was scored from 1 to 5. Data was available on 6,406 cows, in parity 1 to 6 and located on 387 herds. Results showed that differences between herds explained up to 23% of the total phenotypic variance in behaviour traits. This might be due to inconsistent scoring by farmers. Aggressiveness at parturition had higher heritability (0.18) and genetic coefficient of variation (CV_g=17%) than aggressiveness during gestation (h^2=0.06 and CV_g=8%) and matenal care (h^2=0.02 and CV_g=2%). Heritabilities for udder traits (0.14 to 0.21) and teat traits (0.17 to 0.35) were higher than for feet and leg traits (0.03 to 0.19). Genetic coefficients of variation for udder traits and teat traits were also higher (up to 18%) than for feet and leg traits (up to 11%). Strong genetic correlations were found between behaviour traits (0.71 to 0.98), indicating difficulty to improve simultaneously maternal care and reduce aggressiveness. Implementing selection against aggressiveness at parturition and for udder and teat traits is possible using a simple on-farm recording system. Opportunity to select for maternal care and feet and leg traits using this recording system is limited.

The T3811>G3811 mutation in the Blonde d'Aquitaine myostatin gene: effects on beef traits

G. Renand[1], A. Vinet[1], G. Caste[2], B. Picard[3], I. Cassar-Malek[3], M. Bonnet[3], C. Bouyer[4], V. Blanquet[4] and A. Oulmouden[4]
[1]INRA-AgroParisTech, UMR 1313, 78350 Jouy en Josas, France, [2]INRA, UE 0333, 81400 Blaye les Mines, France, [3]INRA-VetAgroSup, UMR 1213, 63122 Saint-Genès-Champanelle, France, [4]INRA-Université de Limoges, UMR 1061, 87060 Limoges, France; gilles.renand@jouy.inra.fr

The mutation (T3811>G3811) recently discovered in the myostatin gene of Blond d'Aquitaine cattle is suspected to result in a reduction in functional Myostatin and consequently to explain the high muscularity in this breed. In order to evaluate the actual effect of this mutation on beef traits, experimental crossbred calves were procreated and tested. Three Blond d'Aquitaine sires were mated to Holstein cows to obtain 3 F1 sires and 9 F1 dams. F2 embryos were implanted into beef heifers. Backcross calves were also obtained mating of F1 sires to Holstein cows. Fifty six calves (43 F2, 13 backcross) were intensively fattened and slaughtered at 169±10 days of age. Birth weight, daily gain and muscularity score were recorded on live animals. Slaughter yield, carcass length, carcass muscularity and fatness scores were recorded at slaughter. The following muscle characteristics were recorded on both Longissimus thoracis (LT) and Triceps brachii (TB) muscles: Isocitrate dehydrogenase (ICDH: oxidative) and Lactate dehydrogenase (LDH: glycolytic) activity, proportions of myosin heavy chain isoforms (MyHC) I, IIa, IIx. ICDH and LDH activities were also recorded in kidney fat. The genotype at the mutation was determined: 18 T/T, 30 G/T, 8 G/G. Highly significant substitution effects (above +1.3 phenotypic s.d.) were shown on carcass yield and muscularity scores. Birth weight was positively affected by the mutation (+0.8 s.d.) but not growth rate, while carcass length and fatness scores were negatively affected (-0.6 to -0.8 s.d.). The TB muscle of calves with a mutated copy had lower oxidative activity and higher proportion of MyHC IIx (around 0.9 s.d.). The mutation had no effect on the LT muscle characteristics neither on the enzymatic activities of kidney fat.

Genetic growth profiles for carcass traits in steers

T.M. Englishby[1,2], R.D. Evans[3], G. Banos[2], M.P. Coffey[2], K.L. Moore[2] and D.P. Berry[1]
[1]Teagasc, Animal & Grassland Research and Innovation Centre, Moorepark, Fermoy, Cork, Ireland, [2]Scotland Rural College, Roslin Institute Building, Easter Bush, Midlothian EH25 9RG, United Kingdom, [3]Irish Cattle Breeding Federation, Bandon, Cork, Ireland; tanya.englishby@sruc.ac.uk

Livestock mature at different rates depending on their genetic merit; therefore, the optimal age at slaughter for progeny of certain sires may differ. The objective of this study was to generate sire genetic profiles for carcass weight, carcass conformation and carcass fat, in steers of multiple beef and dairy breeds, including crossbreeds. Slaughter records from 124,641 steers, aged between 360 and 1,200 days, from 7,949 sires were available from the Irish national database. Variance components for each trait were generated using sire random regression models that included quadratic polynomials for fixed and random effects; heterogeneous residual variances were assumed across ages. Heritability for carcass weight, conformation and fat score across age at slaughter varied from 0.44 (801 days) to 0.53 (650 days), from 0.40 (650 days) to 0.45 (759 days), and from 0.24 (1,183 days) to 0.31 (673 days), respectively. Genetic correlations within each trait weakened as the interval between ages compared lengthened, suggesting a different genetic background for different ages. Eigenvalues and eigenfunctions of the additive genetic covariance matrix substantiated this but also revealed genetic variation among animals in the shape of the growth profiles for carcass traits. Genetic correlations between carcass weight and conformation were positive, reaching a peak correlation of 0.56 at 824 days. Genetic correlations between carcass conformation and fat score were weak (-0.07 at 650 days to 0.06 at 1,010 days), as were the correlations between carcass weight and fat score, ranging from -0.25 (770 days) to 0.17 (1,010 days). Genetic parameters presented indicate genetic variability in growth rate at different ages exists, which can be exploited through breeding programs and used in decision support tools.

Possibility to improve genetic evaluation for carcass traits using data from dairy cows

A.M. Closter[1], E. Norberg[2], J. Pedersen[1] and M. Kargo[1,2]
[1]SEGES P/S, Agro Food Park 15, 8200 Aarhus N, Denmark, [2]Aarhus University, Dept. Molecular Biology and Genetics, Center for Quantitative Genetics and Genomics, P.O. Box 50, 8830 Tjele, Denmark; acl@seges.dk

Results from a Danish study suggests that it could be beneficial to include slaughter data from dairy cows in the present genetic evaluation for growth, which now only includes data from bull calves. The study was based on records from slaughtered cows and bull calves. The first aim of this project was to estimate heritabilities for carcass traits in slaughtered dairy cows. Secondly, the study aimed to test whether genetic and phenotypic variances differed between slaughtered cows and bulls calves from different breeds. Data on 576,981 Holstein (HF) and 56,550 Danish Red (RDM) calves and 394,484 HF calves, 34,222 RDM and 39,927 Jersey cows slaughtered between 2004 and 2014 was extracted from the Danish Cattle Database. Heritabilities for carcass weight of cows was moderate (0.34-0.40) and higher than heritabilities for Holstein and RDM calves. Genetic correlations between carcass weight for cows and calves were medium (0.51-0.52), suggesting that carcass weight in cows and calves to some extent are under different genetic control. Slaughter weight in calves depends on the growth rate from birth to 10-12 months of age, whereas weight in cows depends largely on genetic disposition for mature stature and to a smaller extends on the growth rate. Heritabilities for carcass conformation were moderate (0.23-0.25) for slaughtered cows and somewhat higher for calves (0.35-0.39). Lower heritabilities found for cows could partly be due to gender differences and that parturition of energy resources in a cow largely benefits milk production instead of growth, as calves mainly use energy for growth of the muscles. Genetic correlations between slaughtered cows and calves were medium to high for carcass conformation (0.55-0.62) as for carcass weight. Use of data from both gender can be used to maximize the total improvement for both cows and calves traits.

Development of the Meat Standards Australia (MSA) index – improving carcass feedback for producers

P. Mcgilchrist[1], R.J. Polkinghorne[2], A.J. Ball[2] and J.M. Thompson[2]
[1]Murdoch University, School of Veterinary & Life Sciences, South St, Murdoch, 6150 WA, Australia, [2]Meat & Livestock Australia, Trevanna Rd, University of New England, Armidale, 2351 NSW, Australia; p.mcgilchrist@murdoch.edu.au

The objective of this study was to produce a single number that depicts the eating quality of a beef carcass. The Meat Standards Australia (MSA) grading model accurately predicts the eating quality of 39 individual cuts in a beef carcass. Each cut receives a meat quality score (MQ4) between 0 and 100, based on a prediction and combination of 4 traits; tenderness, juiciness, flavour and overall liking. The MSA model is complex due to the non-linear impact of different model inputs between cuts and the diversity of the Australian cattle demographics and production systems. The potential eating quality of a carcass is of interest to producers as they can use feedback to evaluate the improvement in eating quality due to various factors like: investment in new genetics; different suppliers of feeder cattle; difference between seasons or years and other production factors. However to date, carcass feedback to producers is in the form of individual measurements for carcass traits like carcass weight, ossification, MSA marbling score, rib fat, hump height or Bos indicus content, ultimate pH, gender, hormonal growth promotant, milk fed vealer and saleyard status. The impacts of all these factors on the eating quality of 39 different muscles in the body are not linear due to the complexity of muscle biology. The non-linearity makes it impossible for producers to assess individual carcass traits to evaluate the eating quality of a carcass. Hence the MSA index was created, which is a single number for each carcass and an average of the MQ4 scores for the 39 cuts in the carcass with a fixed weighting for their proportion of the total cut weight. The MSA index generally ranges between 40 and 70 and can be utilised to analyse eating quality over time within a production system, across production systems and to benchmark producers.

For good beef, gender is more important than breed

S.P.F. Bonny[1], D.W. Pethick[1], I. Legrand[2], J. Wierzbicki[3], P. Allen[4], L.J. Farmer[5], R.J. Polkinghorne[6], J.F. Hocquette[7] and G.E. Gardner[1]
[1]Murdoch University, School of Veterinary and Life Sciences, 60 South St, Murdoch 6150, Australia, [2]Institut de l'Elevage, Service Qualite des Viandes, Limoges Cedex 2, 87060, France, [3]Polish beef association, Ul. Smulikowskiego 4, 00-389 Warszawa, Poland, [4]Teagasac, Food and Research Centre, Ashtown, Dublin 15, Ireland, [5]Agri-Food and Biosciences Institute, Newforge Ln., Belfast BT9 5PX, United Kingdom, [6]Polkinghornes, 431 Timor Rd, Murrurundi 2338, Australia, [7]INRA, VetAgro Sup, 1213 Theix, Saint Genes Champanelle 63122, France; sbonny@murdoch.edu.au

Variable eating quality is a major factor in declining beef consumption. The relative accuracy of the Meat Standards Australia (MSA) prediction system, which uses carcass traits to guarantee consumer eating quality, was evaluated within Europe. The MSA model was developed on beef steers and females, so it was expected to predict bulls and dairy breeds less accurately. In total 6,852 muscle samples from 482 carcasses from France, Poland, Ireland and Northern Ireland were evaluated by untrained consumers, according to MSA protocols. The scores were combined to form a Meat Quality score (MQ4). Carcasses were MSA graded during processing. There were four breed categories; British beef breeds, Continental beef breeds, dairy breeds and crosses. The difference between the actual and the MSA predicted MQ4 scores were analysed using a linear mixed effects model including the effects; carcass hang method, cook type, muscle type, sex, country, breed category and post mortem ageing period, with animal identification, grader and kill group as random terms. As expected the predicted MQ4 of females and steers was more accurate than for bulls. Accuracy varied between breed categories, but only for a minority of muscles and the effects were often contradictory. Therefore in a European eating quality prediction system similar to MSA a separate adjustment for bulls is required. Any differences in breed category are already adequately explained by factors currently present in the MSA model.

Environmental impact of beef – the food chain from farm to fork

L. Mogensen[1], T.L.T. Nguyen[1], N.T. Madsen[2], O. Pontoppidan[2], T. Preda[1] and J.E. Hermansen[1]
[1]Aarhus University, Department of Agroecology, Foulum, 8830 Tjele, Denmark, [2]DMRI, Danish Technological Institute, Taastrup, 2630, Denmark; lisbeth.mogensen@agro.au.dk

The environmental impact of different types of beef was analyzed in the whole chain from farm to fork focusing on global warming potential (GWP), land use, biodiversity, fossil energy and eutrophication. The functional unit was 'kg beef products used for human nutrition', i.e. the sum of meat products and edible by-products that are used in human nutrition. The environmental impact from the slaughtering process was based on data on energy use and output of meat, hides and by-products, where the utilization of by-products was taken into account. The largest environmental impact of meat was related to the production at the farm, whereas, the impact related to slaughtering was basically offset by the recovery from the use of by-products from the slaughtering process. The degree to which the living cattle are translated into edible products has huge impact. The share of edible products from dairy breeds and Highland cattle was between 45 and 50% of live weight (LW) depending on age and sex, whereas the share of edible products from Limousine was between 53 and 57%. Furthermore, it was estimated which by-products that at the moment are used for something else than human consumption, but has a potential for use for human consumption at a global marked in the future. Thereby, there was a potential for a 12-15% point higher utilization (e.g. for the Holstein calf the present utilization is 50% of LW, which could potentially be increased to 63% of LW). This would reduce the GWP of the meat by 17-23%. In conclusion, the major environmental burden is related to the farm level stage and innovations to reduce impact should be given high attention. The slaughtering process itself is very energy and resource efficient. The main innovation to reduce environmental impact of beef will be to ensure a higher utilization of the live animal into new edible products not conventionally produced.

Beef production, supply and quality from farm to fork in Europe
K. De Roest
Research Center for Animal Production (CRPA), Viale Timavo 43/2, 42121 Reggio Emila, Italy;
k.de.roest@crpa.it

Europe has a wide variety of beef farming systems, based either on the specialized beef cattle breeds or on the dairy herd. The differences concern breeds, diets, feeding and housing systems which interest both the cow calf as well as the beef finishing farms. Beef meat can be derived from bulls, heifers, bullocks and cull cows which are slaughtered at different ages and at different slaughter weights. One objective of this paper is first of all to provide a statistical description of the different beef production systems in the main Member States of the EU giving details about their development over the last ten years. After a macro analysis of the characteristics of beef production in Europe, detailed information will be given about the latest technical and economic performance of representative beef farms using the Agribenchmark database of typical farms. In the last decades the consumption of beef in Europe declined significantly either because of price competition from pig and poultry meat or because of inadequate levels of beef quality offered on the market. The huge differences in beef production systems have their repercussions on the quality of beef at consumer level. The quality of beef varies considerably from country to country and is only partially reflected by the SEUROP classification, as the eating quality of beef is poorly correlated to this EU carcass classification. Nevertheless of interest are the significant differences reflected by the results of the carcass classification. The second objective of this paper is to present the tendencies in the classification of cattle carcasses and the final output prices for the main EU Member States over the last ten years. These data will be compared with the production costs of the representative beef production systems in these countries which will enable an analysis of the capacity to remunerate labour and capital invested on these farms and of their future economic viability.

Alternative finishing strategies for Holstein-Friesian bulls slaughtered at 19 months of age
B. Murphy[1,2], A.K. Kelly[1] and R. Prendiville[2]
[1]University College Dublin, Belfield, Dublin 4, Ireland, [2]Teagasc, Animal Grassland Research and Innovation Centre, Grange, Dunsany, Co. Meath, Ireland; brian.murphy@teagasc.ie

The aim of this study was to compare the performance of Holstein-Friesian bulls on alternative finishing strategies slaughtered at 19 months of age. Data were available from 58 bulls in a 2 finishing strategy (indoors on concentrates *ad libitum* (AL) or at pasture offered 5 kg concentrate dry matter (DM) per head daily (PC)) × 2 levels of rumen protected fat (RPF) in the finishing diet (control (C) and 5% RPF (T)) factorial arrangement of treatments. Bulls were at pasture for 58 days during the second season. The AL group was adapted to concentrates *ad libitum* over a 21 day period. Straw was offered *ad libitum*. Concentrates were weighed-back twice weekly. The PC group was adapted to 5 kg DM of concentrate per head daily over a 10 day period. Bulls were finished for 102 days. Live-weights were recorded fortnightly. Data were analysed using Proc MIXED of SAS. Live-weight at slaughter tended ($P=0.0745$) to be greater for AL compared to PC (580 and 551 kg, respectively). Carcass weight (CW) and conformation score (CS) were similar; but kill-out proportion (KOP) was greater ($P<0.001$) for PC compared to AL (534 and 516 g/kg, respectively). Fat score (FS), average daily gain during finishing and concentrate DM intake (DMI) were greater ($P<0.001$) for AL compared to PC (6.43 and 5.11, 2.09 and 1.42 kg/d and 1,004 and 453 kg, respectively). Concentrate DMI was lower ($P<0.001$) for T compared to C (716 and 741 kg, respectively). Live-weight at slaughter, CW, CS and FS were similar but KOP ($P<0.01$) was greater for T compared to C (529 and 521 g/kg, respectively). In conclusion, finishing dairy bulls at pasture with limited concentrate input, compared to indoors on concentrates *ad libitum*, resulted in similar live-weights, CW and CS, greater KOP with lower FS. Results also suggest that the system that utilises pasture during finishing is more profitable. Aside from concentrate DMI and KOP the inclusion of RPF had no effect on animal performance.

Comparisons of performances between Holstein and Brown Swiss cattle in a feedlot beef system

Y. Bozkurt, C.G. Tuzun and C. Dogan
Suleyman Demirel University, Department of Animal Science, Faculty of Agriculture, 32260, Isparta, Turkey; yalcinbozkurt@sdu.edu.tr

In this study, the performances of Holstein and Brown Swiss male cattle kept during 12 months in an intensive beef system under the Mediterranean climatic conditions were compared in order to make recommendations for the beef producers in this region. The experiment was conducted in 2012 in Isparta province located in the West Mediterranean part of Turkey. For this purpose, 20 Holstein and 20 Brown Swiss male cattle with an average age of 6 months were assigned to two feedlot paddocks evenly and fed with a similar diet for 12 months. The average initial weights of Holstein and Brown Swiss animals were 158 and 132 kg respectively. The General Linear Model (GLM) procedure was used for the statistical analysis of the data with initial weight and age taken as covariates to eliminate the weight differences at the beginning of the experiment. It was found that at the end of the experiment, final body weights were not statistically significant ($P>0.05$). Holstein and Brown Swiss animals reached average final weights of 502 and 493 kg, respectively. Furthermore, there were no significant ($P>0.05$) differences in mean total gains (344 vs 361 kg) and average daily live weight gains (0.985 vs 1.028 kg). These results indicate that under the Mediterranean climate conditions both animal breeds performed similarly, and there was no superiority of any breed over each other although Brown Swiss cattle tended to show better performances. Therefore, both breeds can be recommended to be reared during 12 months in a feedlot beef system by the beef producers in this region of the Mediterranean.

Meat quality of Dexter cattle kept on alpine pastures compared to Charolais crossbred calves

T. Zehnder[1,2], I.D.M. Gangnat[2], M. Kreuzer[2], M. Schneider[1] and J. Bérard[2]
[1]Agroscope, Institute for Sustainability Sciences, Forage Production/Grassland Systems, Reckenholzstrasse 191, 8046 Zürich, Switzerland, [2]ETHZ, Institute of Agricultural Sciences, Animal Nutrition, Universitätstrasse 2, 8092 Zurich, Switzerland; tobias.zehnder@usys.ethz.ch

The Dexter breed is very small-framed and versatile. It originates from Ireland. Due to its size and robustness, it is particularly well suited for extensive production systems. Its meat quality is considered high, but scientific evidence for this is still scarce. The small size of the cuts, compared to the size of veal cuts, could make it an alternative to veal provided meat quality is competitive. In this experiment, Dexter cattle (D; n=8; 68±5 weeks of age and 116±16 kg carcass weight) were compared to two groups of Charolais crossbred calves (31±2 weeks of age). Charolais had either the same carcass conformation/fat cover score (1=very lean, 5=obese) at slaughter as D but heavier carcass weight (S; n=8; T+/3 on the CH-TAX classification grid; 191±14 kg carcass weight) or the same carcass weight as the D but a lower carcass conformation score (W; n=8; T-/1; 121±13 kg carcass weight). All animals had been grazing on adjacent Swiss alpine pastures (2,000 m a.s.l.) for 11 weeks before slaughter. S calves were additionally suckling from their dams whereas W calves had been weaned. Meat quality analyses were carried out on the longissimus thoracis muscle after 21 days of ageing. Water-holding capacity and tenderness of the meat from D was similar to that of the meat from S and higher than that of W (Tukey's Test, $P<0.05$). In conclusion, the meat quality found in Dexter cattle was very similar to that of younger and heavier beef calves having the same carcass conformation despite the large differences in age at slaughter. In contrast, beef calves with the same carcass weight expressed an inferior carcass and meat quality. The present experiment confirmed that Dexter cattle can produce meat quality as high as the one which can be obtained with common meat breeds slaughtered at a very young age.

Effect of commercial line and housing condition on carcass, meat and sensory quality of beef steers

M. Vitale[1], E. Mainau[1], M. Gil[1], J. Pallisera[1], P. Rodriguez[1], A. Dalmau[1], G. De Planell[2], M.A. Oliver[1] and A. Velarde[1]
[1]*IRTA, Monells, 17121, Spain, [2]Grup Viñas, Vic, 08500, Spain; mauro.vitale@irta.cat*

Housing conditions during the finishing period in beef steers can affect both animals' behaviour and welfare and, as consequence, the final quality of meat. The objective of this trial was to evaluate the effect of the commercial line and the housing conditions on carcass characteristics, meat and sensory quality of beef steers. The two groups of steers used for this trial (n=56) were crossbreeds from two commercial lines and finished in the same farm: (1) pasture steers (PS, n=20), with calves staying on pasture with the mother until 7-8 months of age and then finished with concentrate and (2) indoor steers (IS, n=36), with calves fed with hay and concentrate after weaning. During the finishing period, PS and IS were fed the same diet and divided according to two housing systems: concrete slat (CS, n=10PS, 19IS) and straw (ST, n=10PS, 17IS). The animals were slaughtered at the same age and similar body weight. Instrumental quality of fresh meat was evaluated on the Longissimus lumborum muscle, sampled at 48 h post mortem (pm). Warner-Bratzler shear force was measured at 48 h pm and after 7 days of aging. Sensory analysis was performed by a six member trained panel. For the statistical analysis, differences were considered significant for $P<0.05$. The commercial line affected intramuscular fat content (IMF, %) and instrumental texture of the meat samples. In comparison to IS, PS meat resulted in lower IMF (2.3% and 1.7%) and lower shear force (7.3 and 6.0 kgf/cm^2). The housing system affected meat colour (L*, b*) and instrumental texture, CS animals having higher L* and b* and lower meat shear force than ST ones (WBSF=6.5 and 8.0 kgf/cm^2). Housing conditions affected both hardness and chewiness, according to the trained panel. It is concluded that both commercial line and housing condition have a significant effect on meat quality.

Method for simultaneous identification of bovine and porcine proteins in feed mixtures

M. Natonek-Wiśniewska and P. Krzyścin
National Research Institute of Animal Production, Department of Animal Cytogenetics and Molecular Genetics, 1, Krakowska Street, 32-083 Balice, Poland; malgorzata.natonek@izoo.krakow.pl

Since 2013 the European Commission has eased the law concerning the ban on using processed animal proteins in feeds and in some cases allowed the use of non-ruminant meals. At the same time, the Commission presented detailed requirements concerning the methods of their identification. According to the Commission, the techniques used for this purpose should be based on DNA analysis and be highly accurate. We propose a method for simultaneous identification of bovine and porcine components in feed mixtures. The study was performed using reference samples with known quantitative and qualitative composition. The samples consisted of bovine and porcine meal in a percentage ratio of 0:100, 1:99, 10:90, 50:50, 90:10, 99:1 and 100:0. These samples were used to plot the standard curve and were also treated as samples with unknown concentration. The samples were independently subjected to a three-time duplex qPCR reaction. The study used primers and MGB probes that amplify a COX1 fragment of several dozen bp. Analysis was also made of the commercial samples of feed and pet food (20 samples). The results obtained show that the reaction is repeatable for the whole concentration range. The method is repeatable, with relative standard deviation (RSD) for measurements in separate reactions being ≤8.53% and with relative error (D_{Rel}) of ≤10%. Analysis of the commercial samples showed that the method is suitable for determining their species composition. Using the above standard curve, their percentage content can be identified with lower accuracy ($12.3 \leq D_{Rel} \leq 35.6\%$). The method for simultaneous identification of bovine and porcine protein is fast, sensitive and efficient. To perform quantitative analysis, it is necessary to use the standard curve made from samples whose processing method was as close as possible to the analysed samples.

Using self-feeders for supplementing early-weaned beef calves grazing temperate pastures
V. Beretta, A. Simeone, A. Henderson, R. Iribarne and M.B. Silveira
Facultad de Agronomía-Universidad de la República, Paysandu, 60000, Uruguay; beretta@fagro.edu.uy

An experiment was conducted to evaluate the effect of supplement delivery system (SDS) on post weaning performance of early-weaned female calves grazing temperate pastures (29 ha *Trifolium pratense* and *Cichorium intibus*) during summer (January 3 to March 28, 2014). Twenty-four Hereford calves (64 ± 16 days old; 77 ± 10 kg) were randomly allocated to 6 paddocks to receive 1 of 3 treatments (T) (n=2 paddocks/T): daily supplementation (DS) with 1 kg DM/100 kg liveweight (LW) of a commercial concentrate (18% CP, 80% TDN); self-fed supplementation with same concentrate (SF) or with a concentrate mixture including 17% NaCl for intake regulation at 1% LW (SFS). Supplement in SF and SFS was always available. Pasture was rotative grazed in 7-days strips, with a forage allowance of 8% LW. Calves were weighed every 14 days without fasting. Supplement dry matter intake (SDMI), pasture utilization (PU) and forage intake (FDMI) were estimated weekly. Variables with repeated measures were analysed using the Mixed Procedure, while feed to gain ratio (F:G) was analysed through the GLM procedure (SAS Inst. Inc., Cary, NC). Statistical model included T effect and the initial LW as covariate. Pre-grazing biomass did not differ between T (P>0.05). SDS affected LW gain (DS: 0.631[c]; SFS: 0.859[b]; SF: 1.114[a] kg/d; SE 0.05; P<0.01), SDMI (DS: 0.97[b], SFS: 1.15[b]; SF: 2.70%[a] LW; SE 0.04; P<0.01) and daily water intake (9.5[c], 15.9[b] and 21.0[a] l/d, for DS, SF, SFS, respectively, P<0.05). SF calves showed the lowest PU rate (27.8%, P<0.01) compared to SFS (64.4%) and DS (57.8%) which did not differ (P>0.05). No differences were observed in total DMI between SDS (P>0.05); SF showed the lowest F:G (P<0.05), with no differences between SFS and DS (P>0.05). Results suggest that using self-feeders for supplementing early-weaned calves is a viable alternative. Including salt for supplement intake regulation would probably redound in a more efficient supplement conversion efficiency, as PU is increased.

Correlation between color parameters measured on hot and cold beef carcasses
A. Ivanković and M. Konjačić
Faculty of Agriculture, Department of Animal Science and Technology, Svetošimunska cesta 25, 10000 Zagreb, Croatia; aivankovic@agr.hr

Meat color is one of the important parameter for beef quality. The usual procedure for measuring meat color is performed on cold carcasses after its stabilization. The aim of the study is to determine the color of the meat at two positions on hot carcasses and two positions on the cold carcasses, and to determine their correlations. The study was conducted on 340 beef carcasses. Meat color in hot beef carcasses was measured on m. rectus abdominis and m. gracilis, immediately after processing of carcasses. In cold beef carcasses, meat color (24 h post mortem) was measured on m. gracilis and m. longissimus dorsi 60 min after blooming time (exposing to air). Meat color was measured with a Minolta Chroma Meter CR-410 colorimeter with a 50-mm-diameter measurement area using D65 illumination. The L* value parameter in hot carcasses of m. rectus abdominis (36.62) was not significantly different from the value observed in the cold carcasses of m. gracilis (36.20), but L* value of m. rectus abdominis was significantly lower than the value observed in m. longissimus dorsi (40.93; P<0.01). The L* value observed on hot and cold carcasses on m. gracilis (31.87 vs 36.20) differed significantly (p <0.01). The b* value for meat color in hot carcasses in m. rectus abdominis and m. gracilis (-0.48; 1.88) significantly differed (P<0.01) from the value observed in the cold carcasses in m. gracilisand m. longissimus dorsi (7.80; 8.67). The value for a* parameter in hot carcasses on m. rectus abdominis and m. gracilis (15.80; 16:42) also significantly (P<0.01) differed from the value observed on the cold carcasses in m. longissimus dorsi and m. gracilis (24.87; 23.61). Observed correlations between L*, a* and b* parameters indicated that reliable measurement of the meat color in the hot carcasses has to be done on m. rectus abdominis rather than on m. gracilis. Mentioned meat color parameters, along with pH value can be used as primary indicators for the meat quality.

Effect of dry-aging time on instrumental and sensory quality of cow beef meat
M. Vitale[1], A. Planagumà[2], A. Planagumà[2], M.A. Oliver[1] and N. Panella-Riera[1]
[1]IRTA, Monells, 17121, Spain, [2]Carnisseria Planagumà, Olot, 17800, Spain; mauro.vitale@irta.cat

Aging is crucial for improving beef meat tenderness. Dry aging for more than 20 days can be used to further enhance the flavour of meat, obtaining products with very specific sensory characteristics. The objective of this study was to assess the effect of two dry aging times (30 and 60 days) on instrumental and sensory quality of cow beef meat. Longissimus thoracis (LT) samples from nine Rubia Gallega pure breed cows were used for the trial. The LT samples were divided in two portions, each one assigned to a different dry-aging time: 30 or 60 days. After dry-aging, the LT samples were sliced. Meat instrumental quality was assessed by determining intramuscular fat, colour, expressible juice, cooking loss (CL) and Warner Bratzler shear force. Sensory evaluation was assessed by a six-member trained panel on samples prepared following the indications of an expert chef: pre-heated until 40 °C on a hot plate (set at 200 °C) and then cooked in a convection oven (set at 180 °C) until 55 °C of meat core temperature. The scores given by the panellists were expressed on a 10 cm semi-structured scale. For statistical analysis, differences were considered significant at $P<0.05$. No significant effect of aging time was observed for instrumental quality except for CL, where 30 d samples lost more weight during cooking than 60 d samples (33.6 and 28.4%, respectively). Aging time affected cooked meat lactic-fermented odour (Od) and taste (Ts): as indicated by the sensory panel, 60 d meat presented odour and taste characteristics more similar to a cured product than 30 d meat ($Od_{60d}=6.0$ and $Od_{30d}=4.2$; $Ts_{60d}=6.2$ and $Ts_{30d}=3.7$). Both 30 and 60 d samples presented the same degree of sensory hardness (2.6) and juiciness (7.0 and 6.8, respectively) but they differed in chewiness (3.8 and 2.9, respectively). It is concluded that, in the conditions of this study, dry-aging time of more than 30 days can further modify meat odour and taste, although it does not further improve hardness or juiciness.

Genetic parameters for stillbirth in French beef cattle breeds
R. Lefebvre[1], H. Leclerc[2], F. Phocas[1] and S. Mattalia[2]
[1]INRA, UMR 1313 GABI, Domaine de Vilvert, 78352 Jouy en Josas, France, [2]Institut de l'Elevage, Genetic Department, Domaine de Vilvert, 78352 Jouy en Josas, France; rachel.lefebvre@jouy.inra.fr

Until now in France, selection of beef bulls for stillbirth has been considered only through breeding values for birth weight and calving ease. A new evaluation is under development, based on death of progenies within two days after birth date. Its incidence in the nine breeds evaluated in France ranges from 1.6 to 5.8%. Analyze of a binary trait with low incidence should fit well with a threshold model, as used in routine evaluation of stillbirth in dairy breeds. However it implies to consider contemporary groups (CG) as random in the statistical modeling of environmental effects. Because of the large part of natural service, French beef herds are not as connected as dairy ones and genetic levels are heterogeneous across herds. Therefore, considering CG as random effect may induce some bias in the genetic evaluation. The aim of this study was to compare three models based on data collected from large herds: 1- a threshold model considering CG as fixed effect (reference model), 2- a threshold model considering CG as random effect and 3- a linear model considering CG as fixed effect. In a first step, genetic parameters were estimated for the three models in Limousine (national breed) and Parthenaise (regional breed) populations. Limousine data included 334,285 calf records with 2.5% of stillbirth and Parthenaise data included 32,844 records with 4.1% of stillbirth. Heritabilities on the observed scale ranged from 1 to 3%. In a second step, comparison of breeding values will allow choosing the adequate model for stillbirth, by computing correlations with the breeding values derived from the reference model.

Blood components and immune response of heifers relating to feed efficiency and genomic clusters

E. Crane[1], A. Fontoura[1], R. Ventura[2], J. Munro[1], S. Bourgon[1], K. Swanson[3], A. Fredeen[1], N. Karrow[4] and Y. Montanholi[1,4]
[1]*Dalhousie University, Truro, NS, B2N 5E3, Canada,* [2]*Beef Improvement Opportunities, Elora, ON, N1K 1E5, Canada,* [3]*North Dakota State University, Fargo, ND, 58108-6050, USA,* [4]*University of Guelph, Guelph,ON, N1G 2W1, Canada; emcrane@dal.ca*

Phenotyping for indirect assessments of feed efficiency (residual feed intake; RFI; kg/d), reproduction and health traits are important to spread the genetic improvement for feed efficiency and also to identify novel biomarkers to optimize genomics. Our objectives were to characterize phenotypes of immune response, reproduction and post-absorptive metabolism to identify indicators of complex traits and to identify evidence of genomic associations with the targeted phenotypes. Replacement beef heifers (n=107; body weight: 304±40 kg) were fed a grass-silage based diet and participated in a 112 d performance evaluation including body weight, and body composition ultrasound for RFI calculation. Blood was collected every 28 d over the performance evaluation and also at the occurrence of estrous behavior. Specific antibody response to ovalbumin (IgM) was assessed 10 d after secondary immunization. Heifers were clustered based on genomic similarity (C1=87, C2=14, C3=6). The clusters differed on leukocytes count, IgM response and RFI values. For instance, cluster C3 showed a better feed efficiency than the other two clusters, representing a 3.0 kg/d less feed intake than other two clusters. These genomic-phenotype relationships may serve as basis of further studies on large populations. Heifers were also grouped based on RFI into high-efficiency (HE) and low-efficiency (LE). HE heifers showed lower mean cell hemoglobin (16.1 vs 16.5; P=0.01); and tended to have greater secondary IgM response (0.35 vs 0.23; P<0.10). HE heifers may have decreased demand of oxygen transport and greater immune response. Ongoing data analysis will further investigate sexual and metabolic hormones, as well as, metabolites and respective ties with feed efficiency and genomics.

Changes in CLA and n-3 acids content in milk from cows fed with TMR supplemented with lipid complex

R. Bodkowski, K. Czyz, P. Nowakowski and B. Patkowska-Sokola
Wroclaw University of Environmental and Life Sciences, Institute of Animal Breeding, Chelmonskiego 38c, 51-630 Wroclaw, Poland; robert.bodkowski@up.wroc.pl

An increased interest of consumers and producers in functional food has been noted last years. The study investigated an effect of lipid complex (LC) addition to cows feeding dose on dynamics in the changes of cis-9,trans-11 and trans-10,cis-12 C18:2 (CLA), as well as EPA and DHA acids in milk. The study was conducted on 30 HF cows of red-white variety (2 groups, 15 heads in each). The supplements (Humokarbowit in control group and Humokarbowit with LC in amount of 400 g/head/d in experimental group) were administered individually in TMR dose. LC was based on grapeseed oil with synthesized CLA (alkaline isomerization and urea crystallization) and fish oil enriched in n-3 PUFA (alkaline hydrolysis and urea complexation). The main components of LC were CLA isomers in amounts of about 38% (c9,t11, t10,c12 and c11,t13) as well as EPA and DHA in amounts of about 36.5%. Milk samples were collected at day 1, 7, 14, and 30. In the LC supplemented group, the concentration of CLA c9,t11 increased up to 2.23 g/100 g FA after one week, and was at a similar level until the end of the experiment (2.15 g/100 g FA). The level of c9,t11 in the control group was about 0.62 g/100 g FA for the whole experimental period. In the case of CLA t10,c12, the highest content, i.e. 0.10 g/100 g FA, was noted after 2 weeks of LC application, and after that the level of this isomer underwent a slight decrease towards the end of the experiment, reaching 0.08 g/100 g FA. The content of t10,c12 in the control group was 0.03 g/100 g FA during the whole experiment. The addition of LC also increased n-3 FA in milk fat, in subsequent samplings (week 1, 2, 4), EPA content was 0.21, 0.18 and 0.31 g/100 g FA, and DHA was 0.11, 0.16 and 0.13 g/100 g FA, respectively. In the control group, EPA content was 0.03 g/100 g FA, while DHA 0.01 g/100 g FA or below detection limit for the whole experiment.

Ultra low density genotype panels for breed assignment in Aberdeen Angus cattle
M.M. Judge[1,2] and D.P. Berry[2]
[1]*CIT, Bishopstown, Co. Cork, Ireland,* [2]*Teagasc, Fermoy, Co. Cork, Ireland; donagh.berry@teagasc.ie*

Angus beef is marketed internationally for apparently superior meat quality attributes. Breed composition of carcasses destined for these value-added markets are rarely subjected to quality control measures. Genomic information has previously been used to differentiate between breeds; therefore, the objective of this study was to develop an ultra-low density genotype panel, to quantify the Angus proportion in meat samples. A total of 4,402 (n=440 Angus) purebred animals with high density genotypes were available to determine the genomic characterisation of Angus ancestry. This was undertaken using 13,936 SNPs common to the Illumina high density genotype panel and the commercially available Irish Dairy Beef genotype panel. Panels were developed with between 100 and 1000 SNPs, increasing in increments of 100 SNPs. A combination of the fixation index and Delta scores (while simultaneously selecting SNPs that were relatively evenly dispersed across the genome) were used to select breed-differentiating SNPs and the best 1000 of these SNPs were used to validate the selection process. The SNPs were validated on a population of 1,330 animals with recorded Angus ancestry; the sires of these animals were validated using genomic information. The correlations between the predicted Angus proportions generated by the 1000 SNP chip using 100, 200, 300, 400 or 500 SNPs were 0.966, 0.980, 0.988, 0.990 and 0.992, respectively. A minimum of 300 SNPs are required to accurately determine the breed proportion differentiate in the Angus breed with a standard error of 0.043.

Comparison of carcass quality traits in swiss beef cattle breeds
A. Burren[1], S. Hulst[1], M. Buergisser[2], A. Iten[2], S. Probst[1] and P. Kunz[1]
[1]*Bern University of Applied Sciences BFH, School of Agricultural, Forest and Food Sciences HAFL, Länggasse 85, 3052 Zollikofen, Switzerland,* [2]*Mutterkuh Schweiz, Stapferstrasse 2, 5201 Brugg, Switzerland; stefan.probst@bfh.ch*

The herd book numbers of the Swiss association 'Mutterkuh Schweiz' show that the numbers of beef cattle breeds and crossbred beef cattle have steadily increased in recent years. The question to be studied concerns any relationship between genotype (breed or cross) and carcass quality. In order to answer this question, Mutterkuh Schweiz provided the CH-TAX data for 34,662 animals from 23 different genotypes from the period 2010-2013. The study examined the characteristics of carcass weight, conformation and fat cover. For the analysis, the three variables were corrected with a linear mixed model for sex, birth month, slaughter year and establishment, and standardised to 305 days (mean age at slaughter). Pure Charolais animals achieved the (significantly) highest carcass weight of 248 kg, followed by the crosses Charolais × Brown Swiss (246.3 kg) and Simmental × Charolais (244.5 kg). There was no significant difference for the Limousin × Charolais cross (241.2 kg), and this in turn was not significantly different from the crosses Original Brown Swiss × Limousin (236.2 kg) or Simmental × Limousin (238.5 kg) with respect to the mean standardised carcass weight. The (significantly) highest mean standardised carcass conformation scores were found for pure Limousin (5.82), Limousin × Charolais (5.79), pure Charolais (5.75), pure Aubrac (5.71) and Limousin × Aubrac (5.64) animals. Smaller differences were found for fat cover. Seventeen breeds / crosses had a mean standardised fat cover of 3.00 to 3.35. The following breeds had fat cover that was significantly lower, but still very close to the desirable uniform value: pure Salers (2.84), Limousin × Charolais (2.8), pure Aubrac (2.61), pure Charolais (2.6), pure Limousin (2.59) and Limousin × Aubrac (2.58). The results show that the choice of the genotype according to its fattening potential can significantly affect carcass quality.

Calibrating a dual X-ray absorptiometer for estimating carcase composition at abattoir chain-speed

G.E. Gardner[1], R. Glendenning[2], O. Brumby[2], S. Starling[2] and A. Williams[1]
[1]Murdoch University, School of Veterinary and Life Sciences, Murdoch, 6150, Australia, [2]Scott Technology Ltd., 630 Kaikorai Rd, Dunedin, New Zealand; g.gardner@murdoch.edu.au

A robotic boning system has been developed for use in lamb abattoirs (Scott Technology Ltd.), operating at chain speed and making use of 2D x-ray images to identify cutting lines. This x-ray system has been replaced with a dual energy x-ray absorptiometry (DEXA) system which meets robotic image requirements while also enabling the determination of body composition. This study details the calibration of this prototype DEXA system with the hypothesis that it will provide reliable estimates of body composition which are not influenced by scanning time, or spray chilling. Forty-eight mixed sex lamb carcasses with weights ranging between 17 and 30 kg were scanned using the prototype DEXA system to derive an estimate of lean composition at 1, 12, 24, 60, and 72 hours post-mortem. During this period half of the carcases were spray chilled and half normally chilled. The carcasses were then scanned using Computed Tomography (CT) to determine fat, lean and bone weights which were then expressed as a percentage of carcase weight. Composition of these carcases varied between 12.7-32.2% for CT fat, 52.7-68.6% for CT lean, and 15.0-22.0% for CT bone. The DEXA estimate of lean composition was then used to predict CT fat, lean, and bone percentage, and in support of our hypothesis demonstrated a precision (R^2) of 0.77, 0.63, and 0.60. There was no effect of spray chilling on DEXA values, however in contrast to our hypothesis DEXA values reduced by about 8% over the 72 h scanning period. These results demonstrate the capacity of this prototype DEXA system to estimate body composition and lean meat yield in lamb carcasses at abattoir chain speed.

Analysis of genetic structure within beef cattle breeds using SNPs in four candidate genes

A. Trakovická, N. Moravčíková and R. Nádaský
Slovak University of Agriculture in Nitra, Department of Animal Genetics and Breeding Biology, Tr. A. Hlinku 2, 94976, Slovak Republic; nina.moravcikova@uniag.sk

The aim of this study was to determine and evaluate genetic structure differences in Limousin and Charolais cattle populations using SNPs in four candidate genes associated with economically important traits. Genomic DNA samples were obtained in total from 76 Limousin and 69 Charolais bulls. Animals were genotyped using PCR-RFLP method for each of the following genes: growth hormone receptor (GHR, AF140284:g.257A>G), melanocortin 4 receptor (MC4R, AF265221:g.1069C>G), estrogen receptor (ER, AY332655:g.1213A>G) and cytochrome P450 aromatase (CYP19, Z69241:g.1044A>G). In Limousin and Charolais cattle populations were detected alleles with frequency: 0.54/0.56 and 0.46/0.44 for A and G GHR variants, 0.43/0.55 and 0.57/0.45 for C and G MC4R variants, 0.91/0.94 and 0.09/0.06 for A and G ER variants and 0.68/0.83 and 0.32/0.17 for A and G CYP19 variants. The highest frequency of heterozygous GHR[AG] and MC4R[CG] genotypes were observed in both cattle populations. For ER and CYP19 genes the homozygous GG and AA genotype was predominant. Across loci was observed in analysed populations comparable level of heterozygosity (0.38/0.34) that was also transformed to the median level of loci polymorphic information content (in average 0.31/0.27). Within population variance analysis F_{IS} values close to zero indicated sufficient proportion of heterozygotes. The analysis of molecular variance showed that only 0.5% of the total genetic variability was distributed between populations. The highest proportion of genetic variance was explained by differences within analysed individuals. Our results are in accordance with previously reported data for beef cattle breeds. In the future the association analysis of SNPs effect on production traits will be prepared in order to validation markers utility in gene assisted selection programs for enhancement of cattle production traits.

Effects of beta agonists and immunocastration on fiber frequency in bovine muscle

M.R. Mazon[1], D.S. Antonelo[1], K. Nubiato[1], H.B. Silva[1], T.J.T. Ricci[2], N.T. Bem[2], D.M.C. Pesce[2], P.R. Leme[1] and S.L. Silva[1]
[1]Universidade de São Paulo, Zootecnia, Pirassununga, 13635900, Brazil, [2]Pontifícia Universidade Católica de Minas Gerais, Medicina Veterinária, Poços de Caldas, 37701355, Brazil; sauloluz@usp.br

Beta agonists (BAA) have been used to improve carcass yielding, by promoting skeletal muscle accretion and decreasing protein degradation. However, little is known about the effects of BAA associated with immunocastration (IM) on *Bos indicus* cattle. Therefore, we evaluated the effects of BAA and IM on muscle fiber frequency of Nellore cattle. Ninety-six males (409 ± 50 kg LW; 20-month-old) were divided in two groups and half of the animals received two doses of IM vaccine (Bopriva®) within 30 days interval. Animals were fed during 70 days a common diet containing 76% concentrate and 24% corn silage. Each of these groups were then split in three groups and fed during 30 days one of the following treatments: control diet without BAA (CON); CON diet plus 80 mg/day zilpaterol hydrochloride (Zilmax®) (ZIL); CON diet plus 300 mg/day of ractopamine hydrochloride (Optaflexx®) (RAC). After 100 days, animals were slaughtered and a sample of longissimus muscle was collected between the 12/13th ribs for fiber frequency and sarcomere length determinations. Data was analyzed by ANOVA as randomized complete block (live weight) design in 2×3 factorial arrangement. Compared to non-castrated males, IM males showed higher percentages of SO (slow-twitch oxidative) fibres (10.8 vs 9.1; $P=0.0432$), FG (fast-twitch glycolytic) fibres (64.2 vs 61.8; $P=0.0180$), whereas non-castrated males had higher FOG (fast-twitch oxidative-glycolytic) fibres (29.1 vs 24.9; $P=0.0001$).Treatments CON and RAC induced higher values than ZIL for FG fibres (66.2 and 66.1 vs 56.6; $P<0.0001$) whereas ZIL treatment had higher values than CON and RAC for SO fibres (15.6 vs 7.5 and 6.8; $P<0.0001$). The sarcomere length was not affected by BAA or IM. In conclusion, sex condition and BAA can change muscle fiber frequency with potential effects on meat quality parameters of *B. indicus* cattle.

Identification of SNPs in genes associated with meat tenderness in Nellore cattle

C.U. Braz, R. Espigolan, L. Takada, L.G. Albuquerque, D.G.M. Gordo, F. Baldi, L.A.L. Chardulo and H.N. Oliveira
FCAV-UNESP, Animal Science, Via de Acesso Prof. Paulo Donato Castellane s/n, 14884-900 Jaboticabal, SP, Brazil; holiveira@fcav.unesp.br

The objective of this work was to identify polymorphisms significantly associated with meat tenderness in Nellore beef cattle. A total of 1,694 males with phenotypes and genotypes, from three large animal breeding programs in Brazil, were used. All animals were genotyped with Illumina Bovine HD panel, which contains around 777,000 SNPs distributed throughout the genome. SNPs with call rate<93% and MAF<5% were excluded from analysis. Data were analyzed using MIXED procedure from SAS 9.3. Statistical model included fixed effects of contemporary groups (herd and year of birth, management group at yearling, year and month of slaughter) and number of alleles in each haplotype as covariable (linear effect) and, as random, the sire effect. A total of 52 candidate genes for meat tenderness were analyzed. According to Bonferroni test, haplotypes located in 3 of them (PPARGC1A, ASAP1, CAPN1) showed significant ($P<0.10$) effects on meat tenderness. In order to look for new SNPs, six exon regions, located close to the haplotypes, from 25 non-related animals were sequenced. These regions included exons 14 to 21 from CAPN1 gene and exon 4 from ASAP1 gene. PPARGC1A gene was not sequenced since the significant haplotype was in a promoter region, far from exon 1. There were found 20 new SNPs, located in exon and intron regions. In order to validate these 20 SNPs in the population, a larger number of animals will be sequenced and their association with meat tenderness will be studied. FAPESP # 2013/00035-9 and #2009/16118-5.

Performance of cross-bred cattle in confinement in Brazil

A. Berndt[1], L.S. Sakamoto[1], F.B. Ferrari[2], H. Borba[2], E.D.M. Mendes[1], R.R. Tullio[1] and M.M. Alencar[1]
[1]Embrapa Southeast Livestock, Rod. Washington Luiz, km 234, PB 339, 13560970 Sao Carlos, SP, Brazil,
[2]University of Sao Paulo State, Via Paulo Donato Castellane s/n, 14884900 Jaboticabal, SP, Brazil;
alexandre.berndt@embrapa.br

With the trend towards reducing the length of the productive cycle and increasing demand for higher quality meat, there is a need to adjust some points within the growth curve to produce an animal which meets weight and finishing requirements for slaughter at a younger age. Cattle production relies on a combination of good management, good feed and above all, the use of animals with a genetic makeup which allows a rapid development with major weight gain and good feed efficiency. The objective of this study was to assess different genetic groups of cross-bred cattle and their influence on the performance of these animals in confinement. Recorded variables were daily weight gain (DWG), feed conversion (FC), feed efficiency (FE) and daily consumption of dry matter (DCDM). Steers offspring of Brangus, Canchim (synthetic breed 5/8 Charolais) or Bonsmara bulls and Nellore, ½ Angus + ½ Nellore or ½ Senepol + ½ Nellore cows, reared on pasture and finished in feedlot were evaluated. The animals were confined in collective stalls equipped with GrowSafe troughs, grouped according to weight and the genetic group of the mother and father. The diet was based on maize (corn) silage, ground maize (corn), soybean bran and wheat bran with 51.8% of DM, 13.1% of CP, 71.0% of TDN and 3.2% of EE, provided twice daily, ensuring free consumption. The duration of the confinement period for each animal was variable to allow for a similar finished carcass amongst the animals. Data was analyzed using the PROC MEANS and MIXED procedure of SAS and averages were compared using Tukey's test with significant differences at $P<0.05$. Statistical differences were not found between the genetic groups assessed for the variables DWG, FC, FE and DCDM. The average values obtained were 1.79±0.30, 6.67±0.75, 15.17±1.66 and 11.80±1.29 kg/d, respectively.

Effects of seam (kernel) fat on carcass unit price in Japanese Black cattle

K. Sakoda, S. Maeda, R. Asa and K. Kuchida
Obihiro University of Agriculture and Veterinary Medicine, Department of animal science, 2-11 Inada-cho Obihiro-shi Hokkaido, 0808555, Japan; s23117@st.obihiro.ac.jp

In Japan, beef carcasses are cut and graded at the level of the 6th and 7th rib section. Seam (kernel) fat (SF) is the intermuscular fat surrounded by M. semispinalis capitis, M. semispinalis dorsi and M. longissimus dorsi. According to the Japanese grading rules, if a carcass has over 12 cm^2 SF area then the yield grade is downgraded. Thus, purchasers of carcasses and consumers tend not to favor and subsequently discount carcasses with higher SF. The objective of this study was to investigate the effect of SF on carcass grading traits, image analysis traits and carcass unit price in Japanese Black cattle. Carcass data were collected from carcasses marketed between April 2009 and March 2013 in Hokkaido, Japan. The numbers of records of steer and heifer carcasses were 5,889 and 2,068, respectively. The image analysis traits were calculated from images taken at the 6th and 7th rib section. Correlation coefficients were calculated in order to investigate the relationship between SF and each trait. Analysis of variance was performed by each marbling score (BMS: 2 to 12) to investigate the effect of SF on carcass unit price. The average SF for steer and heifer carcasses was 5.64±2.39 cm^2 and 9.16±3.04 cm^2 respectively. This difference between SF for steer and heifer carcasses is statistically significant ($P<0.01$). The correlation coefficient of each sex between SF and Coarseness index of marbling was 0.25 and 0.27 respectively. However, the correlation coefficients between SF and subcutaneous fat thickness was low for each sex (0.09~0.14). The analysis of variance revealed that carcass unit price was reduced significantly when the SF was higher than 12 cm^2 ($P<0.05$: ex: marbling score BMS 6, small SF 1,582 JPY/kg, middle SF 1,571 JPY/kg, large SF 1,540 JPY/kg). This result indicates that SF has an undesirable effect on the carcass unit price and also leads to coarser (larger) marbling particle size.

Species identification of bovine meat, pork and horse in foods using triplex qPCR
M. Natonek-Wiśniewska and P. Krzyścin
National Research Institute of Animal Production, Department of Animal Cytogenetics and Molecular
Genetics, 1, Krakowska Street, 32-083 Balice, Poland; malgorzata.natonek@izoo.krakow.pl

Food adulteration may have enormous consequences for health and the economy. In recent years in Europe, beef adulteration with pork and horse meat has caused most problems. Efficient laboratory methods are needed to test foods for adulteration. One of these is our method for simultaneous identification of these three components based on mtDNA analysis. Three pairs of primers and MGB probes were designed to amplify short (below 100 bp) fragments in the gene encoding D-loop of the identified species. The study covered model samples of single- and two-species meat (pork/beef and horse meat/beef) containing 0.3, 0.5, 5 and 50%, respectively, pork or horse meat in beef, in several independent replications of qPCR. In addition, 20 commercial samples of processed foods in the form of scalded and/or smoked sausage as well as samples of raw ground meat were investigated. Validation of the method showed that it is effective in detecting beef adulterated with pork and horse meat in the range from 0.5 to 100%. The method is repeatable with relative standard deviation (RSD) for the measurements made in separate reactions being $\leq 4.45\%$, and with $\leq 5\%$ accuracy defined by relative error (D_{Rel}). The usefulness of the method for analysing the composition of commercial food samples was tested using standard curve, produced from the model samples. The results obtained during the analysis of raw meat samples did not differ from those obtained during the analysis of the model samples. For processed foods D_{Rel} ranged from 6.82 to 10.69%, depending on the sample processing method. The method's parameters indicate that it is highly efficient for the species identification of food. The poorer parameters of the method when the processed samples were analysed may suggest the need to use the calibration curve from samples of similar type and processing degree as the analysed samples.

Marbling characteristics at the 6-7th rib and the sirloin in Japanese Black cattle by image analysis
M. Takeo, S. Maeda, R. Asa and K. Kuchida
Obihiro University of Agriculture and Veterinary Medicine, Animal Science, 2-11 Inada-cho Obihiro-shi,
080-8555 Hokkaido, Japan; s22152@st.obihiro.ac.jp

In Japan, beef cattle are graded by Japanese Meat Grading Association (JMGA) by visual appraisal of the cut surface at the 6-7th rib section. The most important economic trait determining carcass price in Japan is the degree of marbling in the rib eye. The purpose of this study was to compare marbling characteristics of rib eye at the 6-7th rib section with that of the sirloin at the 12-13th rib section. The materials of this study were the digital images of 6-7th rib section which were taken at grading and the images of sliced sirloin which were taken after sub-primal meat processing from 4 steers of Japanese Black cattle. Photographs of 1 mm slices of the sirloin were taken at the 12-13th rib section. Up to 17 digital images were taken from each carcass. An investigation of the transition of marbling characteristics between collection sites by image analysis was undertaken for the following traits; Marbling Percent (MP), Coarseness Index (CI), Fineness Index (FI) and New Fineness Index (NFI). The results of the comparison between the 6-7th rib section and sirloin showed: MPs at the sirloin were lower (46.2~49.2%) than at the 6-7th section (54.3~59.4%) for the 4 Japanese Black cattle. The coefficients of variation (CV) of NFI (6.0~9.9%) were smaller than those of CI (15.7~18.9%). Incidentally, CVs of MP were 3.5~6.8%. Thus, these results suggest that the degree of fine (smaller) marbling particles was similar throughout the sirloin, conversely the coarse (larger) marbling particles occurred irregularly and less predictably in the longissimus muscle. Moreover, correlation coefficients of FI (r=0.99) and NFI (r=0.99) between the 6-7th rib section and the sirloin were higher than that of CI (r=0.85). From the above results, a good prediction of marbling characteristics including marbling particle size of the sirloin is possible by using photographs and image analysis results of the 6-7th rib section.

Effects of rice bran, flax seed and sunflower seed on sensory evaluations of Korean Cattle
C. Choi[1], S. Kim[2], H. Kwon[3], U. Yang[4], J. Lee[4] and E. Park[4]
[1]*Yeungnam University, School of Biotechnology, 280 Daehak-ro, 712-749, Gyeongsan, Korea, South,*
[2]*Gyeongbuk Provincial College, Department of Animal Science, 114 Dolibdaehak gil, 757-807, Yecheon,*
Korea, South, [3]*Yeungnam University, Department of Food and Nutrition, 280 Daehak-ro, 712-749,*
Gyeongsan, Korea, South, [4]*Woosung Feed Co., Ltd., 1027 Hanbathdaero, 306-785, Daejeon, Korea,*
South; cbchoi@yu.ac.kr

The objectives of this study were to determine if supplementing rice bran, flax seed or sunflower seed to finishing Korean cattle (Hawoo) steers would affect: (1) carcass characteristics; (2) fatty acid composition; (3) free amino acid contents; and (4) sensory evaluation of longissimus muscle. A total of thirty-nine Hanwoo steers (on average 22.2±2.3-month-old, and 552.2±32.8 kg of body weight) were allocated into either Control, rice bran (RB), flax seed (FS), or sunflower seed (SS) groups considering age and body weight. For RB, FS, and SS groups, 5% of concentrates were replaced with respective ingredients. The animals were fed for 273 days until they reached on average the age of 31.2±2.3 months. Final BW (in kg) and ADG (kg/day) for Control, RS, FS, and SS group were 768.2±36.2 and 0.79±0.1, 785.8±48.2 and 0.85±0.1, 786.2±18.4 and 0.82±0.1, and 789.0±73.3 and 0.84±0.2, respectively, with no statistical differences (P>0.05). Oleic acid (C18:1) was the highest (P<0.05) in Control group (47.73±2.66%). Linolenic acid (C18:3) was 2.3 times higher (P<0.05) in FS group than in the other groups. Methionine, a sulfur containing amino acid, was significantly (P<0.05) higher in FS (1.67±0.41 mg/100 g) and SS (1.17±0.30 mg/100 g) groups than in Control and RB groups. Glutamic acid and alpha-aminoadipic acid (α-AAA) were remarkably (P<0.05) higher in FS group than in the other groups. FS group showed significantly (P<0.05) higher sensory scores for flavor, umami and overall palatability. The results of the current study imply that supplementation flax seed into finishing diets of Hanwoo steers significantly improved beef flavor.

Performance of crossbred Holstein bulls and heifer calves slaughtered at 12-months of age
M. Vestergaard[1], A. Mikkelsen[1], M. Bjerring[1], P. Spleth[2] and M. Kargo[1,2]
[1]*Aarhus University, Foulum, 8830 Tjele, Denmark,* [2]*SEGES, Cattle, Skejby, 8210 Aarhus N, Denmark;*
morten.kargo@mbg.au.dk

Use of sexed semen to produce Holstein heifer calves on the genetically best cows as replacements has increased in Denmark and allows for crossbreeding with beef breeds to the less superior Holstein cows in the dairy herds. In 2014, 4.4% of beef breed semen was used on Holstein females. The crossbred offspring will likely be half bulls and half heifers. The objective of this study was to compare growth and carcass characteristics of beef × HOL bull (BULX) and beef × HOL heifers (HEIX) with that of Holstein bulls (HOL) when slaughtered at 12 months of age. Two beef bulls (one Limousine and one Belgian Blue) were used as sires and half the calves were from each sire. Calves were purchased at 3-4 wk of age. A total of 10 HOL, 10 BULX and 10 HEIX were included in the experiment. Calves were fed milk replacer (850 g/d), grass hay and concentrate pellets until weaning at 8 wk of age. After weaning, calves were gradually changed to a high-energy TMR based on concentrate pellets, corn-cob silage (29% of DM in TMR), ground barley, sugar beet pulp, and soy bean- and canola meal fed *ad libitum* until 200 kg LW where the concentrate pellets where taken out of the TMR (now with 42% of DM from corn-cob silage in TMR) and used until slaughter at 12 mo of age. Individual feed intake was recorded from 4 to 12 months. Average daily Net Energy intake (NEI) and gain from 4 to 12 months were 7.7, 8.0, and 7.3 Scandinavian Feed Units (SFU; P<0.05) and 1.41, 1.63, and 1.33 kg/d (P<0.01) for HOL, BULX, and HEIX, respectively. Thus, the feed conversion efficiency was 5.4, 5.2, and 5.8 SFU/kg gain (P<0.01) for HOL, BULX and HEIX. Carcass weight and EUROP carcass conformation were 258, 325, and 260 (P<0.01), and 3.6, 8.2, and 6.4 (P<0.01) for HOL, BULX and HEIX, respectively. EUROP fatness score was close to 3 and meat and tallow colour was 3.0 for all groups. Based on current beef prices, there was an extra payment of 337 € for BULX and 76 € for HEIX carcasses compared with HOL.

Effect of age at slaughter for early maturing dairy cross heifers

R. Prendiville[1], B. Swan[2] and P. French[3]
[1]*Teagasc, Animal and Grassland Research and Innovation Centre, Grange, Dunsany, Co. Meath, N/A, Ireland,* [2]*Teagasc, Crops, Environment and Land Use Programme, Johnstown Castle, Wexford, N/A, Ireland,* [3]*Teagasc, Animal and Grassland Research and Innovation Centre, Moorepark, Fermoy, Co. Cork, N/A, Ireland; robert.prendiville@teagasc.ie*

The aim of this study was to evaluate alternative systems for April born early maturing dairy cross heifers. Data were available from 91 heifers. Animals were assigned to a 2 (breed group; Aberdeen Angus (AA) and Hereford (HE)) × 2 (age at slaughter, 19 (19MO) and 21 months (21MO)) factorial arrangement of treatments. Heifers in 19MO were slaughtered off pasture at the end of the second grazing season while those in 21MO were finished during the second winter. Animals in 19MO were allocated 2.5 kg of concentrate dry matter (DM) per head daily for the final 51 days. Heifers in the 21MO were offered grass silage ad-libitum with 5 kg concentrates DM/head daily for 72 days pre-slaughter. Live weight at slaughter (LW) was 449 and 518 kg for 19MO and 21MO, respectively (P<0.001). Carcass weight (CW) was greater (P<0.001) for 21MO than 19MO; 258 and 231 kg, respectively. Conformation score (CS) was similar for both groups (5.6 and 5.9 for the 19MO and 21MO, respectively). Kill out proportion (KOP) was greater (P<0.001) for 19MO than 21MO (515 and 497 g/kg, respectively) but fat score (FS) was greater (P<0.001) for 21MO than 19MO (9.7 and 7.3, respectively). Average daily gain during finishing (ADG) was greater (P<0.001) for 21MO (1.06 kg/d) than 19MO (0.76 kg/d). The LW (P<0.05) and CW (P<0.05) were greater for HE (494 and 250 kg, respectively) than AA (473 and 239 kg, respectively). No difference in KOP, CS and ADG was observed between the breeds; 505 g/kg, 5.6 and 0.90 kg/d for AA and 507 g/kg, 5.9 and 0.92 kg/d for HE, respectively. Fat score tended (P=0.573) to be greater for HE (8.9) than AA (8.2). This study showed that finishing early maturing dairy cross heifers indoors resulted in greater CW and FS than off pasture and HE were heavier at slaughter and tended to have a greater FS than AA heifers.

Identifying ideal beef and dairy crossbreeds to optimise slaughter yields

A. Mueller, A. Burren and H. Joerg
Bern University of Applied Sciences BFH, School of Agricultural, Forest and Food Sciences HAFL, Länggasse 85, 3052 Zollikofen, Switzerland; alexander.burren@bfh.ch

This study aims to demonstrate which swiss crossbreeds produce the best returns in terms of carcass weight, conformation and fat cover. The data set consisted of 601,669 individuals being born between 2000 and 2012. The animals were sorted out by sex and age into the following slaughter categories: calves (KV); bulls, no permanent incisors (MT); steers, up to 4 permanent incisors (OB); heifers, up to 4 permanent incisors (RG), and cattle, up to 5 permanent incisors (RV). To compare carcass weights between the crossbreeds, a linear mixed model was used to adjust weights for birth month, slaughter year, sex, litter size, geographical zone, carcass fat cover and conformation. Blonde d'Aquitaine (BA) and Charolais (CH) are suitable breeding partners for Braunvieh (BR) in all slaughter categories (carcass weight [kg] BA × BR = 96.1 (KV), 207.0 (MT), 175.8 (OB), 158.5 (RG) and 180.0 (RV); carcass weight [kg] CH × BR = 88.7 (KV), 198.7 (MT), 187.9 (OB), 172.5 (RG) and 191.1 (RV)) while Belgian Blue (BB) crosses with Braunvieh are characterised by high carcass weights and excellent conformation in the fattening calf and bull slaughter categories (carcass weight [kg] BB × BR = 89.1 (KV) and 204.7 (MT)). Fleckvieh (FT) and Holstein Friesian (HO) crossed with Blonde d'Aquitaine (BA) and Charolais individuals produce good results across all slaughter categories (carcass weight [kg] FT × BA/CH = 93.5/86.0 (KV), 207.0/198.7 (MT), 175.8/187.9 (OB), 158.5/172.5 (RG) and 180.0/191.1 (RV); carcass weight [kg] HO × BA/CH = 95.3/87.9 (KV), 208.8/201.7 (MT), 152.8/170.8 (OB), 138.1/145.5 (RG)). In summary, we found that breeding partners could be found for each of the dairy breeds that produced greater carcass weights, better carcass conformation and better fat coverage.

Supplementation of Caspurea in Crossbred Cattle Diets

P. Paengkoum[1], K. Bunnakit[1] and S. Paengkoum[2]
[1]Suranaree University of Technology, School of Animal Production Technology, Institute of Agricultural Technology, Muang, Nakhon Ratchasima, 30000, Thailand, [2]Nakhon Ratchasima Rajabhat University, Faculty of Science and Technology, Muang, Nakhon Ratchasima, 30000, Thailand; pramote@sut.ac.th

The effects of an extrusion-processed mixture of cassava pulp and urea with protein and pelleting at high temperatures (Caspurea) as a protein source on productive performance in Thai Native × Brahman beef cattle have been assessed. Eight Thai native-Brahman beef cattle with liveweight of 195.5±24.4 kg were used in double 4×4 Latin square arrangement of treatments. The treatments were Caspurea with different protein contents (extrusion-processed mixture of cassava pulp and urea with natural plant protein and pelleting at high temperatures) were as follows: (1) Control Caspurea (45%CP); (2) Caspurea+leucaena leaf; (3) Caspurea+cassava leaf; and (4) Caspurea+soybean meal (SBM). All animals were fed *ad libitum* with urea treated rice straw as roughage. The results showed that total dry matter intake and DM digestibility were not significantly different (P>0.05) among dietary treatments. Ruminal ammonia-N in animals fed the diet containing Caspurea+SBM was lower (P<0.05) than other treatments. Moreover, the results showed that total volatile fatty acid was not differing among dietary treatments. Blood urea nitrogen was affected by the natural plant protein sources and was lowest (P<0.05) in Caspurea+soybean meal group. Body weight change in Caspurea+soybean meal group was higher (P<0.05) than normal Caspurea (Control) and Caspurea+leucaena groups. These results indicated that Caspurea+plant protein has positive effects on crossbred Native × Brahman beef cattle production.

Effects of selection for yearling weight on feed efficiency traits in Nellore cattle

J.N.S.G. Cyrillo, M.E.Z. Mercadante, S.F.M. Bonilha, T. Pivaro, R.H. Branco and F.M. Monteiro
Instituto de Zootecnia, Centro APTA Bovinos de Corte, Rodovia Carlos Tonani km 94, 14174000 Sertãozinho, SP, Brazil; cyrillo@iz.sp.gov.br

This study aimed to evaluate indirect effects of selection for yearling weight (YW) (weight at 378 days for males and at 550 days for females) on feed efficiency traits of 375 Nellore males and females, born from 2004 to 2007, 2011 and 2012, at 280.35±38.12 days of age at the beginning of the performance test (83±15 days long), from the Breeding Program of Instituto de Zootecnia, São Paulo, Brazil. Selection experiment began in 1976, when two selection lines were established: Selection Nellore (NeS), based on maximum selection differential on YW and Control Nellore (NeC), selected for the contemporary group mean of YW. Analyses were performed using Mixed procedure of SAS program. Model statement contained the fixed effects of contemporary group (year, sex and pen), selection line (NeC and NeS) and age of the animal as a linear covariate. Significant differences (P<0.05) between NeC and NeS animals were detected for final test body weight (242±2.7 vs 310±1.9 kg), average daily gain (0.80±0.01 vs 1.08±0.01 kg/d) and dry matter intake (5.18±0.08 vs 6.83±0.05 kg/d), respectively. No differences (P>0.05) were observed for feed conversion ratio (FCR) (6.62±0.10 vs 6.48±0.07 kg DM/g gain), feed efficiency (FE) (0.16±0.01 vs 0.16±0.01 g gain/kg) and residual intake and gain (RIG) (-0.07±0.15 vs 0.11±0.10) between NeC e NeS, respectively. Significant differences (P<0.05) were detected between NeC and NeS animals for residual feed intake (RFI) (-0.206±0.051 vs 0 045±0.036 kg/d) and residual average daily gain (RG) (0.02±0.01 vs 0.015±0.001 kg/d). The results of this study showed that direct selection for post weaning weight did not induce any change in FCR, FE and RIG, but affected significantly RFI and also RG, suggesting that, for this sample, these traits are not independent of growth and size of the animals.

Body condition score and probability of pregnancy of primiparous beef cows grazing grassland
P. Soca[1], M. Carriquiry[1], M. Claramunt[1] and A. Meikle[2]
[1]University of Republic, Animal Science and Pasture, Ruta 3 Km 363, 60000. Paysandú, Uruguay,
[2]University of Republic, Faculty of Veterinary, Lasplaces 1550, Montevideo, Uruguay; psoca@fagro.edu.uy

The objective of this experiment was to analyze the effect of body condition score (BCS) at calving on probability of cycling, early and total pregnancy of primiparous beef cows grazing grassland. Primiparous (3-year-old) Hereford cows (n=153) with normal calving and classified by body condition at calving in (low ≤3.5=L, moderate ≥4=M; 1-8 visual scale) and submitted to suckling management during 12 days and immediately after suckling cows receiving or not (flushing = F vs control = NF) 2 kg/d of whole-rice middling for 22 d. At the beginning of suckling-restriction (55±10 days postpartum [DPP]), all cows were in anoestrus as confirmed from two consecutive weekly plasma samples, with progesterone (P4) concentration being <1 ng/ml. The P4 determination was employed to estimate duration of postpartum anoestrus (PPA) and probability of cycling (PC). The probability of early (EP) and total pregnancy (TP) were estimated by ultrasound scanner with 5 MHz rectal transducers. The duration of PPA was analysed using the MIXED and PC, EP and TP (number of cows pregnant/total cows) were fitted using the GLIMMIX procedure, with the binomial distribution and the logit as the link function specified. The M group increased PC at 67 (L=0 vr M=0.23; P<0.001), 87 (L=0 vr M=0.31: P<0.001) and 117 (L=0.3 vr M=0.6: P<0.005) DPP and reduced PPA (L=123 vs M=98 DPP; P<0.0004). With each unit of increment in BCS at calving, the duration of PPA was reduced by 41 days (P<0.001). Increased EP (L=0.46 vs M=0.84 P<0.01) and TP (L=0.6 vs M=1; P<0.0001) were observed in moderate BCS cows. These results confirmed the relevance of BCS at calving as a link between energetic nutrition- metabolic and reproduction processes in primiparous beef cows grazing native pasture.

Performance of Nelore steers in different Brazil livestock grazing systems
R.R.S. Corte, S.L. Silva, A.F. Pedroso, T.C. Alves and P.P.A. Oliveira
Universidade de Sao Paulo, Av. Duque de Caxias Norte, 13635900, Brazil; rscorte@usp.br

To assess the effects of pasture management and season on the performance of Nelore steers with a mean initial weight and age of, respectively, 290 kg and 15 months, were allotted in four livestock grazing systems with 2 area replications (blocks): (1) IHS: Intensive irrigated with high stocking rate (5.9 AU/ha; *Panicum maximum*); (2) DHS: intensive dryland with high stocking (4.9 AU/ha; *P. maximum*); (3) DMS: Dryland with moderate stocking rate (3,4 AU/ha; *Brachiaria brizantha*); and (4) DP: degraded pasture (1,1 AU/ha; *Brachiaria decumbens*), at an experimental station of the Brazilian Agricultural Research Corporation (EMBRAPA), Southeast of Brazil. Each paddock was grazed by three Nellore (*Bos indicus*) steers as testers and regulating animals that were used to adjust the sward heights. The average daily gain (ADG) were taken for a year according to pasture cycle. The design was a randomized block and the data were analyzed using PROC MIXED. There was observed an interaction among livestock grazing systems and season effects (P<0.001). In autumn, the animals presented similar ADG (IHS:0.766; DHS:0.625; DMS:0.646; DEG:0.576±0.08 kg/d). However, for winter season, the ADG (kg/d) was lower for steers grazing DP (P<0.001) than those grazed the IHS, DHS and DMS area (0.606; 0.682; 0.823; vs 0.344±0.08, respectively) as expected. ADG from steers grazing IHS area was higher (P=0.0259) on spring than DHS, DMS and DP (0.523; 0.502; 0.308; vs 0.795±0.08 kg/d) and steers from DP area presented the worst ADG (P<0.001). At summer season, animals grazing DP area presented higher ADG (P<0.001) than the others (IHS:0.541; DHS:0.614; DMS:0.578; DP:0.826±0.08 kg/d), probably due to a compensatory gain at the past season (spring). Considering the end of the dry season (winter), with limitations of sunlight, temperature and rain, degraded pasture did not allow animals to express their production potential, but the IHS, DHS and DMS system promoted higher average daily gain.

Bayesian analysis for detection of signatures selection in Spanish autochthonous beef cattle breeds

A. González-Rodríguez[1], S. Munilla[1,2], E.F. Mouresan[1], J.J. Cañas-Álvarez[3], J.A. Baro[4], A. Molina[5], C. Díaz[6], J. Piedrafita[3], J. Altarriba[1] and L. Varona[1]
[1]Universidad de Zaragoza, Miguel Servet 177, 50013, Spain, [2]Universidad de Buenos Aires, Viamonte 430, 1053, Argentina, [3]Universitat Autònoma de Barcelona, Edificio V Cerdañola del Valles, 08193, Spain, [4]Universidad de Valladolid, LeónCampus de Vegazana, 24071, Spain, [5]Universidad de Córdoba, Campo Madre de Dios, 0, Córdoba, 14002, Spain, [6]INIA, Crta. de la Coruña, km. 7,5 Madrid, 28040, Spain; mouresan@unizar.es

The BovineHD 770K BeadChip was used on 171 sire/dam/offspring triplets from seven Spanish local beef cattle breeds (Asturiana de los Valles, Avileña-Negra Ibérica, Bruna dels Pirineus, Morucha, Pirenaica, Retinta and Rubia Gallega). The parents were chosen to be as unrelated as possible. We performed a Bayesian analysis with the software SELESTIM to identify loci targeted by a selection process. Further, we selected genomic regions with at least 25 SNP markers, with a signal over an empirical threshold that corresponds to the top 0.1%. Five genomic regions were associated with these strong signals of selection. Three of these regions harbor a clearly candidate gene: BTA2 (MTSN; Myostatin), BTA5 (KITLG; KITLigand) and B18 (MC1R; melanocortin1 receptor); whereas in the other two, located at BTA6 and BTA11, several potential candidate genes can be identified. In BTA6, we identified LAP3 (leucine aminopeptidase3), NCAPG (non-SMC condensing I complex, subunit G), LOCRL (ligand dependent nuclear receptor corepressor-like), ABCG2 (ATP-binding cassette sub-family G member 2) and PPARCG1A (peroxisome proliferator-activated receptor gamma, coactivator 1 alpha); and in BTA11: PROKR1 (prokineticin receptor 1), GFPT1 (glutamine-fructose-6-phosphate transaminase 1), GMCL1 (germ cell-less spermatogenesis associated 1), PCBP1 (poly (rC) binding protein 1) and EHD3 (EH-domain containing 3).

Challenges in dairy breeding under genomic selection

T.J. Lawlor[1], I. I. Misztal[2], Y. Masuda[2] and S. Tsuruta[2]
[1]Holstein Association, USA, 1 Holstein Place, Brattleboro, VT 05302, USA, [2]Department of Animal and Dairy Science, University of Georgia, Athens, GA 30602, USA; tlawlor@holstein.com

The breeding of dairy cattle is often presented as a prime example of a successful implementation and large benefits of genomic selection. In turn, this rapid uptake and wide usage of genomic selection in the dairy industry is presenting researchers with additional challenges due to the growing number of genotyped animals, selection on early genomic predictions and a changing data structure. A major breakthrough termed algorithm for proven and young animals (APY), based on recursion for a large number of animals conditioned on a smaller subset, makes it possible to obtain the inverse of a large genomic relationship matrix. Studies on the national datasets for production and conformation traits in Jerseys and Holsteins, respectively, involving over 11 million animals and 570,000 genotypes have been successfully conducted. Comparisons involving alternative choices of the initial subset of animals used in APY shows the robustness of this approach. Validation studies were used to study the sensitivity of using alternative parameters in the model. Number of proven bulls and the amount that the genomic information is regressed has an impact on both accuracy and estimation of genetic trend. Implementation of SS-GBLUP on a national scale is now possible but it must be coalesced within the routine delivery system of national and international genetic evaluations.

Accuracy of genomic prediction using whole genome sequence data in White egg layer chickens

M. Heidaritabar[1], M.P.L. Calus[1], H.-J. Megens[1], M.A.M. Groenen[1], A. Vereijken[2] and J.W.M. Bastiaansen[1]
[1]Wageningen University, Animal Breeding and Genomics Center, Wageningen, 6700 AH Wageningen, the Netherlands, [2]Hendrix Genetics Research, Hendrix Genetics Research, Technology & Services, Boxmeer, 5830 AC Boxmeer, the Netherlands; marzieh.heidaritabar@wur.nl

There is an increasing interest in using whole genome sequence data in genomic selection breeding programs. Prediction of breeding values is expected to be more accurate when whole genome sequence is used, since the causal mutations are assumed to be in the data. We report genomic prediction for number of eggs in White egg layer chickens using whole genome re-sequence data including more than 4.5 million SNPs. We compared the prediction accuracies based on sequence data with accuracies from the 60k SNP chip data. We applied Genomic Best Linear Unbiased Prediction (GBLUP) and BayesC prediction. Moreover, we evaluated the prediction accuracy using different types of variants (Synonymous, non-synonymous and non-coding SNPs). Genomic prediction using 60k, resulted in prediction accuracy of 0.74 when GBLUP was applied. With sequence data, there was a small increase (~1%) in prediction accuracy over the 60k genotypes. With both 60k and sequence data, GBLUP slightly outperformed BayesC in predicting the breeding values. Selection of SNPs more likely to affect the phenotype (i.e. non-synonymous SNPs) did not improve accuracy of genomic prediction. The small number of sequenced animals in this study, which affects the accuracy of imputed sequence data, may have limited the potential to improve the prediction accuracy. We expect, however, that the limited improvement is because the 60k SNPs are sufficient to accurately determine the relationships between animals.

Usage of genomic information to distinguish between full sibs in layer breeding schemes

M. Erbe[1,2], D. Cavero[3], R. Preisinger[3] and H. Simianer[2]
[1]Institute of Animal Breeding, Bavarian Research Centre for Agriculture, Grub, Germany, [2]Georg-August-University, Animal Breeding and Genetics Group, Goettingen, Germany, [3]Lohmann Tierzucht GmbH, Am Seedeich 9-11, Cuxhaven, Germany; malena.erbe@lfl.bayern.de

In layer breeding, selection of male candidates out of their full sib groups has to be made at a time when no progeny information is available. Thus, distinguishing between full sib males is not possible based on conventional breeding values. With the availability of genomic information, unique direct genomic values (DGVs) can be predicted for each member of a full sib group and be used for directional selection. To study the usefulness of genomic prediction, we used a data set of over 1,200 brown layer individuals from different generations genotyped with the Affymetrix Axiom® Genome-Wide Chicken Array (580k) or with a customized low density SNP chip (50k). Individuals genotyped with the 50k were imputed up to 580k. For four traits (laying rate, egg weight, egg shell strength, egg color) a two-step genomic BLUP (GBLUP) approach with a forward prediction scenario was used to assess the predictive ability of DGVs in a set of full-sib groups. Predictive ability was reasonable (0.35-0.47) when phenotypic information up to the parent generation was available and clearly higher (0.53-0.62) when information of full sib hens was available. Selection differential was always higher when using DGVs instead of random selection. Performance of GBLUP did not decrease when only SNPs from the lower density panel (50k) were used which provides a promising way of cost reduction when a lot of candidates have to be genotyped. Availability of extensive pedigree and phenotypic information made it possible to also study the performance of a single step BLUP for predicting DGVs. Predictive ability was higher (up to +0.1) than with the two step approach. In conclusion, genomic prediction in order to distinguish between full sibs provides a very valuable tool to increase the precision of selection processes in layer breeding programs.

Realized accuracies for males and females with genomic information on males, females, or both

D.A.L. Lourenco[1], B.O. Fragomeni[1], S. Tsuruta[1], I. Aguilar[2], B. Zumbach[3], R. Hawken[3], A. Legarra[4] and I. Misztal[1]
[1]University of Georgia, Animal and Dairy Science, 425 River Rd, 30602 Athens, GA, USA, [2]INIA, Km 10 Rincon del Colorado, 90200 Canelones, Uruguay, [3]Cobb-Vantress Inc., U.S. 412, 72761 Siloam Springs, AR, USA, [4]INRA, 24 Chemin de Borde Rouge, 31326 Castanet Tolosan, France; danilino@uga.edu

Phenotypes were available on 4 production traits recorded for up to 196,613 broiler chickens. A total of 15,723 birds were genotyped for 39,102 SNP. Single-step genomic BLUP (ssGBLUP) was used for genomic evaluation, in a multiple-trait model, with three different reference sets with only males (4,648), only females (8,100), and both sexes (12,748). Realized accuracy of genomic EBV (GEBV) of young genotyped males (1,501), females (1,474), and both sexes (2,975) was used to evaluate the inclusion of genotypes for different reference sets. Using male genotypes as reference, the increase in accuracy of GEBV over EBV for males and females was 12 and 1 percentage point, respectively. When only female genotypes were in the reference, the increase for males and females was 1 and 18 percentage points, respectively. Using genotypes on both sexes as reference increased accuracy by 19 points for males and 20 points for females. For one trait, (G) EBV accuracies for females were lower than for males. For another trait with similar heritability, accuracies were higher, and females had higher accuracy than males. For validation animals GEBV$\approx$w$_1$PA+w$_2$DGV, where PA is parent average, DGV is genomic prediction. When the number of genotyped animals is high, w$_2$ is higher. When an animal is genotyped, the increase in accuracy comes mainly from the DGV portion of GEBV and marginally from improved PA. For non-genotyped animals there is no improvement in accuracy due to DGV. Accuracies for animals of one sex increase with genotypes of the other sex when that sex has independent phenotypic information and when the evaluation methodology avoids double counting. Analysis of realized accuracies reveals different selection pressure for traits and sex.

Genetic evaluation for three way crossbreeding

O.F. Christensen[1], A. Legarra[2], M.S. Lund[1] and G. Su[1]
[1]Aarhus University, Center for Quantitative Genetics and Genomics, Blichers Alle 20, P.O. Box 50, 8830 Tjele, Denmark, [2]INRA, UMR 1388 GenPhySE, BP52627, 31326 Castanet Tolosan, France; olef.christensen@mbg.au.dk

Commercial pig producers generally use a terminal crossbreeding system with three breeds. Genetic evaluation is usually done within each of these breeds based on recorded phenotypes on purebred animals. However, ideally genetic evaluation of purebreds should incorporate phenotypes of interest observed on crossbreds, and breeding values of purebreds for performance in the three-way cross should be computed. The aim of this work is to develop models that handle a combination of pedigree-based and marker-based relationships for single-step genomic evaluation in the three breed crossing system. Wei and Van der Werf presented a model for genetic evaluation using information from both purebred and crossbred animals for a two-breed crossbreeding system, and Christensen et al. reformulated that model and extended it to incorporate a combination of pedigree-based and marker-based relationships. The model provides breeding values for both purebred and crossbred performances for purebred animals. Here, extensions of this model to three-breed terminal crosses are presented. The essential methodology needed for the model is the specification of the additive genetic relationships between genetic values for crossbred performance. Additive genetic relationships in multi-breed populations were derived in Lo, Fernando and Grossman. Gene substitution effects can be defined within populations or across populations, which here are called 'partial genetic' and 'common genetic' approaches. These two approaches provide two different ways of constructing relationships, either partial relationships or common relationships across breeds. Based on these two types of additive genetic relationships, the model for three-way crossbreeding can be formulated using Kronecker matrix products and therefore fitted using standard animal breeding software.

Pedigree and genomic evaluation of pigs using a terminal cross model

L. Tusell[1], H. Gilbert[1], J. Riquet[1], M.J. Mercat[2], A. Legarra[1] and C. Larzul[1]
[1]INRA, Genphyse, UMR 1388 GenPhySE, 31326 Castanet-Tolosan cedex, France, [2]IFIP, Pôle Génétique,
La Motte au Vicomte, 35651 Le Rheu cedex, France; catherine.larzul@toulouse.inra.fr

This study presents a single-step terminal-cross model (GEN) to estimate genetic parameters of growth rate and pH of longissimus dorsi in pigs. The model is compared in terms of parameter estimates and breeding value accuracies with a pedigree-based terminal-cross model (PED) and 2 univariate single-step models (GEN_UNI) for purebred (PB) and crossbred (CB) performance. Ninety Piétrain sires were mated with 306 Piétrain and 306 Large White dams leading a total of 654 PB and 716 CB male piglets. Sires and PB offspring were genotyped using the 60k SNP chip. PB and CB performances were jointly analyzed as 2 traits. The PB animals were accounted for through an animal model, whereas the additive genetic effect of a CB individual was decomposed into its sire and dam allelic contribution effects plus a Mendelian sampling confounded with the residual. Genetic correlation between the PB and the sire contribution for CB performance was estimated. The inverse of a matrix combining both genomic and pedigree relationship matrices was used in the mixed model equations for the Piétrain line as in a single-step procedure. The PED model was of the same form as GEN but accounted only for pedigree information. The GEN_UNI models contained same effects as the GEN model for either the PB or CB performance. (Co)variance components were estimated by Gibbs sampling. Genetic correlations [HPD95%] between PB and CB traits obtained with the GEN model were close to unity: 0.84 [0.45, 1.00] and 0.97 [0.83, 1.00] for growth rate and pH, respectively, suggesting that PB line selection is already successful to improve CB performance. Genotyped animals obtained higher breeding value accuracies with the GEN model than with the PED and GEN_UNI models. Accounting for PB and CB information together with genomic information improves the precision of the genetic evaluation in breeding programs based on crossbreeding.

Comparison of different Marker-Assisted BLUP models for a new French genomic evaluation

P. Croiseau[1], A. Baur[2], D. Jonas[1,2,3], C. Hozé[2], J. Promp[4], D. Boichard[1], S. Fritz[2] and V. Ducrocq[1]
[1]INRA, UMR1313, GABI, 78350 Jouy en Josas, France, [2]ALLICE, 149 rue de Bercy, 75012 Paris, France,
[3]AGROPARISTECH, 16 rue Claude Bernard, 75231 Paris, France, [4]IDELE, 149 rue de Bercy, 75012 Paris,
France; pascal.croiseau@jouy.inra.fr

French genomic evaluations have been based on a Marker-Assisted BLUP (MABLUP) approach where QTL were detected using a variable selection method. The evaluation software was modified to remove the existing constraint on number of QTL included. A set of new MABLUP models were tested using a set of 2574 Montbéliard bulls genotyped on the Illumina Bovine SNP50 BeadChip®. From 1000 to 6000 QTL detected with a BayesCpi approach were included. A model based on SNP only was compared with models based on haplotypes of 3, 4 or 5 SNP. The best SNP corresponding to a QTL position was completed with 2, 3 or 4 flanking SNP to build a haplotype. An alternative to this strategy was developed to select the haplotype with the largest number of 'predictable' alleles from a window of 10 SNP around the QTL-SNP, where observed haplotype allele frequencies were used as indicators of 'good predictability' in genomic evaluation. In the MABLUP model, a polygenic residual component explaining 10 to 50% of the genetic variance was also included in the model and was estimated either through the pedigree or from the SNP of the Illumina Bovine LD chip. The impact of all these scenarios on the correlation between observed daughter yield deviations and genomic estimated breeding values and on the regression slope in a regular validation test was measured and compared to other genomic selection approaches. The best combinations of these scenarios led to an average gain in correlation between 1.5 and 3% with a significant improvement of the regression slope. A reasonable final model for most traits includes 3k QTL haplotypes of 4 SNP and a residual polygenic component based on the LD SNP explaining 20 to 30% of the genetic variance. These results were used to improve the new French national genomic evaluation.

Genomics, sexed semen: changes in reproduction choices in French dairy herds
P. Le Mézec[1], M. Benoit[2], S. Moureaux[1] and C. Patry[2]
[1]Institut de l'Elevage, 149, rue de Bercy, 75595 Paris, France, [2]Allice, 149, rue de Bercy, 75595 Paris, France; pascale.lemezec@idele.fr

The aim of this paper is to explore the mutations in the use of AI bulls in French dairy herds from 2009, first year of genomic evaluation and sexed semen offer, to 2014. Evolution of AI practices during the last six years was examined through data recorded in the National Genetic Information System, which gathers all AI of cattle done by breeding companies or farmers. About 3,500,000 inseminated dairy heifers and cows per year were included in the statistical analyses. The use of young bulls (without progeny) for AI was compared with the use of proven bulls (with more than 40 daughters) according to the type of genetic evaluation the bull benefited in France: genomic evaluation without progeny information, genomic evaluation including progeny information, classical genetic evaluation. The mean genetic level by year and by dairy breed was computed, accounting for the number of 1st AI per bull. From 2009 to 2014, in the three main breeds, the use of young bulls raised from 8% to 67%, whereas proven bulls dropped from 75 to 21%. In the same time, AI with sexed semen increased up to 37% for heifers. Genomics for females also became more and more attractive for breeders: 50,000 young heifers were genotyped on commercial dairy farms in 2014. Depending on breeds, regions and herds, these new tools were used differently. Changes in selection goals combined with the genomics contributed to significant changes in the mean genetic levels. For example, in the Holstein breed, the major progresses of the genetic level of AI were observed for type and functional traits: from 2009 to 2013, in genetic standard deviation of the trait, +1.0 for overall type, +0.7 for fertility, +0.7 for udder health while -0.1 for milk. French breeders and AI organizations are confident about technological innovations. Genomic tools in combination with sexed semen were widely and quickly adopted. It participated to changes in the way of working and in the improvement of performances.

Validation of single step genomic BLUP in field data
S. Wijga, A.M.F.M. Sprangers, R. Bergsma and E.F. Knol
Topigs Norsvin Research Center B.V., P.O. Box 43, 6640 AA Beuningen, the Netherlands; susan.wijga@topigsnorsvin.com

By adding genomic information to traditional pedigree information, single step aims to give a more accurate representation of the genetic relations within the pedigreed population and as such provide more accurate breeding values. Added value of combining genomic information with the traditional pedigree can be tested and quantified though validation. Few examples are available so far on validation of single step in field data and as such this study aimed to validate the approach in the trait litter birth weight. The total dataset was divided in training data (588,187 records) and (blinded) validation data (53,103 records). Validation data consisted of first parity gilts from 5 purebred lines and was selected to contain 10% of the genotyped animals per line. Data on additional parities of these gilts were not in the dataset. Validation data contained 2,941 genotyped animals and training data contained 21,744 genotyped animals. Breeding values were estimated using the training data with single step and traditional pedigree information in the MiXBLUP software package, where the single step relationship matrix was created with the software package calc_grm. Next, a general linear model was applied to the validation data where the phenotype was corrected for the same fixed effects as used in the breeding value estimation and an additional fixed effect estimating either the traditional breeding value or single step. The regression coefficients were 0.88 and 0.99. The correlation between the corrected phenotype and the breeding value increased from 0.17 to 0.24. To reach accuracies correlations were divided by the square root of the heritability of 0.41; the accuracy of selection increased with 42% when single step was applied. In conclusion, analyses from this study based on field data show that use of single step has added value, as single step based models better fit the data. Single step allows for more accurate breeding values, which increase the genetic progress.

Approaches to improve genomic predictions in Danish Jersey

G. Su[1], P. Ma[1], U.S. Nielsen[2], G.P. Aamand[3] and M.S. Lund[1]
[1]Aarhus University, Center for Quantitative Genetics and Genomics, Department of Molecular Biology and Genetics, AU-Foulum, Blichers Alle 20, 8830 Tjele, Denmark, [2]Seges, Agro Food Park 15, Skejby, 8200 Aarhus, Denmark, [3]Nordic Cattle Genetic Evaluation, Agro Food Park 15, Skejby, 8200 Aarhus, Denmark; guosheng.su@mbg.au.dk

Size of reference population is a key factor affecting accuracy of genomic prediction. In dairy cattle, reference population usually consisted of progeny-tested bulls. The small size of reference population is a limitation to the accuracy of genomic prediction in numerically small breeds, such like Danish Jersey. The objective of this study is to investigate various approaches to improve genomic prediction in Danish Jersey. Three approaches were investigated in this study: (1) genotyping cows to be used as reference animals; 2) including North American Jersey bulls into Danish Jersey reference population; (3) integrating information of non-genotyped animals into genomic prediction using single-step approach. The data sets contained about 1,300 Danish Jersey bulls, 1,200 Nordic American Jersey bulls and 8,000 Danish Jersey cows with both genotype and phenotype information, and 610,000 Danish Jersey cows with phenotype information. Deregressed proofs (DRP) were used as response variables in all approaches. Genomic best linear unbiased prediction (GBLUP) model and single-step model were used to predict breeding values. Danish bulls born in 2005 and later were used as validation data. The gain in prediction reliability from the three approaches was assessed by comparing to the GBLUP model with Danish bull reference population. The results show that all approaches can lead to a considerable improvement of genomic predictions for Danish Jersey population. It indicates that genomic selection is feasible in numerically small population.

Genotyping of cows improves the accuracy of genomic breeding values for Norwegian Red cattle

Ø. Nordbø[1,2], T.R. Solberg[2] and T. Meuwissen[3]
[1]Topigs Norsvin, P.O. Box 504, 2304 Hamar 2304, Norway, [2]GENO Breeding and AI Association, Holsetgata 22, 2317 Hamar, Norway, [3]Norwegian University of Life Sciences, Institute of Animal and Aquacultural Sciences, Universitetstunet 3, 1430 Ås, Norway; trygve.solberg@geno.no

To improve the accuracy of genomic breeding values on Norwegian Red cattle, an increased reference population is required. However, all progeny tested Norwegian red bulls are already genotyped, so the next step is to genotype cows. In this study, cows were added to the reference population by including their genotype in the genomic relationship matrix and we utilized a 'one-step' genomic selection method to predict genomic breeding values. For a set of cows, the observed phenotypes for traits including milk production, protein and fat content as well as somatic cell count were masked. We predicted genomic breeding values with and without cows included in the genomic relationship matrix, and utilized the solutions to estimate the phenotype for the validation set. The correlation between observed and estimated phenotype and the accuracy of the GEBV increased for all traits when including cows in the reference population. These results showed that including cows in the genomic relationship matrix in a 'one-step' approach is beneficial for GENO in routine genomic evaluations.

Genomic testing of cows and heifers: An industry perspective

S.A.E. Eaglen[1], G.C.B. Schopen[1], L.I. Stoffelen[1], C. Schrooten[1], M.P.L. Calus[2], A.P.W. De Roos[1] and C. Van Der Linde[1]
[1]CRV BV, P.O. Box 454, 6800 AL Arnhem, the Netherlands, [2]Animal Breeding and Genomics Center, Wageningen UR, P.O. Box 338, 6700 AH Wageningen, the Netherlands; sophie.eaglen@crv4all.com

Genomic breeding values (GEBV) have revolutionized the dairy cattle industry and resulted in big changes in management of many breeding programmes. As industry gets accustomed to genomic selection, doors open for various other industry opportunities with GEBVs leading to new developments including genomic testing of females. By genotyping their herd, farmers can make sharper decisions about young-stock sales and selection as well as mating strategies to increase genetic progress and herd profitability. Genomic testing is expected to play an increasingly large role on the dairy farm, eventually facilitating so-called precision farming. In addition to a new tool for farmers, genomic testing may also warrant both phenotypic as well as genomic data collection for the industry. Considering this, CRV offers Dutch and Flemish herds to join an initiative named DataPlus. On contracted DataPlus herds CRV genotypes all female animals at a reduced price and collects conventional and novel phenotypes. Genotypes of 46,000 females (200 herds) have so far been collected and used to increase GEBV reliabilities. DataPlus data give unprecedented information on: (1) the value of cow genotypes for GEBV reliabilities; (2) the importance of genomic data for herd management and; (3) farmers perception towards genomic testing. In a recent genomic validation study, ~10,000 female genotypes were added to a reference population of 20,000 bulls (subset of the EuroGenomics population). Reliabilities across 23 conformation traits plus hoof health (h^2 0.11-0.52) increased on average by 2%, with a max. of 8%. The aim of DataPlus is to reach 100,000 cow genotypes and increase reliabilities by 10%. Farmer surveys as well as field analyses on genomic herd management are ongoing and results shall be presented.

Overview of beef cattle national genomic evaluation in France

R. Saintilan[1], T. Tribout[2], M. Barbat[1], M.-N. Fouilloux[3], E. Venot[2] and F. Phocas[2]
[1]ALLICE, UMR1313 GABI, 78352 Jouy en Josas cedex, France, [2]INRA, UMR1313 GABI, 78352 Jouy en Josas cedex, France, [3]Idele, UMR1313 GABI, 78352 Jouy en Josas cedex, France; romain.saintilan@jouy.inra.fr

Genomic evaluations have been recently implemented in the three main beef breeds in France, i.e. Charolais, Limousin and Blonde d'Aquitaine. Five direct and two maternal traits routinely evaluated at birth and weaning are currently considered. Genomic estimated breeding values (GEBV) are calculated via a two-step method. First, direct genomic breeding values (DGV) of genotyped animals are estimated using Bayes-C methodology, considering a reference population (RP) constituted by 1,029 to 5,181 animals in Charolais, 606 to 3,995 animals in Limousin, and 1,645 to 4,282 animals in Blonde d'Aquitaine populations, depending on the considered trait. RP animals and candidates were genotyped on the BovineSNP50 BeadChip or imputed from the EuroG10K BeadChip using Fimpute software. RP individual phenotypes are their deregressed national polygenic estimated breeding values (EBV) whose parental information has been previously removed. Between 20% and 40% of RP animals have very accurate deregressed EBV (reliabilities over 0.70). In a second step, the DGV of each genotyped animal is blended with its EBV, taking into account the partial redundancy of information between these 2 breeding values. To date, implementation of genomic evaluations for direct traits resulted in significantly more reliable breeding values for young candidates in Charolais, but the increase in reliability was more moderate in Limousin and Blonde d'Aquitaine. Concerning maternal traits, GEBV and EBV have similar reliabilities, because of the current limited size of the RP for these traits. New traits, as post weaning growth and carcass traits, will be included in national genomic evaluation during 2015, for the three populations. New breeds will also be included in the French genomic evaluation process in 2016 as soon as sufficient reference populations will be available.

Genome-wide association study for carcass traits in Nellore cattle

G.C. Venturini[1], R.M.O. Silva[1], D.G.M. Gordo[1], G.A. Fernandes Junior[1], B.O. Fragomeni[2], D.F. Diercles[1], F. Baldi[1], R. Espigolan[1], J.N.S.G. Cyrillo[3], M.E.Z. Mercadante[3] and L.G. Albuquerque[1]
[1]Univ Estadual Paulista, UNESP, Via de Acesso Prof. Paulo Donato Castellane s/n, Jaboticabal, SP, 14884-900, Brazil, [2]University of Georgia, Athens, GA 30602, USA, [3]Instituto de Zootecnia, Centro APTA Bovinos de Corte, Rodovia Carlos Tonani, km 94, Sertaozinho, 14160970, Brazil; mercadante@iz.sp.gov.br

The aim of this study was to identify genomic regions which could explain part of the genetic variation in carcass traits in a Nellore cattle population. The data set contained 606 ultrasound records for longissimus muscle area (LMA) and subcutaneous rump fat thickness (RFT), and 605 for subcutaneous fat thickness in loin (LFT). Animals were genotyped using BovineHDBeadChip panel with 700k SNP markers. After genomic data quality control, there were available 437,197 SNPs for 498 animals for LMA and RFT, and 497 for LTF. SNP solutions were estimated by genome-wide association study using a single-step BLUP approach (ssGWAS). Variances were calculated for windows of 200 SNP. Fixed effects in the model included contemporary group (sex, year of birth, and pen), plus age of animal as covariable. The results showed the SNP window located from 24.4 to 25.4 mb on chromosome 14 (BTA14) explained 0.62, 2.92 and 1.60% of genetic variances for LMA, RFT and LFT, respectively. On this region are located the following genes: XKR4, TRNAT-AGU, TMEM68, TGS1, LYN, RPS20, MOS, PLAG1, CHCHD7, SDR16C5, SDR16C6, and PENK. There are evidences in the literature that these genes are related to carcass traits in beef cattle. The XKR4 gene have been previously reported to be associated to RFT. In the region from 22.07 to 31.11 mb, on BTA14 were observed 8 adjacent SNP windows which explained from 0.38 to 2.92% of RFT genetic variance. These results indicated that there are genomic regions on chromossome 14 associated to the expression of carcass traits, including several genes which can be related to these traits. São Paulo Research Foundation (FAPESP) grant #2013/01228-5 and #2009/16118-5.

Genome wide association study of reproductive traits in Nellore heifers using Bayesian inference

R.B. Costa[1], G.M.F. Camargo[1], I.D.P.S. Diaz[1], N. Irano[1], M.M. Diaz[1], R. Carvalheiro[1], A.A. Boligon[1], F. Baldi[1,2], H.N. Oliveira[1,2], H. Tonhati[1,2] and L.G. Albuquerque[1,2]
[1]Sao Paulo State University, Animal Science, Via de Acesso Prof Paulo Donato Castelane s/n, 14870-000 Jaboticabal SP, Brazil, [2]CNPq, SHIS QI 1 Conjunto B, Blocos A, B, C e D, 71605-001 Brasilia-DF, Brazil; oliveira.hn@gmail.com

The aim of this study was to identify regions of the bovine genome that significantly influence the traits age at first calving (AFC), and heifers rebreeding (HR) in Nellore females using a high density chip of SNPs. Data from 1,853 females from a large animal breeding program were used. Genotyping was performed using the high density BovineHD BeadChip. After genomic quality control 305,348 SNPs remained. BAYESCπ methodology was used to estimate the effects of SNPs. For HR, the contemporary group (CG) was defined by farm, year and season of birth of the heifer, and sex of its previous calf. For AFC, the CG consisted of farm, year and management group at birth, weaning and yearling. For both traits the systematic effect of GC was considered. For HR, it was considered as a covariate, the linear effect of the resting period (number of days postpartum until the beginning of the second breeding season). SNPs with Bayes Factor (BF) greater than 150 were considered significant. The use of Bayes factor to determine the significance of SNPs allowed the identification of a set of 42 significant SNPs for HR, which explains 11.35% of the phenotypic variance in this trait, and a set of 19 SNPS explaining 6.42% of the phenotypic variance of AFC. All significant SNPs for age at first calving were also significant for heifer rebreeding. Those 42 SNPs signalized 35 genes, distributed across 18 chromosomes, and among those, 27 are protein encoding genes, six are pseudogenes and two are miscellaneous RNA (miscRNA). These SNPs provide relevant information about HR and AFC that can help elucidate which genes are affecting those traits. Sao Paulo State Foundation (FAPESP), grants #2009/16118-5 and 2011/01811-7.

Genomic evaluation of milk production and fertility traits in Russian Holstein cattle

A.A. Sermyagin[1], E.A. Gladyr[1], S.N. Kharitonov[1], A.N. Ermilov[1,2], I.N. Yanchukov[2], N.I. Strekozov[1] and N.A. Zinovieva[1]
[1]L.K. Ernst Institute of Animal Husbandry, Dubrovitsy, 142132 Moscow, Russian Federation, [2]Regional Information selection center Mosplem, Zakharovo, 142403 Moscow, Russian Federation; alex_sermyagin85@mail.ru

The aim of the present study was to evaluate the genomic breeding values (GEBVs) of Holstein bulls in the Russian reference population. Genomic information was obtained by genotyping 256 bulls using the Illumina Bovine SNP50 v2 BeadChip. The dataset included 195 proven (246 daughters per sire on average) and 61 young bulls. After SNP's quality control (by PLINK) for further analysis was used 41,798 markers. GEBVs were calculated as combination residual polygenic effects with the direct genomic values by GBLUP approach. Breeding values estimated by BLUP AM were deregressed (drgEBV) as proposed J.Přibyl for 305-d milk yield (MY),fat percent (FP),protein percentage (PP),milk fat yield (FY),milk protein yield (PY),age at first calving (AC),calving ease (CE),days open (DO) and number of inseminations per conception (NC). Reliability of genomic prediction was based on the square of correlation between drgEBV and GEBV. Heritability coefficients ranged from 0.25 to 0.40 for milk production and from 0.04 to 0.22 for reproduction traits. The GEBV reliabilities of milk production traits for proven and young bulls of the common dataset were 0.61 for MY,0.69 for FP,0.71 for PP,0.69 for FY and 0.63 for PY. The values of predictability of fertility features were 0.67 for AC,0.69 for CE,0.49 for DO and 0.37 for NC that has revealed a possibilities for improving the selection of fitness traits in studied population by genomic selection. The values of the inbreeding calculated by the genomic relationship matrix were within 0.934 to 1.275 and a standard deviation was 0.041. The small reference population of Russian Holsteins allows to be widely used the genomic evaluation of young tested bulls in the early stages of the selection process. Supported by the Russian ministry of education and science,number of the project RFMEFI60414X0062.

Comparison of simulated dairy cattle breeding schemes for residual feed intake

S.E. Wallén and T.H.E. Meuwissen
Norwegian University of Life Sci, Anim. and Aquacult. Sci, P.O. Box 5003, 1432 Ås, Norway; sini.wallen@nmbu.no

Genomic selection (GS) can improve the accuracy of selection for an individual without a phenotype. It is particularly effective for traits that are difficult to measure such as feed efficiency. The objective of the present study was to compare different dairy cattle breeding schemes using stochastic simulations and to quantify genetic gain for residual feed intake (RFI). The accuracy of the breeding values was treated as being dependent on the structure of the breeding scheme simulated. The population structure, including the selection intensity and the number of selected sires, was chosen to reflect the Norwegian dairy cattle population. Achievable genetic gain for RFI was estimated using either traditional ABLUP (Best Linear Unbiased Prediction using pedigree based relationship matrix) schemes or GS schemes. For each scheme, 50 replicates were run and the duration of simulation was 20 years. Two different (2-tier and 3-tier) simulations were used. In the 3-tier simulation, females had 13,000 phenotypic records for RFI which was 1000 records more than in 2-tier simulation. In the 3-tier GS simulation, those 1000 females were also genotyped. Heritability for milk yield and RFI was 0.3 and 0.15, respectively. Compared to the ABLUP method (for the same tier), genetic merit for RFI (σ_g^2=14.2) at year 20 was 44.8 units greater in 2-tier simulation (P<0.05) and 34.9 units greater in 3-tier simulation (P<0.05) for the GS scheme. Genetic merit for milk yield (σ_g^2=15.0) at year 20 was 18.8 units greater in GS method (2-tier: P<0.05) and 36.3 units greater in 3-tier simulation (P<0.05) compared to the ABLUP method. Accuracy for males was approximately 0.4 units greater for both tiers in GS method (2-tier: P<0.05; 3-tier: P<0.05) and accuracy for females 0.1 units greater for 2-tier (P<0.05) and 0.02 units greater for 3-tier (P<0.05) compared to the ABLUP method. However, no difference was found for genetic merit at year 20 for either trait between 2-tier and 3-tier GS simulations.

Predictability of parametric and semi-parametric models for meat tenderness in Nellore cattle

R. Espigolan[1], D.A. Garcia[1], D.G.M. Gordo[1], R.L. Tonussi[1], A.F.B. Magalhães[1], T. Bresolin[1], C.U. Braz[1], G.A. Fernandes Júnior[1], W.B.F. Andrade[1], R.M.O. Silva[1], L. Takada[1], L.A.L. Chardulo[1], R. Carvalheiro[1], F. Baldi[1] and L.G. Albuquerque[1,2]
[1]UNESP, Road Prof. Paulo Donato Castellane, 14884-900, Brazil, [2]CNPq, Brasilia, DF, Brazil; galvao.albuquerque@gmail.com

The aim of this study was to compare the predictive ability of parametric and semi-parametric models using genomic information for meat tenderness (MT) in Nellore cattle. Animals were genotyped using BovineHDBeadChip panel, with 777,962 SNP markers. Quality control criteria were: SNPs with minor allele frequency $\leq 2\%$, call rate $\leq 98\%$ (CR), $P < 10^{-5}$ for H-W equilibrium and correlation between adjacent SNPs > 0.995, were excluded. Only samples with CR $\geq 98\%$ were considered. A total of 369,722 SNPs and 1,624 males remained. Contemporary groups (CG: farm, year of birth, and yearling management group) with less than 3 animals were excluded. Two classes of age at slaughtering (AS) and days in feedlot (DF), nested in farm and year of birth, were defined. The CG, AS, and DF effects were estimated applying a polygenic model and their solutions were used to adjust MT phenotypes (CMT), which were considered as response variable in genomic evaluations. Models were: (1) Genomic BLUP (GM); (2) a combination of polygenic and genomic models (AGM); (3) RKHS regression with a grid for bandwidth parameter (h) equal to 0.2, 0.5, or from 1 to 10 (KM$_h$); (4) Kernel averaging approach, with h being 0.2, 4, and 10 (KAM). A ten-fold cross-validation was carried out and models were compared by their predictive mean square error (PMSE) and correlation (r) between observed and predicted CMT. The AGM model showed the highest r (0.45) and the lowest PMSE (0.60). Estimates of r and PMSE were 0.44 and 0.61 (GM); 0.41 and 0.63 (KM$_{0.2}$); 0.40 and 0.63 (KAM), respectively. Although semi-parametric RKHS models showed estimates of PMSE and r slightly worse than AGM, they were largely advantageous in terms of computational time. Sao Paulo Research Foundation (FAPESP) grant #2009/16118-5 and #2014/00779-0.

Genome-wide association study for carcass weight in Nelore cattle

G.A. Fernandes Júnior[1], R.B. Costa[1], D.A. Garcia[1], R. Espigolan[1], R. Carvalheiro[1], F. Baldi[1], G.J.M. Rosa[2], G.M.F. Camargo[1], D.G.M. Gordo[1], R.L. Tonussi[1], A.F.B. Magalhães[1], T. Bresolin[1], L.A.L. Chardulo[1] and L.G. Albuquerque[1]
[1]UNESP, Via de Acesso Prof. Paulo Donato Castellani, 14884900, Brazil, [2]University of Wisconsin, 1675 Observatory Drive, 53706, USA; lgalb@fcav.unesp.br

Carcass weight (CW) is an economically relevant trait in beef cattle, and the aim of this study was to investigate potential genomic regions harboring genes with major effects for CW in Nelore cattle. The dataset used in this study was composed of 1,566 Nelore bulls with phenotype records. The slaughter was carried out when the animals were around two years of age. Contemporary groups (CG) had at least three animals, and were defined by combining year and farm of birth, and management group at yearling. A subset of 1,409 animals had both genotypic and phenotypic information. Genotypes were generated based on a panel with 777,962 SNPs from the Illumina® Bovine HD chip. Samples with call rates <0.90 were excluded from the analysis. After genotype quality control, 450,971 SNPs were kept. The additive genetic model with fixed effects of CG and slaughter animal age was used to obtain mixed model equations solutions using a pedigree-based relationship matrix modified by contributions from the genomic relationship matrix. The genetic analysis was carried out by two-trait analysis, including weaning weight with the objective to correct for selection effect at weaning. The SNP effects, weighted according both frequencies and allelic substitution effect, were estimated from the gEBVs obtained. A Manhattan plot of the variance explained by 100 adjacent SNP windows was used to assess potential genome regions with major effects on CW. Important signal was observed on window chr14:24.80-25.35 Mb that explained 1.82% of the variance. There are evidence, in the literature, that this region are located in putative QTL affecting carcass weight in beef cattle. Fapesp Grant # 2009/16118-5.

Identification of genomic regions associated with reproductive traits in Holstein-Friesian bulls

T.R. Carthy[1,2], K.E. Kemper[3], D.P. Berry[2], R.D. Evans[4] and J.E. Pryce[3]
[1]University College Ireland, Belfield, Dublin, Ireland, [2]Teagasc, Animal and Grassland and Research Innovation Centre, Moorepark, Co. Cork, Ireland, [3]Agribio, Centre for Agribioscience, Bundoora, Victoria, Australia, [4]Irish Cattle Breeding Federation, Bandon, Cork, Ireland; donagh.berry@teagasc.ie

Genome-wide association studies (GWAS) have provided limited insights into the genetic control of complex traits such as reproductive traits. Regional heritability mapping (RHM) aims to capture the genetic variation associated with genomic regions by combining the effects of multiple single nucleotide polymorphisms (SNP), that each explain too little of the genetic variance to be detected by GWAS. The aim of this study was to calculate regional variance components for reproductive traits in Irish dairy cattle. Illumina Bovine50 beadchip genotypes were available on 3,484 Holstein-Friesian bulls. After quality control, 43,304 SNPs remained. Deregressed breeding values were calculated for the reproductive traits cyclicity, multiple ovulations, cystic structures, embryo loss, and uterine score as well as from the Irish national genetic evaluations for calving interval. The proportion of genetic variance explained was calculated as $\sigma^2_u/\sigma^2_u+\sigma^2_e$ for the whole genome and $\sigma^2_w/\sigma^2_u+\sigma^2_e$ for the genomic region under investigation, where σ^2_u, σ^2_w, σ^2_e are whole genome additive genetic variance, regional genomic additive variance, and residual variance, respectively. All SNPs combined explained between 24% (cyclicity) and 79% (calving interval) of the total genetic variation in the traits. A large proportion (i.e. 70 to 94%) of the genomic regions explained <1% of the genetic variation in the traits, whereas a small proportion (0.5 to 2%) explained >5% of the genetic variation. One region on chromosome 14 explained >7% of the genetic variation in calving interval whereas a region on chromosome 8 explained >7% of the genetic variation in cyclicity. The RHM method may prove to be more informative than conventional GWAS at identifying genomic regions particularly for complex polygenic traits.

Analyses on the impact of foreign breed contribution on genomic breeding value estimation

A. Bunz, G. Thaller and D. Hinrichs
Institute of Animal Breeding and Husbandry, Christian-Albrechts-University, Hermann-Rodewald-Straße 6, 24118 Kiel, Germany; dhinrichs@tierzucht.uni-kiel.de

This study analyzes the impact of Red Holstein (RH) blood on the accuracy of genomic breeding value estimation in the German Simmental population. Daughter yield deviations for milk yield, fat yield, protein yield, and somatic cell score where available from 3,289 bulls genotyped with the Illumina Bovine 50 k array, as well as the information of the proportion of RH blood from these bulls. Based on the proportion of RH blood the data set was split into three different combinations of training and validation data sets (a), (b), and (c), i.e. bulls with RH blood (n=2,516) and without RH blood (n=773), bulls with RH blood below 2% (n=1,026) and bulls with more than 2% RH blood (n=2,263), and bulls with less than 4% RH blood (n=1,991) and bulls with more than 4% RH blood (n=1,298), respectively. In addition, the bulls were allocated randomly, i.e. independently from the proportion of RH blood, to training and validation data sets of similar size (1,991 and 1,298 bulls), to avoid differences due to different numbers of bulls in training and validation data sets. This procedure was repeated three times. The correlation between training and validation in data set (a) was 0.57, 0.53, 0.56, and 0.54 for milk yield, fat yield, protein yield, and SCS, respectively. Similar correlations were observed in data set (b), whereas correlations in data set (c) were slightly lower with 0.44, 0.36, 0.42, and 0.42 for milk yield, fat yield, protein yield, and SCS, respectively. If bulls were allocated randomly to training and validation the average correlation between training and validation was 0.60, 0.54, 0.56, and 0.54 for milk yield, fat yield, protein yield, and SCS, respectively. This study showed that the proportion of RH blood has a small effect on the accuracy of genomic breeding values. For all traits correlations between training and validation were lower (0.04 to 0.16) if bulls were allocated based on the proportion of Red Holstein blood.

GWAS on conception rate and milk production in different stages of lactation for first three parities in US Holsteins

S. Tsuruta[1], D.A.L. Lourenco[1], I. Aguilar[2] and I. Misztal[1]
[1]University of Georgia, Animal and Dairy Science, 425 River Rd Athens, GA 30602, USA, [2]INIA, Las Brujas, Ruta 48 Km. 10. Rincón del Colorado, Canelones, 90.200, Uruguay; shogo@uga.edu

The objectives of this study were to conduct genome-wide association studies (GWAS) on conception rate (CR), production traits, and SCS and determine if the genetic architecture of these traits is different in three stages of lactation for the first three parities in US Holsteins. The data were split into three sets: early (DIM<14 wk), middle (14 wk≤DIM≤29 wk), and end (29 wk<DIM) lactation stages for each one of the first three parities. The SNP file contained 42,503 SNP markers for 34,506 bulls obtained from USDA-AGIL. The DHI data in New York State consisted of 4.5, 3.7, and 2.4M test-day records for cows that calved between 2000 and 2009 in the first three parities, respectively. The multi-trait threshold-linear model included fixed effects: herd-year, age, days to service (CR) or DIM, service month (CR), calving month, and random service sire (CR) and random effects: additive genetic, permanent environmental and residual. Genomic EBV were estimated with a single-step GBLUP via THRGIBBS1F90, and SNP marker variances were estimated via POSTGSF90. Heritability estimates for CR were lowest (0.03) in the middle and highest (0.05 to 0.08) in the end of lactation in all parities. Genetic correlations of CR with other production traits were low and all negative (-0.1 to -0.5). The SNP marker effects were divided into equal segments of 30 SNP. A segment on chromosome 14 was associated with CR only in early and middle stages of lactation in the third parity; the proportion of the total genetic variance explained by this segment for CR were 2.7% and 2.5%, respectively. The proportions for milk and fat yields were highest in the middle of lactation (7% for milk and 8% for fat), whereas those for protein and SCS were low (<2%). The results suggest that gene expression for CR and milk production traits is stronger in early and middle lactation stages and similar over the lactations.

Database management of single nucleotide polymorphisms for use in Polish genomic evaluation

K. Żukowski[1,2], J. Makarski[3], K. Mazanek[3] and A. Prokowski[3]
[1]Dalhousie University, Faculty of Computer Science, Halifax, Canada, [2]National Research Institute of Animal Production, Department of Animal Genetics and Breeding, Balice, Poland, [3]National Research Institute of Animal Production, IT Department, Balice, Poland; kacper.zukowski@izoo.krakow.pl

The full Polish (Genomika Polska) and EuroGenomics partnership in 2014, allows to increased number of deposited genotypes tenfold, from over 3,000 to 30,000 Holstein-Friesian bulls. To facilitate routing genomic evaluation of this this rapidly expanding set of genotyped bulls and cows, we have been built in the National Research Institute of Animal Production database system to store genotypes and support their routine imputation process. The infrastructure of the system is based on dedicated computational servers and matrix storage systems. The database system is built on Microsoft SQL Server 2014, while software is developed using the .NET Framework/Mono platform implemented in the programming language C#. The dedicated interface works with Microsoft Windows (multi-tab graphical application) and Linux (plink like command-line application), which allowed for full management, importing and exporting of genotype. The system also provides database backup and user authentication. The flexible database infrastructure allows the import and export of data with different file parameters such as format (Illumina TOP/AB), chip and pedigree structure, providing standardization of genotypes across all genotyping platforms and sources. Genotypes from all low and high density chips were used in the initial quantification process to assess the call rate, allele frequencies and heterozygosity of the X chromosome. Procedures to check parent-progeny conflicts and other inconsistencies were developed. The low-density genotypes requiring imputation are exported and analyzed automatically, then reimported to the database or used directly as input data in genomic enhanced breeding (GEBV) evaluation.

Performance of pedigree- and genome-based coancestries on selection programmes

S.T. Rodríguez-Ramilo[1], L.A. García-Cortés[1] and M.A.R. De Cara[2]
[1]Instituto Nacional de Investigación y Tecnología Agraria y Alimentaria, Departamento Mejora Genética Animal, Crta. A Coruña Km. 7,5, 28040 Madrid, Spain, [2]Museum National d'Histoire Naturelle, Laboratoire d'Eco-anthropologie et Ethnobiologie, 17 place du Trocadéro, 75116 Paris, France; rodriguez.silvia@inia.es

Estimated breeding values (EBVs) obtained from high-density SNP chips have a higher accuracy than EBVs obtained from pedigree information. Additionally, these genome-based EBVs lead to lower increases in inbreeding compared with EBVs based on pedigree information. Using simulated data, we performed a pedigree-based BLUP and genomic BLUPs based on three different genomic coancestries: firstly, an estimate based on shared segments of homozygosity; secondly, an approach based on SNP-by-SNP count corrected by allelic frequencies; and thirdly, we used the identity by state methodology. We simulated a truncation selection scheme involving different population sizes, different number of genomic markers and different heritabilities for a quantitative trait. The performance of the four measures of coancestry is evaluated in terms of genetic gain, changes in coancestry and maintained diversity. Our results show that the genetic gain is very similar for all evaluated coancestries. However, genomic-based BLUPs are superior to BLUP based on pedigree information in terms of maintaining diversity. Furthermore, the coancestry measure based on shared segments seems to provide slightly better results in terms of genetic gain under some scenarios and the increase in inbreeding and loss in diversity is only slightly larger with selection based on this coancestry measure than in the other genome-based measures. Our results show that genome-based information can maintain more diversity without losing genetic gains, and thus secure the future success of selection programmes.

Identification of epistasis by comparing SNP effects estimated based on EBVs and on phenotypes

T. Suchocki[1], A. Zarnecki[2] and J. Szyda[1]
[1]Wrocław University of Environmental and Life Sciences, Department of Genetics, Kożuchowska 7, 51-631 Wrocław, Poland, [2]National Research Institute of Animal Production, Institute of Animal Breeding and Genetics, Krakowska 1, 32-083 Kraków, Poland; tomasz.suchocki@up.wroc.pl

In genomic selection of dairy cattle pseudo-phenotypes in form of deregressed breeding values are widely used for the estimation of SNP effects. From the nature of the dependent variable use in the model, SNP estimates largely express the additive genetic contributions of the genomic regions marked by them, which is plausible from the breeding perspective. However, from the genetic perspective, such estimates are biased since they lack the dominance and epistatic components. A reliable detection of dominance and epistasis using conventional statistical models suffers from lack of power and a very high dimentionality due to a very large number of possible SNP pairs (in case of a two-locus epistasis) to consider. Therefore, the major goal of this study is the comparison of SNP effects estimated for the Polish Holstein-Friesian dairy cattle using bulls' EBVs and using cows' phenotypes. Differences in effects estimated by both data sets indicate the presence of dominance and/or epistasis within the genomic regions, which can then be formally tested using model comparison cirteria. The data set consisted of 3,438 proven bulls and 2,172 cows. Each animal was genotyped using 54K Illumina Bead Chip. Genotypic data was edited based on technical quality of the chip by removing single nucleotide polymorphisms with call rate lower than 95%. In our analysis milk, fat and protein yield, somatic cell score and the non-return rate of heifers were considered. To estimate of the additive effects of SNPs a SNP-BLUP model was used. Using this model two evaluations separately for bulls and cows were carried out. Finally, statistical significance of the particular SNP estimates and genomic region estimates were compared between these two data sets in order to detect regions of nonadditive contribution to the analysed phenotypes.

Strategies to reduce the bias of model reliability in genomic prediction

P. Ma, G. Su and M.S. Lund
Aarhus University, Department of Molecular Biology and Genetics, Center for Quantitative Genetics and Genomics, Blichers Allé 20, Postboks 50, 8830 Tjele, Denmark; peipei.ma@mbg.au.dk

Model reliability (i.e. reliability estimated from model) in genomic evaluations is important since it could be used as weights in international genomic evaluations and in a selection index. It has been found that the model reliability is generally higher than the realized reliability in genomic evaluations. The reason could be the strong selections of the studied population and the imperfect linkage disequilibrium (LD) between markers and QTL. The objective of this study is to find solutions to correct the bias of model reliability. The data used in this study were simulated data generated using software QMSim. The whole data covered 20 generations. The animals born in 16th generation and after were used as validation dataset. The reference animals consisted of breeding sires. Three factors were investigated: (1) selecting vs random mating population; (2) span of generations for reference animals; starting from 1st, 5th or 10th generation and ending in 15th generation; and (3) genomic relationship matrix (G-matrix): being built using either neutral markers, quantitative trait loci (QTL), or QTL with square of QTL effect as weight. The results showed there was no difference between model reliability and realized reliability when the population was not a selected population. The reference dataset including individuals from 1st generation to 15th generation resulted in larger difference between model reliability and realized reliability than the scenario with fewer generations. QTL-based weighted G-matrix led to smallest difference between the model reliability and realized reliability, while the neutral marker-based G-matrix resulted in the largest difference. The results indicated that in practical genomic selection scheme, improving G-matrix is one of the solutions to reduce bias of model reliability.

Implementation of a genomic selection program for Australian Merino Breed in Uruguay

G. Ciappesoni[1], F. Macedo[2], V. Goldberg[1] and E.A. Navajas[1]
[1]INIA Uruguay, Ruta 48 km 10, Rincón del Colorado, 90200, Uruguay, [2]Facultad de Veterinaria UdelaR, Lasplaces 1550, 11600, Uruguay; gciappesoni@inia.org.uy

Genomic selection (GS) is being implemented in sheep breeding programs in several countries (e.g. Australia, New Zealand and France). Uruguayan breeders are evaluating the possibility of implement GS in the Merino breeding program, which includes as main selection criteria: clean fleece weight (CFW), fibre diameter (FD), yearling weight (YW) and faecal egg count (FEC). For these traits, the genetic progress could be improved using GS by enhancing the accuracy of the expected progeny difference (EPDacc). This work presents the preliminary evaluation of the impact of GS in Merino based on a two-step selection decision using panel of different densities. The EPDacc was computed based on the utilization of a low density chip (500SNP) in 1015 animals from stud-flocks and the Ultrafine Selection Nucleus (NUG). The GS program costs were calculated considering the percentages of selected animals reported in Australia (same breed and similar production systems) and SNP panels currently available (Illumina OvineSNP50 and 500SNP) and under development (ISGC LD15K). Accuracy increased by 8.7, 17.0, 20.5 and 18.5% for FD, CFW, YW and FEC, respectively. The incremental cost for stud breeders for genotyping 20% of the top rams progeny with LD15K and all rams with the 50K panel would be 1500 USD/year. For the NUG, the cost of genotyping the 50% top ram progeny with LD15K, all rams with the 50K panel and all ewe lambs for parentage identification (500 SNP) would be 4,310 USD/year. In 2015-2018, genotyping costs for building the training population (n=3,000) will be funded by research projects. Later, GS will depend on commercial stud breeders decision of paying genotyping service. The preliminary results indicated that GS incremental cost for stud breeders is similar to the price of three average rams, making it cost-effective: breeders will benefit from more accurate selection, higher genetic gain and ram premium prices.

Impact of fitting dominance effects on accuracy of genomic prediction in layer chickens

M. Heidaritabar[1,2], A. Wolc[2,3], J. Arango[3], P. Settar[3], J.E. Fulton[3], N.P. O'sullivan[3], J.W.M. Bastiaansen[1], J. Zeng[2], R. Fernando[2], D.J. Garrick[2] and J. Dekkers[2]
[1]Wageningen University, Animal Breeding and Genomics Centre, Wageningen, 6700 AH, the Netherlands, [2]Iowa State University, Department of Animal Science, Ames, IA 50011-3150, USA, [3]Hy-Line International, Dallas Center, IA 50063, USA; marzieh.heidaritabar@wur.nl

Thus far, most genomic prediction studies fit only additive effects in models to derive genomic breeding values (GEBVs). However, if dominance is an important source of genetic variation for complex traits, accounting for it may improve accuracies of genomic prediction. We investigated the effect of fitting dominance in addition to additive effects on accuracy of GEBVs for 9 traits: egg production, age at sexual maturity, average egg weight, weight of the first 3 eggs, shell quality, albumen height, egg colour, egg colour of the first 3 eggs and yolk weight in a purebred line of brown-egg layer chickens using 42k SNP chip data. Phenotypes were corrected for hatch-date. Additive and dominance variances and SNP effects were estimated using the BayesC method. The GEBVs were predicted using 2 models: one model included both additive and dominance effects and another model included only additive effects. The training population consisted of 1,854 birds hatched between 2004 to 2009, whereas the 302 youngest animals hatched in 2010 were used for validation. Accuracies were computed as the correlation between the phenotypes and GEBVs divided by the squared root of heritability. Dominance effects accounted for 3 to 5% of phenotypic variance, depending on the trait. Prediction accuracies ranged from 0 to 0.60. No improvement in the accuracies of genomic prediction were observed in the models with both additive and dominance effects compared to the models with only additive effects. Moreover, GEBVs (additive component) and total genotypic values had similar predictive ability. We concluded that in this population fitting dominance effect did not have substantial impact on accuracy of genomic prediction.

Genetic parameters for carcass and meat traits using genomic information in Nellore cattle

D. Gordo, R. Espigolan, R. Tonussi, A. Magalhães, G.A. Fernandes Júnior, T. Bresolin, W. Andrade, F. Baldi, R. Carvalheiro, L. Chardulo and L. Albuquerque
Unesp, Via de Acesso P. D. Castellane, 14884900, Brazil; danielmansangordo@gmail.com

Genetic parameter estimates are important tools for animal breeding programs in order to support selection decisions. The aim of this study was to estimate genetic parameters for carcass (CARC) and meat quality (MQ) traits as well as their genetic associations in Nellore cattle. A total of 1,759, 1,753, 1,570, 1,827 and 1,826 records of 12th-13th rib Longissimus muscle area (LMA), backfat thickness (BF), hot carcass weight (HCW), marbling score (MS) and shear force (SF), respectively, were used. Animals were genotyped with Illumina HD panel, which contains about 777k SNP. After quality control, 321,633 SNP were left. The animal model included fixed effects of contemporary group (farm and year of birth, and management group at yearling) and age of animal at slaughtering as covariate (linear). The (co)variance components were estimated by Bayesian inference using a multitrait ssGBLUP analysis. Heritability estimates (SD) for LMA, BF, HCW, MS and SF were 0.31 (0.06), 0.21 (0.06), 0.20 (0.05), 0.09 (0.02) and 0.13 (0.04), respectively. The estimated genetic correlations between LMA and MQ were positive and high with MS (0.62 ± 0.13) and low and negative with SF (-0.10 ± 0.21). Low genetic correlations were estimated between BF and SF (0.24 ± 0.27) and between HCW and MS (-0.23 ± 0.16), indicating that these traits are not controlled by the same set or linked genes. The moderate negative genetic correlation between BF and MS (-0.47 ± 0.16) suggests the need for simultaneous attention for both traits. Similar result was obtained for the estimated genetic correlation between HCW and SF (-0.57 ± 0.20). CARC traits have enough genetic variation for selection purposes. In general, CARC were genetically associated to MQ, however, due to the low heritability estimated for MQ, genetic correlated response will be low. São Paulo Research Foundation Grant # 2009/16118-5 and 2014/11537-8.

Animal feed and human food: competition and synergies

H. Steinfeld and A. Mottet
FAO, Vialle delle Terme di Caracalla, 00153, Italy; henning.steinfeld@fao.org

Livestock are estimated to use about 30% of global crop production. Because of growing population and incomes, especially in urban areas, global demand for livestock products is rapidly increasing and so is the sector's consumption of grain. This trend is particularly strong in developing countries where most of the current sector's growth takes place. Globally, livestock production is shifting from extensive systems to large-scale industrial ruminant systems, and towards monogastrics that rely on concentrate feed. This transition has been enabled by a surplus of grain in parts of the world, largely as a consequence of improvements in crop productivity. However, with rising costs of grain for food, the past decade has seen a growing public debate on the use of food-grain for the production of high value animal protein, especially for ruminant production where the efficiency of feed is low. Increases in grain prices affect poor consumers whose diets are made up of a larger proportion of grains compared to diets of an expanding urban middle-class. The competition for grains also raises the issue of the displacement of crop production with that of feed-crops – a trend particularly visible in Latin America. Other driving forces, especially the rise in oil prices and the diversion of grain to bio-energy production, also play a key role in shaping the food-feed competition. We estimate that feed crops account for about 20% of the global livestock feed rations in dry matter; grass has the highest share (40%) and crop residues, such as straws and stover, account for about 25%. Agricultural by-products, such as brans, pulp, molasses and cakes, contribute 8% to total feed. Livestock therefore directly contribute to food security as a transformer of non-edible resources into food and high nutritional value protein. In addition, livestock is a pillar of resilience, playing the role of a natural buffer during periods of shortfall or extreme events, by adjusting to changes in supply and hence maintaining food consumption in the case of supply shortfalls.

How much meat should we eat – the environmental benefit of feeding food waste to pigs

H.H.E. Van Zanten[1,2], B.G. Meerburg[1], P. Bikker[1], H. Mollenhorst[2] and I.J.M. De Boer[2]
[1]Wageningen University and Research Centre, De elst 1, 6708 WD Wageningen, the Netherlands,
[2]Wageningen University, Department of Animal Sciences, De elst 1, 6708 WD Wageningen, the Netherlands;
hannah.vanzanten@wur.nl

In the years to come global pressure on land use will intensify. Livestock production currently uses about 70% of the agricultural land, mainly for pasture and production of feed crops. Feeding nine billion people the required protein content of 57 g/day in 2050, is only possible when resources are used efficiently. From a land use perspective, this implies that we should ensure that the production of animal proteins results in the same amount or more human digestible protein produced per m^2 compared to crops. This can be achieved by feeding waste products from human diets to pigs. The aim of this study was two-fold: to assess how much pork can be produced based on waste products from a vegan diet and to determine the environmental benefit related to this. Based on a vegan diet of an average European person, 158 kg waste is available per year, for example from the production of margarine or bread. A pig diet based on waste products only did not meet the nutritional requirements of the pig, and therefore, the pig diet was supplemented to prevent health and welfare issues. The amount of waste products in the pig diet was optimized such that the amount of human digestible pork protein produced per m^2 is similar to crop cultivation (i.e. 0.18 m^2), resulting in a pig diet with 70% waste products. These waste products were used in the fraction of availability originating from the human vegan diet. Preliminary results showed that 114 g meat, containing 36 g of protein, can be consumed per day. The production of the daily required 57 g protein based on these waste-fed pigs, resulted in a global warming potential of 0.56 g CO_2-e and a land use of 0.18 m^2, whereas based on plant protein this was 38 g CO_2-e and 0.18 m^2. We, therefore, concluded that pigs fed with human food waste can have a role in future protein supply in human diets.

Updating and renovating the INRA-AFZ multispecies feed tables

G. Tran[1], V. Heuzé[1], P. Chapoutot[1,2,3] and D. Sauvant[1,2,3]
[1]AFZ (Association Française de Zootechnie), 16 rue Claude Bernard, 75231 Paris Cedex 05, France,
[2]AgroParisTech UMR 791 MoSAR, SVS, 16 rue Claude Bernard, 75231 Paris Cedex 05, France, [3]INRA
UMR 791 MoSAR, Phase, 16 rue Claude Bernard, 75231 Paris Cedex 05, France; gilles.tran@zootechnie.fr

The INRA-AFZ 'Tables of composition and nutritional values of feed materials (ruminants, pigs, poultry, rabbits, horses and fish)' (2002-2004, available in French, English, Spanish and Chinese) are being updated and renovated. This update is driven by: (1) the demand of animal feed professionals; (2) the increase in knowledge in animal nutrition, which requires new documents for synthesising such information; and (3) the experience gained from Feedipedia (www.feedipedia.org), the online encyclopaedia on feeds developed by AFZ, INRA, CIRAD and FAO. Feedipedia is helpful to identify the feeds commonly used in Europe and elsewhere, and demonstrates the need for new chemical and nutritional values for a more substantial range of feeds. The new Tables, published by INRA, CIRAD and AFZ (with the support of Ajinomoto Eurolysine), will be available on-line (in French and English) and in print. The innovations introduced in 2002 will be carried over to the 2015 edition. (1) Each feed material will have nutritive values for the main livestock species. (2) Feed compositions will be calculated from the database maintained by AFZ, and values will be corrected to provide consistent compositions. (3) The nutritive values will be calculated from regressions determined for each family of feed materials. The regressions used in 2002 will be updated and the new regressions will be made available to all users. The INRA-CIRAD-AFZ Tables will offer new types of information, including nutritional values for ruminants derived from the INRA Systali programme, and possibly values from other feed unit systems developed in Europe thus making the Tables a first step towards the harmonisation of European feed units.

Protein digestion in broiler: what are the specificities induced by the protein source in the diet?

E. Recoules[1], H. Sabboh-Jourdan[1], A. Narcy[1], M. Lessire[1], G. Harichaux[2], M. Duclos[1] and S. Réhault-Godbert[1]
[1]INRA, UR0083, Recherches Avicoles, Centre Val de Loire, 37380 Nouzilly, France, [2]INRA, Plate-forme
d'Analyse Intégrative des Biomolécules, Laboratoire de Spectrométrie de Masse, Centre Val de Loire, 37380
Nouzilly, France; emilie.recoules@tours.inra.fr

Improving the use of unconventional (different from soybean meal) protein sources in broiler diets requires a better understanding of protein digestion. This study aims to characterize protein digestion using proteomic approaches in broilers (n=72) fed with soybean meal (S), rapeseed meal (R), pea (P) or corn distiller's dried grain with solubles (C) as the unique protein source in the diet between 7 and 21 days. At day 21, birds were euthanized and contents of digestive compartments (gizzard, duodenum, jejunum, ileum) were collected (n=12 per diet). The pH of digestive contents was measured and soluble proteins were analyzed by SDS-PAGE. Results indicated that the diet affected the pH of the digestive contents only in the gizzard (P: 2.48±0.43 vs S: 4.61±0.22; P<0.05). From the SDS-PAGE analysis, 5 bands of 25, 26, 27, 36 and 55 kDa were present in all the digestive compartments of the 4 diets and 3 bands were diet-specific (S: 18 kDa; P: 16 and 24 kDa). Protein composition of these bands was analyzed in the jejunum by mass spectrometry analysis. Forty-two Gallus gallus proteins and 17 plant proteins were identified. Among the chicken proteins, 18 were common to all diets, 8 were identified in either two or three diets and 16 were diet-specific (S: 13 and P: 3). This study allowed a better characterization of the main actors involved in protein digestion, a confirmation of the existence of proteins that were by now only predicted and the identification of protease inhibitors in S and P diets. Therefore, better knowledge about proteases involved in digestion and specificities related to protein sources in the diet would be a way to identify levers of action to improve the efficiency of utilization of unconventional protein sources by broilers.

Yield and amino acid composition of pulp and protein extracted and recovered from legumes and grass
V.K. Damborg[1], L. Stødkilde-Jørgensen[1], S.K. Jensen[1] and A.P.S. Adamsen[2]
[1]Aarhus University, Animal Science, Blichers Alle 20, 8830 Tjele, Denmark, [2]Aarhus University, Engineering, Hangøvej 2, 8200 Aarhus N, Denmark; vinni.damborg@anis.au.dk

Green biomasses such as forage legumes and grasses contain protein in an amount and with an amino acid profile mostly complementary to animal requirements. However, green biomasses contain a large amount of water (80-90%), and a large proportion (30-70%) of the dry matter is composed of fibres indigestible for monogastric animals. Green biomasses are a relatively cheap and readily available alternative to the current sources of protein. The purpose of this study was to examine the recovery of protein from juice extracted from white clover, red clover, lucerne and ryegrass, and compare the four crops in terms of protein composition and extractability. The legumes and the grass were all separated into a liquid fraction (juice) and a solid fraction (pulp) using a commercial laboratory scale screw press. Protein was recovered from the juice using two techniques; a two-step heat precipitation and pH-induced precipitation. The heat precipitation resulted in two protein fractions – green and white. The green protein fraction recovered at 60 °C had true protein contents of 25-41%, corresponding to 64-79% of the protein in the juice. The white fraction was recovered from the residue at 80 °C and had true protein contents of 22-48% DM. However, the recovered white fraction was small; less than 1% of the input mass, corresponding to 3-4% of the protein in the juice. pH-induced precipitation resulted in a single protein fraction. Low pH-values (3-4) gave the highest crude protein contents of the resulting fraction (26-41% DM). In general, the highest protein content in DM was found in white clover and lucerne juice. Ryegrass was found to have the lowest protein content. The results indicate that high quality protein in terms of amino acid composition and content for monogastric animals can be extracted from green forages.

Nonstarch polysaccharide composition influences the energy value of grains and co-products
N.W. Jaworski and H.H. Stein
University of Illinois, 1207 Gregory Dr, Urbana, IL, USA; njawors3@illinois.edu

Three experiments investigated aspects of nonstarch polysaccharide (NSP) composition and digestibility in 6 ingredients commonly fed to pigs. The 6 ingredients included 3 grains (corn, sorghum, and wheat) and 3 co-products [corn distillers dried grains with solubles (DDGS), sorghum DDGS, and wheat bran]. In Exp. 1, the NSP composition of the 6 ingredients was determined. Results indicated that cellulose, arabinoxylans, and other hemicelluloses made up 21.6, 48.6, and 29.8%, respectively, of the NSP in corn, 22.9, 44.3 and 32.8%, respectively, of the NSP in sorghum, and 14.3, 63.0 and 22.8%, respectively, of the NSP in wheat. The NSP make up in the grain co-products was similar to the parent grain, but, NSP concentration was greater in co-products, indicating that NSP composition is not influenced by processing. In Exp. 2, *in vitro* digestibility of NSP in the ingredients was determined. *In vitro* total tract digestibility of NSP was 33.6, 32.7, 12.9, 26.9, 44.9 and 20.6% in corn, corn DDGS, sorghum, sorghum DDGS, wheat, and wheat bran, respectively, indicating that NSP composition influences the extent of NSP digestibility. In Exp. 3, the effects of including 0, 15, or 30% wheat bran in a corn-soybean meal based diet fed to 18 pigs (initial BW: 54.4±4.3 kg) was determined using indirect-calorimetry. Results indicated that the apparent total tract digestibility of DM, GE, ADF, and NDF linearly decreased ($P<0.01$) as wheat bran inclusion increased. The daily O_2 consumption and CO_2 and CH_4 production by pigs linearly decreased ($P<0.01$) resulting in a linear decrease ($P<0.05$) in heat production as dietary wheat bran increased. The NE (1,808, 1,575 and 1,458 kcal/kg) of diets linearly decreased ($P<0.01$) as wheat bran inclusion increased. In conclusion, the amount and type of NSP in ingredients and diets plays an important role in determining the extent of NSP fermentation; therefore, NSP composition influences the energy value of ingredients and diets.

Alternative protein sources for monogastrics: composition and functional assessment
S.K. Kar[1], A.J.M. Jansman[2], L. Kruijt[2], E.H. Stolte[1], N. Benis[1], D. Schokker[2] and M.A. Smits[1,2]
[1]WUR, Host-Microbe Interactomics, De Elst 1, 6708 WD Wageningen, the Netherlands, [2]WUR LR, De Elst 1, 6708 WD Wageningen, the Netherlands; soumya.kar@wur.nl

This study was designed to address key questions concerning the use of alternative protein sources for animal feeds and addresses aspects such as their nutrient composition and impact on gut function, the immune system and systemic physiology. We used casein (CAS), partially delactosed whey powder (DWP), spray dried porcine plasma (SDPP), soybean meal (SBM), wheat gluten meal (WGM) and yellow meal worm (YMW) as protein sources. Advanced proteomic and in silico procedures were applied to approximate the amino acid composition and potential bioactive properties of the selected protein sources. We detected and semi-quantified 70, 130, 210, 748, 586 and 43 different individual proteins in CAS, DWP, SDPP, SBM, WGM and YMW, respectively. Considering the proteins that constitute 90% of the total protein (3, 3, 25, 68, 74 and 13 proteins in CAS, DWP, SDPP, SBM, WGM and YMW, respectively) we concluded that these protein sources are rich in bioactive peptides, in particular of peptides with angiotensin-converting enzyme inhibitor and antioxidative properties. Mouse and pig experiments were performed to evaluate effects of diets containing these and some other alternative protein ingredients such as rape seed meal and black solider fly protein. To discern changes in intestinal mucosal gene expression, intestinal microbiota composition and systemic immunity, genome-wide gene expression profiling in intestinal tissues, 16S rRNA gene (rDNA) sequencing of intestinal microbiota, and multiplex detection of cytokines and chemokines in serum were used. Overall responsiveness of these diets shows different effects on the expression of several gene sets in the ileal mucosa, on the diversity of ileal microbiota, and minor effects on blood immune parameters. The results of these studies will be presented and discussed with regard to the potential use of these alternative protein sources in animal feeds.

Nutritional value of seaweed for ruminants
M.R. Weisbjerg[1], M.Y. Roleda[2] and M. Novoa-Garrido[2]
[1]Aarhus University, Animal Science, 8830, Denmark, [2]Bioforsk, Bodø, 1430 Ås, Norway; martin.weisbjerg@agrsci.dk

Traditionally, seaweeds were used as feed for ruminants in coastal areas especially during periods of scare feed supply. However, this practice has been abandoned and nearly forgotten except in island communities where grazing animals feed on beach-cast seaweeds. The growing need for both feed and alternative protein sources, the wish for locally produced feeds, and the potential for using underutilized biomass from the sea renewed the interest in using seaweed as animal feedstock. But very little is known on the feed value of seaweed, both on chemical composition and digestibility, and also regarding variation between species and between seasons. In this study, seven seaweed species (*Mastocarpus stellatus*, *Porphyra* sp., *Alaria esculenta*, *Palmaria palmata*, *Pelvetia canaliculata*, *Acrosiphonia* sp., *Laminaria* spp.) were sampled in spring (March) and autumn (October-November) 2014 at the coast of Bodø in north Norway, and were analyzed for chemical composition and *in vitro* digestibility. Sampling and analyses will also be conducted in 2015; here, only the 2014 data are shown. *In vitro* organic matter (OM) digestibility (rumen fluid method) varied considerable between species (P<0.0001) but not between seasons; highest *in vitro* OM digestibility was found for P. palmata (85.1%) and lowest for P. caniliculata (34.6%). Ash content in dry matter (DM) was generally high (overall mean 19.1% in DM) and varied considerably, both between species (P=0.03) and between seasons (P=0.02). CP concentration in OM varied both between species (P=0.0003) and seasons (P=0.003). Highest CP in OM was found for *Porphyra* sp. (39.8%), and lowest for *P. caniliculata* (11.5%). In spring samples, CP in OM was 7%-units higher than in autumn samples. These preliminary results shows a huge variation in composition and digestibility of seaweed species, and that protein concentration is lower in autumn than in spring harvest. Several of the investigated species could be interesting as energy and protein feeds for ruminants.

Are byproduct feed systems appropriate for high producing dairy cows?
M. March, M.G.G. Chagunda, J. Flockhart and D.J. Roberts
SRUC, Edinburgh, EH9 3JG, United Kingdom; maggie.march@sruc.ac.uk

Feeding non human edible by-products to dairy cows to produce palatable protein could be construed as a sustainable feeding strategy as farmers can utilise surplus foodstuffs, however the appropriateness of this feeding system for high yielding cows hasn't been quantified. Suitability was examined by evaluating cow health and fertility traits, and milk quality characteristics, of two genetic lines of Holstein Friesian cows fed a diet consisting solely of by-product (BP) feeds. Data were obtained from the SRUC systems study. Genetic lines are selected for milk fat and protein, either the top 1% of UK genetics (S) or chosen to remain close to average genetics (C). Feeds consisted of straw, sugar beet pulp, breakfast cereal, biscuit meal, distillers' grains, molasses and soya bean meal. Approx. 50 cows in the two genetic groups were continuously housed and managed to calve all year round. The regime was developed to meet the criteria of a complete by-product diet and could be considered as an extreme because farmers traditionally use these feedstuffs, to complement forages or other imported feeds. Results show that S cows achieved a 26% higher avg. yield than C cows, 10,602 litres for S vs 8,447 for C. Daily yields differed significantly between groups (P<0.001). Select cows consumed more food at 327 tons/year than the C group which averaged 296 tons. Daily DM intakes were significantly different (P<0.001) with means of 20.3 and 22.69 for C and S groups. Milk fat and protein levels were significantly different between groups (P<0.001), with S group means of 3.67 for fat, and 3.17 for protein, whereas the C group generated means of 3.34 for fat and 3.02 for protein. Milk from C cows was of poor quality, as penalties apply at levels of fat below 3.5%, and protein below 3%. Body condition scores and locomotion scores differed significantly between groups (P<0.001). Results show that dairy cows of diverse genetic merits have the ability to produce higher than average yields solely from BP however milk quality was compromised for animals of average merit.

Testing a moist co-product for dairy cows consuming grass silage based diets
A.S. Chaudhry
Newcastle University, Agriculture, Food and Rural Development, Agriculture Building, NE1 7RQ, Newcastle upon Tyne, United Kingdom; abdul.chaudhry@ncl.ac.uk

Moist co-products are valuable feeds when fresh or conserved forages are limited for ruminants. Their nutritive value can vary and so they should be evaluated against dry feeds. We tested the effect of feeding two concentrates containing either a wheat-based moist feed (Treatment) or rolled wheat (Control) as the test ingredients on yield and composition of milk from dairy cows. Each concentrate was daily mixed with ryegrass silage at 32:68 ratio as total mixed ration (TMR) with 210 g CP and 12.5 MJ ME per kg DM. Seventy-two Holstein-Friesian cows were divided into Treatment and Control groups and loose housed indoors. Each TMR was fed once daily to a cow groups also receiving 2 kg of Distillers' grains per cow in milking parlour. Daily milk yield and composition was recorded over 4 winter months. Mean daily dry matter intake (DMI) of each TMR/cow was uniform (20.19 vs 20.15 kg for Treatment and Control group respectively) for both groups. Daily milk yield and total cell counts per cow did not vary (P>0.05) between groups during various months. While, milk fat and protein contents were greater in Treatment than the Control cows per each month, the differences were significant (P<0.05) only for November and December for fat and January for protein. Overall, Treatment cows tended to have a non-significant increase (P>0.05) in mean daily milk yield by 0.144 kg than the Control cows. The fat (46.2 vs 43.7) and protein (34.5 vs 33.5) contents (g/kg) of milk were also increased significantly (P<0.001) in Treatment compared with Control cows. Mean cell counts remained within acceptable limits (P>0.05). The cows consuming moist feed based TMR remained in good health as reflected by their intake, production, cell counts and general appearance. It would appear that this moist feed can replace rolled wheat in TMR. However, it would be essential to consider the storage, economic and environmental impacts of using such moist feeds in silage based dairy rations.

Systool Web, a new one-line application for the French INRA

P. Chapoutot[1,2,3], O. Martin[1], P. Nozière[4] and D. Sauvant[2]
[1]INRA UMR 791, rue Claude Bernard, 75005 Paris, France, [2]AFZ, rue Claude Bernard, 75005 Paris, France, [3]AgroParisTech UMR 791, rue Claude Bernard, 75005 Paris, France, [4]INRA UMR 123, Theix, 63122 Saint-Genès-Champanelle, France; patrick.chapoutot@agroparistech.fr

The new INRA feed unit system for Ruminants ('Systali') integrates the effects of feeding level (FL), proportion of concentrate (PCO) and rumen protein balance (RPB) on digestive processes. A new on-line tool, Systool Web (SW) (www.systool.fr), was developed by AFZ in order to (1) precisely explain this new model, (2) easily estimate the renewed feed and ration nutritive values and nutrient flows, and (3) allow comparisons with the actual INRA 2007 system. The e-learning module of SW precisely describes all the equations of the Systali model and illustrates all the chaining variables in neighborhood-, predecessor-, and successor-diagrams. The calculation module of SW works with uploaded and downloaded files. The input user-created files include animal characteristics (species, body weight, dry matter intake) and ration description (ingredients, diet composition, etc.). Users can either choose feeds from the 2007 INRA Table or define their own feeds in a specific user-table. SW calculates 'Table' feed values in the Systali system, with a reference value for FL and the hypothesis of PCO=0 and RPB=0. For a given ration, the actual FL and PCO values, as input variables, give initial estimates of digestive interactions. Then, RPB, calculated as an output variable, modulates these interactions through an iterative calculating process, which rapidly converges toward final values of feeds, when included in the rations ('Ration' feed values). Then, the nutritive values of the rations, and the nutrient flows, are calculated according to the feed proportions in the diet. SW allows to match feed and ration values in both INRA 2007 and new Systali models, and also to compare them to published data or predicted values from other unit systems. So, SW could contribute to harmonize European feed unit systems.

Recovery of cheese whey, a by-product from the dairy industry for use as an animal feed

E. El-Tanboly and M. El-Hofy
National Research Center, Dairy Science, Dokki, Tahrir Street, Cairo, 1344 Dokki, Egypt; tanboly1951@yahoo.com

The objective of this research was to determine whether whey, a by-product from a cheese manufacturing process, could be used as an animal Feed. Considering that over 190×106 tons of whey is produced worldwide annually, the desire for new methods to utilize whey can be appreciated. Substantial effort was devoted to the applicability of recovering whey for ruminant feed and the methodology of demonstrating its suitability. Sheep were judged as a suitable alternate ruminant to carry out preliminary trials to allow scientific investigation of the suitability and the value of the whey. Eight week experiments for a six pen trial of five sheep per pen were undertaken in which sheep 8-10 months old, weighing 37-39 kg were fed free-choice six different liquid feeds next to ration components of roughage (berseem hay, Dry Matter (DM): 87.4%, Crude Protein (CP): 13.6%) and mix feed (cotton seed meal; yellow corn; wheat bran; salt; limestone, DM: 89.5%, CP: 17.3%). Benefits and achievements results it could be concluded that for farms: Whey replaces 100% of water intake at the farm, Whey provides a low-cost alternative to liquid feeds, at a fraction of the cost (less than 10% of molasses), 19 litres of liquid whey permeate can replace the same amount of energy and protein as provided by 2.4 kg of a 88% crude protein feed mix/roughage, Roughage intake per kg gain can be reduced from 3 to 1 kg (75% weighing (200 kg/head) being fattened to 400 kg, an additional income of $30/head will be achieved, Whey can improve the feed palatability, texture, and dust control of feedlot rations. It provides a balanced nutrition of energy, protein, minerals, and a safety factor to compensate for poor or variable quality diets, Being a pumpable supplement, whey can save on feeding overheads as it requires less labour and feeding and mixing equipment, and can provide an economic and convenient method to feed urea supplements, vitamins, minerals and feed additives.

Prediction of cold-pressed rapeseed cakes chemical composition using NIRS

I. Zubiria, E. Alonso, A. Garcia-Rodriguez and L. Rincón
Neiker-Tecnalia, Granja Modelo de Arkaute, E-01080, Spain; aserg@neiker.net

With the increasing cost of feed energy and decreasing supply of traditional feedstuffs, a great deal of effort has been directed towards exploring the usefulness of locally alternative feedstuffs in feed formulations. As such, cold-pressed rapeseed cake is a byproduct of biodiesel production that could be used as a valuable source of energy and protein. However, the nutritive value of these cakes depends largely on the processing conditions under which it is performed. Near infrared reflectance spectroscopy (NIRS) can predict chemical values for different feedstuffs. The objective was, therefore, to determine the nutritive value of cold-pressed rapeseed cake using NIRS. Samples (n=90) were collected after mechanically pressing 10 different varieties of rapeseed using 3 different screw rotation speeds combined with 3 different pressure washers. A FOSS NIRSystems model 6500 was used to measure reflectance spectra from 400 to 2,498 nm, every 2 nm. The modified partial least squares (MPLS) regression method was used to obtain NIR calibration equation using full cross-validation. NIRS prediction accuracy was tested after separating the sample dataset into validation (n=10 samples), and in calibration (n=80 samples) datasets. No outlier selection was used. The samples represented a wide range of chemical composition. The dry matter (DM) content ranged from 904 g/kg DM to 916 g/kg DM, the crude protein (CP) content from 232 to 307 g/kg DM, the ether extract (EE) from 186 to 343 g/kg DM, the neutral detergent fiber (NDF) from 181 to 248 g/kg DM and acid detergent fiber (ADF) from 130 to 190 g/kg DM. MPLS regression models accounted for 99.3, 95.4, 93.7 and 87.5% of the variation existing in the validation set on CP, ADF, EE and NDF, respectively. The prediction equations yielded a SEP of 1.8, 3.9, 5.6 and 11.8 g/kg DM for CP, ADF, NDF and EE, respectively. In conclusion the equations for CP, ADF and NDF of cold-pressed rapeseed cakes showed excellent capacity for prediction whereas EE was not so well predicted.

Insect or legume based protein sources to replace soybean cake in a broiler diet

F. Leiber[1], T. Gelencsér[2], H. Ayrle[1], Z. Amsler[1], A. Stamer[1], B. Früh[1] and V. Maurer[1]
[1]Research Institute of Organic Agriculture, Ackerstrasse 113, 5070 Frick, Switzerland, [2]ETH Zurich, Universitaetstrasse 2, 8092 Zurich, Switzerland; florian.leiber@fibl.org

Demands for soybean as a protein source in poultry production are high, and alternatives are needed in Europe. Potential options are grain legumes, conservates of forage legumes like alfalfa, and insect-based products. Combinations of these components were tested in experimental broiler diets against a commercial feed for organic broilers containing 255 g/kg soybean cake, 200 g/kg crude protein [CP] in the diet. In each experimental diet 130 g/kg soybean cake were replaced. Diet HermAlf contained 78 g/kg Hermetia meal (dried pre-pupae of the black soldier fly, Hermetia illucens, raised on food industry by-products) and 52 g/kg alfalfa meal (CP: 200 g/kg in the diet). Diet HermPea contained 78 g/kg Hermetia meal and 52 g/kg peas (CP: 200 g/kg in the diet). Diet AlfPea contained 78 g/kg alfalfa meal and 52 g/kg peas (CP: 170 g/kg in the diet). Diet PeaAlf contained 78 g/kg peas and 52 g/kg alfalfa meal (CP: 170 g/kg in the diet). Respectively 15 broilers (Hubbard JA757) were fattened from day 7 to day 82 of life. They received the respective diet, water and wheat grains *ad libitum*. Feed intake was calculated group-wise, daily gains, live weights, carcass weights, carcass and meat quality (Warner-Bratzler shear force and cooking loss after 45 min at 72 °C) were measured individually (n=15 per group). Data were analysed with SPSS© in a general linear model and a post-hoc Tukey's test for multiple comparisons. No differences of any experimental diet from control were found for feed intake, daily weight gain, carcass weight or feed efficiency. However, for each of these parameters some experimental groups differed between each other, always in advantage of the Hermetia diets. Shear force was not affected but cooking loss was increased with HermPea, compared to the control. The results indicate that the compounds tested could serve as replacers for part of soybean products in broilers diets.

Effects of the full fat walnuts waste on the fatty acids composition of the piglet's tissue

M. Habeanu, D. Voicu, N.A. Lefter, A. Gheorghe, M. Ropota, C. Vlad and E. Ghita
National Research Development Institute for Animal Biology and Nutrition (IBNA), Animal Nutrition, Calea Bucuresti, Nr. 1, Ilfov, Balotesti, 077015, Romania; dorica.voicu@ibna.ro

The objective of this study was to investigate the effects of dietary full fat walnuts waste provided by food industry on the fatty acids (FA) composition of the liver and heart of hybrid Topigs piglets. These tissues are known by their role in body lipid metabolism and type, sources and concentration of FAs delivered may affect the health. The trial was conducted on 19 animals, 12±1.12 kg, randomly assigned for 29-d experimental period, to two groups namely a control (C diet), and an experimental (E diet) with the additional inclusion of 50 g walnuts/kg diet. The FA composition of the organs were determined by gas chromatography. The data were submitted to variance analysis with SPSS, Anova – GLM. The full fat walnuts waste used in our trial had 42% fat and 17% protein, 72% polyunsaturated FA (PUFA) of total FA ester methyl (FAME), especially due to the higher level of the C18:2n-6 fatty acids (60.78%). The addition of 5% walnuts waste in E diet had modified FA composition of both liver and heart tissues promoting the deposition of C18:3n-3 and C18:2n-6 mainly to the detriment of C18:1cis-9. However, they did not favor significantly the deposition of saturated FA in the two organs, but had increased the deposition of LC n-3 PUFA, especially of C20:5n-3 (P=0.02). The level of total n-3 FA was more pronounced affected in the liver than in the heart (>3.59 times). We conclude that, the addition of 5% walnuts waste as dietary sources of PUFA to piglets, modified the level of C18:3n-3 and C18:2n-6 the effects being more pronounced in the heart. However, LC n-3 PUFA and the total n-3 PUFA were influenced especially in the liver. Consequently, n-6/n-3 ratio was significantly lower (P<0.0001) in the liver than in the heart (5.59 compared to 11.6) and in E diet compared to C diet whatever the organs (<35.7% in liver and <15.68% in the heart).

Supplementation of *Tenebrio molitor* larva on growth performance and nutrient digestibility in pigs

X.H. Jin, N.S. Ok, J.H. Jeong, P.S. Heo, J.S. Hong and Y.Y. Kim
Seoul National University, Department of Agricultural Biotechnology, College of Animal Life Sciences, 1 Gwanak-ro, Gwanak-gu, 151-742, Seoul, Korea, South; nsomganbc@naver.com

This experiment was conducted to evaluate the effects of dietary *Tenebrio molitor* larva on growth performance, nutrient digestibility and palatability in weaning pigs. A total of 120 weaning pigs (28±3 days and 8.04±0.08 kg of BW) were allotted to one of 5 treatments in 6 replicates with 4 pigs per pen by randomized complete block (RCB) design. Five different levels of *T. molitor* larva (0, 1.5, 3.0, 4.5 or 6.0%) were used as dietary treatments. Two phase feeding programs (phase I for 0-14 day, phase II for 15-35 day) were used in this experiment. Body weight, ADG and ADFI were improved linearly as dietary *T. molitor* larva increased in phase I diet. During phase II, increasing level of *T. molitor* larva in diet improved the ADG (P<0.01) and ADFI (P<0.05). As dietary *T. molitor* larva was higher, nitrogen retention and digestibility of dry matter as well as protein were improved, respectively (P=0.05). In the results of blood profiles, decrease of blood urea nitrogen (P=0.05) and increase of IGF-1 (P=0.03) were observed as level of *T. molitor* larva increased in diet during phase II. Consequently, supplementation of *T. molitor* larva up to 6% in weaning pigs' diet could improve growth performance and nutrient digestibility.

Intake and growth performance of feedlot lambs fed increasing concentrations of babassu meal

E.H.C.B. Van Cleef[1], O.R. Serra[1,2], J.R.S.T. Souza[2], A.L. Lima[2] and J.M.B. Ezequiel[1]
[1]São Paulo State University, Jaboticabal, 14884-900, Brazil, [2]Maranhão State University, São Luiz, 65055-000, Brazil; ericvancleef@gmail.com

With the objective of minimizing costs of feeding, many studies have been conducted to evaluate the use of byproducts as livestock feed ingredients. Babassu (*Orbignya phalerata*) meal has potential to replace either forage or concentrate ingredients because it possesses 68% NDF and 22% CP. The objective of this study was to evaluate increasing concentrations of babassu meal partially replacing Tifton-85 bermudagrass hay on dry matter intake and feedlot performance of crossbred lambs. Animals (n=27, 90 d of age, 19.57±0.41 kg BW) were randomly assigned to one of three experimental treatments. Isoenergetic and isonitrogenous diets were composed of ground corn, soybean meal, mineral premix, Tifton-85 bermudagrass hay, and 0 (T0), 15 (T15), or 30% (T30) babassu meal. Animals were fed once daily (08.00 h) for a 14-d adaptation period and 56-d performance testing period. The feed delivered and orts were daily weighed and animals` weights were recorded every 14 d. Data were analyzed as a completely randomized design with repeated measures by using MIXED procedure of SAS. The animal was the experimental unit. Contrasts were used to determine linear and quadratic effect of babassu meal addition. The intake of DM (T0=1026, T15=996, and T30=1015 g/d), NDF (T0=504, T15=472, and T30=462 g/d), and CP (T0=202, T15=196, and T30=199 g/d) were unaffected by experimental treatments. However, average daily gain and final body weight were linearly increased (P<0.05) when babassu meal was added to the diets. With no changes in DMI and increased weight gain, feed efficiency was improved by 19.8% with inclusion of 15% babassu meal and by 43.4% with inclusion of 30% of the byproduct. In conclusion, babassu meal is a suitable feed ingredient to partially replace Tifton-85 bermudagrass in diets for finishing crossbred lambs, since it does not promote deleterious effects on feed intake and improve animal performance when added up to 30%.

Inclusion of by-product of *Myrtus communis* in the diet of Sarda sheep: effects on ruminal metabolism

F. Correddu, G. Battacone, G. Pulina and A. Nudda
University of Sassari, Dipartimento di Agraria, Viale Italia 39, 07100, Italy; anudda@uniss.it

By-product resulting from myrtle liqueur preparation of *Myrtus communis* berries could represent a suitable source of polyphenolic compounds. The present study aimed to investigate the effect of dietary supplementation of *M. communis* on ruminal metabolism in Sarda dairy ewes. Eighteen sheep were divided into 3 dietary treatment: a control diet (CON) and 2 diets supplemented with 50 and 100 g/d per head of *Myrtus* by-products (M1 and M2, respectively). Samples of rumen liquid were collected at week 3, 7 and 11 of the trial and analyzed for pH, ammonia and volatile fatty acids (VFA). Methane (CH_4) emission was estimated according to the equation: CH_4 (mM) = 0.45 × acetate − 0.275 × propionate + 0.40 × butyrate. Data were analyzed by a completely randomized ANOVA design with the PROC MIXED procedure of SAS. The model included the diet treatment, sampling and their interaction as fixed effects, and the animal nested within the treatment as random effect. When means were significantly different (P<0.05), groups were separated using Tukey's test. Ruminal pH of supplemented groups did not differ from that of CON (P>0.05), whereas M1 was lower than M2 (6.3 v 6.5; P<0.01). The ammonia content decreased in M1 and M2 compared with CON (mean of 11.32 v 14.32 ml/dl, respectively; P<0.01). The total content of VFA tended to decrease in rumen of M2 compared with CON and M1 (P=0.08). The molar proportion of acetate, propionate, butyrate, the ratio Ac/But and the estimated emission of CH_4 was not affected by the *Myrtus* supplementation (P>0.05). Supplementation with *Myrtus* by-product was able to reduce the accumulation of NH_3 in the rumen of ewes, likely due to the ability of polyphenols to bind proteins and reduce the activity and growth of proteolytic bacteria. Neither 50 nor 100 g/d of *Myrtus* were able to affect the molar proportion of VFA and consequently the ruminal production of CH_4. Research supported by Cargill srl (Animal Nutrition Division).

Utilization of cellulolytic enzymes to improve the nutritive value of date kernels

A.M. Kholif[1], E.S.A. Farahat[1], M.A. Hanafy[2], S.M. Kholif[1] and R.R. El-Sayed[2]
[1]National Research Centre, Dairy Science, Bohouth street, Dokki, Giza, 12311, Egypt, [2]Cairo University, Animal Production, El-Gamaa street, Giza, 12613, Egypt; am_kholif@hotmail.com

Two experiments were carried out to evaluate the effects of cellulases supplementation on *in vitro* degradation of date kernels (the first trial) and *in vivo* rumen fermentation and nutrients digestibility by lactating Zaraibi goats (the second trial). In the *in vitro* experiment, dry matter and organic matter disappearance (IVDMD and IVOMD) were determined for date kernels supplemented separately with Asperozym and a commercial cellulolytic enzyme source (Veta-Zyme Plus®) each at 3 levels (15, 30, and 45 U/kg DM) compared with the control. The highest values ($P<0.05$) of IVDMD and IVOMD were observed with Asperozym supplementation level at 45 U/kg DM compared to control. While, Veta – Zyme Plus® gave the highest ($P<0.05$) IVDMD and IVOMD values at 15 U/kg DM compared to control. In the *in vivo* experiment, nine lactating Zaraibi goats after 7 days of parturition were divided into three groups, three animals each, using 3×3 Latin square designs. The first group was fed 37.5% concentrate feed mixture (CFM), 12.5% date kernel and 50% berseem hay (control diet). The second group was fed control diet supplemented with Veta-Zyme Plus® at level 15 U/kg DM (T1). The third group was fed control diet supplemented with Asperozyme at level 45 U/kg DM (T2). The results indicated that Asperozym and Veta-Zyme Plus® supplementation significantly ($P<0.05$) increased nutrients digestibility, nutritive values, ruminal total volatile fatty acids (TVFA's) and ruminal ammonia nitrogen (NH_3-N) for treated groups compared with the control group.

Optim'Al, an economic optimization tool for ruminant ration calculation for better use of byproducts

P. Chapoutot[1], G. Tran[2], P. Gillet[3], D. Bastien[4] and B. Rouillé[4]
[1]AgroParisTech INRA UMR 791 MoSAR, 16 rue Claude Bernard, 75231 Paris Cedex 05, France, [2]Association Française de Zootechnie, 16 rue Claude Bernard, 75231 Paris Cedex 05, France, [3]Chambre d'agriculture 52, 26 avenue du 109 RI, BP 82138, 52905 Chaumont Cedex 9, France, [4]Institut de l'Elevage, Monvoisin, BP 85225, 35652 Le Rheu Cedex, France; benoit.rouille@idele.fr

Byproduct use in ruminant feeding systems depends on their economic interest. But economic optimization is rarely used to calculate rations, while linear programming is largely applied to formulate compound feeds. In fact, it is difficult to meet the linear constraints when the non-additive and non-linear concepts, such as forage/concentrate substitution or digestive interactions, are used in ruminant ration calculation. Optim'Al, realized by AFZ at the request of Institut de l'Elevage, enables least cost formulation of cattle rations while both applying the INRA systems and meeting the linearity and additivity principles. Optim'Al calculates feed intake, energy, nitrogen and mineral requirements according to the animal daily production. Feeds are available for calculation with their nutritive values, but users can also create their own feed tables. Incorporation limits can be applied to the feeds, individually or by groups, to ensure compliance with production specifications. The user selects different forages and concentrates to create the ration, and can impose chemical composition constraints for it. Optim'Al calculates the optimal combination of feeds, which allows to meet the animal requirements while respecting the substitution and digestive interaction principles, as well as the least cost objective. These multiple goals need several iterative calculation processes. The primal analysis gives different characteristics for the optimized ration. The dual approach studies the sensibility of the optimal solution (opportunity prices, binding constraints, invariance ranges, etc.) and allows technical and economic diagnosis to compare economic interest of byproducts and determine their optimal utilization.

Nitrogen degradability of industrial solvent-extracted rapeseed meal for ruminants

P. Chapoutot[1,2], B. Rouillé[3], P. Gillet[4], C. Peyronnet[5], E. Tormo[5], A. Quinsac[6] and J. Aufrère[7]
[1]INRA UMR 791 MoSAR, rue Claude Bernard, 75231 Paris Cedex 05, France, [2]AgroParisTech UMR 791, rue Claude Bernard, 75231 Paris Cedex 05, France, [3]Institut de l'Elevage, Montvoisin, Le Rheu, France, [4]Chambre d'Agriculture 52, avenue du 109 RI, Chaumont, France, [5]ONIDOL, rue de Monceau, Paris, France, [6]CETIOM, rue Monge, Pessac, France, [7]INRA UMR1213, Theix, 63122 Saint-Genes-Champanelle, France; patrick.chapoutot@agroparistech.fr

The solvent-extracted rapeseed meal (RSM), which is produced in France, presents a certain between- and within-factory variability. This variability can modify the nitrogen value for ruminants. The aim of the study, realized in 2012, was to compare in sacco nitrogen effective degradability (NED) and *in vitro* nitrogen enzymatic degradability (DE1) of different RSM to verify the relation between these criteria used for the prediction of nitrogen value for ruminants. In sacco and *in vitro* measures were conducted on 15 samples of RSM, chosen to be representative of the existing variability studied with 35 samples of RSM previously collected by Chapoutot et al. In sacco measurements were done as described by Michalet-Doreau et al. The values of DE1 presently observed on RSM (20.9±2.4%) were much lower than those obtained during the last 2 decades, which were exploited for the elaboration of the prediction model of NED actually used in France. This reduction of the DE1 values can be related to an increase in the temperature of the treatments used in the oil-extraction process. The results of in sacco measures (NED=69.1±2.6%) allowed to confirm the values proposed in the INRA Tables for RSM. It also permitted to validate the accuracy of the prediction model of NED, calculated for this feed on all the samples of this study plus those of Aufrère et al.: NED=62.2(±1.4) + 0.33(±0.04) x DE1 (n=28; R^2=0.74; ETR=2.8). Moreover, these new results lead to increase the precision of the estimation of NED from current DE1 values.

Influence of using alternative sources of feed for fattening cattle on production and meat quality

D. Voicu, M. Hăbeanu, I. Voicu, R.A. Uta and M. Ropota
National Research Development Institute, Animal Nutrition, Calea Bucuresti, Nr. 1, Ilfov, Balotesti, 077015, Romania; dorica.voicu@ibna.ro

A trial on 30 Romanian Black Spotted fattening steers was designed to monitor the effects of using sorghum grains and dry grape marc as alternative feed sources in the compound feed formulations on the production and quality of meat. The animals were assigned to three homogenous groups of 10 animals each. The control group (C) received a compound feed without sorghum grains and dry grape marc, group E1 received 25% sorghum grains and E2 received 20% dry grape marc. The dry grape marc is a by-product from wine making, or from table grape processing. The use of the new feed sources for groups E1 and E2 didn't influence the intake of the complete diet (CF+alfalfa haylage), while the weight gains were comparable between the three groups of animals (1,306 g for C, 1,266 g for E1 and 1,301 g/steer/day for E2), the differences not being statistically significant (P≥0.05). The fatty acids (FA) content of the Longissimus dorsi muscle of the steers was analysed on samples collected from the slaughtered animals, which were analysed by gas chromatography using a Perkin Elmer-Clarus 500 fitted with injection system in the capillary (splitting ratio 1:100), and with programmed heating of the column oven. The results have shown changes in the FA profile (% of the total methyl ester FA of the muscle). Thus, the content of omega 3 and omega 6 polyunsaturated fatty acids increased as follows: the concentration of alpha linolenic acid C18:(3n3) increased from 1.09% in group C to 1.25% in group E2 and the concentration of linolenic acid (C18:2n6) increased from 7.35% in group C to 8.90% in group E1. These fatty acids are beneficial to human health. We conclude that the total amount and the proportion of the different FA found in the meat of ruminants can be manipulated by feeding, the rumen microorganims playing an important role in the transformation of the dietary lipids depending on nutrient supply.

DDGS in the diet of finishing double-muscled Belgian Blue cows

K. Goossens, L.O. Fiems, J. De Boever, B. Ampe and S. De Campeneere
ILVO, Animal sciences unit, Scheldeweg 68, 9090 Melle, Belgium; karen.goossens@ilvo.vlaanderen.be

The effect of wheat-based dried distiller grains with solubles (DDGS) on the zootechnical performances of finishing double-muscled Belgian Blue (DMBB) cows was studied. Eighty animals were assigned to one of 4 treatments: 2 control diets with a low (3.3%, LS) or a high (21%, HS) starch content in the concentrate, and 2 diets containing DDGS, either as concentrate ingredient (DC, 63% DDGS on DM basis), or as protein supplement (DS). Diets consisted of concentrate and maize silage (50/50 on DM basis), except DS, where maize silage was supplemented with DDGS (65/35 on DM basis; 34% CP in DDGS) and a mineral-vitamin premix. All diets were fed *ad libitum*. There were no differences in daily total DM intake between the groups (on average 12.4 kg DM/day), although the DM intake of maize silage was higher for DS (P<0.001). The incorporation of DDGS in the diet resulted in a higher crude protein (CP) intake per day (P<0.001) and per kg metabolic weight. The daily CP-intake and daily rumen degradable protein balance (OEB) was highest for DS. Although formulated to be isocaloric, LS and DC diets contained more net energy for fattening (NEF) than HS (8.87 and 8.26 vs 7.38 MJ/kg DM) resulting in a higher daily NEF intake (P<0.001). NEF of DS was intermediate between HS and DC. Nevertheless there was no difference in NEF/kg weight. Daily gain ranged between 1.10 and 1.32 kg and was highest in LS, but not significantly different from the other groups. There were no differences in carcass weight, dressing percentage and carcass classification. These results indicate that DDGS can be used effectively in finishing diets for DMBB cows, without negative effects on growth performance and carcass characteristics. Supplementation of DDGS in the concentrate did not result in a reduction in concentrate or maize silage uptake. Separate supply of DDGS even resulted in a higher maize silage intake. Depending on its market price, DDGS can be used as an alternative protein supplement.

Evaluation of milk production and composition of cows fed hydrolyzed sugarcane

V. Endo, V.A. Vieira, G.C. Magioni, F.C. Gatti, A.N.O. Navarro, B. Maximo, I.M. Estima and M.D.S. Oliveira
Sao Paulo State University, Animal Science, Via de Acesso Prof. Paulo Donato Castellane, s/n, 14884-900, Brazil; endo_vica@hotmail.com

Milk has great importance for human nutrition and its main compounds are proteins, minerals and vitamins. There are advantages of using sugarcane in livestock production, such as a high energy level stimulating productivity. However, sugarcane also has disadvantages, e.g. a low crude protein level, thus requiring supplementation. Hydrolysis allows the storage of sugarcane up to three days, eliminating the need for daily cuts. Several factors affect the hydrolysis, e.g. type of lime; however the influences of straw and lime preparation in the efficiency of sugarcane hydrolysis have not been reported. So, the aim of this study was to evaluate the milk production and composition of Holstein cows under those conditions mentioned above. Five cows were distributed to treatments in a latin square design: CIN, in natura sugarcane with straw; ST0 and ST72, sugarcane hydrolyzed with straw using 0.5% calcium hydroxide ($Ca(OH)_2$) prepared zero and 72 h before hydrolysis (respectively); WST0 and WST72, sugarcane hydrolyzed without straw using 0.5% $Ca(OH)_2$ prepared zero and 72 h before hydrolysis (respectively). Cows were confined in individual stalls and roughage:concentrate ratio was 60:40. Milk production and composition were measured and the data were compared by Tukey test at 5% of significance. Milk production (12.49 kg/d) and fat-corrected milk 3.5% (12.82 kg/d) were constant between treatments. Milk composition was not influenced by the presence of straw on sugarcane, or different times of preparation of the suspension lime (0.5% $Ca(OH)_2$ in 2 l of water per 100 kg of chopped sugarcane). The value obtained for fat (3.67%), protein (3.48%), lactose (4.95%) and defatted milk (8.87%) were not affected by treatments. From the production and compositions point of view, sugarcane can be supplied to dairy cows in any way, i.e. with or without straw, and in natura or hydrolyzed condition.

Inclusion of cold-pressed rapeseed cake in dairy cow diet and its effect on milk production
I. Zubiria, R. Atxaerandio, R. Ruiz and A. Garcia-Rodriguez
Neiker-Tecnalia, Granja Modelo de Arkaute, 01080, Vitoria-Gasteiz, Spain; aserg@neiker.net

As a result of the increasing cost of the traditional energy and protein sources for feeding livestock, a great deal of effort has been directed towards exploring the usefulness of locally alternative feedstuffs obtained from the production of renewable energies in feed formulations. Cold-pressed rapeseed cake (CPRC) is a byproduct of biodiesel production that could be used as source of energy and protein. The objective of this trial was to assess the effect of feeding CPRC in the concentrate in dairy cows on concentrate intake, milk yield and composition. Six Holstein and 10 Brown Swiss dairy cows were used. Cows were paired on the basis of breed, milk yield, parity, and days in milk, after a two-week-covariate-period. Two concentrates were formulated to be isoenergetic and isoproteic. Inclusion of CPRC (20%) in the experimental concentrate reduced the content of soybean meal (20 vs 15%) and corn DDGS (14 vs 1.8%) compared to control. Concentrates were offered individually using automatic feeders. Forage was offered *ad libitum*. The experiment lasted for 10 weeks. Individual milk yields and concentrate intake were recorded daily. Individual milk samples were taken every two weeks and were analyzed for protein, fat and urea contents. Data was analyzed considering repeated measures with concentrate, week and the interaction as fixed effects, and pair and its interaction with concentrate as random effects. A reduction in concentrate intake was observed with CPRC from week 6 to 10. No significant differences were found in terms of milk yield (25.3 vs 23.9 kg/d, P=0.494), but feeding CPRC increased milk fat (4.37 vs 3.96%, P<0,001) and protein (3.43 vs 3.29%, P<0,001) with decreased milk urea (180 vs 221 mg/l, P<0,001) concentrations. In conclusion, feeding CPRC in the concentrate decreased concentrate intake from the sixth week without affecting milk yields, increased milk fat and protein content and decreased milk urea content.

Cold-pressed rapeseed cake inclusion in dairy cow diet and its effects on feeding behavior
I. Zubiria, R. Atxaerandio, R. Ruiz and A. Garcia-Rodriguez
Neiker-tecnalia, Granja modelo de Arkaute, 01080 Vitoria-Gasteiz, Spain; izubiria@neiker.net

With the increasing cost of feed energy and protein with decreasing supply of traditional feedstuffs, a great deal of effort has been directed towards exploring the usefulness of locally alternative feedstuffs obtained for the production of renewable energies. Cold-pressed rapeseed cake (CPRC) is a byproduct of biodiesel production that could be used as a source of energy and protein. Unprotected-fat-rich feedstuffs, such as CPRC, can alter rumen function leading to digestion disorders. There is evidence that discomfort in cattle can be identified through observation of changes in feeding behavior. The objective of study was to investigate the effect of CPRC inclusion in dairy cows concentrate in feeding and rumination behavior. Six Holstein and 10 Brown Swiss dairy cows were used. Cows were paired on the basis of race, milk yield, parity and days in milk. Two concentrates were formulated to be isoenergetic and isoproteic. Inclusion of CPRC (20%) in the experimental concentrate reduced soybean meal (20 vs 15%) and corn DDGs (14 vs 1.8%) formulation compared to control. Concentrates were offered individually using automatic feeders. Forage was offered *ad libitum*. The trial lasted for 10 weeks. Behavior was measured for 48 hours in the 6th week of the trial. Eating and ruminating behaviors were monitored visually. Individual behavior activities were noted every 10 min, assuming that each activity persisted for a 10-min interval. Data was analyzed considering concentrate, week and the interaction as fixed effects, and pair and its interaction with concentrate as random effects. Similar time spent eating (355 vs 359 min/d, P=0.913), ruminating (484 vs 462 min/d, P=0.431), chewing (355 vs 359 min/d, P=0.913), resting (582 vs 596 min/d, P=0.670) lying (298 vs 330 min/d, P=0.278) or standing (284 vs 265 min/d, P=0.426) was found in dairy cows eating both concentrates. In conclusion, formulation of 20% CPRC in dairy cows concentrate did not affect time spent eating or ruminating.

Pasteurized milk sensory characteristics as affected by feeding cows with cold-pressed rapeseed cake
I. Zubiria, R. Atxaerandio, R. Ruiz and A. Garcia-Rodriguez
Neiker-tecnalia, Granja modelo de Arkaute, 01080 Vitoria-Gasteiz, Spain; izubiria@neiker.net

Cold-pressed rapeseed cake (CPRC) is a byproduct of biodiesel production that could be used as a valuable source of energy and protein. Diets fed to cows can change milk composition and hence the sensory characteristics of milk. The objective of this study was to assess the effect of feeding dairy cows with CPRC in the concentrate on milk sensory characteristics. Six Holstein and 10 Brown Swiss dairy cows were paired on the basis of race, milk yield, parity and days in milk. Two concentrates were formulated to be isoenergetic and isoproteic. Inclusion of CPRC (20%) in the experimental concentrate reduced soybean meal (20 vs 15%) and corn DDGs (14 vs 1.8%) formulation compared to control. Concentrates were offered individually using automatic feeders. Forage was offered *ad libitum*. The trial lasted for 10 weeks. Milk was collected individually and pooled in four different canteens according to race and concentrate. Milk was pasteurized at 72 °C for 20 seconds and samples were presented tempered in transparent glasses and 1/3 volume of the vessel. A triangular and an acceptation test were performed with 60 people (men and women aged 38±10 years). Data of the triangular test was analyzed according to ISO 4120:2004. The acceptation test was analyzed using ANOVA considering concentrate and race as fixed effects. Acceptance of milk was on a scale of 1 to 9, with 1 being extremely disgusting and 9 extremely good. In the triangular test the evaluation panel found differences between milk from 2 diets in either Holstein (37/60 correct answers, P<0.001) or Brown Swiss cows (38/60 correct answers, P<0.001). Whereas the Holstein cow milk consumer acceptance was similar regardless of the concentrate fed (6.2 vs 6.3, P=0.998), Brown-Swiss-CPRC-fed milk showed a greater acceptance (6.3 vs 5.4, P=0.016). In conclusion consumers were able to distinguish milk from cows fed with different concentrates, showing in Brown Swiss cows a greater preference for CPRC fed cow milk.

Effect of Amaferm® on dry matter intake and milk production in high-productive dairy cows
K. Mahlkow-Nerge[1] and P.A. Abrahamse[2]
[1]Landwirtschaftskammer Schleswig-Holstein, Futterkamp, 24327 Blekendorf, Germany, [2]Cargill, Cargill Animal Nutrition, Veerlaan 17-23, 3072 AN, the Netherlands; sander_abrahamse@cargill.com

Amaferm® is a zootechnical additive which has proven to improve feed efficiency and milk production in many publications. Early lactation trials in high-producing dairy cows (30-35 kg milk/head/d) showed an increase in milk production of 2.8 kg/head/d. The current experiment was set up to test the effect of feeding Amaferm® to high-producing dairy cows in early lactation on a high-maize silage diet at Futterkamp in Germany. 71 Holstein cows with a high feed efficiency (1.6 kg FPCM/kg DMI) were divided into two groups: 36 cows (on average 116 days in milk) were fed a standard farm mineral (Rupromin Fertil) and 35 (on average 99 days in milk) were fed a similar farm mineral that included 5 g/head/d Amaferm® (Rupromin Fiber) at the advised dose (150 g/head/d). The trial setup was a completely randomized design and lasted 8 wk, with measurements being taken during wk 8. Dry matter intake (DMI), milk and fat-and protein corrected milk (FPCM) production, milk fat, milk protein and BCS were 20.9 and 21.7 kg/head/d; 33.6 and 34.6 kg/head/d; 33.8 and 34.8 kg/head/d; 4.05 and 4.09%; 3.46 and 3.40%; 3.07 and 3.12 for the treatment without and with Amaferm®, respectively. There were no significant differences between both treatments, but milk yield (P=0.5394) and FPCM (0.4286) were numerically 1 kg/head/d higher in cows fed Amaferm®.

Influence of rapeseed meal on milk production: a study in dairy cow farms of Haute-Marne (France)
C. Teinturier[1], P. Chapoutot[2], D. Bouthors[1], P. Gillet[1], C. Peyronnet[3], E. Tormo[3], A. Quinsac[4] and B. Rouillé[5]
[1]Chambre d'agriculture 52, 26 avenue du 109ème R.I, BP 82138, 52905 Chaumont Cedex 09, France,
[2]AgroParisTech INRA UMR 791 MoSAR, 16, rue Claude Bernard, 75231 Paris Cedex 05, France, [3]ONIDOL,
11 rue de Monceau, CS 60003, 75008 Paris, France, [4]CETIOM, Rue Monge, Parc Industriel, 33600 Pessac,
France, [5]Institut de l'Elevage, Monvoisin, BP 85225, 35652 Le Rheu Cedex, France; benoit.rouille@idele.fr

Technical and economic results registered in Haute-Marne department confirm the interest of rapeseed meal (RSM) in dairy rations. The objectives of this study was to measure the influence of high levels of RSM on dairy performances and to analyze the interest of RSM compared with meals. Two groups of 42 farms were made by multivariate analysis. For the RSM group, rations included at least 4 kg of RSM/day (on dry matter, DM, basis). For the NoRSM group, diets contained other nitrogen feeds. Performances, feed cost and feed margin were recorded from January to March 2014. Reproductive performances and health status were collected from the national milk recording. The sociological study was based on surveys to collect the motivations of farmers for RSM. It appeared that the use of a large amount of RSM (92% of the nitrogen corrector) in the ration of RSM group, vs NoRSM group, did not changed DM intake, milk yield and fat content, and significantly increased protein content (+0.3 g/kg, $P<0.05$). Farms of RSM group had a lower ration cost (-0.48 €/cow/day; $P<0.001$), and presented a higher feed margin (+0.42 €/cow/day; $P<0.05$). As a local and non GMO feed with low price, RSM had a good image among dairy farmers and feeding advisers. However, due to its lower energy and protein values, RSM was perceived as a less 'noble' byproduct than soybean meal. This study showed that large amounts of RSM can be included in dairy rations without affecting technical performances. The maintain of cow performances combined with the lower price of RSM compared to other sources of protein leads to an increase of the economic profit for the farmers.

Level of glycerol and parameters of *in vitro* fermentation kinetics of different substrates
M. Bruni, M. Carriquiry and P. Chilibroste
University of the Republic, R. 3 km 363, 60000, Uruguay; mbruni@fagro.edu.uy

The study was conducted to assess the effect of levels of glycerol (0, 3, 6, 9 and 12% of dry mater (DM) on the parameters kinetics of *in vitro* gas production of different substrates (S). Sugarcane residues (SR), silage temperate pasture (STP), temperate pasture (TP) and a mixture (40:30:30 DM basis) of TP, corn silage and concentrate (MxS) were used as S. The *in vitro* fermentation was determined using the gas production technique, with a pressure transducer, manual reading and frequent releases of gas. Samples were incubated in triplicate in at least three different runs. The ruminal inoculum was collected from two cows fed forage-based diets. Data was fitted to the first order equation $GP(t)=b(1-e^{-c(t-L)})$ where GP (ml/g DM) was gas production at time t, (b) potential GP which occurs at the fractional rate (c), and (L) lag time (h). The parameters were analyzed with a mixed model that included the effect of S and type of glycerol (pure or crude) as fixed effects, the lineal (NG) and quadratic (NG^2) effects of level of glycerol and the interactions of S by NG and NG^2, and the run as a random effect. Data are expressed as lsmeans$_{(s.e)}$ and were considered to differ if $P<0.05$. Parameters 'b' and 'c' were affected ($P<0.01$) by S, NG and NG^2 while did not differ due to type of glycerol or by the interactions between S and NG or NG^2. The general model for 'b' was $220.4 + 8.2 NG - 0.499 NG^2$ and 'b' was less for STP than for TP, MxS and SR [175.3$_{(23.0)}$ vs 252.9$_{(23.4)}$, 241.9$_{(31.7)}$ and 232.0$_{(24.6)}$ ml/gDM]. The general model for 'c' was $0.043 + 0.002 NG - 0.00008$ NG2 and 'c' was less for SR than STP, TP and MxS [0.017$_{(0.004)}$ vs 0.035$_{(0.004)}$, 0.040$_{(0.004)}$ and 0.044$_{(0.004)}$ h^{-1}], being as expected less in STP than TP. 'L' was not affected by S, NG or NG2 and averaged 0.996$_{(0.22)}$ The kinetic of gas production characterized by parameters 'b and c', increased with the level of glycerol up to 8 or 9% of DM as a result of increased availability of fermentable substrate.

Important factors to consider when applying and interviewing for academic and industry jobs

L.J. Spicer[1] and M.J. Pearce[2]
[1]Oklahoma State University, Department of Animal Sciences, 310 North Monroe, Stillwater, OK 74078, USA, [2]Zoetis, Mercuriusstraat 20, 1930 Zaventem, Belgium; leon.spicer@okstate.edu; michael.c.pearce@pfizer.com

Much has been written about creating the perfect curriculum vitae (CV) and how to best prepare for a job interview. This workshop will focus on preparation for applying and interviewing for entry-level positions in higher education and industry. Differences between CV and resumes as well as structure and content of cover letters will be discussed. Both CVs and resumes should be carefully prepared to reflect relevant experiences, successes and talents. Jobs in academia and industry are very diverse and may be as much as 100% teaching, research or management – or, a mix of these. A CV should be tailored to the responsibilities of the job. Experiences necessary to gain jobs in academia and industry will be outlined and compared. For example, publications are important for both academic and industry jobs, so it is critical that a candidate demonstrate the ability to publish research in a timely manner. For academic jobs with a teaching component, teaching experience is important and may come in a variety of forms, from teaching medical students how to conduct research to serving as a teaching assistant for a class. Information regarding maximizing both research and teaching experiences will be discussed. Obtaining that first interview is always exciting. The dos and don'ts of an interview seminar and teaching presentation will be discussed. Expectations of new faculty in the realm of teaching and research versus industry-related jobs will be compared and discussed.

Genetic selection of dairy cattle for improved immunity

B.A. Mallard
University of Guelph, Pathobiology Ontario Veterinary College, 50 Stone Rd, N1G2W1, Canada; bmallard@uoguelph.ca

The immune system provides protection from a diverse range of pathogens, and has the ability to adapt its response in accordance with the nature of the ever changing pathogenic landscape. Identifying dairy cows with the innate ability to make superior immune responses can reduce disease occurrence, improve milk quality, and farm profitability. Healthier animals also may be expected to demonstrate improvements in other traits, including reproductive fitness. Using the University of Guelph's patented high immune response (HIR) technology it is possible to classify animals as high, average, or low responders based on their estimated breeding value (EBV) for immune responsiveness. High responders have the inherent ability to produce more balanced and robust immune responses compared with average or low responders. High responders essentially have about one-half the disease occurrence of low responders, and can pass their superior immune response genes on to future generations thereby accumulating health benefits within the dairy herd. The Semex Alliance, Canada's largest dairy genetics company obtained an exclusive license from the University of Guelph to utilize the HIR procedure to identify sires with the high immune response classification. These sires are designated as Immunity+, marking their enhanced capacity to make protective immune responses. The immune response traits used in establishing HIR EBVs are moderately highly heritable having heritability estimates of approximately 0.25 to 0.35, which is in the same range as those for milk production traits. To date, we have tested more than 1000 Holstein sires and dams, and have not seen any substantial negative impact of selecting for enhanced immune response on production traits or reproductive traits. This indicates that it possible to genetically improve immunity with minor if any impact on other economically important traits. A GWAS based on the HIR phenotypes is currently under investigation with preliminary results suggesting a genomics test for IR is possible.

Unravelling the genetic background for endocrine fertility traits in dairy cows

A.M.M. Tenghe[1,2], A.C. Bouwman[2], B. Berglund[3], E. Strandberg[3] and R.F. Veerkamp[1,2]
[1]Animal Breeding and Genomics Centre, Wageningen University, P.O. Box 338, 6700 AH Wageningen, the Netherlands, [2]Animal Breeding and Genomics Centre, Wageningen UR Livestock Research, P.O. Box 338, 6700 AH Wageningen, the Netherlands, [3]Department of Animal Breeding and Genetics, Swedish University of Agricultural Sciences, P.O. Box 7023, 750 07 Uppsala, Sweden; amabel.tenghe@wur.nl

Fertility quantitative trait loci (QTL) are of high interest in dairy cows, since insemination failure has increased in some breeds such as Holstein. In this study, 5,563 lactations of 2,492 dairy cows combined from 4 experimental and 12 commercial herds were used to detect genomic regions associated to endocrine fertility phenotypes derived from milk progesterone concentration records. The phenotypes investigated were interval from calving to commencement of luteal activity (CLA) and proportion of samples between 25 and 60 days in milk with luteal activity (PLA). Cows were genotyped with 50k or 80k SNP chip, and imputed to 100k SNPs. Genetic parameters were estimated using a mixed linear animal model, and SNP effects using two models in ASREML: (1) a linear mixed model with each SNP fitted as fixed effect, along with a random polygenetic effect (using pedigree), and a permanent environmental variance; (2) genomic best linear unbiased prediction (GBLUP) model which fits all SNPs simultaneously. Fixed effects in the model were parity, herd-year-season, calving age, and breed proportion. Heritability estimates were 0.14 ± 0.04 for CLA and 0.15 ± 0.04 for PLA. The top 10 significant SNPs ($P<0.001$) for CLA were found on chromosomes 2, 3, 5, 17, 19, 20, 22, and 28 and on chromosomes 5, 10 and 20 for PLA. Six of the top ten top significant SNPs for CLA and 4 for PLA equally had the largest SNP effects from GBLUP; of which a region at 44 Mb for CLA and two regions for PLA (26 and 36 Mb) were found on chromosome 20. In the next step, the regions with significant SNPs will be imputed to full sequence and fine-mapped.

Resumption of luteal activity in first lactation cows is mainly affected by genetic characteristics

N. Bedere[1], L. Delaby[1], V. Ducrocq[2], S. Leurent-Colette[3] and C. Disenhaus[1]
[1]INRA-Agrocampus Ouest, UMR 1348 PEGASE, Domaine de la Prise, 35590 Saint-Gilles, France, [2]INRA, UMR 1313 GABI, Domaine de Vilvert, 78352 Jouy-en-Josas, France, [3]INRA, UE 326 Domaine Expérimental du Pin-au-Haras, 61310 Exmes, France; catherine.disenhaus@agrocampus-ouest.fr

Milk genetic merit is known to impact commencement of luteal activity (CLA) in dairy cows. This effect is usually considered to be related to energy exported in milk. The present study aimed to identify and quantify the effects of genetic characteristics (breed and estimated breeding value (EBV) for milk yield and solids content) and feeding system on CLA of primiparous cows. From 2006 to 2013, an experiment was conducted on 194 primiparous dairy (Holstein) and dual purpose (Normande) cows at the INRA farm of Le Pin-au-Haras. Within breeds, 2 groups were created based on EBV: cows with relatively high milk yield EBV (M) and cows with relatively high fat and protein contents EBV (C). Within breeds, exported energy in milk and weight loss were similar for both genetic groups. Two grazing based strategies were used, a High (H) feeding system (maize silage in winter and grazing plus concentrate) and Low (L) feeding system one (grass silage in winter and grazing with no concentrate). CLA was studied performing survival analyses (Weibull regression). Milk yields and body condition were studied using logistic regression, Chi[2]-test and ANCOVA. H cows produced more milk (+1,690 kg, $P<0.001$) and lost less body weight (BW) from week 1 to 14 pp (+3.8 kg/wk, $P<0.001$) than the L ones. Holstein cows produced more milk (+1,450 kg, $P<0.001$) and lost more BW (-1.4 kg/wk, $P<0.01$) than Normande ones. As expected, Normande cows had a shorter median CLA than Holstein ones (30 and 33.5 days respectively). However, C cows had shorter CLA (associated RR=2.0, P=0.001) than M ones. No effect of milk yield or feeding system on CLA was found. In conclusion, CLA was only affected by genetic characteristics. Beyond breed, the genetic merit to export energy through milk solids had a beneficial effect on post-partum ovarian recovery.

Epigenetic differences in cytokine genes of dairy cows identified with immune response biases

M.A. Paibomesai[1,2] and B.A. Mallard[2,3]
[1]Ontario Ministry of Agriculture, Food and Rural Affairs, Elora, ON, Canada, [2]University of Guelph, Pathobiology Department, Guelph, ON, Canada, [3]University of Guelph, Centre of Genetic Improvement of Livestock, Guelph, ON, Canada; paibomem@uoguelph.ca

Epigenetics, represents level of gene regulation that has been sparsely investigated in cattle. In the past decade epigenetics has been implicated to control disease susceptibility in numerous species, including humans and mice. Dairy cattle undergo periods throughout their production cycle that leave them susceptible to disease with a majority of incidences occurring in early lactation. The identification of cattle that inherently possess a balanced robust innate and adaptive immune response may decrease the incidence of disease through this sensitive period, having both health and economic benefits. CD4+ T-cells are the mediators of the antibody and cell mediated immune response. DNA methylation, an epigenetic modification, has been extensively studied in CD4+ T-cell immune responses (IR) and it has been shown that specific DNA methylation sites within cytokine promoters influence cytokine production in other species. The objective of this study was to determine the role of DNA methylation at cytokine promoters in CD4+ T-cell cytokine profiles of high (H) antibody- (AMIR) or H cell- (CMIR) mediated immune response cows. CD4+ T-cells were isolated from H-AMIR (n=12) and H-CMIR (n=11) cows at three time points before and after calving (-21, +4, and +21 days). DNA was collected for bisulfite pyrosequencing to determine DNA methylation patterns. T-cells from H-CMIR cows produced more IL-4 and IFN-γ than cells from H-AMIR cows at +21 days from calving. Interestingly, H-CMIR cows had lower CpG methylation overall at the IFNG promoter than H-AMIR, which correlated with high IFNγ production. Suggesting that epigenetic modifications, such as DNA methylation, play a role in cytokine expression of CD4+ T-cells from cattle.

Breeding for improved fertility and udder health – requirements

M. Kargo[1,2], G.P. Aamand[1,3] and J. Pedersen[1]
[1]SEGES, Agro Food Park 15, 8200 Aarhus N, Denmark, [2]Aarhus University, Dept. of Molecular Biology and Genetics, Blichers Allé 20, Postboks 50, 8830 Tjele, Denmark, [3]Nordic Cattle Genetic Evaluation, Agro Food Park 15, 8200 Aarhus N, Denmark; morten.kargo@mbg.au.dk

The basic for genetic improvements of fertility and udder health are proper phenotypic data. Furthermore, appropriate economic weights (ew) must be given to the traits in the total merit index. These weights should in principle be given to the core traits, which for female fertility are the ability to show heat and the ability to become pregnant. In Denmark, Sweden, and Finland (DSF), the female fertility traits are AI registrations and these traits are given ew to calculate the combined EBV for female fertility. For udder health, the core trait is resistance to the different types of mastitis, while the number of mastitis treatments is given an ew, and SCS and two udder conformation traits are used as indicator traits. For the future it is urgent to define the core traits correctly and estimate genetic correlations between these traits and possible indicator traits. In genomic selection schemes less but precise registrations of the core traits / indicator traits are needed. Cows with these registrations make the cow reference group, where high quality registrations are needed as the accuracy of EBVs, and thereby the possibility for increased genetic gain, rely on the quality of the phenotypic data. Therefore, breeding organizations in the future might have to pay (more) for these registrations. In Denmark, the EBVs for female fertility and udder health have been published and included in the TMI since 1982 and 1992, respectively, and since 2006 and 2005 as common DSF EBVs. This has resulted in considerable gain for female fertility and udder health in the DSF dairy cattle populations. This is expected to increase in the future since genomic breeding schemes will result in relatively larger genetic gain for fertility and udder health, because the ratio r_{AI} for functional traits over r_{AI} for production traits is increased in these schemes.

Extension programming to address somatic cell count challenges and opportunities
J.M. Bewley
University of Kentucky, Animal and Food Sciences, 407 WP Garrigus Building, 40546-0215, USA;
jbewley@uky.edu

In Kentucky, we have implemented a multi-faceted approach to extension programming for SCC reduction. The University of Kentucky Dairy Extension team has worked closely with the Kentucky Dairy Development Council in a farm-based program entitled M.I.L.K. Counts. The intent of the program is to provide direct, on-farm technical assistance to producers struggling with SCC through evaluation of DHIA records, milking procedures, management protocols, animal hygiene and housing, and dry cow treatment and handling. The M.I.L.K. Counts program incorporates a team based problem solving approach with emphasis on the economic impact of resulting recommendations. Microbiological culturing is performed by the University of Kentucky Veterinary Diagnostic Livestock Laboratory. YouTube videos were developed (http://www.youtube.com/user/UKAgriculture) to demonstrate recommended milking procedures and to provide virtual tours of farms that consistently maintain low SCC. A visual analytics dashboard (http://tinyurl.com/UKMilkBonus) was created to illustrate the potential for increased income through SCC reductions. The Southeast Quality Milk Initiative is a collaborative outreach, educational, and applied research program on mastitis control by milk quality professionals from six Land-Grant Universities in the southeastern US. In this program, we strive to develop cost effective mastitis prevention and control strategies for the southeast region for: higher milk quality, increased milk production, and improved profitability.

Detection of haplotypes responsible for prenatal death in cattle
X. Wu, B. Guldbrandtsen, M.S. Lund and G. Sahana
Center for Quantitative Genetics and Genomics, Molecular Biology and Genetics, Aarhus University, Blichers alle 20, 8830, Tjele, Denmark; xiaoping.wu@mbg.au.dk

Widespread use of a limited number of elite sires in dairy cattle breeding increases the risk of deleterious allelic variants being autozygous. Genomic data is used to search haplotypes that are common in the population, but never occur as homotypes state in live animals. Using this approach, a number of putative recessive lethal-carrying haplotypes causing embryonic death have been identified in cattle. The aim of this study was to detect the recessive lethal haplotypes in three cattle populations and to identify casual genetic factors underlying these haplotypes. The genotypes from Illumina Bovine 50k BeadChip for 26,312, 19,309 and 4,291 animals from Danish Holstein, Nordic Red and Danish Jersey breed were used in this study. The genotypes were phased. Haplotypes which were either never occurred as homotypes or deviated strongly from Hardy-Weinberg proportions were identified. We shortlisted 18, 44, and 17 candidate haplotypes with 11 to 50 expected homotypes, but where only zero or a few homotypes were observed in Danish Holstein, Nordic Red, and Jersey breed. Carrier sire by carrier dam mating are compared to other matings for increased rates of reproductive failure. Whole genome sequence variants will concordance with carrier vs non-carrier status to shortlist candidate causative variants.

Milk and fertility performance of Holstein, Jersey and Holstein × Jersey cows on Irish farms

E.L. Coffey[1,2], B. Horan[2], R.D. Evans[3], K.M. Pierce[1] and D.P. Berry[2]
[1]*University College Dublin, School of Agriculture and Food Science, Dublin 4, Ireland,* [2]*Teagasc Moorepark, Animal & Grassland, Fermoy, Co Cork, Ireland,* [3]*Irish Cattle Breeding Federation, Bandon, Co Cork, Ireland; emmalouise.coffey@teagasc.ie*

Crossbreeding offers the potential to improve milk production, fertility and economic efficiency. Previous research experiments have shown additional gains derived from crossbreeding, capitalising on both additive and non-additive genetic effects. The objective of this study was to compare the biological performance of Holstein, Jersey and Holstein × Jersey crossbred cows using commercial dairy herds practicing crossbreeding over a 5 year period. Milk production and fertility information from the national database on 11,808 cows from 40 spring calving dairy herds that adopted crossbreeding between Holstein and Jersey breeds from 2008 to 2012 inclusive were available. Obvious data errors were removed. Least square means for traits of interest were estimated for purebred and crossbred animals using linear mixed models. Holstein-Jersey first cross cows produced more ($P<0.001$) milk solids (heterosis=6.5%), calved earlier ($P<0.001$) as heifers (heterosis=1.6%), had a shorter ($P<0.001$) calving interval (heterosis=2.1%) and had a higher ($P<0.05$) submission rates in the first 21 days of the breeding season (heterosis=4.5%) relative to their purebred parent breeds. Results observed corroborated with findings from previous component experiments. Breed complementarity and heterosis attainable from crossbreeding resulted in superior performance in crossbreds relative to their purebred parent breeds.

Associations of feed efficiency with reproductive development and semen quality in young bulls

S. Bourgon[1], M. Diel De Amorim[2], S. Lam[3], J. Munro[1], R. Foster[3], T. Chenier[3], S. Miller[3,4] and Y. Montanholi[1,3]
[1]*Dalhousie University, Truro, NS, Canada,* [2]*University of Saskatchewan, Saskatoon, SK, Canada,* [3]*University of Guelph, Guelph, ON, Canada,* [4]*AgResearch Limited, Invermay, Agricultural Centre, New Zealand; stephanie.bourgon@dal.ca*

Improving feed efficiency has positive economic and environmental implications for the beef industry. However, an inverse relationship between improved feed efficiency and fertility related traits has been suggested in beef bulls. Our objective was to describe sexual development and reproductive profile of young bulls in the context of feed efficiency. Crossbred and purebred bulls (n=176; 374±68 kg, 282±28 d) were submitted to feed intake and performance tests. Residual feed intake (RFI) was calculated using daily feed intake, average daily gain, bodyweight (BW) and ultrasound traits through a linear regression model. Reproductive profile assessments consisted of scrotal circumference, blood plasma metabolites and hormones, testis ultrasound for pixel intensity, scrotum infrared images and semen quality. Reproductive tissues were collected from a subpopulation and processed for histomorphometry. The univariate procedure of SAS was used to obtain descriptive statistics. RFI will be used as a classificatory variable (low- and high-efficiency) for means comparison and as a continuous variable for regression analysis. Preliminary results include initial scrotal circumference (30.4±3.0 cm), initial testes pixel intensity (142.6±15.8 pixels), scrotal surface temperature gradient (4.1±1.5 °C), sperm progressive motility (68.8±13.0%), sperm viability (81.4±10.4%), normal sperm morphology (71.3±7.7%), testes weight (53.7±11.2 g/100 kg of BW) and seminal vesicles weight (6.1±1.2 g/100 kg of BW). Further analyses using the complete dataset in the context of feed efficiency are in progress. Comparison of testis morphology with sexual and metabolic hormones and sperm parameters may culminate in the identification of biomarkers with practical application for sexual development, fertility and feed efficiency.

Decision support model for dairy cow management

J. Makulska, A.T. Gawełczyk and A. Węglarz
University of Agriculture, al. Mickiewicza 24/28, 30-059 Kraków, Poland; rzmakuls@cyf-kr.edu.pl

A high milk performance is often associated with lower cow fertility and health problems resulting in decreased longevity and herd profitability. To support the economically optimal decisions concerning management of cow reproduction and production, a dynamic programming model was developed. The optimization problem was formulated as a multi-level hierarchic Markov decision process. Biological, technical and economic parameters of the model were derived on the basis of the data from Polish dairy herds and the literature review. The criterion of optimality was assumed as a maximization of expected discounted net revenues per cow. The model has a structure of 4-level Markov process in which decisions are taken at 3 levels. At the 0 level a founder process was defined with stages corresponding to the lifetime of cows classified according to the age at first calving (3 classes). Child level 1 is a representation of subsequent calving intervals (max. 10) of a cow. Child level 2 represents an open period with subsequent oestrus cycles (max. 11). At child level 3 stages correspond to pregnancy period including dry period (max. 12 weeks). Uncertainty in the reproduction and production processes is taken into account by calculating relevant transition probabilities. Structure of the model, transition probabilities and parameters were programmed as a plug-in to the Multi-Level Hierarchic Markov Processes software. The Markov chain simulations were carried out to calculate key figures characterizing the optimal policy on cow insemination, drying-off and replacement and to provide the information for the critical components of the model. The outcomes from the baseline scenario were compared with those obtained by the sensitivity analyses performed at different values of input parameters. Due to the possibility of alteration in inputs the model can be used to determine economically optimal strategy of dairy cow management in various production conditions. This study was supported by the Polish Ministry of Science and Higher Education (project no. DS-3258/14).

Economic importance of health traits in dual-purpose cattle

Z. Krupová[1], E. Krupa[1], M. Michaličková[2] and L. Zavadilová[1]
[1]Institute of Animal Science, P.O. Box 1, 104 01 Prague, Czech Republic, [2]NAFC-Research Institute for Animal Production Nitra, Hlohovecká 2, 951 41 Lužianky, Slovak Republic; krupova.zuzana@vuzv.cz

Health traits as a part of functional traits have a large impact on the farm profitability through effective use of inputs, reduction of veterinary costs along with improved price, quality and health security of products. Functional traits have been included in the breeding goals and selection schemes of many cattle breeds for several years. Economic value of clinical mastitis and claw-disease incidence in dairy population of Slovak Pinzgau breed was calculated for the first time using the bio-economic model of the program package ECOWEIGHT. To evaluate economic value the trait of interest, the impact on other evaluated traits (e.g. somatic cell count, milk yield during the rest of lactation and cow losses) was not considered to avoid double counting. Under the production and economic conditions of the year 2013, the marginal economic value of clinical mastitis and claw disease incidence was -68.53 € and -27.05 € per case per cow and year, respectively. Their relative economic importance from the all of functional traits (productive lifetime, fertility, somatic cell score, losses of calves) which were evaluated simultaneously, contributed by 3 and 1%, respectively. Lower economic value of the claw-disease incidence corresponds to relative good health status of the breed farmed extensive with the regular access to pasture. For somatic cell score as the trait also related to udder health the high marginal (-241 € per score per cow and year) and relative (26%) economic value was calculated. Based on above mentioned, clinical mastitis incidence and/or somatic cells score are of notable impact on the farm profitability and therefore should be included in the breeding schemes of the local dairy population. Study was supported by project QJ1510217 and MZERO0714 of the Czech Republic and Ministry of Agriculture and Rural Development of the Slovak Republic.

Effectiveness of teat disinfection formulations containing polymeric biguanide compounds
D. Gleeson and B. O'brien
Teagasc, Livestock Systems, Animal & Grassland Research and Innovation Centre, Moorepark, Fermoy, Co. Cork, Ireland; bernadette.obrien@teagasc.ie

The formulation of a teat disinfectant product applied post-milking may have varying degrees of success in reducing bacterial counts on teat skin. The objective of this study was to investigate the efficacy of post milking teat disinfectant products with polymeric biguanide formulations compared to products containing more traditional formulations on reducing bacterial numbers on teat skin. Nine disinfectant products containing formulations of either iodine, chlorohexidine, ammonium lauryl sulphate, quaternary ammonium compounds, lactic acid, or three containing polymeric biguanide compounds were applied to cow's teats post milking. Teats of 20 cows were swabbed for *Staphylococcus* and *Streptococcus* bacteria before disinfection and at 5 (RF & LH teats) and 60 minutes (LF & RH teats) after disinfection (n=800). The statistical analysis (NLMIXED procedure of SAS 9.2, 2009) used a factorial design to examine main effects and interaction of time and treatment by measuring the reduction in bacterial numbers before disinfection compared to 5 and 60 minutes after disinfection. Staphylococcal and Streptococcal bacteria were present on 94% and 69% of teats, respectively, after cluster removal (before disinfection). All disinfectant products reduced bacterial counts on teats at both 5 and 60 minutes after disinfection (P<0.001). The product containing poly hexamethylene biguanide resulted in lower Staphylococcal counts at 5 and 60 minutes compared to lactic acid and iodine products (P<0.05). The quaternary product had the highest percentage of teats with no Staphylococcal bacteria present at 5 (89%) and 60 minutes (60%). Iodine, lactic acid and biguanide products eliminated Streptococcal bacteria on all teats after 5 minutes. Efficacy of all products was reduced at 60 minutes compared to 5 minutes. This study highlighted that disinfectant products containing biguanides are equally as effective as chlorohexidine, iodine and lactic acid in reducing bacterial counts on teat skin.

Bacterial proteoliposomes – new immunological tool for control of mammary pathogens
J. Quiroga, H. Reyes, M. Molina, S. Vidal, L. Santis, L. Sáenz and M. Maino
Veterinary Vaccines Laboratory, Universidad de Chile, Santa Rosa 11735 La Pintana, Santiago, 8820808, Chile; mmaino@uchile.cl

The overall impact of mastitis on the quality and quantity of milk produced for human consumption has provided the need to develop new strategies to enhance disease resistance through immunoregulation. At field level, commercial vaccines against mastitis pathogens have proved to be ineffective in the generation of a protective immune response, in part because of the wide variety of strains and species involved with the infection. The use of bacterial proteoliposomes emerges as an interesting alternative for the development of multivalent vaccines against mastitis. Objectives: To evaluate the specific immune response in dairy cattle immunized with a polyvalent vaccine based on proteoliposomes against *Escherichia coli*, *Staphylococcus aureus* and *Streptococcus uberis*. 60 Holstein cows were randomly divided into 2 groups. Cows in the vaccinated group were immunized 3 times with 500 µg of each bacterial proteoliposomes emulsified with 8 mg of aluminum hydroxide in a total volume of 2 ml. Cows in the control group received 8 mg of aluminum hydroxide in 2 ml. Vaccines were administered subcutaneously, 45 and 15 days before parturition and 45 days postpartum. To evaluate presence of specific antibodies by indirect ELISA, blood samples were collected monthly from the tail vein. Statistical analysis: Specific IgG levels between groups were compared using 2-way ANOVA. Cows did not show any signs of abnormal sensitivity to the vaccine and calves born healthy. After immunization, levels of specific IgG against *E. coli*, *S. aureus* and *S. uberis* were significantly higher in the vaccinated group (P<0.001) compared to the control group, remained significantly higher until 135 days postpartum (P<0.001). Immunization with a multivalent vaccine based on bacterial nanovesicles proves to be a safe and effective tool for generation of specific humoral immune response against mastitis pathogens.

Global DNA methylation in bovine peripheral blood mononuclear cells and in milk somatic cells

M. Gasselin[1], A. Prezelin[1], M. Boutinaud[2], P. Debournoux[2], A. Neveux[1], M. Fargetton[2], J.P. Renard[1], H. Kiefer[1] and H. Jammes[1]
[1]INRA, UMR1198, Domaine de vilvert, 78352 Jouy en Josas, France, [2]INRA, UMR1348, AGROCAMPUS Ouest, 35590 Saint-Gilles, France; maxime.gasselin@jouy.inra.fr

Epigenetic modifications such as DNA methylation play a role in regulating gene expression and consequently in biological processes, such as those involved in health and disease. In human, studies provide evidences that the methylome in peripheral blood mononuclear cells (PBMC) can reflect some disease susceptibility. In dairy cows, the post calving period is characterized by profound changes associated with an immunosuppression increasing the susceptibility to diseases. This study aims to quantify the global DNA methylation levels (dMe) by LUminometric Methylation Assay in bovine PBMC and in milk leukocytes and epithelial cells. Parity, milk production and quality, health and diet were reported from 52 Holstein housed at INRA's experimental dairy farms. Genomic DNAs were extracted from 94 PBMC samples collected at D15 and D60 of lactation. PBMC-dMe varied from 72 to 86%. The proportion of different blood cell types were determined by cytometry and were found stable during the lactation period analyzed. dMe variations do not correlate with blood cell populations. No effect of lactation period, parity or breeding was observed. Moreover, dMe of PBMC was not significantly altered in pathological cows (mastitis or uterine post-partum disease, n=6) in comparison with healthy cows (n=18). For the 24 cows at the two lactation stages, milk leukocytes (ML) and epithelial cells (MEC) were also purified after milk centrifugation and immunomagnetic binding and gDNA extracted. Using the paired samples (PBMC-ML-MEC), dMe was found to be cell type specific: 77.08±1.8% in PBMC vs 65.6±4.1% in ML (significantly different, P<0.001) and 67.39±7.9% in MEC. Further studies are needed to identify the specific changes in DNA methylation using genome-wide analysis. CIFRE fellowship from Pilardière & Codelia.

Effect of fetal membrane release induced by oxoETE injection on the postpartum dairy cow performance

H. Kamada[1] and Y. Matsui[2]
[1]NARO Institute of Livestock and Grassland Science, Animal Reproduction, Ikenodai-2, Tsukuba, Ibaraki, 305-0044, Japan, [2]Hokkaido Research Organization, Konsen Agricultural Experiment Station, Asahigaoka-2, Nakashibetsu, Hokkaido, 086-1135, Japan; kama8@affrc.go.jp

The mechanism of fetus release at delivery is well understood; however, there is little information about the process of fetal membrane (FM) release. We discovered that 12-oxoETE is a candidate of signal for FM release after delivery and succeeded in FM release by its injection to PG-induced delivery cows (retained FM: RFM). The present study showed the effect of FM release induced by 12-oxoETE on milk yield, postpartum plasma P4 level and biochemical parameters of blood plasma. Treatments were as follows; normal delivery (ND: nonRFM), PG-induced delivery (PG: RFM), natural occurred RFM (NR: RFM) and PG+oxoETE treatment (PO: nonRFM). Milk yield of RFM cows in the early lactation period was lower than ND cows; however, FM release induction by 12-oxoETE injection (PO) recovered its decrease. Postpartum plasma P4 elevation in PG cows delayed about 2 weeks compared to ND cows; however, P4 increase in 12-oxoETE treatment cows (PO) started about 14 days after delivery on average as same as ND cows. These results suggested that FM release induced by 12-oxoETE is available physiologically in cows.

Determination of complex vertebral malformation genetic disorder in Holstein cows in Antalya region
M.S. Balcioglu, T. Karsli and E. Sahin
Akdeniz University, Animal Science, Akdeniz University Faculty of Agriculture Department of Animal Science, 07058 Antalya, Turkey; msoner@akdeniz.edu.tr

Complex vertebral malformation is an autosomal recessive genetic disease in Holstein cattle breed. CVM is caused by a point mutation (G-T) at nucleotide position 559 of the solute carrier family 35 member 3 (SLC35A3) gene on chromosome 3. The mutation in homozygous recessive is lethal and lead to the abortion of most of the affected calves before gestation day 260. In this study, was investigated the presence of CVM disease in Holstein cows reared in the Antalya region of Turkey by using allele specific polymerase chain reaction (AS-PCR) method. Blood samples were obtained from a total of 314 Holstein cows randomly selected from farms in different parts of Antalya for this study. In examined 314 samples, homozygous recessive CVM carrier was not detected but eight samples were found to be heterozygous CVM carriers. As a result, prevalence of CVM genetic disorder was calculated as approximately 2.5% in Antalya region of Turkey.

Equine practice into science – what does this mean?
N. Miraglia
Molise University, Agriculture, Environment and Foods, Via De Sanctis, 86100 Campobasso, Italy; miraglia@unimol.it

Equine practice into science – what does this mean? Generally we use to say 'from science to practice' but now, after many years of research activity in the horse field, maybe it is more appropriate to 'invert' the question; in fact the transfer of knowledge from science to practice demonstrated many limits and the most relevant weaknesses stays in the difficulty to identify the exigencies coming from practitioners and in the necessity to give compromises between research and practical applications in the field. Many aspects determined this condition. Over the last years there has been a considerable worldwide changes in the horse sector. Equine Industry is facing to new challenges: the social demand is new and the economic pressure is higher. Moreover, the considerable development of leisure riding, the diversification of usage of horses and the increasing role of horses in the use of territories are of high concern for European and local socio-economy. At a scientific level an important discussion was destined to identify the research needs coming from the modern society. But the gap between the two sectors – research and practice – seemed to increase. It arises that the research approach should be designed and implemented across the scientific disciplines considering that, in practice, the questions which have been raised are multifactorial and are combined throughout consistent breeding systems where the impact of new knowledge and technologies should be integrated. Different attempts have been implemented successfully to initiate cooperation for research between researchers and professionals. So far updated should be implemented to promote new knowledge in equine industry. A closer cooperation between researchers and breeders associations is important to establish a more permanent dialogue: 'Equine practice into science' becomes of high concern to assure the harmonization of knowledge and to stimulate a better understanding and the use of new technologies in the field.

Develpment of horse industry
F. Clément
IFCE, Terrefort, 49411 Saumur, France; francoise.clement@ifce.fr

A 2012 study about the french horse industry to 2030 has described 4 constrasted long term scenarios: (1) everyone on horseback (horses in the leisure market, buoyed by a wide variety of businesses); (2) the high society horse (limited number of users in a socially-divided society); (3) the civic horse (the horse in public and collective action, the link between humans, the land and nature; (4) the companion horse (from exploiting to caring for animals, the quest for animal welfare). Since 2012, some general trends are observed: (1) the economic crisis affects the purchasing power (-8% of the spending for leisures per family in 2014 vs 2013); (2) the increasing concern of society for animal welfare (-1.3% of meat consumption in 2011 vs 2013; change in the legal status of animal as a sentient being); (3) the decline of government support: decreased public fundings (-24% in 2008 vs 2013), transfer of trade services to private sector. Data about the french horse industry show: (1) decreased customers in sports (-2% of licence holders at the equestrian federation from 2012 to 2013), races betting (-3% turnover from 2012 to 2014); (2) decreased breeders (-22% of registered foals from 2003 to 2013), trades (-4.3% of sold horses from 2012 to 2013), employees (-2% from 2012 to 2013); (3) but a steady increase of competitions (+42% of events over the period 2009-2014). So, the probability of achieving the scenario 'the high society horse' increases at the expense of the scenario 'everyone on horseback' mainly because of economic crisis. The scenario 'the companion horses' grows up with potential negative impact on economic efficiency of the industry and on the scenario 'everyone on horse back'. In addition, the initiatives concerning the scenario 'the civic horse' remain scattered. In conclusion, big efforts on the economic efficiency of the horse companies must be made and the government must create an enabling business environment (sanitary prevention, access to land, etc.) to develop a private sector able to satisfy the strong interest in horses.

Equi-ressources: keys to understanding the job market's dynamic in the french horse industry
S. Doaré
Ifce, équi-ressources, Haras national du Pin, 61310 Le Pin au Haras, France; sylvie.doare@ifce.fr

7 years ago, the french horse industry worked to create a unique job center to help to connect employers and employees in the french equine industry. The partnership established between public and private members of the equine sector has the ambition to federate all the segments of the french horse industry. During 2014, 1,800 offers were published on www.equi-ressources.fr. This free service is provided by advisors employed trough the french institute for horse and horse riding (ifce), through various fundings (as FSE). Equi-ressources has endowed itself with an observatory for 4 years. It is an instrument for analysis and prospective dedicated to the equine sector. The equi-ressources observatory works on 3 axis: (1) the link between equine industry and job market in France; (2) the evolution of the vocational courses; and (3) the match between courses and employment. By realizing national reports and thematic studies, equi-ressources became a necessary tool for the policy-makers. Last year was published the second national report about work, trades and courses in France. The french horse industry employed 179,392 persons in 2012, which corresponds to 0.7% of the total domestic employment. The analysis of the datas, issued from the offers managed by the equi-ressources advisors, allowed to map-out the professions where the most job opportunities are. The status of the women is evolving in courses and jobs, and the behaviour of the employers is also changing. The observatory gives guidelines to describe how the equine vocational courses help people to connect or reconnect firstly with school and secondly with the job market. However, the data analysis drawn from the offers shows how the needs expressed by the employment market do not always correspond to the profile of the applicants. So, tensions appear on the market, and the stakeholders use these informations to adapt the courses provided.

Innovation and research in equine science and practice
V. Korpa[1], M. Saastamoinen[2] and L. Rantamäki-Lahtinen[2]
[1]Latvia University of Agriculture, Liela street 2, Jelgava, 3001, Latvia, [2]Natural Research Institute Finland,
Opistontie 10a1, 1, 32100 Ypäjä, Finland; viola.korpa@llu.lv

The proposed paper explores innovation in the Equine Science and practice. An interpretation of innovations, openness to novelties and possibilities to innovate in equine businesses are directly linked with understanding of what the equine sector is and what kind of services or products it can offer to contemporary society. If the equine sector is defined in a very traditional way as the economic activity related solely to the agriculture (horse breeding) and equestrian and other horse sports, the scope of potentially new services and practices more likely will be limited to some novelties of technical improvements and practices related to horse breeding, horse welfare or environmental issues, whereas broader vision on the industry requires thinking beyond the box and allows perceiving opportunities for innovations in horse related activities involving cooperation also with other sectors of local economy, for example, tourism, health care, and recreation. The paper gives examples of practices and ideas that work in particular place and time in relation to horse welfare, diversity of services, different customer groups and customer oriented communication, as well as collaboration and networking as important driving forces for innovation. Innovations are often based on assimilation of research implications, why it is important that research results are delivered effectively. It is also necessary that there is good communication between the horse sector and researchers. The paper presents practices and inspiration identified within a framework of the international project INNOEQUINE 'Equine Industries Promoting Economically Competitive and Innovative Regions' (2011-2013).

The analysis of differentially methylated regions between equine sarcoids and healthy skin
E. Semik[1], T. Ząbek[1], A. Fornal[2], A. Gurgul[1], T. Szmatoła[1], K. Żukowski[3], M. Wnuk[4] and
M. Bugno-Poniewierska[1]
[1]National Research Institute of Animal Production, The Laboratory of Genomics, Krakowska 1, 32-083
Balice, Poland, [2]National Research Institute of Animal Production, Department of Animal Cytogenetics and
Molecular Genetics, Krakowska 1, 32-083 Balice, Poland, [3]National Research Institute of Animal Production,
Department of Animal Genetics and Breeding, Krakowska 1, 32-083 Balice, Poland, [4]University of Rzeszow,
Department of Genetics, Sokołowska 26, 36-100 Kolbuszowa, Poland; ewelina.semik@izoo.krakow.pl

The aim of this study was to identify differentially methylated regions (DMRs) in equine sarcoids – a locally invasive skin tumor of equids, which is considered to be the most common equine skin neoplasm. DNA was extracted from neoplastic and healthy tissue obtained from the same individual. We implemented Reduced Representation Bisulfite Sequencing (RRBS) method in combination with pair-end Illumina sequencing to quantify DNA methylation levels across genome. A total of 17 DMRs were identified, which were located within introns, exons as well as in genes vicinity. The region showing the largest differences in DNA methylation level was located in a possible promoter of the gene SPARC which is involved in regulation of cell growth. Further analysis of the region aimed to determine the level of DNA methylation in greater number of sarcoids and skin samples using bisulfite sequencing of PCR products (BS-PCR) and sequencing of cloned BS-PCR products. The results confirmed that DNA methylation differs between neoplastic and healthy tissue samples in 4 out of 7 of the analyzed animals. Furthermore, sequencing results of cloned products show that the analyzed region in skin without neoplastic changes exhibited higher level of methylation (64.2%) compared to the sarcoids in which the degree of methylation was 14.5%. These results may be helpful to assess epigenetic changes characteristic for sarcoid progression and can be useful to find particular genomic regions involved in molecular alterations specific for the development of equine sarcoids.

The application of aCGH technique to analyse equine sarcoids

K. Pawlina[1], A. Gurgul[1], J. Klukowska-Rötzler[2], C. Koch[3], K. Mählmann[3] and M. Bugno-Poniewierska[1]
[1]National Research Institute of Animal Producttion, Laboratory of Genomics, Krakowska 1, 32-083 Balice, Poland, [2]University of Bern, Department of Clinical Research, Murtenstrasse 35, 3010 Bern, Switzerland, [3]University of Bern, Department of Clinical Veterinary Medicine, Postfach 8466, 3001 Bern, Switzerland; klaudia.pawlina@izoo.krakow.pl

The aim of our study was to give insight into aberrations occurring at the genome level in horse sarcoid, which is the most frequent skin neoplasm in horses. It is classified as locally malignant, fibroblast benign tumor of fibrous tissue. Despite the fact that it does not metastasize, it may recur if treatment is not carried out properly. One of methods which may shed light on molecular characteristics of sarcoid is aCGH technique. This technique is the combination of CGH (comparative genomic hybridization) with microarray technology, which increased the resolution and made it possible to detect even micro rearrangements. In this study, the research material consisted of DNA isolated from eight samples of horse sarcoid tissue as well as healthy, control skin tissue. Firstly, we carried out DNA labeling using SureTag DNA Labeling Kit (Agilent). The labeled DNA was then subjected to 40 h hybridization onto 2×400k custom equine microarrays (Agilent). The washed microarrays were scanned with SureScan G2565CA Scanner (Agilent). The obtained raw data were analyzed using Agilent Genomic Workbench (7.0). The preliminary analysis showed great diversity between single tumors, which manifested not only as the number and type of aberrations but also their size or presence of genes. Moreover, the analysis of pathways and biological processes of genes localized in the aberrant regions revealed that the vast majority of them is engaged in the functioning of the immune system, cell processes, communication and signaling. In conclusion, horse sarcoids appear as a non-uniform group, which may stem from various causes, like heterogeneity of a tumor or different clinical types of the examined tumors.

Influence of relaxing massage on emotional state in Arabian race horses

S. Kowalik[1], I. Janczarek[2], W. Kędzierski[1], A. Stachurska[2] and I. Wilk[2]
[1]University of Life Sciences in Lublin, Akademicka 12, 20-033 Lublin, Poland, [2]University of Life Sciences in Lublin, Department of Horse Breeding and Use, Akademicka 13, 20-950 Lublin, Poland; witold.kedzierski@up.lublin.pl

The objective of the study was to assess the effect of relaxing massage on the heart rate (HR) and heart rate variability (HRV) in young race horses, during their first racing season. In the study, 72 Purebred Arabian race horses were included. The horses from control group (24) and experimental group (48) were included in regular race training six days a week. The horses from experimental group were additionally subject to the relaxing massage for 25-30 min three days a week during the whole study. HR and HRV were assumed as indicators of the emotional state of the horses. The following HRV variables were quantified: time domain analysis (the square root of the mean of the squared successive differences in R-R intervals – RMSSD) and frequency domain analysis (low frequency component – LF, high frequency component – HF and low to high frequency ratio LF/HF). The measurements were taken six times, once 4-5 weeks, during training in the whole race season. The parameters were measured for 10 min at rest, 10 min during grooming and saddling the horse and 10 min of warm-up walking under rider. The changes of the parameters throughout the season suggest that the relaxing massage may be effectively used to make the racehorses more relaxed and calm. The massage may be beneficial not only to the horse's health but also can be a valuable tool for the trainers in an early prevention of the overtraining syndrome. The study was financially supported by the National Centre for Research and Development, Poland (N180061 grant).

Frequently recorded sensor data may correctly provide health status of cows
P. Løvendahl and L.P. Sørensen
Center for Quantitative Genetics and Genomics, Dept. Molecular Biology and Genetics, Aarhus University,
AU-Foulum, 8830 Tjele, Denmark; peter.lovendahl@mbg.au.dk

The implementation of sensor based decision support in commercial dairy herds is highly dependent on having reliable systems. This is also a requirement in obtaining reliable phenotypes for genetic or genomic selection. Problems with sensors give missing and noisy data hampering their use. Also, the presentation of results needs to be in a form which is simple and useful. These issues were addressed using a mastitis sensor and decision support as example. This study aim at providing and evaluating a modular system applicable to the pipeline from sensor to decision support. The case of mastitis was chosen as it is of economic importance and also affect welfare of cows, and because we have worked with a commercial sensor. The problems with sensors causing missing data and noise is described and a range of filtering and monitoring modules are shown to be important to make systems functional for herd management purposes. On top of this a solid method need to be used to interpret and present data to end users, in terms of easy to read categories. Filtering and pre-adjustments of raw data are important in making algorithms robust and reliable for daily use. Re-definition of traits is needed going from traditional few groups to continuous definitions, and then to new action oriented health classes. Also, for this case focusing on mastitis, assignment to 'permanently sick' groups can be helpful in keeping focus on new acute cases. The combined used of filtering, fix-up routines and time series models leading into action oriented categories is needed to provide simple and robust decision support. The systems may be vastly improved by opening for transmission of data between user groups and to common databases – also with a view to use data in genetic selection. For that purpose, the continuous variables are advantageous.

Data from automatic milking systems used in genetic evaluations of temperament and milkability
K.A. Bakke[1] and B. Heringstad[1,2]
[1]*Norwegian University of Life Sciences, Department of Animal and Aquacultural Sciences, P.O. Box 5003, 1430 Ås, Norway,* [2]*Geno Breeding and A.I. Association, 1432 Ås, Norway; bjorg.heringstad@nmbu.no*

The objective was to examine whether data routinely recorded in automatic milking systems (AMS) can be used to define new behavior- and milkability traits, and to estimate genetic correlations between these new traits and the current subjectively scored temperament, milking speed, and leakage. The analyzed AMS data came from 46 herds with DeLaval milking robots, had information from about 6,000 cows and more than 2 million daily records. Milkability was defined as average milk yield per total time spent in the milking robot; kg milk per minute 'box time'. This is a combined measure of milk flow/milking speed and how efficient the cow visit the milking unit. The proportion of milkings with kick-offs and the proportion of incomplete milkings during a lactation were used as behavior traits. Information on temperament, milking speed and leakage of about 330,000 Norwegian Red cows were extracted from the Norwegian Dairy Herd Recording System. Dairy farmers routinely score their first lactation cows for these 3 traits, each in 3 categories. The estimated genetic correlation between milkability from AMS and subjectively scored milking speed of 0.88 imply that these are similar traits genetically. The genetic correlation with leakage was unfavorable and stronger from subjectively scored milking speed (0.84) than from AMS milkability (0.53). Moderate genetic correlation of temperament with proportion kick offs (0.54) and with incomplete milkings (0.27), respectively, suggest that these may be potential traits to consider as alternative measures of temperament. The unfavorable genetic correlation between AMS milkability and temperament (-0.22) imply genetic association between difficult temperament and slower milkability. New traits from AMS can supplement or replace current traits in genetic evaluations.

Air leakage in automatic milk flow recording – an indicator trait for temperament?
D. Krogmeier and K.-U. Götz
Institute for Animal Breeding, Bavarian State Research Center for Agricuture, Prof.-Dürrwaechter-Platz 1, 85586 Grub, Germany; dieter.krogmeier@lfl.bayern.de

In the state of Bavaria, Germany about 80% of milk recording is done using a LactoCorder as the recording device. A LactoCorder records milk yield as well as information about milk flow. One parameter of the milk flow curve is 'air leakage', which is in many cases caused by the cow knocking off the milking cluster. Therefore, air leakage could be an indicator trait for cow temperament. Air leakage can be objectively measured and a large amount of data is already available. An overall of 191,170 Simmental cows with conformation scores were used for a variance component analysis. For these cows first lactation milk flow curves, temperament classification by farmers and linear scores for udder traits were available. Heritabilities and genetic correlations were estimated with DMU based on more than 70,000 cows from farms with at least 20 observations. Mean frequency of air leakage in first lactation was 6.3% and was declining in the course of the lactation. Air leakage showed the highest incidence (9.6%) during the first milk recording. Effects affecting the frequency of air leakage were herd, days in milk, year-month, daytime of milking, milk yield and milkability. Heritability estimates for air leakage were low $h^2=0.015$ with linear and $h^2=0.059$ with a threshold model, respectively. Genetic correlation with temperament classification was moderate ($r_g=+0.45$) and genetic correlations with different udder traits moderate to low but favorable (overall udder score $r_g=-0.32$, teat placement $r_g=-0.31$, teat diameter $r_g=-0.29$). The possible use of air leakage as an indicator trait in a breeding value estimation for temperament based on 'temperament classification by farmers' will be examined in further investigations.

Circadian metabolomic profile of blood plasma from heifers in the context of feed efficiency
A. Macdonald[1], I. Burton[2], T. Karakach[2], S. Lam[1], A. Fontoura[1], S. Bourgon[3], S. Miller[1,4] and Y. Montanholi[1,3]
[1]University of Guelph, Guelph, ON, Canada, [2]National Research Council Canada, Halifax, NS, Canada, [3]Dalhousie University, Truro, NS, Canada, [4]Invermay Agriculture Centre, AgResearch Limited, Mosgiel, New Zealand; alaina@mail.uoguelph.ca

The practical limitations of direct assessment of feed efficiency (residual feed intake; RFI) in the bovine justify the search for biomarkers of production efficiency. This study used metabolomics to investigate biomarkers of feed efficiency over circadian periods and across physiological states. Thirty-six *Bos taurus* crossbred heifers were evaluated for productive efficiency (body weight and body composition via ultrasound). Blood plasma was collected hourly over 24 hours at three sampling occasions that represented open (367±15 d old), early-gestation (542±23 d old; 82±24 d in pregnancy) and late-gestation (704±25 d old; 244±4 d in pregnancy) stages. Samples were stored in liquid nitrogen and thawed for 30 min at room temperature immediately prior to preparation for NMR analysis. Aliquots (300 µl) of plasma were placed in 5 mm NMR tubes, to which 300 µl of 2 mM sodium 3-trimethylsilyl-2,2,3,3-$_d$4-propionate (TMSP) in D_2O was added and mixed. The spectra (1D 1H-NMR) were acquired at 298 K using a 5 mm TCI CryoProbe™ with automatic tuning and matching and a z-axis gradient amplifier and digital lock on a Bruker Avance III spectrometer operating at 700 MHz proton resonance frequency. Spectra were acquired with a pulse-sequence emphasizing lower molecular weight compounds. The spectra were pre-processed before being exported to MATLAB® for advanced multivariate statistical analysis. Prominent small molecular weight metabolites including acetate, free choline, and α-glucose, as well as glycoproteins, influence the patterns of sample spectral aggregation characteristic of the circadian periods. The spectra of blood plasma metabolites is a promising investigation for bovine production research, as further characterization may reveal phenotypic differences across cattle of varying feed efficiencies and physiological states.

Genetic associations between feed intake and conformation traits in two populations

C.I.V. Manzanilla P.[1], R.F. Veerkamp[1], R.J. Tempelman[2], K.A. Weigel[3] and Y. De Haas[1]
[1]Wageningen UR Livestock Research, Animal Breeding and Genomics Centre, P.O. Box 338, 6700 AH Wageningen, the Netherlands, [2]Michigan State University, Department of Animal Science, East Lansing, MI 488284, USA, [3]University of Winsconsin, Department of Dairy Science, Madison, WI 53706, USA; coralia.manzanillapech@wur.nl

To use conformation traits as predictors of feed intake, it is necessary to obtain accurate estimates of genetic correlations between feed intake and conformation traits, however those correlations can vary across countries. The objective was to estimate genetic correlations between two feed intake traits (dry matter intake (DMI) and residual feed intake (RFI)), and conformation traits within United States (US) and the Netherlands (NL). Records were available on feed intake from 83 nutritional experiments conducted in NL (n=1665, 1991-2011) and US (n=1920, 2007-2013), and on conformation records from those cows plus their relatives (NL=37,241 and US=28,809). The model included fixed effect corrections for location × experiment × ration, age of cow at calving, location × year × season, and days in milk for feed intake traits, and for herd × classification-date, age of cow and lactation stage at classification for conformation. The included random effects were the additive genetic effect and residual. Heritability estimates for DMI were 0.32 in NL and 0.29 in US, whereas heritability estimates for RFI were 0.25 in NL and 0.22 in US. Heritability estimates for conformation traits varied between 0.25 and 0.60 in NL, and between 0.17 and 0.40 in US. The highest estimated genetic correlation was between DMI and chest width (CW) in both countries (NL=0.45 and US=0.61), followed by stature (ST; NL=0.33 and US=0.57),body depth (BD; NL=0.26 and US=0.49) and body condition score (BCS; NL=0.24 and US=0.46). Genetic correlation estimates between feed intake traits and conformation traits were slightly higher in US than in NL, but the trend was similar in both countries. Conformation traits (i.e. ST, CW and BD) showed potential to be used as predictors of DMI.

Direct multitrait selection gives the highest genetic response in a ratio trait

L. Zetouni[1], A.C. Sørensen[1], M. Henryon[2] and J. Lassen[1]
[1]Aarhus University, Molecular Biology and Genetics, Blichers Allé 20, Postboks 50, 8830, Tjele, Denmark, [2]SEGES, Danish Pig Research Centre, Breeding & Genetics, Axeltorv 3, 1609 Copenhagen V, Denmark; lzetouni@mbg.au.dk

For a number of traits the phenotype considered to be the goal trait is a combination of two or more traits, like methane (CH_4) emission (CH_4/milk production). Direct selection on CH_4 emission defined as a ratio is problematic, because it is uncertain whether the improvement comes from an improvement in milk production, a decrease in CH_4 production or both. The goal was to test different strategies on selecting for two antagonistic traits – improving milk production while decreasing methane production. The hypothesis was that in order to maximize genetic gain for a ratio trait, the best approach is to select directly for the two traits rather than using a ratio trait or a trait where one trait is corrected for the other as the selection criteria. Stochastic simulation was used to mimic a dairy cattle population without genomic selection. Three scenarios were tested, which differed in selection criteria but all selecting for increased milk production: (1) the ratio of methane to milk; (2) gross methane phenotypically corrected for milk; and (3) selection based on a multitrait approach using the correlation structure between the two traits. A genetic improvement of 0.14 genetic standard deviation per year was obtained in all scenarios. An improvement in milk production was achieved in all scenarios, but on scenario 1 methane yield increased, with a genetic response of 0.00027 genetic standard deviation for the ratio, while in scenarios 2 and 3 milk methane yields decreased, with genetic responses for the ratio of were -0.00145 and -0.00147. These results confirm the hypothesis that in order to obtain the highest genetic gain a multitrait selection is a better approach than selection based on the ratio. The results are exemplified for a methane and milk production situation but can be generalized to other situations where combined traits need to be improved.

Recording of claw disorders in dairy cattle: overview and prospects of international harmonization

K.F. Stock[1,2], A.-M. Christen[3,4], J. Burgstaller[4], N. Charfeddine[4], A. Fiedler[4], B. Heringstad[1], J. Kofler[4], K. Müller[4], P. Nielsen[4], E. Oakes[4], C. Ødegard[4], J.E. Pryce[1], G. Thomas[4] and C. Egger-Danner[1,5]
[1]ICAR WG Functional Traits, Via Savoia 78, sc. A int. 3, 00198 Rome, Italy, [2]vit, Heideweg 1, 27283 Verden, Germany, [3]Valacta, 555 Bvrd des Anciens-Combattants Sainte-Anne-de-Bellevue, H9X 3R4 Quebec, Canada, [4]International claw experts / ICAR WG FT, Via Savoia 78, sc. A int. 3, 00198 Rome, Italy, [5]ZuchtData EDV-Dienstleistungen GmbH, Dresdner Straße 89/19, 1200 Vienna, Austria; friederike.katharina.stock@vit.de

Several countries have recently started routine recording of claw health data in dairy cattle. Documentation of the health status of feet and legs during regular claw trimming has been identified as a valuable source of information for management and breeding. However, heterogeneous documentation practices complicate systematic collection and use of data. This motivated the ICAR Working Group for Functional Traits (ICAR WGFT) to survey the current role of recording and use of claw data in dairy cattle and coordinate an initiative of international harmonization. Responses from 18 countries showed that half of them have single national keys for recording of claw and foot disorders. Information is collected on 6 to 20 different conditions, with digital dermatitis, white line disease, sole ulcer, interdigital phlegmon, corns, and sole hemorrhage being most frequently considered. Optional severity grading and specification of affection sites are common. Across countries, professional claw trimmers are the main source of claw health data, often using mobile recording devices and customized software. To support monitoring and improvement programs, ICAR WGFT has started interdisciplinary collaborative work involving international claw health experts, with the aim to provide harmonized definitions for major claw disorders along with representative pictures. The descriptive approach will benefit usability of the new ICAR claw health atlas as valuable tool for improved management and breeding of dairy cattle.

Examinations on the genetics of digital dermatitis based on improved definitions of clinical status

K. Schöpke[1], A. Gomez[2], K.A. Dunbar[2], H.H. Swalve[1] and D. Döpfer[2]
[1]Martin-Luther University Halle-Wittenberg, Institute of Agricultural and Nutritional Sciences, Theodor-Lieser-Str. 11, 06120 Halle, Germany, [2]University of Wisconsin, School of Veterinary Medicine, 2015 Linden Drive, Madison, WI 53706-1102, USA; hermann.swalve@landw.uni-halle.de

Bovine digital dermatitis (DD) is an increasing problem in all cattle production systems worldwide. The objective of this study was to evaluate the use of an improved scoring of the clinical status for DD via M-scores accounting for the dynamics of the disease, i.e. the transitions from one stage to another. Newly defined traits were then subjected to a genetic analysis for the assessment of a possible genetic background for the susceptibility to DD. Data consisted of 6,444 clinical observations from 729 Holstein heifers in a commercial dairy herd, collected applying the M-score system. M-scores were converted into phenotypic traits deriving new DD trait definitions with different complexity. Linear mixed models and logistic models were used to identify fixed environmental effects and to estimate variance components. In total, 68% of all observations showed a healthy status while 11% were scored as infectious for and affected by DD, and 23% of all observations exhibited an affected but non-infectious status. For all traits, the probability of occurrence and the clinical status were associated with age at observation and time of observation. Risk of getting infected increased with age. Month of observation significantly affected all traits. Estimates of heritabilities of the traits studied ranged between 0.19 ± 0.11 and 0.52 ± 0.17 revealing a tendency for higher values for more complex trait definitions. In terms of genetic selection, all trait definitions concordantly identified the best, i.e. most resistant animals, but only the new trait definitions were able to distinguish between animals with average and high pre-disposition for DD.

Utilizing claw health information in traditional and genomic selection for Norwegian Red

C. Ødegård[1,2], M. Svendsen[2] and B. Heringstad[1,2]
[1]Norwegian University of Life Sciences, Department of Animal and Aquacultural Sciences, P.O. Box 5003, 1432 Ås, Norway, [2]Geno Breeding and AI Association, P.O. Box 5003, 1432 Ås, Norway; cecilie.odegard@nmbu.no

Claw health is important for economic and animal welfare reasons. The aim was to evaluate efficient ways of utilizing data from claw trimming to improve claw health in Norwegian Red cows, applying traditional and genomic selection. Data from 2004 was available for the genetic analyses. A new claw health index was included in the routine genetic evaluation of Norwegian Red in September 2014. The traits included were: corkscrew claw (CSC); infectious claw disorder (INF), consisting of dermatitis, heel horn erosion and interdigital phlegmon; and laminitis related claw disorder (LAM), consisting of sole ulcer, white line disorders and hemorrhage of sole and white line. The number of cows with claw health records increase each year, and have since 2011 been approximately 65,000 cows per year. Claw disorders are low incidence traits, and genetic analyses using a multivariate animal model show heritabilities of 0.05 for CSC and 0.03 for INF and LAM. The daughter groups of sires at first official proofs are below 10 and the challenge is to obtain reliable breeding values. One possible solution is genomic selection. However, claw disorders are novel traits with few animals in the reference population. A two-step approach was conducted and genomic breeding values were validated using 10-fold cross-validation. The mean predictive correlations of GEBV for CSC, INF and LAM were 0.35, 0.32 and 0.33, respectively. Increasing the reference population by including genetic correlated traits in the analyses gave a slight increase in predictive correlation of GEBV for CSC. To further increase the predictive correlation of GEBV, including genotyped females in the reference population and use of one-step approach is of interest. This will be the next step for efficient utilizing of claw health data.

Estimation of genetic parameters for well-established and new milk yield performance test traits and

D. Hinrichs[1], T.J. Boysen[2] and G. Thaller[1]
[1]Institute of Animal Breeding and Husbandry, Christian-Albrechts-University, Hermann-Rodewald-Straße 6, 24118 Kiel, Germany, [2]Institute of Clinical Molecular Biology, Christian-Albrechts-University, University Hospital Schleswig Holstein Campus Kiel, Schittenhelmstr. 12, 24105 Kiel, Germany; dhinrichs@tierzucht.uni-kiel.de

The aim of this study was to estimate genetic parameters for new milk yield performance test traits and their genetic correlations with clinical mastitis (CM). Data were recorded from February 1998 to November 2008 on 3 commercial milk farms with an average herd size of 3,200 cows. Within this time period 55,536 medical treatment records for CM and 324,423 monthly milk performance tests were gathered for 16,545 cows. Animals which not temporal complete recorded lactations and with less than 5 performance tests in the lactation were rejected. Furthermore the data was restricted to 305 days in lactation. The resulting data basis comprises 13,450 animals, 255,784 milk performance tests and 43,183 mastitis records. The median and the variance of milk yield, fat- and protein-percentage, fat-protein ratio, and somatic cell score were calculated based on milk performance test data for different lactations. CM was either treated as a binary trait (BT) or as the number of treatments per lactation. Heritabilities vary between 0.04 (mastitis BT) and 0.76 (fat percentage) if the median was used for the estimation. As expected the highest genetic correlation was observed between the two mastitis measurements (0.97) and somatic cell score (0.68). However, there were also moderate genetic correlations between mastitis and milk yield, fat- and protein-percentage, and fat-protein ratio, varying between -0.24 and 0.18. Standard errors of the estimates fall in the interval between 0.01 and 0.06. If variances were used for the estimation, heritability estimates were lower (0.02 to 0.20) and also the genetic correlations between yield traits and mastitis measurements were lower (0.52 to -0.02) and the standard errors were higher (0.01 to 0.10).

An investigation on determining factors affecting body condition score in first parity Holstein cows

A. Galic[1] and N. Karslioglu Kara[2]
[1]Akdeniz University, Animal Science, Akdeniz University Faculty of Agriculture Department of Animal Science, 07058 Antalya, Turkey, [2]Uludag University, Animal Science, Uludag University Faculty of Agriculture Department of Animal Science, 16059 Bursa, Turkey; galic@akdeniz.edu.tr

Body condition scoring is very important for monitoring of feeding programs in an effective manner. In addition, the relationship between the various traits increases the importance of body condition score. This study was conducted to determine the effects of lactation period, calving age and calving season on body condition score in first parity Holstein cows. Fort this aim, body conditions of 107 cows were used as material. Animals were selected from a farm that is a member of the Cattle Breeders' Association of Antalya. Animals were scored on the same day using the scale of 1-9. The averages of first calving age and body condition score were determined as 26.81 ± 0.52 months and 5.20 ± 0.11, respectively. Also, it was found that averages of body condition scores at different periods of lactation were ranged from 3.78 to 7.75 which were increased with advancing lactation period. In addition, results showed that body conditions were affected by the lactation period significantly, while the effects of calving age and calving season on body condition score were not significant.

Association of STAT1 gene with milk production traits in Iranian Holstein dairy cattle

M. Hosseinpour Mashhadi[1] and N. Tabasi[2]
[1]Department of Animal Science, Mashhad Branch, Islamic Azad University, Mashhad, Iran, 91735-413-, Iran, [2]Bu-Ali Research Institute, Immunology Research Center, School of Medicine, Mashhad University of Med, Mashhad, 99, Iran; mojtaba_h_m@yahoo.com

Identification of polymorphism of candidate genes and their association with milk production traits provides appropriate information for animal breeding specialists. The STAT1 gene is a candidate gene. This study was conducted for investigating the single nucleotide polymorphism of gene (C/T) and association between genotypes of gene with milk production traits in Holstein dairy cattle. The blood samples of 266 cows were used for DNA extraction. DNA was extracted with Cinagene kit (PR881612C). Fragment of 314 bp of STAT1 gene was amplified by PCR and digested by Pag1 restriction enzyme. Genotypes frequencies were calculated by PopGene software (version 1.31). The effects of father (as random effect), HYS, lactation and genotype (as fixed effects) on milk production traits was studied by GLM procedure of SAS package. Duncan multiple range test was used for compare means of traits. Frequency of TT, TC and CC genotypes were 0.1098, 0.4422 and 0.4489 respectively. The T and C allele frequency were 0.33 and 0.67 respectively. Result of Chi-square showed that the sample was not in the Hardy Weinberg equilibrium ($P<0.01$). The mean of genotypes for milk yield, fat yield and protein percentage respectively were 8,324 (CC), 7,938 kg (TC) and 7,318 (TT); 275.6 kg (CC), 265.4 kg (CT) and 249.9 kg (TT); 3.42% (CC), 3.43% (CT) and 3.65% (TT) and differences between these means was significant ($P<0.05$). The genotypes mean of fat percentage and protein yield respectively were 3.31% (CC), 3.35% (CT) and 3.36% (TT); 285.5% (CC), 274 kg (CT) and 268.2 kg (TT) and differences between these means wasn't significant ($P>0.05$). The results show the positive effect of allele C on milk production traits.

Genetic parameters for dry matter intake in primiparous Holstein, Nordic Red and Jersey

B. Li[1,2], W.F. Fikse[2], J. Lassen[1], M.H. Lidauer[3], P. Løvendahl[1] and B. Berglund[2]
[1]Aarhus University, Department of Molecular Biology and Genetics, Blichers Allé 20, 8830 Tjele, Denmark, [2]Swedish University of Agricultural Sciences, Department of Animal Breeding and Genetics, Ulls väg 26, 75007 Uppsala, Sweden, [3]Natural Resources Institute Finland, Green Technology, Humppilantie 7, 31600 Jokioinen, Finland; bingjie.li@slu.se

Dry matter intake (DMI) is an important component of feed efficiency in dairy cattle. This study estimated the genetic parameters for DMI over the first 24 lactation weeks in 3 dairy cattle breeds including Holstein, Nordic Red and Jersey. In total, 1,751 primiparous cows (771 Holstein, 696 Nordic Red and 284 Jersey) from Denmark, Finland and Sweden were included. Genetic parameters were estimated for a combined data set with all cows from 3 breeds as well as for each single breed. Variance components, heritability and repeatability for DMI within the first 24 lactation weeks were estimated based on a linear mixed animal model. Heritability for DMI ranged between 0.25-0.37 in the combined data set, 0.18-0.47 in Holstein, 0.23-0.39 in Nordic Red, and 0.17-0.44 in Jersey; the differences in heritability between breeds were not statistically significant. Genetic variance and the total phenotypic variance increased along the lactation in the combined data set. Nordic Red had a significantly higher total phenotypic variance than the other two breeds and tended to have a higher genetic variance than Holstein and Jersey. Genetic correlations for DMI within the 24 lactation weeks were generally high in all three breeds, especially among lactation weeks 5 to 24. DMI at early lactation (lactation weeks 1 to 4) was genetically different from the middle lactation (lactation weeks 5 to 24). Due to the high genetic correlation among lactation weeks 5 to 24, this period of lactation could be recommended for recording of DMI.

Relationships among feed efficiency, growth and carcass traits in beef cattle

T.M. Ceacero, M.E.Z. Mercadante, J.N.S.G. Cyrillo, A.L. Guimarães, R.C. Canesin and S.F.M. Bonilha
Instituto de Zootecnia, Centro APTA Bovinos de Corte, P.O. Box 63, Sertãozinho SP, Brazil; mercadante@iz.sp.gov.br

This study aimed to estimate genetic (r_g) and phenotypic (r_p) correlations among yearling feed efficiency traits [gross feed efficiency (G:F), residual feed intake (RFI), residual feed intake adjusted for backfat thickness (RFI_{bf}) and adjusted for ultrasound backfat thickness and rump fat thickness (RFI_{ft}), residual average daily gain (RAG) and residual intake and gain (RIG)] and yearling body weight (YW), hip height (HH), chest girth (CG), ultrasound longissimus muscle area (LMA), and ultrasound backfat (BFT) and rump fat thickness (RFT) in Nellore cattle. Dry matter intake (DMI) and average daily gain (ADG) of 955 animals were obtained in 21 performance tests of 83±15 days long. Covariance components were estimated in two-trait animal models by restricted maximum likelihood. Heritability for G:F, RFI, RFI_{bf}, RFI_{ft}, RG and RIG were 0.14, 0.24, 0.20, 0.22, 0.19 and 0.15. All r_p among studied traits were close to zero, excepting some of them between feed efficiency traits and DMI (-0.10 to 0.73); and ADG (0.02 to 0.68). The r_g of DMI and ADG with growth traits were strong and positive (higher than 0.61), whereas with carcass traits were weak to medium and also positive (0.15 to 0.48). Among feed efficiency traits, RAG showed the highest and positive r_g with YW, HH and CG (0.34, 0.25 and 0.34) and the lowest r_g with BFT and RFT (-0.17 and 0.18). RFI, RFI_{bf} and RFI_{ft} showed unfavorable r_g with YW (0.17, 0.23 and 0.22), BFT (0.37, 0.33 and 0.33) and RFT (0.30, 0.31 and 0.32). The r_g of G:F and RIG with growth traits were weak and favorable (0.08 to 0.22), however they were medium and unfavorable with BFT and RFT (-0.22 to -0.38). BFT and RFT inclusion in RFI calculation model did not decrease genetic relationship between RFI and carcass fat thickness. Selection for improving feed efficiency in growing Nellore cattle is expected to decrease carcass fat deposition, with little changing in growth traits.

A simulation study on selection to lower the fat to protein ratio in the early lactation of Holstein

A. Nishiura[1], O. Sasaki[1], M. Aihara[2], H. Takeda[1] and M. Satoh[1]
[1]NARO Institute of Livestock and Grassland Science, Tsukuba, Ibaraki, 305-0901, Japan, [2]Livestock Improvement Association of Japan, Koto-ku, Tokyo, 135-0041, Japan; akinishi@affrc.go.jp

The high fat to protein ratio of milk (FPR) values in the early lactation indicate poor energy status of lactating cow. The objective of this study was to investigate the response to the selection to lower FPR in the early lactation by a simulation study using the field test data. The data consisted of test-day milk records of Japanese Holstein cows from 2001 to 2010. All lactations in the first three parities were required 10 records from 6 to 305 days in milk (DIM) from herds with over 100 cows. The data set included 9,779,580 records of 655,355 cows in 5,212 herds. We divided the data set into 10 subsets at random. We estimated genetic parameters of FPR for one of these 10 subsets using random regression test-day animal model. The breeding values of FPR from 6 to 105 DIM in 2001-2005 for the other nine subsets were estimated based on these estimated parameters. In the simulation, we selected the cows with the top 10% estimated breeding values (EBV). Then we estimated breeding values of FPR in 2006 and selected the cows with the top 10% EBV. We repeated this selection to 2010. In the control, the EBV were estimated from the whole data in 2001-2010 for the same nine subsets. In our results, the average EBV of FPR by test year of cows were getting lower year by year. The EBV of FPR were lower in the simulation than in the control, but the differences between them were small. The EBV of fat % showed the similar tendency to FPR. The EBV of protein % were almost constant in 2001-2010. The EBV of milk yield were upward to 2006, and then were constant to 2010. There were little differences between the simulation and the control in the EBV of milk yield. It supposed that cows with high FPR values in the early lactation were culled not only in the simulation datasets, but also in the control datasets.

Genetic analysis of lactation persistency and conformation traits in Polish Holstein-Friesian cows

A. Otwinowska-Mindur, E. Ptak and W. Jagusiak
University of Agriculture in Krakow, al. Mickiewicza 24/28, 30-059 Krakow, Poland; rzmindur@cyf-kr.edu.pl

The objective of this study was to estimate genetic relationships between lactation persistency and conformation traits of Polish Holstein-Friesian cows. Data consisted of 5 descriptive type traits (scored from 50 to 100), stature, 16 linearly scored (1-9 scale) traits and 3 measures of lactation persistency. The first definition ($P_{2:1}$) was milk yield in the second 100 days in milk (DIM) divided by yield in the first 100 DIM. The second definition ($P_{3:1}$) was milk yield in the third 100 DIM to yield in the first 100 DIM, and the third definition (P_d) was milk yield at 280 DIM divided by milk yield at 60 DIM. All 3 measures of persistency were calculated based on lactation curves fitted to TD milk yields from 18,216 first lactations of cows calved in 2003-2007. The MT-REML method was used for (co)variance component estimation. The linear model for persistency included fixed effect of herd-year-season of calving, fixed effect of age of calving class, and random animal effect; the linear model for type traits contained fixed effect of herd-year-season of calving-classifier, fixed effect of age of calving class, fixed effect of lactation stage and random animal effect. All 3 persistency measures showed higher genetic correlations with descriptive type traits and stature than with linearly scored type traits. Among 5 descriptive traits, udder had the highest correlation with lactation persistency (0.75-0.80), and type & conformation the lowest (0.02 to 0.14). Negative and moderate genetic correlations were estimated between persistency and stature (0.64 to 0.76). Linearly scored traits were weakly or not genetically correlated with lactation persistency (0.01-0.40, ignoring sign). A thorough examination of the estimated genetic correlations indicated that some improvement in lactation persistency could be achieved by indirect selection for udder, rear udder height and udder width, especially when P_d was used as the measure of lactation persistency.

Differences in novel traits between genetic groups of dairy cows in pasture-based production systems

K. Brügemann[1], M. Jaeger[1], U. Von Borstel[2] and S. König[1]
[1]University of Kassel, Dept. Animal Breeding, Nordbahnhofstr. 1a, 37213 Witzenhausen, Germany; [2]University of Göttingen, DNTW, Albrecht-Thaer-Weg 3, 37075 Göttingen, Germany; kerstin.bruegemann@uni-kassel.de

Breeding goals of Holstein dairy cattle in New Zealand (NZL) focus on traits reflecting adaptation to pasture-based systems. Due to high prices for concentrates, and advantages for cattle health and welfare, low input grazing systems become important in regions of North-West Germany. The present study addresses the hypothesis that progeny of NZL-Holstein sires are superior in 30 German grassland farms compared to offspring of Holstein sires being progeny tested in German indoor systems. A research design was implemented to create three different genetic groups (F1 generation) within herds on the basis of a German Holstein cow (GHC) population: Group 1=GHC × NZL sires, group 2=GHC × GH sires, group 3=GHC × GH_pasture sires. GH_pasture sires represent bulls of German origin with high breeding values for the traits being important in NZL. Group comparison focused on the full set of production and functional traits from official recording systems, and in addition on further novel traits reflecting cow health and welfare: Scores for body condition (BCS), locomotion (LOCS), hock lesions (HLS), and total hygiene (THS). Linear and generalized linear mixed models were applied to estimate group differences for a longitudinal data structure (2998 observations). In parity 1, daughters of NZL sires had the highest LSMean for BCS (2.56), and lowest LSMean for LOCS (0.10), HLS (0.25) and THS (3.67), reflecting an enhanced robustness of HF strains from NZL. Within breed genetic analyses simultaneously considering all genetic groups revealed a moderate genetic background with the following heritabilities: BCS (0.25), LOCS (0.12), HLS (0.04), and THS (0.17). Genotype by environment interactions between pasture-based and indoor systems were studied by applying multiple trait models. For all traits, r_g were smaller than 0.80, especially indicating G×E interactions for low heritability functional traits.

Lactation persistency and environmental factors affecting on this characteristic in dairy cattle

M. Elahi Torshizi
Department of Animal Science, Mashhad Branch, Islamic Azad University, Mashhad, Iran; elahi222@gmail.com

In this study the data set contained 435,390 test day records of first lactation Holstein cows corresponding to 48955 cows calved between 2001 and 2011. Four different measures of persistency were milk produced between 101-200 days divided into the first 100 days (pers1), milk produced between 201-305 days divided into the first 100 days (pers2), milk produced between 201-305 days divided into 101-200 days (pers3), milk produced between 5-305 days divided into 5-82 days (pers4) and calculation of persistency by Kamidi formula. Fitting individual lactation curve was done by the Wood function. The average of pers1, pers2, pers3 and pers4 were 103.16±0.044, 91.38±0.07, 87.89±0.04 and 3.67±0.001 and positive phenotypic correlation among these measures and 305 d milk yield were 0.43, 0.98, 0.96 and 0.81 respectively. Low heritabilities estimated using animal model for these characteristics (0.084, 0.080, 0.062 and 0.079). The phenotypic trend of persistency was 0.5%/year for pers1 followed by 0.72%/year, 0.28%/year for pers2 and pers3 respectively. The most important environmental factors influences on persistency were calving year, season of production and season of calving. Persistency was highest and lowest in cows calved in summer and spring respectively. Moreover, persistency calculated by Kamidi model were 99.98 in cows calved in summer. According to the results, the highest persistency of milk yield is related to pers1 followed by Kamdid formula, pers2, pers4 and pers3 respectively and it seems that pers1 is an optimum criterion for selection of cows with the most persistency.

Circadian profile of methane and heat production at different physiological states in beef heifers

Y. Montanholi[1,2], B. Smith[2], K. Colliver[2], A. Fontoura[2] and S. Miller[2,3]
[1]Dalhousie University, Truro, NS, Canada, [2]University of Guelph, Guelph, ON, Canada, [3]AgResearch Limited, Puddle Alley, Mosgiel, New Zealand; yuri.montanholi@dal.ca

The beef industry has been emphasizing the improvement of feed efficiency. Calorimetric assessments are fundamental for identifying biomarkers and for assessing response to selection for improved feed efficiency. This study aims to characterize the circadian calorimetry profile of beef heifers with known feed efficiency (RFI; kg/d) at different developmental stages, as well as, verify associations between heat and methane production with infrared images from different body locations. Thirty-six *Bos taurus* crossbred heifers were evaluated for productive efficiency (body weight and body composition via ultrasound) and individual feed intake assessment to calculate RFI. Heifers were sampled using a four-head-chamber-open-indirect-calorimeter during 24 h continuously at three sampling occasions that represented open (367 ± 15 d old), early-gestation (542 ± 23 d old; 82 ± 24 d in pregnancy) and late-gestation (704 ± 25 d old; 244 ± 4 d in pregnancy) stages. The calorimeter was set to perform 6 min of sampling in each chamber plus the reference channel for a full cycle every ½ h over 24 h of breath gases sampling (CO_2, O_2 and CH_4). Heifers were concomitantly infrared imaged on an hourly basis, with images taken from eye, snout, feet, hind area and flanks. Mixed procedure of SAS was used to analyze the data. Preliminary results, from the late-gestation stage, indicate that more feed efficient heifers have greater oxygen consumption (819 vs 743 W, P=0.01) and produce more methane (153 vs 124 ml/min, P=0.02) over the circadian period. Infrared images indicated lower infrared radiation emissions in more feed efficient cattle at the eye-globe, caudal view of the feet (i.e. 31.8 vs 32.9 oC, P=0.03) and both flank regions. These preliminary results provide evidence to support that improved feed efficiency in the bovine is associated with increased metabolic rate and diminished radiant heat loss.

Genetic diversity in Swiss cattle breeds

H. Signer-Hasler[1], B. Gredler[2], M. Neuditschko[3], A. Burren[1], C. Baes[1,2], B. Bapst[2], D. Garrick[4], C. Stricker[5] and C. Flury[1]
[1]Bern University of Applied Sciences, School of Agricultural, Forest and Food Sciences HAFL, Länggasse 85, 3052 Zollikofen, Switzerland, [2]Qualitas AG, Chamerstrasse 56, 6300 Zug, Switzerland, [3]Agroscope, Swiss National Stud Farm, Les Longs-Prés, 1580 Avenches, Switzerland, [4]Iowa State University, Animal Science, Kildee Hall, Ames, IA 50011, USA, [5]agn Genetics, Börtjistrasse 8b, 7260 Davos, Switzerland; christine.flury@bfh.ch

In this study 50k SNP genotypes of more than 9,000 sires were used to assess genetic diversity and population structure of Brown Swiss (BS), Braunvieh (BV), Original Braunvieh (OB), Holstein (HO), Red Holstein (RH), Swiss Fleckvieh (SF), Simmental (SI), Eringer (ER) and Evolèner (EV) cattle. In total 28,948 autosomal SNPs with call rates $\geq$90% and minor allele frequency >1% in all populations were used in the analyses. The pairwise F_{ST}-values ranged from 0.007 (BV/BS) to 0.156 (HO/BS) indicating, that differentiation is largest for HO and BS and lowest for BV and BS. The average observed heterozygosity varied between 35.3% (BS) and 39.3% (SF). The average genomic relationship was lowest for SF (4.5%) and largest for BS (15.4%), respectively. In addition fine-scale population structure of each breed is investigated. The presented parameters allow to assess the current genetic diversity within and between Swiss cattle breeds. Selection signatures highlight genomic regions under selection which led to the differentiation between breeds.

How performance impairs implementation: a review of performance of sensor-based automatic monitoring

K.N. Dominiak
University of Copenhagen, Department of Large Animal Sciences, Groennegaardsvej 2, 1870 Frederiksberg C, Denmark; knd@sund.ku.dk

High performance in early warning systems is crucial for minimizing the amount of false positive alarms. Inadequate performance decreases the effect and the reliability of the early warning system and thereby implementation in modern livestock production is prevented. Through a thorough review of sensor-based automatic monitoring systems in livestock production it is revealed that a consistent challenge is to optimize the performance to a level that enables implementation in commercial production. In the discussion a majority of the papers mentions challenges with the performance of the methods causing their early warning systems to be non-implementable in the current form. Although thoughts on how to optimize the performance is presented in several of the papers, they are typically both vaguely expressed and described as being beyond the scope of the presented studies. Only few papers focus on ranking or prioritization of alarms generated by sensor-based monitoring. Two of those papers aim to rank or prioritize an alert-list of dairy cows likely to have Clinical Mastitis (CM) by combining sensor-based data from Automatic Milking Systems (AMS) with non-AMS information like CM-history, parity or day in milk in different ways. Even though the methods presented reduced the number of cows that needed further visual checking the authors concluded that it was not to a level implementable in commercial herds. Hence the conclusion of the review of the performance of sensor-based automatic monitoring systems in livestock production is that the majority of papers recognizes the importance of prioritization of alarms and discusses in vague terms how to improve the performance. Accordingly it is clear that the topic is highly underexposed in scientific literature and future focus on the task of implementation is of highest importance. This major challenge for precision livestock farming will be addressed in an ongoing international research project.

Bayesian prediction of mastitis using sensor data routinely collected in dairy herds

D.B. Jensen[1] and A. Devries[2]
[1]Copenhagen University, Department of Large Animal Sciences, Grønnegårdsvej 2, 1870 Frederiksberg C, Denmark, [2]University of Florida, Department of Animal Sciences, 2250 Shealy Drive, Gainesville, FL 32669, USA; daj@sund.ku.dk

In modern dairy cattle herds, several types of data is routinely collected for monitoring purposes and to raise alarms to check for undesired health conditions, e.g. mastitis. However, a meaningful way of combining the different types of information is still lacking, and alarms raised from the values in individual data series seem to universally require an unacceptable compromise between sensitivity and specificity. Here we demonstrate the use of a naïve Bayesian classifier as a simple yet effective method for combining eight types of continuous sensor information (milk yield, milk conductivity, milk protein- fat- lactose- and blood percentage, individual cow weight and activity) and four types of categorical information about the cow and its environment (cow parity, cow mastitis history, somatic cell count level, and current season). We use data from 1,275 individual cows, all from the same herd, collected over a 6 year period. Half of these were randomly selected to form the learning set, while the other half formed the testing set. From the learning set, likelihoods were calculated for each category of the categorical data, conditioned on the health state of the cows (mastitis or no mastitis). The values of the continuous variables were categorized as low, lower-middle, upper-middle or high, compared to the daily mean and standard deviation found for the learning set. Likelihoods of observing each of variables within each category, given the health state, were calculated. Using these likelihoods, the posterior probability of mastitis was calculated for each day after calving for each cow in the test set, and the predicted health state classified accordingly. Thus mastitis could be classified in the test set with a sensitivity of 0.816 when the specificity is kept at 0.8.

Using acceleration data to detect automatically the beginning of farrowing in sows
I. Traulsen[1], W. Auer[2], K. Müller[3] and J. Krieter[1]
[1]Institute of Animal Breeding and Husbandry, Christian-Albrechts University, Olshausenstr. 40, 24098 Kiel, Germany, [2]MKW electronics GmbH, Jutogasse 3, 4675 Weibern, Austria, [3]Chamber of Agriculture Schleswig-Holstein, Gutshof, 24327 Blekendorf, Germany; itaulsen@tierzucht.uni-kiel.de

Growing numbers of livestock per farm increases management effort for farmers and stockpersons. Technical equipment can be beneficial to support management. The aim of the present study was to use accelerations measurements to detect automatically the beginning of farrowing in sows. From December 2013 until March 2014 sows in the farrowing unit of the Futterkamp research farm were equipped with an ear sensor to sample acceleration (1 Hz). As reference video recordings of 24 sows were used to determine the beginning of farrowing defined as birth of the first piglet. To implement an early warning system a self-starting CUSUM chart using an acceleration index based on the sensors' acceleration measurements was set up staring four days before the calculated date of farrowing. Before the birth of the first piglet the acceleration increased for all sows up to a higher level than the days before farrowing. The CUSUM statistic found this increase and gave an alert when control limits were exceeded. Sow individual modeling of the CUSUM chart increased the detection rate. In a first step the alert was given approximately eight hours before the beginning of farrowing. A second alert three to four hours before start might be beneficial to the farmer. A targeted monitoring of the sows can improve the working flow and reduce losses during farrowing. Summarizing the results showed that acceleration measurement of an ear sensor can be used to detect the beginning of farrowing in sows.

Monitoring growth in finishers by weighing selected groups of pigs – a dynamic approach
A.H. Stygar and A.R. Kristensen
University of Copenhagen, Faculty of Health and Medical Sciences, Department of Large Animal Sciences, Grønnegårdsvej 2, 1870 Frederiksberg C, Denmark; as@sund.ku.dk

In an ongoing project on improving welfare and productivity in growing pigs using advanced ITC methods (the PigIT project) the end goal is to become able to monitor growth automatically by vision based methods. Such automatic monitoring systems for body weight are not yet common in commercial herds. Instead, some farmers use control groups for growth monitoring where a chosen subset of pigs is regularly weighed manually and the resulting information is used as a basis for monitoring, forecasting and decision support. The objective of this study was to evaluate the use of manually weighed control groups for monitoring and forecasting. All pigs (around 500) in a section were individually identified by RFID ear tags and weighed at insertion and at delivery of first pigs. Additionally, in four pens (18 pigs each) the body weight has been measured on weekly basis. A linear mixed-effects model with contributions from fixed effect of time, random intercept and slope for pen and random effects with repeated measurements for pigs was fitted to the data on body weight. The two pen effects (intercept and slope) were assumed to be independent. Additionally, a Gaussian correlation structure was used to model within-pig correlations. Obtained estimates for variance components, autocorrelation and the mean values for slope and intercept were implemented in a multivariate dynamic linear model (DLM). The resulting monitoring tool was evaluated on its precision in forecasting body weights. Furthermore, the differences between observed and predicted values of body weight from the DLM were monitored on weekly basis using a Control Chart for detection of impaired growth. This study will provide an evaluation of the manual body weight monitoring, but it is also expected that the described model could be used in developing a monitoring system based on automatic weight assessment by vision methods.

Real time measurement of reticular temperature for the prediction of parturition and estrus in cows

J. Gasteiner[1], J. Wolfthaler[2], W. Zollitsch[2], M. Horn[2] and A. Steinwidder[1]
[1]AREC Raumberg-Gumpenstein, Raumberg 38, 8952 Irdning, Austria, [2]BOKU-University of Natural Resources and Life Sciences, Gregor-Mendel-Straße 33, 1180 Vienna, Austria; marco.horn@boku.ac.at

The suitability of the reticular temperature (RT) as an indicator for expected parturition and estrus of dairy cows was investigated. 25 parturitions and 43 estruses were recorded. Estrus was confirmed by frequent measurement of milk-progesterone and, retrospectively, by a successful artificial insemination. The RT was measured continuously every ten minutes with indwelling reticular sensors and data were read out by telemetry. The average ambient temperature during the study period was 4.43 ± 7.86 °C and the mean RT 39.23 ± 0.33 °C. In this study the averages of RT-day, 5 days before up to 2 days after estrus, the RT-4 hour averages from 48 hours up to 20 hours after the temperature maximum at estrus and the RT-day averages 10 days before up to 10 days after calving, were analyzed. Time of day, feeding, breed and lactation number were also considered. RT was sign. influenced by time of day and ambient temperature. RT was also sign. affected by the occurrence of estrus. The mean RT on the day of estrus was 0.15 °C higher than the day before. The maximum RT-4-hour average on the day of estrus (39.71 °C) was also increased sign. The results for heat detection showed an area under curve (AUC) of 0.81. A sign. effect of parturition on the RT was also found. 48 hours prior to calving RT decreased sign. by 0.43 °C. No sign. difference was found between one day before parturition and the day of parturition. Up to a temperature threshold of ≥0.40 °C, 100% of the parturitions were detected by RT within 24 up to 48 hours, with a specificity of up to 93%. The prediction of a parturition within 24 and 48 hours showed an AUC of 0.99. We conclude that continuous RT measurement as used herein is highly suitable for detecting upcoming parturitions and, to a lesser extent, to indentify cows in heat.

Application of GPS to monitor cattle behaviour and pasture use in European Alpine regions

J. Maxa, S. Thurner, H. Wirl and G. Wendl
Bavarian State Research Centre for Agriculture, Institute of Agricultural Engineering and Animal Husbandry, Voettinger Str. 36, 85354 Freising, Germany; jan.maxa@lfl.bayern.de

Animal monitoring based on various techniques has been applied since many decades and became more common since global positioning systems (GPS) can be used for civilian purposes. High labour workload together with difficult relief conditions on alpine farms resulted in a decrease of livestock units and succession processes in many regions of the Alps. Therefore, the main aim of this study was to analyse cattle behaviour and pasture use based on data from GPS tracking collars together with an analysis of workload of herdsmen with focus on possible optimization of the management of alpine farms. Two independent trials were conducted in 2013 and 2014 on alpine farms in Austria and Germany, one for analyses focusing on distances walked by cattle from successive GPS locations and autocorrelation of recorded information and second on monitoring of cattle behaviour using the information from GPS collars. Time-sampling method was conducted to score the cows behaviours which were used to classify the GPS data. Turn angle, distances walked by cattle together with statistical analyses (Spearman correlation for autocorrelation analysis and Wilcoxon rank sum test for identifying the differences among the classes of behaviour scores) of collected data were calculated using R software. The results showed an antagonism among time intervals of successive locations needed to calculate distances travelled and to correctly interpret cattle grazing patterns. Walking, grazing and standing could be distinguished based on GPS data basis (P<0.05) but additional information from other sensors such as an accelerometer are needed to recognize a wider spectrum of behavioural data. Overall the study concluded the potential of use of cattle tracking system for optimizing the workload of herdsmen and pasture management in the Alps and revealed information necessary for developing of classification algorithms for behavioural recognition.

Development of a multi-Kinect-monitoring system using complete 3D information of walking cows
J. Salau[1], J.H. Haas[1], G. Thaller[1], M. Leisen[2] and W. Junge[1]
[1]Kiel University, Institute of Animal Breeding and Husbandry, Olshausenstrasse 40, 24098, Germany, [2]Rinderzucht Schleswig-Holstein eG, Rendsburger Str. 178, 24537 Neumünster, Germany; jsalau@tierzucht.uni-kiel.de

Every additional technological system comes with additional costs or efforts for the dairy farmer and potential interfacing problems with commonly applied herd management software. Automated systems that reliably deliver data concerning several monitoring purposes therefore help to reduce technical overload and data traffic. In this study a system which uses methods of computer vision to not only analyze dairy cows' gait but also assess body characteristics is developed. Six Microsoft Kinect 3D cameras are mounted in a framework with ≈2 m in both passage height and width. They are facing the passage from both sides and allow the collection of 3D information on the complete cow surface during several steps. Software for recording, synchronizing, sorting and segmenting images, and claw determination had already been implemented. The system worked stably during the recording of a six hours development data set that consisted of Holstein Friesian cows led by rope. To describe the animals' gait, trajectories of front claws have been extracted. The mean step height and length calculated to 25.9 cm and 74.6 cm. -3.5° were determined as mean angle of deviation from moving the claw on a straight line parallel to the cow's motion direction. Monitoring these measures over time enables the detection of changes in gait. For object recognition, 2D-wavelet transforms had been applied and binary classifiers based on the local high frequencies have been implemented to decide whether a pixel belongs to the image foreground. It performed very well according to the Area Under the receiver operation characteristic Curve (AUC>0.8). Additionally, wavelet applications were used to analyze the cow surface's curvature with reference to changes in body condition. Simultaneously addressing multiple components of herd monitoring, the aspired system could be a promising holistic tool.

Precision dairy cattle nutrition: statistics of rumen pH in commercial cattle
T.T.F. Mottram
Royal Agricultural University, School of Agriculture Food and Environment, Cirencester, GL7 6SJ, United Kingdom; toby.mottram@rau.ac.uk

The development of the rumen telemetry bolus now permits precision in determining the risk of acidosis for commercial cattle. At the EAAP 2014 Mottram et al. presented case studies of the effects of husbandry on reticulo-rumen pH. In this paper we analyse 450 rumen pH records up to 150 days in length from commercial and research herds. Where metadata was available the rumen profiles were categorised by husbandry (robotic milking, three times a day milking, twice a day milking) and feeding type (grazing, TMR, grass silage and concentrate). Statistics will be presented of features of the rumen pH profiles under each husbandry treatment the mean, minimum, maximum, hours below threshold of 5.8 pH, mean daily range, numbers of drops per day. For the rumen temperature profiles, the mean, mode, maximum, minimum and the number of drinking events per day will be presented. Seasonal variations in pH and temeprature will also be discussed.

Using a dairy cow model to interpret *in vivo* individual data and to upscale results at herd level

O. Martin[1,2], C. Gaillard[3], J. Sehested[3] and N.C. Friggens[1,2]
[1]AgroParisTech, UMR 791 MoSAR, 75005 Paris, France, [2]INRA, UMR 791 MoSAR, 75005 Paris, France, [3]Aarhus University, Foulum, 8830 Tjele, Denmark; olivier.martin@agroparistech.fr

Models are appropriate tools to help understanding the functioning of biological systems and many of them are used to forecast and do simulations. Precision farming has primarily been developed through the automation of data acquisition but interpretative tools to capitalize on this raw material are lacking. In this context, animal models can be used as translators of individual time series data on animal performance into phenotypic information providing quantification on variability and further useful benchmarks for decision support. In this study, we propose a fitting procedure on experimental data to synthesize records on individual cows and a demonstration of the use of this model-based interpretation to upscale results at herd level. We used a modified version of the GARUNS model of dairy cow lifetime performance proposed by Martin and Sauvant in 2010, and data from an experimental trial on extended lactation conducted at the Danish Cattle Research Centre in Aarhus University (Denmark, 2012-2015). The data concerned insemination and parturition times, diet energy and dry matter content, body weight, body condition score, dry matter intake, and milk yield and composition. The model was fitted on individual cow data with a step-by-step fitting procedure. Each cow (n=40) was thus characterized by an adjusted version of the model with a specific set of 12 parameters. The variability of these parameters is then used to design and simulate a herd of individual virtual cows managed with different strategies for extended lactation. The present communication is intended to describe the model fitting procedure, to present the results of the fitting at animal level and the results of the in silico experiment at herd level, and to put into perspective the use of this method and more generally of model-based approaches as management tools in the context of precision farming.

A tool to assess nitrogen efficiency, autonomy and excretion at dairy herd level: CowNex

P. Faverdin[1], C. Baratte[1], R. Perbost[2], S. Thomas[2], E. Ramat[2] and J.-L. Peyraud[1]
[1]INRA, Agrocampus Ouest, UMR1348 PEGASE, Domaine de la Prise, 35590 Saint-Gilles, France, [2]Univ Lille Nord de France, ULCO, LISIC, BP 719, 62228 CALAIS, France; philippe.faverdin@rennes.inra.fr

The efficiency of nitrogen (N) use in dairy herds is an important challenge due to the environmental impacts of the N cascade and to the cost of protein resources. The assessment of global N efficiency and N excretion of dairy herds is complex due to the large diversity of feeding management during the year (several diets and several feeding groups). CowNex is a new web application (http://www.cownex-record.inra.fr/) (Project RedNex UE FP7 KBB-2007-1) that facilitates the calculation of dry matter and nitrogen use in dairy herd according to the management, with a special attention to feeding management. It is a subpart of the whole farm MELODIE model using the Record modelling platform. It simulates the daily intake, production and excretion of the different categories of animal in a dairy herd according to the feeding management described by the users. CowNex enables the assessment of existing farms and the simulations of changes in feeding management both on production and excretion. Applied to a large diversity of dairy systems encountered in France, CowNex was used to estimate the actual situation and to test some mitigation options to reduce NH_3 emissions and associated N_2O emissions from manure. The estimation of these impact was calculated using the EMEP-EEA equations with the CowNex results. The reduction of protein supplementation to reach a crude protein content of 14% with maize silage based diets is a simple and efficient solution to highly reduce ammonia emissions and decrease GHG emissions between 10 and 20%, with a good choice of protein sources, but without changing any manure management options. In many systems, there is both economic and environmental benefit to pay attention to the protein supplementation.

Influence of several factors in the parameters of quality of Manchega sheep milk

L. Jiménez Sobrino
Cersyra Valdepeñas, Dairy Research Laboratory, Avenida del Vino, 10, 13300 Valdepeñas (Ciudad Real),
Spain; lorenaj@jccm.es

The dairy sheep sector has traditionally an important implantation in the European Mediterranean area. Specifically, in La Mancha region it is supported one of the most important indigenous breeds of dairy aptitude in Spain: Manchego sheep breed. The milk of this breed is used to produce a high quality of differential product like Manchego Cheese with Origin Designation (PDO). Therefore, the control of overall quality of milk from primary production is an issue of great importance to ensure the quality of final product. In this context, the present study aims to determinate the influence of several factors, like the year station, size farm and being member of National Breed Association of Manchega Sheep (AGRAMA) over certain physic-chemical and technological characteristics of Manchego sheep milk. The study was conducted in 80 Manchego sheep herds, where it have been taken four tank milk samples, from each farm during the years 2012 to 2013, one in every season. At the same time, there has been a survey to find productive, social, economic and environmental structure of farms. The results indicate the influence of the season over most of physic-chemical characteristics (fat, protein, dry matter, total casein, urea content, pH, colorimetric: red light and index) and technology (clotting time, hardening rate of curd, average curd hardness of thirty (A30) and sixty minutes (A60), and the curd yield). On the other hand, farm size factor only appears to affect significantly on a colorimetric component, the rate of red. Finally, being a member of National Breed Association of Manchega Sheep (AGRAMA) has significant influence over most of the physicochemical characteristics of milk except urea content, although paradoxically, only a technological component A60. Knowledge of the characteristics of Manchega sheep milk and its factors of variation could help to sector operators to make a decision in order to optimize the production of PDO Manchego Sheep.

Using technology to study relationships between behaviour, feed efficiency and growth of beef steers

B.N. Marsiglio-Sarout[1], M. Haskell[2], A. Waterhouse[2], C. Umstatter[3] and C.A. Duthie[2]
[1]Federal University of Rio Grande do Sul, Porto Alegre, 91540-000, Brazil, [2]SRUC, Edinburgh, EH9 3JG,
United Kingdom, [3]Agroscope, Tänikon 191, 1580 Avenches, Switzerland; bruna.sarout@sruc.ac.uk

The study was carried out to evaluate the use of technologies to monitor livestock. Automated feed intake equipment and accelerometer-based activity sensors were used to collect detailed information on feed intake behaviour, feed intake, and cattle activity, alongside animal performance characteristics. Analysis included circadian rhythms of activity and feeding behaviour. Two contrasting diets and two breeds were used to increase the variation between animals and gain information of potential breed by diet interactions. A 2×2 factorial design was used, conducted with 40 crossbred Charolais (CHx) and 40 purebred Luing (LU) assigned to *ad libitum* diets: (1) Forage based (FO) with (DM basis) 28.6% whole crop barley, 20.5% grass silage, 39.7% barley, 10.3% maize dark grains, 0.9 minerals; (2) Concentrate based (CO) with 72% barley, 17.3% maize dark grains, 7.6% barley straw, 2.0% molasses, 1.1% minerals. Feding behaviour was evaluated over a 8 week period: daily Bunker Visits (BV), events/d; daily Feeding Duration (FD), minutes/d; daily Feeding Rate (FR), DM grams/d. Statistical analyses were conducted using Spearman correlation and mixed procedure of SAS with day within week as the repeated measurement. All feeding behaviour traits and Dry Matter Intake (DMI) were affected by breed and diet ($P<0.01$) and the interaction between breed and diet was significant for BV, DMI and FR ($P<0.01$). LW gain (LWG) was not affected by diet, however, was greater for CHx compared to LU. Residual Feed Intake (RFI) was not affected by diet and CHx were more efficient (lower RFI) than LU ($P<0.01$). There were correlations between some feeding behaviour parameters and DMI, LWG and RFI ($P<0.01$). Circadian rhythms will be investigated further. These results suggest that behaviour data collectable by technology can be related to feed efficiency and growth.

Assessing beef production chain data quality

F. Maroto Molina[1], R. Santos Alcudia[2], J. Gómez Rodríguez[2] and A. Gómez Cabrera[1]
[1]*University of Cordoba, Department of Animal Production, Ctra. Madrid-Cádiz km 396, 14013 Cordoba, Spain,* [2]*Cooperativa Andaluza Ganadera del Valle de los Pedroches, Ctra. Industrial Dehesa Boyal, 14400 Pozoblanco, Spain; g02mamof@uco.es*

Every day, a number of data are recorded throughout the beef production chain: farmers, feedlots, slaughterhouses, distribution and sales. These data are recorded mainly for administrative and traceability purposes, but they can be used to answer scientific and technical questions if they are pre-processed correctly. The objective of this paper is to evaluate the quality of beef production chain real-world data. A database containing 22 variables and more than 600,000 data (identifications, dates, breeds, weights, scores, etc.) on roughly 30,000 yearling bulls and heifers slaughtered between October 2010 and October 2014 at COVAP, a big agri-cooperative in the south of Spain, was used. The three main components of data quality were assessed: outliers, usefulness and completeness. Both univariate and multivariate outliers were detected. The last are the most important outliers regarding scientific and technical purposes, as they affect relationships between data and the values of some technical indexes (e.g. dressing percentage as the rate between hot carcass weight and live weight). On the other hand, no usable data due to their granularity (e.g. weight on leaving the feedlot, recorded as the mean weight of all animals transported in the same lorry) or their consistency (e.g. breed, recorded under different descriptors by farmers and feedlots) were located. Finally, some data needed to support important technical decisions (e.g. feed consumption for determining optimal slaughter endpoint) or commercial decisions (e.g. meat color) were not recorded. Based on these results, we highlight the need to improve the quality of some data, e.g. avoiding manual handling and standardizing descriptors, and to use Precision Livestock Farming concepts and tools to obtain more frequently recorded and new data.

Temperature as a predictor of fouling and diarrhea in slaughter pigs

D.B. Jensen[1], N. Toft[2] and A.R. Kristensen[1]
[1]*Copenhagen University, Department of Large Animal Sciences, Grønnegårdsvej 2, 1870 Frederiksberg C, Denmark,* [2]*Technical University of Denmark, National Veterinary Institute, Bülowsvej 27, 1870 Frederiksberg C, Denmark; daj@sund.ku.dk*

The PigIT Project aims at improving welfare and production of slaughter pigs by integration of various sensor systems for alarm purposes. Here we present an exploratory analysis to assess the predictive value of temperature sensor data with respect to pen fouling and diarrhea. We recorded the temperature at four locations in two double-pens (by the drinking nipples and by the corridor) between November 2013 and December 2014. Logistic regression models were made to express the probability of fouling and diarrhea per day, and were reduced via backwards elimination. Furthermore, fitting the models was attempted with the raw temperature data as well as data averaged over 10, 15, 30 and 60 minutes. The predictive performances were evaluated with Matthews Correlation Coefficient (MCC). For diarrhea, the minimal and maximal temperatures at the water nipple and the corridor, as well as the maximal rate of temperature decrease, were found to be either significant or borderline significant. The same factors, with the addition of maximum rate in temperature increase, were found to be significant or borderline significant predictors for pen fouling. Both conditions were consistently detected at better than randomly (MCC between 0.422 and 0.557 for diarrhea, and between 0.386 and 0.560 for fouling). Thus, temperature information seems to contain predictive value in relation to fouling and diarrhea, but not enough to stand alone. It would thus be meaningful to combine this information with other available data to achieve an optimal predictive power.

Use of specter data for breeding and feeding

T. Ådnøy
Department of Animal and Aquacultural Sciences (IHA), Norwegian University of Life Sciences (NMBU),
IHA, NMBU, N1432, Norway; tormod.adnoy@nmbu.no

Specter data are used to estimate milk composition (fat%). Similar multivariate info appears in many fields. We have been trying to advocate the use of mixed models to have better use of multivariate data. We know (CR Henderson 1975) that when $y = Xb + Zu + e$ with $var(u) = G$ and $var(e) = R$ the Best Linear Unbiased Predictor for random explanatory variables (u) is given by the Mixed Model Equations. The G may include covariance structures for additive breeding values, permanent environment of animal effects, and herd-testday. This theory also holds for the multivariate situation where y is expanded to a long vector, and the rest of the model is accommodated correspondingly. A challenge is to find the G and R in multivariate cases. There are limitations to how many random variables may be included when estimating covariance components with available computer programs. For specter data we therefore propose to reduce their dimension by Principle Components or Factor Analysis and then analyze the factors and thus find covariance structures for the population in question based on them. Knowing covariance estimates for the random components of the mixed model we may then use MME for BLUP prediction of random variables for new data coming from such a population. We showed that this may give 2-4%? better prediction of breeding values for milk content in dairy goats compared to established methods where content phenotype (fat%) is first found and then using univariate analysis in the MME. We are also trying to verify whether this multivariate specter method gives a better prediction of ketosis diagnosis in a Polish dairy cow data set. The residual deviation of a cow's milk specter values from her predicted milk specter values based on BLUP is postulated to give a better diagnosis of her ketosis.

Lameness detection in gestation kept sows from ear tag- sampled acceleration data

C. Scheel[1], I. Traulsen[1], W. Auer[2], E. Stamer[3] and J. Krieter[1]
[1]Kiel University, Institute of Animal Breeding and Husbandry, Hermann-Rodewald-Straße 6, 24118 Kiel, Germany, [2]MKW electronics GmbH, Jutogasse 3, 4675 Weibern, Austria, [3]TiDa Tier und Daten GmbH, Bosseer Str. 4c, 24259 Westensee/Brux, Germany; cscheel@tierzucht.uni-kiel.de

To implement an automated system for lameness detection, the use of acceleration data is suggestive. Acceleration data is a reasonable proxy for the behavioural patterns of movement an animal exhibits, and modern acceleration sensors can be integrated in an ear tag. Our research seeks to develop a method to interpret time series of acceleration data obtained from such ear tags with the aim of detecting lameness early and reliably. The method developed splits each sows' individual time series of acceleration s into a past record s_p and current part s_c. On the assumption that s_p represents the sow in a healthy state, it then compares s_p to s_c, and seeks for deviations of s_c from s_p larger than those observed within s_p. To implement the comparison numerically, a collection of localized features is extracted and a feature representation is assigned to each day in the record. It consists of pairwise similarity measures between days, based on to scale autocorrelation measures from the time series representation as well as coherence measures from the wavelet representation of the signal. The method was tested on data from an experimental run of a monitoring system supplied by MKW Electronics, installed at the Futterkamp agriculture research farm in Schleswig-Holstein, Germany. A total of 196 Sows were fitted with ear tags, sampling at 1 Hz resolution, over a time span of 18 months. In a validation sample of 7 lame and 10 healthy sows set aside for development, with a record of 14 days each, the method is able to reveal a higher degree of irregularity in the lame samples for the last days of their records, the very last day being the day of a diagnosed lameness. Currently a measurement is being developed with the aim to condense this irregularity into a reliable single number which can be used as a warning for lameness.

Genetics of male reproductive performance in White Leghorns

A. Wolc[1,2], J. Arango[1], P. Settar[1], J.E. Fulton[1], N.P. O'sullivan[1], R.L. Fernando[2], D.J. Garrick[2] and J.C.M. Dekkers[2]
[1]Hy-Line International, Dallas Center, IA 50063, USA, [2]Iowa State University, Department of Animal Science, Ames, IA 50011, USA; awolc@up.poznan.pl

Ability to produce viable progeny is a complex trait involving male and female components. In poultry, birds are maintained in groups with a sex ratio of approximately 1 male to 8-12 females, and selection in males is much more stringent then in females. Thus, the impact of male reproductive failure is much larger than that of a female. In this study, genetic determination of reproductive performance, by natural mating and artificial insemination, was evaluated. Semen quality was studied in 1575 preselected White Leghorn males from nine generations. Genetic parameters for fertility, hatch of fertile, hatch of set, sperm motility, sperm count and fertility using artificial insemination were estimated using single and multi-trait animal models, with test (generation) as fixed effect. Selected birds were genotyped using the 600k Affymetrix SNP chip. Genomic data were analyzed with the BayesB method. Fertility and hatchability were highly genetically (r_G=0.82-0.99) and phenotypically (r_P=0.28-0.99) correlated, but the correlations with semen quality traits were not strong and varied between tests (-0.13-0.14). Birds used for fertility and hatchability test were preselected based on sperm motility and sperm count, which could contribute to the lack of strong correlations between these traits. Low to moderate heritability was estimated for reproductive traits using pedigree (h^2=0.08-0.21). Markers explained a low proportion of phenotypic variance, probably due to stringent selection of genotyped individuals (h^2_m=0.04-0.15). No genes with large effects (QTL) were identified. Genomic breeding values were more accurate than pedigree-based EBV only for hatch of fertile and fertility using AI. Despite low estimates of accuracy in validation, genetic trends were positive for all analyzed traits. Continued selection can result in genetic improvement of reproductive performance of White Leghorns.

Genetic parameters of body surface temperature in laying hens exposed to chronic heat

T. Loyau[1], T.B. Rodenburg[2], J. Fablet[3], M. Tixier-Boichard[4], J. David[1], M.-H. Pinard-Van Der Laan[4], T. Zerjal[4] and S. Mignon-Grasteau[1]
[1]INRA, UR83 Recherches Avicoles, 37380 Nouzilly, France, [2]Behavioural Ecology Group, Wageningen University P.O. Box 338, 6700 AH Wageningen, the Netherlands, [3]Institut de Sélection Animale, S.A.S, Hendrix Genetics, 1 rue Jean Rostand, 22440 Ploufragan, France, [4]INRA, UMR1313 GABI, Domaine de Vilvert, 78352 Jouy en Josas, France; thomas.loyau@tours.inra.fr

The exposure of laying hens to chronic heat stress induces egg production losses. Therefore selecting more robust hens would be interesting, but measuring birds' sensitivity through internal temperature is laborious and has a poor precision. In this study we used infrared thermography to measure the hens' capacity to dissipate heat in two commercial lines of laying hens cyclically subjected to either neutral (N, 19.6 °C) or high (H, 28.4 °C) ambient temperatures. Mean body temperatures (BT) were estimated from wing, comb and shank (WT, CT, ST) infrared pictures (n=5,588 in line A, 9,355 in line B). Genetic parameters were estimated combining the data from N and H temperatures in line A (due to lower number of birds) and for each temperature separately in line B. WT had a low heritability in both lines (0.00 to 0.09), consistent with the fact it reflects mainly the environmental temperature. CT heritability was higher (0.08 in line A, 0.15 and 0.19 in line B in N and H conditions, resp.). Finally, ST presented the highest heritability with values ranging from 0.12 in line A to 0.22 and 0.20 in line B in N and H conditions, which demonstrates that heat dissipation is partly under genetic control. Plumage condition showed negative genetic correlations with ST and CT, indicating that birds with a poor plumage condition have the possibility to dissipate heat also through bare skin zones. CT and ST were correlated with egg size and weight, yolk brightness and yellowness and Haugh units only under H conditions. On the contrary, shell color was correlated with leg BT only at thermo-neutrality.

Bivariate analysis of individual and pooled data on social interaction traits

K. Peeters[1,2], P. Bijma[2] and J. Visscher[3]
[1]Hendrix Genetics, Research and Technology Centre, Spoorstraat 69, 5831 CK Boxmeer, the Netherlands,
[2]Wageningen University, Animal Breeding and Genomics Centre, Droevndaalsesteeg 1, 6708 PB Wageningen,
the Netherlands, [3]Hendrix Genetics, Institut de Sélection Animale B.V., Spoorstraat 69, 5831 CK Boxmeer,
the Netherlands; katrijn.peeters@wur.nl

Collecting individual data on group-housed animals can be difficult and expensive. Alternatively, data collection can be done at group level, e.g. egg production traits or feed intake. Thus, phenotypic records on an animal can come from different levels. Bivariate analysis of individual and pooled data is complex, in particular for social interaction traits. In the direct-indirect genetic model for individual data, direct and indirect environmental effects are captured by two model terms: the random group effect and the residual. However, for pooled data, direct and indirect environmental effects are captured by the residual only. Most statistical software programs cannot model the correlation between the random group effect of the individual trait and the residual of the pooled trait. Disregarding this correlation can result in biased genetic parameter estimates. We have found that this problem can be solved by adding a random group effect for the pooled trait and correlating it with the random group effect of the individual trait. The variance of the random group effect of the pooled trait needs to be set to a fixed value to avoid over-parameterization. A simulation study was conducted and showed that the adjusted model, unlike the unadjusted model, resulted in unbiased genetic parameter estimates. The adjusted model was used to estimate the genetic correlation between survival time (individual records) and early egg production (pooled records) in crossbred laying hens. A slightly negative (-0.09), but non-significant, genetic correlation was found between both traits. This differs from the correlation found with the unadjusted model (0.14), demonstrating the need for the adjusted model.

New genomic regions associated with litter size and its variation in pigs

E. Sell-Kubiak[1], M.S. Lopes[2], N. Duijvesteijn[2], E.F. Knol[2], P. Bijma[1] and H.A. Mulder[1]
[1]Animal Breeding and Genomics Centre, Wageningen University, P.O. Box 338, 6700 AH Wageningen, the Netherlands, [2]Topigs Norsvin Reaserch Centre B.V., P.O. Box 43, 6640 AA Beuningen, the Netherlands; ewa.sell-kubiak@wur.nl

One of the main breeding goals in pig breeding is increasing total number born in a litter (TNB). Variation in this trait is large and increasing TNB in a population could exceed the physiological capacity of sows to provide for a litter pre- and post-farrowing. Thus there is a desire to breed for increased mean TNB while at the same time reducing its variability. Here we study the variation of TNB in a Large White pig population by applying Double Hierarchical GLM (DHGLM) and a genome-wide association study (GWAS). The residual variance of TNB (varTNB) and its variance components were estimated with DHGLM in ASReml. For this step, 263,088 observations on TNB were available. Estimated breeding values (EBV) obtained with DHGLM were used to calculate deregressed EBV (dEBV) for 2,351 sows and boars genotyped with 64k chip. The GWAS was performed with a Bayesian Variable Selection method in Bayz. SNP's were considered significant if their Bayes Factor was above 30. Genetic coefficient of variation of the standard deviation for varTNB was estimated as 0.09, indicating good opportunities for improvement of uniformity by selection. Genetic correlation between additive genetic effects onTNB and on its variation was 0.5. This indicates that the selection for TNB increases the variation of TNB. Ten SNPs were detected for TNB and nine SNPs for varTNB. The most significant SNP explained 0.4% of genetic variance in TNB (SSC11) and 0.5% in varTNB (SSC7). Possible candidate genes for varTNB on SSC7 are: heat shock protein (HSPCB alpha), vascular endothelial growth factor (VEGFA), and CUL9 protein regulating p53 function. This is the first study reporting SNPs and candidate genes associated with varTNB in pigs.

Effects of selection on the efficiency and variability of sow reproduction and maternal ability

P. Silalahi[1,2], T. Tribout[2], J. Gogué[3], Y. Billon[4] and J.P. Bidanel[2]
[1]Bogor Agricultural University, Jl. Lingkar Kampus, Dramaga, 16680 Bogor, Indonesia, [2]INRA, AgroParisTech, UMR1313 GABI, 78350 Jouy-en-Josas, France, [3]INRA, UE332 Domaine expérimental de Bourges, 18520 Avord, France, [4]INRA, UE1372 GENESI, 17700 Surgères, France; parsaoran.silalahi@jouy.inra.fr

The objective of this study was to estimate the effects of selection on the efficiency and the variability reproductive and nursing performance of French Large White (LW) sows along their career (6 parities). LW sows were inseminated with stored frozen semen of LW boars born in 1977 and 1998, leading to 2 experimental groups of pigs (D7 and D8). A second generation was produced by inter se mating within D7 and D8 groups. Crossfostering was practised shortly after birth to obtain mixed D7-D8 litters to investigate direct and maternal effects on piglet growth. Traits investigated included the number of piglets born in total (TNB) and born alive (NBA) per litter, litter weight at birth (LWB), sow weight loss from birth to weaning (SWL), weaning to first oestrus interval (WEI), individual and average piglet weight at birth and at 21 days of age. The data were analysed using mixed linear models. The variance of performance across parities was computed from mixed models residual values. All traits were significantly affected by selection, with an increase in litter size and weight (+0.12 and +0.09 piglet/year and +0.22 kg/year for TNB, NBA and LWB, respectively), a higher SWL (+0.52 kg/year) and a shorter WEI (-0.09 d/year). The results also showed that the variability of sow performance increased over time for TNB, NBA and SWL and decreased for WEI. Similarly, the variability of birth weight and of average daily gain from birth to 21 d of age was higher in D8 than in D7 piglets. This higher variability might be an indicator of a lower sow and piglet robustness.

Genetic background of longitudinal feed efficiency and feeding behaviour traits in Maxgro pigs

M. Shirali[1], P.F. Varley[2] and J. Jensen[1]
[1]Center for Quantitative Genetics and Genomics, Aarhus University, 8830, Tjele, Denmark, [2]Hermitage Genetics, Sion Road, Kilkenny, Ireland; mahmoud.shirali@mbg.au.dk

The aim of this study was to determine the genetic characteristics of feed efficiency and feeding behaviour traits at different stages of growth in pigs. Data were available on 3850 purebred Maxgro pigs from 52 kg body weight (BW) (11 kg, standard deviation (SD)) to 110 kg BW (11 kg, SD) in 7 weeks. Feed efficiency was estimated by using residual feed intake (RFI) as daily feed intake per each week of growth adjusted phenotypically for average daily gain (ADG) in the test period and lean meat percentage (LMP) at the end of test. Furthermore, average daily occupation time in automatic feeder (OT), number of visit (NV) and feeding rate (FR) obtained as feeding behaviour traits. All traits were expressed as weekly averages. Feed efficiency and feeding behaviour traits were analysed using random regression model containing animal genetic effect, permanent environmental effect of each pig and contemporary group using Legendre polynomials of days on test along with heterogeneous residual variances. Heritability of weekly RFI increases by age ranging from 0.07 to 0.17. The heritability estimates of weekly OT (0.26 to 0.34), NV (0.16 to 0.21) and FR (0.26 to 0.31) were consistent at different stages of growth. Feed efficiency at early stages of growth (week 1 and 2) had substantially low genetic correlation with late stages of growth (week 7) ranging from 0.03 to 0.35. The genetic relationship of RFI at each week of growth with feeding behaviour traits of the same week varied depending on the trait OT (-0.02 to 0.50), NV (-0.23 to 0.33) and FR (0.25 to 0.01). Feeding behaviour traits can help to understand the feeding characteristic of efficient pigs and possibly can be used for selection to further improve feed efficiency. Genetic selection for feed efficiency should consider different stages of growth. Selection on curve of feed efficiency can be carried out to improve efficiency of lean meat production.

Estimating the genetic correlation between resistance and tolerance in sheep

H. Rashidi[1], H.A. Mulder[1], L. Matthews[2,3], J.A.M. Van Arendonk[1] and M.J. Stear[2,3]
[1]*Wageningen University and Research Centre, Animal Breeding and Genomics Centre, Droevendaalsesteeg 1, 6708 PB Wageningen, the Netherlands,* [2]*University of Glasgow, The Boyd Orr Centre for Population and Ecosystem Health, Jarrett Building, G61 1QH, Glasgow, United Kingdom,* [3]*University of Glasgow, Institute of BAHCM, Jarrett Building, G61 1QH, Glasgow, United Kingdom; hamed.rashidi@wur.nl*

Resistance and tolerance are the two main strategies of host's immune response against infections. Resistance is the ability to resist pathogen by, for example, preventing the pathogen from entering the body or controlling the replication of the pathogen within the host. Tolerance is the ability to minimize the damage caused by pathogen. Breeding for resistance is now widespread in the livestock industries. However, there is uncertainty about whether it is better to breed for resistance or tolerance. Though the genetics of resistance to infection has been widely investigated, the genetics of tolerance to infection and its relationship with resistance remains poorly understood. From a commercial flock of Scottish Blackface sheep 962 pedigreed lambs were studied. Lambs were born outside and were continuously exposed to natural mixed nematode infection by grazing on pasture. We applied a reaction norm model to quantify the change in lambs' bodyweight towards changes in their faecal nematode egg count. We observed significant genetic variation for tolerance as the slope of the reaction norms, indicating the possibility of improving tolerance by selective breeding. We also applied a bivariate model to study the genetic correlation between resistance (measured as increased IgA and decreased FEC) and tolerance. A negative genetic correlation was observed between tolerance and resistance, indicating that animals that are genetically more resistant are less tolerant. These findings indicate that unless both traits are carefully included in breeding programs, breeding for increased resistance may decrease tolerance.

Including caseine αs1 major gene effect on genetic and genomic evaluations of French dairy goats

C. Carillier-Jacquin, H. Larroque and C. Robert-Granié
INRA, INPT ENSAT, INPT ENVT, UMR1388 Génétique, Physiologie et Systèmes d'Elevage, INRA, 24 Chemin de Borde Rouge, Auzeville CS 52627, 31326 Castanet Tolosan cedex, France; celine.carillier@toulouse.inra.fr

Casein αs1 gene has a major effect on protein content of dairy goat milk. All French bucks (Alpine and Saanen breeds) used for artificial insemination, 2,145 testing bucks, and 2,983 dams of these bucks, have been genotyped for this gene since 1990. The casein αs1 gene is a multi-allelic gene with 6 different alleles. The aim of this study is to investigate how to include the casein αs1 gene effect in genetic or genomic evaluation currently based on a polygenic model. In French dairy goats, casein αs1 genotype has a significant effect on protein content, milk yield and fat content. Genetic evaluations based on daughter yield deviations of bucks were computed in Alpine and Saanen breed separately, including casein αs1 genotype as fixed or random effect. Validation correlations estimated on the 252 youngest bucks were slightly improved by considering the casein αs1 gene as fixed or random effect in Saanen breed (+18% for protein content). Adding the 50k bead chip genotypes of bucks (471 Alpine and 354 Saanen bucks) in the model already including casein αs1 genotype did not improve validation correlations. The most of females used for genetic evaluation were not genotyped for casein αs1 gene. Including casein αs1 gene effect in genetic evaluation, based on female performances, requires to predict casein αs1 genotypes for all these females. Probabilities of each 19 possible genotypes were estimated for each female, from pedigree information and true genotypes of the subset of the population. Several models were tested to include these probabilities in the model as random or fixed effect. The first results have shown no improvement of validation correlations compared to a model not taking into account the probabilities of genotype. This could be linked to a non-optimal genotype prediction. Other methods or models are investigated to avoid the problem of imputation.

Relationships between immune traits found in the blood and milk of Holstein-Friesian dairy cows

S.J. Denholm[1], T.N. Mcneilly[2], G. Banos[1], M.P. Coffey[1], G.C. Russell[2], A. Bagnall[1], M.C. Mitchell[2] and E. Wall[1]
[1]SRUC, Scotland's Rural College, Edinburgh, EH25 9RG, United Kingdom, [2]Moredun Research Institute, Edinburgh, EH25 0PZ, United Kingdom; scott.denholm@sruc.ac.uk

Immune traits in blood have been shown to be associated with health, productivity and reproduction in dairy cows. The aim of the present study was to determine if any associations exist between immune traits in blood and those in milk, the latter of which is routinely collected and in a less invasive manner. All animals were Holstein-Friesians from the Langhill research herd (n=288) housed at the SRUC Dairy Research centre (Scotland). This research herd has been subject to long-term selection for genotype (control and select) and diet (by-product and homegrown) in a 4 by 4 approach. Samples were collected on 11 separate occasions (bi-monthly April-October 2013; monthly January-July 2014) and serum and milk analysed by ELISA for natural antibodies (NAb) haptoglobin and Tumour Necrosis Factor Alpha (TNF-α). Data from blood-only samples collected from the same herd (n=251) between Jul 2010 and Mar 2011 were also available. Data were analysed using a bivariate mixed linear model in ASReml version 3. Fixed effects included diet group, genetic group, lactation week, assay, year by month of calving, and lactation number by age at calving. Random effects included additive genetic effect of cow, permanent environmental effect to account for repeated measures, and a random residual effect. Significant (P<0.01) positive genetic correlations were found to exist between blood and milk NAb (from 0.69-0.95) and blood and milk haptoglobin (0.12). Strong phenotypic correlations were found between blood and milk NAb (from 0.30-0.72). Analyses also yielded significant environmental and residual correlations. Results from the present study show that immune traits found in blood and milk are strongly correlated and highlights the potential of using routinely collected milk samples as a less invasive resource of data for predictive modelling of animal immune traits.

A new model for an imprinting analysis of Brown Swiss slaughterhouse data

I. Blunk[1], M. Mayer[1], H. Hamann[2] and N. Reinsch[1]
[1]Leibniz-Institute for Farm Animal Biology (FBN), Department of Genetics and Biometry, Wilhelm-Stahl-Allee 2, 18196 Dummerstorf, Germany, [2]Landesamt für Geoinformation und Landentwicklung Baden-Württemberg (LGL), Stuttgarter Straße 161, 70806 Kornwestheim, Germany; blunk@fbn-dummerstorf.de

Genomic imprinting is a phenomenon that arises when the expression of genes depends on the parental origin of their alleles. It is caused by epigenetic mechanisms and is newly established during each gametogenesis depending on the respective sex of animal. Several studies demonstrated the importance of genomic imprinting for the expression of agricultural important traits in farm animals and lacking attention in animal breeding programs leads to biased estimates of genetic parameters. The aim of this study was to investigate the influence of imprinting on value-determining slaughter traits in Brown Swiss cattle. Data sets of net body weight gain, killing out percentage, EUROP class and fat score were available for up to 247,883 fattening bulls. Pedigrees contained up to 490,464 animals representing the entire Brown Swiss population of Austria and Germany. To test for significant imprinting variances, we applied a linear BLUP model with two additive genetic effects per animal that accounts for all variants of imprinting. Moreover, we developed an equivalent model allowing direct estimation of imprinting effects and standard errors. Results of analyses proofed both traits to be significantly imprinted with respective imprinting variances accounting for 9% of the additive genetic variance. Moreover, the maternal gamete showed major influence contributing up to 93% to the imprinting variance. This study presents new evidence demonstrating the significant influence of imprinting on value-determining beef traits in Brown Swiss and proposes a new model to directly estimate imprinting effects and standard errors.

Genomic prediction for feed efficiency in pigs based on crossbred performance

C.A. Sevillano[1,2], J.W.M. Bastiaansen[2], J. Vandenplas[3], R. Bergsma[1] and M.P.L. Calus[3]
[1]Topigs Norsvin Research Center B.V, P.O. Box 43, 6640 AA Beuningen, the Netherlands, [2]Wageningen University, Animal Breeding and Genomics Centre, P.O. Box 338, 6700 AH Wageningen, the Netherlands, [3]Wageningen UR Livestock Research, Animal Breeding and Genomics Centre, P.O. Box 338, 6700 AH Wageningen, the Netherlands; claudia.sevillanodelaguila@topigsnorsvin.com

Traits related to body maintenance processes, such as feed intake and residual feed intake, show low influence of selection at the nucleus level on the rate of genetic change in the production level. Therefore, optimal selection strategies for feed efficiency should be based on feed intake data recorded in commercial conditions measured on crossbred pig. When using crossbred animal as reference population we have to take into account that QTL allele in breed A can behave different that in breed B. To find these differences, and employ them in genomic prediction, a model with line origin of allele will be implemented. First SNP genotypes for crossbred pigs will be phased and compared to haplotypes observed in the parental purebred lines to derive which allele originates from each parental line. Second, this information will be used to construct breed-specific partial relationship matrices. With this approach the effect of a SNP allele from breed A when it is present in a crossbred AB animal can be differentiated from the effect of a SNP allele from breed B. The data used to test this methodology included 1.3k three-way-crossbred pigs (A(BC)), with individual daily feed intake observations from 25 kg until 120 kg and genotyped with a 10k SNP chip and imputed to 60k. In addition, we had 1.3k two-way-crossbred pigs (BC) and 9.7k purebred pigs (1.7k A line, 3.5k B line and 4.5k C line), all genotyped with a 60k SNP chip. We expect that this methodology will allow us to better estimate SNP effects and therefore, provide a tool to better select purebred animals based on crossbred performance for feed efficiency.

Regions with high persistency of linkage disequilibrium in seven Spanish beef cattle populations

E.F. Mouresan[1], A. González-Rodríguez[1], S. Munilla[1,2], C. Moreno[1], J. Altarriba[1], C. Díaz[3], J.A. Baro[4], J. Piedrafita[5], A. Molina[6], J.J. Cañas-Álvarez[5] and L. Varona[1]
[1]Facultad de Veterinaria, Universidad de Zaragoza, Unidad de Mejora Genetica, c/ Miguel Servet 177, 50013 Zaragoza, Spain, [2]Facultad de Agronomia, Universidad de Buenos Aires, Producción Animal, Av. San Martín 4453, 1417, CAB, Argentina, [3]Instituto Nacional De Tecnologia Agraria Y Alimentaria, INIA, Mejora Genética Animal, Carretera La Coruña, km 7,5, 28040 Madrid, Spain, [4]Universidad de Valladolid, Ciencias Agroforestales Producción Animal, Pza. del Campus Universitario, s/n, 47011 Valladolid, Spain, [5]Universitat Autónoma de Barcelona, Ciencia Animal y de los Alimentos, Campus UAB, 08193 Cerdanyola del Vallès, Barcelona, Spain, [6]Universidad de Córdoba, Genética, Edificio Gregor Mendel (C-5). Campus de Rabanales, 14071 Córdoba, Spain; mouresan@unizar.es

This study analyzed the persistency of the linkage disequilibrium (LD) between 7 Spanish beef cattle populations. Using the populations in pairs, the mutually segregating markers were separated and regions of 1 Mb were defined along the genome. Then, the LD between all markers of each region was calculated and the persistency of LD between the populations was calculated as the correlation between the LD values of the markers of each region. The values of the persistency per region were all positive and ranged between 0.127 (Mo-Pi) and 0.999 (ANI-BP and BP-RG). The mean persistency along the genome ranged between 0.632 (AV – BP) and 0.480 (Pi – Re). 42 regions in total showed persistency values higher than the 99.9% of the empirical distribution and 10 of those regions were found to be consistent in more than 10 of the 21 comparison cases. Using the tools of www.biomart.org, these 42 regions were found to accommodate around 220 genes associated with biological processes (cell adhesion), molecular functions (translation repressor activity) and cell components (connexon complex).

Genetic structure of Slovak Pinzgau cattle and related breeds

V. Šidlová[1], I. Curik[2], M. Ferenčaković[2], N. Moravčíková[1], A. Trakovická[1] and R. Kasarda[1]
[1]Slovak University of Agriculture in Nitra, Department of Animal Genetics and Breeding Biology, Tr. A. Hlinku 2, 949 76 Nitra, Slovak Republic, [2]University of Zagreb, Department of Animal Science, Svetošimunska cesta 25, 10000 Zagreb, Croatia; veron.sidlova@gmail.com

The aim of this study was utilization of high-throughput genotypes for definition of the population structure of Slovak Pinzgau cattle. To provide preliminary investigation of 19 bulls the similarity among individuals belonging to the populations that historically contributed to the breed origin was evaluated. Cluster algorithms were used to identify groups of related individuals without reference to prior information of the genetic subdivision. Possible number of clusters increased from 1 to 3, after which it balanced. Based on logarithm of probability, two clusters have been chosen as a possible division of 6 observed populations. We have investigated the affiliation of individual breeds to cattle production types. Based on this, division to 2 main populations was observed, Holstein population as a representative of milk production type and Simmental population as a representative of dual-purpose type. Systematic selection due to increasing of milk production is evident based on genetic insight in the Slovak Pinzgau cattle population. Additional monitoring and preserving of this endangered breed in original phenotype and therefore dual-purpose production type is necessary.

Determination of the purebred origin of alleles in crossbred animals

J. Vandenplas[1], M.P.L. Calus[1], C.A. Sevillano[2,3], J.J. Windig[1] and J.W.M. Bastiaansen[2]
[1]Wageningen UR Livestock Research, Animal Breeding and Genomics Centre, 6700 AH, Wageningen, the Netherlands, [2]Wageningen University, Animal Breeding and Genomics Centre, 6700 AH Wageningen, the Netherlands, [3]Topigs Norsvin Research Center B.V., 6640 AA Beuningen, the Netherlands; jeremie.vandenplas@wur.nl

Several production systems, mainly for pig and chicken, are based on crossbreeding. However, one limitation of associated breeding programs is that selection is performed on purebreds while the aim is to improve crossbred performance. Also, in crossbred populations, single nucleotide polymorphism (SNP) effects may differ between purebred and crossbred populations due to different genetic backgrounds and environments, and, thus, may be line (breed) specific. Assuming that purebred origin of alleles in crossbreds is known, different implementations for genomic selection of purebreds for crossbred performance were recently proposed. Results from simulations showed that genomic models that consider line (breed) specific SNP effects could outperform current genomic models that assume the same SNP effect across breeds. Estimation of breed specific SNP effects in real datasets will require phasing of genotypes in crossbreds and assigning breed origin of their alleles. Therefore, an approach was developed that applies long range phasing to assign the purebred origin of alleles of crossbreds. Long range phasing methods can be applied in the absence of pedigree and with the presence of haplotypes within a Linkage Disequilibrium block common between breeds. The aim of this study was to determine the accuracy of allelic assignment of the proposed approach on datasets simulating crosses between closely related, distantly related and unrelated breeds. Effects of different factors and parameters, like the presence of a haplotype in another pure breed that would preclude the assignment to the first pure breed, were also tested. An application on real pig genotype data was performed. Determining the allelic origin of crossbreds was feasible without pedigree information.

Temperate and tropical conditions impacts on production and thermoregulatory traits in growing pigs

R. Rosé[1], H. Gilbert[2], D. Renaudeau[3], J. Riquet[2], M. Giorgi[4], Y. Billon[5], N. Mandonnet[1] and J.-L. Gourdine[1]
[1]INRA, UR143 URZ, Petit-Bourg, France, 97170, Guadeloupe, [2]INRA, UMR1388, GenPhySE, Castanet Tolosan, 31326, France, [3]INRA, UMR1348, PEGASE, St Gilles, 35590, France, [4]INRA, UE 503 PTEA, Petit-Bourg, 97170, Guadeloupe, [5]INRA, UE1372 GenESI, Surgères, 17700, France; roseline.rose@antilles.inra.fr

The aim of the study was to evaluate the effects of two environments (temperate TEMP vs tropical humid TROP) on production traits and thermoregulatory response of genetically linked growing pigs. Large white (LW) pigs were crossed with tropical pigs from the Creole breed (CR), which is less productive but more thermotolerant than LW. A total of 1,296 half-sib backcross, from 10 (LW × CR) boars and about 65 LW sows in each environment, were reared in TROP (n=667) and in TEMP (n=629). TROP was characterized by an average daily ambient temperature (T) of 26.0±0.3 °C and an average daily relative humidity (RH) of 84.5±0.5%. The corresponding values for TEMP were 25.1±0.6 °C and 61.2±3.8%, respectively. Live body weight (BW) was measured at the beginning of test (week 11), at weeks 19 and 21, and at the end of test (week 23). Backfat thickness (BT) and cutaneous temperature (CT) were measured at weeks 19 and 23. Rectal temperature (RT) was recorded at weeks 19, 21 and 23. Under TROP, the average daily gain (ADG) during the test and BT were lower than under TEMP (718 vs 813 g/d and 14 vs 18 mm, P<0.001). CT and RT declined from week 19 to week 23 (-0.5 and -0.2 °C, respectively, P<0.001), and they were higher at all stages in TROP than in TEMP (35.9 vs 34.8 °C and 39.5 vs 39.4 °C, respectively, P<0.001). A moderate phenotypic correlation was obtained between CT and RT (rp=0.3). Correlations were low between ADG from week 19 to week 23 and thermoregulatory traits (rp=0.1 with CT and rp=-0.03 with RT). Furthermore, BW, ADG, RT and BT showed significant sire × environment interactions. Further research will focus on genetic approach to quantify the level of genetic by environment interactions.

Haplotype analysis of mtDNA in Iranian Sheep breeds: new insights on the history of sheep evolution

M.H. Moradi[1], S.H. Phua[2], N. Hedayat[3] and M. Khodayi-Motlagh[1]
[1]Arak University, Arak, Iran, [2]AgResearch Invermay, Mosgiel, New Zealand, [3]University of Mohaghegh Ardabili, Ardabil, Iran; hoseinmoradi@ut.ac.ir

Archaeological evidence suggests that sheep (*Ovis aries*) is the first grazing animal known to have been domesticated, approximately 9000 years ago, in northern Iraq and nearby regions in Iran. Therefore, Iran is one of the first origins of sheep domestication in the world. The aim of the present study was to investigate the diversity of mitochondrial haplotypes in indigenous Iranian sheep breeds. Using the Sequenom MassARRAY platform, a multiplex of ten mitochondrial SNP markers were used to genotype 257 animals from two sheep breeds; 107 of Lori-Bakhtiari breed and 150 of Zel breed. Phylogenetic analysis classified all animals into two mitochondrial haplogroups, A and B. The two haplogroups were found in both breeds. The presence of two haplogroups suggests that Iranian sheep came from two sources. The results of haplotype analysis led to the identification of nine different haplotypes in the animals. Two haplotypes, designated H1 and H3, had higher frequencies in both breeds. Haplotype H2 was identified as population-specific in Lori-Bakhtiari breed, and H5, H6, H7, H8 and H9 in Zel breed. The existence of two common haplotypes (H1 and H3) in the two Iranian breeds, which came from strikingly different geographical regions, suggest that these haplotypes could be the early haplotypes, with population specific haplotypes developing later. The Tajimma's test was positive for both Iranian sheep breeds, suggesting that the population size has declined in recent years for both breeds and that the balancing selection has been occurred. Analysis of variation (ANOVA) revealed that around 97% of the total genetic diversity distributed within breeds that in general, it appears due to the high level of gene flow between breeds in the past and the lack of systematic breeding programs in the recent years, the high genetic differentiation has not been formed yet between these breeds.

Evaluation of genetic diversity in Sri Lankan Indigenous chicken using microsatellite markers

P.B.A.I.K. Bulumulla[1], H.A.M. Wickramasinghe[2], P. Silva[2] and H. Jianlin[3]
[1]Uva Wellassa University of Sri Lanka, Badulla, 90000, Sri Lanka, [2]University of Peradeniya, Peradeniya, 20400, Sri Lanka, [3]International Livestock Research Institute, Nairobi, 0100, Kenya; pbaikbulumulla@gmail.com

Sri Lanka is rich in indigenous chicken (IC) genetic resources which are adapted to wide range of production systems with varying climatic and natural environments. Although they play a significant role in rural poultry production systems, and also they harboring a wealth of genetic diversity, an adequate attention is not been paid due the absence of significant production standards. Hence, systematic evaluation and characterization have so far been limited to few studies. Here, we investigate the genetic diversity of IC from 5 distinct regions of the country. A total of 150 birds were genotyped with 20 microsatellite markers on 11 chromosomes. There were 188 alleles observed and number of allele ranged from 5 to 18 with a mean of 9.4. Heterozygosity (H) values of 5 populations were more than 0.5. The average H value and polymorphism information content (PIC=0.663) suggested that Sri Lankan IC population possesses a high genetic diversity compared to many reported to-date. The high average gene diversity (0.676) and the presence of a high number of population specific alleles (privet alleles=36) also prove genetic distinctiveness of IC in Sri Lanka. The fixation coefficients of subpopulations within total population (FST) for 20 loci were 0.05 with moderate genetic differentiation. Estimates of Nei's genetic distance and Neighbor-joining method revealed that, the IC populations have been clustered according to their geographic distribution. High genetic diversity of Sri Lankan IC is in agreement with high phenotypic diversity exhibited by them, and also supported by the existing breeding system at village level. The genetic diversity and relationships among populations estimated may be useful as a guide for designing future investigations and conservation strategies for Sri Lankan Indigenous Chicken genetic resources.

SNPs in candidate genes for intramuscular fat in Parda de Montaña and Pirenaica beef cattle breeds

L.P. Iguácel[1], J.H. Calvo[1,2], P. Sarto[1], G. Ripoll[1], D. Villalba[3], I. Casasús[1], M. Serrano[4] and M. Blanco[1]
[1]CITA, A. montañana, 50059, Spain, [2]ARAID, C. María de Luna, 50018, Spain, [3]U. Lleida, A. Alcalde Rovira Roure, 25198, Spain, [4]INIA, C. la Coruña, 28040, Spain; lpiguacel@cita-aragon.es

Intramuscular fat (IMF) content influences sensory quality traits, such as tenderness, taste and flavor. Due to its interest, there is a search for genetic markers associated to IMF deposition. The aim of this study was to evaluate the association with IMF content of 9 SNPs located at TG (BTA14: g.9509309C>T), SCD1 (BTA26:g.21144730C>T), LEP (BTA4:g.93261931T>C), RORC (BTA3:g.19010079T>G), FASN(BTA19:g.5140203G>A), CAST(BTA7:g.98535683A>G) and CAPN1(BTA29:g.1827088G>C; g.1843665G>A; g.1845653T>C) genes in Parda de Montaña (n=225) and Pirenaica (n=68) beef cattle breeds. IMF was determined following the Ankom procedure in Longissimus thoracis muscle and all SNPs were genotyped by PCR-RFLPs. All SNPs were in Hardy-Weinberg equilibrium in both breeds. In Pirenaica breed, TG gene was the only gene that affected IMF content (P=0.03). The CT genotype had greater IMF content than the CC one (1.50 vs 0.70%, respectively; P=0.03), after the Bonferroni adjustment. These results are in agreement with other authors that found similar results in German Holstein and Charolais breeds. In Parda de Montaña breed, IMF content was affected by 2 SNPs in CAPN1gene: g.1845653 T>C (P=0.03) and g.1843665 G>A (P=0.05). IMF content of TT was greater than that of CT genotype (1.91 vs 1.49%, respectively; P=0.02) for g.1845653 T>C. In g.1843665 G>A, AA genotype tended to have greater IMF content than AG and GG genotypes (2.09, 1.65 and 1.65%, respectively; P>0.10). These results are consistent with those found by others authors suggesting an association between the CAPN1 gene and IMF and marbling in several beef breeds. The lack of consistent candidate gene effects between breeds could indicate that association between markers and IMF content could be influenced by different genomic backgrounds. Thus, different polymorphisms should be used to predict IMF content in each breed.

Estimated genetic variances due to QTL candidate regions for carcass traits in Japanese Black cattle

S. Ogawa[1], H. Matsuda[1], Y. Taniguchi[1], T. Watanabe[2], A. Takasuga[2], Y. Sugimoto[3] and H. Iwaisaki[1]
[1]Graduate School of Agriculture, Kyoto University, Kitashirakawa Oiwake-cho, Sakyo-ku, Kyoto 606-8502, Japan, [2]National Livestock Breeding Center, 1 Odakurahawa, Odakura, Nishigo, Fukushima 961-8511, Japan, [3]Shirakawa Institute of Animal Genetics, 1-83 Odakurahara, Odakura, Nishigo, Fukushima 961-8061, Japan; sogawa@kais.kyoto-u.ac.jp

Databases on bovine genome could provide information about the genetic architecture of traits of interest that may be useful for successful genomic prediction. In this study, the proportion of genetic variance explained by single nucleotide polymorphism (SNP) markers located on the pre-reported quantitative trait loci (QTL) candidate regions to the total genetic variance was estimated for marbling score (MS) and carcass weight (CW) in Japanese Black cattle. Phenotypic data about 872 fattened steers with genotype information on approximately 40,000 autosomal SNPs (Btau4.0) were used. Information about the physical location of 144 and 119 QTL candidate regions for MS and CW, respectively, was obtained from the Animal QTLdb database (http://www.animalgenome.org/cgi-bin/QTLdb/index). The SNPs were divided into two groups located inside and outside the given candidate region(s); variance components were then estimated using the statistical model containing different terms for the two SNP groups. The proportion of genetic variance explained by SNPs located inside all candidate regions, approximately 40% and 30% of all SNPs considered here, was estimated to be 50% and 40% for MS and CW, respectively. For CW, 20% of genetic variance was estimated to be explained by SNPs located inside the candidate regions reported in Japanese Black cattle (only 2.4% of all SNPs), many of which were on BTA6 and BTA14. However, these proportions are likely to be an overestimate and of limited value. Therefore, further careful investigation is necessary to select SNPs useful for the valid genomic prediction of carcass traits in Japanese Black cattle.

Genetic characteristics of Sokolski and Sztumski horses based on microsatellite polymorphism

A. Fornal[1] and G.M. Polak[2]
[1]National Research Institute of Animal Production, Department of Animal Cytogenetics and Molecular Genetics, Krakowska 1, 32-083 Balice, Poland, [2]National Research Institute of Animal Production, National Focal Point, Wspolna 30, 00-930 Warsaw, Poland; agnieszka.fornal@izoo.krakow.pl

Sztumski and Sokolski horses are local breeds of Poland. Sztumski horse was developed in the first half of the 19th century as a result of crossing among the local mares and Coldblood stallions (i.e. Ardennes and Belgian) in northern Poland. Sokolski horse was derived from local mares and imported Ardennes and Breton stallions in first half of the 20th century, climate and environmental conditions of Podlaski region were crucial for developing this breed. A total of 180 animals (105 Sokolski horse and 75 Sztumski horse) were generically characterized by using 17 STR markers recommended by ISAG for identification and pedigree analysis of horses. Individuals were genotyped and mean number of alleles per locus was estimated: 7.235 ± 0.489 and 6.765 ± 0.546 respectively. All the DNA microsatellite markers were highly polymorphic. The most polymorphic loci were ASB17 with 14 variants, the least polymorphic with 4 allelic variants were HTG6 (Sokolski horse) and HTG7 (Sztumski horse). The average observed heterozygosity H_o and expected H_e were estimated: $H_o=0.693$ and $H_e=0.705$ for Sokolski horse, $H_o=0.729$ and $H_e=0.705$ for Sztumski horse. The mean polymorphic information content was estimated (about 0.7) for seventeen microsatellite markers and indicates usefulness of this set of markers in parentage testing of Sokolski and Sztumski horses despite the relatively small sizes of both populations.

RyR1 single nucleotide polymorphisms in *Equus caballus*
A. Fornal, A. Piestrzyńska-Kajtoch and A. Radko
National Research Institute of Animal Production, Department of Animal Cytogenetics and Molecular Genetics, Krakowska 1, 32-083 Balice, Poland; agnieszka.fornal@izoo.krakow.pl

The ryanodine receptor type 1 (RyR1) gene, located on the equine chromosome 10, encodes one of three isoforms of ryanodine receptor. It is probably associated with equine malignant hyperthermia. Rapid release of calcium (as a result of calcium channel disorders) has great influence on intracellular metabolism in endoplasmic reticulum in skeletal muscle and leads to activating metabolic processes, then hypermetabolism and even death. The mechanism of this equine disorder has not been known enough. The cause of the dysfunction is probably single missense point mutation – substitution in exon 46 of RyR1 gene. The aim of the study was to analyse RyR1 gene sequences in Anglo-Arab, Thoroughbred and Małopolski horse breed (45, 46 and 7 individuals, respectively). The fragment, which could be related with malignant hyperthermia disease, containing part of exon 45, part of exon 46 and intron 45 (according to GeneBank AH015510.2) was amplified and sequenced. Three novel polymorphic sites (SNPs) were identified in the whole population studied: A9554637G (AA frequency – 92.86%, AG – 7,.%), C9554835T (CC – 75.51%, CT – 21.43%, TT – 3.06%), C9554701T (CC – 84.69%, CT – 14.29%, TT – 1.02%). Although RyR1 exons aren't still annotated to equine chromosome 10 (GeneBank NC_009153.2), according to interspecies alignment and GeneBank AH015510.2 sequence SNPs C9554835T and C9554701T are probably located in exon sequence and SNP A9554637G is probably located in intron. Analysis on the basis of GeneBank AH015510.2 sequence also revealed, that C9554835T SNP is silent mutation. A9554637G in Thoroughbred was not identified as well as C9554701T in Małopolski horse. TT variant of C9554701T was not observed in Anglo-Arab breed. All SNPs seemed not to be breed-specific.

Association and functional impact of SNPs in 3'UTR CAST gene with tenderness in cattle
L.P. Iguácel[1], A. Bolado-Carrancio[2], G. Ripoll[1], P. Sarto[1], D. Villalba[3], M. Serrano[4], J.C. Rodríguez-Rey[2], M. Blanco[1], I. Casasús[1] and J.H. Calvo[1]
[1]CITA, A. montañana, 50059, Spain, [2]U. Cantabria, A. Cardenal Herrera Oria, 39011, Spain, [3]U. Lleida, A. Alcalde Rovira Roure, 25198, Spain, [4]INIA, C. la Coruña, 28040, Spain; lpiguacel@cita-aragon.es

The system calpain-calpastatin (CAPN1-CAST) regulates post-mortem proteolysis and affects beef tenderness. Some SNPs in CAST gene have been associated with meat tenderness, including the SNP BTA7:g.98579663A>G (UMD 3.1) in 3'UTR. Association results of this SNP are variable across breeds. The aim of this study was to find out the SNPs in the 3'UTR region of the CAST gene and evaluate their effect on meat tenderness in Parda de Montaña breed (n=147), as well as the functional consequences using luciferase assays. In total, 8 polymorphisms were found in this region. The majority of polymorphisms occurred as multiSNP combinations for individual subjects. Only the g.98579663A>G SNP was associated with meat tenderness at 7 days post-mortem. The AA genotype was more tender than AG genotype (P<0.05). Haplotype analysis identified 4 main haplotypes, which were not associated with meat tenderness. In silico analysis using Microinspector software showed that 6 SNPs modify putative target sites of three bovine miRNA. The SNP g.98579663A>G modified a putative target site for bta-miR-542-5p. In order to assess the activity of the 3'UTR of CAST gene, luciferase assay within C2C12 cells was performed. A 749 bp fragment of the 3'UTR of CAST gene for each main haplotype was cloned. There were no differences between haplotypes in the activity of the luciferase, but their signal was approximately 30% lower than that of the cells transfected with empty pmirGLO vector. These findings suggest that the 3'UTR of CAST gene is an active zone. Perhaps, different miRNA binding activities among haplotypes could be found in other conditions such as other types of cell cultures, growth media or using some medium additives.

Association of polymorphism c.145 A>G in rabbit LEP gene with meat colour of crossbreed rabbits

Ł. Migdał[1], O. Derewicka[1], W. Migdał[1], D. Maj[1], T. Ząbek[2], S. Pałka[1], A. Otwinowska-Mindur[1], K. Kozioł[1], M. Kmiecik[1], A. Migdał[1] and J. Bieniek[1]
[1]University of Agriculture in Krakow, al. Mickiewicza 24/28, 30-059 Krakow, Poland, [2]National Research Institute of Animal Production, ul. Krakowska 1, 32-083, Poland; l.migdal@ur.krakow.pl

This study examined the associations of single nucleotide polymorphisms in the rabbit leptin (LEP) gene with colour parameters of meat of New Zealand White and Belgian Giant Grey crossbred rabbits. Meat colour of the m. longissimus lumborum (L*45'l, a*45'l, b*45'l) and hind leg (L*45'h, a*45'h, b*45'h) muscle were recorded 45 minutes post-mortem and after 24 h storage at 4 °C – m. longissimus lumborum (L*24l, a*24l, b*24l) and hind leg (L*24h, a*24h, b*24h). Meat colour was assessed by the L* (lightness), a* (redness) and b* (yellowness) system (CIELab) using a Konica Minolta CR-400. Polymorphism was analysed using MlyI restriction enzyme (A=202 bp; G=137+65 bp). The effects of sex, genotype, interaction between gender and genotype and linear covariate of litter size were investigated by analysis of variance. Polymorphisms c.145 A>G in rabbits leptin gene affects (P<0,05) a* value of hind leg after 24 h storage (12.12±1.69 for AA and 13.91±2.03 for GG genotype, respectively). Polymorphism in the rabbit leptin gene influences meat traits like colour parameters of NZW × BGG crossbreeds.

Genetic analysis of phosphorus utilization in Japanese quails using structural equation models

P. Beck[1,2], R. Wellmann[2], H.-P. Piepho[3], M. Rodehutscord[1] and J. Bennewitz[2]
[1]University of Hohenheim, Institute of Animal Science, Animal Nutrition, Emil-Wolff-Str. 10, 70599 Stuttgart, Germany, [2]University of Hohenheim, Institute of Animal Science, Farm Animal Genetics and Breeding, Garbenstraße 17, 70599 Stuttgart, Germany, [3]University of Hohenheim, Institute of Crop Science, Biostatistics, Fruwirthstraße 23, 70599 Stuttgart, Germany; r.wellmann@uni-hohenheim.de

The current study was undertaken to estimate the genetic parameters of phosphorus utilization (PU), feed efficiency (FE) and bodyweight gain (BWG) in an F_2-population of Japanese quails. These are complex traits putatively affecting each other. Structural equation models (SEMs) can quantify causal relationships among different traits, are able to distinguish between directly and indirectly acting genetic effects, and may help to understand complex biological and physiological relations between traits. Therefore the data were analysed with SEMs as well as with classical multitrait models (MTM). A number of 920 young Japanese quails of the F_2-generation was phenotyped in a standard balance trial. Hatchlings were wing-banded at the first day of life and raised in groups in floor pens. In the first 5 days of life they were fed with a commercial starter diet and from day 6 to day 15 they were fed with a P low experimental diet (4.0 g/kg dry matter) which was based on corn and soybean-meal without supplementing phytase or mineral P. Between day 10 and day 15 growth and feed intake was measured and total excreta was individual sampled in metabolic cages. PU was determined individually using the difference in P intake and total P excretion. Average PU was 71% ranging from 22 to 87%, FE was 1.78 g/g and BWG was 24.5 g, each with high variation. Heritabilities were estimated using a SEM and were 0.14 (SE=0.06) for PU, 0.12 (SE=0.06) for FE and 0.09 (SE=0.14) for BWG. Heritabilities were similar to heritabilities estimated with a MTM. Genetic correlations differed if estimated with a SEM or a MTM, which shows that indirect genetic effects are important for the genetic correlations.

Genetic evaluation models in Holstein cows: genetic parameters for test-day and 305-day milk yield

Y. Ressaissi and M. Ben Hamouda
National Institute of Agricultural Research of Tunisia, Animal and fodder production, Rue Hédi Karray-Tunis, 2049 Ariana, Tunisia; yos.re@hotmail.fr

The study has focused on comparing genetic evaluation models commonly used in dairy cattle in order to appreciate the model that permit a better genetic analysis of the raw performances recorded on the Holstein cows in Tunisia. Herds were divided into groups according to their sizes and for each group the genetic parameters were estimated and the breeding values were predicted for 305-days cumulative milk yield (L305) and test-day milk yield (TDM). Two animal models applied to the method of restricted maximum likelihood were used and the analyses counted the random additive genetic effect and the permanent environmental effect. The contemporary groups were defined as Herd × Calving year for L305 and Herd*Control year for TDM. Spearman rank coefficients (ρ) were established to compare the breeding values. Heritabilities ranged respectively from 0.002 and 0.07 and from 0.02 to 0.16 for L305 and TDM and repeatabilities have varied respectively from 0.20 to 0.38 and from 0.34 to 0.47. Breeding values have fluctuated between -608 kg and 519 kg for L305 and between -4.62 kg to 6 kg for TDM; however the contribution of the permanent environment components have respectively oscillated between -2525 to 2590 and between -15.27 to 13.76. Genetic parameters of milk yield were low and the variance components of the permanent environment were higher which suggests a limited genetic variability between herds. The environment effect seems to be fundamentally implicated in milk yield character; therefore the expression of the adequate genetic abilities in Tunisian Holstein herds remains restricted, especially in small herds. The ranking of cows through TDM and L305 were approximately similar; nevertheless the genetic contrast between herds and the estimation of genetic parameters were more explanatory under TDM.

Polymorphism and expression of genes connected with neurodegeneration in sheep

A. Piestrzynska-Kajtoch, G. Smołucha, M. Oczkowicz, A. Fornal and B. Rejduch[†]
National Research Institute of Animal Production, Sarego 2, 31-047 Kraków, Poland; agata.kajtoch@izoo.krakow.pl

Prion protein (PrP), putative infectious agent of scrapie (fatal prion disease), is encoded by the host PRNP gene. The disease is connected with the PrP expression. However, other genes involved in PrP metabolism may be linked to scrapie occurrence and neurodegeneration. The aim of the project was to analyze the polymorphism and the expression of several genes connected to PrP metabolism, proliferation, differentiation and apoptosis of neuronal cells. Expression profiles of PRNP, RPSA, MGRN1, PSEN1, PSEN2, BAX, CASP3 and TP53 were investigated in 13 different ovine tissues (5 sheep). The RNA was isolated, normalized, reverse transcribed, amplified with gene specific primers, electrophoresed and sequenced. Polymorphism of the PRNP (DNA) and MGRN1 (cDNA) genes was studied by sequencing in two sheep breeds (45 sheep; Romanov, Polish Merino). Sequences were analyzed in BioEdit Sequence Alignment Editor. PRNP, RPSA, MGRN1, PSEN1, BAX, CASP3 and TP53 transcripts were found in all tissue analyzed. PSEN2 transcripts were absent in pituitary gland, heart and skeletal muscle. As the method was half-quantitative, tissue specific differences in the expression levels were observed for all genes except for the RPSA. The highest expression for PRNP was in the brain stem. Some primer pairs gave more than one PCR product suggesting more transcript variants for BAX, TP53, PSEN1 and PSEN2. Next, the Real-time PCR experiment will be performed to confirm and precise these results. Seven PRNP genotypes were found in the studied group: ALRH/ALRQ (2,2%), ALRQ/AFRQ (2,2%), ALRQ/ALRQ (20%), ALRR/AFRQ (8,9%), ALRR/ALHQ (6,7%), ALRR/ALRQ (42,2%) and ALRR/ALRR (17,8%). Additionally, other SNPs were observed: H143R, E224K and 2 silent mutations: R231R, L237L. 66 bp indel was found in MGRN1 cDNA sequence, what suggest presence of two transcripts in cerebellum. Apart from that, 2 SNPs were noticed in MGRN1: T>C and G>A. The study, financed by grant project no. N N311 406339, has been continued.

Assessment of genetic diversity and population structure in six brown layer lines by STR markers

T. Karsli and M.S. Balcioglu
Akdeniz University, Animal Science, Akdeniz University Faculty of Agriculture Department of Animal Science, 07058 Antalya, Turkey; takikarsli@akdeniz.edu.tr

The aim of this study was to determine genetic diversity and population structure in six brown layer chicken lines [two Rhode Island Red (RIRI and RIRII) lines, two Barred Ploymouth Rock (BARI and BARII) lines, Colombian Rock (COL) and Line-54 (L-54)] by using twenty two microsatellite markers. A total of 180 chickens including 30 from each purebred line were used. A total of 233 alleles were observed and all microsatellite loci were polymorphic. The mean of number alleles per locus was 10.59 ± 4.89. The observed heterozygosities ranged from 0.309 (RIRII) to 0.496 (BARII), the expected heterozygosities ranged from 0.540 (RIRII) to 0.645 (RIRI), the inbreeding coefficients ranged from 0.167 (BARII) to 0.443 (RIRII) and PIC values ranged from 0.487 (RIRII) to 0.587 (RIRI) per line. These findings indicated homozygous excess and severe inbreeding in six lines. Pairwise Fst value between pairs of lines ranged from 0.115 to 0.352. The lowest pairwise Fst value (0.115) was between BARI and BARII lines. These values showed the genetic differentiation between studied populations. Phylogenetic tree (NJ) constructed using genetic distance (D), Bayesian model-based clustering on multilocus genotypes of individuals and Factorial correspondence analysis (FCA) indicated four genetic clusters consistent with genetic origin of lines (first group RIRI and RIRII lines; second group BARI and BARII lines; third group L-54 line and the last group COL line). Results from this study showed that microsatellite markers were useful for determination of genetic diversity and populations structure in purebred chicken lines.

Compost bedded pack barns as a lactating cow housing system

J.M. Bewley[1], R.A. Black[1], F.A. Damasceno[2], E.A. Eckelkamp[1], G.B. Day[2] and J.L. Taraba[2]
[1]University of Kentucky, Animal and Food Sciences, 407 WP Garrigus Building, 40546-0215, USA,
[2]University of Kentucky, Biosystems and Agricultural Engineering, 128 C.E. Barnhart Building, 40546-0276, USA; jbewley@uky.edu

A compost bedded pack barn is a lactating dairy cow housing system consisting of a large, open resting area, usually bedded with sawdust or dry, fine wood shavings. Bedding material is composted in place, along with manure, when mechanically stirred on a regular basis. Recently, the popularity of compost bedded pack barns has unquestionably increased in the Southeast (at least 80 compost bedded pack barns have been constructed in Kentucky. Producers report reduced incidence of lameness and improved hoof health resulting from greater lying times and a softer, drier surface for standing. Cows may be more likely to exhibit signs of estrus because of improved footing on a softer surface, leading to improved heat detection rates. Compost bedded pack barns reduce the need for liquid based manure storage systems and provide producers with the option to economically transport nutrients in a dry, concentrated form to areas where there is an off-farm demand for nutrients. The initial investment costs of a compost bedded pack barn are lower than for traditional freestall or tie-stall barns, because less concrete and fewer internal structures (stall loops, mattresses) are needed. Proper composting increases the bedding temperature and decreases the bedding moisture by increasing the drying rate. Keeping the top layer of bedding material dry is the most important part of managing a compost bedded pack barn. The pack should be stirred at least two times per day. Stirring is typically accomplished while the cows are being milked, using various types of cultivators or roto-tillers. Proper management of compost bedded pack barns includes facility design, ventilation, timely addition of fresh, dry bedding, frequent and deep stirring, and avoidance of overcrowding.

Barn design, sustainability and economics of bedded back barns in the Netherlands
P.J. Galama[1], H.J. Van Dooren[1], H. De Boer[1], W. Ouweltjes[1], G. Kasper[1] and F. Driehuis[2]
[1]Wageningen University and Research, Livestock Research, De Elst 1, 6708 WD Wageningen, the Netherlands,
[2]NIZO food research BV, Kernhemseweg 2, 6718 ZB Ede, the Netherlands; paul.galama@wur.nl

Since 2009, about 50 compost bedded pack barns are built in the Netherlands. Ten of those farms are involved in our research project, one more intensive than the other. They differ in farm design, type and management of the bedding and aerating system. Five farms use wood chips as bedding material. They compost these in the stable together with the faeces and urine of the cows. The heat of the composting process stimulates the evaporation of moisture. Four farmers control the composting process with an aerating system by blowing or suckling air through the bedding. If they manage the composting process well this is a promising sustainable housing system from the point of view of animal welfare, longevity, environment (emissions of ammonia, greenhouse gasses and odor), manure quality (soil improver), milk quality and economics. The other four farmers used composted green waste from compost factories as bedding and one uses straw to absorb moisture. Since 1/1/2015 this is prohibited by the Dairy Industry due to an increased concentration of spores of Thermophilic Aerobe Sporeformers bacteria (TAS) in raw milk from cows kept on such bedding, which may lead to decay of sterilized dairy products. Therefore these farmers are testing alternative bedding materials. Some alternative bedding materials like straw, miscanthus, wood chips, wood fiber, cocos fiber, peat and humus from worms are evaluated on lab scale for suitability for composting and absorptive based on the criteria C/N ratio, demolition, porosity and water absorption. An economic evaluation is made based on experiences with composting of wood chips. The yearly costs for such a bedded pack barn with 15 m^2 bedding area per cow are higher than a cubicle stable, but depends a lot on the price of wood chips. However with 10% lower cow replacement the yearly costs are about the same.

Ammonia and nitrous oxide emissions from organic beddings in bedded pack dairy barns
H.J.C. Van Dooren, J.M.G. Hol, K. Blanken, H.C. De Boer and P.J. Galama
Wageningen UR Livestock Research, P.O. Box 338, 6700 AH Wageningen, the Netherlands;
hendrikjan.vandooren@wur.nl

Bedded pack dairy barns (BPBs) are an alternative for loose housing systems (reference system) that have cubicles for lying, concrete floors for walking and slurry storage underneath. Dutch farmers use either compost from municipal organic waste or wood chips as bedding material in BPBs. The objective of this study was to get insight in the emission levels of ammonia (NH_3) and nitrous oxide (N_2O) of both bedding materials used in BPBs. Emissions were measured twice for 24 hour in each of two naturally ventilated dairy barns, one bedded with wood chips (WCB) and the other with compost (CB). SF_6 was used as a tracer gas and injected at a constant and known rate at the air inlet over the whole length of the barn. SF_6 concentrations were measured with a gas chromatograph (GC). NH_3 concentrations were measured using an open path laser and two photo-acoustic multi gas monitors. Concentrations of N_2O were measured in 24 hour samples using the lung method and were analysed in the lab gasses using a GC. Emissions were calculated according to Ogink et al. Average NH_3 emission was 16.5 and 42.4 kg/animal/year for WCB and CB respectively. Average N_2O emission was 2.7 and 2.6 kg/animal/year for WCB and CB respectively. NH_3 emissions from the WCB were considerably lower than from the CB. However, ammonia and nitrous oxide emissions of both barns were higher than the emission of a reference barn (14.4 kg NH_3 and 0.23 kg N_2O per animal per year) reported by Mosquera et al. One of the possible explanations is the larger area per cow (17.5 m^2/cow at the WCB and 16.5 m^2/cow at the CB) compared to 7 m^2/cow in a typical reference system. Also availability of oxygen may explain the higher nitrous oxide emissions. Emissions expressed as N-loss were 127% and 305% compared to the reference system for WCBs and CBs respectively. Optimization of WCBs may lead to acceptable ammonia emissions.

Compost bedded loose housing dairy barn for sustainable dairy production

J.L. Taraba[1] and J.M. Bewley[2]
[1]University of Kentucky, Biosystems & Agricultural Engng, CE Barnhart Bldg, Lexington, KY 40546, USA, [2]University of Kentucky, Animal & Food Sci, WP Garrigus Bldg, Lexington, KY 40546, USA; joseph.taraba@uky.edu

In the US, compost bedded loose housing (CBP) dairy barns were developed by Virginia dairy producers in the 1980's to increase cow comfort and longevity. The key component of CBP barn consists of a large, open resting area, usually bedded with dry, fine sawdust. Bedding material and manure is composted in place when mechanically stirred on a regular basis to support aerobic composting. Studies in Minnesota in the early 2000's built a knowledge base which researchers in Kentucky utilized during the past 5 years as the foundation for research and extension activities concerning the management and construction of CBP barns, herds housed within them, and compost fertility. CBP barns fit within goals of sustainable agriculture because of benefits to the cow (space, rest, exercise, and social interaction), farmer (low investment, labor-extensive, reduced manure storage costs), and environment (reduced NH_3, odor, dust and GHG emissions). CBP barns reduce the size of liquid based manure system and allow producers to economically transport dry, concentrated nutrients to areas of need on and off-farm. Active aerobic composting increases the bedding temperature and decreases bedding moisture by increasing the drying rate. A dry top layer of bedding material is the most important part of managing a CBP barn for cow welfare and productivity. The pack should be stirred at least 2X per day using a cultivator or rotary tiller. Poor bed management leads to undesirable compost bed conditions, dirty cows, and increased clinical mastitis incidence. Instead of using the cow hygiene scores or bed temperature, moisture content >60% is viewed as the primary, leading indicator of impending poor bed performance. Success of CBP barns depends on facility design, ventilation, timely addition of fresh, dry bedding, frequent and deep stirring, and avoidance of overcrowding. Research results of completed and on-going CBP barn projects are presented.

Low N loss from a bedded pack barn with active composting of woodchip bedding

H.C. De Boer, H.J.C. Van Dooren and P.J. Galama
Wageningen UR, Livestock Research, Postbus 338, 6700 AH Wageningen, the Netherlands; herman.deboer@wur.nl

A better cow welfare stimulates an increasing number of Dutch dairy farmers to change their barn type from a free stall barn with a concrete slatted floor to a bedded pack barn with organic bedding in addition to a slatted floor. Changes in floor characteristics have consequences for the barn's environmental impact, most notably on different forms of gaseous N loss from the barn and after manure application to the field. The level of total gaseous N loss from a barn is a general indication of its environmental impact and can be derived from the barn N balance. Initial calculation of N balances for different bedded pack barns in the Netherlands indicated a high level of N loss, ranging between 19 and 63% of excreted N. N loss from barns with composting woodchip beddings was at the lower end of the range. Further research therefore focused on barns with this bedding type. A bedded pack barn with active composting of woodchips at 40 to 50 °C was visited every three weeks to collect information for calculation of the N balance and the development of N loss over time. After eight months, cumulative N loss from this barn was 8.7% of excreted N, lower than the 10.6% lost from a reference freestall barn. Further analysis indicated that cumulative N loss was 18% of N excreted on the bedding and 3% of N excreted on the slatted floor. Initial N loss from the barn was negative, which indicates that the composting bedding acted as a biofilter and fixed N from the air. Over time, N loss from the bedding increased and surpassed the level of N loss from the slatted floor. Linear regression revealed an inverse relationship ($P<0.001$; $R^2_{adj.}=87\%$) between N loss from the barn and bedding C/N ratio. Above a C/N ratio of 34, there was no net N loss from the barn. The results of our study indicate that active composting of woodchip bedding can result in a low level of N loss from bedded pack barns, with potential to avoid N loss when bedding C/N ratio is maintained above a threshold level.

Model-based tool to estimate the NH$_3$ emission reduction potential of adapted dairy housing systems

L.B. Mendes[1,2], P. Demeyer[2], E. Brusselman[2], N.W.M. Ogink[3] and J. Pieters[4]
[1]International Institute for Applied Systems Analysis, Schlossplatz 1, 2361 Laxenburg, Austria, [2]Institute for Agricultural and Fisheries Research, Burg. Van Gansberghelaan 115, 9820 Merelbeke, Belgium, [3]Wageningen UR Livestock Research, De Elst 1, 6700 AH Wageningen, the Netherlands, [4]Ghent University, Coupure Links 653, 9000 Ghent, Belgium; mrmendes2010@gmail.com

In order to address the environmental issues raised within the Natura2000 EU policy, an action plan titled 'Integrated Approach to Nitrogen (acronym in Dutch PAS)' was launched in Flanders (Belgium). ILVO provides policy support creating a scientific framework to develop, test and validate new ammonia (NH$_3$) emission mitigation strategies. The specific PAS task which this project is about refers to the development of an algorithm that can estimate the potential NH$_3$ emission reduction capacity of mitigation techniques for dairy cattle (DC) housing systems. The algorithm is based on peer reviewed literature and mechanistic modelling of NH$_3$ emission processes normally occurring in DC housing systems. The model relies on three main physicochemical processes: (1) enzymatic degradation of urinal urea into ammonium (NH$_4^+$); (2) pH-driven dissociation of NH$_4^+$ into NH$_3$; and (3) convective mass transfer of volatilized NH$_3$. The calculation tool features aspects such emissions from slatted floor and underlying manure storage and includes effects of: (1) different types of floor; (2) scraping manure and scraping frequency; (3) flushing with water and/or buffer solutions and (4) acidification of manure. New or adapted dairy cattle systems were compared to a reference system comprising slatted floor and manure cleaning once a day, with estimated NH$_3$ emissions of 12 to 13 kg/animal-place/year. NH$_3$ emission reduction factors obtained with the algorithm were validated against data from peer reviewed literature. The tool will be used by the Scientific Team that advises the Flemish Ministry of Environment in order to attribute estimated NH$_3$ emission reduction factors to new or adapted systems.

The compost barn – an innovative housing system for dairy cows

E. Ofner-Schroeck[1], M. Zaehner[2], G. Huber[1], K. Guldimann[2], T. Guggenberger[1] and J. Gasteiner[1]
[1]HBLFA Raumberg-Gumpenstein, Institute for Animal Welfare and Animal Health, Raumberg 38, 8952 Irdning, Austria, [2]Institute for Sustainability Sciences (ISS), Tänikon 1, 8356 Ettenhausen, Switzerland; elfriede.ofner-schroeck@raumberg-gumpenstein.at

Compost barns for dairy cattle are showing increased popularity also in Central Europe. The compost barn typically consists of a large bedded lying area and a solid feeding alley. The lying area is mostly bedded with sawdust or dry fine wood shavings or wood chips and has to be stirred twice a day. Stirring aerates and mixes manure and urine into the bedding material. The mixture decomposes by means of aerobic microorganisms. In a joint research project between the Agricultural Research and Education Centre Raumberg-Gumpenstein (HBLFA) and the ISS Tänikon among others, the topics skin lesions, cleanliness, lying behaviour and the current lameness situation of animals in compost barns were analyzed. The investigations were conducted on five Austrian dairy farms keeping a total of 138 cows in compost barns. All cows were visually scored and animal behaviour was observed by data loggers as well as by direct observation. Concerning lying behaviour cows showed no differences between times of day and temperatures. Large differences in lying behaviour were evident between farms. The dirtiness of animals averaged 0.44, while the udder was the cleanest and the lower leg the dirtiest area. Only a few lesions in carpal and tarsal joints could be found. In lameness assessments 25% of cows were scored to be lame in compost barns. This percentage is significantly (P<0.001) lower than a series of results on cubicle housing systems (31-46%). From the present results, the compost barn can be seen as an animal-friendly system. Further investigations are desirable to analyze other factors affecting animal health and to resolve any outstanding issues concerning economy and alternative litter materials.

Through-flow patterns in naturally ventilated dairy barns – three methods, one complex approach

S. Hempel[1], L. Wiedemann[1], D. Janke[1], C. Ammon[1], M. Fiedler[1], C. Saha[1,2], C. Loebsin[3], J. Fischer[4], W. Berg[1], R. Brunsch[1] and T. Amon[1,5]
[1]Leibniz Institute for Agricultural Engineering Potsdam-Bornim, Engineering for Livestock Management, Max-Eyth-Allee 100, 14469 Potsdam-Bornim, Germany, [2]Bangladesh Agricultural University, Department of Farm Power and Machinery, BAU Main Road, 2202 Mymensingh, Bangladesh, [3]Landesforschungsanstalt für Landwirtschaft und Fischerei Mecklenburg-Vorpommern, Wilhelm-Stahl-Allee 2, 18196 Dummerstorf, Germany, [4]Universität Hamburg, Neue Rabenstr. 13, 20354 Hamburg, Germany, [5]Freie Universität Berlin, Department of Veterinary Medicine/Fachbereich Veterinärmedizin, Robert-von-Ostertag-Str. 7-13, 14163 Berlin, Germany; tamon@atb-potsdam.de

The highly variable flow patterns in naturally ventilated barns determine the air exchange rate and are crucial for the transport of pollutants, humidity and heat. Understanding the processes in detail and relating those to typical emission rates is essential. The through-flow processes of naturally ventilated barns are highly complex and turbulent. Even for symmetric buildings the flow is spatially heterogeneous and not stationary. The observed patterns depend strongly on the inflow conditions and the building design. In most of cases, those cannot be changed to study the effect on the flow in detail. It is sensible to complement field data with data from validated models. Studies in the boundary layer wind tunnel and numeric simulations with an orthogonal inflow, supported by the field data, revealed a meandering flow. Wind speed in the front half of the barn is in the animal occupied and the emission active zone typically much higher than in the back half. Close to the roof a reverse flow is observed. Physical and numeric modeling is essential to obtain sufficiently high resolved data and to study the influence of boundary conditions. Field measurements are crucial to validate models and assess the flow patterns regarding their effect on emission rates.

Comparison of mastitis, its indicators, and lameness in compost bedded pack and sand freestall farms

E.A. Eckelkamp[1], J.L. Taraba[2], R.J. Harmon[1], K.A. Akers[1] and J.M. Bewley[1]
[1]University of Kentucky, Department of Animal and Food Sciences, 400 W.P. Garrigus Building, Lexington, KY 40546, USA, [2]University of Kentucky, Department of Biosystems and Agricultural Engineering, 128 CE Barnhart Building, Lexington, KY 40546-0276, USA; eaec223@uky.edu

The objective of this research was to describe the relationships among SCC, high SCC prevalence (HSP, percent of cows with SCC>200,000 cells/ml), cow cleanliness, and locomotion status in 8 compost bedded pack (CB) and 7 sand freestall (SF) farms in Kentucky from May 2013 to January 2014. The same observer evaluated cow hygiene scores (HS) and locomotion scores (LS) bi-weekly for 50 cows per herd. Throughout the study, producers identified quarters displaying clinical mastitis signs (reported clinical mastitis incidence, RCMI). Test-day SCC and HSP were obtained from DHIA. The MIXED procedure of SAS (SAS Institute, Inc., Cary, NC) was used to assess fixed effects of barn type (BT), maximum temperature humidity index (MT), and HS on SCS, MIP, and RCMI. The MIXED procedure of SAS was used to assess fixed effects of BT and MT on LS and HS. Stepwise backward elimination was used to remove non-significant interactions (P≥0.05) with main effects remaining in the model regardless of significance. A χ^2 analysis was conducted using the FREQ procedure of SAS to determine different LS groups and mastitis severity scores between BT. No differences were detected between BT for HS (P=0.38), SCC (P=0.69), HSP (P=0.43), RCMI (P=0.90), or LS (P=0.57). No differences were observed in LS groups between BT (P≥0.05). A difference was noted between BT (P<0.05) in mastitis severity score with greater score 1 (abnormal milk only; 67.2 vs 54.8%) and lower score 3 (systemic signs; 3.4 vs 15.3%) in CB vs SF. In this study, cows housed in CBP and SF performed similarly.

3D-head acceleration used for lameness detection in dairy cows

Y. Link[1], W. Auer[2], S. Karsten[3] and J. Krieter[1]
[1]Institute of Animal Breeding and Husbandry, Christian-Albrechts-University, Olshausenstr. 40, 24098 Kiel, Germany, [2]MKW electronics GmbH, Jutogass 3, 4657 Weibern, Austria, [3]TiDaTier und Daten GmbH, Bosseer Str. 4c, 24259 Westensee, Germany; ylink@tierzucht.uni-kiel.de

Nowadays, acceleration data is used in dairy husbandry mainly for heat detection. There are also attempts to detect lameness or other behaviours, such as rumination, on a regular basis with acceleration. In this study, 3D-head acceleration were analysed in order to detect lameness in dairy cows. In total, 69 cows were equipped with an ear tag sensor measuring the acceleration with 1 Hz and 10 Hz transmission rates (20 cows: 1 Hz, 30 cows: 10 Hz) and were rated three times a week according to a five point locomotion scoring system (1=not lame, 5=severe lame) between February 2014 and December 2014. Cows were classified as lame with a locomotion score (LS) greater than 2. The sensors measured the acceleration of three axes (x, y, z), which were combined to a total value. The results showed that 10 Hz sensors provided a higher daily average total acceleration with $1,039.10\pm0.34$ mm/s2 than 1 Hz sensors with $1,027.75\pm0.39$ mm/s2. The 3D-head acceleration of cows is influenced by many different factors, e.g. age and lactation state. The biggest impact has the single individual. Thus, the acceleration data of each cow can only be compared with itself to detect variations in the activity patterns and to potentially predict lameness. First results showed that the daily mean and standard deviation of the daily values decrease with LS greater than 2. The distribution of the data also changed during 'days of lameness'. Kurtosis and skewness increased, which implies that during 'days of lameness' more extreme values occurred. Before detecting lameness the skewness decreased. Therefore, changes in the data distribution might be a way to detect lameness.

Effect of milking behaviour of dairy cows on lifetime production efficiency

W. Neja, A. Sawa, M. Jankowska, M. Bogucki and S. Krężel-Czopek
University of Science and Technology, Cattle Breeding, ul. Mazowiecka 28, 85-084 Bydgoszcz, Poland; nejaw@utp.edu.pl

The aim of the study was to analyse the effect of milking behaviour of cows on their lifetime production efficiency. The animals represented the active population in Pomorze and Kujawy regions, first calved in 2006, were assessed for milking behaviour, and were used or culled until the end of 2012. The following milking behaviour categories were accounted for, based on the methodology of the Polish Federation of Cattle Breeders and Dairy Farmers: 1: calm; 2: normal; 3: excitable or aggressive. Most cows showed a normal temperament (89.52%). Over successive years of the evaluation, there were increases in the proportion of cows with a calm (from 2.98% to 6.85%) and excitable temperament (from 7.50% to 8.27%). Temperament was found to have a highly significant effect on first lactation milk yield, first lactation daily milk yield, milk yield per day of age, and milk yield per day of productive life. In cows with a calm temperament, first lactation milk yield was 621 kg higher than in cows with a normal temperament and 329 kg higher compared to excitable (aggressive) cows. Large individual differences in animal behaviour and their interpretation pose a considerable difficulty in cattle behaviour research regardless of the housing system, environmental conditions and the natural patterns of animal behaviour. On the other hand, behavioural observations in cattle are increasingly accurate and their analysis is increasingly taken into consideration when developing animal production technologies. It is concluded that milk yield, first lactation daily milk yield, and milk yield per day of age and day of productive life all depend on the cow's temperament. Most of the cows in the analysed population exhibited a normal temperament, but the proportion of animals with a calm and excitable temperament increased over successive years.

Cattle welfare analysis in dairy cattle farms based on new legal regulation and veterinary inspection

E.A. Bauer, J. Żychlinska-Buczek and D. Praglowski
University of Agriculture in Krakow, Animal Science, Al. Mickiewicza 24/28, 31-120, Poland;
e.bauer@ur.krakow.pl

Welfare concerns for intensive cattle production have often been raised, but on farm welfare assessment studies are rare. The aim of the study was to apply the legal regulations welfare assessment system for dairy cattle on farms to evaluate the state of welfare of the level of Polish Veterinary Inspection measure of and of aggregated scores, as well as overall classification. An animal welfare is a condition of the physical and mental health of animals achieved in conditions of a profound sympathy with the surrounding environment. Official animal welfare inspections in Poland primary control compliance with animal welfare legislation based on resource measures (e.g. housing system) and usually do not regard animal response parameters (e.g. clinical and behavioural observation). The legal regulations assessing the degree of the cattle welfare were described, which the Polish Veterinary Inspection. The aim of the study was to determine on the basis of breeding documentation and control protocol SPIWET animal welfare conditions in accordance with the Regulation of the Minister of Agriculture and Rural Development from date 28 June 2010. On minimum conditions for the maintenance of livestock species. Types of dairy barn buildings, its equipping and the microclimate, maintenance systems, and also kinds of ground were described. At the work to degree of achievement of the cattle welfare was described in the polish public farms, district in 2008-2013 years. As a main contacts result of the inspection conducted of ten dairy farms enough a soundness of adapting buildings and the equipment to requirements of cattle management according to Polish law regulations was stated.

SWOT analysis of different housing systems for dairy cows

M. Klopčič
University of Ljubljana, Biotechnical Faculty, Department of Animal Science, Groblje 3, 1230 Domžale,
Slovenia; marija.klopcic@bf.uni-lj.si

Dairy cattle, specifically the milking herd, close-up dry cows and young stock, are housed in different types of dairy facilities. A well-designed barn provides a clean, comfortable space for the herd and a pleasant, efficient workplace for the farmer. These facilities can include tie stalls for individual dairy cows, free stalls, loose housing, pasture systems, and more seldom, compost bedded packs. Tie-stalls are still the most common in Slovenia. Dairy farmers know that housing system has a substantial impact on the overall production, health and longevity of their cattle. When farmer selecting the most suitable housing system for their farm, it is important to know construction costs, ongoing management and maintenance costs as well as the animal health and welfare implications. Our objective was to compare different housing systems for dairy cattle and on the base of SWOT analysis to help dairy farmers, which intend to build new barn or renovate old barn for dairy cattle. A study on housing and management of dairy cows were performed on 15 selected dairy farms, which built new barn last two years. All activities related to the housing of dairy cattle on these farms were observed and farmers were interviewed using a questionnaire. Results of these observations and SWOT analysis on the base of depth interview with farmers will be presented. There are advantages and disadvantages to different housing systems.

Horses, humans, and life in the 21st century: new challenges, new opportunities

R. Evans
Hogskulen for landbruk og bygdeutvikling, 213 Postvegen, 4353, Klepp Stasjon, Norway; rhys@hlb.no

Horses have been a part of human life for millennia. A study of horses in human history shows a remarkable versatility in terms of the roles they have served. These range from food to part of the production of food, through industrial and transportation helpmate all the way to therapist, physical activity facilitator and companion. Although horses still fill all those roles somewhere in the world, in the developed world their role is in the process of adapting to meet the needs of increasingly urbanised life in the 21st century. In every case, horses have taken up roles which are produced by the state of human life. Changing economies, changing societies and technological change have also taken away the predominant roles held by horses in human life. Thus for example, we have seen the decline of numbers with the advent of motorization due to the loss of horses working in agriculture, forestry and transportation. Throughout history, the potential of the close relationship possible between humans and horses means that they have been companions, and the focus of leisure interests. Now, however, the leisure sector (including fitness, competition, outdoor recreation, tourism, etc.) takes a new role in contemporary society and the economy. Leisure activities are highly significant generators of wealth, and people are willing to pay significantly for them partly because their working and domestic situations generate a need for the things equine activities provide. This paper will explore how these trends in contemporary society is creating new roles for horses and providing new opportunities for the equine economy. At the same time, it will explore some of the challenges this provides and ask a few significant questions about the state of our knowledge and some of the taken-for-granted ideas which need critical assessment.

How to evaluate the welfare of racing and sport horses?

M. Minero
Università degli Studi di Milano, DIVET, Via Celoria 10, 20133 Milano, Italy; michela.minero@unimi.it

Horse welfare is a cause for concern due to increased public awareness and demand for improved animal welfare and to limitations of the present European legislation. Animal welfare includes both physical and mental states and describes a measurable quality of a living animal at a particular time. It is a scientific concept that requires a multi-dimensional approach and that takes into consideration several aspects that are almost independent. The European Animal Welfare Indicators Project (AWIN) addressed the development, integration and dissemination of animal-based welfare indictors for horses with emphasis on pain assessment and pain recognition. AWIN also translated the welfare assessment protocol into an interactive app to facilitate data collection, data storage and data analysis. Equine stakeholders have been constantly updated and they actively contributed to the development of the protocol outlining potential barriers and identifying possible solutions. The welfare assessment protocol for horses developed by AWIN is grounded on the four welfare principles and twelve criteria developed by Welfare Quality® and is focused on animal-based indicators. Resources-based and management-based indicators are included for those criteria where animal-based indicators cannot provide results in a validated or feasible way. The protocol considers a stepwise strategy of assessment, with a more detailed assessment dependent on the outcome of a smaller number of important first measures. After the assessment, welfare data are entered into the app or in a data set and an objective descriptive output is generated. The aim of the output is to give a visual feedback about welfare of the animals on the farm, highlighting positive conditions and enabling comparison with a reference population. Currently, the reference population considered for the outcome refers to data collected during the AWIN project on fifty farms in Germany and Italy. Long periods of confinement without access to turnout, and a lack of social interaction appear to be areas of concern.

Horse purchases, horse husbandry and riding schools: key quality demands of 'tomorrow's customers'

K. Wiegand, C. Ikinger and A. Spiller
Georg-August University of Goettingen, Department of Agricultural Economics and Rural Development,
Platz der Göttinger Sieben 5, 37073 Goettingen, Germany; katharina.wiegand@agr.uni-goettingen.de

The German horse sector has become very heterogeneous as riding styles and horse breeds from all over the world appear on the market. The percentage of riders preferring nature- and leisure-time experiences has increased. Due to demographic and social changes new target groups have evolved. The present study investigates quality aspects that play a key role for modern horsemen when buying a horse or choosing a stable. An online study among 2,048 equestrians was conducted. The analysis of the data was carried out with the software SPSS 21. With regard to horse purchase the results showed that commonly advertised characteristics such as the level of training, breed and pedigree or competitive success generally only play a secondary role. In contrast the character, behavior, health status and the rearing conditions are regarded as most important aspects. In view of horse husbandry the key quality aspects were an adequate feeding regime, regular access to pasture, the husbandry system, health management and a good working atmosphere. The results concerning riding schools show that the central quality aspects are the health of the school horses and the respective husbandry. The qualification and tournament success of the instructor are regarded as less important. Looking more deeply into the various types of equestrians, there are vast differences regarding the importance of individual quality aspects, e.g. between success and leisure oriented equestrians. Aspects concerning the husbandry and rearing of horses play an increasing role across all areas of the horse business. For horses standing for sale, boarding stables and riding schools efforts made to enhance these aspects should not only be made clear but also well-documented and communicated to the potential client. The present results provide important information for actors in the equestrian field who wish to act customer-orientated in the future.

Equitation science: a research-based approach to improved understanding of horse perspective

C.L. Wickens
University of Florida, Animal Sciences, 2250 Shealy Drive, 32611, Gainesville, FL, USA; cwickens@ufl.edu

The mission of equitation science is to promote and encourage the application of objective research and advanced practice which will ultimately improve the welfare of horses in their associations with humans (www.equitationscience.com). Equitation science is a relatively young scientific field which employs a multidisciplinary, evidence-based approach to investigation of how horses learn and the impact of training, performance, and management on the physical and psychological well-being of the ridden horse. Previous International Society for Equitation Science conference themes have encompassed the physical and psychological aspects of the sustainable athlete, the importance and effects of the daily routine outside the time each horse is worked, equine stress, learning and training. A substantial amount of research has been conducted to assess the soundness of traditional methods and opinions frequently practiced and observed in the horse industry. Recent work has examined the effects of negative and positive reinforcement on equine learning, horse response to rider and equipment attributes, and horse response to various management practices and the stable environment. Measurement and interpretation of behavioral indicators of stress and discomfort as well as the development of novel approaches to the assessment of equine affective state have received increased attention and investigation among equitation science researchers. Findings from these studies are proving helpful in reducing conflict and unwanted behaviors in horses and improving our ability to evaluate horse temperament and to determine horses' preferences for different resources and training modalities. A crucial element of equitation science is to extend the knowledge obtained through research to the equestrian practitioner. This paper will explore current equitation science research with emphasis on understanding the horses' perspective as well as present ways in which equitation science can be incorporated into practice.

Improving horse businesses by introducing the lean production process model – an exploratory study
J. Öwall and A. Herlin
Swedish University of Agricultural Sciences, Biosystems and Technology, P.O. Box 103, S-23053 Alnarp,
Sweden; johannaowall@gmail.com

Methods of improving business operation and competiveness are essentially unknown and rarely described in the horse sector. However, many horse businesses struggle because of e.g. high costs, accident risks, difficulty of attracting skilled personnel on long term, traditional ways of management impeding a progressive development in the everyday operation and business model. There is a need to make the operations cost effective and safer. By streamlining everyday processes, introducing mechanization and the lean production process model, benefits in economics and working conditions can be obtained. The objectives of the study were: (1) Investigate whether lean production is applicable to the horse business sector; (2) Examine which work processes that can be streamlined; and (3) Examine the interest in work efficiency. A total of six horse businesses were studied; three riding schools and three racehorse trainers respectively. The operations were studied by the use of semi structured interviews, observations and the lean production tool value stream mapping exercise. These were used to study the tasks mucking out, sweeping aisles, preparation of feed and walking of horses to and from paddocks. The results indicate that the core value of the businesses contact with customers, horse owners and business development could be increased and given better focus with more work time efficient routines and housing conditions. Streamlining the processes mucking out, distributing feed and sweeping aisles and optimize the logistics of horses, feed and manure will save time. However, interest in methods to improve work efficiency was low and there is a sceptic attitude concerning new ideas and technology. Businesses working with lean production are suggested to benefit by having the less time spent on non-horse activities, more on staff development, the staff staying longer in the occupation, having fewer accidents and fewer days off.

An explorative study on the potentials for an animal welfare label for horse husbandry in Germany
C. Ikinger and A. Spiller
Georg-August University of Goettingen, Department of Agricultural Economics and Rural Development,
Platz der Göttinger Sieben 5, 37073 Goettingen, Germany; cikinge@gwdg.de

A growing concern about the animal-friendliness of current horse management practices and husbandry systems characterizes the German horse industry. However, in the German horse market only few horse husbandry labels exist, and hardly any of them emphasize horse welfare. The aim of the present study is to explore the demand for an innovative animal welfare label for horse husbandry in Germany. For the present purpose, an online study among a total sample of 2,048 equestrians was conducted which encompassed both open and closed questions. The analysis of the data was carried out with the statistics software IBM SPSS 21. The results show that, although the perceived uncertainty when evaluating a horse husbandry system seems to be rather moderate, a great interest and demand does exist for a horse welfare label. Concerning the perceived difficulties within the evaluation of a horse-keeping farm in general most participants were afraid that stable managers can prepare for the usually pre-announced inspections and therefore express the desire for regular, unannounced controls. Another difficulty seems to be the existence of different opinions on the animal-friendliness of horse husbandry. Further difficulties mentioned include the evaluation of the feed quality and management, the selection and objectivity of adequate inspectors or testing institutions and the selection and weighting of appropriate criteria. Concerning potential criteria for an animal welfare label for horse husbandry the three most important criteria were an adequate feeding regime, regular access to pasture and the appropriate riding of horses. The management skills of the manager were rated least important. In summary, one can conclude that a great interest in and also a great potential for both existing and future animal welfare labels for horse husbandry does exist, which carries the potential to improve the level of horse welfare in the long term.

Reserach funding in Norway/Sweden

S. Johanson

HNS (Swedish Horse Industry Foundation), Hästsportens Hus, 161 89 Stockholm, Sweden;
stefan.johanson@nshorse.se

The Swedish/Norwegian Equine Research Foundation (SHF) was founded in 2004 by the Swedish Horse Industry Foundation in collaboration with AB Trav & Galopp (the Swedish horse betting company), Agria (the market leading horse animal insurance company in Sweden) and the Swedish Foundation for Agricultural Research. During 2009 there began a collaboration between Sweden and Norway. Now there is an agreement between SHF and the Norwegian Research Council, which among other things supports to lift collaborative project between the two countries. The stakeholders from Norway and Sweden also use a common research administration to be as cost efficient as possible. For the period 2013-2015 has through agreements been secured a financing of at least SEK 8.1 million per year to the Swedish equine research. The government contributes with SEK 3.0 million through the Swedish Research Council Formas, while the remainder coming from the founders/horse industry (AB Trav & Galopp, SEK 3.6 million and Agria, SEK 1.5 million). From 2016 the founders will increase their contribution with 10%. It is also very positive that the funding of Norwegian equine research is long-term secured. The Norwegian government, through the Norwegian Research Council, contributes with NOK 2 million. The horse industry via Norsk Rikstoto (the Norwegian horse betting company) ensured annual contributions together with the Norwegian agricultural operations, each NOK 2 million per organization. In total NOK 6 million/year. From Norwegian/Swedish aspects there are a great interest to enlarge this collaboration with our neighbors Denmark and Finland, with the aim to create a total Nordic platform for equine research. To maybe implement similar financing models in Denmark/Finland and begin, from a wider scope, to exchange communication between the different team of researchers in these four countries can be the start of very fruitful development processes. Our main point of view is that new competence/knowledge via research is not limited by nation borders.

How to fund the équine research

P. Lekeux

University of Liege, Faculty of Veterinary Medicine, 20 Bvd de Colonster, 4000 Liege, Belgium;
pierre.lekeux@ulg.ac.be

Despite the evident importance of the horse sector, the equine species is considered by most funding bodies and by big industry as a minor species. As a result, research applications related to the equine sector have little chance to be successful, when compared to major species in production animals. Therefore a realistic option for building a serious R&D project is to use the co-funding, as developed in public-private partnership. An example of funding equine research is the Hippolia Foundation based in Normandy (France) and created in 2011. The main objective of Hippolia is to boost the equine sector by promoting R&D and innovation, specially in equine health, welfare and performance. The main strategy of Hippolia is to promote excellence, innovation, development, education, pooling of facilities, collaborative and transnational projects, and, last but not least, exploitation of the results into the private sector. The main funding of Hippolia comes from private-public partnership, ie co-funding from the public sector (State, Region, Department), from the private sector (pharmaceutical industry) and from the research laboratories. Despite the difficult economical context, Hippolia has already supported several successful projects in equine R&D.

AUTOGRASSMILK – combining automatic milking and precision grazing

B. O'brien[1], A. Van Den Pol-Van Dasselaar[2], F. Oudshoorn[3], E. Spörndly[4], V. Brocard[5] and I. Dufrasne[6]
[1]Teagasc, Animal and Grassland Research & Innovation Centre, Livestock Systems Department, Moorepark, Fermoy, Co. Cork, Ireland, [2]Wageningen UR Livestock Research, P.O. Box 65, 8200 AB Lelystad, the Netherlands, [3]Danish Agriculture & Food Council, Agro Food Park 15, 8200 Aarhus N, Denmark, [4]Swedish University of Agricultural Sciences, 753, 23 Uppsala, Sweden, [5]Institut de l'Elevage, BP 85225, 35652 Le Rheu Cedex, France, [6]Université de Liège, 4000, Liege, Belgium; bernadette.obrien@teagasc.ie

Automatic milking (AM) systems have changed dairy herd management and have significantly reduced the physical work involved. Indoor feeding systems have been well adapted to AM, however grazing has not, thus leading to a decrease in grazing on farms with AM. Grazing has many advantages for the economy, environment, animal welfare and product quality. Thus, AM and cow grazing are each highly valued, but have not generally been considered compatible in combination. The 'AUTOGRASSMILK' project (Jan, 2013-Dec, 2015) is working to address this concern by developing and implementing sustainable farming systems that integrate grazing with AM as appropriate to different countries in Europe. This project involves 14 partners across 6 EU countries that include research organizations, SME (small medium enterprise) Associations (e.g. farmer or milk purchaser groups) and end-user farmers. Current research findings on pasture inclusion in cows' diets (10-25%) in Sweden and (80-95%) in Ireland for different cow breeds are being generated. Recommendations on grass management and daily grass allocations are being developed together with guidelines for the operation of mobile and carousel AM systems. AUTOGRASSMILK is also working to provide three web-based decision support tools that will provide information on grazing strategies, sustainability and economic efficiency. Dissemination of the information is firstly to the SME Associations, their members and stakeholders, and thereafter to research scientists, EU dairy farmers and policymakers.

Pasture-based automatic milking systems in Australia

N. Lyons
NSW Department of Primary Industries, Elizabeth Macarthur Agricultural Institute, 2567, Australia; nicolas.lyons@sydney.edu.au

Automatic milking systems (AMS) arrived in a pasture-based system in Australia in 2001. Currently there are 32 farms operating AMS, and at least another 6 being installed. They milk over 8,000 cows producing almost 45 million litres of milk per year. The average AMS farm has 250 cows milked with 4 robots. The dairy industry should expect these numbers to increase exponentially from now onwards. Every dairy region in Australia has farmers that have decided to commission AMS. These installations are from every commercially available type (single box robots, multi box robots and robotic rotary) and operate in an array of farming system types (although over 80% of them operate under grazing with variable levels of supplementation). The experience of the existing commercial operations, together with almost 10 continuous years of research conducted at FutureDairy, has proven that the technology and existing farm management knowledge of AMS can be implemented successfully in Australian dairy farming pasture-based systems. Cows in pasture-based AMS have lower milking frequencies, compared to indoor AMS. Although AMS farmers can achieve similar pasture utilisation than those milking conventionally, pasture management does have an impact on cow traffic. Lower milking frequencies are associated with lower pre-grazing pasture biomass and increasing proportion of pasture in the diet, distance to pasture and pasture allowance. Management practices such as feeding during milking, offering 3 pasture allocations every 24 hours (instead of 2) and offering access to supplement after milking all increase milking frequency. Currently research is focused on the adoption of AMS on herds with over 600 milking cows. Efforts are also concentrated on understanding the impact of AMS on animal behaviour and wellbeing, as well as improving efficiency of individual cows and whole system performance. Work is being conducted to categorise and quantify the costs involved in running and maintaining an AMS as well as a whole financial performance analysis.

Challenging land fragmentation thanks to a mobile milking robot: two cases of implementation

V. Brocard[1], I. Dufrasne[2], I. Lessire[2] and J. Francois[3]
[1]Institut de l'élevage, BP 85225, 35652 Le Rheu Cdx, France, [2]Université de Liège, Faculté Vétérinaire, Chemin de la ferme, 6 B39, 4000 Liège, Belgium, [3]Pôle Herbivores des Chambres d'Agriculture de Bretagne, Ferme expérimentale de Trévarez, 29520 Saint Goazec, France; valerie.brocard@idele.fr

Grazed grass has always been the base of the forage system in many west European regions because of its low production cost and high availability. However, the recent development of milking robots has been followed by a change in the production systems with a lesser resort to grazing. The increase in herd sizes also accounts for a smaller area of grazed grass per cow near the buildings. A mobile automatic milking solution might be the opportunity to better integrate robotic milking and grazing and graze big blocks of paddocks located on the other side of a busy road or far from the current milking parlour. Two experimental farms, Liège in Belgium and Trévarez in France, chose 5 years ago to design a mobile milking robot and to use it inside a building during the winter period, and on a grassland summer site six months per year. This paper will describe the technical data used in the call for tenders to design the prototypes of mobile robots and tanks, the grass management systems tested and their influence on cow flows and milking frequency (one vs two paddocks per 24 h), the grass valorisation and the dairy production, the milk quality parameters, the herdsmen working time and the feeding cost reached compared to the winter 'indoor' period. The two experiences show that it is now possible to design and implement a mobile robot milking solution and that the grass management with two paddocks per 24 h system improves cow flows with limited human interventions. Moreover, the grass valorisation reached is high, and few limited technical problems appeared since the start of the two units. The relatively high investment cost of such solutions, in particular if no infrastructure at all pre-existed on the summer site, remains a restraint that has to be counteracted by a low feeding cost.

Drivers for grazing and barriers to grazing on dairy farms with and without automatic milking

A. Van Den Pol-Van Dasselaar
CAH Vilentum University of Applied Sciences / Wageningen UR Livestock Research, P.O. Box 338, 6700 AA Wageningen, the Netherlands; agnes.vandenpol@wur.nl

In the last decade, grazing of cattle has become a societal issue in the Netherlands. Grazing in combination with automatic milking systems (AMS) is seen as complicated. The aim of this research (part of the FP7 funded project Autograssmilk – 314879) was to study the technical and social factors that affect the extent of grazing on commercial farms with and without AMS. A conceptual framework was developed, based on the Theory of Planned Behavior, in which the extent of grazing is influenced by attitude towards grazing, subjective norms about grazing, perceived behavioural control of grazing and technical possibilities for grazing. An on-line questionnaire was sent to farmers and 212 valid answers were obtained. Factor analysis revealed that attitude towards grazing consisted of two components: the belief of the farmer in the effect of grazing on farm continuity and on grass yield. The subjective norms reflect the social norms in the outside world. The technical possibilities were expressed as milk production ha^{-1}. Combining all these factors in a multiple linear regression model could account for 47% of the variation in the extent of grazing. On average, social norms were seen as driver for grazing; dairy farmers believed that the outside world supports grazing. Grass yield was seen as barrier to grazing; farmers believed that grazing has a negative effect on grass yield. Farmers that grazed associated grazing with farm continuity and perceived behavioural control. Farmers on the other hand that did not graze had the opposite association. This was consistent with their choices in grazing management. The effect of milking system (AMS or no AMS) on all these factors was limited; there was only a significant effect of milking system on perceived behavioural control ($P<0.01$). The perceived behavioural control might change if the infrastructure of the farms can be improved or if the knowledge of the farmer on grazing management increases due to training.

Milk production, milking frequency and rumination time of grazing dairy cows milked by a mobile AMS

F. Lessire[1], J.L. Hornick[1], J. Minet[2] and I. Dufrasne[1]
[1]*FARAH Faculty of Veterinary Medicine University of Liège, Department of Animal Productions, Chemin de la Ferme, 6, 4000 Liège, Belgium,* [2]*University of Liège, DER Sc. et gest. de l'Environnement Agrométéorologie, Arlon Campus avenue de Longwy 185, 6700 Arlon, Belgium; flessire@ulg.ac.be*

In Europe, analysis of meteorological data shows that the average temperature has increased by ~1 °C over the past hundred years. Heat stress periods are thus expected to be more frequent even in temperate areas. The use of an automatic milking system (AMS) implies the need to stimulate cows' traffic to the robot, especially with grazing cows. The aim of this study is to describe how heat stress influenced cows' traffic to the robot. Grazing dairy cows milked by an automatic system (AMS) experienced heat stress (HS) periods, twice during the summer 2013 in July (J) and August (A). The daily temperature humidity index (THI) during these periods was higher than 75. Each HS period was compared with a 'normal period' (N), presenting the same number of cows, similar lactation number, days in milk, distance to come back to the robot and an equal access to water. The first HS period of 5 days with a mean THI of 78.4 was in J, and a second that lasted for six days in A with a THI value of 77.3. Heat stress periods were cut off with the same duration of days with no stress (N) and mean THI <70. Milk production, milkings and refusals to the robot during HS were compared with N periods. Milkings and refusals were significantly more numerous in HS periods in July (HS: 2.54±0.11 vs N: 2.19±0.08, 1.87±0.20 vs 0.72±0.16) but milk production dropped from 21.8±0.6 kg/cow/day during N periods to 18.9±0.8 kg in HS. In August, MY increased slightly during HS. This could be explained by lower ambient temperatures and decreased walking distances inducing less energy expenditure. The increase in milkings and refusals to the robot during HS could be linked to water availability nearby the robot and confirmed previous findings.

Genetic evaluation for automatic milking system traits in the Netherlands

J.J. Vosman, G. De Jong and H. Eding
CRV, P.O. Box 454, 6800 AL Arnhem, the Netherlands; jorien.vosman@crv4all.com

Half of the newly built milking systems in the Netherlands and Flanders are automatic milking systems (AMS). The AMS records a lot of data about individual milkings. This data can be used for developing breeding values, a tool for selection of AMS suitable cows. The traits analysed are AMS efficiency (EF), milking interval (MI), and habituation of heifers (HH). EF is the milk yield in kg per total box time in minutes, MI is the time between two consecutive successful milkings in minutes and HH reflects the time period a heifer needs to get familiar with the AMS. Traits were analysed with a multi-trait animal model. Heritabilities were 0.27 for EF, 0.12 for MI, and 0.07 for HH, with associated genetic standard deviations of 0.20 kg milk per minute for EF, 35.8 minutes for MI, and 20.3 minutes for HH. Based on these genetic parameters, breeding values for bulls and cows were estimated. These breeding values make it possible to select cows that produce more milk per minute AMS time, visit the AMS more frequently and heifers that reach the standard milking interval early after calving. Selecting only bulls with a breeding value of two standard deviations above average will results in more kg of milk per AMS. In other words: to milk more cows per AMS. The amount of extra milk based on an average herd is close to 80,000 kg yearly, or 7 cows extra per AMS based on a production per cow of 30 kg milk per day. A point of attention is the negative genetic correlation between EF and udder health. As of April 2015 these breeding values will be estimated and published routinely in the Netherlands. A summarizing AMS index based on these breeding values is in development to avoid a reduction in udder health.

How reliable are milk sample data from AMS for decision support and regular recording?

P. Løvendahl[1], M. Bjerring[2], K. De Koning[3] and C. Allain[4]
[1]Dept. Molecular Biology and Genetics, Aarhus University, AU-Foulum, 8830 Tjele, Denmark, [2]Dept. Animal Science, Aarhus University, AU-Foulum, 8830 Tjele, Denmark, [3]Dairy Campus, Wageningen UR, Boksumerdyk 11, 9084 AA Leeuwarden, the Netherlands, [4]Institut de l'Elevage, BP 85225, 35652 Le Rheu, France; peter.lovendahl@mbg.au.dk

Automated milking goes hand in hand with automated management and decision support based on automated data recording. This of course involves either in-line sensors or automated milk sampling and assaying of milk samples. For the off-line samples in this presentation, not only traditional components are assayed but also some other test variables such as metabolites, hormones, antigens and DNA fragments. Some of the new assays are extremely sensitive. However, the benefit of the high sensitivity is sacrificed if the basic samples are of improper quality. Unfortunately, samples from AMS tend to be of a lesser quality because of carry-over problems. Here, carry-over means that a given milk sample from the cow milked now contains milk from the cow milked just before. The fractional carry-over in percent has been observed between 2 and 20%. With highly variable traits such as cell counts and test-days with single samples, false positive alerts are obviously possible. The carry-over problem comes from the sum of milk residues sitting in receiver tank, pump house, pipes and tubes and valves between the cow and the final sample cup. With sub-optimal adjustment of settings these reservoirs are not emptied completely. Carry-over has not previously been estimated directly during ICAR approval testing of meters and samplers. A new testing procedure offers an effective way to test for carry-over and is now being implemented in coming ICAR-guidelines. Hopefully, manufacturers will take up this test method, in order to improve quality and robustness of the sampling procedures. In the meantime, we may critically assess the use of AMS milk sample data, and support users with guidelines as to their interpretation and need for repeated sampling.

Successful grazing with automatic milking

A.J. Van Der Kamp, G.L. De Jong, A. Gouw and T. Joosten
Lely International, Farm Management Support, Cornelis van der Lelylaan 1, 3147 PB, the Netherlands; avanderkamp@lely.com

The demand for grazing becomes more prominent in the Netherlands, although a common belief is that increasing farm size and automation is limiting the grazing options on farm. Many different grazing systems are applied and selection gates are used to combine grazing with automatic milking, allowing farmers to be able to manage their farm. In order to find important factors which can secure successful grazing in combination with an automated milking system the management of 200 farms is evaluated. The number of milk visits per cow and the number of refusals are found to be important parameters to evaluate grazing systems in combination with an automatic milking. Different management systems like grazing strategy and the use of a selection gate have influence on the success of the system. The parameters are evaluated both before and after the transition to grazing as well as non-grazing farms are compared to grazing farms. Comparing the differences in milk visits before and after the transition to grazing, a change of -0.1 milking per cow per day was found. Cows averaged 2.66 milk visits per cow per day before grazing and 2.56 milk visits per cow per day while grazing. This has a limited effect on the amount of milk per milk visit which is important for cow health and robot performance. The visits to the milking robot in which the cows were too soon to be milked dropped from 3.43 to 2.77. While grazing cows visit the milking robot less and when they visit they are more likely allowed to be milked, resulting in a minor change in milk visits and a major drop in the visits without milking. With the right farm management, grazing system and selection gate it is possible to realize successful grazing on a farm with automatic milking. The challenge is to change the common belief and show that grazing and automatic milking go together.

Effect of milking frequency on hoof health and locomotion scores of cows in a pasture based AMS

J. Shortall[1,2], K. O'driscoll[2], C. Foley[2], R. Sleator[1] and B. O'brien[2]
[1]Cork Institute of Technology, Department of Biological Sciences, Bishopstown, Co. Cork, Ireland,
[2]Animal and Grassland Research and Innovation Centre, Teagasc Moorepark, Fermoy, Co. Cork, Ireland;
john.shortall@teagasc.ie

The aim of this study was to investigate the effect of two different milking frequencies on hoof health and locomotory ability of cows in a pasture based automatic milking system. A herd of 70 spring calving cows were randomly assigned into one of two treatment groups by breed, parity, days in milk and previous twenty one days milk yield and milking frequency. Cows in group one had a milking frequency of 1.4 milkings per day, while cows in group two had a milking frequency of 1.8 milkings per day. All cows grazed as one herd moving between three grazing blocks during a 24 hour period. Hoof health exams were carried out on a subsample of 41 cows at 14, 44, 85, and 167 days in milk (DIM). Hooves were scored on a 1-5 scale for bruising (B), heel erosion (HE), dermatitis (D) and white line disease (WLD). Locomotion scoring took place on a subsample of 67 cows at 64, 85 and 113 DIM for spine curvature (SC), speed (S), tracking (T), head carriage (HC) and ab/adduction (Ab/Ad). Data were examined using distribution plots to test for normality, then analysed using Proc Glimmix and Proc Mixed statements of SAS. Fixed effects were milking frequency, breed, parity, DIM, exam and interactions, with exam included as the random/repeated effect and the initial exam as a covariate. There was no effect of treatment on any aspect of hoof health or locomotion. However, there was an interaction between treatment and exam for both SC and S ($P<0.001$). Exam also had an effect on SC, S, HC, B ($P<0.001$) and Ab/Ad ($P<0.05$), and their sum ($P<0.01$), with scores generally decreasing over time. These data suggest that reducing milking frequency has no beneficial effect up to 167 DIM on hoof health or locomotory ability of dairy cows milked in a pasture based automatic milking system.

The economic and environmental performance of grazing and zero-grazing systems in a post-quota era

C.W. Klootwijk[1], C.E. Van Middelaar[1], A. Van Den Pol-Van Dasselaar[2], P.B.M. Berentsen[3] and I.J.M. De Boer[1]
[1]Wageningen University, Animal Production Systems, De Elst 1, 6708 WD Wageningen, the Netherlands,
[2]Wageningen University and Research Centre, Livestock Research, De Elst 1, 6708 WD Wageningen, the Netherlands, [3]Wageningen University, Business Economics, Hollandseweg 1, 6706 KN Wageningen, the Netherlands; cindy.klootwijk@wur.nl

Grazing of dairy cattle is decreasing in the Netherlands. This trend is associated with an increase in use of automatic milking systems, herd size and stocking rate. To maintain grazing, more knowledge is required on grazing strategies for future dairy farms. Therefore, we need insight into the economic and environmental consequences of grazing and zero-grazing systems for future dairy farms. A whole-farm linear programming model based on the objective to maximize labour income was used to evaluate the economic consequences of grazing and zero-grazing systems for a Dutch dairy farm using automatic milking after abolishment of the milk quota. In addition, life cycle assessment was used to calculate greenhouse gas (GHG) emissions per ton fat-and-protein-corrected milk (FPCM) for each system. We modelled a dairy farm on sandy soil with 75 hectares of land and compared day and night grazing, day grazing and summerfeeding. Day grazing resulted in the highest labour income (€69,444 per year), followed by day and night grazing (€66,909 per year), and summerfeeding (€46,760 per year). The lower income in case of summerfeeding related mainly to higher feed costs compared to the two grazing systems. Summerfeeding resulted in the lowest GHG emissions per ton FPCM (1002 kg CO_2-equivalents (CO_2-eq)), followed by day grazing (1096 kg CO_2-eq), and day and night grazing (1214 kg CO_2-eq). Results indicate that grazing can contribute to higher economic performance of future dairy farms. To utilize the full economic potential of grazing while at the same time minimize GHG emissions, further fine tuning of grazing strategies is necessary.

Successful combining of robotic milking- and grazing-strategies in the Dutch context

M. Vrolijk[1], A.P. Philipsen[1], J.M.R. Cornelissen[2] and R. Schepers[3]
[1]Wageningen UR, Livestock Research, P.O. Box 338, 6700 AH Wageningen, the Netherlands, [2]Hin Strategic Consultancy, Nieuwe Gracht 3, 2011 NG Haarlem, the Netherlands, [3]Schepers Adviseurs, Rozenhoflaan 12, 7201 AV Zutphen, the Netherlands; maarten.vrolijk@wur.nl

3,475 Dutch dairy farmers use 5,400 robotic milking units. Many farms have high yields per cow and ha and allow cows to walk to the barn for milking or feed, while grazing. Dutch dairy farmers have not enough expertise to realise excellent results with combined grazing- and AMS strategies. To create new solutions ZuivelNL started Robotic Milking and Grazing with 514 dairy farmers with an AMS. -460 practice the Farmwalk, monitor and register grass growth, grazing results and milk production, discuss the results with colleagues, to find new solutions. Farmwalk is a new concept for Dutch farmers. -50 learn in 5 network groups from each other practices and results, in several workshops on different farms. -4 teams make in wintertime an integrated plan with 2015 goals for grazing and robotic milking for 4 farmers. During grazing season the results are shared on monthly basis. The AMS supplier, feed advisor, technical and economic expert, veterinarian, farmer and partner invest their own time. February 2015, provisional results -Experts and advisors, working together in a team, can make an integrated plan when they really participate and learn. -Working together with many actors (farmers, dairy companies, experts and advisors) generates energy and willingness to make new solutions for successful combinations of AMS and grazing. -A matrix with 5 strategies to combine AMS and grazing. Based on skills, entrepreneurial style and farm characteristics, farmers choose the best fitting strategy. Each consisting of rules: what to do, what not, options: possibilities, reflections, tips: helping hands, warning signals. These information is classified in 10 attributes: farmer, number milking times/cow/day, dry matter (dm) intake/cow/grazing day, cow logistics, cow traffic, weather, labour, barn-pasture system, cow behaviour, AMS.

Automatic milking within a grass based dairy farming system

B. O'brien, C. Foley and J. Shortall
TEAGASC, Livestock Systems Department, Animal and Grassland Research & Innovation Centre, Moorepark, Fermoy, Co. Cork, Ireland; bernadette.obrien@teagasc.ie

If automatic milking (AM) is to be considered as a serious alternative to conventional milking (CM) in a grass-based milk production system as standard in Ireland, then it has to incorporate a similar cow nutritional strategy and focus on cow utilization of grass. This approach is also crucial in partial grazing systems in different EU counties if cow grazing is to be optimized on farms with AM systems. The challenge posed is to establish the compatibility of AM with cow grazing. This challenge is being addressed in the EU FP7 project 'AUTOGRASSMILK'. In this scenario, fresh pasture has to be the main incentive for cows to come to the milking unit. Within a trial conducted in Ireland, the land area was divided into three grazing sections. Grass was allocated to cows in each of the three grazing areas during each 24 h period. A herd of 70 cows grazed to a post-grazing sward height of 4.0 cm. All cows received 0.5 kg concentrate feed per 24 h period during the grazing season. A Merlin AMS unit was used. An average milk yield of 4,316±1,167 litres and milk solids yield of 369 kg per cow was achieved over the lactation, which is comparable with a large proportion of Irish dairy farms. The average number of milkings per cow over the lactation was 406±87. The average number of milkings/cow per day was 1.6±0.2. An average milk somatic cell count of 107,000 cells/ml was observed. An extensive range of parameters including cow lameness, milk quality and sustainability measures were also focused on. Successful integration of AM into a grass based system was achieved using the strategy outlined above. However the economic viability of AM will determine how widely the technology will be adopted.

Effect of short-term incentive and priority yard on waiting times in a pasture-based robotic system

V.E. Scott[1], K.L. Kerrisk[1], S.C. Garcia[1] and N.A. Lyons[2]
[1]The University of Sydney, Faculty of Veterinary Science, Brownlow Hill Loop Road, Camden 2570, Australia, [2]Agriculture NSW, NSW Department of Primary Industries, Orange NSW 2800, Australia; nicolas.lyons@dpi.nsw.gov.au

The commercialisation of a high-throughput robotic rotary (RR; Automatic Milking Rotary (AMR); DeLaval, Tumba, Sweden) in 2011 will likely increase adoption of AMS in large (>500 cow) herds. One challenge that arises from this new unit relates to the fact that large numbers of cows need to access the RR through a single entry point (compared to multiple entry points with box robots). A study was conducted over 22 days with 262 cows on a commercial RR in Tasmania (Australia) to investigate the effect of offering and removing a short-term feed incentive (molasses) on the milking platform, and to assess the use of a priority yard, on pre-milking waiting times. A 7 day habituation period allowed cows to adjust to the installed molasses feeder units. Data were then collected over three 5 day periods (Pre-Molasses, Molasses and Post-Molasses). Cows accessed the milking unit either through the main yard or the priority yard; a smaller yard where cows that had extended milking intervals (MI>13 h in the night and >15 h in the day) were able to traffic directly to the front of the queue. Following milking, all cows had access to grain in feeding stations. Data were analysed using linear mixed modelling (REML). The short-term offer of molasses did not reduce voluntary waiting times, nor did voluntary waiting times worsen following the removal of molasses (P>0.05). Cows directed through the priority yard had 50% shorter pre-milking waiting times than cows trafficking through the main yard (P<0.001). Only 48% of cows licked the molasses on offer, however there was no significant difference in waiting times between treatments for these cows. The strong effect of the priority yard suggests that this could be a valuable design feature of an AMS dairy that would allow farmers to assist poor traffickers (e.g. cows that struggle to hold their place in a milking queue), and minimise waiting times of cows that are overdue for milking.

Transient effect of two milking permission levels on milking frequency in an AMS with grazing

C. Foley, J. Shortall and B. O'brien
Teagasc, Livestock Systems, Moorepark, Fermoy, Co. Cork, Ireland; cathriona.foley@teagasc.ie

In Ireland automated milking systems (AMS) have been integrated with grazing and spring calving herds. This system presents a number of challenges, in particular when cows within a herd reach peak milk yield simultaneously. As a result there is less free time in the AMS, potentially leading to increased waiting time pre-milking and therefore, reduced time grazing. However, farmers could alleviate this pressure by reducing milking permission (MP) which subsequently reduces milking frequency (MF). Cows are permitted to milk based on time since last milking and expected milk yield. The number of times the cow voluntarily visits the AMS and is permitted milking is defined as the milking frequency (MF). The objective of this study was to compare the transient effect of two levels of MP on cow MF. Sixty two cows were randomly assigned to one of two groups with a treatment of either low (LP) or high (HP) MP of 2 and 3 times per day, respectively. Groups were balanced for breed, lactation and days in milk. The experiment was divided into 16 time periods: (1) period -1 (10 days), all cows with same MP (2) period 0 (10 days), LP and HP treatments (3) periods 1 to 12 (12×1 weeks), LP and HP treatments, and (4) periods +1 to +2 (2×1 weeks) all cows with same MP. The statistical model used was a repeated measures ANOVA in SAS (PROC MIXED) and Tukey's post-hoc analysis. The fixed effects were period, breed, days in milk and MP. The main effects MP and period, and the interaction between MP and period were significant. Comparing MF between groups, there was no significant difference during periods -1, 0, +1 and +2 and there was a significant difference (P<0.05) during periods 1 to 12. Comparing MF over time and within each group, period -1 was not significantly different to period +2. In conclusion the results of this study would suggest that once MP is set there is a requirement of between 10 and 14 days for milking frequency to adjust regardless of MP level.

Equations to predict grass metabolisable energy in pasture-based systems

S. Stergiadis[1], M. Allen[2], X.J. Chen[1,3], D. Wills[1] and T. Yan[1]
[1]Agri-Food and Biosciences Institute, Large Park, Hillsborough, Co Down, BT26 6DR, United Kingdom,
[2]Agri-Food and Biosciences Institute, 18a Newforge Lane, Belfast, Co Antrim, BT9 5PX, United Kingdom,
[3]Lanzhou University, Lanzhou, 730020, China, P.R.; tianhai.yan@afbini.gov.uk

The aim of this study was to develop predictions of grass metabolisable energy (ME) contents from nutrient concentrations and digestibility, using non-pregnant dry cows fed solely fresh cut grass at maintenance energy level. During three consecutive grazing seasons, 33 cows were fed perennial ryegrass from swards varying in harvest dates, maturity stage, fertiliser input and grass variety. Grass chemical composition and dry matter (DM) and nutrient intakes/outputs were assessed daily over 50 weeks and 464 data were used, as a result of using a 3-day average of 1,392 data. Predictions were developed using residual maximum likelihood analysis in GenStat. Equations using gross energy (GE) or organic matter (OM) digestibilities, either as sole predictors or in combination with nutrient contents, improved prediction accuracy when compared to those using digestibilities of DM or digestible OM in DM. When GE digestibility (GEd) was used as primary predictor, grass nitrogen (N), GE, neutral-detergent fibre (NDF) and acid-detergent fibre (ADF) contents were also identified as significant predictors in the following equation: ME (MJ/kg DM)=1.047+17.46 GEd (MJ/MJ)-31.72 N (kg/kg DM)-0.145 GE (MJ/kg DM)+2.223 NDF (kg/kg DM)-2.735 ADF (kg/kg DM), R^2=0.90. When OM was used as primary predictor, using only N as additional predictor increased the explained variation and showed relatively low prediction error. According to an internal validation that was performed for all developed in the current study equations and those currently used in feeding systems, the new equations showed better accuracy and may be recommended for grazing animals. An under-prediction of ME for high-quality grass was observed in most cases but the range of this error was smaller in the equations developed in the current study.

How historical background influenced size and efficiency of farms in Central and Eastern Europe

P. Polák
National Agricultural and Food Center, Research Institute for Animal Production, Hlohovecká 2, 95141 Lužianky, Slovak Republic; polak@vuzv.sk

Efficiency of agriculture enterprises in any time of human history was affected by combining of many factors from which farm size, market possibilities, technological and social development are important. However the most important are people affected by their knowledge, skills, political system and religion. In Roman Republic basic structure of food production was private – citizen farm powered by work of owner and his family. Later in the Roman Emperor it was 'latifundium' – large farm owned by noble family operated by slaves. Increasing of production was done by increasing of area size and number of workers. Similar it was in medieval time. Important changes in production per farm brought industrial revolution. Establish of industrial centres create market for agricultural products and consequently provoked increasing of agricultural production by technological development. Period of constant war in 17th and 18th centuries led to decrease of industrial development in region on Central and Eastern Europe (CEE). Majority of CEE countries are middle size or rather small. Absence of bigger cities in that smaller countries and agricultural character of the region have had negative effect on agricultural market. However main changes which have had the most important impact in efficiency of agriculture in region were both World Wars and socialistic way of communal agricultural production established in 50's in last century. Communal farming brought negative attitude of people to work on land and reduce relationship to the private land. Privately owned area of agricultural land is rather small and for increasing is necessary to rent land. Small and family farms must behave like business companies and create profit to cover cost for renting. In marginal regions with high unemployment rate is visible tendency for 'microfarming' – keeping small number of animals for self supplying. However efficiency of farming is depended on number of supported production unites.

The vulnerability of goat production in the Mediterranean region

O.F. Godber and R. Wall

University of Bristol, Veterinary Parasitology and EcologyGroup, Bristol Life Sciences Building, University of Bristol, Bristol, BS8 1TQ, United Kingdom; olivia.godber@bristol.ac.uk

Goat production is an important contributor to the national economy, rural livelihood and food security in many Mediterranean countries, particularly within marginal habitats that are unsuitable for crop production. Here, the vulnerability of the goat sector in the Mediterranean region is modelled using vulnerability analysis, to predict the effects of changes in climate, human population and novel disease, using a range of indicators derived from FAOSTAT and World Bank statistics. This modelling approach is used widely to predict impacts on both food and economic security and the results can help inform policy decision-making and direct the targeting of interventions. This model shows that Southern Mediterranean nations are the most vulnerable whilst Greece is the most vulnerable nation of the European Union. The relatively higher adaptive capacity of France, Italy and Spain is shown to reduce their high exposure and overall vulnerability. Despite this model being unable to capture the impact of agricultural policy within the adaptive capacity element, it does serve as a useful starting point for the identification of areas where further research into the bio-economics of goat production might be focused.

Economic and technical structure of the water buffalo breeders: the case of Turkey

M. Gul[1], M.G. Akpinar[2], Y. Tascioglu[2], B. Karli[1] and Y. Bozkurt[3]

[1]Suleyman Demirel University, Agricultural Economics, Süleyman Demirel University, Faculty of Agriculture, Department of Agricultural Economics, Isparta, 32260, Turkey, [2]Akdeniz University, Agricultural Economics, Akdeniz University, Faculty of Agriculture, Department of Agricultural Economics, Antalya, 07058, Turkey, [3]Suleyman Demirel University, Animal Science, Süleyman Demirel University, Faculty of Agriculture, Department of Animal Science, Isparta, 32260, Turkey; mevlutgul@sdu.edu.tr

The world water buffalo husbandry is carried out mainly in Asia. In general, the numbers of water buffalos and products in the world have shown significant increases in the last 2-3 decades. However, the numbers of water buffalos and its products have shown significant declines in Turkey. The decrease in water buffalo numbers has led to the need for the development of specific public policies for water buffalo keeping in 2000s. As a consequence, new agricultural policies in water buffalo farming began to take up a shape in 2008. Water buffalo production is important in Turkey in terms of biodiversity. In this study, the technical and economic structures of water buffalo enterprises in Turkey were examined and current situation of farms and their problems were determined. The data used in this study were collected through questionnaires from the water buffalo farmers in Afyonkarahisar, Bitlis, Diyarbakır, Istanbul, Muş, Samsun and Tokat provinces of Turkey. These provinces account for more than half of Turkey's number of water buffalo and production of milk. Farms were divided into four size groups, ranging as the number of water buffalo <6, 6-15, 16-35 and >36 respectively. A total of 462 water buffalo farms were interviewed for the analysis. The economic and technical structure, production and marketing problems of the farms in the study area were analysed.

Self-sufficient pasture-based farms enhance economic performance and provision of ecosystem services

R. Ripoll-Bosch[1], E. Tello[2], T. Rodriguez-Ortega[2], M. Joy[2], I. Casasús[2] and A. Bernués[2]
[1]Wageningen University, P.O. Box 338, 6700 AH Wageningen, the Netherlands, [2]CITA, Av. Montañana 930, 50059 Zaragoza, Spain; raimon.ripollbosch@wur.nl

Sheep farming systems in the Euro-Mediterranean basin are usually located High Nature Value (HNV) areas and generally imply the use of local breeds. We aimed to unveil possible relationships between key farm technical/economic parameters and management practices, and their potential environmental implications. Firstly, we surveyed 30 mixed cereal-sheep farms rearing a local breed (Ojinegra, in North-West Spain). Data regarding farm structure, management, economic and social aspects were collected through direct interviews to farmers. Secondly, we performed a deliberative research (focus groups, n=5) to identify the perceptions of farmers and other citizens on the most important ecosystem services delivered by livestock production in HNV areas. A principal components and cluster analysis allowed to identify relationships among variables and classify farms into four homogeneous groups: intensive (high use of inputs); feed self-sufficient (great reliance on grazing); specialized (lamb meat as main or unique income); diversified (agriculture as predominant income). Feed self-sufficiency and reliance on natural resources (i.e. grazing) greatly determined the economic profitability of the farms, due to lower variable costs (i.e. feed inputs). The response of ewe's productivity to the intensification of production (i.e. higher variable costs per ewe) was minimal, which can be explained by limited breeding potential of the breed and/or, inefficient management. According to the focus groups, grazing management was perceived as a key agricultural practice, which in combination with other practices, could prevent forest wildfires, contribute to biodiversity conservation, and provide food products with inherent higher quality. We conclude that feed self-sufficiency in these particular conditions should be a target for future research due to its link to the farm economic performance and the provision of ecosystem services.

Drivers of the competing use of land: the case study of the pheripheral zone of the National Park W

C. Tamou, R. Ripoll Bosch and S.J. Oosting
Wageningen University, Animal Production System, Droevendaalsesteeg 4, 6708 PB Wageningen, the Netherlands; charles.tamou@wur.nl

The National Park of W is a nature reserve shared between Burkina Faso, Niger and Benin and where nature, pastoralism and crop production compete for the use of land and water. The present study aims at assessing the extent of the competition for land and water in the zone adjacent to the park and to analyze some important drivers. Data were collected from secondary literature and interviews with pastoralists, farmers and park authorities. It was found that the increase of cotton cultivation put pressure on crop lands and grazing lands for production of food crops and livestock grazing. This was facilitated by the introduction of animal-drawn cultivation and conversion of the banks of the Niger River from communal grazing lands with access to water to crop lands. Traditional grazing lands are also converted to crop lands to feed the growing population. As a result pastoralists moved illegally inside the park to sustain the feed of the animals, as well as some crop farmers. After a reinforcement of the park management laws by rangers, that forbidden grazing inside the park, grazing lands largely diminished and the total number of livestock of pastoralists in the region has declined. Since livestock is still important for livelihood of smallholders and for sustainable food crop production, there is a need for innovative strategies to integrate crop production, livestock production and conservation.

African Chicken Genetic Gains (ACGG) Program: an innovative way of delivering improved genetics

T. Dessie[1], J. Bruno[2], F. Sonaiya[3] and J. Van Arendonk[4]
[1]ILRI, Addis Abeba, EThiopia, P.O. Box 5689, Ethiopia, [2]ILRI, Addis Abeba, Ethiopia, P.O. Box 5689, Ethiopia, [3]Obafemi Awolowo University, fsonaiya1@yahoo.com, fsonaiya1@yahoo.com, Nigeria, [4]Wageningen University, Wageningen University, Wageningen University, the Netherlands; t.dessie@cgiar.org

The African Chicken Genetic Gains (ACGG) is a program for making available well-adapted low-input chickens for productivity growth of small holders in sub-Saharan Africa. The vision of this program is to catalyze public-private partnerships for increasing smallholder chicken production and productivity growth as a pathway out of poverty in sub-Saharan Africa. The aim of the program is to increase production and productivity and reduce poverty, especially for poor women. With the program we aim to empower smallholder farmers, especially women, by providing them access to more productive chicken breeds that they can optimally manage in their specific agro-ecology and production systems. Typically, very low-producing chicken genotypes dominate smallholder production systems in sub-Saharan Africa, largely due to the lack of effective long-term genetic improvement, multiplication, and delivery systems. Through the program we will make pre-vaccinated chicks of different breeds that are adapted to typical low-input systems available to farmers in poor rural communities. We will test different breeds on 2,700 households in the 3 project countries to determine the best breeds for the different production environments. In addition, activities are undertaken to improve breeds to better meet the needs. This proposal details R4D partnership between the International Livestock Research Institute (ILRI), National Agricultural Research Institutes in Ethiopia, Tanzania, and Nigeria, and international research organisations including Wageningen University aimed at developing public-private partnerships that will contribute to this overall vision of change in the chicken sector through increased farmer access to preferred, highly productive birds.

Cross-breeding in developing countries: extent, constraints and opportunities

G. Leroy, B. Scherf, I. Hoffmann, P. Boettcher, D. Pilling and R. Baumung
Food and Agriculture Organization of the United Nations, Animal Production and Health Division, Viale delle Terme di Caracalla, 00153 Rome, Italy; gregoire.leroy@fao.org

Well-planned cross-breeding programmes, including the creation of synthetic breeds, can be considered a potential means of helping to alleviate rural poverty in developing countries by improving the productivity of livestock, while maintaining sufficient robustness to cope with harsh environmental conditions. Nevertheless, the lack of adaptation of exotic breeds to local production environments and logistical and organizational constraints may affect the sustainability of such schemes and limit smallholder's interest in participating in them. Among the 128 country reports submitted in 2014 as part of the preparation of the second report on The State of the World's Animal Genetic Resources for Food and Agriculture, indiscriminate cross-breeding was the factor most frequently reported as a cause of genetic erosion (particularly among African countries). According to various census and studies, the extent of cross-breeding varies widely among species and countries, ranging from almost non-existent for sheep in numerous African countries, to widespread for dairy cattle in Brazil. The sustainable implementation and maintenance of cross-breeding schemes is dependent on a number of perquisites, including long-term access to improved germplasm, a moderate or controlled environment, availability of fodder and feeds, market access and technical, financial and veterinary support. Implementation of planned cross-breeding schemes can be a catalyst for innovation and development for smallholders. We investigate the challenges behind cross-breeding based on an international survey and literature review.

Goat breeding strategies of farmers in Nepal

B. Moser[1], R. Roschinsky[2], C. Manandhar[3], M. Malla[3] and M. Wurzinger[1,2]
[1]BOKU University of Natural Resources and Life Sciences Vienna, Department of Sustainable Agricultural Systems, Gregor Mendel Strasse 33, 1180 Vienna, Austria, Austria, [2]BOKU University of Natural Resources and Life Sciences Vienna, Centre for Development Research, Peter Jordan Strasse 82, 1190 Vienna, Austria, Austria, [3]Caritas Nepal, Dhobighat, Lalitpur, P.O. Box 9571, Kathmandu, Nepal; romana.roschinsky@boku.ac.at

Climate change affects Nepal leaving smallholders with diverse challenges concerning their small, mixed farming system. Goats are an integral part of Nepalese smallholder farms providing income and nutritional security. Crossbreeding of local and exotic breeds is one solution to minimize negative climatic effects as crossbreds might be better adapted to new climatic conditions. The aim of this study was to assess current breeding strategies and the impact of crossbreeding at farm level within the context of climate change. In 31 semi-structured interviews with farmers from Pokhara, Nepal, quantitative and qualitative data was collected in 2014. Performance parameters of local and crossbred animals were estimated. Interviews with livestock experts completed the database. An analysis was conducted to assess if technical training had an impact on breeding and husbandry practices. Participating farmers own diverse farms with various agricultural activities. Goats are mainly used for meat production and manure is used for fertilizing crops. Technical training has a positive influence on selection of breeding bucks, inbreeding prevention, castration practice, feeding and husbandry practices. Farmers consider similar traits in local and crossbred goats, but consider physical characteristics of exotic breeds in the crossbreds. Some farmers value crossbred goats as possibility for adaption to climate change. Benefits of crossbreeding include higher income. Few farmers report challenges resulting from crossbreeding such as higher workload and higher need for veterinary treatment. Respondents stated that they want to continue crossbreeding in the future, introduce improved, exotic breeds, milk production and increase herd sizes.

Meat research for market development for indigenous pork produced by smallholders in Son La, Vietnam

P.C. Muth[1], L.T.T. Huyen[2], A. Markemann[1] and A. Valle Zárate[1]
[1]University of Hohenheim, Animal Husbandry and Breeding in the Tropics and Subtropics, Garbenstr. 17, 70593 Stuttgart, Germany, [2]National Institute of Animal Sciences, Department of Economics and Livestock Farming Systems, Thuy Phuong, Bac Tu Liem, Hanoi, Viet Nam; p_muth@uni-hohenheim.de

Based on 11 years of research in Son La province, northwest Vietnam, a breeding and marketing program for local pig breeds has been transferred to an ethnic minority smallholder community. One goal of the program was to evaluate the potential of the indigenous Ban pig breed for its incorporation into a short food supply chain targeting urban niche markets. Initial market research on the gourmet restaurant and retail sector highlighted the need to define meat quality attributes that allow discrimination of Ban pork from commercial produce and to identify optimum slaughter weights for Ban finishers. In 2013/14, slaughter trials on 81 barrows (61 purebred Ban and 20 commercial crossbreds) raised by smallholders in Son La province were conducted. Mixed model analyses showed that at low slaughter weights of around 15 kg, as preferred by the urban market, the primal cut sizes of Ban were substantially smaller compared to those of commercial crossbreds slaughtered at the same age, allowing the sale of whole cuts. The marbling fat content for Ban loins was also considerably reduced and the fatty acid pattern of meat revealed marked differences between both product categories. With respect to slaughter weight, moderate increases up to around 30 kg would not dramatically change the proportion of fat and lean in Ban carcasses, while profitability for producers and retailers could be improved. In conclusion, meat research in the context of the short food value chain for Ban specialty pork identified opportunities to enhance vertical integration between stakeholders. Improved connectedness of ethnic minority smallholders to pork niche markets could contribute to income diversification and reduce the risk of dependence on crop production.

Diversity of pig farm types in the northern mountain of Vietnam and their sustainability

L.T.T. Huyen[1], N. Hostiou[2], S. Cournut[3], S. Messad[4] and G. Duteurtre[4]
[1]NIAS, Tu Liem, Hanoi, Viet Nam, [2]French National Institute for Agricultural Research, INRA, Paris, France, [3]VetAgro Sup UMR, Métafort, Lempdes, France, [4]CIRAD, Montpellier, Languedoc-Roussillon, France; nhostiou@clermont.inra.fr

Smallholder farming systems in Vietnam are diversified but little is known on this diversity. Due to a national increasing demand for pork, the changes in farms' structure occur clearly in pig production. However, the different livestock farming systems do not show the same ability to be sustainable in their context. The study aims at categorizing farming systems involving pig production, and investigating their sustainability. Data were collected from 160 pig farms in Mai Son district, Son La province (northwest of Vietnam) using stratified random sampling. Results of the multiple factor analysis showed four farm types characterized with economic, social and environmental indicators: small diversified farms with low levels of productivity; small farms with off farm activities and high productivity of land; specialized farms with large pig herd; and large mixed crop-livestock farms. The type 1 included mainly ethnic minorities located in the intermediate highland. They had low productivity of labour and relied mainly on cropping. Pig production was characterized by low level of economic and social sustainability. Farmers of type 2 with reduced land and crops specialized more on livestock production with medium pig herd and with off-farm jobs. They depended less on credit or location, and performed more on environmental issues than the two other types of larger farms. The type 3 seemed more sustainable to economic and social components than the other types but their environmental sustainability was questionable. The type 4 obtained high income but depended more on credit than the other types, and the environmental issue was also an opening question. An industrial pig enterprise is still new in the district. The results help to build foresight scenarios and to support future policy decisions for the development of pig production.

Trajectories of changes of pig farms in Vietnam – comparison between a northern and a southern dist

N. Hostiou[1], L.T.T. Huyen[2], S. Cournut[3] and G. Duteurtre[4]
[1]INRA, UMR Métafort, 63122 Theix, France, [2]NIAS, Tu Liem, Hanoi, Viet Nam, [3]VetAgroSup, Avenue de l'Europe, 63370 Lempdes, France, [4]Cirad, 16, Thuy Khue Street, Hanoi, Viet Nam; nhostiou@clermont.inra.fr

In Vietnam, due to the industrialization of pig value chain, pig family farms are becoming more and more integrated to markets. However, the dynamics of changes are not yet very analyzed in terms of farmers' practices and relation with industry. Our aim was to characterize the transformation of pig farms and to identify internal and external factors explaining those changes. Surveys were carried out in two districts of Vietnam with contrasted pig production changes: Mai Son district in the North of Vietnam and Thong Nhat district in the South. Twenty farmers in each district were selected for interviews to obtain a diversity of pig production types in terms of size, farm location, activities and ethnicity. Information was collected on farm's trajectory. The results show different technical model between the two districts. In Thong Nhat, farmers have adopted a homogenous semi industrial husbandry system since the beginning of their activity. It relies on a strong dependency to industrial companies for know-how, feed and veterinary products provision, improved housing and breeding. In Mai Son the technical model is more diversified in terms of breed (local or exotic), feed (local resources or industrial feed), housing and outlets. In both district we found different paths for pig farms according to the choice of diversification vs specialisation, evolution of workforce available (family or hired labour), size of the herd and relations to industries (contracts or market relations). Farms' trajectories intend to depend heavily on internal factors as the the availability of family labour. External factor is related to the development of the value chain. In the two location, policies had a lower influence according to the perception of farmers themselves. The results could be used to discuss foresights scenarios on the pig sector development with local and national stakeholders.

Goat feeding strategies of smallholders in Nepal in the context of climate change

C. Gerl[1], R. Roschinsky[2], C. Manandhar[3], M. Malla[3], M. Wurzinger[1,2] and W. Zollitsch[2]
[1]BOKU-University of Natural Resources and Life Sciences Vienna, Department of Sustainable Agricultural Systems, Gregor Mendel Strasse 33, 1180 Vienna, Austria, [2]BOKU-University of Natural Resources and Life Sciences Vienna, Centre for Development Research, Peter Jordan Strasse 82, 1190 Vienna, Austria, [3]Caritas Nepal, Dhobighat Lalitpur P.O. Box 9571, Kathmandu, Nepal; c.gerl@students.boku.ac.at

In smallholder farming systems in developing countries like Nepal, goats play an important role by ensuring household food security and often being the only asset for poor families. A current challenge for smallholders in Nepal is climate change, resulting in varying monsoon patterns and increasing periods of droughts. Goats are very tolerant to drought or irregular access to water and survive by browsing woody plants. Therefore they are a possible element of climate change mitigation strategies. The aim of this study was to document and evaluate the goat feeding system on Nepalese smallholder farms in this context. 31 smallholder farmers participating in SAF- BIN project activities in Nepal were interviewed using a semi-structured questionnaire. Individual feeding calendars were generated and fodder samples collected. Quantitative and qualitative data was analysed using SAS software. Results show that goats are important in the present mixed farming systems. A wide range of fodder plants is available and farmers possess a comprehensive traditional knowledge on these plants. This makes it possible for most farmers to bridge the dry season well. An adequate amount of fodder in a good quality is very important to all farmers interviewed. Most farmers recognized a change of the available fodder plants during the last years which may be an effect of climate change. Seasonal aridity is increasing and growing periods are changing. Some farmers introduced new fodder plants supported by training and supplies from NGOs. This leads to the conclusion that goat feeding systems are changing and that goats may play an important role in the adaption process to climate change in rural areas of Nepal.

Sustainable dairy intensification in Kenya: typologies of production systems and breeding practices

S.A. Migose[1,2], B.O. Bebe[1] and S.J. Oosting[2]
[1]Egerton University, Box 526, 20115 Egerton, Kenya, [2]Wageningen University, Box 338, 6700 AH Wageningen, the Netherlands; sally.migose@wur.nl

Increased production of smallholder dairy systems is needed to meet increasing demand for animal source protein. Adoption of innovations depends on farming systems characteristics. We defined three typologies based on interviews and discussions with stakeholders (farmers, input suppliers, service providers and dairy experts). The peri-urban system (PUS) is located in peri-urban regions and is a high input high output-system, using 5.0 ± 2.5 kg of concentrates/herd/day (/h/d) to produce 26.5 ± 23.4 kg of milk/h/d. The rural commercial system (RCS) is located in rural regions and is a medium input medium output-system, using 2.9 ± 1.4 kg of concentrates/h/d to produce 20.2 ± 21.0 kg of milk/h/d. The rural subsistence system (RSS) is located in rural regions and is a low input low output-system, using 2.4 ± 2.9 kg of concentrates/h/d to produce 15.5 ± 7.7 kg of milk/h/d. Production strategies and marketing differed among system: PUS-farmers kept 4.9 ± 1.6 tropical livestock units (TLU) of exotic cattle, used artificial insemination (AI) with local and imported semen, fed on Napier grass and stover and sold 19.4 ± 16.6 kg of milk/h/d to consumers at a price of 0.45 ± 0.04 \$/kg for cash income; RCS-farmers kept 7.1 ± 4.5 TLU of exotic-zebu crosses upgraded using AI with local semen, grazed their cows on open fields and sold 14.9 ± 13.4 kg of milk/h/d to middlemen at a price of 0.32 ± 0.08 \$/kg for cash to support crop production; RSS-farmers kept 7.0 ± 8.3 TLU of zebu crosses upgraded using unproven exotic crossbred bull mating, grazed their cows on paddocks, sold 9.9 ± 6.1 kg of milk/h/d to middlemen at 0.31 ± 0.04 \$/kg and dairy played subsistence and cultural roles. Breeding practices were suited to production strategies determined by resources, household priorities and market situation. Dairy productivity increase through AI with semen of superior breeds may not be suitable in rural systems with poor-functioning markets.

Meat quality of Thai indigenous crossbred chickens kept under free-range raising system

W. Molee, S. Khempaka and A. Molee
Suranaree University of Technology, School of Animal Production Technology, 111 University Avenue, Sub District Suranaree, Muang District, Nakhon Ratchasima 30000, Thailand; wittawat@sut.ac.th

The 'Korat Meat Chickens' represents one of indigenous crossbred chickens developed by the cooperation between Suranaree University of Technology (SUT), Thailand Research Fund (TRF), and Department of Livestock Development (DLD). The aim of this study was to determine the effect of free-range raising system on meat quality of 'Korat Meat Chickens'. A total of 250 birds (4 weeks of age), were randomly allocated into 2 treatments; indoor and free-range treatments. In indoor treatment, the chickens were raised in an indoor pen (8 birds/m^2). In the free-range treatment, the chickens were raised in a similar indoor pen; in addition, they also had a free-range grass area (1 bird/m^2). Each treatment was represented by 5 pens containing 25 birds each. All birds were provided with the same diets and were raised until slaughter at 9 weeks of age. The results showed that breast meat from free-range treatment had lower cooking loss than that of indoor treatment (P<0.05). Thigh meat from free-range treatment had lower drip loss and higher shear force value than that of indoor treatment (P<0.05). Breast skin color from free-range treatment had more red (a*) and yellow (b*) than that of indoor treatment (P<0.05). Thigh skin color from free-range treatment had more lightness (L*) than that of indoor treatment (P<0.05). The proportion of SFA, MUFA, PUFA, total n-3, total n-6 and n-6 to n-3 ratio in breast meat were not different among treatments (P>0.05). However, the proportion of PUFA and total n-6 in thigh meat were higher in free-range treatment than in indoor treatment (P<0.05). There was no difference between treatments in cholesterol content in breast and thigh meat (P>0.05).

Factors affecting the sustainability of the production system of the Greek buffalo (*Bubalus bubalis*)

C. Ligda[1], D. Chatziplis[2], E. Komninou[3], S. Aggelopoulos[2], V.A. Bampidis[2] and A. Georgoudis[4]
[1]Veterinary Research Institute, P.O. Box 60272, 57001 Thessaloniki, Greece, [2]Alexander Technological Educational Institute, P.O. Box 141, 57400 Thessaloniki, Greece, [3]Ministry of Rural Development and Food, N. Mesimvria, 57011 Thessaloniki, Greece, [4]Aristotle University of Thessaloniki, Thessaloniki, 54124 Thessaloniki, Greece; chligda@otenet.gr

The Greek buffalo (*Bubalus bubalis*) is a local autochthonous breed and the study of its production system is of particular interest for the high-quality traditional foods produced and the protection and management of its natural living environment. The zootechnical, economic, social and environmental factors affecting the viability of the production system was studied using a questionnaire and interviews of 27 buffalo breeders. The data collected concern the general description of the farm, the livestock kept, the management system (nutrition, selection and breeding), the social position of the breeders, the sales and marketing of the products and the financial results of the holdings. The productivity is sufficiently high to ensure the sustainability of the system, but the limitation of available pasturelands in the surrounding wetlands requires the frequent movement of the animals and the use of supplementary feed that increase the production costs and decrease the income, with possible negative consequences on the future of buffalo farming business. A preservation plan and a scheme for sustainable conservation of the Greek buffalo will be developed, including proposals for a population management plan adapted to the existing conditions for buffalo farming, with the aim to ensure the sustainability of the system. This research has been co-financed by the European Union (European Social Fund – ESF) and Greek national funds through the Operational Program 'Education and Lifelong Learning' of the National Strategic Reference Framework (NSRF) – Research Funding Program: ARCHIMEDES III. Investing in knowledge society through the European Social Fund.

Effects of the zootechnical parameters on the financial result: the case of Greek buffalo farms

S. Aggelopoulos[1], D. Chatziplis[1], C. Bogas[2], A. Georgoudis[3] and V.A. Bampidis[1]
[1]Alexander Technological Educational Institute, P.O. Box 141, 57400 Thessaloniki, Greece, [2]Technological Educational Institute of Central Macedonia, Serres, 62124 Serres, Greece, [3]Aristotle University of Thessaloniki, Thessaloniki, 54124 Thessaloniki, Greece; chatz@ap.teithe.gr

Greek buffalo (*Bubalus bubalis*) farming is a traditional livestock sector in Greece. The aim of this study was to determine policy and financial measures of Greek buffalo farms, to improve productivity and their competitiveness, after developing a typology, based mainly in the zootechnical behavior and in specific economic parameters. Zootechnical and economic survey data were gathered through appropriate structured questionnaire, filled with personal interviews of 27 Greek buffalo farmers. In order to develop the typology of the farms, Hierarchical Cluster Analysis was used. The formulation of the clusters was made using Ward's criterion, while the squared Euclidean distance was used to measure the (dis)similarity between the farms. The farms that achieved better zootechnical management had the best prices in the financial parameters that were selected. The improvement of zootechnical parameters of the holdings was related to the improvement of reproduction parameters of the farms. In Greece, as a measure policy, the preparation and the financing of an efficient performance recording program is suggested, not only for the improvement of the existing genetic material, but also for its proper management, which can lead to the decrease of the inbreeding coefficient and, consequently, to the improvement of the reproduction parameters. The vertical integration of the commercial system of the produced products is also suggested. This research has been co-financed by the European Union (European Social Fund – ESF) and Greek national funds through the Operational Program 'Education and Lifelong Learning' of the National Strategic Reference Framework (NSRF) – Research Funding Program: ARCHIMEDES III. Investing in knowledge society through the European Social Fund.

Mastitis incidences in Greek buffalo farms

E. Palla, K. Mazaraki, A. Founta, I. Mitsopoulos, V. Lagka and V.A. Bampidis
Alexander Technological Educational Institute, P.O. Box 141, 57400 Thessaloniki, Greece; gmitsop@ap.teithe.gr

Milk samples from 58 buffalo cows (28 in April and 30 in May 2014) of the Greek buffalo breed (*Bubalus bubalis*), from 6 farms located in proximity to Lake Kerkini, Central Macedonia, Greece, were microbiologically analysed and examined for mastitis. 14 and 12 milk samples, respectively, were positive in California Mastitis Test (CMT), while Total Mesophilic Flora (TMF) ranged from 43 to 100×0^3 colony forming units (cfu)/ml and from 75 to 100×0^3 cfu/ml, respectively. All milk samples positive to CMT were Gram stained and cultivated in Nutrient agar, MacConkey agar, Mannitol Salt agar and Blood agar, while biochemical tests (Coagulase test, IMVIC test, etc.) for the identification of the microorganisms isolated were performed. In April, *Escherichia coli* (4 samples, 28.6%), *Klebsiella* spp. (2 samples, 14.3%), *Staphylococcus aureus* (5 samples, 35.7%), *Staphylococcus epidermidis* (2 samples, 14.3%) and fungi (1 sample, 7.1%) were isolated. In May, *E. coli* (4 samples, 33.3%), *Klebsiella* spp. (2 samples, 16.6%), *S. aureus* (4 samples, 33.3%) and *S. epidermidis* (2 samples, 16.6%), and no fungi, were isolated. In the isolated microorganisms, an antibiotic sensitivity test was performed using the Kirby-Bauer method, against tetracycline, oxytetracycline and streptomycin. *E. coli* and *S. aureus* were resistant to tetracycline and streptomycin and sensitive to oxytetracycline, while *Klebsiella* spp. and *S. epidermidis* were sensitive against all examined antibiotics. The results of this study indicated that pathogenic microorganisms were isolated in the mammary gland of a relative small number of buffalo cows, while mastitis was subclinical and could be controlled with antibiotics. This research has been co-financed by the European Union (European Social Fund – ESF) and Greek national funds through the Operational Program 'Education and Lifelong Learning' of the National Strategic Reference Framework (NSRF) – Research Funding Program: ARCHIMEDES III. Investing in knowledge society through the European Social Fund.

Effect of forage type on milk production and quality of Greek buffalo (*Bubalus bubalis*)

I. Mitsopoulos[1], V. Christodoulou[2], B. Skapetas[1], E. Nistor[3], K. Vasileiadis[1], D. Nitas[1] and V.A. Bampidis[1]
[1]Alexander Technological Educational Institute, P.O. Box 141, 57400 Thessaloniki, Greece, [2]Hellenic Agricultural Organization – Demeter, Paralimni, 58100 Giannitsa, Greece, [3]Banat's University of Agricultural Sciences and Veterinary Medicine 'King Michael I of Romania', Timişoara, 300645 Timişoara, Romania; gmitsop@ap.teithe.gr

In an experiment with twenty four lactating Greek buffalo (*Bubalus bubalis*) cows, effects of total replacement of alfalfa hay (AH) with corn silage (CS) on productivity and milk composition were determined. In the experiment, which started on week 10 postpartum, buffalo cows were allocated, after equal distribution relative to milk yield and lactation number (i.e. 2 or 3), into 2 treatments being AH and CS of 12 buffalo cows each. For a period of 8 weeks (i.e. weeks 10-18 postpartum), buffalo cows were individually fed a concentrate mixture (7.3 kg dry matter – DM/buffalo cow/day), wheat straw (1.8 kg DM/buffalo cow/day) and alfalfa hay (treatment AH; 5.4 kg DM/buffalo cow/day) or corn silage (treatment CS; 5.6 kg DM/buffalo cow/day). In the 8-week experimental period, there were no differences (P>0.05) between AH and CS treatments in milk fat (81.0 g/kg), protein (46.1 g/kg), lactose (51.2 g/kg) or ash (8.1 g/kg) contents, as well as in somatic cell counts (SSC; 108.8 ×1000 SSC/ml) and colony forming units (cfu; 53.6 ×1000 cfu/ml) of milk. Moreover, average milk yield (5.8 kg/day), and fat (0.47 kg/day), protein (0.26 kg/day), lactose (0.29 kg/day) and ash (0.047 kg/day) yields, was not affected (P>0.05). Thus, total replacement of alfalfa hay with corn silage did not affect the productive performance of lactating Greek buffalo cows. This research has been co-financed by the European Union (European Social Fund – ESF) and Greek national funds through the Operational Program 'Education and Lifelong Learning' of the National Strategic Reference Framework (NSRF) – Research Funding Program: ARCHIMEDES III. Investing in knowledge society through the European Social Fund.

An assessment of factors affecting water buffalo breeding decisions of the enterprises in Turkey

M.G. Akpinar[1], M. Gul[2], Y. Tascioglu[1], B. Karli[2] and Y. Bozkurt[3]
[1]Akdeniz University, Agricultural Economics, Akdeniz University, Faculty of Agriculture, Department of Agricultural Economics, Antalya, 07058, Turkey, [2]Suleyman Demirel University, Agricultural Economics, Süleyman Demirel University, Faculty of Agriculture, Department of Agricultural Economics, Isparta, 32260, Turkey, [3]Suleyman Demirel University, Animal Science, Süleyman Demirel University, Faculty of Agriculture, Department of Animal Science, Isparta, 32260, Turkey; mgoksel@akdeniz.edu.tr

According to Turkish Statistical Institute (TUIK) data, Turkey had a 1.04 million water buffaloes in 1980. The water buffalo's inventory dropped to 107,435 head by 2013. The average carcass weight of water buffalo has increased from 112.3 kg in 1980 to 152.3 in 2013, an average of growth of nearly 0.93% annually. It is commonly accepted that per- water buffalo milk yields in Turkey have also increased the 1980s, an average of growth of nearly 0.7% annually. Despite the increase in carcass and milk yields, the productivity increases have not been sufficient to prevent reductions in output potential caused by declining water buffalo inventories. The prices of meat have been increased due to the failure in red meat production to satisfy the demand. Recently, subsidies in livestock policies of the current government have been introduced to encourage the new entrepreneurs to invest and improve animal production sector especially beef, water buffalo, sheep and goat in the country. The goal of this research was to assess the factors affecting the water buffalo enterprises in Turkey. For this purpose, a survey was conducted with 462 farmers by face to face interviews as a sample of rural areas in Afyonkarahisar, Bitlis, Diyarbakır, Istanbul, Muş, Samsun and Tokat provinces of Turkey. The factor analysis was used for the evaluation of the data. In this study, it was focused on identifying the attitudes towards water buffalo husbandry and the factors that affect these attitudes.

Characterization of farming systems in land reform farms of the Waterberg District Municipality, RSA

A.J. Netshipale
Wageningen University, Animal Production Systems, P.O. Box 338, 6700 AH Wageningen, the Netherlands;
avhafunani13@gmail.com

Land reform in South Africa brought changes to farms in the aspects of resource endowment, production orientation and farm operation styles. This study characterized the farming systems in land reform farms on the basis of activities being practiced, production orientation, scale of production and farm operation style. Fifty-one farms that represented the diversity in land reform programmes, and in production orientations and operation styles were selected. Ten percent of the active households per farm in the sampled farms 87 were selected for interviews by means of a semi-structured questionnaire. Focus group discussions were also held to acquire information regarding the farms as enterprises. Using Categorical Principal Component Analysis (CATPCA) and Two-Step Cluster analysis, eleven distinctive farming clusters were identified. The clusters distinguished between livestock, crop, mixed operations at high, medium, low input scale, and operation style entity, individual and a combination of these. From the cluster analysis results, it was clear that land reform has broadened the farming systems. Relevant stakeholders should align their support with farm needs in accordance with the apparent farming systems and their viability for business/livelihoods sustainability for improvement purposes.

Meta-analysis of GWAS of bovine stature with >50,000 animals imputed to whole-genome sequence

A.C. Bouwman[1], H. Pausch[2], A. Govignon-Gion[3], C. Hoze[3], M.P. Sanchez[3], M. Boussaha[3], D. Boichard[3], G. Sahana[4], R.F. Brøndum[4], B. Guldbrandtsen[4], M.S. Lund[4], J. Vilkki[5], M. Sargolzaei[6], F.S. Schenkel[6], J.F. Taylor[7], J.L. Hoff[7], R.D. Schnabel[7], R.F. Veerkamp[1], M.E. Goddard[8], B.J. Hayes and The 1000 Bull Genomes Consortium[8]
[1]Animal Breeding & Genomics Centre, WUR, Wageningen, the Netherlands, [2]Lehrstuhl fuer Tierzucht, Technische Universitaet Muenchen, Freising, Germany, [3]INRA, Génétique Animale et Biologie Intégrative, Jouy-en-Josas, France, [4]Center for Quantitative Genetics and Genomics, Aarhus University, Tjele, Denmark, [5]MTT, Biotechnology and Food Research, Jokioinen, Finland, [6]Centre for Genetic Improvement of Livestock, University of Guelph, Ontario, Canada, [7]Division of Animal Sciences, University of Missouri, Columbia, MO, USA, [8]Biosciences Research Division, DEPI, Bundoora, Australia; aniek.bouwman@wur.nl

Extensive meta analysis of GWAS in humans has identified 697 significant SNP, however these SNP explain only 20% the total genetic variation. In order to compare the genetic architecture of stature in humans to stature in cattle, we performed a large meta-analysis using imputed sequence data. The 1000 Bull Genomes project provided a multi-breed reference population of 1,147 sequenced animals to impute SNP-chip genotypes up to whole genome sequence for 15 populations. The populations from Australia, Canada, Denmark, Finland, France, Germany, the Netherlands, and the USA represented the Angus, Fleckvieh, Holstein, Jersey, Montbeliarde, Normande, and Nordic Red Dairy Cattle breeds. Genome-wide association studies were performed on stature phenotypes for each of the populations. Individual GWAS studies revealed many QTL regions and several regions harboured good candidate genes, e.g. PLAG1, IGF2. Results from these GWAS studies were combined in a meta-analysis to increase the power for QTL detection and to refine QTL regions exploiting the different patterns of LD among the breeds. Results of this meta-analysis will be validated in an independent population to determine how much of the variation in stature can be explained by the significant SNP.

Multi-breed GWAS and meta-analysis using sequences of five dairy cattle breeds improve QTL mapping
I. Van Den Berg[1,2,3], D. Boichard[2] and M.S. Lund[3]
[1]*AgroParisTech, 16 rue Claude Bernard, 75231 Paris, France,* [2]*INRA, UMR1313 GABI, Domaine de Vilvert, 78350 Jouy-en-Josas, France,* [3]*Aarhus University, Center for Quantitative Genetics and Genomics, Department of Molecular Biology and Genetics, Blichers Allé 20, 8830 Tjele, Denmark; irene.vandenberg@mbg.au.dk*

GWAS in dairy cattle often result in a large number of associated variants due to the high levels of linkage disequilibrium (LD) within breed. Furthermore, with sequence based GWAS, high detection thresholds are necessary to avoid excessive numbers of false positives. The sample size in smaller breeds might not be sufficient to overcome such thresholds to detect QTL. A multi-breed GWAS could improve these issues. With less LD shared across breeds, fewer variants are expected to be associated with the same causal variant. Additionally, jointly analyzing multiple breeds increases the sample size and thereby detection power, aiding the detection of smaller QTL that are missed within breed. Improved GWAS results can subsequently benefit both in the detection of causative mutations and the selection of variants used for genomic prediction. It is, however, not always possible to access all data necessary for a full multi-breed GWAS. In human genetics, it is common practice to combine results of several GWAS studies by meta-analysis. Combining within breed GWAS by a meta-analysis would make a multi-breed analysis possible when the data for a full joint analysis is not available. The objective of our study was to compare a multi-breed GWAS with within breed GWAS and three meta-analysis models for the fine mapping of variants associated with milk, fat and protein yield, using imputed sequences of 4,993 Danish Holstein, 984 Jersey, 768 Danish Red, 5,626 French Holstein, 1,935 Montbéliarde and 1,725 Normande bulls. Our findings show that multi-breed GWAS is more precise that single-breed GWAS and that results obtained by the weighted Z-score meta-analysis were the most similar to the multi-breed GWAS.

Association studies for growth index in three dairy cattle breeds using whole-genome sequence data
X. Mao[1,2], G. Sahana[2], D.J. De Koning[1] and B. Guldbrandtsen[2]
[1]*Swedish University of Agricultural Sciences, Department of Animal Breeding and Genetics, Ulls väg 26, 75007 Uppsala, Sweden,* [2]*Aarhus University, Department of Molecular Biology and Genetics, Blichers Allé 20, 8830 Tjele, Denmark; xiaowei.mao@mbg.au.dk*

Beef production from dairy cows is important for overall farm profitability, and faster growth is important for reproduction of young heifers. However, few gene mapping studies have been performed for growth traits for dairy breeds. The objectives of this study were to perform association analyses for growth traits in Holstein, Jersey and Nordic Red cattle (RDC) using whole-genome sequence data, and to increase power of QTL detection by applying meta-analysis on these dairy breeds. Linear mixed model with a single marker approach was applied on growth index in Holstein (5,519), Jersey (1,231), and RDC (4,411) separately. The meta-analysis was carried out on single-breed results. Within-breed association analyses identified six genome-wide significant QTL in Holstein located on *Bos taurus* autosome (BTA) 7, 10, 21, 23, 28, and 29 and three for RDC on BTAs 6, 10 and 14. No association in Jersey reached the genome-wide significant threshold. Meta-analysis identified seven genome-wide significant QTL on BTAs 6, 10, 14, 23, 28, and 29. Significant variants at BTA10 were common between Holstein and RDC, and meta-analysis increased the significance of these variants. Genes and variants within the associated regions were prioritized, and their biological relevance to the traits was interpreted through bioinformatics analyses. Genes AP4E1 (BTA10: 59,578,464-59,644,086), C-MOS (BTA14: 24,975,950-24,976,948), PLAG1 (BTA14: 25,007,291-25,009,296), and CHCHD7 (BTA14: 25,052,885-25,058,779) were considered as candidate genes. Two missense mutations, variants at BTA10:59,615,446 and at BTA14:24,976,045, were proposed as candidate causal mutations. This study has identified major QTLs underlying growth index in dairy cattle and a meta-analysis was useful to increase the significance through combining information across breeds.

Screening for selection signatures in Norwegian Red cattle

B. Hillestad[1], J.A. Woolliams[1,2], S.A. Boison[3], D.I. Våge[1], T. Meuwissen[1] and G. Klemetsdal[1]
[1]Norwegian University of Life Science, Department of Animal and Aquacultural Sciences (IHA), P.O. Box 5003, 1432 Ås, Norway, [2]University of Edinburgh, The Roslin Institute and Royal (Dick) School of Veterinary Studies, Easter Bush, EH25 9RG, United Kingdom, [3]University of Natural Resources and Life Sciences Vienna, Department of Sustainable Agricultural Systems, Greogor Mendel Str. 33, 1180 Vienna, Austria; borghild@salmobreed.no

The objective of this study was to search for selection signatures using inbreeding measurements from runs of homozygosity (ROH) in Norwegian Red cattle. The data consisted of 381 Norwegian Red bulls born between 1971 and 2004, genotyped with the Illumina HD-panel gaining 708k SNP-markers after genotyping quality controls. Positional F-values for each SNP were made from ROH and used to define SNP autozygosity, for which changes over time were estimated by logistic regression; being most expressed on BTA 6, 24, 5 and 12, in descending order. Further evidence came from plotting individual ROH by chromosome, demonstrating a convincing sweep on BTA 6; precisely at the spot where QTLs are known to segregate in dairy cattle, affecting milk and mastitis. These results support that the chromosomal regions have been under a high selection pressure. Thus, access to dense markers for animals born over a longer time span can be used to search for selection signatures. This method is independent of phenotypic data, and can therefore potentially reveal any trait exposed to directional selection.

Use of Wright's fixation index (Fst) to detect potential selective sweeps in Nelore Cattle

D.F. Cardoso[1], G.C. Venturini[1], M.E.Z. Mercadante[2], J.N.S.G. Cyrillo[2], M. Erbe[3], C. Reimer[3], H. Simianer[3], L.G. Albuquerque[1] and H. Tonhati[1]
[1]São Paulo State University, Jaboticabal, SP, Brazil, [2]Instituto de Zootecnia, Centro APTA Bovinos de Corte, Sertãozinho, SP, Brazil, [3]Animal Breeding and Genetics Group, Georg-August-University, Göttingen, Germany; dcardos@gwdg.de

The aim of this study was to detect differentiation in genomic regions among three experimental lines of Nelore cattle, which has been kept under selection since 1980 in the Instituto de Zootecnia, Sertãozinho, SP, Brazil. Control Line (NeC, n=60) is a closed line, where stabilizing selection to yearling weight is performed. Selection Line (NeS, n=120) is another closed line selected for high yearling weight. Traditional Line (NeT, n=170) is selected for highest yearling weight as well, but occasionally received sires from NeC, NeS or from commercial herds. 89, (189, 485) animals from NeC (NeS, NeT), were genotyped using the Illumina BovineHD Genotyping BeadChip (529k autosomal SNPs after quality control). Differentiation among lines, was estimated with Wright's fixation index (F_{st}) per SNP in each pairwise combination of lines, and averaged values in windows of 100 kb (75 kb overlap) were considered. The means of these values in NeC vs NeS, NeC vs NeT and NeS vs NeT were 0.03, 0.02, and 0.01, respectively. As expected, highest differentiation was observed between closed lines. F_{st} values different from zero observed in NeS vs NeT were probably caused by neutral processes, since the same selection criterion has been applied to both lines. Thus, the maximum value (0.096) observed in NeS vs NeT was used as empirical threshold to define F_{st} values caused by selection in NeC vs NeS and NeC vs NeT. We identified 82 genomic regions above this threshold, shared in NeC vs NeS and NeC vs NeT. The chromosomes with highest percentage of size considered under selection were 6, 9, 10 and 24. The largest candidate selective sweep (425 kb) was detected on chromosome 18, containing 21 genes, including AKT1S1, which is a potential regulator of insulin sensitivity.

How selection shaped the Italian Large White pig genome: a tale of short term artificial evolution

G. Schiavo[1], G. Galimberti[2], D.G. Calò[2], A.B. Samorè[1], F. Bertolini[1], V. Russo[1], M. Gallo[3], L. Buttazzoni[4] and L. Fontanesi[1]
[1]Bologna University, Department of Agricultural and Food Sciences, Division of Animal Sciences, viale Fanin 46, 40127, Bologna, Italy, [2]Bologna University, Department of Statistical Sciences 'Paolo Fortunati', Via Belle Arti, 40126, Bologna, Italy, [3]Associazione Nazionale Allevatori Suini, Via L. Spallanzani 4, 00161, Roma, Italy, [4]Centro di Ricerca per la Produzione delle Carni e il Miglioramento Genetico, Monterotondo, 00015, Roma, Italy; luca.fontanesi@unibo.it

In this work we investigated how the genome of Italian Large White pig breed has been shaped over the last two decades of artificial directional selection based on boar genetic evaluation obtained with a classical BLUP animal model. Boars with estimated breeding value (EBV) reliability >0.85, thus the most influencing boars of this breed (n. 192), born from 1992 (the beginning of the selection program for this breed) to 2012, were genotyped with the Illumina Porcine SNP60 BeadChip. Boars were grouped in eight classes according to their year of birth; then filtered single nucleotide polymorphisms (SNPs) were used to evaluate the influence of time on allele frequency changes using multinomial logistic regression models. A total of 493 SNPs had a Bonferroni corrected P-value<0.10. The largest proportion of these SNPs was on chromosomes SSC7, SSC2, SSC8 and SSC18. These SNPs belonged to 204 haploblocks. Functional annotations of genomic regions including the 493 SNPs reported few Gene Ontology terms that might indicate the biological processes that contributed to increased performances of the pigs over the 20 years. These results show that directional selection in pigs using methodologies assuming the infinitesimal model has led to allele frequency changes and has shaped the genome of the Italian Large White pigs.

The use of imputed haplotypes to identify lethal recessive effects in pigs

D.M. Howard[1], R. Pong-Wong[1], P.W. Knap[2], N. Deeb[3] and J.A. Woolliams[1]
[1]The Roslin Institute & R(D)SVS, The University of Edinburgh, Midlothian, United Kingdom, [2]Genus PIC, Ratsteich 31, 24837 Schleswig, Germany, [3]Genus PLC, 100 Bluegrass Commons Blvd, Suite 2200, Hendersonville, TN 37075, USA; david.howard@roslin.ed.ac.uk

The moderate effective population sizes of livestock can lead to lethal recessive alleles drifting more rapidly to higher frequencies, impacting upon reproductive success. The identification of lethal recessive alleles using a haplotype approach has yet to be reported in pigs. This study sought to detect putative lethal recessive haplotypes, which negatively impacted on the total number born (TNB) or the number born alive as a proportion of the TNB (NBA/TNB). This study also compared the efficacy of using imputed data with non-imputed data and also to a SNP-by-SNP approach. A large commercial line was analysed, consisting of 5,660 individuals genotyped using the PorcineSNP60 BeadChip, with a further 12,534 animals genotyped for 402 SNPs. The population was phased and imputed and following quality control, resulted in a total of 23,327 individuals, haplotyped for 47,704 SNPs. Candidate haplotypes were identified based on their frequency in the population or the lack of homozygotic offspring produced from matings between carriers and then examined for their effect on TNB or NBA/TNB, to identify putative lethal recessive haplotypes. Evidence of putative lethal recessive haplotypes with an effect on the TNB were found within 7 regions. These regions relate to at least 5 different putative lethals, each located on a different chromosome. SSC1, SSC6 and SSC14 contained strong evidence of an effect whilst SSC10 and SSC15 provided tentative evidence of an effect. Imputed haplotypes were found to be superior for detecting effects compared to non-imputed data alone and a SNP-by-SNP approach. No putative lethal recessive haplotypes were detected with an effect on the NBA/TNB. The study developed a robust statistical framework for allowing animal breeders to optimise reproductive performance in a timely and cost-effective manner.

Analysis of copy number variations in 32 Polish Holstein-Friesian cow genomes based on NGS data

M. Mielczarek[1], M. Frąszczak[1], R. Giannico[2], G. Minozzi[2,3], E.L. Nicolazzi[2], K. Wojdak-Maksymiec[4] and J. Szyda[1]
[1]*Wroclaw University of Environmental and Life Sciences, Department of Genetics, Kozuchowska 7, 51-631 Wrocław, Poland,* [2]*Parco Tecnologico Padano, Via Einstein Albert, 26900 Lodi, Italy,* [3]*University of Milan, Via Celoria 10, 20133 Milan, Italy,* [4]*West Pomeranian University of Technology, Aleja Piastow 17, 70-310 Szczecin, Poland; magda.mielczarek@up.wroc.pl*

Copy number variations (CNVs) are an important source of genetic diversity and they are defined as gains and losses of long DNA fragments from 50 bp to several Mbp. Recent advances in next-generation sequencing (NGS) provide a new approach to identify the high number not only common, but also rare, not known CNVs. Whole genome DNA sequences were determined for 32 cows representing Polish Holstein-Friesian breed. The total number of raw reads generated for a single animal varied between 164,984,147 and 472,265,620 and the average coverage along the genome was 14.03. Alignment to the reference genome was carried out using BWA-MEM, and CNV calling was done by CNVnator software. Minimum and maximum number of duplications per individual was 2,310 and 5,929, while minimum and maximum number of deletions per cow was equal to 14,122 and 24,156, respectively. The total number of duplications shared between all cows was 6,434. The fewest of them (50) were located on BTA20, and the most (536) on BTA14. The total number of shared deletions was 49,690. The fewest (904) were located on the BTA25, and the most (2,934) on the BTA1. The distribution of the number of duplications and deletions per chromosome differed between animals. Moreover, the correlation between the number of duplications and the number of deletions was highly negative. The lengths of CNVs varied as well. The shortest deletion was 200 bp and the longest 724,000 bp, while the shortest duplication was 200 bp and the longest 439,300 bp. In conclusion, individual genomes vary widely in terms of the number, distribution and length of CNVs, even within the same breed.

Predicting the accuracy of multi-population genomic prediction

Y.C.J. Wientjes[1,2], P. Bijma[1], R.F. Veerkamp[1,2] and M.P.L. Calus[2]
[1]*Wageningen University, Animal Breeding and Genomics Centre, P.O. Box 338, 6700 AH Wageningen, the Netherlands,* [2]*Wageningen UR Livestock Research, Animal Breeding and Genomics Centre, P.O. Box 338, 6700 AH Wageningen, the Netherlands; yvonne.wientjes@wur.nl*

For designing breeding programs, knowing the accuracy of genomic prediction of different designs is essential for maximising response to selection. Accuracy of genomic prediction when using a single reference population can be predicted by the deterministic equation derived by Daetwyler et al. Our aim was to derive a deterministic equation to predict the accuracy when multiple populations are used for genomic prediction. The equation was derived based on a selection index approach for a scenario with population A and B in the reference population (RP), and using population C as selection candidates (SC). It was assumed that n_g independent quantitative trait loci (QTL) were affecting the trait, that were the same across populations, and each QTL explained the same part of the genetic variance. The derived equation contains population parameters, such as n_g, heritability and size of the populations in RP, and genetic correlations between populations. Due to linkage disequilibrium, however, loci are not independent. Therefore, n_g was replaced by the number of effective chromosome segments, as was described for the equation for single population scenarios. The derived equation can also be applied *in situ*ations with SC from population A or B included in RP and *in situ*ations with more than two populations in RP. The equation was validated using a GBLUP model with simulated phenotypes and real genotypes of one Holstein Friesian population. This population was split in two populations for RP (population A and B) and one population for SC (population C), using different values for heritability and genetic correlations between the populations. Three different cattle breeds with real genotypes and simulated phenotypes, assuming different genetic correlations, were also used for validation.

Can multi-subpopulation reference sets improve the genomic predictive ability for pigs?

A. Fangmann[1], S. Bergfelder[2], E. Tholen[2], H. Simianer[1] and M. Erbe[1]
[1]Georg-August-University Goettingen, Animal Breeding and Genetics Group, 37075 Goettingen, Germany, [2]University of Bonn, Group of Animal Breeding and Husbandry, 53115 Bonn, Germany; anna.fangmann@agr.uni-goettingen.de

Separate breeding work of different pig breeding organizations in Germany for decades has led to stratified subpopulations in the breed German Large White. Due to this fact and the limited number of genotyped Large White animals available in each organization, there was an urgent need to assess whether within-breed, but multi-subpopulation genomic prediction is advantageous in pigs. Genotypes (Illumina Porcine 60K SNP Beadchip) from over 2,000 German Large White individuals from five different pig breeding organizations were available. With a genomic BLUP (GBLUP) model, direct genomic breeding values (DGVs) were estimated for the trait 'number of piglets born alive' based on different within- or multi-subpopulation reference sets. To test the predictive ability of within- or multi-subpopulation reference sets, a random five-fold cross-validation with 20 replicates was performed. Predictive ability for a specific subpopulation was measured as correlation between quasi-phenotypes (conventional estimated breeding values, EBVs) and DGVs for the individuals of the respective subpopulation in the validation set. Results were similar for all subpopulations; thus, only results for subpopulation 1 are presented here. The predictive ability within subpopulation 1 (0.76) was relatively high. In general, the predictive ability for subpopulation 1 did not increase significantly by enlarging the reference population with individuals of other closely related subpopulations (0.76-0.77) while it slightly decreased when distantly related subpopulations were added (e.g. 0.74 with all subpopulations). When adding additional female individuals of the same subpopulation the predictive ability increased remarkably (0.63 to 0.76). Thus, available resources should be preferably used to enlarge the reference set within subpopulation.

Accuracy of genomic predictions for dairy traits in Gyr cattle (*Bos indicus*)

S.A. Boison[1], A.T.H. Utsunomiya[2], D.J.A. Santos[2], H.H.R. Neves[2], G. Mészáros[1], R. Carvalheiro[2], J.F. Garcia[3], M.V.G.B. Silva[4] and J. Sölkner[1]
[1]University of Natural Resources and Life Sciences, Vienna, Gregor Mendel Strasse 33, 1180, Austria, [2]UNESP, Jaboticabal, 14884-900, Brazil, [3]UNESP, Aracatuba, 16015-050, Brazil, [4]Embrapa Dairy Cattle, Juiz de Fora, 36038-330, Brazil; solomon.boison@students.boku.ac.at

Accelerating genetic gain with the use of genomic information in *Bos indicus* breeds is of great interest. We present results of genomic predictions in Gyr for milk yield (MY, kg), fat yield (FY, kg), protein yield (PY, kg) and age at first calving (AFC) using information from bulls and cows. Additionally, four different SNP panels were studied. A total of 440 bulls and 1,597 cows were genotyped with the Illumina BovineHD (HD) and Bovine50K (50K) chip, respectively. Genotypes of cows were imputed to HD using FImpute. SNPs from the GeneSeek 20Ki, GeneSeek 75Ki and Illumina 50K in common with HD were subset and used to assess the impact of lower SNP density on accuracy of prediction. Deregressed breeding values (dEBV) were used as pseudo-phenotypes. Data were split into training (TR) and validation (VAL) set to mimic a forward prediction scheme. TR consisted of sires (TR1, n=282) or a combination of sires and dams (TR2, n=1,879), while VAL were only young bulls (n=158). SNP-BLUP model was used to estimate SNP effects and genomic breeding values (GEBV). Predictive ability (R^2) was defined as the weighted squared correlation between dEBV and GEBV. Using TR1, estimate of R^2 ranged from 0.50 (MY) to 0.78 (FY). Extension of the TR with cow genotypes (TR2) increased R^2 for almost all traits except for FY. The average R^2 were 0.59 (20Ki) and 0.61 for 50K, 75Ki and HD using only sires in TR. Reduction in prediction accuracy was observed when imputed cow genotypes were added to the TR. R^2 of VAL was observed to be biased upwards when reliability of dEBV were close to parent averages, suggesting a need for more accurate dEBV of young bulls to be able to fully quantify the gain in R^2 of GEBVs over parent averages.

Genomic prediction of performance in pigs using additive and dominance effects

M.S. Lopes[1,2], J.W.M. Bastiaansen[1], L. Janns[3], E.F. Knol[2] and H. Bovenhuis[1]
[1]Wageningen University, Droevendaalsesteeg 1, 6708 PB Wageningen, the Netherlands, [2]Topigs Norsvin, Schoenaker 6, 6641 SZ Beuningen, the Netherlands, [3]Aarhus University, Blichers Allé 20, 8830 Tjele, Denmark; marcos.lopes@topigsnorsvin.com

Using prediction based on either pedigree or genomic information, the focus of animal breeding has been on additive genetic effects translated as breeding values. However, if the aim is to predict performance rather than breeding values of an animal, models that simultaneously account for additive and dominance effects are expected to be more effective. The objectives of this study were to estimate the contribution of additive and dominance effects to the total phenotypic variation, and to assess the prediction accuracy of performance for life-time daily gain (DG) in a Pietrain pig population accounting for additive and dominance effects. Animals (n=1,424) were genotyped using the Illumina SNP60K Beadchip and they were divided into a training dataset to estimate the genetic parameters and the SNP effects, and a validation dataset (20% youngest animals) used to assess the prediction accuracy. We evaluated two models that were based on random regression on SNP genotypes, implemented in the program BayZ. The model MA accounted only for additive effects and the model MAD accounted for both additive and dominance effects. Additive heritability estimates were 0.26±0.04 using MA and 0.21±0.04 using MAD. Dominance heritability estimate was 0.19±0.07. Prediction accuracy was 0.18 using MA and 0.22 using MAD. Dominance effects make an important contribution to the genetic variation of DG in the population evaluated, accounting for 47% of the total genetic variation. In this study, we evaluated a purebred population and already found prediction to be improved. Using the same method to predict crossbred performance is expected to be even more interesting as dominance levels are expected to be higher. The prediction accuracy using this method in a population of finishing pigs (n=1,408) is being tested.

Equivalence of genomic breeding values and reliabilities estimated with SNP-BLUP and GBLUP

L. Plieschke[1], C. Edel[1], E. Pimentel[1], R. Emmerling[1], J. Bennewitz[2] and K.-U. Goetz[1]
[1]Bavarian State Research Center for Agriculture, Institute of Animal Breeding, Prof.-Dürrwaechter-Platz 1, 85586 Poing-Grub, Germany, [2]University Hohenheim, Institute of Animal Husbandry and Breeding, Garbenstraße 17, 70599 Stuttgart, Germany; laura.plieschke@lfl.bayern.de

Genomic evaluation models can define coefficient matrices of order equal to the number of animals (GBLUP) or to the number of markers (SNP-BLUP). Different studies have shown that GBLUP solutions are equivalent to the ones from SNP-BLUP. For practical applications it would also be important to know how to obtain equivalent reliabilities of DGVs from both approaches. In this study we use real data to show under which conditions GBLUP and SNP-BLUP lead to equivalent results for DGVs and reliabilities. Data used comprised 11,852 Fleckvieh bulls genotyped at 41,266 SNPs. The DGVs obtained with both models were the same, apart from rounding errors, which confirms the equivalence between the two models. Reliabilities of DGVs obtained from both models were also the same, when genomic inbreeding was taken into account. Neglecting the genomic inbreeding coefficient in the calculation of reliabilities in the SNP-BLUP resulted in some differences for lower reliabilities despite a correlation of 0.999. Therefore, a prerequisite is that the error variance of the mean is consistently taken into account as well as the genomic inbreeding coefficient. In addition, a coding that is consistent with quantitative genetics theory should be used. In our data set, in which the number of markers by far exceeds the number of animals, GBLUP has significant advantages with respect to computation time and storage requirement. However, in a scenario where the number of animals exceeds the number of markers SNP-BLUP will have computational advantages. Another advantage of SNP-BLUP is the easier way to calculate breeding values for candidates. The choice of model can therefore be based solely on the structure of the dataset, i.e. the relation between number of markers and number of genotyped animals.

Genomic predictions in Norwegian Red Cattle: comparison of methods

O.O.M. Iheshiulor[1], J.A. Woolliams[1,2], X. Yu[1] and T.H.E. Meuwissen[1]
[1]Norwegian University of Life Sciences, Department of Animal and Aquacultural Sciences, Arboretveien 6, 1432, Ås, Norway, [2]University of Edinburgh, The Roslin Institute (Edinburgh), Royal (DICK) School of Veterinary Studies, Midlothian, EH25 9RG, Scotland, United Kingdom; oscar.iheshiulor@nmbu.no

Genomic selection (GS) is increasingly widely adopted and implemented in most livestock breeding programs with the dairy cattle sector taking the lead. For the estimation of genomic breeding values (GEBV), plethora's of methods are available and based on assumptions about SNP (single-nucleotide polymorphisms) effects; these methods can broadly be classified into two groups (i.e. G-/RR- BLUP and variable selection methods). Here we investigated the accuracy of genomic predictions based on three methods: best linear unbiased prediction (GBLUP), GblupBayesC (GBC, a deterministic method similar to BayesC but fits gblup next to the SNP effects), and MixP (based on Pareto Principles and uses mixture of two normal distribution). A data set of 3,244 Norwegian Red (NRF) bulls, which were genotyped with 54k SNP panel, was used. After quality control, a total of 48,249 SNPs remained. Progeny-based daughter yield deviation (DYD) was used as the response variable and the study was performed on four traits (milk yield, fat yield, protein yield, and somatic cell count). Data was split into reference and validation set corresponding to forward prediction scheme. Predictability was defined as cor(DYD,GEBV) within the validation set. Accuracies ranged between 0.532-0.620 (GBLUP), 0.536-0.632 (GBC), and 0.552-0.648 (MixP), with the lowest accuracies observed in somatic cell count. For the four traits respectively, using GBC increased accuracy by 0.004, 0.012, 0.021, and 0.003, while with MixP accuracy increased by 0.020, 0.028, 0.014, and 0.003, in comparison to GBLUP. MixP gave the highest accuracies in three of the traits, followed by GBC. We conclude that GBC and MixP are viable tools for routine evaluations.

Overview of possibilities and challenges of the use of infrared spectrometry in cattle breeding

N. Gengler[1], H. Soyeurt[1], F. Dehareng[2], C. Bastin[1], F.G. Colinet[1], H. Hammami[1] and P. Dardenne[2]
[1]ULg, GxABT, Passage des Déportés 2, 5030 Gembloux, Belgium, [2]CRA-W, Chée de Namur, 24, 5030 Gembloux, Belgium; nicolas.gengler@ulg.ac.be

Near or mid-infrared (NIR or MIR) spectrometry is a versatile and cost-efficient technology used in cattle production to trace the chemical composition of gases, liquids and solid matters. Recent research showed the potential of MIR spectrometry in milk to predict many different milk components but also status and well-being of the cows, quality of their products, their efficiency and their environmental impact. Under changing socio-economic circumstances, novels traits could help to select for enlarged breeding objectives. But the following challenges need to be overcome: (1) access to and harmonization of MIR data; (2) availability of reference values representing the variability to be described, also highlighting the importance of international collaborations; (3) difficulties to obtain, but also to transfer prediction equations between instruments; (4) modeling of the massive longitudinal data generated; (5) estimation of parameters to assess phenotypic and genetic variability and links with other traits leading to the; (6) assessment of the position of novel traits in breeding objectives. Recent research reported how to address these issues for traits close to routine use including fatty acids and methane. Expected future developments include direct use of MIR data and multivariate modeling of novel traits. Similarly, genomic prediction for novel traits, which are limited by the availability of phenotyped reference populations, will also benefit from the use of correlated, MIR predicted, traits. Currently, MIR instruments can only be used in the frame of milk recording and not on-farm. But recent research showed that NIR is closing the gap thereby allowing advances in precise on-farm phenotyping and giving new opportunities for breeding, but also management. Possibilities for the use of infrared technologies for other trait groups such as meat composition and quality should allow cross-fostering of developments.

Novel metabolic signals in health risk assessment in dairy cows

K. Huber[1], S. Dänicke[2], J. Rehage[1], H. Sauerwein[3], W. Otto[4], U. Rolle-Kampczyk[4] and M. Von Bergen[4]
[1]University of Veterinary Medicine, Bischofsholer Damm 15, 30173 Hannover, Germany, [2]Federal Research Institute for Animal Health, Bundesallee 50, 38116 Braunschweig, Germany, [3]Institute for Animal Science, Katzenburgweg 7-9, 53115 Bonn, Germany, [4]Helmholtz Centre for Environmental Research, Permoser Strasse 15, 04318 Leipzig, Germany; korinna.huber@uni-hohenheim

At the onset of lactation, modern dairy cows are at risk for metabolic disturbances and consequently, production diseases. However, not all cows are equally prone to disease. Predisposition to metabolic disturbances might match with a distinct plasma metabolome, most likely reflecting the origin of the failure. For identifying cows at risk early, blood variables such as NEFA, BHBA and glucose/insulin indices were used but the predictive value is low. This study aimed to perform a targeted metabolomics approach to identify new metabolites for assessing health risks in dairy cows. Plasma of 25 periparturient dairy cows was obtained at day -42, 3, 21 and 100 relative to calving; and 180 metabolites (AbsoluteIDQ p180 kit, Biocrates, Austria) were quantitatively determined by HPLC/mass spectrometry. All data were statistically evaluated by Two Way ANOVA for the factors time and group. Groups were assigned according to the post partum health status of the cows: clinically sick, healthy but left herd within current lactation (group A) and healthy (group B). Comparing A and B, plasma NEFA, BHBA, glucose and insulin were not different. However, the metabolomics approach revealed distinct differences in metabolites involved in oxidative lipid metabolism and inflammatory pathways. Group A had less acylcarnitines (AC) but higher kynurenine/tryptophan (kyn/trp) ratio compared to group B. High plasma AC reflect an efficient mechanism of mitochondrial protection, thus the cows of group B were metabolically more healthy and in a less inflammatory state as indicated by a low kyn/trp ratio. This suggests high predictive values of AC and kyn/trp ratio for health risk assessment in dairy cows.

Milk biomarkers to detect ketosis and negative energy balance using MIR spectrometry

C. Grelet[1], C. Bastin[2], M. Gelé[3], J.B. Davière[4], R. Reding[5], A. Werner[6], C. Darimont[1], F. Dehareng[1], N. Gengler[2] and P. Dardenne[1]
[1]CRA-W, 5030, Gembloux, Belgium, [2]ULg,GxABT, 5030, Gembloux, Belgium, [3]IDELE, 49105, Angers, France, [4]CLASEL, 53942, St Berthevin, France, [5]Convis, 9085, Ettelbruck, France, [6]LKV-BW, 70190, Stuttgart, Germany; c.grelet@cra.wallonie.be

In order to manage negative energy balance and ketosis in dairy farms, rapid and cost-effective detection is needed. Among the milk biomarkers, citrate was recently identified as an early indicator of negative energy balance and acetone and BHB are of particular interest regarding ketosis. The objective of this study was to evaluate the ability of Mid-Infrared (MIR) spectrometry to predict these biomarkers as this technology can routinely provide rapid and cost-effective predictions. A total of 566 milk samples were collected in commercial and experimental farms in Luxembourg, France and Germany. Acetone, BHB, and citrate contents were determined by flow injection analysis. Milk MIR spectra were recorded and standardized for all samples. Acetone content ranged from 20 to 3,355 µmol/l with an average of 103 µmol/l; BHB content ranged from 21.3 to 1,595.6 µmol/l with an average of 215.4 µmol/l; and citrate content ranged from 4.5 to 15.5 mmol/l with an average of 8.9 mmol/l. Acetone and BHB contents were log-transformed to approach a normal distribution. Prediction equations were developed using PLS. The R^2 of calibration was 0.73 for acetone, 0.75 for BHB and 0.90 for citrate with RMSE (root mean square error) of 87.7 µmol/l, 86.5 µmol/l and 0.75 mmol/l respectively. An external validation was performed and RMSE of validation was 45.2 µmol/l for acetone, 65.33 µmol/l for BHB and 0.80 mmol/l for citrates. Although the practical usefulness of the equations developed should be verified with field data, results from this study demonstrated the potential of MIR spectrometry to predict citrate content with good accuracy and to supply indicative contents of BHB and acetone in milk, providing consequently detection tools of ketosis and negative energy balance in dairy farms.

Effectiveness of mid-infrared spectroscopy to predict bovine milk minerals
P. Gottardo, M. De Marchi, G. Niero, G. Visentin and M. Penasa
University of Padova, Department of Agronomy, Food, Natural resources, Animals and Environment
(DAFNAE), Viale dell'Università 16, 35020 Legnaro (PD), Italy; paolo.gottardo.1@studenti.unipd.it

The importance of milk mineral composition for human health is well known and accepted. Because of that, a rapid and cost effective monitoring of milk mineral composition is an important goal to achieve. This study aimed to investigate the potential of mid-infrared spectroscopy to predict calcium, potassium, sodium, magnesium and phosphorus in milk through the use of Uninformative Variables Elimination partial least squares (UVE-PLS) procedure. Individual samples (n=246) of Holstein-Friesian, Brown Swiss, Simmental and Alpine Grey cows from different stages of lactation and parities were collected from single-breed herds. Reference analysis was undertaken with inductively coupled plasma optical emission spectrometry (ICP-OES) in accordance with standardized methods. For each milk sample MIRS analysis in the range of 900 to 5,000 cm^{-1} was performed. Prediction models were developed using PLS regression after UVE for each considered trait and prediction accuracy was based on leave one out cross validation (246 segments). The coefficient of determination in cross validation was greatest for phosphorus (0.72), followed by potassium (0.70), calcium (0.68) and magnesium (0.67). Sodium prediction was poor, with a very low coefficient of determination (0.46). All the prediction models were unbiased (P<0.05). The ratio of performance deviation suggested that prediction models for P can be used for analytical purposes. Calcium and potassium exhibited RPD close to the threshold and thus they could be considered for analytical purposes as well. Results from the present study demonstrated that MIRS, combined with PLS regression is useful for the acquirement of milk mineral phenotypes at population level and the use of UVE combined with PLS represent a valid approach to improve the accuracy of prediction models.

Genetic parameters for milk calcium content predicted by MIR spectroscopy in three French breeds
A. Govignon-Gion[1], S. Minery[2], M. Wald[2], M. Brochard[2], M. Gelé[3], B. Rouillé[4], D. Boichard[1],
M. Ferrand-Calmels[2] and C. Hurtaud[5]
[1]INRA, UMR 1313 GABI, 78350 Jouy-en-Josas, France, [2]Institut de l'Elevage, 149, rue de Bercy, 75595
Paris, France, [3]Institut de l'Elevage, CS 70510, 49105 Angers, France, [4]Institut de l'Elevage, BP 85225,
35652 Le Rheu, France, [5]INRA, UMR PEGASE, 35590 Saint-Gilles, France; didier.boichard@jouy.inra.fr

The aims of this study were to develop an equation to estimate calcium content (Ca) in bovine milk, using mid-infrared (MIR) spectroscopy and to determine Ca genetic parameters. To develop the Ca equation, 300 milk samples were selected from PhénoFinlait milkbank to cover a large range of breeding practices (3 breeds, different areas, seasons, lactation numbers, diets, etc.). Those samples were both analyzed by MIR and by atomic absorption spectrometry which is the reference method for Ca measurement. 210 out of the 300 samples were used as calibration dataset and the remaining 90 were used as independent validation set. The determination coefficient of validation of the equation (Rv^2) reached 0.79 and its residual standard deviation (sy,x) was 4%. Genetic parameters of Ca were estimated for the three French major dairy breeds (Prim'holstein (HOL), Montbéliarde (MON), Normande (NOR)). Ca equation was applied to 35,326 spectral records collected from 6,723 first lactation HOL cows, 28,508 spectral records collected from 5,590 first lactation NOR cows and 50,505 spectral records collected from 6,330 first lactation MON cows. Three different models were used to estimate genetic parameters (1) an individual test-day repeatability model, (2) a lactation model, where the trait is the average of test-day records and (3) a test-day random regression model. The heritabilities of Ca estimated with lactation model were 0.44 in HOL, 0.74 in NOR and 0.70 in MON. The coefficients of genetic variation were 3.6, 4.3 and 4.2 in HOL, NOR and MON respectively. And data from more than 8,000 cows in the 3 breeds will be used for the next step: analysis of genomic sequences to identify causal mutations for Ca.

Estimating genetic parameters for predicted energy traits from mid-infrared spectroscopy on milk

S.L. Smith[1], S.J. Denholm[1], V. Hicks[2], M. Coffey[1], S. Mcparland[3] and E. Wall[1]

[1]SRUC, Animal and Veterinary Sciences, Roslin Institute Building, Easter Bush, Midlothian, EH25 9RG, Edinburgh, Scotland, United Kingdom, [2]NMR, Fox Talbot House, Greenways Business Park, Bellinger Close, Chippenham, SN15 1BN, Chippenham, England, United Kingdom, [3]Teagasc, Animal & Grassland Reserach and Innovation Centre, Moorepark, Fermoy, Co. Cork, Ireland, Co. Cork, Ireland, Ireland; stephanie.smith@sruc.ac.uk

The balance of energy across lactation can have health and fertility consequences for the dairy cow. Mid-infrared analysis (MIR) of milk has been shown to provide a robust prediction of a cow's energy balance, using calibration equations developed by McParland et al. This study aimed to create a larger 'calibration' dataset of body energy traits from UK cows with their concurrent MIR spectral data to generate a prediction tool for use on national herds. Phenotypic data from 922 cows in the Langhill research herd, at Crichton Royal Farm (Scotland), were used to generate the following energy traits using the equations of Banos and Coffey (2010); energy balance (EB), body energy content (EC) and effective energy intake (EI), all in MJ/day. Pre-treatment of spectral data involved 'standardisation' according to files generated as part of an InterReg-funded project (Optimir), to account for drift over time and machines. Alignment of energy estimates with their spectral data produced a calibration dataset of 13683 animal-testdates with the resultant prediction equation, generated using partial-least square analysis, with an accuracy of c.0.85 for EB. These prediction equations were applied to national spectra on over 117,000 animals (over 1 million testdates) from 355 farms, across 2013-2015 and genetic parameters estimated using an animal model in ASReml. Heritability estimates for EB, EC, and EI were 0.11, 0.25 and 0.10. MIR-based energy trait predictions from routinely collected national data have potential use in genetic improvement of livestock with regards to optimal and sustainable energy profiles.

Heritability and genetic correlations for health and survival in Norwegian Red calves

K. Haugaard[1] and B. Heringstad[1,2]

[1]Norwegian University of Life Sciences, Department of Animal and Aquacultural Sciences, P.O. Box 5003, 1432 Ås, Norway, [2]GENO Breeding and A.I. Association, P.O. Box 5003, 1432 Ås, Norway; katrine.haugaard@nmbu.no

The aim of this study was to estimate heritability for and genetic correlations among survival, respiratory disease, gastritis/enteritis and arthritis in Norwegian Red calves. Information from 606,447 calves, both males and females born from 2004 to 2013 and offspring of 1,320 Norwegian Red AI sires was extracted from the Norwegian Dairy Herd Recording System and used for the analyses. The traits were defined as binary, where for each of the diseases 1 = veterinary treatment and 0 = no treatment for that disease and for survival 1 = alive and 0 = dead at 6 months of age. The prevalence of disease was 3.32% for respiratory disease, 2.31% for gastritis/enteritis and 2.31% for arthritis, while 4.43% of the calves died before 6 months of age. A multivariate threshold sire model using Gibbs sampling was applied for estimation of genetic parameters. A total chain length of 500,000 iterations, after 50,000 iterations burn-in, was used for posterior inference. The posterior means (SD) of heritability of liability was 0.04 (0.01) for survival, 0.09 (0.01) for respiratory disease, 0.05 (0.01) for gastritis/enteritis and 0.08 (0.01) for arthritis. This is in the range of heritability for other health traits. The genetic correlations of survival to the diseases was negative and strong (i.e. less disease = higher survival) and ranged from -0.61 (arthritis) to -0.68 (respiratory disease), while the genetic correlations among the three diseases were positive and moderate, ranging from 0.23 (arthritis/respiratory disease) to 0.33 (arthritis/gastritis/enteritis). An advantage for calfhood diseases is that information is available early and on both sexes, yielding highly accurate breeding values early in the bulls life.

Loci associated with adult stature also effect calf birth survival in cattle
G. Sahana, J.K. Höglund, B. Guldbrandtsen and M.S. Lund
Aarhus University, Center for Quantitative Genetics and Genomics, Molecular Biology and Genetics,
Blichers Alle 20, 8830, Denmark; goutam.sahana@mbg.au.dk

Understanding the underlying pleiotropic relationships among quantitative traits is integral to predict correlated responses to artificial selection. The availability of large-scale next-generation sequence data in cattle has provided an opportunity to examine whether pleiotropy is responsible for overlapping QTLs in multiple economic traits. In the present study, we examined QTLs effecting cattle stillbirth, calf size, and adult stature located in the same genomic region. A genome scan with whole sequence variants revealed one QTL with large effects on the service sire calving index (SCI), and body conformation index (BCI) at the same location (~39 Mb) on chromosome 6 in Nordic Red cattle. The targeted region was analyzed for SCI and BCI component traits. The QTL peak included LCORL/NCAPG genes, which were earlier reported to influence fetal growth, and adult stature in several species. The QTL exhibited large effects on calf size and stature in Nordic Red cattle. Two deviant haplotypes (HAP1 and HAP2) were resolved, which increased calf size at birth, and effected adult body conformation. However, the haplotypes also resulted in increased calving difficulties and calf mortality due to increased calf size at birth. Haplotype locations overlapped, however linkage disequilibrium (LD) between the sites was low, suggesting two independent mutations responsible for similar effects. The difference in prevalence between the two haplotypes in Nordic Red subpopulations suggested independent origins in different populations. Results of our study identified QTLs with large effects on body conformation, and service sire calving traits on chromosome 6 in cattle. We present robust evidence that variation at the LCORL/NCAPG locus effects calf size at birth and adult stature. We suggest the two deviant haplotypes within the QTL were due to two independent mutations.

Fine-mapping of a QTL region on BTA18 affecting non-coagulating milk in Swedish Red cows
S.I. Duchemin[1,2], M. Glantz[3], D.J. De Koning[2], M. Paulsson[3] and W.F. Fikse[2]
[1]Wageningen University, P.O. Box 338, 6700 AH,Wageningen, the Netherlands, [2]Swedish University of Agricultural Sciences, P.O. Box 7023, 750 07, Uppsala, Sweden, [3]Lund University, P.O. Box 124, 221 00, Lund, Sweden; sandrine.duchemin@wur.nl

Milk with poor coagulation properties is undesirable within cheese production, and up to 18% of the Swedish Red (SR) cow population was observed to produce non-coagulating (NC) milk. The aims of this study were to perform a GWAS for NC milk and to fine map a QTL region on BTA18 using sequences from the 1000 Bull Genomes project. Phenotypes were available on 382 SR cows belonging to 20 herds in the southern part of Sweden. Milk samples that had not started to coagulate within 40 min after rennet addition were defined as NC milk. Based on 777k SNP genotypes available for all cows, a GWAS identified several genomic regions associated with NC milk in SR cows. The most promising region was found on BTA18 between 0-30 Mbp. Whole-genome sequences were available for 427 bulls and for 2 cows from 15 different breeds (Run 3 of the 1000 Bull Genomes consortium). Among these sequences, 33 belonged to SR and Finnish Ayrshire bulls with a large impact in the SR cow population. The 382 cows were imputed from 777k genotypes to sequence level for half of BTA18 using Beagle. Three different scenarios of imputation were run to account for the nature of the different variants: nordic-red-specific, dairy-breeds, and common variants. For each variant, the imputed genotype with the highest imputation accuracy across the three scenarios was selected. After imputation, the number of variants on half of BTA18 increased from 7,869 SNP to 564,552 variants, of which 133,277 were polymorphic. Single-SNP analyses were run using a linear animal model in ASReml. A total of 130 variants were significantly associated with NC milk in SR cows. In comparison with the 777k SNP genotypes, the use of imputed sequence data yielded lower P-values for associations with NC milk.

PTR-ToF-MS to study the effect of dairy system and cow traits on the volatile fingerprint of cheeses

M. Bergamaschi[1,2], F. Biasioli[1], A. Cecchinato[2], C. Cipolat-Gotet[2], A. Cornu[3], F. Gasperi[1], B. Martin[3] and G. Bittante[2]
[1]Fondazione Edmund Mach, Department of Food Quality and Nutrition, via E. Mach, 1, 38010 San Michele all'Adige (TN), Italy, [2]University of Padova, DAFNAE, viale dell'università, 16, 35020 Legnaro (PD), Italy, [3]INRA, UR 1213 Herbivores, 63122 Saint Genès Champanelle, 63122 Saint Genès Champanelle, France; matteo.bergamaschi@studenti.unipd.it

The aim of this work was to investigate the volatile fingerprint of 1,075 model cheeses through a rapid and non-invasive analysis: Proton Transfer Reaction Time of Flight Mass Spectrometry (PTR-ToF-MS). Cheeses were produced using individual milk from Brown Swiss cows reared in 72 herds belonging to 5 dairy systems ranging from traditional to modern ones and were ripened for 60 d. PTR-ToF-MS analysis of the sample headspace reveled more than 500 spectrometric peaks, among which the 240 most intense (above 1 ppb_v) were used for statistical analysis. The data obtained were compared with those yielded on a subset of 150 cheese samples with the SPME/GC-MS. A principal component (PC) analysis was applied to compress the multiple responses of PTR-ToF-MS spectra into 5 synthetic variables representing together 62% of the total variance (from 28% for PC1 to 6% for PC5). Dairy system influenced 57 peaks ($P<0.05$) and correlate with the 2nd PC. Days in milk affected 139 peaks ($P<0.05$) and the 1st, 2nd, 4th and 5th PC. Parity of cow affected 22 peaks ($P<0.05$). Lastly, the daily milk yield affected 31 peaks ($P<0.05$) and correlate with the 3rd PC. Since many volatile compounds (VOC) are involved in cheese flavor, it may be anticipated that the farming characteristics that affect VOC profiles also affect cheese sensory properties. The VOC fingerprint of cheese obtained by PTR-ToF-MS proved to be useful for discriminating samples according to dairy system. Individual cheese-making procedure associated with this spectrometric technique open new avenues for the study of cheese quality, the selection of dairy cattle and the authentication of milk production conditions.

Genome-wide association study for cheese yield and curd nutrients recovery in bovine milk

C. Dadousis[1], C. Cipolat-Gotet[1], S. Biffani[2], E.L. Nicolazzi[3], G. Rosa[4,5], D. Gianola[4,5], G. Bittante[1] and A. Cecchinato[1]
[1]University of Padova, Department of Agronomy, Food, Natural Resources, Animals and Environment (DAFNAE), Viale dell'Università 16, 35020 Legnaro, Italy, [2]IBBA, CNR, Via Einstein, Loc. Cascina Codazza, 26900 Lodi, Italy, [3]PTP, Department of Bioinformatics, Via Einstein- Loc. Cascina Codazza, 26900 Lodi, Italy, [4]University of Wisconsin-Madison, Department of Biostatistics and Medical Informatics, Madison Wisconsin, 53706, USA, [5]University of Wisconsin-Madison, Department of Animal Sciences, Madison Wisconsin, 53706, USA; christos.dadousis@studenti.unipd.it

A genome wide association study was conducted with cheese yield (CY) and milk nutrient and energy recoveries (REC) in the curd measured in individual cows. Three CY traits expressing the weight (wt) of fresh curd ($\%CY_{CURD}$), curd solids ($\%CY_{SOLIDS}$), and curd moisture ($\%CY_{WATER}$) as percentage of wt of milk processed, and 4 REC (REC_{FAT}, $REC_{PROTEIN}$, REC_{SOLIDS}, and REC_{ENERGY} calculated as the % ratio between the nutrient in curd and the corresponding nutrient in processed milk) were analyzed. Milk and blood samples were collected from 1,264 Italian Brown Swiss cows from 85 herds. Animals were genotyped with the Illumina SNP50 Beadchip v.2. A single marker regression was fitted using the GenABEL R package (GRAMMAR-GC). Days in milk, parity and herd were considered as non-genetic effects. In total, 103 significant associations (88 SNP) were identified ($P<5\times10^{-5}$) in 10 chromosomes (2, 6, 9, 11, 12, 14, 18, 19, 2, 7, 28). For REC_{FAT} and $REC_{PROTEIN}$, highly significant associations were identified in *Bos taurus* autosomes (BTA) 6 and BTA11, respectively. SNP ARS-BFGL-NGS-104610 was highly associated to $REC_{PROTEIN}$ ($P=6.07\times10^{-36}$) and Hapmap52348-rs29024684, closely located to the casein genes on BTA6, to REC_{FAT} ($P=1.91\times10^{-15}$). Genomic regions identified may enhance marker assisted selection in bovine cheese breeding. Acknowledgements: Trento Province (Italy), Italian Brown Swiss Cattle Breeders Association (ANARB, Verona, Italy) and Superbrown Consortium of Bolzano and Trento.

Phenotypic analysis of low molecular weight thiols in milk of dairy and dual-purpose cattle breeds

G. Niero, A. Masi and M. Cassandro
University of Padova, Department of Agronomy, Food, Natural resources, Animals and Environment, Viale dell'Università 16, 35020 Legnaro (PD), Italy; giovanni.niero@studenti.unipd.it

Free radicals (FR) are unstable molecules with adverse effects on animal and human cells. Antioxidants, acting against FR, are essential for cell homeostasis maintenance. Antioxidants are molecules that compete with other oxidizable substrates, preventing the oxidation of the substrates themselves. This definition encloses a large number of molecules including glutathione (GSH), that belongs to the class of low molecular weight (LMW) thiols. GSH and LMW thiols are involved in FR scavenge with the formation of disulphide bound (GS-SG). This study aimed to characterize LMW thiols of bovine milk. Twenty-four individual milk samples from each of 4 dairy cattle breeds, Brown Swiss, Holstein-Friesian, Alpine Grey and Simmental, were collected in 8 herds. After sampling, milks were added with preservative and stored at -20 °C. Low molecular weight thiols were extracted after thawing, from the soluble fraction of milk sample, obtained removing fat and proteins. Thiols were detected in Reverse Phase (RP)-HPLC, after the application of a derivatization protocol, that aimed to mark target molecules with a die (SBDF), detectable with a fluorimetric relevator. Six thiols species were detected and only two of them were identified. The mean (standard deviation) of Cys-Gly and GSH were 0.087 (0.068) and 0.067 (0.064) μM, respectively. Average concentration of Cys-Gly was greater than that of GSH across all breeds. Milk from dual-purpose breeds (SI and AG) was richer in Cys-Gly and GSH concentration than milk from HF and BS cows. Great variation of both thiols was found among herds. Cysteine Glicine exhibited a moderately low Pearson correlation with protein (0.290; $P<0.001$) and casein (0.271; $P<0.001$), while GSH did not show any correlation with protein and casein. Glutathione and Cys-Gly, closely linked in biosynthesis pathway, were quite strongly correlated (0.653; $P<0.001$). Negligible effects of days in milk and parity were detected.

Epigenetics of bovine semen: tools to analyse DNA methylation

J.P. Perrier[1], A. Prezelin[1], L. Jouneau[1], E. Sellem[2], S. Fritz[2], J.P. Renard[1], D. Boichard[1], L. Schibler[2], H. Jammes[1] and H. Kiefer[1]
[1]INRA, Domaine de Vilvert, 78352 Jouy-en-Josas, France, [2]ALLICE, Rue de Bercy, 75595 Paris, France; jpperrier@jouy.inra.fr

Prediction of male fertility remains a major goal to promote animal insemination (AI). The combination of several markers of sperm quality provides promising results but still remains unsatisfactory. The crucial role of epigenetics in sperm functions and fertilization efficiency is now recognized. A better understanding of epigenetic mechanisms such as DNA methylation may highlight new key criteria that could improve prediction models. Here, we present tools that can be used to monitor sperm DNA methylation (dMe) at different scales and to identify individual and temporal variations in the methylation contents and patterns. Global quantification by LUminometric Methylation Assay (LUMA) in bovine, porcine, ovine, human and mouse suggests a species-specific behaviour of sperm dMe. Indeed, dMe was lower in bovine semen than in other species, and this result was not affected by cryoconservation. In contrast, somatic cells showed similar content of dMe in all species. In bovine, subtle breed-dependent variations were also detected by LUMA, suggesting underlying genetic factors in the determination of dMe content. To map dMe at a genome-wide scale, we used immunoprecipitation of methylated DNA followed by hybridization on a microarray (MeDIP-chip). The microarray targets the upstream region (-2,000 to +1,360 bp) of 21,416 bovine genes (UMD3.1 assembly). Methylation pattern in sperm was compared with liver and fibroblasts, and Differentially Methylated Regions between tissues (tDMRs) were identified, that are enriched in pathways potentially important for sperm physiology. Pyrosequencing validation of these tDMRs is now on-going and will be presented. In conclusion, we developed relevant tools to analyze dMe in bovine semen. These tools could now be used to identify and monitor dMe variations related to environmental changes affecting the quality of gametes and the fertility of AI bulls.

Milk mammary epithelial cells as a non-invasive indicator of lactation persistency in dairy cows?

M. Boutinaud[1,2], V. Lollivier[1,2], L. Yart[1,2], J. Angulo Arizala[3], P. Lacasse[4], F. Dessauge[1,2] and H. Quesnel[1,2]
[1]INRA, UMR1348 PEGASE, 35590 Saint-Gilles, France, [2]Agrocampus Ouest, UMR1348 PEGASE, 35000 Rennes, France, [3]University of Antioquia, AA 1126, Medellin, Colombia, [4]Agriculture and Agri-Food Canada, Dairy and Swine R & D Centre, Sherbrooke J1M0C8, Canada; helene.quesnel@rennes.inra.fr

Mammary epithelial cells (MEC), that produce milk, are shed into milk during the lactation process. We propose the estimation of MEC exfoliation rate into milk as an original trait to characterize the lactation persistency of dairy cows. We developed a non-invasive method to determine this rate through milk MEC purification by an immuno-magnetic method. The rate of MEC exfoliation into milk was determined in two trials where lactation persistency of Prim'Holstein cows was affected. In the first study, 14 cows were ovariectomized or sham-operated around 60 days in milk. Milk was collected at 5, 21, 37, 47 and 54 weeks of lactation to estimate MEC exfoliation rate. Milk yield decreased as lactation progressed (P<0.05) and the number of MEC daily exfoliated into milk tended to increase (P<0.10). Ovariectomy reduced by 8.1% the decline of milk yield (P<0.05) and decreased MEC loss in milk at 47 weeks (P<0.05). In the second study, 9 cows were assigned to 3 treatments during 3 5-d periods: (1) daily injections of Quinagolide (Quin), a prolactin-release inhibitor; (2) daily injections of Quin and injections of prolactin (PRL); (3) daily injections of vehicles as control. Quin decreased milk yield (P<0.05) and increased the MEC loss into milk (P<0.05). PRL injections did not restore milk yield but tended to reverse the effect of Quin by increasing milk protein content (P<0.10) and decreasing the milk MEC content (P=0.10). Our results showed that less MEC exfoliation into milk was associated with a greater lactation persistency in ovariectomized cows and that PRL could regulate MEC exfoliation. A large scale study is now necessary to determine if MEC exfoliation rate into milk is a phenotype that could characterize the ability of cows for a high lactation persistency.

Temperament traits and stress responsiveness in livestock: a developmental perspective

M.J. Haskell, S.P. Turner and K.M.D. Rutherford
SRUC, West Mains Road, Edinburgh EH9 3JG, United Kingdom; marie.haskell@sruc.ac.uk

An animal's behavioural and physiological response to stressful events is affected by a number of key influences. The animal's genotype and its exposure to pre-natal stress are important factors. Thereafter, the challenges of the post-natal environment and the animal's learning experiences further modify its response. Using data from cattle and pigs, this paper will explore the relevance of these different factors using specific examples of how animals respond to interactions with humans or other animals. Animals are consistent in the way they respond to human handling and to social challenge from conspecifics, and there is variation between individuals. An excitable behavioural response is associated with a higher stress response. The influence of genotype is shown by the moderate heritability estimates shown for response to handling in cattle, and for conspecific-directed aggression in cattle and pigs. Pre-natal stress also influences behaviour and stress responsiveness. Research on pigs has shown that it affects maternal behaviour, physiological responses to social challenge and neurotransmitter systems. In cattle, maternal stress during pregnancy has been shown to result in increased physiological stress reactions in calves. Learning from experience further modifies the response. Poor quality handling in cattle results in chronic stress and a poorer subsequent response to handling. In other species, the experience of winning and losing interactions influences the propensity to engage in further interactions. Higher excitability is associated with poor meat quality, growth, efficiency and health. We conclude that prenatal effects modify the outcome from the genetic template and in combination with postnatal learning, will dictate how animals respond to challenge. Understanding the interaction of genotype, pre-natal and post-natal stress and learning across the animal's lifetime are important in determining how animal welfare and productivity can be improved.

Genetic and molecular background of cattle behaviour and its effects on milk production and welfare
J. Friedrich[1], B. Brand[2], J. Knaust[2], C. Kühn[2], F. Hadlich[2], K.L. Graunke[3], J. Langbein[3], S. Ponsuksili[2] and M. Schwerin[1,2]
[1]University of Rostock, Institute for Farm Animal Research and Technology, Justus-von-Liebig-Weg 8, 18059 Rostock, Germany, [2]Leibniz Institute for Farm Animal Biology, Institute for Genome Biology, Wilhelm-Stahl-Allee 2, 18196 Dummerstorf, Germany, [3]Leibniz Institute for Farm Animal Biology, Institute for Behavioural Physiology, Wilhelm-Stahl-Allee 2, 18196 Dummerstorf, Germany; juliane.friedrich@uni-rostock.de

Breeding efforts for cattle temperament are becoming more relevant due the effect of behaviour characteristics on production traits and animal welfare, but knowledge about underlying genetic and biological mechanisms is still limited. In this study, a genome-wide association study was conducted (1) to identify genomic regions with impact on cattle behaviour traits assessed in early life at the age of 90 days and (2) to test for their associations to milk production traits. In total, 41 single nucleotide polymorphisms (SNPs) distributed over 21 chromosomes were identified to be significantly associated with active, inactive and exploratory behaviour in open-field and novel-object tests. Of the 9 SNPs which are simultaneously significant for behaviours and milk production, all showed competitive genotype effects for agitated behaviours and milk production. Further, gene expression profiles of the bovine adrenal cortex of cows slaughtered in the second lactation were analysed to identify transcripts with relevance for behaviour by applying a partial least squares regression analysis. 2,888 adrenocortical transcripts, significantly enriched in glucocorticoid receptor signalling and catecholamine biosynthesis, were detected to be associated with behaviour traits assessed in early life, indicating a consistency of stress responsiveness across time and situations. Thus, our results indicate a genetic background of cattle behaviour and we identified physiological mechanisms of the stress response to be a major behaviour determining factor in challenging situations in cattle.

Effects of late gestation heat stress on the physiology of dam and daughter
G.E. Dahl
University of Florida, Animal Sciences, P.O. Box 110910, 2250 Shealy Dr., Gainesville, FL 32607, USA; gdahl@ufl.edu

In lactation, heat stress (HS) effects are well known and include lower dry matter intake (DMI) and metabolic shifts that reduce milk production efficiency. When dry cows are HS they reduce DMI, but maintain normal metabolic profiles despite lower DMI. Late gestation HS reduces mammary growth, possibly due to suboptimal placental function. Dry period HS alters immune function and those impacts persist into lactation, resulting in a poorer transition outcome. Thus, a relatively brief duration of heat stress at a specific phase of the production cycle can have dramatic, negative outcomes on multiple physiological systems and overall productive efficiency. A recent focus has been on the effects of late gestation heat stress on calf survival and performance, with a series of studies to examine pre-weaning growth and health, and later reproductive and productive responses, to quantify acute and persistent impacts of in utero HS. Calves born to dams HS when dry have lower birth weight and remain lighter up to 12 mo of age vs calves from dams that are cooled (CL) when dry. Calves HS in utero have lower immune status compared with those from CL dams, beginning with poorer apparent efficiency of IgG absorption and extending to lower survival rates. Relative to CL calves, HS calves also have shifts in metabolism that lead to greater peripheral accumulation of energy and less lean growth vs those from CL dams. Comparing reproductive performance in HS vs CL calves, we observe that the CL heifers require fewer services to attain pregnancy and are pregnant at an earlier age. Finally, milk yield in calves HS vs those CL in utero reveals a 5 kg/d reduction in yield through 35 wk of lactation, despite similar bodyweight at calving. These observations indicate that a relatively brief period of HS in late gestation dramatically alters the health, growth, and performance of calves. Thus, environmental factors can program shifts in physiological systems in a sustained manner to the detriment of productive efficiency.

Betaine improves milk yield under thermoneutral conditions and after heat stress in dairy cows

J.J. Cottrell[1], M.L. Sullivan[2], J.B. Gaughan[2], K. Digiacomo[1], B.J. Leury[1], P. Celi[3], I.J. Clarke[4], R.J. Collier[5] and F.R. Dunshea[1]
[1]The University of Melbourne, Faculty of Veterinary and Agricultural Sciences, Royal Pde, Parkville 3010, Australia, [2]The University of Queensland, School of Agriculture and Food Sciences, Warrego Hwy, Gatton 4343, Australia, [3]The University of Sydney, Faculty of Veterinary Science, Private Bag 4003, Narellan 2567, Australia, [4]Monash University, Faculty of Medicine, Nursing and Health Sciences, Wellington Road, Clayton 3800, Australia, [5]The University of Arizona, Department of Animal Sciences, Shantz 205, P.O. Box 210038, Tucson, AZ 85721-0038, USA; fdunshea@unimelb.edu.au

Betaine is an organic osmolyte sourced from plant material that is accumulated in cells during osmotic stress. Since accumulation of betaine lowers energy expenditure and therefore metabolic heat production, the aim of this experiment was to investigate if betaine supplementation ameliorated heat stress in lactating dairy cows. Sixteen Friesian × Holstein cows (ca. 120 DIM) were allocated to control (0, n=6), 12.5 (n=5) or 25 g/d (n=5) natural betaine and 7 d Thermoneutral (TN, 18 C), 7 d Heat Treatment (HT, 8 h 35 °C/28 °C) then 5 d recovery (18 °C). Milk yield, feed intake and FCE were measured daily. Head and flank skin temperatures and respiration rate were counted every 2 h for one day during TN and recovery and daily during HS. Results were analysed for the effects of betaine dose, climate and the interaction using an ANOVA in Genstat. HT increased indices of heat stress, marked by increased temperature (head and flank) and RR (P<0.001 for all). Moreover HT reduced feed intake and milk yield by 38% compared to TN, while FCE was not affected. No effect of betaine was observed during HT, however betaine fed cows had increased RR and cooler temperatures during recovery (P<0.001). Betaine improved milk yield in TN (12.5 g/d), recovery (both, P=0.003) and FCE (25 g/d, P=0.016) by ~13%. In conclusion, addition of natural betaine up to 25 g/d improves the recovery from heat stress in dairy cows, improving welfare and milk production.

Cardiovascular function and anatomy in relation to feed efficiency in beef cattle

J. Munro[1], S. Lam[2], A. Macdonald[2], P. Physick-Sheard[2], F. Schenkel[2], S. Miller[3] and Y. Montanholi[1,2]
[1]Dalhousie University, Truro, NS, Canada, [2]University of Guelph, Guelph, ON, Canada, [3]AgResearch Limited, Invermay, Agricultural Centre, New Zealand; munroj@dal.ca

Oxygen consumption, heart rate (HR; bpm) and heart weight are associated with metabolic rate and thus with basal energy expenditure in the bovine. This suggests cardiovascular parameters are potential biomarkers of feed efficiency (residual feed intake, RFI; kg/d). Our objective was to characterize differences in cardiovascular function and anatomy associated with feed efficiency in beef cattle. Yearling beef bulls and steers (weight; 670±79 kg) from three populations (SS 16, SB 16, FS 16) were subjected to a performance test (bodyweight, body composition) and classified by RFI value into two groups (RFI$_{LOW}$, RFI$_{HIGH}$). Blood volume (BV; ml/kg; indocyanine dilution technique), HR and the indicator dilution curve will be used in the estimation of stroke volume. One day prior to slaughter animals were equipped with an electrode based HR recorder. HR values were averaged over 30 s into lowest HR during rest (HR$_{REST}$), transport (HR$_{TRANS}$) and stunning (HR$_{STUN}$). At slaughter myocardial tissue was collected for histomorphometrics and hearts were dissected and components weighed. Statistics were obtained using proc UNIVARIATE and means compared using proc GLM in SAS. In preliminary analysis means for BV, HR$_{REST}$, HR$_{TRANS}$ and HR$_{STUN}$ were; 94±28.4 ml/kg, 59.6±7.3, 83.3±12.3 and 114±28 bpm. Total ventricular weight (% heart weight; %HW) was higher in RFI$_{LOW}$ animals than in RFI$_{HIGH}$ animals (86.6 vs 85.9%HW; P=0.05). RFI$_{LOW}$ animals had a heavier right ventricle (22.0 vs 21.3%HW; P=0.04) than RFI$_{HIGH}$ animals with no difference in left ventricle weight. Differences in component weights between feed efficiency groups propose that cardiac structural capacity is associated with bovine energy metabolism. Estimation of stroke volume and completion of histomorphometrics will further our understanding of bovine energy metabolism and may identify new industry applicable phenotypes of feed efficiency.

Changes of NFkB activity in hyperglycemic piglet adipose tissue
J. Zubel-Łojek, K. Pierzchała-Koziec, E. Ocłoń and A. Latacz
University of Agriculture, Department of Animal Physiology and Endocrinology, Al. Mickiewicza 24/28,
30-059 Kraków, Poland; rzkoziec@cyf-kr.edu.pl

One of the most important regulators of inflammation is the transcription factor, nuclear factor kB (NFkB) that changes the expression of many cytokine genes involved in inflammatory responses. Inflammation leads to the development and progression of many diseases such as obesity, diabetes, cardiovascular disease and cancer. Obesity is associated with a low-grade inflammation of white adipose tissue (WAT) resulting from chronic activation of the immune system and which can subsequently lead to insulin resistance. Hyperglycemia in diabetes is associated with elevated levels of pro-inflammatory factors such as circulating cytokines. The aim of the experiment was to estimate the effect of hyperglycemia on the expression, concentration and secretion of NFkB from visceral adipose tissue after pro- and anti-inflammatory factors treatment. The experiment was carried out on the 10 weeks old piglets (Polish Landrace, n=18) divided into groups: control, hyperglycemia I (streptozotocin-treated, STZ) and hyperglycemia II (dexamethasone treatment, DEX). Total RNA was extracted from adipose tissue and quantitative PCR analysis was performed. NFkB levels were measured in tissue homogenates and incubation media by ELISA kit. The obtained results shown increase of NFkB activity in visceral adipose tissue during hyperglycemia in piglets (P<0.01). The augmentation of NFkB expression after streptozotocin by 428% and after glucocorticoid treatment by 856% was observed Concentrations of NFkB in tissue from hyperglycemic groups were higher approximately 2-fold compared to control animals. Interestingly, *in vitro* secretion of transcription factor was increased in STZ (pro-inflammatory) treated group whereas decreased after DEX (anti-inflammatory) treatment. The data suggest, that NFkB is an important factor connecting inflammation and metabolic diseases in WAT. Supported by grant: 12006406, NN 311227138 and DS 3243/KFiEZ.

The physiology of coping with and without water
C.H. Knight
University of Copenhagen, SUND IKVH, Dyrlægevej 100, 1870 Frederiksberg C, Denmark; chkn@sund.ku.dk

In the few days before parturition, a dairy cow might be drinking in the region of 15 litres of water a day and excreting 5 litres as urine. Within the first 24 or 48 hours of lactation, this will become around 125 litres drunk and maybe 30 litres excreted (it is remarkably difficult to find good data on the latter). Milk water will constitute a further flux of maybe 30 litres per day, so if we regard water drunk as a valuable resource, the 'efficiency' is better pre-partum (33% loss in urine) compared to post-partum (48% loss in urine and milk combined). The reason is simple; biologically, the mother is drinking for two, and the offspring need water just as much as she does. Despite many years of feed efficiency research, the concept of water flux having an efficiency is almost unheard of. We are increasingly familiar with the notion that agriculture accounts for a major proportion of the global water footprint, but we (correctly) assume that the animal's immediate needs are only a small proportion of this. Is this being realistic, or complacent? What would happen if the essential few percent of supply was not available to the lactating animal at the correct time? The answer is quite obvious, animals would die, and the fact that female mammals have adopted a large variety of physiological coping strategies to ensure that they and their offspring can maintain water homeostasis during lactation is testimony to the importance of water physiology. Curious, then, that water efficiency is un-researched, the relationships (if any) between mammary water flux and kidney water flux are unknown and the exact mechanism whereby water moves into milk is uncertain. This review will attempt to shed light on what we do know and what we need to know if we are to maximize water efficiency of our production animals, with emphasis on lactating ruminants.

A mechanistic model of water partitioning in dairy cows; water use under feeding and climate changes
J.A.D.R.N. Appuhamy and E. Kebreab
University of California, Department of Animal Science, One Shield Avenue, 95616, USA;
jaappuhamy@ucdavis.edu

Reliable estimates of water kinetics of dairy cows assists in determining water use efficiency of dairy cows accurately. The objective of the present study was to construct a mechanistic model to quantify water partitioning in individual lactating dairy cows. The model contains four body water pools: reticulo-rumen (Q_{RR}), post-reticulo-rumen (Q_{PR}), extracellular (Q_{EC}), and intracellular (Q_{IC}). Dry matter intake (DMI), dietary forage, DM, ADF, and ash contents, milk yield and milk fat and CP contents, days in milk, and body weight (BW) are input variables to the model. Drinking, feed, and saliva water inputs and fractional water passage from Q_{RR} to Q_{PR} are estimated within the model using a set of empirical equations. Water transfer via the rumen wall is adjusted for changes in Q_{EC} (e. g. dehydration) and total water input to Q_{RR}. Post-reticulo-rumen water passage is adjusted for DMI. Metabolic water production and respiratory cutaneous water losses are estimated with functions of heat production, which is estimated within the model. Water loss in urine is modelled to be driven by absorbed N left after being removed via milk. Model parameters were estimated using data (n=670) related to thermo-neutral conditions. Model was evaluated with an extra set of data (n=377) related to similar conditions. Model predicted drinking water intake, and fecal, urinary and total manure water output with root mean square prediction errors (RMSPE) 18.1, 15.4, 30.7, and 14.4% of average observed values, respectively. When evaluated with data (n=135) related to a summer in California, the respective RMSPE were 21.6, 15.3, 31.2, and 14.8%. Modifying the model by including the impact of environmental temperature and individual electrolyte balance would improve accuracy of the predictions, particularly drinking water intake and urine excretion under different feeding and climate conditions.

Relationships between water stable isotopes and electrolytes in dairy cattle body fluids
F. Abeni[1], F. Petrera[1], M. Capelletti[1], A. Dal Prà[1], L. Bontempo[2], A. Tonon[2] and F. Camin[2]
[1]Consiglio per la ricerca in agricoltura e l'analisi dell'economia agraria (CRA), Centro di ricerca per le produzioni foraggere e lattiero-casearie, Via Porcellasco 7, 26100 Cremona, Italy, [2]Fondazione Edmund Mach – Centro di Ricerca Istituto Agrario San Michele all'Adige, Dipartimento Qualità Alimentare e Nutrizione, Via E. Mach 1, 38010 San Michele all'Adige (TN), Italy; fabiopalmiro.abeni@entecra.it

The aim of this study was to approach water stable isotope changes as possible markers of heat stress in dairy cattle linking them to plasma and urinary changes across seasons. Preliminary data were collected in a farm in Po valley on Italian Friesian cattle from 3 different reproductive stages (RS): gestating heifers, primiparous lactating, and pluriparous lactating cows. Body fluids (fecal water, blood plasma, urine, and milk) were collected in winter and summer from 5 head/RS. All the samples were analysed to determine d^2H and $d^{18}O$; blood plasma and urine were analysed for mineral components (Na, K, Cl), creatinine (crea), protein, albumin. From stable isotope analysis, deuterium excess (d) in each sample was calculated. From urine analysis, the ratio between each mineral and creatinine was calculated to adjust the raw mineral urine data from a spot sampling. The urine:plasma creatinine ratio (UC:PC) was also calculated. Data from plasma and urine metabolites were processed by ANOVA to assess the effect of season and RS. Simple Pearson's correlation coefficients were calculated between the obtained values within each season. During summer, the $d^{18}O$ values in fecal water, plasma, and urine were correlated (r>0.850, P<0.001) with plasma creatinine level, and the d values in plasma and urine were negatively correlated with plasma creatinine. Urine Na:crea, K:crea, and Cl:crea were lower in summer for all RS. No significant correlations were evidenced between water isotopes data and these ratios in urine, nor with UC:PC ratio.

Betaine improves milk yield in grazing dairy cows supplemented with concentrates in summer
F.R. Dunshea, K. Oluboyede, K. Digiacomo, B.J. Leury and J.J. Cottrell
The University of Melbourne, Faculty of Veterinary and Agricultural Sciences, Royal Parade, Parkville
3010, Australia; fdunshea@unimelb.edu.au

Betaine is an organic osmolyte sourced from sugar beet that is accumulated in cells undergoing osmotic stress. Since accumulation of betaine lowers energy requirements and therefore metabolic heat production, the aim of this experiment was to investigate if betaine supplementation improved milk yield in grazing dairy cows in summer. One hundred and eighteen Friesian × Holstein cows were paired on days in milk and within each pair randomly allocated to a concentrate supplement containing either 0 or 2 g/kg natural betaine per day for 4 weeks during February/March 2015. The mean maximum February temperature was 30 °C. Cows were allocated approximately 14 kg dry matter pasture and 7.5 kg of concentrate pellets (fed in the dairy) per cow per day. and were milked through an automatic milking system three times per day. Results were analysed for the effects of betaine and day of treatment and the interaction using an ANOVA in Genstat v.15 and pre-treatment values as covariates. Betaine supplementation increased average daily milk yield by 5% (22.6 vs 23.7 kg/d, P<0.001) with the response increasing as the study progressed as indicated by the interaction (P<0.001) between betaine and day. Milk protein (P=0.98) and fat (P=0.46) content were unchanged by dietary betaine. However, betaine supplementation increased milk protein (706 vs 741 g/d, P<0.001) and fat (903 vs 943 g/d, P<0.001) yield with responses again being more pronounced (P=0.003) as the study progressed. In conclusion, dietary betaine supplementation during late summer increased milk and milk component yield in grazing dairy cows.

Relationship between the intensity of ruminal mat stratification and rumination in dairy cow
K. Izumi
Rakuno Gakuen University, Department of Sustainable Agriculture, 582, Bunkyodai Midorimachi, Ebetsu,
069-8501, Hokkaido, Japan; izmken@rakuno.ac.jp

Rumination is induced by tactile stimulation of ruminal wall, and it is thought that extensively stratified ruminal mat plays the main role in this stimulation. In this study, the product of consistency and thickness of the ruminal mat in cow was defined as ruminal mat stratification index (RMSI). The aim of this research was to examine the relationship between rumination activity and RMSI in a dairy cow. Five experiments were carried out on rumen-cannulated Holstein cows. In experiment 1, four lactating cows were fed either the control total mixed ration (TMR) or the twice-chopped TMR, containing mainly corn and grass silage, in a crossover design. In experiment 2, four lactating cows were provided each two diets in a crossover design. Diets were designed using two forage-to-concentrate ratios (low forage, 40:60; high forage, 60:40). In experiment 3, red bean hulls (RBH) were the non-forage fiber source, and four non-lactating cows were fed a control diet of 60.1% forage and an RBH diet containing 51.6% forage and 9.4% RBH in a crossover design. In experiment 4, three non-lactating cows were fed sake cake and grass hay (GH) at a ratio of 35:65 or 65:35. In experiment 5, four non-lactating cows were fed four different diets: alfalfa hay and beet pulp (BP) at 8:2 (dry matter basis, A8B2) and 2:8 ratios and GH and BP at 8:2 and 2:8 ratios in a 4×4 Latin square design. The consistency (q_c value, N/cm^2) and thickness (cm) of the ruminal mat and rumination activity were measured in all experiments. Ruminal mat characteristics were measured using the penetration resistance tests. A broken line correlation at the intersection point was noted between the RMSI (N/cm^2•cm) and rumination time. Increasing the RMSI up to 707.8 N/cm^2•cm significantly increased the rumination time (r=0.920; P<0.001; n=10). The correlation reached the asymptotic plateau over the intersection point at the rumination time 522 min/d. In conclusion, RMSI may be considered a factor that regulates rumination activity.

Effect of music on emotional state in Arabian race horses

A. Stachurska[1], I. Janczarek[1], I. Wilk[1] and W. Kędzierski[2]
[1]University of Life Sciences, Department of Horse Breeding and Use, Akademicka 13 str., 20-950 Lublin, Poland, [2]University of Life Sciences, Department of Biochemistry, Akademicka 12 str., 20-033 Lublin, Poland; anna.stachurska@up.lublin.pl

Race horses subject to training and races undergo strong stress. The emotional stress is expressed by among others a change of heart rate (HR) and heart rate variability (HRV). Physiological and psychological benefits from listening to the music, such as pain relief, lower anxiety, blood pressure and HR, are known in humans and some animal species. The question is whether the music may be also used in enrichment of the stable environment. The objective of the study was to determine the effect of music featured in the stable, on the emotional state of race horses. Three-year-old Purebred Arabian horses were studied in their first race season. The experimental group included 40 horses and control group 30 horses. The groups were placed in separate stables. For the experimental group, specifically composed music was featured in the stable for five hours a day during the whole study. The emotional state in the horses was assessed at rest, saddling and warm-up walk under rider. The measurements were taken six times, once 30-35 days from the beginning of the experiment. The horse's emotional state was assessed by HR and HRV. The cardiac activity variables were compared with repeated measures design. The music positively affected the emotional state in race horses. The influence was noticeable already after first month of featuring the music and increased in the second and third months. Later, the variables began to return to initial levels. The results suggest that the music may be featured in the stable, preferably for two to three months as a means of improving the welfare of race horses. The study was financially supported by the National Centre for Research and Development, Poland (N180061 grant).

Effects of *Spirulina* spp. when supplemented to dairy cows' ration, on animal heat stress

M.A. Karatzia and E.N. Sossidou
Hellenic Agricultural Organization-DEMETER, Directorate General of Agriculture Research, Veterinary, P.O. Box 60272, 57001 Thermi, Thessaloniki, Greece; mkaratz@vet.auth.gr

Heat stress (HS) in dairy cows is directly related with reduced dry matter intake (-56% DMI), decreased milk yield (-38%) and differentiation in behavioral traits, while severely suppressing farm profits (-417.3€/cow/year). In Greece, where cows are exposed to HS-inducing weather conditions for 5 months annually, the above mentioned consequences have been recently confirmed (significantly lower milk yield and DMI). Previous research indicated diverse feeding practices and management adjustments in order to minimize the effects of HS. *Spirulina* is a microalga, rich in proteins ($\approx$70% Dry Matter), vitamins and polyunsaturated fatty acids. The aim of this study is to investigate the effect of *Spirulina* powder, when incorporated into total mixed rations, on the alleviation of the consequences induced by HS on productivity, milk fatty acid profile and parameters determining animal welfare and behavior. Two experiments (one during winter and one during summer) will be carried out at a typical Greek Holstein dairy farm. Measurements will include milk composition and fatty acid profile, hematological profile, albumin, urea, cortisol and oxidative stress levels. Body condition score will be evaluated, rumination, breathing rate and behavior will be monitored. Results are expected to clarify whether *Spirulina* powder can serve as a nutritive tool to minimize the adverse effects that HS has on dairy cows, whilst supporting their health, productivity and welfare, as well as the economic profitability of farms. This research project is funded under the Action 'Research & Technology Development Innovation Projects'-AgroETAK, MIS 453350, in the framework of the Operational Program 'Human Resources Development'. It is co-funded by the European Social Fund through the National Strategic Reference Framework (Research Funding Program 2007-2013) coordinated by the Hellenic Agricultural Organization – DEMETER.

Behavioural response of lambs weaned from ewes that grazed in two pasture allowances in pregnancy
A. Freitas-De-Melo, R. Ungerfeld, M.J. Abud and R. Pérez-Clariget
Universidad de la República, Montevideo, 11600, Uruguay; alinefreitasdemelo@hotmail.com

The aim of this study was to compare the behavioural response to abrupt weaning of lambs born from ewes that grazed on high or low native pasture allowance (NPA) from 23 days before mating until 122 days of gestation. Twenty-four multiparous single-lambing Corriedale ewes were assigned to two NPA: (1) high (group HPA: n=12): the ewes grazed on 10 to 12 kg dry matter kg/100 kg of body weight (BW) per day; and (2) low (group LPA; n=12): the ewes grazed on 5 to 8 of dry matter kg/100 kg of BW per day. Three paddocks divided in two by electrical fences were used; therefore each treatment had 3 replications. From 122 days of gestation until 3 days before parturition all ewes grazed on Festuca arundinacea (14 kg of DM/100 kg of BW per day), and from then on, on native pastures with unlimited availability (12 to 15 kg of DM/100 kg BW per day) until parturition. From 116 days of gestation until lambing ewes were daily supplemented with 200 g/animal of rice bran and 50 ml of crude glycerine/animal. After lambing ewes were management as a single group and grazed on native pasture. Lambs were weaned 65 days after birth (day 0). The frequency of pacing, walking, standing, lying, grazing and ruminating were recorded on days 0 and 1, using 10 min scan sampling. Vocalizations were recorded for a 30 s period every 10 min using 0/1 sampling. The frequency of behaviours were compared between treatments on each day by ANOVA (standing, lying, ruminating and grazing) or Mann-Whitney test (pacing, walking and vocalizing). On day 0, HPA lambs vocalized more than LPA lambs (14.9±2.9 vs 6.4±2.1%; P=0.02), but this difference disappeared on day 1. On day 0 and 1, the frequency of other behaviours did not differ between groups. The greater frequency of vocalization at weaning in lambs from the HPA group, which is a main behavioural indicator of a stress response, suggests that those lambs were more stressed after weaning.

Digestion and metabolic parameters of domestic sheep and their hybrids with Argali
N.V. Bogoliubova, V.N. Romanov, V.A. Devyatkin, I.V. Gusev, R.A. Rykov and V.A. Bagirov
L.K. Ernst Institute for Animal Husbandry, Animal feeding and feed technology, Moscow region, Dubrovitsy, 142132, Russian Federation; 652202@mail.ru

The promising way to increase the biodiversity and to introduce the new properties in farm animal species is their hybridization with wild animals. The objective of our research was to compare the digestion processes and metabolism status of domestic sheep and their hybrids with Argali. The experiments were carried out on tree groups of animals with a rumen fistula: (1) pure bred Romanov sheep (n=3); (2) Romanov sheep with 12.5% of Argali blood; (3) Edilbaj sheep (50%), Romanov sheep (37.5%) and Argali (12.5%). The animals were fed on diet consisting of silage, haylage, hay, concentrates and salt balanced for sheep. The production of volatile fatty acids (VFAs), amylolytic activity and microbial mass in rumen have been measured and the metabolic parameters of blood have been determined. The increased level of VFAs production in rumen of hybrid animals comparing to Romanov sheep was observed (+8.7 vs +17.6%, P<0.05), but the ammonia production was higher in Romanov sheep (+8.4 vs +27.7%, P<0.05). The increased amylolytic activity was measured in the rumen of hybrid sheep of group 2 comparing to Romanov sheep (+2.04 U/mg) as result of higher content of the microbial mass, mainly bacteria in their rumen. The analysis of blood metabolic profiles revealed the higher concentrations of glucose (+16.0 vs +19.5%, P<0.001), ALT (+11.8%, P<0.05) and AST enzymes (+11.5 vs +15.5%, P<0.05) and reduced cholesterol level (-12.2 vs -20.4%, P<0.01) in group 3 of hybrid animals compared to Romanov sheep. The hybrid animals of group 2 were characterized by the higher creatinine level (+11.2 vs +15.5 m/l, P<0.05) and lower cholesterol level (-13.9 vs -16.7%, P<0.05). Our results indicated the differences in digestion and metabolic processes between pure bred and hybrid sheep and will be used to develop the optimal diets for hybrid animals. The study was carried out with financial support of the Russian Science Foundation within Project no. 14-36-00039.

Effects of prepartum housing environment on the gestation length and parturition performance in sows

J. Yun, K. Swan, O. Peltoniemi, C. Oliviero and A. Valros
Univeristy of Helsinki, Department of Production Animal Medicine, P.O. Box 57, Koetilantie 2, 00790, Helsinki, Finland; jinhyeon.yun@helsinki.fi

Studies suggested that parturition is implemented by hormonal changes, and that the gestation length is associated with litter size and piglet mortality. We investigated whether prepartum environments affect the gestation length in sows. We also studied whether the gestation length is associated with litter size, and prepartum maternal behaviour and hormones in sows. Sows were transferred from group-housing gestation pens to farrowing units 7 days before the expected farrowing date. We allocated 33 sows in: (1) CRATE: the farrowing crate closed (210×80 cm), with provision of a bucketful of sawdust; (2) PEN: the farrowing crate opened, with provision of a bucketful of sawdust; (3) NEST: the farrowing crate opened, with provision of abundant nest-building materials. Sows were video-recorded from 18 h prior to parturition until birth of the first piglet for analysing prepartum nest-building and bar-biting behaviour. We collected sow blood samples for oxytocin and prolactin assays via catheters on days -3, -2, -1 from parturition twice a day. Oxytocin and prolactin concentrations were analysed with a mixed model using repeated measures. Litter size, still born, and born alive piglets were recorded. A multiple comparison procedure, according to a randomized complete block design, was used for analysis of the gestation length. Correlations between the gestation length and prepartum oxytocin and prolactin, bar-biting and nest-building behaviour, and litter performance were tested with Spearman rank correlation coefficients (r_s).The gestation length was negatively correlated with litter size (r_s=-0.39, P<0.0001), as well as with the number of total born alive piglets (r_s=-0.28, P<0.01). However, the gestation length did not differ among treatments (P>0.10), and had no correlations with prepartum oxytocin and prolactin concentrations, or bar-biting and nest-building behaviour in sows.

Employment of the Wolff Chaikoff phenomenon to mitigate acute heat stress-induced hyperthermia

H.J. Al-Tamimi and Z. Mahasnih
Jordan University of Science and Technology, Animal Science, College of Agriculture, 22110 Irbid, Jordan; hosamt@gmail.com

Some of the key changes linked to global warming are the pronounced sudden heat waves. This is often associated with severe hyperthermia, especially in grazing livestock. In response to heat stress (HS), animal's metabolic heat production defensively falls, through lowering calorigenic hormones, primarily thyroid. However, thyroid suppression requires prolonged period. High iodine treatment induces rapid hypothyroidal state – Wolff Chaikoff effect (WCE). We aimed to investigate the WCE efficacy in delineating exacerbated thermophysiological responses to acute HS. Twelve radiotelemetered rats, intensively (at 15-minute intervals) emitting body temperature (Tcore), locomotive activity (LA) and heart rate (HR) data (split plot in time), were exposed to thermoneutrality (TNZ; 23 °C) for 7 days, followed by an acute HS bout (36 °C; 5 hours) and finally resumption of TNZ for another 7 days. On the HS day, male Wistar rats (n=12) either received 0 (CON) or 0.5% potassium iodide (KI) in drinking water, and blood samples to quantify circulating free triiodothyronine (FT3) were collected immediately prior to return to the final TNZ period. Feed and water intakes alongside body weights were recorded daily. Successfully, KI induced (P<0.01) hypothyroidal response (2.26 and 1.66±0.12 pg/ml FT3), for the CON and KI groups, respectively. This momentary goitrogenic response enhanced the thermal state (0.75±0.12 °C drop in Tcore; P<0.05) during HS. Yet, WCE exerted an instant and delayed twofold suppression in LA, but a delayed elevation by nearly 25% in daytime HR. A rebound in Tcore – likely due to an 'escape phenomenon' – occurred within 24 hours from KI withdrawal. No adverse effects (P>0.05) were noticed in growth performance data due to KI treatment. Results implicate the efficacy of the WCE to overcome acute HS in rats. Further research is needed to assess efficacy of such protocol in alleviating the impact of acute HS on farm animals.

Progesterone administration reduces the behavioural response of ewes at weaning

A. Freitas-De-Melo, R. Ungerfeld and R. Pérez-Clariget
Universidad de la República, Montevideo, 11600, Uruguay; alinefreitasdemelo@hotmail.com

The aim of this study was to determine if the injection of oil-based progesterone (P4) immediately before abrupt weaning of lambs reduces the behavioural response of ewes. The study was performed during the non-breeding season with 24 multiparous single-lambing Corriedale ewes. Twelve ewes received 50 mg of oil-based P4 (MAD-4, Laboratorio Río de Janeiro, Argentina) sc immediately before weaning (group GP), and the other 12 ewes remained as an untreated controls (group GC). Lambs were weaned 66 days after birth (day 0). The frequency of pacing, walking, standing, lying, grazing and ruminating were recorded from day -3 to 0 and day 2, using 10 min scan sampling. Vocalizations were recorded for a 30 s period every 10 min using 0/1 sampling. The frequency of behaviours was compared between treatments using ANOVA (standing, ruminating and grazing) or Mann-Whitney test (pacing, walking, lying and vocalizing). The frequency of each behaviour recorded before weaning did not differ between treatments. On Day 0, GP ewes had lower frequency of pacing in GP (only one ewe paced once) than in GC ($2.6\pm0.8\%$; $P=0.01$). GP ewes walked less (1.7 ± 0.6 vs $10.6\pm1.4\%$; $P<0.0001$) and ruminated and standed more than GC ewes (15.7 ± 1.5 vs $11.0\pm0.9\%$; $P=0.01$; and $48.7\pm2.2\%$ vs $41.7\pm1.7\%$; $P<0.02$ respectively). The frequency of grazing tended to be lower in GP than GC ewes (27.3 ± 2.7 vs $32.8\pm1.5\%$; $P=0.08$); without difference in lying. On day 2, GP ewes had lower frequency of standing than GC ewes (73.2 ± 1.9 vs $82.2\pm1.7\%$; $P=0.002$ respectively); the frequency of lying was greater in GP than in GC ewes (26.7 ± 1.9 vs $10.9\pm1.5\%$; $P<0.0001$). The differences in walking and ruminating frequencies were still present ($P<0.0001$ and 0.0004 respectively). The frequency of other behaviours did not differ between groups. Oil-based progesterone injected immediately before abrupt weaning of lambs reduced the behavioural response of ewes to weaning.

Neuroendocrine responses to prebiotic treatment in lambs

K. Pierzchała-Koziec[1], J. Zubel-Łojek[1], E. Ocłoń[1], A. Latacz[1], K. Pałka[1] and S. Kwiatkowski[2]
[1]University of Agriculture, Department of Animal Physiology and Endocrinology, Al. Mickiewicza 24/28, 30-059 Kraków, Poland, [2]University of Kentucky, Department of Cellular and Molecular Biochemistry, 138 Leader Ave., Lexington, KY 40506, USA; rzkoziec@cyf-kr.edu.pl

Mannan oligosaccharides, extracted from the yeast wall of *Saccharomyces cerevisiae*, are widely used as prebiotic feed additives. However, little is known about their impact on the physiology of nervous and endocrine systems. Thus, we tested the hypothesis that mannan perturbs brain and peripheral hormones levels and disrupts neuroendocrine interactions essential for stress response and appetite in lambs. The experiments were carried out on 48 female lambs, 30 days old, divided into two controls and six experimental groups injected 5 times, every 6 days with ghrelin, naltrexone and glucocorticoid. Half of the animals were fed with regular feed and the others were fed with addition of mannan (0.1%) for 30 days. Blood, pituitary and adrenals were taken out at the end of experiment. The levels of ACTH, cortisol, ghrelin, leptin and Met-enkephalin were estimated by RIA, transcript levels of some genes were measured by PCR-RT. Injections of hormones and opioid receptor antagonist, naltrexone, significantly changed the levels of all tested parameters. Mannan treatment reduced the plasma levels of ACTH, cortisol, ghrelin, leptin and Met-enkephalin compared to lambs fed without prebiotic. Prebiotic caused elevation of the native Met-enkephalin level in pituitary and affected the transcript levels of genes involved in stress and appetite elevation. Cortisol concentration and releasing from the adrenal were decreased by mannan treatment. Collectively, these results showed that mannan, despite its positive effect on the gastrointestinal physiology, is a neuroendocrine disruptor, impacting the stress and feeding responses in lambs. Supported by NCBR grant 12006406.

A preliminary study on the changes of FT3 concentration during prepubertal period in red deer hinds
J. Kuba, B. Błaszczyk, T. Stankiewicz and J. Udała
West Pomeranian University of Technology in Szczecin, Department of Animal Reproduction Biotechnology
and Environmental Hygiene, Judyma 6, 71-466 Szczecin, Poland; jan.udala@zut.edu.pl

The aim of the study was to analyze the annual changes in concentration of free triiodothyronine (FT3) in prepubertal red deer females maintained in an ecological farm. The study was performed on eight red deer (*Cervus elaphus elaphus*) females which in the beginning of the experiment were 6 months old. The experiment continued for the 12 months that preceded and partially involved pubertal period of animals. Samples were collected monthly from November to October the following year. To evaluate the length of the prepubertal period, the concentration of progesterone in the hinds serum was monitored. Statistical analysis included the repeated-measure ANOVA and the significance of differences between means was determined by Duncan's test. Analyses were performed with significance level of $P<0.01$. The concentration of progesterone was at a low and stable mean level (0.98 ± 0.41 ng/ml) from November to August. In the next two months the concentration of P4 significantly ($P<0.01$) increased, reaching 2.91 ± 0.68 ng/ml in September and 3.04 ± 0.72 ng/ml in October. The profile of changes in FT3 in females was characterized by significantly ($P<0.01$) higher levels during the winter-spring (January to June) period than in the summer-autumn (July to December). The highest concentration of FT3 was observed in April at 8.06 ± 1.34 pmol/l and the lowest in October, at 3.74 ± 1.17 pmol/l. The results suggest that in red deer hinds puberty, pronounced by an elevated progesterone level indicating the luteal activity, appears during the second autumn of life, which corresponds to the age of 16 months. The dynamics of changes in FT3 secretion shows the annual rhythm which responds to seasonal variations, aside from sexual maturity.

Effects of xylanase on nutrient digestibility and blood and urine xylose concentrations in pigs
M.C. Walsh[1] and E. Kiarie[1,2]
[1]DuPont Industrial Biosciences, Danisco Animal Nutrition, Marlborough, United Kingdom, [2]University of
Manitoba, Department of Animal Sciences, Winnipeg, Canada; maria.walsh@dupont.com

The use of xylanase to target the insoluble fiber fraction of cereal by-products is well known to mitigate some of the negative effects fiber has on the nutrient digestibility of the diet. Delivering the optimum dose of in-feed xylanase appears to be of fundamental importance as high inclusion levels have been thought to result in excessive production of pentose sugars (xylose and arabinose) which are known to be a poor source of energy for pigs. This study was conducted to examine the efficacy of a high dose of 3 xylanases (20,000 U/kg) on both nutrient digestibility and the release of xylose in the blood and urine of pigs fed corn/soybean meal diets (PC) or diets containing 70% PC + 30% corn DDGs w/w (NC) with or without xylanase A, B or C. All diets contained supplemental microbial phytase (500 FTU/kg) and were formulated to meet nutrient requirements of growing pigs (20-50 kg BW, NRC 2012). Sixty pigs (BW, 21 kg) were individually housed in metabolism crates and randomly allotted to 1 of 5 diets in a completely randomized design over two treatment periods (6 pens/diet/period). Treatment period included 4 days of adaption followed by 3 days of faecal and urine collection. Pigs were restricted fed at 3% of BW which was offered twice daily. Blood samples were collected following a 12 h fast at the end of each period. All 3 xylanases tested increased the faecal digestibility of fat compared to the NC diet ($P<0.05$). Xylanase C increased the ADF digestibility compared to the NC diet ($P<0.05$). There were no differences in blood xylose concentration between treatments. The NC diet increased ($P<0.05$) urinary excretion of xylose compared to the PC diet but this increase was not enhanced by the addition of any of the xylanases tested. The present data indicates that excessive xylose production may not be a constraint to feeding high doses of xylanase to pigs.

Open-field response in fish subjected to high and low stocking density

W.M. Rauw[1], L.A. García Cortes[1], A.M. Larrán[2], J. Fernández[1], J. Pinedo[2], M. Villarroel[3], M.A. Toro[3], C. Tomás[2] and L. Gómez Raya[1]
[1]*Instituto Nacional de Investigaciones Agroalimentarias, Carretera de la Coruña km 7, 28040, Spain,* [2]*Instituto Tecnológico Agrario (Junta de Castilla y León), Centro de Investigación en Acuicultura, Ctra Arévalo s/n, 40196 Zamarramala, Spain,* [3]*E.T.S.I. AGRÓNOMOS (Universidad Politécnica de Madrid), Departamento de Producción Animal, Senda del Rey s/n, 28040 Madrid, Spain; gomez.luis@inia.es*

Reducing stress and its harmful effects on fish production is a goal among aquatic farmers. The stocking densities used in fish production are a welfare concern. Sustained levels of stress result in significant changes in swimming patterns. The aim of this study is to investigate behavioral alterations in rainbow trout resulting from different stocking densities. A total of 2,000 rainbow trout (*Onchorynchus mykiss*) were randomly allocated to either a high (37 kg/m^3; HD) or a low (6 kg/m^3; LD) stocking density treatment. After six weeks, HD fish were reallocated to a low stocking density. On day 0, 14, 42 (T42), 61 (T62), and 78, 40 fish per treatment were randomly sampled, weighed, and subjected for 5 minutes to a novel aquarium of 25 l. Movements were recorded by means of an overhead camera. Distance travelled was estimated with the software Smart© version 3.0 (Panlab Harvard Apparatus®) between 15-30 (P15-30), 30-60 (P30-60), and 60-120 (P60-120) seconds. Results are reported for T42 and T62 only. HD fish weighed less (72.5±2.4 and 92.0±2.4 g) than LD fish (88.0±2.4 and 111.0±2.4 g) at T42 and T61, respectively (P<0.0001). No differences were observed for distance reached in any of the time periods at T42. However, at T62, when HD fish had been allocated to a low stocking density, LD fish swam a significantly larger distance than HD fish, which was significant in P30-60 (205±25 vs 156±26 cm, respectively; P<0.01). Differences in body weight did not significantly influence these observations. The results suggest that stocking density has affected swimming patterns and possibly coping styles in trout.

Evolution of body condition score, insulin and IGF-I during pre and postcalving

P. Soca[1], M. Carriquiry[1], M. Claramunt[1] and A. Meikle[2]
[1]*University of Republic, Animal Science and Pasture, Ruta 3 Km 363, 60000. Paysandú, Uruguay,* [2]*University of Republic, Faculty of Veterinary, Lasplaces 1550, Montevideo, Uruguay; psoca@fagro.edu.uy*

The objective of this experiment was to analyze the effect of body condition score (BCS) at calving on BCS, insulin and IGF-I during pre and postcalving of primiparous beef cows grazing grassland. During pre (-40 postpartum days DPP) and postcalving (100 DPP) we employed (n=56) primiparous cows classified by BCS at calving (low ≤3.5=L, moderate ≥4=M; 1-8 visual scale). At 55±10 DPP all cows were in anestrous as confirmed from two consecutive weekly plasma samples with progesterone concentration <1 ng/ml and submitted to suckling management during 12 days and immediately after suckling cows receiving or not (flushing = F vs control = NF) 2 kg/d of whole-rice middling for 22 d. Each 10 days BCS, insulin and IGF-I blood samples were registered. The effect of BCS at calving and DPP on BCS, insulin and IGF-I were analyzed by a repeated-measures using MIXED procedure with DPP postpartum as the repeated effect and first-order autoregressive as the covariance structure. The BCS and insulin was affected by DPP (P<0.001) and BCS at calving (P<0.001) while BCS at calving × DPP (P<0.001) controlled IGF-I patterns. The difference in BCS at -40 DPP and calving (L=4.0±0.3 vr M=4.5±0.5) was not generated experimentally. Loss of BCS and IGF-I during -40-+20 DPP were superior in M group. Insulin and IGF-I concentration were 17 (L=7.4 vr M=8.7 µI, P<0.03) and 55(L=48.9 vr M=71.6 µI P<0.0001) per cent superior in M. Increased in insulin and IGF-I during flushing (67-89 DPP) were relevant in M (P<0.001). Small difference en BCS at calving reflect cow´s differences in efficiency of energy use and suggests that M moderate would have greater metabolic plasticity and the importance of metabolic memory in endocrine response to flushing.

Effects of vitamin E supplements on growth and blood parameters in hybrid catfish
S. Boonanuntanasarn
Suranaree University of Technology, Institute of Agricultural Technology, 111 University Avenue, Muang,
Nakhon Ratchasima, 30000, Thailand; surinton@sut.ac.th

The present study investigated the effects of supplementary vitamin E in a practical diet for hybrid catfish (*Clarias macrocephalus* × *Clarias gariepinus*) on growth, hematological and immune parameters. Four experimental diets were used: the basal diet (non-supplemented), and the basal diet supplemented with vitamin E equivalents t of 125, 250 and 500 mg/kg of feed. In this study, the α-tocopherol present in the feed ingredients was detected in all treatment diets and ranged from 8.42 to 10.22 ppm. In addition, the analyzed tocopheryl acetate in treatment diets reflects the supplementary levels. The survival rate and feed conversion ratios were similar in all groups. Growth performance, including weight gain, specific growth rate and feed conversion ratio, appeared to be similar for all treatment groups. Blood sampling and hematological and immune parameters were assessed after week 4 and week 8 of the experimental period. No significant differences in red blood cell number, hematocrit and hemoglobin concentration were observed among hybrid catfish that were fed various level of dietary vitamin E. In addition, different levels of vitamin E intake had no effect on lysozyme, plasma protein and total immunoglobulin. Therefore, a low level of vitamin E supplementation would be sufficient to prevent physiological disorder, as well as to maintain a non-specific immune response in hybrid catfish. Further experiments would be required to investigate the supplementary level that would be benefit to fish under stressful condition.

Effect of group housing gilts with ESF on physiological and reproductive responses
J.C. Jang, S.S. Jin, J.S. Hong, S.O. Nam, H.B. Yoo and Y.Y. Kim
Seoul National University, Department of Agricultural Biotechnology, College of Animal Life Sciences, 1
Gwanak-ro, Gwanak-gu, 151-742, Seoul, Korea, South; yooykim@snu.ac.kr

An experiment was conducted to assess the welfare of gestating gilts in groups with Electronic Sow Feeding (ESF) system compared to conventional stalls. A total of 84 gilts (Yorkshire × Landrace) was housed in individual stalls to be artificially inseminated. Gilts with confirmed pregnancy were introduced to conventional stalls (ST) or groups with ESF system (ESF) treatment. All gilts were taken to the farrowing crates 1 week prior to their expected farrowing date. In gestation period, gilts housed in group tended to higher live body weight and backfat thickness compared to ST treatment. Moreover body muscle content of pregnant gilts was also higher in ESF treatment at 110 days of gestation (P<0.01). Reproductive performance was not changed by treatment and parturition time tended to decrease when gilts were reared with ESF system but piglet mortality was increased in ESF treatment (P<0.01). In blood profiles, ST gilts showed higher cortisol level at 110 days of gestation than that of ESF gilts (P<0.01). Weaning to estrus interval was shorter in ESF treatment than that of ST group (P<0.01). Higher locomotion score and incidence of scratches were observed in ESF treatment particularly at 36 days of gestation because of competition of gilts for ranking in group (P<0.05). Consequently this experiment demonstrated that gilts in ESF system are negatively influenced in terms of piglet mortality and incidence of scratches, but have higher locomotion. ESF gilts showed higher health status, farrowing time was shortened compared to ST treatment.

How does cows activity change after feeding bin change?

M. Soonberg and D. Arney
Estonian University of Life Sciences, Kreutzwaldi 1A, 51006, Estonia; maria.soonberg@emu.ee

Understanding the feeding behaviour of dairy cattle in different indoor housing systems is important to optimise production and welfare. Outdoors,grazing cattle walk about 4 km/day, graze for about 4-14 hours within a 24-hour period and lie down for about 9-12 hours. Monitoring the locomotion of cow can be used to predict oestrus and lameness. And the same activity monitors can be used to estimate activity and feeding visits by cows. In a system in which cows are grouped and given different access to feeding bins with different rations, and these groups change over time, it is important to find out how a change in the ration, and a change in the feeding bin, affects the cow's feeding behaviour, and if so, for how long. Ice tag activity monitors were attached to the right hind leg of six cows. Walking, standing, lying data and health records were used to record changes before and after a change in the feed ration / feeding bin.

Udder characteristics affects teat utilisation in three pig breed

M. Ocepek[1], I. Andersen-Ranberg[2] and I.L. Andersen[1]
[1]Norwegian University of Life Sciences, Department of Animal and Aquacultural Sciences, P.O. Box 5003, 1430, Norway, [2]Topigs Norsvin, P.O. Box 504, 2304 Hamar, Norway; marko.ocepek@nmbu.no

The ongoing selection of maternal sow lines has resulted in larger litters, greater piglet weight gain and an increased number of functional teats. Still, competition for teats causes the loss of several live born piglets and some preliminary findings suggest that this could be related to an impaired udder morphology reducing teat utility early after birth. The aim of the present study was to investigate differences in udder characteristics (length and diameter of functional teats and teat pair distance) and teat utilisation (teats being used) in two different breeds (Norsvin Duroc; n=12, and Norsvin Landrace; n=12) and one crossbreed (Norsvin Landrace × Yorkshire; n=14). The differences in udder characteristics and teat utilisation were analysed in SAS. There were no significant difference between the three breeds in teat pair distance and teat diameter. In all breeds, multiparous sows had greater teat pair distance, teat length, and teat diameter (P<0.0001). In Norsvin Landrace, teat pair distance, teat length and diameter increased with an increasing number of functional teats. Teats were less used on day 1 postpartum than later in lactation period (P<0.0001). On day 1, teat utilisation declined from anterior to posterior position and this bias was greater in the lower teat row, where approximately half of the middle and posterior teats were not used. Use of lower middle teats declined when teat pair distance became greater than 16 cm, in particular to longer teats (P=0.0090). Utilisation of lower posterior teats reduces when teat pair distance exceeded 14 cm. With further surge in teat pair distance for around 10%, use of teats in upper teat row declined, especially to shorter teats. Fewer posterior and anterior teats were used as parity increased. The present results suggest that teat pair distance is an important trait to include in the breeding programme to ensure that the piglets can access and use the teats.

Effect of heat stress on age at first calving and calving interval of Wagyu (Japanese Black) cattle

T. Oikawa
University of the Ryukyus, 1 Senbaru, Nishihara-cho, Okinawa, 903-0312, Japan; tkroikawa@gmail.com

Prolonged calving interval has been a problem to be improved in Japanese Black cows. One of a factor of the prolongation is considered to be heat stress. Average temperature of Japan is rising considerably in recent 30 years because of global warming. Okinawa is located in sub-tropical zone; 700 km south of main island of Japan, where half of a year is in warm season, and the rest is in neutral season. The objective of this study is to evaluate an effect of heat stress on reproductive traits; age at first calving (AFC) and calving intervals (CI) in Okinawa. CI in study was three CIs, first to third CI (CI1, CI2 and CI3). The data set consisted of 7961 to 9,279 reproductive records of cows after eliminating records of small farms and minor sires. As a result, number of farms was 35, and number of sires was 50. Also reproductive records were limited so as to be within a certain range. Statistical analysis was conducted by REG, GLM and NLIN procedure of SAS 9.3. Because of sub-tropical climate of Okinawa, average THI (temperature humidity index) was above 72 in half a year. Least square means of calving months were low in summer, however, high in winter in these reproductive traits. The result was opposite to previous results studied on dairy cattle. Thus, in Okinawa, nutritional condition was considered to be a limiting factor for reproduction. Then effect of THI was analyzed on days from 31 before a service to 61 after a service. THI with maximum temperature and minimum humidity was chosen to be evaluated for this analysis according to AIC value. Significant days of THI were -7, -2 and +31 for AFC; -3 and +33 for CI1; -20, +3 and +45 for CI2; -25, -17, -12 and -8 for CI3. The regressions were in a range of linear to quadratic and downward to upward, depending on a day of THI. These results are graphically presented.

The effect of plant extracts on rumen fermentation and the rumen microbiota

C.J. Newbold[1], G. De La Fuente[1,2], A. Belanche[1], K.J. Hart[1], E. Pinloche[1], T. Wilkinson[1], E.R. Saetnan[1] and E. Ramos Morales[1]
[1]Aberystwyth University, IBERS, Aberystwyth, SY23 3AL, United Kingdom, [2]Universitat de Lleida, Lleida, 25003, Spain; cjn@aber.ac.uk

Microbial fermentation plays a central role in the ability of ruminants to utilize fibrous substrates; however rumen fermentation also has potential deleterious environmental consequences as it ultimately leads to the emission of greenhouse gases and breakdown of dietary protein leading to excessive N excretion in faeces and urine. The removal of antibiotic growth-promoters and consumer preferences for more 'natural' production systems has led to an increased interest in alternative means of manipulating rumen fermentation. There is a significant body of work that shows plant extracts can modify rumen fermentation. However, many studies have been carried out *in vitro* and there is a lack long term *in vivo* and production based studies. Nevertheless, it is becoming clear that the main classes of plant extracts used, organosulphurous compounds, essential oils, tannins and saponins, differ in their mode of action within the rumen. Indeed even within classes of additive it is apparent that the action of individual compounds varies dependent on their chemical structures and that mixes of compounds often have actions that cannot be accurately predicted by actions of individual compounds studied in isolation. Molecular techniques based on amplification of ribosomal genes have allowed both quantitative and qualitative studies on the response of microbial populations in the rumen to plant extracts to be carried out. The ability to target archaea, prokaryotes and eukaryotes using quantitative PCR and massively parallel amplicon sequencing is bringing a new level of understanding as to the effect of plant extracts on the rumen microbiota. Whilst the developments in the study of functional genes (metagenomics) and their expression (metatranscriptomics) promises to greatly enhance our understanding of how plant extracts affect the activity of individual microbial groups in the rumen.

Comparison of effects of forage legumes containing condensed tannins on milk and cheese quality

M. Girard[1,2], F. Dohme-Meier[1], D. Wechsler[3], M. Kreuzer[2] and G. Bee[1]
[1]Agroscope, ILS, Tioleyre, 1725 Posieux, Switzerland, [2]ETHZ, ILS, Universitätstrasse 2, 8092 Zurich, Switzerland, [3]Agroscope, IFS, Schwarzenburgstrasse 161, 3003 Bern, Switzerland; marion.girard@agroscope.admin.ch

An experiment was conducted to evaluate whether diets containing condensed tannins (CT) from different forage plants can affect milk and cheese quality by increasing polyunsaturated fatty acids without affecting negatively their sensory properties. Twenty-four Holstein cows were assigned to four treatments for 26 d (period E) after having passed a 26-d control (C) period, where a basal diet composed of hay, maize silage, linseed:concentrate and lucerne (LU) pellets was fed (45:25:5:7:18%). In E, in 3 of the 4 groups LU was replaced by either sainfoin (SF; 19% CT) or 2 cultivars of birdsfoot trefoil (polom, BP, 3% CT; bull, BB, 5% CT). At the end of the C and E, milk was collected on 3 consecutive days and analysed for milk fatty acid profile and processed to Gruyère-type cheese. In C and E, neither feed intake nor milk yield differed among treatment groups. From C to E, milk urea concentration was reduced (P<0.001) by 23% with SF, remained unchanged with BP and tended to increase (P< 0.10) with LU and BB. The odor of the fresh BB milk was judged to be different (P<0.05) from LU milk. From C to E, the 18:3n-3 proportion of total fatty acids increased by 17% in milk (P=0.07) and cheese (P<0.001) with SF, increased (P=0.04) by 3% in BP cheese and tended (P=0.07) to decrease in BB cheese. From C to E, the 20:5n-3 and 22:5n-3 proportions tended (P<0.10) to increase in SF cheese. Compared to LU cheeses, CT cheeses were judged harder (P<0.05) and tended to be less adhesive (P=0.08) on the palate. In addition, SF and BP cheeses had less rind (P<0.05). In conclusion, feeding SF seems preferable from a dietetic point of view and this at unchanged sensory quality and similar cheese properties compared to the birdsfoots trefoil. Interestingly, despite similar CT content these 2 cultivars had opposite effects on milk urea and 18:3n-3 deposition.

Effect of sainfoin (*Onobrychis viciifolia*) silage on feed digestibility and methane emission in cows

N.T. Huyen[1], O. Desrues[2], W.H. Hendriks[1,3] and W.F. Pellikaan[1]
[1]Animal Nutrition Group, Wageningen University, P.O. Box 338, 6700 AH Wageningen, the Netherlands, [2]Parasitology and Aquatic Diseases, University of Copenhagen, 1870 Frederiksberg C, Denmark, [3]Department of Farm Animal Health, Utrecht University, P.O. Box 80.163, 3508 TD Utrecht, the Netherlands; huyen.nguyen@wur.nl

Sainfoin is a tanniniferous, leguminous plant that has potential nutritional and health beneficial effects on preventing bloating, reducing nematode larval establishment, improving N utilization and reducing greenhouse gas emission. However, the use of sainfoin as a fodder crop for dairy cows in the Netherlands is still limited. The objective of this study was to evaluate the effect of sainfoin silage on feed digestibility, animal performance, rumen functioning and methane (CH_4) production. Six rumen cannulated lactating dairy cows housed individually, received a grass and maize silage based control diet. Animals were first allowed to adapt to the control diet for 10 days prior to the start of the trial. Subsequently, they were assigned to their diet treatments, either remaining on the control diet or receiving a sainfoin based diet in a cross over design. In the sainfoin diet, 50% of grass silage in TMR was exchanged by sainfoin silage. The animals remained on their diet treatment for 4 weeks, successively placed on a control diet for 10 days, and changed for diet treatment for a second 4-week period. During the 4-week periods, the cows were transferred for three days per week to climate controlled respiration chambers to determine feed intake, digestibility, CH_4 and milk production. All data were analyzed using the mixed procedure of SAS. Total daily DM (18.3 vs 17.1 kg DM) and ADF (4.1 vs 3.7 kg ADF) intake of the sainfoin diet was higher (P≤0.04) than the control diet. The digestibility of DM, OM, NDF and ADF was similar among the two diets. Methane production per kg milk production was lowest (P=0.03) on the sainfoin diet. Based on the results, it was concluded that sainfoin silage improved feed intake and reduced CH_4 per kg milk production.

Effects of quercetin on liver health in dairy cows during periparturient period

A. Stoldt[1], M. Derno[1], G. Nürnberg[1], A. Starke[2], S. Wolffram[3] and C.C. Metges[1]
[1]Leibniz Institute for Farm Animal Biology (FBN), Wilhelm-Stahl-Allee 2, 18196, Dummerstorf, Germany,
[2]University of Leipzig, An den Tierkliniken 11, 04103, Leipzig, Germany, [3]Christian-Albrechts-University
of Kiel, H.-Rodewald-Straße 9, 24118, Kiel, Germany; metges@fbn-dummerstorf.de

Flavonoids, such as quercetin (Q), have hepatoprotective potential and reduce hepatic lipid accumulation in rodents, and may prevent high yielding dairy cows from fatty liver during the periparturient period. Because Q is largely degraded in the rumen, we investigated effects of intraduodenally administered Q on metabolic parameters, energy metabolism, and liver status in periparturient cows. Holstein cows, equipped with a duodenal fistula, were fed grass silage ante partum (ap) and a TMR *ad libitum* post partum (pp). Starting from 3 wk ap, cows received either daily 100 mg of Q-dihydrate/kg BW duodenally (n=5) or NaCl as control (CTR; n=5) for 6 wk. At 3 and 2 wk ap as well as at 2 wk pp, energy expenditure (EE), energy balance (EB), and fat oxidation were measured via indirect calorimetry. Plasma metabolites and enzymes were analyzed and fat content was determined in liver biopsies ap and pp. Q-cows had higher plasma flavonoid levels than CTR cows (167±23 vs 5±24 nmol/l; P<0.05). Q-cows tended to have higher plasma flavonoid levels (P=0.09) pp than ap, possibly due to reduced Q metabolism. Plasma metabolites and EB did not differ between Q and CTR cows, whereas Q reduced pp plasma aspartate aminotransferase (AST) activity by 80% (P<0.05) and lowered pp plasma glutamate dehydrogenase (GLDH) numerically by 50% (P=0.15).There were no Q effects on EE and EB, but cows showed a decline in respiratory quotient (P<0.05) from ap to pp, and fat oxidation peaked after calving, indicating the higher energy supply from fatty acids derived from lipomobilization. Liver fat content tended (P=0.10) to be lower in Q cows pp. It is concluded that Q supplementation might be beneficial in periparturient cows because of lowered plasma AST and GLDH activities and reduced liver fat.

Effects of phytogenic additives associated with acidifiers on intestinal morphology of broiler chick

F. Foroudi[1], H. Rahnama[2] and Y. Almasi[2]
[1]Islamic Azad University, Animal Science, Varamin-Pishva Branch, Varamin, Tehran, 3371857554,
Iran, [2]Behparvar Producing Group, no. 21, Eskandari shomali, 4th floor, Tehran, 1419744465, Iran;
f.foroudi@iauvaramin.ac.ir

A study was undertaken to examine the synergic effects of a phytogenic feed additive (ENTX) containing cinnamon, fenugreek and garlic with an acidifier including butyric acid (C4) on performance, carcass characteristics and intestinal morphology of broiler chickens. The experiment was conducted in a completely randomized design with a 2×2 factorial arrangement of treatments (with 0.05% or without ENTX × with 0.15% or without C4). A total of 400 one day old Ross 308 broiler chicks were equally distributed to 4 dietary treatments with 25 birds per pen and 4 replicates per treatment. Two birds per replicate (8 birds per treatment) with body weights close to the pen average were selected for carcass and intestinal morphology evaluation. Results show that weight gain, feed intake and feed conversion ratio were not significantly influenced by using ENTX and C4 isolated or associated in broiler diets. However, the use of ENTX isolated or associated with C4 in broiler diets decreased the fat deposits (P<0.05). Isolated or associated ENTX and C4 decreased the epithelium thickness and increased the number of goblet cells (P<0.05) in small intestine of broiler chicks. In conclusion, isolated or associated phytogenic additives and organic acids provided better carcass characteristics and better small intestine condition in broiler chicks.

EU regulatory update on botanical substances and preparations in animal nutrition

A. Holthausen
DELACON Biotechnik GmbH, Weissenwolffstrasse 14, 4221 Steyregg, Austria; antje.holthausen@delacon.com

Botanical substances and preparations are widely used in animal nutrition for different purposes. The most basic application in a legal sense is the use of minimally processed plants, providing micronutrients and aromatic notes, classified as feed ingredients and subject to the feed regulation (Reg. 767/2009). More sophisticated botanical preparations, such as essential oils, are defined as natural feed flavourings, subject to the feed additive regulation (Reg. 1831/2003), and requiring evaluation by EFSA (European Food Safety Authority). Industry has submitted 20 group dossiers covering 268 botanical substances now undergoing EFSA scrutiny as sensory feed additives – flavouring compounds. Many plant preparations exert bacteriostatic effects, have anti-oxidative or anti-inflammatory properties, impact digestive physiology or provide other beneficial effects on animal health and well-being. A few botanical preparations have succeeded in obtaining approval under this last category, as zootechnical feed additives that enhance animal performance. Botanical substances and preparations encompass a huge variety of plant families with many thousands of different constituents, depending on the plant genotype, origin, environmental conditions, extraction procedure, solvents, etc. These factors influence the composition and concentration of desirable botanical substances, but may also affect the presence and amount of naturally-occurring substances of toxicological concern. Hence every botanical preparation must be specified carefully to allow a targeted safety assessment. The EFSA Feed Unit has recently stated that the intentional addition of substances with genotoxic-carcinogenic properties to the food chain via feed additives is not acceptable, including naturally-occurring substances of botanical origin. This stance is a challenge to companies involved in developing and marketing natural botanical preparations, and underlines the importance of selecting and standardising botanical ingredients that meet EFSA's safety criteria.

Reflection of cinnamon, garlic and juniper oils supplement on rabbits performance

A.A. Abedo, Y.A. El-Nomeary, M.I. Mohamed, H.H. El-Rahman and F.M. Salman
Animal Production Department, 33 El Bohouth st. (former El Tahrir st.), Dokki, 12622 Giza, Egypt; abedoaa@yahoo.com

This study aimed to evaluate effect of supplement some essential oils (cinnamon, garlic and juniper) to rabbits diets on its performance. Twenty-four growing male white New-Zealand rabbits weighed 818 ± 37.61 g in average were divided into four groups and fed four experimental diets, the first as control and the other diets were supplemented with 0.5 ml/kg feed cinnamon, garlic and juniper oil, respectively. The diets were fed to cover the growing requirements for rabbits according to NRC (1977) recommendation, and the feeding trials were extended to 45 days. The results indicated that the average body weight gain for rabbits fed diet supplemented with garlic oil was significant ($P\leq0.05$) increased compared with control and the other diets. Essential oils supplement ($P\leq0.01$) improved feed conversion compared with control diet and juniper supplement achieved the best value. All nutrients digestibility were ($P\geq0.05$) increased with garlic oil supplement, and were no significant difference in digestibility were observed between different diets, except crude protein digestibility was significant ($P\leq0.01$) decreased with juniper oil supplement compared with the other diets. Oil supplement ($P\leq0.01$) decreased feed intake, especially cinnamon and juniper, and the feeding value as digestible crude protein was take the same trend. Garlic oil was significant ($P\leq0.01$) increased cecum total microbial, fungi, actinomyces and cellulolytic bacteria counts compared with the other diets. It can be concluded that garlic oil supplement was improved average daily gain, feed conversion and cecum microorganisms of rabbits.

The effect of dietary sea buckthorn leaf extract on the health in neonatal calves with diarrhoea
L. Liepa, M. Viduza, E. Zolnere and I. Dūrītis
LUA, Faculty of Veterinary Medicine, Helmana 8, 3004 Jelgava, Latvia; laima.liepa@llu.lv

The aim of the study was to investigate the effect of sea buckhorn leaf (SBL) extract on the health parameters in newborn calves. SBL have attracted interest during past few years as the most promising source of active compounds after berries. Materials and methods. The experiment was performed in a calf shelter of 275 dairy cow herd in March-November, 2014. Cryptosporidiosis and Corona virus induced – diarrhoea was recognised in newborn calves. In the control group (C) 8 calves were included and in the experimental group (E) – 14 calves. The nutritional and holding conditions were equal in both groups. None of the calves got antibiotic therapy in the period of diarrhoea. In group E the extract of SBL was given orally before feeding milk in increasing doses from 1-10 ml twice a day, starting from the day of birth (D0) till day 15 (D15). The health indices were controlled every day, but weight gain was checked and blood samples for biochemical and haematological analyses were collected on D1, D10 and D15. The data were statistically analysed by software 'SPSS STATISTICS 22'. Results. In group E calves had a longer diarrhoea period than in group C, but group E calves had better appetites and fewer days with a body temperature above 39,5 °C at this time. In group E animals had significantly lower (P<0.01) TNF-alfa concentration than in group C- on D10 10.4±7.1 pg/ml and 58.1±25.1 pg/ml; on D15 27.6±8.1 pg/ml and 108.5±35.1 pg/ml, respectively. On D10 and D15 calves in group E had a tendency of higher blood concentration of glucose, cholesterol and number of segment nuclear leukocytes, and lower serum haptoglobin, IL-6 and number of band leukocytes (P>0.05). Conclusions. The SBL extract reduced inflammatory cytokine TNF-alfa in the serum of newborn calves with diarrhoea, but it is necessary to have more experiments with modified SBL extracts and treatment strategies for reducing clinical signs of different kinds of diarrhoea.

The effect of different level of Pennyroyal (*Mentha pulegium* L.) extract on carcass traits and serum
J. Nasr[1], R. Shamlo[1] and F. Kheiri[2]
[1]Save Branch, Islamic Azad University, Department of Animal Science, Faculty of Agriculture, Save Branch, Islamic Azad University, Saveh, Iran, [2]Sharekord Branch, Islamic Azad University, Department of Animal Science, Faculty of Agriculture, Sharekord Branch, Islamic Azad University, Shahrekord, Iran; Javadnasr@iau-saveh.ac.ir

This study was conducted to investigate the effects of pennyroyal (*Mentha pulegium* L.) extract on carcass yield and serum cholesterol in broiler chickens. A total of 320 one-day old broilers (Ross 308) were randomly allocated to 4 treatments, 4 replicates with 20 birds in each in a completely randomized design. Treatments included were control and addition of various levels of pennyroyal extracts at 50, 100, and 150 ppm in diets. A completely randomized design was employed to data analysis. One-way analysis of variance was performed using the GLM procedure of SAS software. Duncan's multiple range test was used to find the significance difference (P<0.05). The results showed that inclusion of 150 ppm pennyroyal extract in diets increased live body weight, carcass relative weight, carcass efficiency and heart relative weight (P<0.05). In addition, inclusion of pennyroyal extract at levels of 100 or 150 ppm in diets increased relative weights of breast and wings (P<0.05). Furthermore, the lowest serum cholesterol and triglyceride and low density lipoprotein levels were obtained in birds fed 150 ppm pennyroyal group (P<0.05). The findings of present study revealed that pennyroyal inclusion in diet at the level up to 150 ppm improved body growth and serum cholesterol profile in broilers.

Effect of chloroform on *in vitro* methane production of diets supplemented with nitrate or saponins

L. Maccarana[1], J.B. Veneman[2,3], E. Ramos Morales[2], L. Bailoni[1] and C.J. Newbold[2]
[1]University of Padua, Dept. BCA, Viale dell'Università 16, 35020 Legnaro, Italy, [2]Aberystwyth University, Dept. IBERS, Penglais, SY23 3AL, United Kingdom, [3]Cargill Animal Nutrition Innovation Center, Veilingweg 23, 5334 LD Velddriel, the Netherlands; laura.maccarana@studenti.unipd.it

This study investigated the possible synergistic effect of feed supplements, with different modes of action, on gas (GP) and methane (CH_4) production, using a semi-continuous flow system (RUSITEC®). A diet based on grass hay (CP=7.1%; NDF=52.8%; lipids=1.3%, on DM), was incubated alone (basal diet; B) or supplemented with nitrate (N; 31.5 g/kg) or saponins from ivy extract (S; 50 g/kg). These 3 diets were incubated without (Ch-) or with chloroform (Ch+; 2 µl/l), obtaining 6 treatments (BCh-; BCh+; NCh-; NCh+; SCh-; SCh+). Four RUSITEC units were used, each containing all treatments, for a total of 24 vessels. Each vessel was inoculated with 800 ml of buffered rumen fluid. Whole incubation lasted 21 d. During the last 5 d, GP was measured by a dry test gas meter. Gas was sampled from air-tight bags for the analysis of CH_4 concentration by GC. Data were analyzed using a 3×2 factorial design (PROC MIXED, SAS) including diets, chloroform addition and their interaction as fixed factors, and RUSITEC unit as random blocking factor. Total GP (on average 2.56 l/d) was not affected by diet or chloroform addition. Daily CH_4 production was influenced by diet (199, 69, and 151 ml/d for B, N, and S, resp.; P=0.0024), and by chloroform addition (266 and 14 ml/d for Ch- and Ch+, resp.; P<0.0001). Statistical significance of the interaction was P=0.003. Specifically, CH_4 reduction, due to the chloroform addition, was significant in the BCh+ and SCh+ treatments compared to the BCh- and SCh- (-96% and -94%, resp.), but not in the NCh+ treatment compared to NCh-. Chloroform and nitrate, singularly, reduced CH_4 emissions significantly, with the largest magnitude of effect observed with chloroform. However, none synergistic effect emerged between chloroform and hydrogen sink (N) or antiprotozoal agent (S).

Effect of *Echium* oil supplementation on milk fatty acid profile in lactating dairy goats

M. Renna, C. Lussiana, P. Cornale, L.M. Battaglini, R. Fortina and A. Mimosi
University of Torino, Department of Agricultural, Forest and Food Sciences, L. go P. Braccini 2, 10095 Grugliasco (TO), Italy; manuela.renna@unito.it

Plant seeds of the genus *Echium* (Boraginaceae) contain high amounts of n3 polyunsaturated fatty acids (PUFA), predominantly α-linolenic and stearidonic acids. They may therefore be considered promising to increase, in ruminant products, the content of long-chain n3 PUFA (EPA and DHA), well known for their ability to lower cardiovascular disease risk in humans. This study aimed to assess the effects of dietary supplementation of *Echium plantagineum* (Ep) oil (unprotected against ruminal biohydrogenation) on the fatty acid (FA) profile of goat milk. Twenty-four Camosciata goats were divided into two groups and fed for 44 days a 70:30 forage:concentrate diet supplemented (Ep group, *Echium*) or not (C group, control) with 40 ml Ep oil. Individual milk samples were collected at 11, 22, 33, and 44 days following supplementation and analysed for their FA profile. Data were statistically treated as a repeated measure design. *Echium* oil supplementation decreased (P≤0.001) the n6/n3 FA ratio and the concentrations of odd-chain, branched-chain and total saturated FA, including lauric (-33%), myristic (-24%) and palmitic (-20%) acids, thus significantly diminishing the atherogenicity index of milk. The Ep diet increased milk content (g/100 g fat) of total monounsaturated FA (25.42 vs 21.16; P≤0.001), total polyunsaturated FA (6.30 vs 3.76; P≤0.001), total C18:1 trans FA (7.76 vs 1.99; P≤0.001), total C18:2 trans FA (3.66 vs 1.19; P≤0.001), total conjugated linoleic acids (2.22 vs 0.67; P≤0.001) and total n3 FA (1.33 vs 0.68; P≤0.001). EPA was higher in Ep than C milk (0.024 vs 0.017; P≤0.001) while DHA was undetectable in all analyzed milk samples. Data suggest that dietary supplementation of ruminally unprotected Ep oil can significantly ameliorate the nutritional quality of lipids in goat milk. However, one serving (250 mg) of Ep milk just provided 1.84 mg of beneficial long-chain n3 PUFA (EPA+DHA).

Effect of *Calendula officinalis* on the immune system of broiler chickens exposed to heat stress

Y. Jafariahangari[1], A. Alinezhad[1], S.R. Hashemi[1] and A. Akhlaghi[2]
[1]Faculty of Animal Science, Gorgan University of Agricultural Science and Natural Resources, 49189-43464, Gorgan, Iran, [2]Department of Animal Science, Shiraz University, 71441-65186, Shiraz, Iran;
yjahangari@yahoo.co.uk

A total of 160 day-old broiler chicks (Ross 308) were randomly assigned to four dietary treatments with four replicate pens per treatment (10 birds/pen) in a completely randomized design using factorial arrangement. Chickens received one of four experimental diets including: basal diet as control, basal diet supplemented with 1.5 or 3% *Calendula officinalis* powder, and basal diet supplemented with 300 mg vitamin E per kg diet. The birds had a free access to feed and water and the lighting regimen and ventilation were continuously monitored from d 1 to 42. Commencing from d 35, half of the broiler chickens were exposed to heat stress for 6 h/d (34±1 °C and 75% RH). Blood samples were collected on days 35 and 42 of experiment in order to evaluate the antibody response to sheep red blood cells (SRBC) and Newcastle Disease Virus (NDV). The results showed that there was no significant difference among the treatments for antibody titer against NDV on the days studied, although a higher antibody titer observed for diets supplemented with Vit. E and 3% *C. officinalis*. However, the lowest concentration of antibody titer against SRBC was found for chicken in control group under heat stress and the highest concentration was recorded for the birds receiving 3% *C. officinalis* in their diets with or without heat stress conditions. Results showed that *C. officinalis* powder improved the immune response to SRBC and also increasing total number of WBCs as well as improving the heterophil to lymphocyte ratio in broiler chickens exposed to heat stress.

Rumen fatty acid metabolism in lambs fed silages containing bioactive forage legumes

P.G. Toral[1], L. Campidonico[2], G. Copani[3], G. Hervás[1], G. Luciano[2], C. Ginane[3], A. Priolo[2], P. Frutos[1] and V. Niderkorn[3]
[1]Instituto de Ganadería de Montaña, CSIC-ULE, Finca Marzanas, 24346 Grulleros, León, Spain, [2]Di3A, University of Catania, Via Valdisavoia 5, 95123 Catania, Italy, [3]INRA, UMR1213 Herbivores, 63122 Saint-Genès-Champanelle, France; pablo.toral@csic.es

Inclusion of forage legumes, such as red clover (RC) or sainfoin (SF), in grass silage might affect rumen lipid metabolism. Polyphenol oxidase in RC is known to alter lipolysis, and tannins in SF modify biohydrogenation, which may subsequently modulate the fatty acid (FA) profile of ruminant-derived products. To compare the effects of these fodder legumes on rumen FA composition, 40 lambs (initial body weight: 31±0.3 kg) were allocated to 5 experimental groups (n=8) and fed *ad libitum* one of the following five silages: timothy grass (T, control, without bioactive components), T+SF (1:1), T+RC (1:1), T+SF+RC (2:1:1), or SF+RC (1:1). Forages were mixed on a DM basis. In addition to the silages, animals received daily a restricted amount of barley and straw. After 10 weeks on treatments, lambs were slaughtered and rumen digesta samples were collected for FA determinations by gas chromatography. Data were analyzed by 1-way ANOVA. Replacement of T with forage legumes was associated with greater rumen concentrations of polyunsaturated FA and lower of monounsaturated FA (P<0.05), mainly due to changes in 18:3n-3 and t11-18:1, respectively. However, the concentrations of other bioactive FA, such as c9-18:1, t10-18:1, c9t11-18:2, and t10c12-18:2, were not affected by diet (P>0.10). Only subtle variations between the effects of RC and SF would suggest that both forage legumes are able to similarly decrease the extent of ruminal FA metabolism, without altering its major pathways, while greater responses to SF+RC might indicate cumulative effects. The fact that RC and SF differently affected rumen odd- and branched-chain FA, which are commonly used as microbial markers, would most probably be related to the ruminal mechanisms underlying the response to each fodder legume.

Effect of concentration and source of condensed tannins on ruminal fatty acid biohydrogenation

A. Grosse Brinkhaus[1], G. Bee[1], W.F. Pellikaan[2] and F. Dohme-Meier[1]
[1]Agroscope, Institute for Livestock Sciences, Tioleyre 4, 1725 Posieux, Switzerland, [2]Wageningen University, Institute of Animal Science, Droevendaalsesteeg 4, 6708 Wageningen, the Netherlands; frigga.dohme-meier@agroscope.admin.ch

Ruminal biohydrogenation (BH) of polyunsaturated fatty acids (PUFA) might be decreased by condensed tannins (CT). However, the impact of the source and concentration of CT on BH is unclear. For the *in vitro* trail, 0.5 g of a hay:maize silage:linseed mixture (60:33:7) was weighed into glass flasks. This mixture was then supplemented with either 0 mg (CON), 2.5 mg birdsfoot trefoil (BT5), 2.5 mg sainfoin (SF5) or 14 mg sainfoin CT extract (SF28). Subsequently, 50 ml of a rumen fluid buffer mixture (1:4 v/v) was added to each flask, which were then incubated at 39 °C for 24 h. The incubation was stopped after 3, 6, 12 and 24 h. Each treatment and time point combination was incubated in triplicate in 2 runs (n=6) and all residues were analysed for fatty acid composition. Data were evaluated by analysis of variance for each time point separately using treatment as fixed effect. After 3 h of incubation, PUFA profile was similar among treatments. At the other time points the greatest (P<0.001) 18:2-n6 and 18:3-n3 concentrations were observed in SF28 followed by BT5, SF5 and CON. After 6 and 12 h of incubation, the concentration of t11-18:1 was lower (P<0.001) in SF28 compared to CON with intermediate values for BT5 and SF5. These differences among treatments disappeared after 24 h of incubation. The concentration of 18:0 after 6, 12 and 24 h of incubation was always lower (P<0.01) with SF28 and BT5 compared to CON. With increasing incubation time, the concentration of the c9t11 CLA isomer increased linearly with the SF28 treatment and tended to be greater (P=0.06) compared to CON after 24 h of incubation. In conclusion, supplementation of CT lowered ruminal BH of PUFA and this effect was more pronounced with increasing CT concentration. Differences between CT sources at an added amount of 2.5 mg extract could not be detected.

Silages containing bioactive forage legumes: a promising protein-rich food source for growing lambs

G. Copani, V. Niderkorn, F. Anglard, A. Quereuil and C. Ginane
INRA, UMR1213 Herbivores, 63122, Saint-Genès-Champanelle, France; giuseppe.copani@clermont.inra.fr

Growing lambs require high levels of protein, especially during the early stage of growth. In ruminant nutrition, forage legumes are of great interest as they can provide protein in a sustainable way. Some legumes contain bioactive compounds such as condensed tannins (CT) in sainfoin (SF), known to preserve protein degradation during the ensiling process and ruminal digestion; or polyphenol oxidase (PPO) in red clover (RC) known for similarly effects on protein preservation. We investigated the effects of these bioactive legumes on intake and performances of growing lambs, by feeding forty 4-month old male Romane lambs (initial BW 30.7±0.3 kg), *ad libitum*, with 5 different diets (n=8 per group) for 10 weeks: T diet (timothy grass, control, without bioactive compounds), T+SF (1:1), T+RC (1:1), T+SF+RC (2:1:1), or SF+RC (1:1). All the mixtures were prepared on a DM basis. Lambs also received daily a limited fixed amount of barley and straw. Data were analysed using the Mixed Procedure of SAS. Daily silage intake was greater (+160 g DM on average) in lambs fed RC-containing silages than in lambs fed T or T+SF silages (P<0.01). Consistently, lambs fed RC-containing silages showed better performances (final BW=43.9 vs 39.5 kg for lambs fed T or T+SF silages, P<0.0001). When expressed relative to metabolic weight, silage intakes ranked the same even if differences between treatments decreased through weeks, excepted for SF+RC which maintained at high level (P<0.0001). The positive effect of RC and differences between RC and SF could be due to the slightly better nutritive value of RC-containing silages and to their different profile in secondary compounds that may have impacted the ensiling process, the lambs' feeding motivation and their digestive efficiency. Furthermore, ensiling mixtures containing RC is a promising way to provide animals a protein-rich food allowing to combine high animal performances and reduced environmental impacts.

Use of chitosan and ivy saponins as natural products to decrease rumen methanogenesis *in vitro*

A. Belanche, E. Ramos-Morales and C.J. Newbold
IBERS, Aberystwyth University, SY23 3DA, United Kingdom; aib@aber.ac.uk

Decreasing methane emissions from ruminants without compromising their ability to digest fibrous material is one of the biggest challenges in ruminant nutrition. This experiment investigates the effect of a soluble chitosan (CHI) and ivy saponins (IVY) on the main factors which determine rumen methanogenesis based on three *in vitro* trials. Using 14C-labelled bacteria to study the bacterial breakdown; it was observed that CHI and IVY linearly decreased the rumen protozoal activity at concentrations above 1 g/l. Using rumen batch cultures it was observed that CHI promote a shift from butyrate to propionate, while IVY shifted fermentation from acetate to propionate. A rumen simulation technique was also used to further investigate the mode of action of both additives when used at 5% inclusion rate. This study confirmed that CHI and IVY promoted a similar decrease in rumen methanogenesis (-42 and -40%, respectively) without affecting feed digestibility and rumen levels of bacteria and methanogens. CHI had a multifactorial anti-methanogenic effect based on: (1) lower stoichiometric H2 production as a result of decreased acetate and butyrate and increased propionate; (2) increased lactate production as an alternative H2 sink; (3) CHI promoted a shift in the structure of the methanogens population; and (4) modified structure of the whole bacterial community promoting a lower bacterial diversity and firmicutes/bacteroidetes ratio. All this resulted in more efficient rumen fermentation in terms of NDF degradability and microbial protein synthesis. The effect of IVY was more specific since it did not modify the general structure of the bacteria and methanogens populations. Instead IVY lowered methanogen biodiversity and the abundance of two genuses within the family Methanomassiliicoccaceae. This was accompanied by decreased gas production, acetate molar proportion and microbial protein synthesis. In conclusion, CHI and IVY demonstrated to have promising anti-methanogenic properties *in vitro* and should be further investigated *in vivo*.

Goats browse preference in relation to chemical composition and tannin content of Ethiopian browse

G. Mengistu[1,2], W.F. Pellikaan[2], M. Bezabih[3] and W.H. Hendriks[2,4]
[1]Mekelle University, Animal,Rangeland and Wildlife Sciences, Mekelle University, P.O. Box 231, Ethiopia, [2]Wageningen University, Animal Nutrition Group, P.O. Box 338, 6700 AH Wageningen, the Netherlands, [3]International Livestock Research Institute, Addis Ababa, P.O. Box 5689, Ethiopia, [4]Utrecht University, Animal Nutrition Division, P.O. Box 80.163, 3508 TD Utrecht, the Netherlands; genet.mengistu@wur.nl

Preference of browse species by goats, dry matter intake (DMI) and browse chemical composition were evaluated in a cafeteria trial. Four mature goats with browsing experience (14.4±1.07 kg live weight) were used in a Latin square design. In trial 1, 25 g of air-dried *Capparis tomentosa, Dichrostchys cinerea, Dodonea angustifolia, Cadaba farinosa, Euclea schimperi, Rhus natalensis, Maeura angolesnsis, Acacia etbaica, Maytenus senegalensis* and *Senna sanguinea* were offered for 30 minutes for 10 days with daily sample collection. In trial 2, a similar procedure was followed except 25 g polyethyleneglycol (PEG)4000 included. Chemical composition and condensed tannin (CT) were determined using Near Infrared Spectroscopy. GLM procedure of SAS was used to determine mean differences and Pearson correlation analysis to compare intake and browse composition. In both trials DMI differed significantly ($P<0.0001$) between browses. The ranking order for DMI was *D. cinerea>R. natalensis>A. etbaica>M. senegalensis>E. schimperi>C. tomentosa>S. sanguine>M. angolesnsis>C. farinosa>D. angustifolia*. Inclusion of PEG increased intake for all browses, whilst the order changed only for the first three browses; *A. etbaica>D. cinera>R. natalensis*. The DMI was positively related with CT without PEG ($r^2=0.45;P<0.0001$) and negatively correlated with crude protein without PEG ($r^2=-0.39;P<0.0001$). Intake was negatively correlated with NDF ($r^2=-0.30;P<0.0001$) and ADF ($r^2=-0.36;P<0.0001$) fractions without PEG. Our observations showed that goats with browsing experience have a preference for browse with low fiber fractions, apparently associated with lower crude protein and high CT contents.

The effect of grape seed extract and vitamin C feed supplementation on some blood parameters

H. Hajati, A. Hassanabadi, G. Golian, H. Nassiri-Moghaddam and M.R. Nassiry
Ferdowsi University of Mashhad, Animal Science, Ferdowsi University of Mashhad (FUM) campus, Azadi
Sq., Mashhad, Khorasan Razavi, 9177948974, Iran; hassanabadi@um.ac.ir

In this experiment, the effect of hydroalcoholic grape seed extract (GSE) and vitamin C feed supplementation on some blood parameters of broiler chickens suffered from heat stress was investigated. Experimental diets included control diet (with no additive), 3 levels of grape seed extract (150, 300, 450 mg/kg diet), and one level of vitamin C (300 mg/kg diet). Each of the five diets was fed to 5 replicates of 12 male chicks each, from d 1 to 42. The birds were under chronic heat stress under 34 ± 1 °C temperature with 65-70% relative humidity for 5 hours from 29 up to 42 days of age. Results showed that grape seed supplementation at the level of 300 mg/kg increased body weight of broilers both at 28 d (pre-heat stress condition), and at 42 d (at the end of heat stress condition). Also, birds fed with 300 mg GSE/kg diet had higher European production efficiency factor before heat stress condition, during heat stress condition, and whole period of the experiment. GSE supplementation decreased the concentration of serum glucose, however, dietary treatments did not affect serum triglyceride, cholesterol, high- density lipoprotein (HDL), low- density lipoprotein (LDL), very low- density lipoprotein (VLDL) and uric acid at 28 d (pre heat stress condition). GSE supplementation decreased glucose concentration of serum blood, also GSE supplementation at 450 mg/kg diet decreased cholesterol, triglyceride, LDL and VLDL concentration of serum blood at 35 d (during heat stress condition). Vitamin C supplementation decreased serum cholesterol concentration of broilers suffered from heat stress. Heat shock protein 70 (HSP70) gene expression in heart and liver of broilers reduced by grape seed extract and vitamin C supplementation pre and during chronic heat stress condition.

Effects of tannin supplementation on duodenum morphology and cell proliferation in fattening boars

D. Bilić-Šobot[1], M. Čandek-Potokar[2], V. Kubale[3] and D. Škorjanc[1]
[1]University of Maribor, Faculty of Agriculture and Life Sciences, Department of Animal Science, Pivola 10, 2311 Hoče, Slovenia, [2]Agricultural Institute of Slovenia, Animal Production Department, Pig breeding, Hacquetova ulica 17, 1000 Ljubljana, Slovenia, [3]University of Ljubljana, Veterinary Faculty, Institute of Anatomy, Histology and Embryology, Gerbičeva ulica 60, 1000 Ljubljana, Slovenia; diana.bilic@outlook.com

The effect of tannins supplementation on intestinal morphology and cell proliferation in duodenum of fattening boars was studied. A total of 24 boars (Landrace × Large white) were assigned to four treatment groups, a control (T0 fed mixture with 13.2 MJ/kg, 15.6% crude proteins) and three experimental groups for which T0 feed was supplemented with 1% (T1), 2% (T2) and 3% (T3) of sweet chestnut extract of tannins. Boars were kept individually with *ad libitum* access to feed and water and slaughtered at 193 days of age and 122 ± 10 kg live weight. Morphological alteration of duodenum was analysed. No significant effect of tannin supplementation was observed on duodenum villi width, crypt depth and cell proliferation, whereas an effect was observed on villus height, perimeter and mucosal thickness. Our results indicate potential beneficial effect of tannins supplementation in boars diet on duodenum small intestinal morphology.

How to solve a conflict without getting into a fight

S.P. Turner[1], G. Arnott[2], M. Farish[1] and I. Camerlink[1]
[1]Scotland's Rural College (SRUC), West Mains Rd, Edinburgh, EH9 3JG, United Kingdom, [2]Queen's University Belfast, School of Biological Sciences, Belfast, BT9 7BL, United Kingdom; simon.turner@sruc.ac.uk

Excessive aggression is an important welfare issue in pig husbandry and mainly occurs when unfamiliar pigs meet. In nature, the agonistic behaviour of pigs comprises threat displays, to which withdrawal may follow. In this way, dominance relationships can be established and maintained without escalation. Commercial housing may impede this process and consequently provoke escalated fighting. We studied the importance of the full expression of agonistic behaviours on the time and strategy to settle conflicts. Contests (n=52) were staged between unfamiliar pairs of pigs of similar age (10 wk) and body weight, in an arena measuring 2.9×3.8 m. Contests lasted until a clear winner was present. Behaviour was observed from video. Contests lasted on average 5½ min (339 ± 19 s) with 87 ± 6 s of the contest spent on display behaviour (e.g. parallel walking), 35 ± 6 s on pushing, and 54 ± 6 s on mutual fighting. Pairs showing more display had a longer contest duration (b=2.4 ± 0.3 s/sec display; P<0.001), but did not differ in fight duration (P=0.96). In 28% of contests, pigs reached an outcome (winner/loser) without fighting. In these contests there was 53% more non-damaging investigation of each other (P=0.06), 46% more parallel walking (P=0.01), and 64% less pushing (P=0.04). However, bullying increased 2.8 fold in contests without a fight (P<0.001), which might be due to more energy reserves or a heightened need to affirm the outcome. Pigs which invest more time in display behaviour, and have space to do so, seem able to resolve conflicts without escalated fights. Negative consequences of fighting which impair welfare and productivity (e.g. injuries, reduced food intake, increased energy expenditure) might be reduced when pigs are given more opportunity to signal their intent. Space for conflict resolution should therefore not be regarded as an unnecessary luxury.

Does group size have an impact on welfare indicators in fattening pigs?

S. Meyer-Hamme[1], C. Lambertz[2] and M. Gauly[2]
[1]Georg-August-University, Department of Animal Science, Albrecht-Thaer-Weg 3, 37075 Goettingen, Germany, [2]Free University of Bolzano, Faculty of Science and Technology, Universitätsplatz 5, 39100 Bolzano, Italy; sophie.meyer-hamme@agr.uni-goettingen.de

Production systems for fattening pigs have been characterized over the last two decades by rising units and increasing group sizes, resulting in a serious discussion regarding animal welfare and health. Even though public discussions focus very much on these factors, it is still unknown whether they are really related to animal welfare. Therefore, the aim of this study was to describe the animal welfare status on 60 conventional pig fattening farms with different group sizes (small: <15 pigs/pen, n=207; medium: 15 to 30 pigs/pen, n=257; large: >30 pigs/pen, n=136) in Germany using animal- and pen-based measures of the Welfare Quality® protocol. Moderate bursitis (35%) was found as the most prevalent indicator of welfare-related problems. However, group size did not affect its incidence (P>0.05). In contrast, moderate and severe manure was found more often in medium-sized (17.7 and 11.5%) and large-sized (19.1 and 11.8%) when compared with small-sized groups (14.4 and 9.4%) (P<0.05). With increasing group size, the incidence of moderate wounds rose from 8.7% in small to 17.6% in large groups (P<0.05). Tail biting was observed at very low rates of 1.9%. The human-animal relationship was improved in large-sized in comparison to small-sized groups. In conclusion, group size affected some aspects of welfare in different directions, although there was no overall benefit of a particular group size.

Does housing influence maternal behaviour in sows?

C.G.E. Grimberg[1], K. Büttner[1], C. Meyer[2] and J. Krieter[1]
[1]Institute of Animal Breeding and Husbandry, Christian-Albrechts-University, Olshausenstr. 40, 24098 Kiel, Germany, [2]Chamber of Agriculture Schleswig-Holstein, Gutshof 1, 24327 Blekendorf, Germany; cgrimberg@tierzucht.uni-kiel.de

Fixation of sows during lactation in farrowing crates limits the opportunity of the sows to move freely and to interact naturally with their piglets. The aim of this study was to compare sows (n=23) in a group housing system (GH) and sows (n=24) in a conventional single housing system (SH) during lactation with regard to maternal behaviour. The GH sows were only fixed 3 days ante partum and 1 day post partum. Data were collected in 4 batches with 6 sows per batch in each housing system. All sows were observed in week 2 and 4 of lactation in 6 successive tests concerning maternal behaviour. The sows' reaction to handling (e.g. lifting piglets), separation and reunion of their piglets was tested both in their home pen (HP) and in a test arena (TA) for 5 min. The TA (14,4 m^2) had a piglet nest in a corner. The sows were only able to hear and smell their piglets. In the HP, GH sows showed stronger reactions to piglet handling compared to SH sows (P<0.05). However, in the TA, SH sows remained more frequent near their handled piglet (P<0.05) and vocalized more (P<0.05). During separation in the HP, GH sows showed more stress behaviour (P<0.05) and less resting positions (P<0.05). However, in the TA, SH sows were more active (P<0.05) and visited more frequent the piglet nest (P<0.05), while GH sows explored more frequent the floor (P<0.05). During reunion in the HP, GH sows vocalized more frequent (P<0.05) and showed less resting positions (P<0.05). In the reunion test in the TA, no difference was found between GH and SH sows. Regarding total piglet losses (e.g. crushing, underweight) GH sows had significant lower losses (P<0.05). To conclude, GH sows showed a stronger reaction in the HP and SH sows in the TA. Thus, the housing system has an effect on maternal behaviour. Nevertheless, GH sows had significantly lower piglet losses which could also indicate good maternal behaviour.

Environmental enrichment in piglet rearing: elevated platform with different enrichment materials

F. Lüthje, M. Fels and N. Kemper
University of Veterinary Medicine Hannover, Foundation, Institute for Animal Hygiene, Animal Welfare and Farm Animal Behaviour, Bischofsholer Damm 15, 30173 Hannover, Germany; nicole.kemper@tiho-hannover.de

The aim of this study was to test a modified conventional rearing system for weaned piglets, which was characterized by the construction of an elevated platform in the pen where various enrichment materials were offered to create a special area for activity and playing. In each of four rounds, 40 piglets were housed together after weaning for 33 days in an enriched pen equipped with an elevated platform. Space allowance was 0.6 m^2 per piglet. There were nine different enrichment materials available on the platform. Under the platform there was space to retreat and rest. Video recordings were performed every week for two days, respectively. For video analyzing each day was divided into three parts (morning, afternoon, night) and the number of standing and lying piglets was counted in four different areas of the pen as well as the number of piglets which were using the enrichment materials on the platform. For statistical analysis, univariate analyses of variance followed by SNK test were carried out with SPSS Statistics. Behavioural observations revealed that in the afternoon more piglets were on the platform than in the morning or at night (7.2 vs 4.9 and 0.6, P<0.05). The area under the platform was more preferred in the morning and at night than in the afternoon (18.5 and 21.6 vs 12.6, P<0.05). On average 10 piglets were lying and 2 piglets were standing under the platform. Enrichment materials were used more in the afternoon than in the morning (1.5 vs 0.7 contacts). From week 1 to 4 the contacts of piglets with the enrichment materials increased (0.9 vs 1.1) followed by a decrease in week 5 (0.9). This study shows that the area on and under the platform as well as the enrichment materials were used during the whole rearing period. The new system enables the farmer to increase space allowance and animal welfare in existing barns by minor structural changes.

Relation between orientation reaction and saliva testosterone levels in pigs
P. Juhas, O. Bucko, J. Petrak, P. Strapak, K. Vavrisinova and O. Debreceni
Slovak University of Agriculture, Faculty of Agrobiology and Food Resources, Department of Animal Husbandry, Tr. A. Hlinku 2, 94901 NItra, Slovak Republic; peter.juhas@uniag.sk

The aim of this research was to compare behavior of pigs with different testosterone levels (TSL) placed in an unfamiliar room. 29 pigs weighing from 18.5 to 27.5 kg were tested (14 females and 15 males). Males were castrated during the 2nd week after birth. Behavior in the unfamiliar space was tested in an empty room, using an open field test. Orientation was evaluated by comparing time spent near doors and in other part of the room. Animals were placed in the unfamiliar room for 20 minutes. Behaviors were recorded and analyzed from records using Noldus Observer XT software. Saliva samples for testosterone level were sampled before (calm state) and immediately after testing. TSL were measured using a commercially available ELISA kit (DiaMetra Testosterone saliva). Optical absorbance was measured by Microplate Reader (Model DV 990BV4, UniEquip Deutschland). TSL varied from 78.91 to 1,131.06 pg/ml in the calm state, and from 93.12 to 1,589.36 pg/ml after testing. Time spent near the door (TD door) ranged from 123.11 to 892.64 sec. Difference in TD door was tested by Mann-Whitney test. We found significant differences in TD door between animals from the 1st quartile and 3rd quartile of TSL during the calm state (U=7, P=0.025). Mean duration of time spent near door was higher in animals with higher TSL after testing (r=0.432, P=0.019). The research was supported by project VEGA 1/0364/15.

The effect of Tellington Ttouch® method on the horse behaviour in daily tasks
T. Majerle and K. Potočnik
University of Ljubljana, Biotechnical Faculty, Department of Animal Science, Groblje 3, 1230 Domžale, Slovenia; klemen.potocnik@bf.uni-lj.si

Working with horses needs awareness of possible dangerous situations, because of confidence lack between horse and human. The main objective of the study was to examine the effect of Tellington Ttouch® method (TT) on the horse behaviour during daily tasks. TT is a collection of different circles done with hands and fingers over various parts of horse's body to enhance trust, improve health, performance, to accelerate a horse's ability to learn, more willingly to give us their attention, etc. Two trials were performed. In the first study, 6 Lipizzan horses were included and the trail lasts two weeks. In the 1st week, traditional method while in the 2nd week, TT was used. During the hoof care, the horses' behaviour were observed and following parameters were measured: time needed for hoof care, number of attempts to remove each foot, number of actual foot remove, horses heart rate and optical communication (OC). OC describes horse body language, as score of horse-human relationship from cooperating to aggressive behaviour on scale of 5 to 1. Data were analysed with programme SAS/STAT. It was realised that TT was not significantly affected on the horse behaviour during every day work. Horses were calm and unproblematic during the hoof care in the time of whole experimental period. On the other hand, in the second case trial which lasted two months, one problematic horse was included, which was kicking when its legs were touched as well as showing threats toward its new owner in the pasture. After using the TT for 4 weeks at the pasture the horse became relaxed and willingly accepted touching with its head and neck. The next 4 weeks using the TT during the hoof care. Heart rate was measured to determinate if the aggressive behaviour was a result of fear and optical communication was observed. The heart rate was decreased from 56 to 42.5 beats/min and so we can conclude that the horse feel fear. After two months of TT using, the trust between horse and human were enhanced.

Effects of reduced feeding space on feeding, comfort and agonistic behaviour of goats

G. Das[1], M. Yildirir[2], C. Lambertz[3] and M. Gauly[3]
[1]Leibniz Institute for Farm Animal Biology (FBN), Wilhelm-Stahl-Allee 2, 18196, Dummerstorf, Germany, [2]Animal Genetic Resources Research Working Group, Tarım Kampüsü İstanbul Yolu Üzeri, 38, 06171,Yenimahalle, Ankara, Turkey, [3]Free University of Bolzano, Faculty of Science and Technology, Universitätsplatz 5, 39100, Bolzano, Italy; christian.lambertz-1@agr.uni-goettingen.de

Competition for resources can negatively affect welfare of individuals. We determined effects of a reduced number of feeding places on feeding, comfort and agonistic behaviour of Boer goats (n=40). Goats received either an individual feeding place (FP-1) or there was 0.5 feeding place per goat (FP-0.5). Behaviours were video-recorded and evaluated once a week for 24 h with the continuous sampling method over a 6-week period. A dominance index was used to estimate rank orders. Data were analysed with repeated measures ANOVA. The FP-0.5 increased frequency of feeding bouts (P<0.001) and decreased the duration of feeding (P<0.001). Also, FP-0.5 elevated duration and frequency of competition behaviours (P<0.001), which increased with decreasing rank order (P≤0.001). High-ranked goats had longer lying durations than medium- and low-ranked goats in FP-0.5 (P=0.003). However, all rank categories had similar durations of comfort behaviours in FP-1 (P>0.05). The frequencies of aggressive behaviour were also higher in FP-0.5 than in FP-1 (P≤0.01). Aggressive behaviour increased particularly at feeding (P<0.001). The interactions between group and rank on certain aggressive behaviours indicated that low-ranking goats were exposed to more aggressions in FP-0.5 than those in FP-1 (P≤0.05). It is concluded that reducing the number of feeding places decreases the time spent for feeding and comfort behaviour, but increases competition and agonistic interactions in Boer goats. Goats with restricted feeding place compensated the effect of lower feeding duration by increasing the frequency of feeding bouts. Therefore, low-ranking individuals may suffer from competition and aggression resulting from the reduced feeding place.

Non genetic factors affecting hunting ability in Italian Maremma Scent Hound dog

S. Riganelli[1], S. Antonini[1], M. Gubbiotti[2], A. De Cosmo[1], A. Valbonesi[1] and C. Renieri[1]
[1]University of Camerino, Via Gentile III da Varana s/c, 62032 Camerino, Italy, [2]Guglielmo Marconi University, Via Plinio 44, 00193 Roma, Italy; stefania.riganelli@unicam.it

The aptitude for wild boar hunting by Maremma Italian Scent Hound is estimated by scoring 5 traits: search, approach, tracking of prey, standstill barking and physical skills. A total of 763 Maremma Scent Hound (488 males and 275 females) were studied. The data consisted of 1147 results from competitions held in North-Central Italy which took place between 2010 and 2011. Dog was tested as individual, pairs and pack (6 to 12 components). The effect of five non-genetic factors (sex, coat colour, judges, type of competition, field of trial) was carried out by ANOVA using SPSS12.0 predictive analysis software. Correlations between variables were estimated using the Pearson correlation coefficient. The results indicate that there is no effect on traits for coat colour. A significant difference between males and females is observed only for search (P<0.001). Type of competition (individual, pair and pack) has a significant effect on search (individual vs pair; P<0.05) and approach (all three types; P<0.05). Field of trial (site of boar hunting and climatic conditions) has a significant effect on search (P<0.05) and approach (P<0.05). The 'judges' factor is highly significant for physical skills (P<0.001) and standstill barking (P<0.001). There is a highly significant positive phenotypic correlation between tracking of prey and approach (P<0.01). Physical skills are positively correlated with both search (P<0.01) and approach (P<0.01). Approach is highly positively correlated with tracking of prey (P<0.01). Standstill barking shows no correlation with any traits. Sex, judges, type of competition and field of trial affect the five tested traits in the Maremma Italian Scent Hound. These data are the basis to improve our knowledge about the values of variability in considered hunting traits. Also, they provide genetic criteria to breeders to achieve more stringent selective choices.

Stability of milking rank in a newly formed group of goats
P. Cornale, M. Renna, C. Lussiana, L.M. Battaglini, R. Fortina and A. Mimosi
University of Torino, Department of Agricultural, Forest and Food Sciences, L. go P. Braccini 2, 10095
Grugliasco (TO), Italy; manuela.renna@unito.it

The aim of the present study was to assess: (1) if milking order of a new group of goats constituted for experimental reasons was stable over period; and (2) if milking rank was affected by age, milk yield, milk quality, and morphometric measurements. In a dairy goat farm (NW Italy), 21 Camosciata goats were housed indoors and fed a diet based on fresh grass and concentrate. Goats were machine-milked twice a day in a 12-stall milking parlour and their milking order was recorded. Eleven goats were selected from the flock on the basis of their stage of lactation, milk yield and composition. They were moved to a pen and fresh grass was replaced with mixed hay. The goats were daily monitored at morning milking for 17 days after group formation. The entrance order in the milking parlour, milk yields, and milk samples (for determining fat, protein and somatic cell count) were collected everyday. Body weight, trunk length, height at withers, rump width, chest girth, hornedness, length and maximum diameter of horns of each goat were measured at the end of experimental period. The consistency of milking order of the goats was calculated using the Kendall's concordance coefficient. The correlations between mean milking order vs age, days in milk, milk yield, milk traits, and morphometric traits were calculated using the Spearman correlation coefficient. The Kendall's coefficient showed a significant consistency of entrance in the milking parlour (W=0.69; P≤0.001). Milking order was significantly correlated with rump width (R=0.70; P≤0.01) and trunk length (R=0.59; P≤0.05). The results demonstrated that milking order of goats was not random and it was consistent during the trial. Moreover, the milking rank showed in the original flock was preserved in the experimental group. These results suggest that goats, including subjects involved in experimental trials, should always freely enter the milking parlour in order to avoid possible stress.

Grazing behaviour of early-weaned calves under different supplement delivery systems
V. Beretta, A. Simeone, A. Henderson, R. Iribarne and M.B. Silveira
Facultad de Agronomía-Universidad de la República, Paysandu, 60000, Uruguay; beretta@fagro.edu.uy

An experiment was conducted to evaluate the effect of supplement delivery system (SDS) on grazing behaviour of early-weaned calves grazing temperate pastures (29 ha *Trifolium pratense* and *Cichorium intibus*) during summer (12 weeks). Twenty-four Hereford calves (64±16 days old; 77±10 kg) were randomly allocated to 6 paddocks and 3 treatments: daily supplementation (DS) with 1 kg DM/100 kg liveweight (LW) of a commercial concentrate (18% CP, 80% TDN); self-fed supplementation with same concentrate (SF) or with a concentrate mixture including 17% NaCl for intake regulation at 1% LW (SFS). Supplement in SF and SFS was always available. Pasture was grazed in 7-days strips, with a forage allowance of 8% LW. During week (W) 2 and 9, on days (D) 2, 4 and 6 after entering a new strip, animals were observed individually (0900-2100 h), recording every 15 minutes, grazing (G), supplement intake (S) or ruminating (R) activities. Bite rate (BR) was measured every 3 h recording number of bites in 1 minute. To characterized daily defoliation pattern, sward height was recorded every 24 hs. Behaviour data was submitted to LOGIT transformation assuming binomial distribution and analysed as the probability to find an animal on a specific activity: $Ln(p/(1-p))=m+Ti+Wj+Dk+(TW)ij+(TD)ik+(TWD)ijk+eijkl$. As calves were older, independent of SDS, G (W2 0.32; W9 0.49) and S (W2 0.03; W9 0.06) increased (P<0.01) with no changes in R. Grazing and BR were affected by SDS (G: DS 0.47[a], SFS 0.46[a], SF 0.29[b]; BR: DS 12[ab], SFS 14[a], SF 10[b] bites/min, P<0.01). An interaction TxD was observed for G, with no differences due to SDS in D2, but with G increasing in DS and SFS on D4 and D6, while in SF it remained stable. This response was consistent with sward defoliation pattern. Results show that even under same grazing management, SDS determines changes in grazing behaviour. This should be taken into account as pasture/concentrate ratio in the diet might be affected, as well as rumen kinetics and fermentation pattern.

Effect of elevated concentrations of crude glycerin on feeding behavior of crossbred lambs

E.H.C.B. Van Cleef, M.T.C. Almeida, J.R. Paschoaloto, F.O. Scarpino-Van Cleef, E.S. Castro Filho, J. Mangolini, V.H.C. Rau, A. Queiroz, I. Monsignati and J.M.B. Ezequiel
FAPESP, São Paulo State University, Via de Acesso Paulo Donato Castellane, 14884-900, Brazil; ericvancleef@gmail.com

Forty crossbred (Santa Ines × Dorper) male lambs (90 d of age and 18±2.4 kg BW) were used to evaluate the effect of increasing concentrations of crude glycerin on their feeding behavior. Animals were blocked by initial body weights and randomly assigned to 40 individual pens (1.5 m^2), where they received for 70 d four experimental diets composed of corn silage (40%) and concentrate (60%) containing soybean meal, soybean hulls, minerals, and corn grain or 0 (CON), 10 (G10), 20 (G20), or 30% (G30) crude glycerin in diets dry matter. In treatment G30, corn was totally replaced by crude glycerin, which contained 83% glycerol. Behavior observations were performed by 6 observers who recorded, each 5 min during three period of 12 h, the following activities: interaction with feed bunk (IB), interaction with waterers (IW), ruminating standing (RS), ruminating laying (RL), standing still (SS), laying (LA), stereotypies (ST), and other activities (OA). Furthermore, the chewing activity was evaluated with observations of number of chews per bolus (NB), and time of chewing per bolus (TB). Data were analyzed as a randomized complete block design by using the mixed models. Animal was the experimental unit, and model effects included block and treatment. Orthogonal contrasts were used to determine the linear, quadratic, and cubic effects of glycerin and 0% glycerin vs glycerin treatment. No interaction of day of observation and treatments was observed. NB and TB were increased in G10 and G20 with crude glycerin inclusion (quadratic, P=0.004, P=0.006, respectively). The activities observed (IB, IW, RS, RL, SS, LA, ST and OA) were not affected by dietary treatments (P>0.01). Crude glycerin at up to 30% can be used, totally replacing corn, without adversely affecting behavior variables of crossbred lambs, except for time of chewing and number of chewing per feed bolus.

Levels change of testosterone in individual neuroreflection types of pigs after mental load

J. Petrák, O. Debrecéni, O. Bučko and P. Juhás
Slovak University of Agriculture, Nitra, Department of Animal Husbandry, Tr. A Hlinku 2, 949 76, Slovak Republic; peter.juhas@uniag.sk

The dynamics of a reaction to physiologial stress (load) can be assessed on the grounds of excitability of the nervous system pigs. Neuroreflection types in terms of arousal of the nervous system are divided into highly excitable (EHb+), medium excitable (EHb°) and low excitable (EHb-) types. Increasing concentrations of testosterone after load indicates its neuroprotective effect, which becomes apparent on activity of brain-derived neurotrophic factor – BDNF. The aim of this work was to monitor the change in concentration of testosterone within different excitatory type of pigs. Animals were assigned to excitatory type (EHb+, EHb°, EHb-) based on the amount of movement in an unfamiliar chamber using a habituation test. We tested 77 Large White pigs (barrows and gilts). All neuroreflection types were divided into two subgroups in which the concentration of testosterone after load increased (EHb+↑, EHb°↑, EHb-↑) or decreased (EHb+↓, EHb°↓, EHb-↓). Psychological load was applied by a 20 minute stay in the unfamiliar chamber. There were 16 EHb+ types, 44 EHb° type and 17 EHb- type pigs. The concentration of testosterone in saliva was determined by ELISA. Type EHb+: Concentration of testosterone was increased after load compared to before for the subtyp EHb+↑, (P≤0,001). Mean concentration of testosterone decreased after load for subtype EHb+↓ (P≤0,01). Type EHb°: The average concentration of testosterone in subtype EHb°↑ increased after load compared before load (P≤0,05). For subtype EHb°↓ we are observed decreased concentration after load (P≤0,01). Type EHb-: In the subtype EHb-↑, average concentration increased after load without statistical significance. Subtype EHb-↓ decreased concentration of testosterone after load (P≤0,05). Our results indicate that highest level of testosterone in these animals ensures its neuroprotective function. This work was supported by projects VEGA 1/0364/15.

Genomic selection breeding programs

M. Lillehammer[1], A.K. Sonesson[1] and T.H.E. Meuwissen[2]
[1]Nofima, P.O. Box 210, 1431 Ås, Norway, [2]Norwegian University of Life Sciences, Department of Animal and Aquacultural Sciences, P.O. Box 5003, 1432 Ås, Norway; marie.lillehammer@nofima.no

When designing genomic selection breeding programs, choices about when and who to genotype, how to obtain phenotypes, and when and how to perform selection will affect genetic gain, as well as rate of inbreeding, accuracy of selection and contribution of each trait to genetic gain. To ensure a sustainable breeding program that gives good response both in a long and short term perspective requires careful planning of the collection of phenotypes and genotypes as well as the steps of selection. We will present results from simulations of cattle, pig and salmon breeding programs to show how the design of the breeding programs affect multiple result parameters. Presented strategies will include a genomic pre-selection program for dairy cattle, with similar genetic gain and rate of inbreeding as using genomic selection directly, but with a higher accuracy of selection, pig breeding schemes carefully designed to ensure genetic gain for all traits under selection, to be sustainable long lasting breeding programs, and salmon breeding programs where pedigree and genomic information are combined to give high genetic gain for all traits at acceptable costs. General for all programs, a key issue is updating the reference population regularly for all traits. Too few new animals with records and genotypes every generation will cause reduced selection accuracy over time. Few reference animals for some of the traits under selection will shift the genetic gain towards other traits. Another important factor is the economic weight of each trait, as a low economic weight will give low progress for that trait, irrespective of the genotyping strategy. We will show examples of when increasing the size of the reference populations does not affect genetic gain, because of lack of weight of this trait in the breeding goal or because of that the information is already captured by including parents of the same animals in the reference population.

Stochastic simulation of breeding plans in mink: evaluation of genomic selection

K. Meier, J. Thirstrup, M.S. Lund and A.C. Sørensen
Aarhus University, Center for Quantitative Genetics and Genomics, Department of Molecular Biology and Genetics, Blichers Allé 20, Postboks 50, 8830 Tjele, Denmark; kristian.meier@mbg.au.dk

The aim of this study is to evaluate the effect of genomic selection in mink breeding. In Danish mink production, breeding is focused on improving pelt quality, obtaining larger litters and increase pelt length. These traits differ in complexity, and current genetic progress differs between the traits. Pelt length has increased due to selection on animal weight, but breeding for litter size and pelt quality is more complex. Litter size has a low heritability, is only measured on females, and is negatively correlated with animal size. Selection for pelt quality is indirect as grading is done on live animals making it less effective. Current breeding practice allows individual famers to use breeding values estimated from pedigree and phenotypic records. However, selection based on phenotypes alone is also common practise, making breeding for complex traits less effective. One solution to improve genetic gain for these complex traits is to use genomic selection. With the use of stochastic simulations we compared current breeding methods with different scenarios of genomic selection based on different accuracies of genomic predictions and on different genotyping strategies. We evaluated genetic gain and total economic gain for the traits analysed. Our results suggest that use of genomic selection has the potential to improve total economic gain substantially compared to traditional breeding methods. The improvement is dependent on accuracy and genotyping strategy, but even with low accuracy and low genotyping effort total economic gain is improved compared to traditional breeding methods. Our results also suggest that it is possible to increase genetic gain for litter size and pelt quality, as these traits benefit the most from genomic selection.

Stochastic simulation of alternative future blue fox breeding strategies

J. Peura[1], A.C. Sørensen[2], K. Meier[2] and L. Rydhmer[1]
[1]SLU, Department of animal breeding and genetics, Box 7023, 7023 Uppsala, Sweden, [2]Aarhus university, Center for Quantitative Genetics and Genomics, Department of Molecular Biology and Genetics, Blichers Allé 20, 8830 Tjele, Denmark; jussi.peura@slu.se

In Finnish blue fox, selection is currently done within farm using BLUP breeding values. The most important goal traits in breeding scheme are litter size, pelt quality and animal size. In Finland, there is, however, increasing interest to include feed efficiency and leg conformation in the breeding goal. There is also increasing discussion if genomic selection should be used in blue fox farms. The aim of this study is to compared three blue fox breeding strategies: (1) selection based on traditional BLUP evaluation and old selection criteria; (2) selection based on traditional BLUP evaluation and new selection criteria; and (3) selection based on genomic selection and new selection criteria. Old selection criteria include animal size, pelt size and litter size. New selection criteria include animal size, pelt quality, litter size, front leg conformation and feed efficiency. In all three scenarios, the goal traits were animal size, pelt quality, litter size, front leg conformation and feed efficiency. ADAM software was used for stochastic simulation. Genetic evaluation was done using multitrait breeding value evaluation with DMU software. Economic values and genetic parameters were collected from literature. Genomic selection was simulated using pseudo-genomic simulation and three different levels of accuracies of genomic breeding values were tested (low, medium and high). The genomic selection scenario resulted in higher total economic genetic gain than scenarios based on pedigree information. Also inclusion of records for feed efficiency and front leg conformation increased total economic genetic gain. For some traits genetic change was unfavorable and therefore adjustment of economic weights is needed if genetic progress in all traits is preferred.

Genomic selection for feed efficiency in pig breeding programs

K.G. Nirea and T. Meuwissen
Norwegian University of Life Sciences, Department of Animal and Aquacltural Sciences, Universitetstunet 3, 1432, Norway; kahsay.nirea@nmbu.no

Production in modern pig breeding is based on high quality feed resources. In the future, food prices will increase and access to the present feed resources will be limited. This calls for robust animals adapted to local feed resources. Breeding for feed efficiency is challenging and individual feed intake is expensive to measure. However, genomic selection reduces cost because animals can be selected at an early age without observation on themselves. In addition, environmental sensitivity between nucleus and production herds could be taken into account. At present, performance testing of boars is based on records in the nucleus. However, the nucleus herds are fed high quality concentrates to which the production herds do not have access. Thus, the best animals in the nucleus might not be the best in the production environment. Our aim is to design a breeding scheme combining records from the nucleus and production herds. We simulated a typical pig breeding program containing nucleus and production herds. The production herds were a cross between nucleus herds and other breeds. Genomic breeding value of the selection candidates were estimated for two genetically correlated traits (rg=0.5) with a heritability of 0.4 and 0.3 recorded in the nucleus and production environment respectively. In one approach genomic breeding value was estimated including nucleus animals in the training population (N). In the second approach, the training population was a mixture of nucleus and production animals (N+P). In the third approach the training population only consisted of production herds (P). In the production environment, ΔG was 0.064, 0.115 and 0.128 for N, P and N+P respectively. This is an increase in ΔG by 44 and 50% for P and N+P as compared to N. Therefore, when using genomic selection for feed efficiency traits, inclusion of production animals in the training population would significantly increase the rates of genetic gain in the production environment.

Using loci with identified dominance effects to improve the prediction of heterosis

E.N. Amuzu-Aweh[1,2], P. Bijma[1], H. Bovenhuis[1] and D.J. De Koning[2]
[1]*Wageningen University, Animal Breeding and Genomics Centre, P.O. Box 338, 6700 AH Wageningen, the Netherlands, [2]Swedish University of Agricultural Sciences, Animal Breeding and Genetics, P.O. Box 7023, 75007 Uppsala, Sweden; piter.bijma@wur.nl*

The availability of high-density single nucleotide polymorphism (SNP) panels creates opportunities for the genomic prediction of heterosis, which would increase the efficiency of commercial crossbreeding programs. In a previous study, the genome-wide average squared difference in allele frequency (SDAF) between parental pure-lines was used to predict heterosis in egg number and egg weight. This assumed that all SNPs had an equal contribution to heterosis. Heterosis was predicted with an accuracy of ~0.5, and results suggested that dominance plays a major role in heterosis. Therefore, in this study our aim was to identify SNPs with dominance effects and to specifically use these SNPs to calculate a weighted SDAF for predicting heterosis. We hypothesised that predicting heterosis with an emphasis on SNPs with dominance effects would increase the accuracy of prediction. We used 60k SNP genotypes from 3,427 sires, allele frequencies from 9 pure-lines, and egg production records from 16 crosses between those lines, representing ~210,000 crossbred hens. We calculated genotype probabilities for the crossbred offspring of each sire and from that derived additive and dominance coefficients. We estimated additive and dominance effects of all SNPs by regressing offspring performance on their additive and dominance coefficients. Next, using the reciprocal of the variance of the estimated dominance effects, we calculated a weighted SDAF and used that as a predictor of heterosis. Preliminary results indicate that dominance effects are more important for egg number than for egg weight: we identified 37 SNPs for egg number and 5 SNPs for egg weight that had significant dominance effects (10% false discovery rate). Results will reveal whether weighing SNPs by their estimated dominance effects can improve the prediction of heterosis.

Optimal contribution selection in breeding schemes with multiple selection stages

B.S. Dagnachew and T.H.E. Meuwissen
Norwegian University of Life Sciences, Department of Animal and Aquacultural Sciences, Arboretveien 6, 1432, Aas, Norway; theo.meuwissen@nmbu.no

Optimal contribution (OC) selection aims at finding an optimal balance between rate of inbreeding and genetic gain in the long-term. Implementing OC selection has focused mainly on the final stage of selection (i.e. selection of elite animals). In practice, however, the ultimate selection of the best animals consists of many pre-selection steps. For instance, in dairy cattle breeding, male calves are preselected to enter a progeny testing schemes, and the final selection of sires comes after progeny testing of the candidate bulls. When it comes to management of inbreeding, these pre-selection steps may be as important as the final selection step. Thus, it is important that these pre-selection steps also accounted for inbreeding, instead of only controlling inbreeding in the last selection step. One way to achieve this goal is by using the OC selection scheme for overlapping generations to optimize the pre-selection and the final selection stages simultaneously. In this approach, the young animals are added as an extra group of selection candidate to the optimization process. This takes into account a decision made in stage one when making a selection decision in stage two. Hence, it will simultaneously optimize both selection stages. An OC selection algorithm, was amended to handle the extra group of selection candidates. Two scenarios were compared: one scenario with pre-selection using truncation selection and OC selection for the final selection stage. In the second scenario, multiple OC selection was used to optimize both selection steps simultaneously. Results show that simultaneously optimizing multiple selection stages resulted in lower average relationship at the same level of genetic gain. If one of the selection step is completed earlier in time than the other, which is the case in most practical breeding programs, one might re-optimise accounting for the actual selections performed in the previous selection step.

Genome-wide inbreeding and coancestry patterns in dairy cattle

D. Kleinman-Ruiz[1], B. Villanueva[1], J. Fernández[1], M.A. Toro[2], L.A. García-Cortés[1] and S.T. Rodríguez-Ramilo[1]
[1]Instituto Nacional de Investigación y Tecnología Agraria y Alimentaria, Departamento Mejora Genética Animal, Crta. A Coruña Km. 7,5, 28040 Madrid, Spain, [2]Universidad Politécnica de Madrid, Departamento Producción Animal, Escuela Técnica Superior de Ingenieros Agrónomos, 28040 Madrid, Spain; rodriguez.silvia@inia.es

Taking advantage of the large numbers of SNPs currently available, both inbreeding and coancestry can be estimated in detail over the genome. In this study, intra-chromosomal patterns of inbreeding and coancestry were obtained for the Spanish Holstein population. Animals were genotyped with the Illumina® BovineSNP50 BeadChip. After applying filtering criteria, the genomic dataset included 36,693 autosomal SNPs and 10,569 animals. SNP-based inbreeding and coancestry were calculated using 5 Mb sliding windows. The results indicate that SNP-based measures show different patterns of inbreeding and coancestry on specific chromosome regions. These intra-chromosomal patterns could provide a more detailed picture for the management of populations in conservation and selection programmes. In addition, differences on intra-chromosomal patterns of inbreeding and coancestry between individuals born in 1960-1979 and individuals born in 2013 were also detected. These differences could be useful for detecting selection signatures.

Mating strategies using genomic information reduce rates of inbreeding

H. Liu[1], M. Henryon[2,3] and A.C. Sørensen[1]
[1]Aarhus University, Department of Molecular Biology and Genetics, Blichers Allé 20, 8830 Tjele, Denmark, [2]University of Western Australia, School of Animal Biology, 35 Stirling Highway, Crawley WA 6009, Australia, [3]Danish Agriculture and Food Council, Pig Research Centre, Axelborg, Axeltorv 3, 1609 Copenhagen V, Denmark; lhmsai007@yahoo.com

Appropriate non-random mating strategies can help animal breeding schemes reduce rates of inbreeding by improving family structure and decreasing progeny homozygosity. The purpose of this study is to develop two existing mating strategies, i.e. minimum-coancestry (MC) and minimized covariance of ancestral contribution mating (MCAC) by including genomic information, and to test whether use of genomic information in mating can realize lower rates of inbreeding than pedigree. We conducted stochastic simulations to estimate rates of inbreeding and cumulative genetic gains realized by MC and MCAC using pedigree or genomic information in five breeding schemes with truncation selection for 20 discrete generations. Selection was based on the genomic breeding values predicted by a ridge regression model (GEBVs). Breeding schemes differed in family structure or heritability. Also, we calculated the number of ancestors that made a genetic contribution to the offspring in generation 20. The results show that rates of inbreeding generated by marker-based MCAC were 8.7 to 16.2% lower than pedigree-based MCAC, and similarly, marker-based MC generated 6.4-21.4% lower rates of inbreeding than pedigree-based MC in all breeding schemes. Moreover, more ancestors made a genetic contribution to the offspring in generation 20 with marker-based mating, explaining lower rates of inbreeding realized. The differences in the cumulative genetic gain between all mating strategies were minor. But marker-based mating maintained higher genetic variance than pedigree-based mating at generation 20. It was concluded that genomic information, instead of pedigree information, should be used in mating strategies when GEBVs are used for genetic evaluation to control inbreeding regardless of family structure and heritability.

Strategic genotyping of cow groups to improve reliability of genomic predictions
C. Edel, E. Pimentel, L. Plieschke, R. Emmerling and K.U. Götz
Institute of Animal Breeding, Bavarian State Research Center for Agriculture, Prof.-Dürrwaechter-Platz 1, 85586 Poing-Grub, Germany; christian.edel@lfl.bayern.de

For dairy cattle populations using genomic breeding value estimation, it can be assumed that the available bulls have already been genotyped. A further increase in the reliability of genomic predictions can be achieved by international collaboration and the exchange of bull genotypes. Additionally, enlarging the reference population by genotyping cows with phenotypes is an obvious consideration. At present, a complete genotyping and integration of all cows under milk recording is not within reach. Furthermore, deliberately including potentially preselected cows, genotyped by farmers, may introduce a bias to genomic breeding value estimation. In this study we investigate the advantage of enlarging the reference population by routinely genotyping a random sample of the first-crop daughters of every AI bull of the breeding program. To this end we analyzed small nuclear pedigrees, each consisting of a genotyped selection candidate and three generations of genotyped male ancestors. Genotypes were taken from the genomic routine evaluation of Fleckvieh cattle. The phenotypic information of the daughters of any male in each of these pedigrees was either assumed to be averaged to the DYD of the corresponding sire or was assumed to be individually linked to a genotype. Daughter genotypes in this case were simulated from phased haplotypes of their sires and random maternal gametes drawn from a haplotype library. We measured the gain from genotyping daughters as the increase in model-based theoretical reliability of a putative selection candidate. Results are encouraging especially for traits with higher heritabilities and higher number of daughters assumed to be genotyped. However, more refined follow-up investigations will have to clarify, if the results can be confirmed in large scale standard validation scenarios and whether additional benefits with respect to precision and unbiasedness of predictions are possible.

SNPs associated with traits of economic importance of Pinzgau cattle
R. Kasarda, N. Moravčíková, J. Candrák, A. Trakovická and O. Kadlečík
Slovak University of Agriculture in Nitra, Department of Animal Genetics and Breeding Biology, Tr. a. Hlinku 2, 94976 Nitra, Slovak Republic; radovan.kasarda@uniag.sk

Genome wide association study was made to identify genetic markers (single nucleotide polymorphisms; SNP) associated with milk production as well as SCS of cattle. The association study was performed on sample population of Pinzgau cattle in Slovakia. SNP data were obtained from Illumina BovineSNP50K BeadChip and BovineHD BeadChip, respectively. After data quality control SNP's were selected with call rate higher than 0.95, minor allele frequency>0.01 and Hardy-Weinberg equilibrium under Fisher exact test with P<0.001. Average call rate of selected SNP's was 0.995 ± 0.005 and 35,851 to 36,359 SNP's per individual were obtained. Phenotypic data were result of 8,430 daughters of 40 genotyped sires of Slovak as well as Austrian origin. Additive regression model was used for identification of significant associations. In total, with single-trait genome-wide association study, 12 genome-wide significant SNP's associated with milk production and 15 with somatic cell score were detected (P<0.0001). Our results are in agreement with previously published GWAS studies, which indicated their perspective use in selection on economically important traits in dairy populations.

Estimation social (indirect) genetic effects for growth rate according to breeding line

J.K. Hong, N.R. Song, D.W. Kim and Y.M. Kim
National Institute of Animal Science, Swine Science Division, 114, Sinbang 1-St, Seonghwan-eup, Seobuk-gu, Cheonan-city, Chungcheongnam-do, 331-801, Korea, South; john8604@korea.kr

Social interactions can have favourable or unfavourable effects on productivity and welfare. Several studies have shown that social interactions affect the genetic variation in a trait. The objective of this study was to estimate social (Indirect) genetic effects for average daily gain according to breeding line in Korea. A total of 14,371 records were used. There were 5,180 records from sire line and 9,191 from dam line. Total heritable variance expressed relative to phenotypic variance was 28% for sire line and 104% for dam line. This value of dam line clearly exceed the usual range of heritability for their traits. But that of sire line was not because the social variance is very small. Results indicate that efficiency of social effects differed among breeding lines. Further research on social effect including more lines and more breeds would clarify efficiency according to characteristics of population.

Genetic parameters of body weight at different stages of growth in male and female broiler chicken

W. Shibabaw[1], M. Shirali[1], P. Madsen[1], R. Hawken[2] and J. Jensen[1]
[1]Aarhus University, Molecular Biology and Genetics, Blichers Alle, 8830, Denmark, [2]Cobb Vantress Inc., Siloam Springs, AR 72761-1030, USA; wossenie.mebratie@mbg.au.dk

In this study, data from 54 selection rounds for growth rate from COBB breeding company was used for analysis with the objective of estimating genetic parameters for body weight in male and female broiler chicken at three different stages of growth. The numbers of animals with body weight and pedigree data were 646,703 and 649,483, respectively, and the pedigree covers about 10 generations back from the youngest animals. Body weight was measured at three different stages of growth in different animals (t, t+3 and t+7 days) of both sexes. For the first 39 selection rounds body weight was recorded at t+7 days, however, as selection continues the birds start to grow faster and the weighing age was changed and body was recorded at t+3 days from 7 selection rounds and then at t days from the 8 last selection rounds. Multivariate animal model was used to estimate genetic parameters in males and females using REML. In males, heritabilities of body weight were found to be 0.37, 0.33 and 0.29 at t, t+3 and t+7 days of age, respectively, whereas, in females, heritabilities of body weight were found to be 0.40, 0.38 and 0.38, at t, t+3 and t+7 days of age respectively. Genetic correlations of body weight between age groups were found to be 0.98 (t and t +3 days), 0.88 (t+3 and t+7 days) and 0.83 (t and t+7 days) which is different from unity. The genetic correlations of body weight between males and females at t, t+3 and t+7 days of age were found to be 0.94, 0.89 and 0.89, respectively. The heritability and genetic correlation estimates in this study suggests that body weight in males and females should be considered as two different traits. Similarly body weight in the three different ages should be considered as three different traits.

Molecular characterization of fibroblast growth factor 5 (fgf5) gene in alpaca
S. Pallotti, D. Pediconi, D. Subramanian, M.G. Molina, M. Antonini, C. Renieri and A. Laterza
University of Camerino, School of Bioscience and Veterinary Medicine, Via Gentile III da Varano, 62032
Camerino (MC), Italy; stefano.pallotti@unicam.it

Two coat phenotypes have been described in alpaca, Huacaya and Suri. The most common Huacaya type, is characterized by a crimped, dense fleeces, while the Suri type is characterized by a non-crimped, straight fleece. The genetic background of the two coat phenotypes has not been yet clearly defined. Results from segregation analysis provide statistical evidences that the Suri trait is possible inherited as a single dominant gene. The aim of our study was to molecularly characterize the FGF5 as a possible candidate gene for hair length phenotype in Suri type because of its role in the regulation of the hair follicle growth cycle. Currently, we have isolated and characterized two different cDNA clones encoding for FGF5 obtained from total RNA purified from skin biopsies of Peruvian Suri type alpaca. Sequence analysis revealed two types of transcripts: a long form (FGF5L) containing an ORF of 498 bp encoding for a putative 166 amino-acid polypeptide, and a a short form (FGF5S) of 375 bp encoding for a putative 125 amino-acid polypeptide. On the basis of the partial FGF5 genomic sequence data retrieved from the low-coverage 2.51X assembly of alpaca genome at the Ensembl database, the two transcript are produced by the alternative splicing of exon 5, which results in the loss of a fragment of 104 bases. Furthermore, RACE approaches specifically devised to characterize the 3' and 5' UTRs of FGF5 transcripts, show that the FGF5S isoform possesses two 3'-UTRs of 713 and 542 bases respectively. Meanwhile a single 3'UTR has been so far identified for the FGF5L isoform. Work is in progress in our laboratory to better characterize the FGF5 isoforms and their expression in the skin of Suri and Huacaya alpacas, and to identify mutations potentially involved in the fleece variations of alpaca.

The effect of genomic and classic selection on selection response in threshold traits
R. Behmaram and R. Seyed Sharifi
University of Mohaghegh Ardabili, Animal Science, Iran- Ardabil- University Street, 5619911367, Iran;
behmaram.reza@yahoo.ca

In this study, a genome consisting 6 chromosomes each with 100 cM in length was simulated. In order to create sufficient linkage disequilibrium and achieve a mutation drift balance after 50 generations of random mating in a finite population (Ne=100), population was expanded to obtain intended population size (500 male and 500 female). Three measures of heritability (0.10, 0.25, and 0.50) and four different numbers of markers (100, 200, 400, and 800) were considered. Each simulation was replicated 20 times and results were averaged across replications. Selection responses were estimated using by genomic and classic methods. With h^2=0.10, selection responses for genomic and classic path ranged from 0.07 to 3.45 and 0.47 to 2.85 respectively. For genomic and classic methods with h^2=0.25, selection responses varied from 0.26 to 12.35 and 0.42 to 5.68 respectively. With h^2=0.50, selection responses for genomic and classic approaches ranged from 0.42 to 15.46 and 0.91 to 6.12 respectively. The results indicated that increasing of markers number and heritability in many cases can be effective for heightening genetic gain in threshold traits, although in both cases of genomic and classic path, adding the number of markers in some cases especially after 400 markers showed a decrease in genetic gain. Selection responses were also relatively high for low heritable traits, implying that genomic selection could be especially beneficial to improve the selection on traits related to health and fertility. Comparing different ratio of males showed that the highest genetic gain occurred when selected males were between 10 and 25%. Our findings indicated that using genomic selection can be useful for threshold traits which include some of important traits in animal breeding, although, we must consider notable decay in selection response after first generations in long term breeding programs.

Genome-wide association study in dam line breeds for reproduction and production traits

S. Bergfelder-Drüing, C. Große-Brinkhaus, K. Schellander and E. Tholen
University of Bonn, Institute of Animal Science, Group of Animal Breeding and Husbandry, Endenicher Allee 15, 53115 Bonn, Germany; cgro@itw.uni-bonn.de

Selection programs in pig breeding are focused on production traits like average daily gain (ADG), lean meat percentage (LMP) as well as reproduction traits like number of piglets born alive (NBA). Of particular interest are antagonistic relationships between both trait complexes. Such conflicts will be induced by pleiotropic acting genes which should be considered particular when genomic based selection strategies are applied. To clarify the genetic background of NBA, ADG and LMP and to identify possible pleiotropic effects genome-wide association studies (GWAS) were performed using 3,496 Large White (LW) and Landrace (LR) pigs from four different commercial breeding organizations. Illumina PorcineSNP60 BeadChip was used for animal genotyping. Data of each breeding organization was analyzed separately with single-trait analyses. The applied mixed model contained the genomic relationship matrix and was extended by the Eigenstrat approach in order to control the distinct population stratification. Detection of pleiotropic SNPs was performed forming principal components reflecting phenotypic variance and covariance of all traits within each breeding organization and analyzed with GWAS. In total, 51 chromosome-wide and one genome-wide significant markers affecting analyzed traits were found in both breeds. Only one significant chromosome area for both breeds was detected on SSC12 affecting NBA and ADG. On SSC8 possible pleiotropic effect was detected for carcass composition traits. In total 23 and 75 significant associations for LW and LR, respectively, were found, when GWAS was performed using PC as phenotype. Many of these QTLs were found for PCs, which had opposite loadings for NBA and ADG or NBA and LMP. These results are indicators for hidden antagonistic effects, which might be important for breeding programs.

Genomic background for increased stillbirth rate in short and long gestating cows

G. Mészáros[1], S. Boison[1], M. Rauter[1], C. Fuerst[2] and J. Sölkner[1]
[1]University of Natural Resources and Life Sciences, Vienna, Augasse 2-6, 1090 Vienna, Austria, [2]ZuchtData EDV Dienstleistungen GmbH, Dresdner Straße 89, 1200 Wien, Austria; gabor.meszaros@boku.ac.at

A genome wide association study was conducted to highlight regions and identify genes with a possible influence on stillbirth rate in cattle. Special attention was devoted to calvings after a short or a prolonged gestation. Genotypes of 6,564 Fleckvieh bulls from the German-Austrian genotype pool were used for the analysis of direct and maternal stillbirth based on deregressed breeding values. The Austrian subset of bulls was used to determine the differences in genomic regions influencing stillbirth after a short or a prolonged gestation. The results for direct stillbirth showed prominent peaks on BTA14 and BTA21, with additional small signals on BTA3, BTA4 and BTA10. The top region at BTA14 was very near to SNAI2 gene associated with congenital heart defects and infant mortality in humans. A deletion in USP1 gene near out region on BTA3 leads to high postnatal mortality in mice. The genes NFE2L3 on BTA4 and NUSAP1 on BTA10 restrict the growth potential of fetuses in humans. Additional links were found to calving ease and gestation length. A larger number of smaller peaks were identified for the maternal stillbirth, with the most notable regions in BTA5, BTA21 and BTA27. Some of these regions were connected to preeclampsia, a disorder with adverse effects on fetal and maternal health in humans. The regions associated with the stillbirth with short and long gestation were different to those identified in the full set, although some were only suggestive, rather than significant. For prolonged gestations regions on BTA14, BTA16 and BTA20 were connected to abnormalities in fetal growth, muscular atrophy and preeclampsia. The top genomic regions for the short gestation subset were harboring genes associated with neonatal mortality, preterm delivery and preeclampsia on multiple chromosomes, although none of these regions reached the significance threshold.

Effect of including feed efficiency and beef production traits in the breeding goal for dairy cattle

P. Hietala and J. Juga
University of Helsinki, Department of Agricultural Sciences, Koetilantie 5, 00014 Helsinki, Finland;
pauliina.hietala@helsinki.fi

Improving the profitability of milk and beef production in a suitable way plays an important role in ensuring domestic self-sufficiency. More efficient feed utilization in dairy cattle could have a substantial effect both on the profitability and on reducing the harmful environmental impacts of dairy production through lower feed costs and emissions from dairy farming. In addition, beef production based on dairy herds has been shown to produce less greenhouse gas emissions per unit of product and to be more profitable under Finnish production conditions than beef production based on suckler cow systems. Several scenarios were used to assess the economic benefits of the inclusion of residual feed intake (RFI) and beef production traits in the breeding goal for Nordic Red dairy cattle using a deterministic simulation program ZPLAN+. In the scenario, where an integration of RFI traits in the breeding goal was studied, three breeding goal traits for RFI were considered (RFI in growing heifers, in fattened animals and in lactating cows). These traits were selected through two indicator traits which were RFI measured on young bulls in test stations and an indicator trait for RFI in lactating cows. For both traits, genomic breeding values were assumed to be available for genotyped bulls. The inclusion of beef production traits resulted in a higher increase in the discounted profit of the breeding program (+4.7%) than the inclusion of RFI traits (+1.2%). Considering the total costs of the breeding program, measuring RFI in bull test stations made only a marginal contribution when focusing on a population-wide perspective. Even though the inclusion of RFI and beef production traits in the breeding goal had a relatively small direct economic impact on the profit of the breeding program, their economic importance will likely increase in the future if economic values taking into account environmental impacts are introduced.

Genetically active dairy cows eat more feed and produce more milk but they are not more efficient

G.F. Difford[1,2], J. Lassen[2] and P. Løvendahl[2]
[1]Wageningen University, Animal Breeding and Genomics Centre, P.O. Box 338, 6700 AH Wageningen, the Netherlands, [2]Aarhus University, Center For Quantitative Genetics and Genomics, Department of Molecular Biology and Genetics, Blichers Alle, 8830 Tjele, Denmark; gareth.difford@mbg.au.dk

More feed efficient dairy cattle are essential for the future profitability and sustainability of the dairy cattle industry. The relationship between the feed efficiency complex and energy expenditure is not yet fully understood. The relative ease with which activity measurements can be acquired from pedometers or activity tags begs the question 'What relationship does energy expended on activity have with efficiency of feed utilization?' The objective of this study was to infer the genetic (co)variances between feed efficiency such as residual feed intake RFI_{koch}, feed efficiency component traits and activity over the entire lactation period. Data for each week of the first lactation (44 weeks) for dry matter intake (DMI), energy corrected milk (ECM), live weight (LW), body condition score (BCS) and activity (ACT) were collected on 460 Holstein and 230 Jersey cows from the Danish Cattle Research Center (Foulum). Univariate repeatability animal models were employed to estimate trait parameters along with pairwise bivariate repeatability animal models to estimate genetic correlations. Moderate heritability in the range of 0.15 to 0.36 was found for RFI with strong positive correlations to DMI 0.88, and weak positive correlations to ACT 0.13. The heritability of ACT ranged from 0.14 to 0.26 with a strong correlation to ECM in the range of 0.62 to 0.93 and moderate correlations with DMI of 0.28. Results of this study indicate that genetically active cows produce more milk and eat more but at the same time end up without being more or less efficient than inactive cows. Although the present study employed small sample sizes, it contributes to elucidating genetic relationships between the feed efficiency complex and activity.

Dissecting a complex trait – towards a balanced breeding goal for a better energy balance

N. Krattenmacher, G. Thaller and J. Tetens
Institute of Animal Breeding and Husbandry, CAU, Hermann-Rodewald-Str. 6, 24118 Kiel, Germany;
nkrattenmacher@tierzucht.uni-kiel.de

The focus of dairy cow breeding has shifted from mainly yield-based towards more balanced approaches that take into account fitness traits, fertility, and longevity. The negative energy balance p.p. is considered to play a decisive role in this context. Feed intake (DMI) is one of the major determinants of energy balance (EB) and recently some efforts have been made to pave the way for a direct inclusion of DMI in breeding goals. However, there is no consensus on how this should be done as there are different requirements in the course of lactation (reducing energy deficit p.p. vs subsequently improving feed efficiency to reduce production costs). The aim of this study was a genetic and genomic dissection of EB and its major determinants (DMI and energy-corrected milk (ECM)) at different lactation stages applying random regression methodology and GWAS to data from 1,174 primiparous cows. Daily heritability estimates ranged from 0.29 to 0.49, 0.26 to 0.37, and 0.58 to 0.68 for EB, DMI, and ECM, respectively, for the first 180 days in milk. Genetic correlations between ECM and DMI were positive, ranging from 0.09 to 0.36 with the lowest values found at the beginning of lactation. However, ECM and EB were negatively correlated (r_g=-0.26 to -0.59). The strongest relationship was found at the onset of lactation, indicating that selection for increased milk yield in this stage will dramatically intensify the energy deficit p.p. To address this risk, breeding for a higher DMI in early lactation seems to be a promising strategy as the relationship between DMI and EB was more pronounced (r_g=0.71 to 0.81) compared to the correlation between ECM and EB. We found some evidence that genetic regulation of ECM, DMI and EB is complex and partially divergent with trait-specific and lactation stage-specific candidate genes suggesting that the trajectories of the analyzed traits can be optimized in a way that satisfies both the economic and welfare issues.

Genetic parameters for carcass weight, age at slaughter and conformation of young bulls

I. Croué[1,2], M.N. Fouilloux[2], R. Saintilan[3] and V. Ducrocq[1]
[1]INRA, UMR1313 GABI, 78352 Jouy-en-Josas Cedex, France, [2]Institut de l'Élevage, UMR1313 GABI, 78352 Jouy-en-Josas Cedex, France, [3]ALLICE, UMR1313 GABI, 78352 Jouy-en-Josas Cedex, France; iola.croue@idele.fr

Meat production has a large economic impact for dual purpose breeds farms, partly through the production of young bulls, males slaughtered between 12 and 24 months. The profitability of young bull production can be increased through genetic improvement of carcass traits, including conformation based on EUROP carcass classification. In order to develop genetic evaluations for carcass traits in the Normande and Montbeliarde breeds, genetic parameters were estimated for each breed separately using a multitrait animal model based respectively on 166,762 and 173,125 records for young Montbeliarde and Normande bulls. Heritabilities were low for age at slaughter (0.05 to 0.10±0.01) and moderate for carcass weight (0.16 to 0.21±0.01 to 0.02) and conformation (0.21 to 0.26±0.01 to 0.02), implying that carcass weight and conformation of young bulls can be improved via selection in these two breeds. Genetic correlations were high and negative (favorable) between age at slaughter and carcass weight (-0.58 to -0.79±0.03 to 0.07), moderate and negative (favorable) between age at slaughter and conformation (-0.12 to -0.40±0.05 to 0.08) and high and positive between conformation and carcass weight (0.48 to 0.56±0.03 to 0.04). These estimates highlight favorable correlations for simultaneous genetic improvement of carcass weight and conformation, linked to a reduction of age at slaughter, in the two breeds. These results will be used for the development of routine genetic and genomic evaluations of carcass weight, conformation and age at slaughter of young bulls in Normande and Montbeliarde breeds in France. This work was performed within the Mixed Technology Unit 'Gestion Génétique et Génomique des populations bovines' (UMT3G).

Breeding for better welfare; feasibility and consequences
S.P. Turner, R. Roehe, J. Conington, S. Desire, I. Camerlink, R.B. D'eath and C.M. Dwyer
Scotland's Rural College (SRUC), West Mains Road, Edinburgh, EH9 3JG, United Kingdom;
simon.turner@sruc.ac.uk

Animals vary in response to the same stressor and this variation is typically partially genetically determined. Selective breeding leads to permanent and cumulative change and could benefit welfare at low cost to producers. Three examples will demonstrate the technical feasibility of breeding to improve welfare (lamb neonatal survival, foot infections ('footrot') in sheep and aggression between pigs). Each is a long-standing and routine welfare problem. Significant moderate heritabilities, higher than some traits currently under selection (lamb vigour which contributes to survival, 0.32; footrot, 0.15-0.25, aggression, 0.43) show that each can be improved by selection. New rapid and robust phenotyping methods have led to uptake (footrot) or commercial interest in selection (survival, aggression). Antagonistic genetic correlations with economic traits suggest that, in some cases, care is needed to avoid economic penalties of selection (e.g. r_g between aggressiveness and growth, 0.29-0.54). Economic weights need to be quantified for welfare traits and exaggerated in some cases to prevent their deterioration. Phenotyping costs remain higher for behavioural traits than traits such as growth and constrain uptake. For aggressiveness, genomic selection is robust (explaining 61-67% of the variance from a polygenic model), whilst selection on associative effects (heritable effects on growth of group-mates) may reduce chronic aggression. Strategic phenotyping via these methods may overcome barriers to selection. Improving welfare traits by selection presents ethical dilemmas that may be more easily resolved where opportunities for management change are few and unintended changes in other welfare traits are unlikely (e.g lamb survival in unpredictable climates compared to aggression). Barriers to selection for improved welfare are diminishing but the wider economic and welfare outcomes need to be understood where selection will involve changes in complex behaviour.

Genetic variability of MIR predicted milk technological properties in Walloon dairy cattle
F.G. Colinet[1], T. Troch[1], V. Baeten[2], F. Dehareng[2], P. Dardenne[2], M. Sindic[1] and N. Gengler[1]
[1]University of Liege, Gembloux Agro-Bio Tech, 5030, Gembloux, Belgium, [2]Walloon Agricultural Research Centre, 5030, Gembloux, Belgium; frederic.colinet@ulg.ac.be

Cheese yield, acidity and coagulation parameters are important technological and economic criteria at the level of farm cheese manufacturing. Since few years, several studies have demonstrated the usefulness of the mid infrared (MIR) spectrometry for the prediction of milk technological properties. Recently, equations for the prediction of Individual Laboratory Cheese Yield (ILCY), acidity and coagulation parameters were developed. The cross-validation coefficients of determination (R^2_{cv}) of those calibration equations ranged from 0.65 for rennet coagulation time to 0.81 for dry ILCY. The objective of this study was to estimate the genetic parameters related to the MIR predicted milk technological properties in first-parity Holstein cows from Walloon Region of Belgium. These equations were applied on the spectral database (FUTUROSPECTRE) generated during the Walloon routine milk recording. For each trait separately, records with standardized Mahalanobis distance value higher than 3 or with a predicted value out of the range of the reference values were discarded. After editing, the dataset includes about 100,000 predicted records, depending of the predicted trait, collected on 15,338 Holstein cows between January 2009 and November 2014. The variances components were estimated by REML using single-trait random regression animal test-day model. Random regressions were performed with modified normalized second order Legendre polynomials. The model took into account the heterogeneity of the residual variances over days in milk (DIM). Estimated daily heritabilities are low to moderate, thereby confirming the potential of selection. E.g. they ranged from 0.20 at 5th DIM to 0.49 at 240th DIM for dry ILCY. Further researches will study phenotypic and genetic correlations between predicted technological properties and milk production traits.

Inclusion of polled geno- or phenotype into breeding goals: impact on genetic gain and inbreeding

C. Scheper, T. Yin and S. König
University of Kassel, Animal Breeding, Nordbahnhofstr. 1a, 37213 Witzenhausen, Germany;
cscheper@uni-kassel.de

Breeding for polled cattle is of increasing importance in Holstein dairy cattle breeding programs and a viable alternative to the practice of dehorning. Currently available polled AI bulls have lower breeding values and higher levels of inbreeding compared to horned AI bulls. Evaluation of different polled breeding strategies in a long term perspective (20 generations) on population wide levels was performed via stochastic simulations. As a novelty, we developed a simulation package allowing for simultaneous consideration of qualitative (e.g. 'polled') and quantitative traits (e.g. milk yield) in breeding goals. A variety of assignment schemes for polled genotypes (i.e. PP and Pp) along with a flexible range of founder allele frequencies depicts genetic architectures in large Holstein as well as in small endangered populations: (1) random assignment; (2) assignment to animals with low estimated breeding values (EBV); (3) assignment to animals with low EBV and close genetic relationships. Intensified selection for polled cattle in ongoing generations was achieved by adding weighting factors (might be interpreted as economic weights) on (1) polled genotypes = selection scheme PO_GENO; and (2) polled phenotype = selection scheme PO_PHENO. Applying PO_GENO, a fixation of the polled allele (i.e. only homozygous polled progeny) was achieved in generation G9. Maximum polled allele frequencies in selection scheme PO_PHENO ranged between 0.75 and 0.85 in the final generation (G20). Genetic gain was substantially reducded in both selection schemes PO_GENO (-8%) and PO_PHENO (-5%) when compared to a base breeding goal without selection for polled. The results imply the need for specific mating designs to generate a broader polled breeding pool, i.e. selective elite mating schemes of superior horned animals (high EBV) with the presently available polled animals (lower EBVs), to prevent a potential loss of genetic gain related to an intensified selection for the polled trait.

Comparison of two approaches to account for G×E interactions in South African Holstein cattle

A. Cadet[1], C. Patry[1], J.B. Van Wyk[2], L. Pienaar[2,3], F.W.C. Neser[2] and V. Ducrocq[1]
[1] *INRA, Génétique Animale et Biologie Intégrative, UMR1313, 78302 Jouy-en-Josas, France,* [2] *University of the Free State, Animal Wildlife and Grassland Sciences, P.O. Box 339, 9300 Bloemfontein, South Africa,* [3] *ARC, P Bag X2, 0002 Irene, South Africa; neserfw@ufs.ac.za*

If not accounted for, Genotype × Environment (G×E) interactions can decrease the accuracy of genetic evaluations and decrease the efficiency of selection schemes. In countries characterized by a substantial heterogeneity of production systems, G×E is likely to exist within country. The purpose of the study was to assess the existence of G×E interactions for production traits in Holstein in South Africa using two different models. Production data from 378,782 first lactation cows were used. First, different regions were considered as separate traits. A multivariate animal model was fitted to obtain heritabilities for each region and genetic correlations between them. An imposed rank reduction of the genetic correlation matrix was also tested. Alternatively, a random regression model making use of herd climatic variables was used. Models with a heterogeneous within region residual variance and/or full or reduced rank multivariate analyses substantially improved goodness of fit. Strong differences in heritabilities were found between regions (from 0.23 to 0.36 from milk yield) and relative low genetic correlations were found (0.75 to 0.87), leading to substantial bull re-rankings. In terms of goodness of fit, the reaction norm model including average rainfall at herd level was as good as the full rank multivariate model by region. A possible interpretation is that average rainfall may be seen as a proxy of herd production systems (pasture vs total mixed rations). G×E interactions should be considered when genetic evaluations are performed for Holstein in South Africa.

Direct health traits and rearing losses in the total merit index of Fleckvieh cattle

B. Fuerst-Waltl[1], C. Fuerst[2], W. Obritzhauser[3] and C. Egger-Danner[2]
[1]University of Natural Resources and Life Sciences Vienna (BOKU), Dep. Sust. Agric. Syst., Gregor Mendel-Str.33, 1180 Vienna, Austria, [2]ZuchtData EDV-Dienstleistungen GmbH, Dresdner Str. 89/19, 1200 Vienn, Austria, [3]University of Veterinary Medicine Vienna, Institute of Veterinary Public Health, Veterinärplatz 1, 1210 Vienna, Austria; birgit.fuerst-waltl@boku.ac.at

Routine genetic evaluations for the direct health traits mastitis, early reproductive disorders, ovarian cysts and milk fever were introduced for Fleckvieh cattle in 2010. Since 2013, udder and fertility health traits have been included in the total merit index (TMI) through the udder health index and the fertility index, respectively. For rearing losses, a routine genetic evaluation is currently under development. For the definition of the future TMI, economic values were (re)estimated and the index calculation will be optimized. In order to analyze the genetic gain particularly for fitness and health traits, different weightings of direct health traits were compared. According to the derived values, the relative economic weights per genetic standard deviation would sum up to 19% for all health traits including mastitis, SCC, early reproductive disorders, cysts, milk fever and ketosis. For rearing losses, the relative economic weight is approximately 4.4%. If all traits from the current TMI and the fitness traits ketosis, milk fever and rearing losses were considered, the relative weights of dairy:beef:fitness were 37:13:50. Genetic gains for most traits would be above zero. For fertility, the index weight needs to be increased to achieve noticeable positive genetic gain. As high genetic gains may be obtained for dairy traits due to genomic selection, their weights could be slightly decreased in favour of fertility or other functional traits. This follows breeders' as well as consumers' demands.

Dairy breeding programs focusing on animal welfare and environment – a simulation study

A. Wallenbeck, T. Mirkena, T. Ahlman and L. Rydhmer
Swedish University of Agricultural Sciences, Animal Breeding and Genetics, Box 7023, 750 07 Uppsala, Sweden; anna.wallenbeck@slu.se

Swedish dairy producers were asked if 15 given breeding traits are related to productivity, animal welfare or environmental impact in a web based questionnaire. These traits were chosen so that they represent production, functional, traditional and potential future breeding traits. A trait was categorized as either an animal welfare (calving ease, claw and leg health, longevity, mastitis-, disease- and parasite resistance), environmental load (feed conversion, longevity and methane emission) or productivity related (milk yield, longevity, fertility, calving ease, mastitis- and disease resistance, claw and leg health, temperament) if more than 40% of the 468 producers considered the trait to be much related to welfare, environmental load, or productivity. Three breeding program scenarios reflecting producers' preferences regarding; (1) productivity – modeling the genetic change seen the last two decade; (2) animal welfare – 50% of the economic weight proportionally on productivity traits and 50% on animal welfare traits; (3) Environment – 50% economic weight on productivity traits and 50% on environmental traits were developed. The scenarios were simulated for a population of 140,000 Swe. Red dairy cows and compared using the program ZPLAN. Genetic parameters were collected from literature. Results showed that the breeding program emphasizing animal welfare increased the genetic improvement in the animal welfare traits compared to the productivity scenario but had unfavorable genetic change in milk yield. The simulation also showed favorable correlated genetic responses for fertility, temperament and methane emission. The scenario emphasizing reduced environmental load showed genetic improvement in traits categorized as environment related, with maintained genetic improvement in milk yield comparable with the productivity scenario.

Derivation of economic values for breeding goal traits in conventional and organic dairy production

M. Kargo[1,2], J.F. Ettema[2], L. Hjortø[1], J. Pedersen[1] and S. Østergård[3]
[1]SEGES, Agro Food Park 15, 8200 Aarhus N, Denmark, [2]Aarhus University, Center for Quantitative Genetics and Genomics, Department of Molecular Biology and Genetics, Blichers Allé 20, 8830 Tjele, Denmark, [3]Aahus University, Department of Animal Science, Blichers Allé 20, 8830 Tjele, Denmark; morten.kargo@agrsci.dk

With genomic selection high genetic gain can be achieved in minor populations. This may enable different lines for different production systems. The objective of this study was to derive economic values (EV) for breeding goal traits in different dairy production systems. In order to reflect a real production system we used a stochastic herd simulation model (SimHerd) to derive EV. SimHerd includes complex relationships between traits which make it possible to simulate structural relationships. These latter relationships caused by e.g. changes in age structure of the herd should be included in the derivations of EV, while genetic and environmental correlations should not. Earlier we have shown that these requirements can be handled using multiple regression and mediator variables. We have also shown that this method overcomes the problem with double counting. EV for nearly all traits of economic importance were derived for four different Danish productions systems: Conventional Organic, Hitec with use of modern in-line registration, and Environmental friendly with focus on reduced emission. EV for a selected group of traits in conventional and organic (in parenthesis) production systems are given below. Milk yield: €0.32 (0.35) per kg of ECM; Feed efficiency: €9.3 (11.7) per +0.01 kg ECM/consumed Scand. Feed Unit; Ketosis: €90 (199) per case; and Pregnancy rate: €2.5 (0.3) per % point. Overall, the results showed substantial differences between EV in Conventional and Organic production systems and the general trends were that milk production, feed efficiency and health traits have higher EV and cow fertility traits have lower EV in the organic production system.

Definition of breeding goals for dairy breeds in organic production systems

L. Hjortø[1], A.C. Sørensen[2], J.R. Thomasen[2,3], A. Munk[1], P. Berg[4] and M. Kargo[1,2]
[1]SEGES, Agro Food Park 15, 8200 Aarhus N, Denmark, [2]Aarhus University, Center for Quantitative Genetics and Genomics, Institute for Molecular Biology and Genetics, P.O. Box 50, Foulum, 8830 Tjele, Denmark, [3]Viking Genetics, Ebeltoftvej 16, 8960 Randers SØ, Denmark, [4]NordGen, P.O. Box 115, 1431 Ås, Norway; jorn.rthomasen@mbg.au.dk

In a previous study, economic values for organic dairy production were derived from a stochastic, bio-economic model. In this study, we attempt to assess the Danish organic dairy farmers' acceptance of the results from the objective method. As a null hypothesis, all non-market values were assumed to be zero. In accordance with this, we designed a partial choice survey in a way so farmers were given a number of tasks. In each task the farmer was given the choice between two alternatives. The alternatives differed in means of only two traits. The differences in trait means were calibrated according to the objectively derived economic values, so that the difference in all traits had the same monetary value. Therefore, any deviation from equal priority can be interpreted as a non-zero non-market value. There are approximately 350 organic dairy farmers in Denmark and all of them will have the opportunity to participate in the survey. We included 10 traits corresponding to 45 pair-wise comparisons. The questionnaire and the subsequent analyses were conducted by means of the internet-based software 1000Minds. We expect the results to show that farmers give increased priority to some traits above the objectively derived economic value. The breeding goals defined by organic dairy farmers are to be included in a study designed to examine the possibility of setting up breeding schemes for dairy breeds in organic production systems.

Runs of homozygosity and deletion-type events form different patterns in cattle genome
A. Gurgul, T. Szmatoła, I. Jasielczuk and M. Bugno-Poniewierska
National Research Institute of Animal Production, Laboratory of Genomics, Krakowska 1, 32-083 Balice,
Poland; artur.gurgul@izoo.krakow.pl

Runs of homozygosity (ROH) are continuous segments of the genome where the two haplotypes inherited from the parents are identical. The useful tool to detect ROH are genotyping microarrays which with simple computational tools allow for identification of segments of homozygous genotypes. However, at least some portion of this genotypes may be caused by deletion-type events accounting to the copy number variation (CNV) which is not directly informing about the level of genome autozygosity. To determine the size of this phenomenon with use of BovineSNP50BeadChip we detected and compared ROH and deletion CNV calls in the genome of 292 Polish Red cattle. ROH were identified using Plink software with minimum number of SNPs in ROH set to 15, with no heterozygous genotypes allowed inside the region and minimal ROH size of 1 Mb. CNVs were detected using PennCNV algorithm. LLR for all SNPs were corrected for GC genomic waves and along with BAF and genotypes were analyzed with PennCNV software using the following parameters: LRR SD<0.35, minimum 3 probes within a CNV region. In the studied population we detected 11,341 ROHs. The mean ROH size was estimated at 3.28 Mb (±4.6). On average we detected 38.8 homozygous segments per individual raging from 13 to 64 for individual samples. In 127 (43.5%) of this animals at least one deletion CNV was detected (from 1 to 10 per sample). The mean CNVs size was estimated at 241.3 kb and ranged from 43.8 kb to almost 2.3 Mb. Comparison of ROH and CNV breakpoints within individuals showed only 26 overlaps detected in 23 animals. On average CNVs may be responsible for 0.62% (±0.74) of whole ROH length in separate animals (with the highest value of 2.15%). This suggest that deletion events occurring in cattle genome are not significantly affecting ROH lengths and genomic inbreeding estimates. However, more detailed research on this topic in regard to different cattle breeds should be conducted.

Inbreeding trends in Holstein cattle population
I. Pavlík[1], O. Kadlečík[2], J. Huba[1] and D. Peškovičová[1]
[1]NPPC, Research Institute for Animal Production Nitra, Hlohovská 2, 951 41 Lužianky, Slovak Republic,
[2]Slovak University of Agriculture in Nitra, Department of Animal Genetics and Breeding Biology, Tr. A.
Hlinku 2, 949 76 Nitra, Slovak Republic; ivan.pavlik@vuzv.sk

The objective was to assess the trend of inbreeding in population of Holstein cows (including Red Holsteins) in Slovakia by the methods of genealogical analysis. There were 75,835 cows born between 1997-2012 taken into account. Of those 73.59% were inbred (inbreeding coefficient F higher than 0). The average inbreeding coeffcient in inbred population was 1.29% (average increase in inbreeding $\Delta F=0.34\%$). These values are not alarming but breeders should pay more attention in creating mating plans to avoid inbreeding depression. The trend of F was slightly increasing through the years. The average F of cows born in 1997 was 0.30%($\Delta F=0.096\%$) while in 2012 it was 1.10% ($\Delta F=0.269\%$). The same situation was observed in average relatedness of animals. These facts reflect using limited number of sires with lower blood-line diversity in recent years. Therefore, the evaluation of coancestry and inbreeding is important part of breeding work as well as maintenance of genetic diversity ensured by optimal mating strategy. Genomic selection represents very useful tool to maintain genetic diversity within large commercial populations.

Genomic regions bearing the signatures of differential selection in two breeds of cattle

A. Gurgul, K. Pawlina, I. Jasielczuk, T. Ząbek, T. Szmatoła, M. Frys-Żurek and M. Bugno-Poniewierska
National Research Institute of Animal Production, Laboratory of Genomics, Krakowska 1, 32-083, Balice,
Poland; tomasz.zabek@izoo.krakow.pl

History of individual cattle breeds formation is closely related with domestication and strong artificial selection which have led to the separation of phenotypically different types of cattle, significantly differing in terms of productivity and conformation traits. Such genetic improvement, lasting for many generations, should have left a clear signature in the genome of different breeds of cattle, particularly apparent in the changes of frequency of genes located in regions subject to strong selective pressure. Therefore, in this experiment the genotype data for the 43,315 single nucleotide polymorphisms evenly distributed in the genome were used to indicate the regions with the greatest differences in allele frequency between genetically different breeds of Polish cattle (Polish Red and Holstein-Friesian). Using the sliding window approach for the simultaneous analysis of 10 markers, we identified 19 regions of the genome with an average absolute difference in allele frequency greater than 0.55. In these regions we identified 55 coding genes involved in a wide range of processes, starting from the metabolism of fatty acids and vitamins, by the activity of the immune system to coat color formation. Together, regions included a 9.6 Mb of genomic sequences corresponding to 0.37% of the genome. The study may give rise to the search for genes that have a significant impact on the distinctive characteristics of the investigated breed.

Kappa-casein (CSN3) alleles and genotypes frequencies in different breeds bulls used in Latvia

D. Jonkus and L. Paura
Latvia University of Agriculture, Liela 2, Jelgava, 3001, Latvia; daina.jonkus@llu.lv

The aim of the study was to analyse CSN3 (kappa-casein) gene frequency in artificial insemination used bulls in Latvia. 84 bulls from AI station were included in the study: 19 genetic resources Latvian Brown (LBGR) breed bulls, 29 Red breed bulls (RB), which are presented by German Red (GR), Danish Red (DR), Swedish Red (SR), 9 Holstein Red and White (HRW) and 27 Holstein Black and White (HBW) breed bulls. DNA was obtained from semen of the sires. Genotypes for the CSN3 gene were determined by Polymerase Chain Reaction and Restriction Fragment Length Polymorphism methods (PCR-PFLP). In the LBGR population were found A and B κ-casein alleles. The Allele A was more frequent than the B allele (0.816 vs 0.184). The RB, HRW and HBW breed bulls' had the A, B and E allele. The A allele frequency was 0.694 and 0.593 in RB and HBW breed, respectively, whereas the B allele frequency in RB was 0.116 and in red and black Holstein breed bulls 0.222. In RB breed were found higher frequency of E allele (0.190) and higher frequency of genotypes with this allele. The CSN3 genotype frequencies in LBGR bulls were AA=0.632 and AB=0.368. In RB breed were found six genotypes: the more frequent was AA genotype – 0.462, the BB and EE genotypes' were with frequency 0.017. HRW bulls had 4 genotypes': homozygote AA (0.333) and three heterozygote AB (0.333), AE (0.223) and BE (0.111). Genotype frequencies in HBW bulls were AA=0.370, AB=0.148, AE=0.296, BB=0.111, BE=0.074 and EE=0. In Latvia ongoing research to find CSN3 and β-LG genotypes in LBGR cows population.

Dairy breeding objectives: the farmers' perception

M. Klopčič
University of Ljubljana, Biotechnical Faculty, Department of Animal Science, Groblje 3, 1230 Domžale,
Slovenia; marija.klopcic@bf.uni-lj.si

The main aim of dairy cattle breeding last 50 years focused on increasing milk production. The increase in production has been accompanied by declining ability to reproduce, increasing incidence of health problems, and declining longevity of dairy cows. Functional traits become more important also because of growing concern about animal well-being and consumer demands for healthy and natural products. Dairy cattle breeding is an important area also for Slovenian dairy sector. For this reason, a study was conducted on dairy farms using a questionnaire survey involving more than 250 dairy farms with bull dams Holstein, Brown or Simmental breed. The aim of this study was to identify which traits in breeding goals are the most important for Slovenian dairy farmers and whether variations in their farm system and breed of cows influenced these opinions. A majority of the farmers would like give more emphasis on health and fertility traits including longevity. For farmers with dual-purpose breeds (Simmental and Brown) also beef traits are important. The selection of traits in breeding goals was largely influenced by the farmer's herd breed and breeding objective (the 'profit cow' vs the 'management cow'). Farmers typically selected traits that are recognised as weaknesses of their corresponding herd, e.g. fertility, longevity, health (SCC, metabolic diseases, lameness, etc.). Farmers are aware that they will need to pay in the future more attention to feed efficiency, reduce emissions and minimize the negative effects on the environment.

Survival analysis of productive life in Brazilian Holstein using a Weibull proportional hazard model

E.L. Kern[1,2], J.A. Cobuci[2], C.N. Costa[3] and V. Ducrocq[1]
[1]Inra, UMR1313 Génétique Animale et Biologie Intégrative, Domaine de Vilvert, Jouy-en-Josas, 78352, France, [2]UFRGS, Department of Animal Science, 7712 av. Bento Gonçalves, Porto Alegre, 91540-000, Brazil, [3]Embrapa Gado de Leite, rue Eugênio do Nascimento 610 Dom Bosco, Juiz de Fora, 36038-330, Brazil; elikern@hotmail.com

The objectives of this study were to assess the most important factors that influence productive life (PL) of Brazilian Holstein cows and to estimate genetic parameters for PL using a piecewise Weibull proportional hazard model. Records of PL from first calving to last recording (culling) of 13,292 cows coming from 945 herds were used. They had to have at least one of their first five calvings occurring between 1989 and 2013 and they were daughters of 6804 sires. The model included the time-dependent effects of region within year, class of milk production within herd-year, class of milk production within lactation number, class of fat and protein content within herd and (variation in) herd size as well as the time-independent effects of age at first calving, the random effects of herd-year and sire. All fixed effects had a significant effect on PL ($P<0.001$). The relative risk (RR) for within herd class of milk yield varied from 3.16 for the worst 10% class to 0.41 for the best 10% class. RR also increased as protein and fat decreased, but to a lesser extent compared to milk yield. Significant region-year effects were found on PL, reinforcing the importance of the inclusion of this effect in the analysis. RR increased with age at first calving and with herd size but lower risks were observed when herd size was increasing or decreasing, compared to stable herds. The Weibull shape parameters (and therefore RR) increased with lactation number and with stage of lactation. The sire genetic variance estimate was 0.030 ± 0.002 which corresponds to an effective heritability estimate of 7.8% in absence of censoring. A positive genetic trend of PL was observed. These results will contribute to the development of a genetic evaluation necessary to improve PL of Brazilian Holsteins.

Functional longevity in Polish Simmentals – genetic evaluation

M. Morek-Kopec[1] and A. Zarnecki[2]
[1]University of Agriculture in Krakow, Department of Genetics and Animal Breeding, Al. Mickiewicza 21, 31-120, Poland, [2]National Research Institute of Animal Production, ul. Krakowska 1, 32-083 Balice k. Krakowa, Poland; rzmorek@cyf-kr.edu.pl

The Weibull proportional hazard model was applied to genetic evaluation for functional longevity in Polish Simmentals. Data from the SYMLEK National Milk Recording System consisted of production and disposal records for 12,527 Simmental cows, daughters of 294 sires, calving for the first time from 1995 to 2014 in 286 herds. Length of productive life (LPL) of cows was calculated as number of days from first calving to culling or censoring. Average length of productive life of 4,462 cows with complete (uncensored) survival records was 1,198 days; mean censoring time for the remaining 8,065 cows was 1,093 days. Functional longevity was defined as length of productive life corrected for production. The model included time-independent fixed effect of age at first calving, time-dependent fixed effects of year-season, parity-stage of lactation, annual change in herd size, relative fat yield and protein yield, and random herd-year-season and sire effects. Likelihood ratio tests showed a highly significant impact of all fixed effects on longevity, except for relative fat yield. The estimated sire variance was 0.069, resulting in effective heritability of 0.25. Sire breeding values (EBV) expressed as relative risk of culling ranged from 0.524 to 1.724, with mean 1.024 and standard deviation 0.157. The average reliability of EBVs was 0.47. Moderate heritability indicates some possibility of selection for functional longevity in Polish Simmentals.

Inserting of Interbull MACE values in national RR test day model in single-step genomic evaluation

J. Přibyl[1], J. Bauer[1], P. Pešek[1], J. Šplíchal[2], L. Vostrý[1] and L. Zavadilová[1]
[1]Institute of Animal Science Prague, Pratelstvi 815, 104 00 Uhříněves, Czech Republic, [2]Czech Moravian Breeding Corporation, Hradištko 123, 252 09, Czech Republic; vostry@af.czu.cz

Nation-wide test-day records of milk production from 1995 to 2014 and Interbull file (August 2014) of all available Holstein bulls were combined in single-step procedure (ssGBLUP) in national genomic evaluation for all genotyped and non-genotyped animals in the population. For pedigree relationship matrix, 5 generations of ancestors were included. Pedigree and genomic relationship matrix were weighted in ratio 20 and 80% respectively. In the analysis totally 3,520 genotyped (Illumina 50k chip) proven and young bulls were used. National and Interbull values were included into multi-trait model. National test-day records were included into random regression model by 4 parameter Legendre polynomials. Interbull values consisted of index from three lactations. Genetic covariance matrix covered mutual covariances of regression coefficients and covariances of genetic regression coefficients to the Interbull index value. For all individuals in the national evaluation reliabilities of EBV and GEBV were approximated. The procedure of approximation of reliabilities was based on the contributions from effective number of observations in contemporary groups in herd-test-days, from pedigree relationship and from genomic relationship. All animals in the population (genotyped and non-genotyped) were included on the basis of genetic evaluation into one common ranking. The national and Interbull databases and genotyping of animals notably contributed to the accuracy of estimated breeding values (EBV) and genomic enhanced breeding values (GEBV).

Modelling phosphorous intake, digestion, retention and excretion in pigs of different genotypes

I. Kyriazakis, V. Symeou and I. Leinonen
Newcastle University, School of Agriculture, Food & Rural Development, Newcastle NE1 7RU, United
Kingdom; ilias.kyriazakis@newcastle.ac.uk

A deterministic, dynamic model able to represent phosphorous (P) digestion, retention and excretion in pigs of different genotypes was developed. The model represented the limited ability of pig endogenous phytase (Phy) to dephosphorylate phytate as a function of dietary calcium (Ca). Phytate dephosphorylation in the stomach by exogenous microbial Phy was expressed by first order kinetics. The absorption of non-phytate P from the small intestine was set at 0.8. The net efficiency of using digested P (ep) was set at 0.94 and assumed to be constant across genotypes. P requirements for maintenance and growth were made functions of body protein. Undigested P was assumed to be excreted in the feces in both soluble and insoluble forms. If digestible P exceeded the requirements for P then the excess P was excreted in the urine. Model behaviour was investigated for its predictions of P under different levels of inclusion of microbial phytase and dietary Ca, and different non-phytate P: phytate ratios in the diet. Uncertainties associated with the underlying assumptions of the model were identified. Model outputs were most sensitive to the values of ep and the non-phytate P absorption coefficient from the small intestine; all other model parameters influenced model outputs by <10%. Independent data sets were used to evaluate model performance. They were selected on the basis of the following criteria: they were within the BW range of 20-120 kg, pigs grew in a thermo-neutral environment; and they provided information on P intake, retention and excretion The model predicted satisfactorily pig responses to variation in dietary inorganic P supply, Ca and Phy supplementation. The model performed best with 'conventional', European feed ingredients and poorly with 'less conventional' ones, such as DDGS and canola meal. Explanations for these inconsistencies will be offered; they are expected to lead to further model development and improvement.

Consequences of nutritional strategies on decreasing phosphorous excretion from pig populations

I. Kyriazakis, V. Symeou and I. Leinonen
Newcastle University, School of Agiculture, Food and Rural Development, Newcastle NE1 7RU, United
Kingdom; ilias.kyriazakis@newcastle.ac.uk

A validated deterministic model that predicts phosphorous (P) digestion, retention and metabolism for different pig genotypes was developed into a stochastic one to investigate the consequences of different management strategies on P excretion by pigs growing from 30-120 kg BW. The conversion was associated with several challenges, including the description of the variation and co-variation between the different parameters that describe pig genotype. The strategies investigated were: (1) changing feed composition frequently in order to match more closely pig digestible P (digP) requirements to feed composition (phase feeding); and (2) grouping pigs into light and heavy groups and feeding each group according to the requirements of their group average BW (sorting). Phase feeding reduced P excretion as the number of feeding phases increased. The effect was more pronounced as feeding phases increased from 1 to 2, rather than 2 to 3. Increasing the number of feeding phases increased the % of pigs that met their digP requirements during the early stages of growth and reduced the % of pigs that were supplied less than 85% of their digP requirements at any stage of their growth. Sorting of pigs reduced P excretion to a lesser extent; the reduction was greater as the % of pigs in the light group increased from 10 to 30%. This resulted from an increase in the P excreted by the light group, accompanied by a decrease in the P excreted by the remaining pigs. Sorting increased the % of light pigs that met their dig P requirements, but only slightly decreased the % of remaining pigs that met these requirements at any point of their growth. Exactly the converse was the case as far as the % of pigs that were supplied less than 85% of their digP requirements were concerned. The developed model is flexible and can be used to investigate the effectiveness of other management strategies in reducing P excretion from groups of pigs.

Using a value based method to determine the optimal dose of phytase for gestating and lactating sows

A.L. Wealleans and L.P. Barnard
Danisco Animal Nutrition, DuPont Industrial Biosciences, Marlborough, SN8 1XN, United Kingdom;
alexandra.wealleans@dupont.com

The level at which to include phytase in monogastric diets is a contentious industry issue. This study aimed to use a method based on the law of diminishing marginal utility to estimate the optimal dose of phytase for gestating and lactating sows. 40 commercial hybrid sows were assigned to 5 treatments (8 replications/treatment): a positive control (PC) diet, a negative control (NC) diet based on wheat/barley and deficient in phosphorus and calcium, and 3 graded levels of an *Escherichia coli* phytase expressed in *Schizosaccaromyces pombe* over the negative control (500, 1000, 7,500 FTU/kg). Optimum inclusion was determined based on the law of diminishing marginal utility through the replacement of inorganic P from the ration. Phytase supplementation had no significant effect on feed intake or litter performance ($P>0.05$). Significant increases in P and Ca digestibility were seen in both gestating and lactating sows following phytase supplementation ($P<0.05$). Non-linear regression was conducted on faecal phosphorus digestibility using JMP 11, and optimum phytase dose was determined using the marginal decision rule, assuming the value of incremental P from phytase was equal to the value of the same amount of P from DCP. The optimum dose of phytase under standard price conditions was calculated to be 1,034 FTU/kg for gestating sows and 799 FTU/kg for lactating sows. The difference in optimal dose is likely to be a function of the change in metabolic state post-parturition, and the high demand for phosphorus in lactating sows.

Effect of forage particle size on phosphorus retention and levels in plasma and saliva in maintenance

A. García-Robledo, A.H. Ramírez-Pérez and S.E. Buntinx
Facultad de Medicina Veterinaria y Zootecnia. Universidad Nacional Autónoma de México, Nutrición Animal y Bioquímica, Av. Universidad 3000, 04510, Mexico; rpereza@unam.mx

This study was carried out to determine the effect of forage particle size on digestive availability of dietary phosphorus (P), as well as on plasmatic and salivary P concentrations in maintenance ewes. A digestibility trial was developed using 6 ewes (Pelibuey × East Friesian, 49±2.7 kg BW), in a double, 3×3 Latin Square design with 3 particle sizes of oat hay (S, 2.54 cm.; M, 5.08 cm. and G, long hay coarse particle). Ewes were confined in metabolic crates, fed oat hay (793.2 g/kg DM) and concentrate (206.8 g/kg DM, ground corn, soybean meal, coconut meal and minerals). Ration met NRC 2006 requirements for ewes, which require 981±36.7 g DM/d. The digestibility trial consisted of three 21-d periods, 14 d for adaptation to particle size and 7 d for sample collection. Saliva samples were taken at the parotid duct opening; whereas, blood samples were taken at the jugular vein. Both, saliva and blood were sampled before (600 h) and two hours after the first meal (800 h). The forage particle size did not affect ($P>0.05$) apparent digestibility of DM (542.25±7.61 g/kg), CP (565.8±25.65 g/kg CP), NDF (596.5±16.86 g/kg NDF) and energy (0.53±0.02). Also it did not have an effect on plasma P concentrations either in fasting ewes (52.3±4.5 mg/dl) or two-hours after first meal (52.4±3.7 mg/dl), as well as in saliva concentrations ($P>0.05$, 581.2±24.9 µg/g). In this study, forage particle size did not show conclusive effects ($P=0.08$) on P balance data, which showed negative apparent absorption coefficient values (-20.43±61.6 g/kg P) and P retention (-0.1±0.1 g of P/d) that were accompanied by important standard errors of the mean for particle G. Further research is needed regarding other physiological states of ewes where metabolic challenges are more important.

Effects of structural and chemical changes of soy proteins during thermal processing on proteolysis
S. Salazar-Villanea[1], E.M.A.M. Bruininx[1,2] and A.F.B. Van Der Poel[1]
[1]Wageningen University, P.O. Box 338, 6700 AH Wageningen, the Netherlands, [2]Royal Dutch Agrifirm, P.O. Box 20018, 7302 HA Apeldoorn, the Netherlands; sergio.salazarvillanea@wur.nl

The aim was to analyse the decisive factors for protein modifications during thermal processing and their influence on proteolysis. The experiment consisted of a $3 \times 2 \times 2$ factorial design of heat (not added, 80 or 120 °C), moisture (0 or 25% w/w) and glucose (with or without) addition. A soy protein isolate devoid of soluble sugars was used. Thermal treatments were performed by autoclaving for 30 minutes, with 3 replicates per treatment. Glucose was added at a 2:1 molar ratio of reducing sugars to total lysine level. Each sample was analysed for nitrogen solubility index (NSI) in 0.2% KOH and proteolysis with the pH-STAT method. Proteolysis was analysed during 10 minutes at pH 8 using trypsin, chymotrypsin and peptidase. Statistical model included heat, moisture and glucose addition as main effects and the 2-way and 3-way interactions. The interaction between heat and moisture on the degree of proteolysis was highly significant ($P<0.001$). The degree of proteolysis was highest in the samples heated at 80 and 120 °C without added moisture (12.79 and 11.71%, respectively), in the mid-range for samples without any heat treatment (11.60% with added moisture and 11.39% without added moisture) and lowest for samples heated at 80 and 120 °C with the addition of moisture (10.24 and 8.88%). The 3-way interaction was significant for NSI ($P=0.01$). Samples without any heat treatment and samples heated at 80 °C without added moisture were highly soluble (92-96%), followed by samples heated at 80 °C with added moisture (74-77%), samples heated at 120 °C without added glucose (38-47%), and samples heated at 120 °C with added glucose (5-9%). In conclusion, these results suggest that protein solubility is highly impaired by both structural and chemical changes, whilst proteolysis is mainly reduced by structural modifications of proteins.

Starch degradability and energetic value of maize silage
J. Peyrat[1], E. Meslier[1], A. Le Morvan[2], A. Férard[1], P.V. Protin[1], P. Nozière[2] and R. Baumont[2]
[1]ARVALIS Institut du Végétal, Station expérimentale de la Jaillière, 44370 La Chapelle Saint Sauveur, France, [2]INRA, UMR1213 Herbivores, Site de Theix, 63122 Saint-Genès Champanelle, France; j.peyrat@arvalisinstitutduvegetal.fr

Maize silage provides the main source of energy in the diet of high-yielding ruminants. Thus, accurate evaluation of its energetic value becomes a key economic issue. The nature of available energy for animal from maize silage results in rate and extent of ruminal degradation of starch and cell-wall fractions. In this study, we investigated the effects of maturity stage and type of hybrids grain on ruminal starch degradability of maize forage. Four maize hybrids (F1, F2, FD and D varying from flint to dent grain) were harvested at four maturity stages, over two years. All silage samples were dried 72 h at 60 °C and ground at 4 mm before being nylon-bagged (3 g in 5×10 cm R510 Ankom bags) and incubated in the rumen of three fistulated cows at six incubation times (2, 4, 8, 24, 48 and 72 h). Effective starch degradability was calculated assuming a particulate passage rate of 0.06 h^{-1} (ED6Starch). Effects of maturity stage, hybrid and year of harvest were analysed on ED6Starch using MIXED procedure of SAS. With increasing maturity, the ED6Starch decreased from 0.74 to 0.58. Hybrid D with dent grain has the highest ED6Starch (0.72) and hybrid F1 the lowest (0.59, $P<0.01$). Significant interactions were observed between maturity stage, hybrid and year of harvest on ED6Starch. Negative correlations were observed between ED6Starch and indicators of maturity, i.e. sum of temperature from flowering to harvest ($r=-0.55$) or whole plant DM content at harvest ($r=-0.49$). Taking into account effects of hybrid and year of harvest, ED6Starch is closely related to the DM content at harvest with a quadratic effect ($r^2=0.77$; RSD=4.57). This result indicated that the use of criteria characterizing maturity stage and type of hybrid could be relevant to estimate ED6Starch and their implications for diet formulation.

LCIA results of feedstuffs for French livestock

A. Wilfart[1], S. Dauguet[2], A. Tailleur[3], S. Willmann[3], M. Laustriat[2], M. Magnin[1] and S. Espagnol[4]
[1]INRA, Agrocampus Ouest, UMR 1069 Soil, Agro and hydroSystems, 35000 Rennes, France, [2]CETIOM,
Technical Institute for Oilseeds, 33600 Pessac, France, [3]ARVALIS, Institut du végétal, 44370 la Chapelle
Saint Sauveur, France, [4]IFIP, Institut du porc, 35651 Le Rheu, France; Aurelie.Wilfart@rennes.inra.fr

Animal feed contributes very significantly to the overall environmental impacts of animal products assessed by Life Cycle Assessment (LCA). The current feed formulation only takes into account economical and nutritional constraints. The French project ECOALIM aims at improving the environmental impacts of husbandries by optimizing their feed. This will lead to select differents feedstuffs in the formulation, or to change their ways of production (crop managements, transformation processes). To perform such optimization, LCA data concerning the feedstuffs are needed. 142 LCA data of 69 different feedstuffs were produced. Several scales were treated: national average data representative for France with different poduction processes for transformed feedstuffs, and different managements for the main crops used in feed like cereals and oil seeds ((1) organic fertilization, (2) introduction of intermediate crops, (3) introduction of legumes in the rotation).The LCA were perfomed with a same methodology with a focus made on five impact indicators (energy demand, GHG emissions, acidification, eutrophication, land occupation). Considering the weight of field emissions on environmental impacts of crops, models were preferably adapted to the French context and when possible, special developments were made to estimate more precisely emissions like nitrate leaching. These results represent a step forward to share with the agricultural sector in order to promote environmental evaluation and good farming practices. Thus, this should encourage the first processing and the feeding industries to promote better agricultural practices.

Comparison of urea and slow release urea, supplied in 4 portions a day, in dairy cattle

L. Vandaele, K. Goossens, J. De Boever and S. De Campeneere
ILVO Institute of Agricultural and Fisheries Research, Animal Sciences Unit, Scheldeweg 68, 9090 Melle,
Belgium; karen.goossens@ilvo.vlaanderen.be

Feed urea is used to optimize microbial protein production in cattle diets with a shortage of rumen degradable protein. Slow release urea sources are claimed to realize a better synchronization of energy and protein in the rumen, because of a retarded release of nitrogen, but they are quite expensive. The aim of the present study was to compare a dispersed supply of urea (U) and slow release urea (SRU) (Optigen®, Alltech) in a cross over trial with 2 periods of 4 weeks. The trial was set-up with 24 high producing dairy cows and the diets were formulated according to the energy and protein requirements of each individual cow. Maize silage, prewilted grass silage and pressed beet pulp (60/30/10 on dry matter base) were fed *ad libitum* and were individually supplemented with (protected) soybean meal, wheat and concentrates. To attain a rumen degradable protein balance (RDPB) of 0, each cow daily received at least 100 and max 228 g of one of both feed urea sources in 4 equal portions at 8 am, 12 am, 4 pm and 8 pm. Beside daily registration of dry matter intake (DMI) and milk production, milk composition was determined on 4 consecutive milkings. Intake and performance data were compared using analysis of variance (P<0.05). Cows achieved a higher DMI with SRU (21.3 kg) than with U (20.8 kg). This 0.5 kg difference (P=0.13) did not result in significantly different performances. Cows produced on average 25.50 kg (U) and 25.76 kg (SRU) milk with a fat content of 4.41% for U and 4.39% for SRU; and a protein content of 3.40% for both treatments. The difference in DMI resulted in a different crude protein content (12.9% for U and 12.7% for SRU) and RDPB (64 g/d for U and 35 g/d for SRU). Nitrogen efficiency was however not affected (30.8% for U and 31.3% for SRU). In conclusion, when urea supply in diets with a shortage of degradable protein is spread over 4 times per day the use of slow release urea sources only has minimal benefits on dairy cattle performance.

Reconsidering acquisition and allocation in animal nutrition models to better understand efficiency

L. Puillet[1], D. Reale[2] and N.C. Friggens[1]
[1]INRA, UMR791 Modélisation Systémique Appliquée aux Ruminants, 75005 Paris, France, [2]UQAM, Chaire de Recherche du Canada en Ecologie Comportementale, H3C 3P8 Montréal, Canada; laurence.puillet@agroparistech.fr

Improving livestock production efficiency is a key issue for animal scientists. At the animal level, efficiency relies on resource acquisition and allocation among biological functions. A potential trade-off exists between improving short term efficiency (resource partitioning toward one function) and long term efficiency (involving longevity). Unravelling this trade-off implies a better understanding of how these efficiency criteria are impacted by differing levels of investment in acquisition and by different allocation strategies. Existing nutrition models are based on the concept of production potential (e.g. milk yield, average daily gain). Resource acquisition and allocation are then derived from this input. As a result, these models are not suitable to study and disentangle the role of acquisition, allocation and utilization in animal efficiency. In this study, we developed an animal model that simulates the biological variables that underpin efficiency, depending on inputs representing energy acquisition, allocation and utilization. The dynamic model comprises 4 sub-models. The acquisition sub-model simulates gross energy intake from the environment and the costs associated to its transformation into metabolizable energy. The allocation sub-model simulates proportions of metabolizable energy allocated to growth, survival and reproduction throughout lifespan. Based on acquisition and allocation dynamics, the utilization sub-model achieves energy partitioning among functions and the body mass sub-model simulates body reserves and structural mass changes. A screening technique (Morris method) was used with 26 inputs related to energy acquisition, allocation and utilization to determine their impacts on both short-term and long-term efficiencies, while reducing computational cost.

Data mining of *in vitro* and *in vivo* dietary intervention datasets

D. Schokker[1], I. Wijers[2] and M.M. Hulst[1]
[1]Wageningen Livestock Research, Genomics, Droevendaalsesteeg 1, 6708 PB Wageningen, the Netherlands, [2]Wageningen UR, Animal Nutrition, De Elst 1, 6700 AH Wageningen, the Netherlands; dirkjan.schokker@wur.nl

The objective of this study was to identify overlap in the response between *in vitro* (cultured intestinal epithelial cells of pig jejunum, IPEC-j2) and *in vivo* (pigs, chicken) systems upon (dietary) interventions. We used three different interventions: a high concentration of ZnO, rye, and amoxicillin. In the *in vitro* system we also challenged the cells by administrating *Salmonella* after the dietary exposure to trigger an immune reaction. Our approach was to use bioinformatic tools and data mining to investigate to what extent the *in vitro* system is able to predict the *in vivo* effects by identification of overlapping genes and biological processes affected by the intervention. Here, we focused mainly on genes, pathways, and biological processes involved in intestinal functionality and immune competence. We only found a minor overlap in genes or processes between both systems. For ZnO the main overlapping process was related to the HIF-1α pathway, whereas for rye we observed many overlapping processes related to cell cycle, and for amoxicillin the reactive oxygen species (ROS) was overlapping. ROS activates CREB activity which may lead to apoptosis. For the *in vitro* system we observed an up-regulation of key cytokines (IL6 and IL8) in the control samples. In the *in vitro* system, ZnO and amoxicillin were both able to down-regulate the *Salmonella* induced inflammatory genes. In the *in vivo* system ZnO also regulates genes involved in immune related processes (i.e. TGFβ, cytokine, and chemokine signaling), whereas the amoxicillin treatment led to higher expression of genes involved in the gut barrier function. In conclusion, although we only observed a minor overlap between the *in vitro* and *in vivo* system, this bioinformatics and data mining approach showed that the *in vitro* system gives an impression of the *in vivo* system. Potentially, such an approach may yield new angles for selecting compounds for dietary interventions.

Effect species on the chemical composition and nutritive value of oak tree leaves

A.I. Atalay[1], M. Sahin[2], O. Kurt[2] and E. Kaya[2]
[1]Igdir University, Faculty of Agriculture, Animal Science, Suveren Campus, 76100 Igdir, Turkey,
[2]Kahramanmaras Sutcu Imam University, Faculty of Agriculture, Animal Science, Avsar Campus, 46100
Kahramanmaras, Turkey; aliihsanatalay66@hotmail.com

Oak is one of the most important trees which provides considerable amounts of leaves for ruminant animals to meet their requirements. There are a lot of oak trees from different species in Turkey. The aim of the current experiment was to determine the effect of species on the chemical composition and nutritive value of oak tree leaves. The crude protein content of leaves of *Quercus suber* was significantly higher than the other oak species. The neutral detergent fiber, acid detergent fiber and condensed tannin contents of *Quercus ilex* and *Quercus coccifera* were significantly higher than the others. The ether extract content of *Quercus cerris* was significantly higher than the other oak species. The gas production and metabolisable energy of leaves of *Quercus infectoria* and *Q. cerris* were significantly higher than those of *Q. suber*, *Q. ilex* and *Q. coccifera*. True digestibility of leaves of *Q. infectoria* was significantly higher than for the other oak species. Partitioning factor and microbial mass of *Q. ilex*, *Q. coccifera* and *Q. infectoria* were significantly higher than those of *Q. suber* and *Q. cerris*. As a conclusion, species had a significant effect on the chemical composition and nutritive value of oak tree leaves.

Effect of high N efficiency rations on N excretion in heavy pigs

D. Biagini and C. Lazzaroni
University of Torino, Department of Agricultural, Forest and Food Sciences, largo P. Braccini 2, 10095
Grugliasco, Italy; carla.lazzaroni@unito.it

To spread among farmers the good practices in animal feeding, a study was carried out in heavy pig's Italian commercial farm, adopting rations with high N efficiency. Results obtained in three fattening period (about 6 months each and differing for genetic type) in an heavy pig's farm, located in the north-west plain of Italy (Piemonte region) and producing for the PDO cured pork ham 'Prosciutto di Parma' supply chain are reported. N balance was adopted to evaluate the reduction of N excretion using diets with low protein (LP) content and supplemented with essential amino acids vs ordinary one (HP). The LP diet was formulated reducing soy meal and adding lysine, methionine, threonine and tryptophan, according to NSNG 2010 recommendations, with a feed gross protein reduction of 10% for each feeding phases (3 phases: 25-90, 90-120, 120-170 kg LW). Data collected for the N balance were: feed gross protein (samples collected every two weeks), feed consumption, initial and final pig's weight, carcass classification, lean meat yield. Data were analysed by GLM ANOVA procedure. In the first fattening cycle, the N balance elements (only HP diet) were 10.5, 2.9 and 7.6 kg for N intake, N gain and N excretion respectively, and the N conversion efficiency was 28%, in agreement with heavy pig average efficiency in literature. In the second fattening cycle, the results for N balance elements for HP and LP diet were 10.1 vs 9.2 (P<0.001), 2.9 vs 2.5 (P<0.001) and 7.2 vs 6.7 kg (P<0.001) for N intake, N gain and N excretion respectively. In the third fattening cycle, the results for HP and LP diet were 11.4 vs 10.4 (P<0.001), 2.8 vs 2.7 (P>0.05) and 8.6 vs 7.7 kg (P<0.001) for N intake, N gain and N excretion respectively. No differences were found for live and slaughtering performances (final LW 160-170 kg, ADG 630-670 g/d, and lean meat yield 50-52%, as average). The adoption of high N efficiency rations allowed a 10% reduction in N excretion. (LIFE+ AQUA project).

Effect of dietary crude protein on water intake and manure production by pigs

A.N.T.R. Monteiro[1,2], J.-Y. Dourmad[1], P.A.V. De Oliveira[3] and T.M. Bertol[3]
[1]INRA Agrocampus Ouest, UMR 1348 Pegase, Saint-Gilles, 35590, France, [2]UFRGS, Departamento de Zootecnia, Porto Alegre, 91540-000, Brazil, [3]CNPSA, Embrapa Suínos e Aves, Concórdia, 89700-000, Brazil; a_monteiro@zootecnista.com.br

The aim of this study was to evaluate the effects of a nutritional program with reduced crude protein (CP) content on water use and manure production. Forty gilts and 40 barrows (24.5±1.8 kg) were distributed in a randomized block design with two treatments, 10 replications per treatment and four animals per experimental unit. The feeding program was in four phases. Two diets were formulated for each feeding phase. The first one was adjusted by using InraPorc® model (LP) to adjust nutrient supplies to animal's requirements, and the second was formulated according to standard Brazilian nutritional levels (SP). Water was supplied *ad libitum* and consumption was measured by using a water meter for each pen. The volume of the manure was calculated for every pen by measuring its height in the collector rails with a graduated ruler. The average protein content was 145 and 168 g/kg for LP and SP treatments, respectively. Nutritional adjustment allowed a reduction of protein intake by 13, 20, 14 and 15% in grower I, grower II, finisher I, and finisher II phases, respectively. There was no effect of treatments on water use (6.17 and 5.84 l/d for LP and SP respectively, P=0.78), and volume of manure produced (4.05 and 3.70 l/d for LP and SP respectively, P=0.61). It is reported in literature that, for each percentage point of reduction in CP, water intake is reduced by 2% and slurry volume is reduced by 4%. However, this effect was not observed in this trial, probably because the difference in protein content was not enough to influence water intake. Another reason could be the high ambient temperature (average 22 °C) which is also known to affect water consumption. The results of this trial indicate that water use and volume of manure produced by growing-finishing pigs are not affected in the range of protein levels evaluated in this study, at hight temperature.

Synchronous least-cost ration formulation using nonlinear programming

A.A. Aboamer[1], S.A.H. Abo El-Nor[1], A.M. Kholif[1], H.M. Saleh[2], I.M. Khattab[3], M.M. Khorshed[4], M.S. Khattab[1] and H.M. El-Sayed[4]
[1]National Research Centre, Cairo, 12311, Egypt, [2]Nuclear Research Center, Cairo, 13759, Egypt, [3]Desert Research Center, Cairo, 11753, Egypt, [4]Faculty of Agriculture, Ain Shams University, Kaliobeya, 11241, Egypt; Aboamer@agr.asu.edu.eg

Least-cost procedures do not always produce the most practical rations for ruminants. The aim of this work was to reduce the cost of feeding by achieving the highest efficiency of utilization from the diet at the lowest possible cost through incorporating the concept of synchronizing ruminal energy and protein digestion into the traditional least-cost ration model. In a preliminary study, degradation kinetics of seven commonly used feedstuffs were estimated by the nylon bag technique. The observed patterns of N and OM degradability of feedstuffs were largely varied and greatly depend on the rumen outflow rate (fraction/h). The synchrony index (SI) was calculated based on the OM and N fractions of each ingredient and their degradation rates. A new user-friendly software application 'Lacto-sheep' was being developed based on the C# language and the .NET Framework 4.0 to facilitate the processes of ration formulation. Simplex and Hybrid local search (HLS) solvers in Microsoft Solver Foundation 3.1 have been used to solve both linear and nonlinear programming models, respectively. Example synchronous least-cost ration (SLCR) and least-cost ration (LCR) formulations of 50-kg lactating ewe based on the 1975 NRC recommendation is presented. SLCR and LCR showed two different patterns of N/OM release within a day. SI was about 0.87 and 0.60, respectively. As cost is often a limiting factor in the traditional least cost ration formulation therefore, LCR did not contain any source of protein-rich sources. Synchronous least-cost procedure was more appropriate to produce practical rations.

Phosphorous sources for ruminants

K.L.A. Yokobatake, M.A. Zanetti, J.A. Cunha, R.S.S. Santana and W.L. Souza
FZEA, Sao Paulo University, Animal Science, Rua Duque de Caxias Norte, 225, 13633-900, Brazil;
mzanetti@usp.br

Phosphorus is one of the most important minerals for ruminants, it has several functions in animal body, as well as it high cost. It is important not match up of additional requirements, not only by economic factors, but also to avoid environmental contamination. To study the different phosphorus (P) sources for cattle on P metabolism, nutrient digestibility and animal performance, two trials were conducted at the FFZEA, University of São Paulo, Pirassununga, Brazil. A metabolism and digestibility trial using 4 Nellore in the growth phase (250-300 kg BW), during 76 days, in a Latin square design, 4×4 (four treatments and four replicates), with a total collection of feces and urine in metabolic cages. The second on feedlot performance, with duration of 112 days, using 48 animals in the growth stage (300 kg), divided in four treatments: Control (0.1% of P in the diet) and three dicalcium phosphates (0.2% of P in the diet). Body weight was recorded each 28 days, when blood samples were also sampled for the P and alkaline phosphatase analysis. Data for performance were analyzed as a completely randomized design through PROC MIXEX/SAS. There was no effect of treatments on dry matter digestibility ($P>0.05$). The P apparent absorption was higher for the supplemented animals ($P<0.05$), being 33.56% for the control and 46.71% for the P supplemented animals. The blood P increased significantly for the animals supplemented with 84 days, and alkaline phosphatase increased only for animals receiving the P deficient diet ($P<0.05$). The dry matter intake and weight gain were higher for the supplemented animals ($P<0.05$), but there were no differences among the P sources. Feed conversion was better for the animals supplemented with P related to the diet deficient in phosphorus ($P<0.05$), but there was no difference between P sources ($P>0.05$), although there are differences of up to 9% between sources. It was concluded that there was no difference between the three inorganic phosphate sources used in the experiment.

***In situ* rumen degradation kinetics of four sorghum varieties in Nili Ravi buffalo bulls**

M. Ali, H. Nawaz and U. Ashfaq
Institute of Animal Sciences, University of Agriculture, Faisalabad, 38040, Pakistan;
drmubarak434@gmail.com

Different sorghum varieties are introduced in practice to ensure the provision of good quality fodder for ruminants. Limited information is available on *in situ* rumen degradation characteristics of dietary nutrients of sorghum varieties. This study was planned to evaluate the chemical composition and *in situ* rumen degradation kinetics of four sorghum varieties in Nili Ravi bulls. Four sorghum varieties (Viejin 4003, Viejin 4005, Viejin 4006 and Viejin 4571) were harvested at 80th day after sowing. Samples of fresh fodders were taken and used for the determination of chemical composition. These sorghum varieties were evaluated using *in situ* nylon bag technique. The fresh samples were filled in nylon bags and incubated in the rumen of four rumen cannulated buffalo bulls for 0, 2, 4, 6, 12, 24, 36, 48, 72 and 96 h. Rumen incubated residues were analyzed for dry matter (DM), neutral detergent fiber (NDF) and acid detergent fiber (ADF). Chemical analysis showed that the crude protein and ADF content were found higher ($P<0.05$) for Viejin 4003 compared to other varieties, whereas, the higher values of NDF and hemicellulose contents were found for Viejin 4005. Results showed that significant ($P<0.05$) higher effective rumen degradation (ED) and extent of degradation of DM were found for Viejin 4571. The extent of degradation and ED of NDF and ADF were found higher ($P<0.05$) for Viejin 4006 variety compared to other varieties. It is concluded that Viejin 4006 variety has the most favourable nutritive profile and rumen degradation characteristics.

The energy (oil) sparing effect of LYSOFORTE® on the performance of growing-finishing pigs

Y.J. Ji[1], C.H. Lee[2], J.S. Hong[1], H.B. Yoo[1], S.O. Nam[1] and Y.Y. Kim[1]
[1]Seoul National University, Department of Agricultural Biotechnology, College of Animal Life Sciences, 1 Gwanak-ro, Gwanak-gu, 151-742, Seoul, Korea, South, [2]Kemin®, Department of Technical Sales, D-224, Sigma 2, 164 Tancheonsang-ro, Bundang-gu, Seongnam-si, 463-741, Gyeonggi-do, Korea, South; nsomganbc@naver.com

This study was conducted to evaluate energy sparing effect of LYSOFORTE® on growth performance and productivity in growing-finishing pigs. A total of 140 crossbred ([Yorkshire × Landrace] × Duroc) pigs with averaging 7.29±1.62 kg of initial body weight were randomly allotted to one of four treatments based on sex and initial body weight according to randomized complete block (RCB) design in 5 replicates with 7 pigs per pen. The 2×2 factorial arrangement was used and the first and the second factors were dietary energy level (3,200 kcal ME/kg or 3,300 kcal ME/kg), and supplementation of LYSOFORTE® (supplementation levels: 0 or 0.05%). Feeding program was composed of three phases (Phase I, 0-2 week; Phase II, 3-5 week; Phase III, 6-10 week). In Phase I, any significant improvements of average daily gain (ADG) and average daily feed intake (ADFI) were not observed by dietary treatments. However, gain to feed (G/F) ratio was significantly higher in low energy treatment (P=0.04). In phase II (3-5 week), both dietary energy level and LYSOFORTE® supplementation had no effect on growth performance. In Phase III, increased ADG (P<0.01) and tendency of improving G/F ratio (P=0.09) were observed when LYSOFORTE® was added to diets. Supplementation of LYSOFORTE® improved ADG by 15% and 11% in 5-10 week and 0-10 week, respectively. Also, supplementation of LYSOFORTE® improved G/F ratio by 20% in 5-10 week and 13% in 0-10 week. The feed cost/weight gain was reduced when pigs were fed diets containing LYSOFORTE® during the whole experimental periods except for Phase II. Consequently, LYSOFORTE® supplementation can improve growth performance and productivity in weaning and growing pigs.

Biohydrogenation of linoleic acid by the rumen bacterium *Butyrivibrio fibrisolvens* as affected by pH

M.R.G. Maia[1], R.J.B. Bessa[2], A.R.J. Cabrita[1], A.J.M. Fonseca[1] and R.J. Wallace[3]
[1]REQUIMTE, LAQV, ICBAS, Universidade do Porto, Rua de Jorge Viterbo Ferreira 228, 4050-313 Porto, Portugal, [2]CIISA, FMV, Universidade de Lisboa, Av. da Universidade Técnica, 1300-477 Lisboa, Portugal, [3]Rowett Institute of Nutrition and Health, University of Aberdeen, Bucksburn, Aberdeen AB21 9SB, United Kingdom; mrmaia@icbas.up.pt

Although the ruminal biohydrogenation (BH) of unsaturated fatty acids has been shown to be a detoxification process, the diversity of BH intermediates and proportions found in the rumen might be a result of changes in the microbial population or of metabolic adaptations to environmental stress stimuli. Our aim was to evaluate if *Butyrivibrio fibrisolvens*, a biohydrogenating bacterium, might modulate the linoleic acid BH pattern in face of environmental stress (low pH). The *B. fibrisolvens* was inoculated into Hungate tubes with the liquid form of Hobson's M2 medium adjusted at pH 5.5 or 6.5, and supplemented with 50 mg/l linoleic acid. Tubes were incubated at 39 °C for 0, 1, 2, 3, 6, 9, 12, 16, 20, 24, 30 and 36 h in triplicate. At each time point, growth was determined by turbidity at 650 nm, pH measured and fatty acids analysed by GC-FID equipped with a 100 m capillary column. Growth of *B. fibrisolvens* was strongly affected by pH. The lag phase varied from 6 to 20 h at pH medium of 6.5 and 5.5, respectively. Additionally, the maximum optical density measured was three times higher at pH 6.5 (1.8) compared to pH 5.5 (0.6). pH had no effect on the linoleic acid BH pathway. Linoleic acid was isomerised to rumenic acid and saturated to vaccenic acid regardless of medium pH, yet the BH rate greatly differed. Growth at pH 5.5 only occurred after all dienoic acids were converted to monoenoic, while at pH 6.5 growth started when the conjugated linoleic acid was still present. In resume, although growth is affected, the pathway of BH by *B. fibrisolvens* is unchanged.

Tracing the potential of Portuguese seaweeds as ruminant feed

A.R.J. Cabrita[1], M.R.G. Maia[1], I. Sousa-Pinto[2], A.A. Almeida[3] and A.J.M. Fonseca[1]
[1]REQUIMTE, LAQV, ICBAS, Universidade do Porto (UP), Rua de Jorge Viterbo Ferreira n.° 228, 4050-313 Porto, Portugal, [2]Coastal Biodiversity, CIIMAR, FC/UP, Rua do Campo Alegre 1021/1055, 4169-007 Porto, Portugal, [3]REQUIMTE, LAQV, FF/UP, Rua de Jorge Viterbo Ferreira n.° 228, 4050-313 Porto, Portugal; arcabrita@icbas.up.pt

The nutritional value of seaweeds as feed supplements for domestic animals has been known for centuries. Based on their known structural and compositional characteristics, it is expected that ruminants are among the most suitable animals for algal use. This study aimed to evaluate the potential of 18 common Portuguese seaweed species belonging to green, brown, and red classes. Variability between species was evident in the proximate composition and energy content. Factor and cluster analysis were performed to distinguish seaweeds according to nutritional composition, and *in vitro* true dry matter digestibility (DMD). *Codium adhaerens*, *Codium tomentosum*, and *Halopteris filicina* were joined through their ash and fibre contents, thus mostly suitable for low producing ruminants. *Codium vermilara*, and *Plocamium cartilagineum* were grouped based on fat content, despite the overall content being low. Ulva sp., *Gigartina* sp., *Saccharina latissima*, and *Gracilaria vermiculopylla* were characterized by higher crude protein content, and DMD, suggesting the possibility of higher dietary inclusion for high productive animals. *Bifucaria bifucarta*, *Cystoseira usneoides*, *Sargassum muticum*, *Sargassum vulgare*, *Laminaria ochroleuca*, *Pelvetia caniculata*, *Fucus spiralis*, *Fucus guiryi*, and *Fucus serratus*, were joined in the same cluster. Despite its gross energy content, these seaweeds presented relatively high contents of low digestible fibre, thus lower DMD, preventing their use in great amounts in diets for high producing animals. A detailed mineral characterization showed a huge different mineral profile between seaweeds. Overall, seaweeds have potential as ruminant feeds, their level of fibre and the presence of toxic minerals dictating the level of inclusion in diets for high producing ruminants.

Impact of a new bacterial phytase on protein, energy and mineral utilization in broilers and piglets

A. Cerisuelo[1], M. Cambra-López[2], E.A. Gómez[1], L. Ródenas[2], P. Ferrer[1], B. Farinós[2], P. Añó[2], R. Marqués[3], I. Salaet[3], R. Aligué[4] and J.J. Pascual[2]
[1]Centro de Investigación de Tecnología Animal (IVIA), Pol. La Esperanza, 100, 12400 Segorbe, Spain, [2]Instituto de Ciencia y Tecnología Animal (UPV), Camino de Vera s/n, 46022 Valencia, Spain, [3]Global Feed S.L.U. (Grupo Tervalis), Avda Francisco Montenegro s/n, 21001 Huelva, Spain, [4]Departamento de Biología Celular (UB), Casanova, 143, 08036 Barcelona, Spain; jupascu@dca.upv.es

The effect of a new phytase [produced by *Pichia pastoris* from a cloned gene of *Serratia odorifera*, with a wider optimal pH range (3.5-5.8) and high resistance to pepsin and trypsin] on nutrient utilization in broiler (Exp.1) and piglet (Exp.2) feeds was evaluated. A digestibility balance was conducted housing 96 male chicks (28-day-old, Exp.1) and 72 male pigs (53-day-old, Exp.2) in metabolic cages. Six diets differing in the mineral content and enzyme doses were used: (1) C+ (recommended levels of Ca and P; without phytase); (2) C- (low levels of P; without phytase); (3) 250 (C- with 250 IU phytase/kg); (4) 500 (C- with 500 IU phytase/kg); and (5) 1000 (C- with 1000 IU phytase/kg). Tibias were analysed for mineral retention. In Exp. 1, treatments 250 and 500 showed the highest (P<0.05) protein (+5 units) and energy (+1.8 units) digestion rates compared to C-. Dietary apparent metabolic energy (kcal/kg) content increased linearly (P<0.08) with phytase inclusion respect to C-. In Exp. 2, protein digestibility was not improved by phytase but dietary digestible energy (kcal/kg) content increased linearly (P<0.05) with the addition of phytase. As expected, Ca and P utilization rates were higher (P<0.05) in phytase treatments, which also showed lower (P<0.05) mineral excretion rates compared with C- in both experiments. In Exp.2, Ca and P in tibia was on average 1.3 and 0.5 g higher in diets containing phytase compared with C-. The supplementation with 250 and 500 IU/kg of new phytase improved energetic value of feeds and mineral utilization in broilers and pigs, but protein utilization was only improved in broilers.

Effects of forage type and protein supplementation on chewing and faecal particle size in sheep
P. Nørgaard[1], S. Hedlund[2], S. Muche[3], A. Arnesson[2] and E. Nadeau[2]
[1]University of Copenhagen, Dept. of Veterinary Clinical and Animal Science, 1870 Frederiksberg, Denmark, [2]Swedish University of Agricultural Sciences, Dept. of Animal Environment and Health, 532 23 Skara, Sweden, [3]LKS mbH, August-Bebel-Straße 6, 09577 Niederwiesa OT Lichtenwalde, Germany; pen@sund.ku.dk

The objectives were to assess effects of forage type and protein supplementation on chewing activity and faecal particle size in sheep. Eight ram lambs (87 ± 5.3 kg) with a body condition score of 3.4 ± 0.2 were assigned to a duplicated 4×4 Latin square design with four different forages with and without supplementation of 150 g rapeseed meal daily during four 4-week periods. The forages were early (EG) and late (LG) harvested grass silage and early (EB) and late (LB) harvested whole-crop barley silage. The sheep were kept in individual pens during the first three weeks of each period; two weeks of adaptation and one week of *ad libitum* intake registrations. The sheep were fed at 80% of *ad libitum* intake in metabolic cages during the last week of each period. The EG, LG, EB and LB had NDF contents (% of DM) of 45, 55, 42 and 46, respectively, and a particle size of 30 mm. Jaw movement oscillations were recorded for 96 h from day 25. Faeces were collected during the same four days, washed, freeze dried and particle DM (PDM) was sieved (1, 0.5, 0.212, 0.106 mm sieves and bottom bowl). Data were statistically analyzed using a mixed model with fixed effects of forage type, protein supplementation and period, and a random effect of sheep nested within protein supplementation. The 80% of *ad libitum* intake of forage DM ($P<0.001$), duration of eating bouts ($P<0.01$), number of ruminating periods ($P<0.03$), basic chewing rate during eating ($P<0.02$), faeces DM content ($P<0.001$) and the proportion of faeces PDM below 0.1 ($P<0.001$), 0.2 ($P<0.001$), 0.5 ($P<0.001$) and 1 mm ($P<0.001$) were affected by forage type. Protein supplementation did not significantly affect intake, eating activity, rumination activity and faeces characteristics.

Different sources of glycerol on fermentation parameters and *in vitro* digestibility of corn silage
J.M.B. Ezequiel, E.S. Castro Filho, M.T.C. Almeida, J.R. Paschoaloto, I. Monsignati and E.H.C.B. Van Cleef
São Paulo State University, Jaboticabal, 14884-900, Brazil; janembe_fcav@yahoo.com.br

This study evaluated the effect of glycerol sources on *in vitro* digestibility of dry matter and neutral detergent fiber, and *in vitro* fermentation parameters of corn silage. The treatments were: (1) corn silage; (2) corn silage + 10% crude glycerin (86% glycerol); (3) corn silage + 10% dry glycerin (68% glycerol); and (4) corn silage + 10% pure glycerin (99% glycerol). Rumen content was collected from 2 ruminally canulated male sheep (68.5 kg BW) fed diet containing 40% corn silage and 60% corn-based concentrate. *In vitro* digestibility of DM and NDF were evaluated using the Ankom Daisy[II] fermenter. Filter bags (6 bags/treatment; 0.5 g/bag) were incubated (72 h, 39 °C) into fermentation jars containing 400 ml rumen fluid, 1,330 ml buffer A and 266 ml buffer B. When the first step of incubation was terminated (48 h), 8 g of pepsin and 40 ml of 6N HCl were added to each jar. The gas production, pH, ammonia nitrogen (NH_3-N) and DM disappearance were evaluated after 24 h of incubation at 39 °C. The substrates (0.2 g) and buffered rumen fluid were placed into 20 bottles (60 ml), purged with helium gas and sealed. All trial procedures were replicated. The addition of glycerol increased IVDMD ($P<0.01$), however there was no difference among glycerol treatments (66.6%). No effect was observed on IVNDFD ($P>0.05$; 39.4%). The 24-h pH (6.1), NH_3-N (40.1 mg/dl), total production of gas (51.2 ml), production of CH_4 (8.8 ml) and CO_2 (39.7 ml) were not different among treatments. The gas production as ml per gram of degraded DM as well as production of CO_2 tended to be greater for corn silage compared to glycerol treatments ($P=0.09$). In conclusion, the incubation of 10% of different sources of glycerol (crude, dry or pure glycerin) with corn silage improve its IVDMD without adversely affecting the IVNDFD. Moreover, glycerol, regardless the source, decreases the total gas and CO_2 production, when the disappearance of DM is taken into account.

Effect of diets containing blue lupin seeds on rabbit doe milk yield and the growth of their litters

L. Uhlířová[1,2], Z. Volek[2], M. Marounek[2], E. Tůmová[1] and V. Skřivanová[2]
[1]Czech University of Life Sciences Prague, Kamýcká 129, CZ-165 21 Prague, Czech Republic, [2]Institute of Animal Science, Přátelství 815, CZ-104 00 Prague, Czech Republic; volek.zdenek@vuzv.cz

White lupin (WLS) is a suitable crude protein (CP) source for growing-fattening rabbits, as well as for lactating rabbit does. To our knowledge no reports have studied the effect of blue lupin on rabbit doe milk production and composition as well as the growth of their litters. Two isonitrogenous and isoenergetic lactation diets were formulated. The first (control) diet contained WLS (25%, cv Amiga) as the main CP source, whereas the second diet was based on blue lupin seeds (BLS, 28.5%, cv Probor). No additional fat was added to any of the diets. A total of 32 Hyplus rabbit does (16 animals per treatment; at the 3rd parturition) were fed 1 of the 2 lactation diets during the entire lactation (35 days). Does were housed in modified cages which allowed controlled suckling (once a day at 7 am) and separate access of does and their litters to feed. The litters were standardized to 9 kits immediately after the birth. Litters were offered weaning diets with the same CP source as in the lactation diet of their mothers from d 17 of lactation. Feed intake, feed efficiency and milk production of does were not affected by dietary treatments, as well as milk efficiency and the growth of their litters. Live weight at weaning (P=0.005), milk dry matter (P=0.028) and fat contents (P=0.016), as well as fat output per kg of metabolic weight (P<0.001) were higher in does fed the WLS diet. It can be concluded that blue lupine seeds could be another suitable dietary CP source for lactating rabbit does. Dietary fat, however, should be added in order to eliminate higher mobilisation of body reserves of does in the later phase of lactation. Supported by project no. MZERO0714.

Effect of lipids source and supplementation frequency on intake of grazing heifers

M.C.A. Santana[1,2], V.C. Modesto[3], R.A. Reis[4], G. Fiorentine[4], J.F.H. Rodrigues[5], J. Cavali[5] and T.T. Berchielli[4]
[1]Faculdade Evangélica de Goianésia, Agronomy, Av. Brasil, n° 1000, Covoa, Goianésia, Goiás, CEP: 76.380-000, Brazil, [2]EMATER, Rua 227A n.331, Setor Leste Universitário Goiânia, Goiás, CEP: 74.610-060, Brazil, [3]UNESP, Animal Science, Avenida Brasil, 56. Ilha Solteira, São Paulo, CEP: 15.385-000, Brazil, [4]UNESP, ANIMAL SCIENCE, Rodovia de Acesso Prof. Paulo Donato Castellane s/n. Jaboticabal, São Paulo, CEP:14884-900, Brazil, [5]UNIR, Animal Science, Presidente Medici, Rondônia, CEP: 76.916-000, Brazil; mcaspaz@yahoo.com.br

The objective of this study was to evaluate effects of soybean oil from different sources [i.e. ground soybean (GS), soybean oil (SO) and calcium salts of soybean oil (CS)] on intake variables of heifers grazing marandu palisadegrass (Brachiaria brizantha cv. Marandu) supplemented in two frequencies [i.e. daily and three days a week (Monday, Wednesday and Friday – MWF)]. The experiment was carried out during five months. Forty two crossbred heifers (1/4 Nellore × 1/4 Santa Gertrudis × 1/2 Braunvieh), 17 month-old, [(297.1 kg±30.5 body weight (BW)], were used. Supplements were offered at 0.75% BW/day. The forage intake was measured by concentrations of n-alkanes in the supplement, forage and feces. The allocation of treatments was done in a completely randomized design and distributed in a factorial arrangement (3×2 – soybean oil from three different sources and two supplementation frequencies). Total dry matter intake (DMI) was greater in August (2.2 kg DM/kg BW) and October than in November (1.5 kg DM/kg BW). There was effect of sources on total DM, herbage and nutrients intake, but no effects were resulted from supplementation frequency (daily 5.6 kg/d and MWF 5,9 kg/d). The animals supplemented with SO (6.3 kg/d) had greater DMI (P<0.05) than the ones fed GS (5.7 kg/d) and CS 5.5 kg/d). Overall, this data indicated that the intake can be influenced by the soybean oil sources and the frequency of supplementation can be reduced to increase the (work) efficiency on grazing system.

Effect of dietary energy level on the zootechnicals performances of rabbits bred in hot climate
Y. Dahmani, H. Ainbaziz, N. Benali, D. Saidj, M. Chirane, G. Ahmil, N. Tazkrit and S. Temim
École nationale supérieur veterinaire, ENSV BP 161 Hacen badi EL-Harrach Alger, 16000, Algeria; Dahmani_vet@yahoo.fr

The aim of this experiment was to determine the effect of dietary energy level on growth performance of rabbit of the local population exposed to heat stress. For this purpose, 48 rabbits of 42 days of age were weighed and allotted to two groups of 24 rabbits each with their mean weight at 855.2±9.1 g. The two lots of rabbits were each fed a 16% iso-proteic tow diet with each containing a different digestible energy level namely: 2,551 kcal/kg for the control group (C), and 2,700 kcal/kg for the group (A). Rabbits of both lots were subjected to high daytime temperatures and humidity of 33 °C and 69% respectively. The animals performances were measured and calculated every week between 42 and 91 days, the carcass yield was determined on 10 rabbits of each group. The results revealed that the zootechnicals performances were not influenced by the dietary energy level in the beginning of the test. However, the results show that the live weight, the daily weight gain and the average daily feed intake between the period (70-91 days) of the young rabbits of the group C is significantly higher than those of the group A (+13.1, +16.4 and +19.1%; $P<0.05\%$) respectively. Moreover, no significant effect was found on the feed conversion throughout the experiment except for the period 56-63 days, which was significantly higher in the animals of group C (+19,35%; $P<0.05\%$). Furthermore, a significant effect was found on the anterior part of the carcass, peritoneal fat and liver; they were lower in group C by (-5, -30 and -41%, respectively), while carcass yield, inter-scapular fat, and intermediate parts were significantly higher in group C (+5, +19 and +33%; $P<0.05$). No significant effect was found on the yield of the other components of the carcass. In conclusion, increasing the dietary energy level does not improve growth performances and carcass yield of rabbits of the local population under chronic heat stress.

Evaluation of reduced nutrient levels in growing-finishing pig diets
A.N.T.R. Monteiro[1,2], J.-Y. Dourmad[1], T.M. Bertol[3], P.A.V. De Oliveira[3] and A.M. Kessler[2]
[1]INRA Agrocampus Ouest, UMR 1348 Pegase, Saint-Gilles, 35590, France, [2]UFRGS, Departamento de Zootecnia, Porto Alegre, 91540-000, Brazil, [3]CNPSA, Embrapa Suínos e Aves, Concórdia, 89700-000, Brazil; a_monteiro@zootecnista.com.br

The aim of this study was to evaluate the effects of a nutritional program with reduced nutrient content, on nutrient ingestion and nitrogen (N) and phosphorus (P) balance in growing-finishing pigs from 25 to 130 kg body weight. Forty gilts and 40 barrows (24.5±1.8 kg) were distributed in a randomized block design with two treatments, 10 replications per treatment, and four animals per experimental unit. The feeding program was in four phases. Two diets were formulated for each feeding phase. One was adjusted using InraPorc® model to minimize the nutrient excess (LN), and the other (SN) was formulated with standard Brazilian recommendations. Nutrient consumption was measured and N and P balance were determined according to intake and retention. Data were subjected to ANOVA with the main effects of treatment, block, sex and interaction between treatment and sex. Over the whole experimental period, there was no statistical difference ($P>0.10$) between treatments for energy intake (mean of 33.3 MJ ME/d) and average daily gain (0.92 kg/d). Dietary treatments affected ($P<0.05$) daily lysine (18.9 vs 21.3 g/d), crude protein (352 vs 419 g/d) and P (11.3 vs 12.6 g/d) intake, with lower values for LN pigs. Nitrogen (28.8 vs 38.0 g/d) and P (5.7 vs 6.5 g/d) excretion were lower ($P<0.05$) for LN pigs, without difference ($P>0.05$) on body retention (n=23.5 and P=4.8 g/d). This resulted in higher nutrient utilization efficiency in animals that were fed with the adjusted diet. The sex affected ($P<0.05$) all variables, with the gilts showing lower intake and excretion values than barrows. According to these results it seems possible to reduce the levels of nutrients that are commonly used in Brazil, without affecting the N and P retention. Moreover, nutritional adjustment proves to be an efficient tool to reduce nutrient excretion and, consequently, environmental impacts.

In vitro methanogenic potential of three cultivars of *Trifolium pratense*

M.D.E.J. Marichal[1], L. Piaggio[2], R. Crespi[1], G. Arias[1], S. Furtado[1], M.J. Cuitiño[3] and M. Rebuffo[3]
[1]Facultad de Agronomía, Garzón 780, 12900 Montevideo, Uruguay, [2]Secretariado Uruguayo de la Lana, Rambla B. Brum 3764, 11800 Montevideo, Uruguay, [3]INIA La Estanzuela, Ruta 50 Km 11, Colonia, Uruguay; mariadejesus.marichal@gmail.com

The aim of this work was to evaluate and compare the *in vitro* methanogenic potential of the cultivars of *Trifolium pratense* INIA Mizar, Estanzuela116 (E116) y Antares (LE113) in late vegetative stage (mean stage by weight: 30.4±1.4, height: 31.3±1.4 cm). Four replicates of each cultivar were seeded in a complete randomized design in the Experimental Research Center INIA La Estanzuela, Uruguay (34° 20´ S, 57° 41' W). An *in vitro* gas production procedure was followed using the rumen contents collected two hours after morning feed of two fistulated whethers fed alfalfa hay. Three batches of 15 bottles (3 bottles by cultivar, 3 with alfalfa as standard, and 3 blanks) were incubated for 24 h. Gas produced from 0 to 8 and 8 to 24 h was collected and CH_4 was measured by gas chromatography. Results of gas and CH_4 were expressed in ml/g organic matter incubated (ml/g iOM). The three cultivars produced similar (P>0.193) volumes of gas at 8, 8 to 24 and 0 to 24 hs of incubation (169.6, 50.4 and 220 ml gas/iOM, respectively). Differences (P<0.041) in CH_4 production were observed. From 0 to 24 h, LE113 produced greater (P=0.0295) and E116 tended to produce greater (P=0.087) CH_4 than INIA Mizar (28.0, 26.2 and 24.4 ml CH_4/g iOM, respectively). At 8 h of incubation, CH_4 was greater (P=0.0409) in E116 and tended to be higher (P=0.0901) in LE113 than in INIA Mizar (19.6, 17.0 and 18.5 ml CH_4/g iOm, respectively). The three cultivars produced similar (P=0.501) CH_4 between 8 and 24 h of incubation (8.9 ml of CH_4/g iOM). The lower production of CH_4 in INIA Mizar could reflect the introgression of american cultivar with E116 in their genetics what produced molecular and physiological modifications resulting in differences in the nature and characteristics of substrates available for microbial fermentation.

Influence of diet type and dietary magnesium on mineral status in lactating dairy cows

P. Schlegel[1], M. Rérat[2], M. Girard[1] and A. Gutzwiller[1]
[1]Agroscope, Tioleyre 4, 1725 Posieux, Switzerland, [2]Federal Food Safety and Veterinary Office, Schwarzenburgstr. 155, 3003 Bern, Switzerland; marion.girard@agroscope.admin.ch

Magnesium (Mg) absorption is influenced by dietary potassium (K) and may be influenced by diet type. Herbage based diets can be rich in K due to organic fertilisation of grassland and Mg availability from spring pasture may be limited due to a rapid rumen passage rate. The aim of this experiment was to assess the Mg status of lactating dairy cows fed two high K diets (30 g/kg DM) based on 24 h spring pasture or on hay and, as control, a low K diet (20 g/kg DM) based on maize-grass silage. All basal diets contained three levels of Mg (2.0, 2.7 or 3.4 g/kg DM). Eighteen multiparous lactating dairy cows (25 d in milk) were fed one of the basal diets over 8 weeks. After 14 days adaptation, cows were assigned to each Mg level for 14 days each in a cross-over design. Samples were collected on d 11 and 14 of each period. Serum Mg was reduced (P<0.001) by dietary Mg of 2.0 g/kg DM and urinary Mg decreased (P<0.001) with decreasing dietary Mg. Serum and urinary Mg were reduced (P<0.001) when fed the hay based diet and the high K diets, respectively. Serum and urinary Mg correlated linearly with dietary Mg in high K diets (R^2=0.69 and 0.62, respectively). Urinary Mg correlated linearly with dietary Mg in the low K diet (R^2=0.69). Serum and urinary Mg correlated quadratically (R^2=0.55). Serum and urinary K were reduced (P<0.001) when fed the low K diet. Serum phosphorus (P) was reduced (P<0.001) when fed the pasture diet. Urinary P content was highest when low K diet was fed and was lowest when fed on pasture (P<0.001). No diet × Mg interaction was observed. The data indicate that 3.0 g Mg/kg DM was required in the high K diets to reach serum Mg plateau obtained from the low K diet. Independent from dietary K, a 24 h spring pasture was not more problematic for an adequate Mg status than was a hay based diet. Finally, the P status was sub-optimal on pasture, although dietary P (4.4 g/kg DM) was sufficient.

Valine deficiency in piglet diets can be partly compensated by β-alanine

A. Cools[1,2], C. De Moor[1], A. Lauwaerts[1], W. Merckx[3] and G.P.J. Janssens[2]
[1]Eastman Chemical Company, Pantserschipstraat 207, 9000 Gent, Belgium, [2]Ghent University, Animal Nutrition, Hiedestraat 19, 9820 Merelbeke, Belgium, [3]Catholic University Leuven, ZTC, Bijzondere Weg 12, 3360 Lovenjoel, Belgium; angelolauwaerts@eastman.com

Research has shown that the non-essential amino acid β-alanine (β-ala) is increasing voluntary feed intake in weaned piglets. Biochemically, pathways of valine (Val) and β-ala are related but this relation is never tested *in vivo*. Therefore, β-ala supplementation in Val deficient piglet diets was tested. At weaning (21 days) 480 piglets were divided over 48 pens of 10 piglets of equal sex and weaning weight. Pens were assigned to very low Val (VL: Val/Lys 0.46 in weanling and 0.47 in starter diet), low Val (L: Val/Lys 0.54 and 0.55), or control (C: Val/Lys 0.62 and 0.63) with equal distribution of sexes and weaning weight. In each diet group, half of the pens was supplemented with 0.5 g β-ala/kg of weaning (day 0 to 14) or starter (day 15 to 42) diet. Growth and feed intake was recorded to calculate average daily gain (ADG), average daily feed intake (ADFI) and feed conversion ratio (FCR). Diet effects on body weight changes were tested with a linear mixed model (piglet within pen and pen as random factors) and effect on ADG, ADFI, and FCR were tested with a linear model including weaning weight, diet (VL, L, C) and β-ala dose. Overall body weight gain was 270 g/day (P<0.001) but was 56 g/day lower in VL compared with C diet (P<0.001). Supplementing VL diet with 0.5 g β-ala/kg increased body weight gain with 31 g/day (P<0.001). Both ADG and ADFI were reduced (P<0.001) for VL compared to C diet (ADG -56 g/day; ADFI -73 g/day). Addition of β-ala to VL diet tended to increase ADG with 34 g/day (P=0.072) and ADFI with 48 g/day (P=0.079). The FCR was only affected by VL (+0.077, P=0.024) and not by β-ala addition. In conclusion, the results indicate that β-ala can partly prevent production losses in Val deficient diets. Further research on this issue is needed to clarify this mechanism and optimize the β-ala dose needed.

Production and caecal microbial counts of broilers fed with wheat-barley based diets and dry whey

C. Pineda-Quiroga, R. Atxaerandio, A. Hurtado, R. Ruíz and A. García-Rodríguez
Neiker-Tecnalia, Campus Agroalimentario de Arkaute, 01080 Vitoria-Gasteiz, Spain; cpineda@neiker.net

Dry whey is a co-product of the dairy cheese industry and is characterized by a high lactose content (approx 70% of dry matter). In poultry, lactose is not digested in the upper intestine but is fermented in the hindgut and can stimulate the growth of the intestinal acid lactic bacteria, preventing the colonisation by potential pathogenic bacteria. The aim of this study was to evaluate the effect of inclusion of dry whey in wheat-barley based diets on productive performance and caecal microbial populations. Six hundred day-old broiler chicks Ross 308 were randomly allocated in pens of 3.75 m^2. Two treatments with 10 replicates of 30 chickens were evaluated: control diet (0% dry powder whey, 0W) and 6% dry powder whey (6W). The experimental diets were supplemented with β-glucanase and β-xylanase were offered in pelleted form. Animals were fed with starter diet from day 0 to 21 and then were switched to a grower-finisher diet until day 42. Feed and water were provided *ad libitum*. Body weight (BW), daily weight gain (DWG), feed intake (FI) and feed conversion ratio (FCR) were calculated on a weekly basis. Mortality rate (M) was calculated at day 42. At day 21, 3 chickens per treatment were randomly selected and slaughtered to remove cecum for *Bifidobacterium* spp., *Lactobacillus* spp., *Clostridium perfringens* and *Escherichia coli* counts. At day 42 animals fed with 0W had higher BW (2,004 vs 1,726 g, P<0.001), DWG (59 vs 49.1 g, P=0.050) and FI (129 vs 116 g, P=0.012). However, differences were not found between treatments in FCR (2.2 vs 2.4, P=0.423) or M (4.1 vs 7.5%, P=0.257). Only an increase in *Lactobacillus* spp. (11.9 vs 8.3 log10 cfu/g caecal content, P=0.007) and *E. coli* counts (12.5 vs 8.9 log10 cfu/g caecal content, P<0.001) was observed in 6W fed animals. In conclusion, although supplying 6% of dry whey in wheat-barley based diets for broilers increased *Lactobacillus* spp., a decrease in final BW was observed, without affecting FCR or M.

Production, duodenal histomorphometry and caecal microbial counts of broilers fed with dry whey
C. Pineda-Quiroga, R. Atxaerandio, A. Hurtado, R. Ruiz and A. García-Rodríguez
Neiker-Tecnalia, Campus Agroalimeario de Arkaute, 01080 Vitoria-Gasteiz, Spain; cpineda@neiker.net

Dry whey is a co-product of the dairy cheese industry characterized by a high lactose content (approx 70% of dry matter). In poultry, lactose is not digested in the upper intestine but is fermented in hindgut causing a modulation in the intestinal flora. The aim of this study was evaluate the effect of inclusion in diet of dry whey on productive performance, duodenal histomorphometry and cecal microbial counts. Six hundred day-old broiler chicks Ross 308 were randomly allocated in pens of 3.75 m^2. Two treatments with 10 replicates of 30 chickens were evaluated: control diet (0% dry whey, 0W) and 6% dry whey (6W). The experimental diets were corn-soybean based and offered in pellet form. Animals were fed with starter diet from day 0 to 21 and then were switched to a grower-finisher diet until day 42. Feed and water were provided *ad libitum*. Body weight (BW), daily weight gain (DWG), feed intake (FI) and feed conversion ratio (FCR) were calculated on a weekly basis. Mortality rate (M) was calculated at day 42. At day 21, 3 chickens per treatment were randomly selected and slaughtered to remove cecum for microbial counts and duodenum to measure villus height (VH), crypt depth (CD) and apparent villus surface area (A). At day 42, non-significant differences were found between treatments in BW (2,616 vs 2,559 g, P=0.148), DWG (62 vs 61 g, P=0.780) neither M (3.3 vs 2.2%, P=0.408). Chickens fed with 6W had lower FI (92 vs103 g, P<0.001) and FCR (1.51 vs 1.66, P<0.001). Counts of *Bifidobacterium* spp. increased in 6W compared to 0W (9.91 vs 6.75 log_{10} cfu/g caecal content, P=0.002) while *Lactobacillus* spp., *Clostridium perfringens* and *Escherichia coli* counts were not different. Finally, VH increased in animals fed with 6W (1,349 vs 1,152 μm, P=0.002), while D and A were not affected. In conclusion, supplying 6% of dry whey in corn-soybean diets for broilers offered in pelleted form promote *Bifidobacterium* spp. growth, increase VH and improve FCR without affecting BW, DWG and M.

Biofortificated milk: selenium and vitamin E in cow's diet to improve nutritional components in milk
M.S.V. Salles[1], A. Saran Netto[2], L.C. Roma Junior[1], M.A. Zanetti[2], F.A. Salles[1] and K. Pfrimer[3]
[1]APTA, Ribeirão Preto, 14030-670, Brazil, [2]FZEA/USP, Pirassununga, 13635-900, Brazil, [3]FMRP/USP, Ribeirão Preto, 14049-900, Brazil; marciasalles@apta.sp.gov.br

Healthy nutrition is a goal for most of the world's population, emphasizing the importance of animal science studies to improve milk nutrient composition. The aim of this was to evaluate the addition of vitamin E and selenium with sunflower oil (SFO) to the diet of lactating cows to improve the nutrient profile of milk. Twenty-eight cows were allocated to four treatments: C (control diet); O (4% of SFO in dry matter (DM) diet); A (3,5 mg/kg DM of organic selenium + 2,000 IU of vitamin E/cow/day); OA (4% of SFO in DM diet + 3,5 mg/kg DM of organic selenium + 2,000 IU of vitamin E/cow/day). Cows were fed with 0.50 of concentrate, 0.42 of corn silage and 0.08 of coast-cross hay (on DM-basis). Blood and milk were taken in the last week of the trial and analyzed for selenium and alpha-tocopherol. Data were analyzed as a Completely Randomized Block Design with a factorial treatment structure (GLM/SAS). The addition of selenium and vitamin E in the cow's diet increased selenium and alpha-tocopherol serum (0.083, 0.085, 0.337 and 0.389 μg/l, P<0.0001, 0.022 SEM of selenium and 4.51, 6.26; 6.92 and 8.97 mg/dl, P=0.0009, 0.66 SEM of alpha-tocopherol for C, O, A and OA, respectively) and milk (0.011, 0.027, 0.235 and 0.358 μg/l, P<0.0001, 0.033 SEM of selenium and 2.27, 1.56; 3.08 and 2.89 mg/l, P=0.0088, 0.36 SEM of alpha-tocopherol for C, O, A and OA, respectively). SFO supplementation increased alpha-tocopherol in cow's blood (P=0.0097). The antioxidants added to the diet of lactating cows improved the nutrient profile of milk. Financial support: FAPESP.

Effects of dietary crude protein and energy levels on carcass trait and blood composition on rabbit

N. Benali, H. Ainbaziz, Y. Dahmani, D. Saidj, S. Bouzidi, N. Kedachi, R. Bellabes, M. Cherrane and S. Temim
High National School Vetrerinary, El Harrach, BP 161, Hacene Badi, El Harrach, 16000 Algiers, Algeria;
na.benali@yahoo.fr

The aim of this experiment was to study the effect of different dietary crude protein and energy levels on the carcass traits and blood composition of local population rabbits. Ninety six (96) male rabbits of 35 days-old were used and allotted to six groups of 16 rabbits each. They were fed with six diets iso crude fiber (13%), but containing different levels of crude protein (CP) and digestible energy (DE): A (16% CP and 2,500 kcal/kg DE), B (18% CP and 2,500 kcal/kg DE), C (20% and 2,500 kcal/kg DE), D (16% CP and 2,700 kcal/kg DE), E (18% and 2,700 kcal/kg DE) and F (20% and 2,700 kcal/kg DE). The rabbits were slaughtered at the end of the experiment at 92 days. Carcass traits and blood composition were determined on 24 rabbits. The results revealed that the high CP (20%) and low DE (2,500 kcal/kg) treatment had a high significant effect on carcass dressing percentage of rabbits compared to group D (+6.45%, $P<0.05$). Moreover, the weight of skin of group B and C respectively was higher than for D (+17.36 and 16.41%; $P<0.05$). Rabbits fed diet F had the highest weight of liver compared to group A and B, respectively (+21.20 and +29.42%; $P<0.05$) and highest weight of kidney fat compared to B and C, respectively (+42.28 and 36%; $P<0, 05$). Interaction between dietary protein (16, 18 and 20%) and low energy (2,500 kcal/kg) showed the best weight of the fore part, intermediate and hind part ($P<0.05$). The plasmatic concentration of glucose, total protein, triglycerides and creatinine were not significantly affected by treatments, except for the serum cholesterol level which was significantly lower in group B compared to group D (1.43 vs 2.25 mmoles/l; $P<0.05$) and urea which was higher on A than group D and E (7.50 vs 5.51 mmoles/l and 6.09 mmoles/l; $P<0.05$). In conclusion, based on the carcass dressing percentage, a diet containing 20% CP and 2,500 kcal/kg DE resulted in optimum meat yield.

Feed conversion and meat quality: bulls or steers?

A. Simeone, V. Beretta, C.J. Caorsi and J. Franco
Facultad de Agronomía-Universidad de la República, Paysandu, 60000, Uruguay; asimeone@adinet.com.uy

Productivity of intensive backgrounding and finishing beef cattle systems, based on pastures and grain feeding, could benefit from raising not castrated calves, without affecting meat quality if animals are slaughtered at young ages. This experiment evaluated lifetime performance, carcass and meat traits of Hereford calves castrated (CAS) or not (NCAS) at birth. Twenty-four calves born in spring 2012 were randomly allotted at birth to each treatment (T). Calves were early weaned (71±10 days old) and managed altogether until slaughter. After weaning they were lot-fed (FL1, summer 2013) and then grazed on improved pastures during autumn, winter and spring (6 kg DM/100 kg liveweight LW). During finishing (summer-autumn 2014) cattle remained grazing but with *ad libitum* access to a self-fed fattening ration (FL2, EM 2.8 Mcal/kg, CP 12%). All animals were slaughtered on Apr 22nd 2014 (526±10 days old). LW was recorded every 28 day (grazing) or 14 days (FL), and LW gain (LWG) estimated by regression. Feed to gain ratio (FG) was estimated for FL phases, and carcass and meat traits were recorded at slaughter. The statistical model included T effect and initial LW as covariate. Pre-weaning LWG did not differ between T (0.780±0.11 kg/d $P>0.05$). During FL1, NCAS showed higher LWG ($P<0.01$) but did not differ in FG (4.9 vs 5.1; $P>0.05$). By the end of the grazing season NCAS were heavier compared to CAS (398 vs 370 kg, $P<0.01$), showed higher LWG in FL2 (1.71 vs 1.31 kg/d; $P<0.01$) and better FG ratio (7.9 vs 9.0; $P<0.01$). NCAS cattle had higher carcass weight and yield (262 vs 235 kg; 56 vs 54%; $P<0.01$), but lower subcutaneous back fat depth (5.1 vs 9.2 mm, $P<0.01$). Meat quality parameters (L. dorsi 48 h pH and color, lightness L* and brightness *a) were better for CAS ($P<0.01$): lower pH (5.9 vs 6.7 $P<0.01$), higher *L (35.9 vs 29.6) and *a values (21.5 vs 14.0). Although NCAS are more efficient under intensive feeding systems, reduced meat quality should be taken into account at decision-making.

Prediction of fatty acid profile on the basis of their content in the feed or blood plasma

A.M. Brzozowska[1], M. Lukaszewicz[1], P. Micek[2] and J.M. Oprzadek[1]
[1]Institute of Genetics and Animal Breeding PAS, Animal Improvement, Jastrzebiec, 05-552, Poland,
[2]University of Agriculture, Animal Sciences, Krakow, 31-120, Poland; j.oprzadek@ighz.pl

The aim of the study was to design a method of prediction of the fatty acid (FA) profile in milk, blood, liver, and adipose tissue of cows. The experiment on 24 Polish Holstein-Friesian cows fed corn or grass silage with protein-energy feed additive was conducted. During the experiment, the samples of feed, milk, liver, blood and adipose tissue were collected, in which the fatty acids were determined. The matrices of regression coefficients were used to predict the fatty acid profiles. Determination coefficients between real milk values and the ones predicted based on FA profile in the ration were satisfactory in case of palmitic, stearic, α-linolenic, conjugated linoleic, eicosanoic and cis-9 eicosanoic acids. In turn, determination coefficients of milk FA predicted based on FA profile in the plasma were decent for α-linolenic acid, C22:1 and eicosapentaenoic acid. The analyses showed that it cannot be determined which source of information is better to predict FA profile in milk. High determination coefficients in the models predicting profile in the liver suggest their accuracy. Determination coefficients in the model for adipose tissue FA prediction were unsatisfactory. The obtained results show that the accuracy of prediction of blood, milk, liver and adipose tissue FA profiles of cows varies for different FA.

Response of early-weaned beef calves to graded levels of sorghum wet distiller grains in the diet

A. Simeone, V. Beretta, M. Acuña, M. Loustau and A. Suárez
Facultad de Agronomía-Universidad de la República, Paysandu, 60000, Uruguay; asimeone@adinet.com.uy

An experiment was conducted to evaluate the effect of level of inclusion of wet distiller grains (WDG) from sorghum in the diet of lot-fed early-weaned beef calves. Twenty Hereford female calves (90.2±17.6 kg) were randomly allocated to 4 total mixed rations (20% alfalfa hay/80% concentrate) differing in the level of WDG in the concentrate: 0, 12, 24 or 36% (on DM basis), in substitution of soybean meal and corn grain in the basal diet (WDG 0%, SBM 23.9%, corn 71.8%, urea 0.58%, molasses 2,0%, mineral and vitamin premix 1.78%). Concentrates were isoenergetic (ME: 3.0 Mcal/kg DM) and isonitrogeneous (CP: 198 g/kg DM) across treatments, increasing fat and NDF content with WDG level. Animals were fed in individual pens outdoor during 12 weeks, with feed offered *ad libitum* in 3 meals. Calves were weighed every 14 days. Dry matter intake (DMI) was determined daily and feed:gain ratio (F:G) calculated based on mean values for the experimental period. Apparent DM digestibility (DMD) was estimated on week 8. The experiment was analyzed according to a randomized plot design with repeated measures, with animal as the experimental unit. Linear and quadratic effects associated to WDG level were tested. Animal LW increased linearly with time (P<0.01) but no differences were observed in LW gain due to WDG level (0% 1.21, 12% 1.19, 24% 1.14, 36% 1.25 kg/d, SE 0.07; P>0.10) or in LW/height ratio (P>0.10). Fresh feed intake increased linearly with WDG level (b=0.12 kg; P<0.01) but no differences were observed in DMI (b=0.016 kg; P>0.10). DMD was also similar between treatments (P>0.10), which altogether was consistent with similar F:G ratio between treatments (0% 3.37, 12% 3.55, 24% 3.64, 36% 3.70 kg/d, SE 0.15; P>0.10). Results suggest that performance of early-weaned calves lot-fed during postweaning would not be affected if sorghum WDG is included up to 30% of ration DM.

Substitution effect of corn by barley on growth performance and digestive tract of broilers

H. Belabbas[1], N. Adili[2] and A. Benkhaled[1]
[1]University of Mohamed Boudiaf, Microbiology and Biochimistry Departement, M'sila, 28000, Algeria,
[2]Institute of veterinary sciences and agricultural sciences, University of El-HadjLakhdar, Veterinary
Medicine Department, Batna, 05000, Algeria; anavet.belabbas@yahoo.fr

The effect of partial substitution of corn by barley on the digestive tract function and the growth performance of broilers was studied. Three diets were tested per trial period; the first one (control group) was based on corn. In the second and third one (groups 2 and 3), we substituted 10% and 20% of corn with barley respectively. These diets were isocaloric, isonitrogenous and fed to broiler chickens (n=40/group) in floury form. The body weight development, size and the weight of three intestine portions (duodenum, jejunum and ileum) were measured during four weeks, as well as the intestinal enzymatic activity which was measured at the 14th day. Results revealed that the inclusion of barley (10 or 20%) in a broiler diet did not have a significant effect (P>0.05) on the body weight development, size and weight of three segments of the intestine throughout the experimental period. On the other hand, there was no significant deference (P>0.05) in the jejunum alkaline phosphatase activity of the incorporated barley diet. Conclusively, the incorporation of barley in a broiler chicken diet at the rate of 20% and less is strongly feasible, without any chicken growth performance alterations.

Effect of rats dietary supplementation on ALA, EPA and DHA content in muscles and intramuscular fat

T. Wysoczanski[1], E. Sokola-Wysoczanska[2], A. Prescha[3], R. Bodkowski[4], B. Patkowska-Sokola[4] and K. Czyz[4]
[1]FLC Pharma Ltd, Muchoborska 18, 54-424 Wrocław, Poland, [2]Lumina Cordis Foundation, Szymanowskiego 2A, 51-609 Wrocław, Poland, [3]Wroclaw Medical University, Department of Food Science and Dietetics, Borowska 211, 50-556 Wrocław, Poland, [4]Wroclaw University of Environmental and Life Sciences, Institute of Animal Breeding, Chelmonskiego 38c, 51-630 Wrocław, Poland; robert.bodkowski@up.wroc.pl

A study on alpha-linolenic acid (ALA) conversin into EPA and DHA was conducted with male Wistar rats. The experiment lasted 8 weeks, the rats were divided into the following groups: I = control; II = linseed oil; III = ethyl esters of linseed oil; IV = fish oil. All the supplements were provided individually, in an amount of 0.04 g/kg of body weight per day. On the last day of the experiment all animals were subject to euthanasia, and samples of muscular tissue and intramuscular fat were collected in order to examine fatty acids profile. The analysis were performed at the Department of Food Science and Dietetics, Wroclaw Medical University, using gas chromatograph 6890 N. Chemstation v. B.04.02 software was used to analyse the results. The results demonstrated a significant increase in ALA content in groups II and III in case of muscular tissue (over 5-fold increase), but what is interesting, in case of intramuscular fat, also the content of EPA and DHA increased significantly in groups III and IV. EPA increase was over 3-fold higher in group III, and over 4-fold higher in group IV, while in case of DHA these values were 14-fold, and 22-fold, respectively. In conclusion, the most profitable was supplementation with ethyl esters of linseed oil, since it causeda significant increase in EPA and DHA content, but moreover an increase in ALA content, which is a kind of 'reservoir' of EPA and DHA for an organism. Linseed oil, being the substrate for ethyl esters manufacturing, demonstrated lower biological activity, which proves that ethyl ester form is definitely more bioavailable.

Effect of a complex-enzyme incorporated in the diet on zootechnical performances of broilers

W. Doumandji[1], H. Ainbaziz[2], L. Sahraoui[2], H. Boudina[3], R. Kaddour[2] and S. Temim[2]
[1]Faculté des Sciences de la Nature et de la Vie et des Sciences de la Terre, Université de Buira, 10000, Algeria, [2]ENSV, Alger, 16200, Algeria, [3]Itelv, Bab Ali, 16200, Algeria; waffadoumandji@yahoo.fr

The aim of this study was to evaluate the impact of dietary supplementation with enzymes (E) on growth performances and some physiological parameters in broilers. A total of 840 one day old chicks (ISA F15 strain) were divided into two experimental groups of homogeneous weight (7 repetitions of 60 subjects): a control group (C) fed with a standard diet and an experimental group (E) fed continuously with the same standard diet supplemented with 0.1% of an complex enzyme (xylanase, α-galactosidase, beta-glucanase, protease, pectinase and amylase). The incorporation of enzymes in the complete feed increased feed intake (+5%; P<0.05) and the cumulative weight gain (+4%; P<0.05) without changing feed conversion rate. Furthermore, the weight of the carcass ready to be cooked and the proportion of abdominal fat were not affected by dietary supplementation with enzymes. The E group had a significant increase in liver proportion (+12%), gizzard (+14%) and the total digestive lactobacilli (+6%) compared to group C. The plasmatic concentrations of cholesterol, triglycerides and total protein were higher in the E group (+14, 19 and 37%; P<0.05, respectively) compared to the C group. We did not find any difference between groups for triglycerides and urea (P>0.05). The rate of creatinine and glucose was lower in the E group (-33 and -27%; P<0.001) in comparison with the C group. The E group showed high villi in the proximal duodenum, jejunum and ileum (+26%; P<0.01). In conclusion, the enzyme supplementation has improved some zootechnical, physiological and digestive performances.

Nutritive value and protein molecular structure of yellow and black canola seed

K. Theodoridou[1] and P. Yu[2]
[1]Queens university belfast, 18-30 Malone Road, BT95BN, Belfast, United Kingdom, [2]University of Saskatchewan, 51 Campus Drive, S7N5A8, Saskatoon, Canada; k.theodoridou@qub.ac.uk

Oilseeds and their products are the most valuable agricultural crops in world trade. Canola includes the yellow and the black-seeded varieties. Apart from chemical composition of oilseed crops, their protein secondary structure profiles may also influence protein quality, nutrient utilization and availability. Although data are rare, it is vital to study protein secondary structure in order to understand crop's digestive behavior and nutritional value. A new approach is the use of Fourier-transformed infrared-vibration spectroscopy (FT/IR), a technique for studying the secondary structural composition, stability and conformational changes. The aim was to evaluate and compare the nutritional value of two canola seed varieties. The items assessed included: (1) chemical and nutrient profiles; (2) rumen degradation kinetics; and (3) protein molecular structures. Statistical analyses were performed using the MIXED procedure of SAS. Yellow (CS-Y) and black (CS-B) canola seeds (B. napus) were used as feed sources and three dry Holstein cows fitted with a rumen cannula for the *in situ* rumen degradation parameters. CS-Y was lower in nitrogen detergent fiber and acid detergent fiber (P<0.05). Crude protein content was not different between the two varieties but the non-protein nitrogen was lower (P<0.05) for CS-Y compared to CS-B. The net energy for lactation, the digestible and metabolisable energy were higher (P<0.05) for CS-Y compared to CS-B. The C20:1 Eicosenoic acid content (omega-9) was lower (P<0.05) while the total polyphenols were tended to be lower (P≤0.10) for the CS-Y than for the CS-B. No significant differences were observed for the effective degradability of protein and the protein's inherent molecular structural make up, between the CS-Y and the CS-B. The findings of the present study indicate that the breeding of CS-Y has the potential to be a promising route to reducing fiber and hull content, while at the same time increasing the level of oil.

Determination of AMEn of different oil and powder fat sources for broilers in starter period
F. Sadi and H. Moravej
Tehran university, Animal Science, Faculty of Agriculture, Karaj, 31587-77871, Iran; hmoraveg@ut.ac.ir

This study was conducted to evaluate the effects of sex, different type and levels of fat and oil sources on metabolisable energy (AMEn) in the starter period of broiler chicks. The experiment had a completely randomized block design in a 6×2 factorial arrangement with 3 blocks (3 levels of each fat source) and 5 birds per replicate. The treatments included three different oil sources (soybean, free fatty acid, and poultry) and 3 different fat powder sources (animal-vegetable blend, vegetable, and vegetable-free fatty acid blend). The levels of different fat sources were 5, 10 and 15%. AMEn value of different fat sources was determined by excreta collection using a marker method (chromic oxide) in the starter period (6-11 days). The results showed a significant difference in AMEn value between the treatments (P<0.05). The AMEn of fats were not influenced (P>0.05) by the sex, and there were no significant interaction effects (P>0.05) between fat sources and sex of broilers for the AMEn. The highest and least AMEn value of different fat sources belong to soybean oil (8,081 kcal/kg) and vegetable fat powder (5,971 kcal/kg), respectively. Furthermore, there were no significant differences (P>0.05) between the AMEn of vegetable-free fatty acid blend (6,599 kcal/kg) and free fatty acid (6,593 kcal/kg). The influence of different levels of fat and oil sources on AMEn were significant (P<0.05). There were no significant differences (P>0.05) between AMEn of 5% (7,063 kcal/kg) and 15% (7,059 kcal/kg) levels of fat and oil sources, while both of them had significant different with 10% (7,218 kcal/kg) (P<0.05).

In vitro rumen fermentation of three varieties of sorghum grain, with or without polyethylene glycol
Z. Amanzougarene, A. Gimeno and M. Fondevila
Universidad de Zaragoza, Producción Animal y Ciencia de los Alimentos, M. Servet 177, 50013, Zaragoza, Spain; mfonde@unizar.es

Sorghum grain is commonly used in ruminant feeding, but some varieties have moderate levels of tannins that negatively affect rumen fermentation and overall nutrient digestion. Tannins bind to fermentable nutrients, reducing starch and N utilization in the rumen. This study aimed to determine the potential differences in ruminal fermentation of grains of three sorghum varieties (a white sorghum, S1, and two brown sorghum, S2 and S3), and to what extent such differences are caused by their tannin content. Four 12 h *in vitro* incubation series were carried out, with pH adjusted to 6.4, to measure the volume of gas produced in the presence or absence of polyethylene glycol (PEG) as tannin binder. Bottles including 500 mg of substrate were incubated in triplicates, and gas production was recorded every 2 h. The comparison of the fermentation pattern of sorghum grains in natural conditions (without PEG) shows that S1 rendered a higher volume of gas than S2 throughout all the incubation period (P<0.05), whereas differences between S1 and S3 were only manifested at 12 h. This response should be, at least in part, related to the different tannin content of sorghum grains, which was highest in S2. Thus, the improvement in gas production in S2 when tannins were inactivated (with PEG) was observed at all times of incubation (P<0.05), whereas it was not manifested for S1 and S3 (P>0.10). After 12 h, gas production when adding PEG was 1.03, 1.07 and 1.26 times that recorded without PEG in S1, S2 and S3. The effect of adding PEG promoted that, if tannin effect is avoided, S2 can be even more fermentable than S3 from 8 h onwards (P>0.05). Rumen fermentation of sorghum grains is constrained by their tannin content in some, but not all, brown varieties, and the potential effect of tannins can be estimated by gas production. Nutrients protected by tannins that are not fermented in the rumen would be available for their utilisation in the lower gut.

Growth performance of growing osimi lambs fed diets containing different levels of carob pulp

S. Abdel-Magid, I. Awadalla and M. Mohamed
National Research Centre, Animal Production, 33 El- Bohouth str., 12622 Dokki, Giza, Egypt;
soha_syd@yahoo.com

In a feeding trial for 16 weeks, three concentrate mixtures replacing 0, 25 and 50% cotton seed meal protein by equivalent amount of carob pulp protein to three groups of growing Osimi meal lambs (5 animals/group), aged 4-5 months old and weighted in average 30.60 kg. Tested rations were manipulated to be isocalorice and isonitorgenous. The results indicated that, inclusion of carob pulp protein at levels of 25 or 50% of cotton seed meal protein decreased (P<0.01) DM, CP, CF digestibilities, and average daily gain, while feed conversion as kg feed/kg gain was increased (P<0.01). It could be concluded that the inclusion of carob pulp in rations of grouing Osimi sheep might not exceed 12.5% of the total dietary protein.

Growth performance of rahmani lambs fed complete feed mixture containing new source of roughage

S. Abdel-Magid, I. Awadalla and M. Mohamed
National Research Centre, Animal Production, 33 El- Bohouth str., 12622 Dokki, Giza, Egypt;
soha_syd@yahoo.com

Fifteen Rahmani weaned lambs, 3-4 months old and 16 kg average body weight, were used, classified into three groups (5 animals each), in a feeding trial to study the effect of using fenugreek straw (FS) and pea straw (PS) instead of berseem hay (BH, control) as a roughage sources in the complete feed mixture on growth performance. The trial lasted for 16 weeks divided into two periods. The first was considered as growing period (8 weeks) during which animals received diets contained the tested roughages at 40% dietary level while the second was considered as finishing period (8 weeks) during which each group of lambs continued on the same tasted roughage in used during the growing period but at a level representing 20% of the diet. Diets within each period were formulated to be isocaloric and isonitrogenous. Another 9 mature rams were used in two digestibility and N-balance trials, one at the end of each period. The results showed that, OM, CP, CF and NFE digestibilities of the diets containing FS at either 40 or 20% levels were lower than those of diets containing PS or BH (control) which showed similar values. Body weights at the end of the trial were similar for groups fed on the control or PS diets, being higher (P<0.05) than those of the group fed on FS diets. Body weight gain and average daily gain showed similar trend. It can be concluded that PS could be successfully used in diets for rahmani lambs as a source of roughage instead of BH.

Effects of different levels of patulin on *in vitro* ruminal microbial fermentation
M. Mojtahedi
Department of Animal Science, University of Birjand, 97175-331, Birjand, Iran; mojtahedi@birjand.ac.ir

Patulin is a toxic mycotoxin produced by certain moulds of the genera Penicillium, Aspergillus and Byssochlamys growing on various food commodities like fermented feeds. Patulin exhibits a number of toxic effects in animals and its presence in feed is undesirable. The objective of this study was to evaluate the effects of patulin on *in vitro* rumen fermentation and gas production parameters. In an anaerobic batch culture system, 50 ml of buffered rumen fluid was dispensed into a 125-ml serum bottle containing 0.5 g dry matter (DM) of the experimental diet. Experimental treatments included four dose levels of patulin (0.0, 30, 60 and 120 µg/ml). All bottles were purged with anaerobic CO_2, sealed and placed in a shaking water bath for 72 h at 38.6 °C. Gas production of each bottle was recorded at 2, 4, 8, 12, 16, 24, 48 and 72 h of the incubation and then gas released. After 24 h incubation a set of bottles were opened and the pH, *in vitro* dry matter digestibility (IVDMD) and VFA production were determined. Results showed that addition of patulin affected the rate and cumulative gas production ($P<0.05$), so, by increasing the level of patulin from 0 to 120 µg/ml, the gas production rate decreased from 0.086 to 0.054 and cumulative gas production decreased from 167.2 to 133.9 ml/g DM, respectively. In addition IVDMD decreased significantly ($P<0.05$) with inclusion of patulin in culture medium, so that the lowest and the highest IVDMD values were observed in treatments with 120 and 0 µg/ml patulin, respectively (0.511 vs 0.574). Total VFA production was reduced significantly with addition of patulin (77.6, 71.4, 62.5 and 48.9 mM for 0.0, 30, 60 and 120 µg/ml patulin, respectively). While the pH and molar proportions of acetate, propionate, butyrate and valerate were not affected by the patulin, significantly ($P>0.05$). Results of this study indicated that addition of different levels of patulin affected *in vitro* fermentation characteristic negatively, as represented in reduced gas production, dry matter digestibility and VFA production.

Challenges related to introducing genomic selection in sport horse breeding
A. Ricard[1,2]
[1]IFCE, Recherche et Innovation, 61310 Exmes, France, [2]INRA, UMR 1313, 78352 Jouy-en-Josas, France; anne.ricard@toulouse.inra.fr

Genomic selection is now widely used in dairy cattle breeding schemes, but not yet in horses. We will discuss reasons and review the state of the art in genomic evaluation in horses with focus on advantages and disadvantages for breeding plans. The list of advantages is as follows: We have rather good linkage disequilibrium at small distance (r^2 about 0.25 for 50 kb). A new HD beadchip is now available. There are good possibilities of imputation from low density beadchip. The genotyping cost is low compared to the price of the animal. The long generation interval may be reduced. Selection for many traits difficult to measure on large scale, e.g. criteria related to health, reproduction, behavior, will benefit of genomic evaluation based on a well phenotyped reference population. Among the challenges, technical and practical issues may be distinguished. Technical points are as follows: the breeding schemes based on own performances measured on both sexes (impossible in dairy cattle) have relatively good efficiency. It is difficult to have a large reference population of animals with breeding values of high accuracy because stallions have less progeny than bulls, and overall population size is much lower. Low numbers of QTLs were found for traits of interest except for DMRT3. A potential subsequent gain of accuracy from theoretical framework was determined, but low actual gain of accuracy was measured in first practical applications. The latter may be explained by the low numbers of sire/son pairs and by the sampling exclusively from stallions. Political points are the difficulty to build a large international reference population and possibly to have support of breeders. However, genomic evaluation may be more accepted by breeders than genetic evaluation because it is also based on molecular data rather than only statistics. Furthermore, genomic evaluation has the potential to become a valuable tool for pre-selection of best horses for sport rather than only to improve future breeding schemes.

Application of genomic selection in the Spanish Arab horse breeding program

I. Cervantes[1] and T.H.E. Meuwissen[2]
*[1]Department of Animal Production, Complutense University of Madrid, E28040, Spain, [2]Department of Animal
Aquacultural Sciences, Norwegian University of Life Sciences, 1432 Ås, Norway; icervantes@vet.ucm.es*

Sport horse breeding programs are characterized by long generation intervals and suboptimal selection intensities. The objective of this study was to compare the genomic selection tools with traditional BLUP by simulation in Arab horse breeding program in Spain. Thirty-two chromosomes of 100 cM were simulated with a mutation rate of 2×10^{-9}. An average of 83,000 SNPs markers were selected using a minimum allele frequency of 0.03. The probability to become a QTL of each marker was 3.7%. We simulated a historical population with effective size of 200 and 2,000 generations. From the last generation the founders of a subset of the real pedigree of Arab Horse were randomly chosen, and finally a simulated genotype for each animal included in the pedigree was obtained. True breeding values and phenotypes were simulated with heritabilities of 0.1, 0.2 and 0.3. Twenty replicates of each scenario were used. The real pedigree contained 3,613 Arab horses. The training set included 2,000 animals and the validation set included 613 young animals. A traditional BLUP evaluation using pedigree and phenotypes and a GBLUP evaluation using phenotypes and genotypes were performed. Both methods were compared by accuracies between predicted and true breeding values. The results showed significant differences between accuracies obtained using GBLUP and traditional evaluation using pedigree. In the validation set the accuracy was 8, 10 and 11% higher using genotypes than pedigrees for the scenarios with heritabilities 0.1, 0.2 and 0.3, respectively. We can conclude that the structure in Spanish Arab Horse population is appropriate to attain reliable breeding values using genomic data. The application of genomic selection could reduce the generation interval in the breed and help preselecting candidate sires.

The importance of the DMRT3 mutation on riding traits and gaits in Standardbred and Nordic trotters

K. Jäderkvist, L.S. Andersson and G. Lindgren
*Swedish University of Agricultural Sciences, Department of Animal Breeding and Genetics, Box 7023,
75007 Uppsala, Sweden; kim.jaderkvist@slu.se*

Previous studies have shown that a single base-pair mutation, a change from cytosine (C) to adenine (A), in the DMRT3 gene affects both the ability to show ambling and lateral gaits in a wide range of horse breeds, as well as racing performance and trotting technique in Standardbred and Nordic trotters. The variant allele is present in gaited breeds but is absent, or found at a very low frequency, in breeds used for Western-European style riding and flat racing, like the Swedish Warmblood and Thoroughbreds as well as in draught horses. This indicates that the variant allele might have a negative effect on certain riding performance traits in horses. This study investigated whether the DMRT3 variant affects canter in Standardbred trotters, and if heterozygous horses (CA) are better suited for Western-European style riding than homozygous horses (AA). Riding traits were analysed in 115 Standardbreds and 55 Nordic trotters. The owners answered a questionnaire and had to score how well the horse performed walk, trot and canter on a scale from 1 to 6, with focus on rhythm, balance and transitions. The survey also included questions on whether the horse ever performed tölt and pace and if it had competed in harness racing, dressage or jumping. The results showed that CA Standardbreds had significantly better balance in both collected and extended canter compared with AA horses. The CA horses also received significantly higher scores for transitions in collected canter. For rhythm in canter we found no significant differences between the genotypes. There were no significant differences between the genotypes for trot. The AA horses performed tölt and pace significantly more often than CA horses. No significant differences were present for canter ability in the Nordic trotters but the mutation was strongly associated with the gaits tölt and pace. The results of this study suggest that CA horses may be better suited for traditional riding disciplines rather than harness racing.

Should DMRT3 be used for estimation of breeding values in the French trotter?
S. Brard[1,2,3] and A. Ricard[4,5]
[1]INRA, GenPhySE (Génétique, Physiologie et Systèmes d'Elevage), 31326 Castanet-Tolosan, France, [2]Université de Toulouse, INP, ENSAT, GenPhySE (Génétique, Physiologie et Systèmes d'Elevage), 31326 Castanet-Tolosan, France, [3]Université de Toulouse, INP, ENVT, GenPhySE (Génétique, Physiologie et Systèmes d'Elevage), 31076 Toulouse, France, [4]IFCE, Recherche et Innovation, 61310 Exmes, France, [5]INRA, UMR 1313, 78352 Jouy-en-Josas, France; sophie.brard@toulouse.inra.fr

An A/C mutation in the DMRT3 gene is responsible for the ability to pace in horses, and allele C is supposed to have a negative effect on performances of trotters. However, the frequency of allele A in French Trotter (FT) is only 77%, and allele C has shown a positive effect on late performances of FT. We aimed to ascertain if the genotype at SNP BIEC2-620109 (linked to DMRT3) could be used for computation of estimated breeding values (EBVs) for FT. We used the genotypes of 630 horses, with 47,711 SNPs retained, and pedigree data of 6,599 horses. The qualification status (trotters need to keep a time limit on a 2,000 m course to begin their career) and earnings at different ages were pre-corrected for fixed effects and evaluated with a multi-trait model. EBVs were computed with and without the genotype at SNP BIEC2-620109 as a fixed effect in the model. The analyses were performed using only the pedigree with BLUP, and using the genotypes with GBLUP. For validation, 3 groups of 100 candidates were alternatively used. Validations were also performed on 50 randomly clustered groups of 126 candidates, and compared to the results of the 3 disjoint groups. For performances on which DMRT3 has a small effect, the coefficients of correlation between the EBVs and the performances were not improved by including the genotype at SNP BIEC2-620109 in the model, but the coefficient of regression was closer to 1. For qualification status and early earnings (strongly related to DMRT3), using the genotype of the SNP BIEC2-620109 improved the coefficients of correlation a lot. We therefore recommend the use of the genotype at DMRT3 for the FT genetic evaluation.

Genomic analysis of hydrocephalus in Friesian horses
A. Schurink[1], B.J. Ducro[1], J.W.M. Bastiaansen[1], I.J.M. Boegheim[2], F.G. Van Steenbeek[2], M. Vos-Loohuis[2], I.J. Nijman[3], G.R. Monroe[3], I. Hellinga[4], B.W. Dibbits[1], W. Back[5,6] and P.A.J. Leegwater[2]
[1]Wageningen University, Animal Breeding and Genomics Centre, Wageningen, the Netherlands, [2]Utrecht University, Department of Clinical Sciences of Companion Animals, Faculty of Veterinary Medicine, Utrecht, the Netherlands, [3]University Medical Center Utrecht, Department of Medical Genetics, Utrecht, the Netherlands, [4]Koninklijke Vereniging 'Het Friesch Paarden-Stamboek', KFPS, Drachten, the Netherlands, [5]Ghent University, Department of Surgery and Anaesthesiology of Domestic Animals, Faculty of Veterinary Medicine, Merelbeke, Belgium, [6]Utrecht University, Department of Equine Sciences, Faculty of Veterinary Medicine, Utrecht, the Netherlands; anouk3.schurink@wur.nl

Hydrocephalus in Friesian horses is a developmental disorder that often results in stillbirth of affected foals and dystocia in their dams. The occurrence is probably related to a founder effect and inbreeding in the population. The aim of our study was to identify the causal mutation for hydrocephalus. From the Illumina EquineSNP50 Genotyping BeadChip, 29,270 SNPs remained after quality control. A genome-wide association study of hydrocephalus in 13 cases and 69 controls indicated the involvement of a region on ECA1 ($P<1.68×10-6$). All cases, and none of the controls, carried 2 copies of a 0.58 Mb haplotype. Next generation DNA sequence analysis was performed of gene exons in this region on 4 cases and 6 controls and revealed a nonsense mutation that was identical to a mutation identified in a human case of muscular dystrophy-dystroglycanopathy with hydrocephalus. All available cases and none of the controls were homozygous for the mutation. Tested dams (n=17) of cases were all heterozygous. Hydrocephalus in Friesian horses has an autosomal recessive mode of inheritance. Currently, Friesian horses are tested for the presence of the disease allele, which allows selection against this allele and thereby reduction in losses and suffering caused by hydrocephalus in the Friesian horse population.

Admixture analysis in the Franches-Montagnes horse breed using whole-genome sequences

M. Frischknecht[1,2,3], M. Neuditschko[1,2], V. Jagannathan[1,3], J. Tetens[4], G. Thaller[4], T. Leeb[1,3] and S. Rieder[1,2]
[1]Swiss Competence Center of Animal Breeding and Genetics, University of Bern, Bern University of Applied Sciences HAFL & Agroscope, Bremgartenstrasse 109a, 3001 Bern, Switzerland, [2]Agroscope, Swiss National Stud Farm, Les Longs-Prés, 1580 Avenches, Switzerland, [3]Institute of Genetics, Vetsuisse Faculty, University of Bern, Bemgartenstrasse 109a, 3012 Bern, Switzerland, [4]Institute of Animal Breeding and Husbandry, Christian-Albrechts-University, Hermann-Rodewald-Straße 6, 24118 Kiel, Germany; mirjam.frischknecht@vetsuisse.unibe.ch

The Franches-Montagnes (FM) is the last indigenous Swiss horse breed. The breed was established in the 19th century by crossing of local mares with Warmblood (WB) Stallions from England and France. Subsequently, horses from other breeds, such as Arabians were also occasionally used. The last introgression with two WB stallions occurred in the 1990s. We collected paired-end whole-genome sequencing data for 30 FM and 14 WB horses. In order to investigate the impact of using different references for calculating the admixture, three different approaches based on pedigree information, genotype information corresponding to the 50 k Illumina SNP chip, and whole genome-sequences were applied. The levels of admixture between FM and WB horses were determined with the software Admixture 1.23. Using pedigree-based ancestry information the 30 FM horses showed on average a level of admixture of 13%. Based on 50 k genotype information and including 14 WB horses in the reference population, the average level of admixture of the 30 FM horses was 18%. Using the whole-genome sequence information of the 30 FM and 14 WB horses the average level of admixture was 11%. The level of admixture is often used to describe the origin of breeds. Thus, it is a very important population parameter for breeding management decisions and to conserve genetic resources. Our results demonstrate that the calculation of admixture depends on the source of ancestry information which must be considered for appropriate interpretation.

Breeding objectives and practices of sport horse studbooks: results of a worldwide inventory

K.F. Stock[1], K. Quinn Brady[2], K. Christiansen[3], Å. Viklund[4], I. Cervantes[5], A. Ricard[6], B.J. Ducro[7] and S. Janssens[8]
[1]vit, Heideweg 1, 27283 Verden (Aller), Germany, [2]Horse Sport Ireland, Millennium Park, Kildare, Ireland, [3]Danish Warmblood, Vilhelmsborg Allé 1, 8320 Maarslet, Denmark, [4]SLU, Box 7023, 75007 Uppsala, Sweden, [5]UCM, Avda. Puerta de Hierro, 28040 Madrid, Spain, [6]INRA, UMR 1313 GABI, Domaine de Vilvert, 78352 Jouy-en-Josas, France, [7]WUR, P.O. Box 338, 6700 AH Wageningen, the Netherlands, [8]KU Leuven, Livestock Genetics, KA30, 3001 Leuven, Belgium; friederike.katharina.stock@vit.de

The breeding of sport horses is carried out by many breeding organizations worldwide. Common breeding goals and well developed logistics for the exchange of genetic material across countries have created an internationalized and highly competitive situation. Recent economic development and structural changes have increased pressure on the studbooks, implying reconsidering and possible adjusting of strategies and practices in the international sport horse scene. This motivated the 2015 studbook survey of the Interstallion working group on key determinants of breeding programs. Questions related to consideration and relative importance of traits, sources of phenotypic data, future developments and strategies for collecting data. Organizations were asked to assess the role of performance tests, national and international competition results and provide information on current and prospective selection methods and their acceptance by breeders. When compared to 2002, there were fewer changes in the trait groups relevant for selection than in the role of individual traits. Major challenges of the studbooks were seen regarding phenotypes and methodology, and the need for refined trait definitions and new traits (health, behavior, international sport performances) for routine genetic and future genomic evaluations was obvious. To meet these demands and comply with the broad interest in using genomic tools, improved international collaboration will be crucial to strengthen and develop sport horse breeding within and across countries.

Linear profiling protocols for 3-year-olds for genetic evaluation of Swedish Warmblood horses
Å. Viklund, S. Bonow and S. Eriksson
Swedish University of Agricultural Sciences, Animal Breeding and Genetics, P.O. Box 7023, 75007 Uppsala,
Sweden; asa.viklund@slu.se

A linear profiling protocol has been introduced in 2013 at performance tests for 3-year-old Swedish Warmblood horses (SWB). In this protocol traits are subjectively described on a linear scale between two extremes. This is a complement to the traditional scoring where the quality of the horse is assessed in relation to the current breeding objective. A linear profile could provide the horse owner with more information about the horse. The aim of this study was to investigate the suitability of the linearly scored traits for genetic evaluation of SWB. Data from 1,889 3-year-old horses tested in 2013 and 2014 were analyzed. The linear protocols from both years consisted of 54 traits in total, including 22 conformation traits, 17 gait traits, 14 jumping traits and one temperament trait. Each trait was recorded on a nine point linear scale. Genetic parameters were estimated with an animal model. For conformation traits the heritabilities ranged from 0.07 for hock joint from behind (cow hocked-bow hocked) to 0.48 for shape of neck (arched-straight). For gait traits the highest heritability (0.62) was estimated for suppleness in trot (supple-stiff) and the lowest (0.05) for energy in walk (hurries-lacks engagement). The heritabilities for jumping traits ranged from 0.19 for taxation skills (secure-insecure) to 0.70 for scope (much-little). Genetic correlations between linear traits and corresponding traditionally scored traits were strong for gaits (-0.70 to -0.97) and jumping traits (-0.62 to -0.99). For gaits the strongest genetic correlation was between the score for trot and stride length (long-short). For jumping the strongest correlation was between the score for jumping technique and angle of haunches (open-tight). The results show that it is possible to use the linear information in genetic evaluation and that it can be a tool for more detailed descriptions of breeding objective traits in the SWB.

Genetic parameters for eventing and their relationship to stallion performance test traits
H. Frevert[1], K. Brügemann[1], U. König V. Borstel[2] and S. König[1]
[1]University of Kassel, Animal Breeding, Nordbahnhofstr. 1a, 37213 Witzenhausen, Germany, [2]University of
Göttingen, Animal Breeding, Albrecht-Thaer-Weg 3, 37075 Göttingen, Germany; uk020548@uni-kassel.de

Breeding values and genetic (co)variance components for eventing were estimated using 6.42 m. equestrian competition results from the years 2005-2013, including 67,665 observations from 14,753 animals for eventing. As a novel approach, results from (German) national and international competitions were defined as two different traits. Bivariate genetic analyses using animal models were carried out in a Bayesian framework for all trait combinations, i.e. results from national and international eventing competitions with ranks for dressage and showjumping competitions, and with results from stationary stallion performance test (PT) traits (n=862 and n=736 stallions tested in 30- and 70-d PTs, respectively, between 2001-2003). Heritabilities for eventing were both at the national and international level h^2=0.05 and comparable with those of dressage and showjumping. A moderate genetic correlation of 0.46 between national and international eventing illustrates a divergent genetic background for horses to be successful in national or international competitions. High genetic correlations between dressage comptitions and eventing at national level (0.78) could be related to the major impact of dressage performance to the overall result of eventing competitions. Pronounced relationships were also found for national eventing with traits reflecting the dressage component of 70-d PT-traits. Surprisingly at the international level, between eventing and 70-d PT traits, including canter and cross-country jumping, genetic correlations were close to zero. Better predictors for international eventing ability were traits assessed in younger stallions during 30-d PTs: r_g=canter: 0.15; rideability: 0.24. Overall, estimated genetic covariance components suggest implementation of direct selection strategies for eventing rather than indirect selection based on dressage or showjumping traits.

The genetic variability of equine whey proteins

J. Brinkmann, G. Thaller, N. Krattenmacher and J. Tetens
Institute of Animal Breeding and Husbandry/CAU, Hermann-Rodewald-Str 6, 24118 Kiel, Germany;
nkrattenmacher@tierzucht.uni-kiel.de

The genetic variability of milk protein genes results in different protein variants, which may influence the nutritive value or processing properties of the milk. Among the milk proteins, α-lactalbumin (LALBA) and β-lactoglobulin (LGB) represent the major whey proteins in equine, but also bovine milk. While LALBA plays an important role in lactose synthesis, several functions of LGB have been discussed. Based on its retinol-binding capacity it has mainly been linked to vitamin-uptake, but its ultimate physiological role remains to be elucidated. LGB belongs to the protein family of lipocalins, known to be common inhalation and food allergens, and is absent from the milk of humans making it a major agent provoking cow's milk protein allergy (CMPA). The milk of horses and donkeys is better tolerated by CMPA patients, and beneficial effects of consumption have been described in the context of atopic dermatitis. Thus, there is an increasing interest in mare's milk and especially equine milk proteins. Studies about the genetic variability of equine milk protein genes are, however, still limited. In the current study, the open reading frames of the milk protein genes LALBA, LGB1 and LGB2 were resequenced in 194 horses belonging to 7 breeds used in milk production. A total of 14 protein variants, 12 of which were previously unknown, were predicted from the variants identified at the DNA level. For these variants a preliminary nomenclature was established. The highest genetic variability was detected in the Icelandic Horse, the Russian Heavy Draft was least variable. The LALBA gene was found to be monomorphic, which is not surprising given the essential function of the protein. Notably, the LGB1 gene was also nearly monomorphic, while most variants were detected in LGB2. Both genes have been shown to be expressed in horses. A possible explanation is that LGB2 has previously been a pseudogene as also found in other species, while LGB1 fulfills an ancestral and important, but unknown function.

Reproductive results of Sokólski mares involved in the conservation programme between 2008 and 2011

G.M. Polak
National Research Institute of Animal Production, National Focal Point, Wspolna Str. 30, 00-930, Poland;
grazyna.polak@izoo.krakow.pl

The genetic resources conservation programmes for coldblood Sokólski horses contributed to an increase in their population. The increase was driven by agri-environmental payments as part of the Rural Development Programme. The conservation programme specifies that mares should only be mated to designated stallions and every mare should give birth to at least one foal during the two years of participation in the programme. The aim of the present study was to evaluate reproductive results of mares and stallions participating in the genetic resources conservation programme for the Sokólski population. The results showed that over a period of 6 years (from 2008 to 2013), 1474 mares took part in the programme, 899 of which weaned 2,599 foals by the designated stallions (2.9 foals/mare on average). Although during the first year of the programme the Polish Horse Breeders Association assigned as many as 755 stallions for mating 320 mares, these sired only 88 foals. In 2013 only 261 stallions were approved for mating, of which 204 were actually used. During the study period, offspring from a total of 479 stallions were born. The average number of foals per stallion over the 6 years was 5.4, with a maximum of three foals per year. The average number of mares mated per year ranged from 2.6 in 2009 to 4.54 in 2013. These results show that the number of foals and the number of mares mated to stallions were very low, and the reproductive results were unsatisfactory.

Genetic distance of Sztumski and Sokolski coldblood horse in Poland

G.M. Polak[1] and A. Fornal[2]
[1]National Research Institute of Animal Production, National Focal Point, Wspolna 30 Str., 00-930 Warsaw, Poland, [2]National Research Institute of Animal Production, Department of Animal Cytogenetics and Molecular Genetics, Krakowska 1, 32-083 Balice, Poland; grazyna.polak@izoo.krakow.pl

Sztumski and Sokolski horses are the local populations of coldblood horses developed in Poland in the 19/20th century. These local populations have been formed by small primitive horses, improved mainly with Belgian and Ardennes horses. The number of purebred local Polish horses decreased dramatically in the second part of 20th century when all local types were crossed among themselves for creating a new breed: Polish Coldblood Horse. In 2008 a genetics recourses conservation program for Sztumski and Sokolski horses has been founded in which 1,046 Sokolski and 930 Sztumski mares are currently enrolled. The aim of our analyses was to estimate the genetic distance between local populations of Sztumski and Sokolski horses and 6 other breeds: Polish Coldblood Horse, Swedish Ardennes, French Ardennes, Belgian Draught, Trakehner and Polish Konik horse (Polish primitive horse). In the analysis of class II markers 17 microsatellites were tested. The results showed that genetic distance between Sztumski and Sokolski horses – (0.050) is greater than between Sokolski and Polish Coldblood Horse (0.016) and Sztumski and Polish Coldblood Horse (0.035), but still smaller than between the other breeds. The largest genetic distance was observed between Trakehner horses and other breeds. These results indicate considerable exchange of genetic material between analyzed breeds, implying the need for more accurate selection.

Association analysis of mutations in KIT and MITF genes with white markings in two horse breeds

S. Negro[1], F. Imsland[2], M. Valera[3], A. Molina[1], M. Solé[3] and L. Andersson[2]
[1]University of Córdoba, Edf. Gregor Mendel. Ctra. Madrid-Córdoba, km 396ª, 14072, Córdoba, Spain, [2]Uppsala University, Biomedical Center, Box 597, 75124, Uppsala, Sweden, [3]University of Seville, Ctra. Utrera km 1, 41013, Seville, Spain; z12neras@uco.es

The genetics of coat color in animals is one of the most popular model traits for breeding. Some horse breed registries have restrictions towards some coat color phenotypes, as the Menorca Purebred horse (PRMe): black coat color with just very little white marks in the head and limbs. The aim of this study was to analyse the association between white markings and mutations in KIT, PAX3 and MITF genes. A total of 140 horses from two horse breeds were analysed: 70 PRMe and 70 Pura Raza Español (PRE). In this study, an accurate standardized scoring system to estimate the extension of white marks was proposed. All horses were genotyped for MC1R and ASIP genes, and we analysed the exon 2 (c.59C>G and c.209G>A) of PAX3; the intron 1 (g.77784972T>C) and the exon 14 (c.2045A>G) of KIT; and the intron 1 (g.20189177T>A), the promoter (g.20117302Tdelins1), the intron 2 (g.20147039C>T) and the exon 5 (c.837_841del5) of MITF gene, using pyrosequencing. After statistical analysis using R software, we found by t-test that the A allele at KIT_exon14 was associated with white marks in head (0.09 in PRE) and total score (0.009 in PRMe). All the animals with AG genotypes had a wide star in the forehead. By Linear Regression analysis, we obtained that the A allele in KIT_exon14 was associated with white marks in head (0.009) and total score (0.02) in PRMe and also with hindlimb score in PRE (0.003). C allele in KIT_77784972 was associated with hindlimbs (0.041) in PRE. The G allele in MITF_20147039 was associated with white marks in forelimbs (0.017) in PRMe. Our preliminary results indicated that the A allele in exon 14 of KIT gene determined white markings in the head. Foreleg markings were driven predominantly by MC1R and the mutation MITF_20147039; and hind leg markings by the mutations in exon 14 and intron 1 of KIT gene.

Microsatellites as a tool for studying diversity of the horse major histocompatibility complex

E. Horecká[1], Č. Horecký[1], E. Jánová[2], P. Hořín[2] and A. Knoll[1]
[1]Mendel University in Brno, Dep. of Animal Morphology, Physiology and Genetics/CEITEC MENDELU, Zemědělská 1, 61300, Brno, Czech Republic, [2]University of Veterinary and Pharmaceutical Sciences Brno, Dep. of Animal Genetics/CEITEC VFU, Palackého třída 1/3, 612 48 Brno, Czech Republic; knolla@seznam.cz

The major histocompatibility complex (MHC) of vertebrates is a genomic region playing a crucial role in immune responses. Due to the complexity of the MHC region, deciphering and routine analysis of genetic variation including copy number variations and genotyping of MHC genes remain difficult. In horses, microsatellites spanning the MHC region have been identified, but with only few markers in the functionally important and complex class I region. The aim of this study was to establish a panel of microsatellite markers, which could be used for studying the genetic diversity of the entire MHC and for analyzing associations with selected diseases. Bioinformatic analysis of genomic nucleotide sequences retrieved from whole genome sequences allowed identifying polymorphic short tandem repeats (microsatellites) located in the three major MHC regions, class I, II and III. For the panel, we selected six microsatellites located in the class I region, four previously described (305-93, ABGe17402, TKY2933, UMN-JH34-2), and two newly detected loci. In addition, six microsatellites (TKY3324, COR112, COR113, UM011, COR114, and one new marker) within the class II region, one unknown marker located between class I and III, and two microsatellites (ABGe9019 and HMS082) located within the class III region were included in the panel. All markers selected for the panel could be genotyped by fragment analysis using genetic analyzer ABI PRISM® 3500. Numbers of alleles of each locus were determined. For markers described previously, a comparison with the original reports was made. The results indicate possible extension of the panel of five microsatellites used in our previous studies to fifteen markers to allow better characterization of the equine MHC. This study was supported by CEITEC CZ.1.05/1.1.00/02.0068.

Pony index: what is new in jumping indexation ?

M. Sabbagh[1], C. Blouin[2], G. Blanc[3], S. Danvy[1] and A. Ricard[4]
[1]IFCE, département R&I, jumenterie du Pin, 61310 EXMES, France, [2]INRA, UMR de Génétique Animale et Biologie Intégrative Equipe BIGE, Domaine de Vilvert, 78352 Jouy-en-Josas Cedex, France, [3]IFCE, DAFAT, jumenterie du Pin, 61310 EXMES, France, [4]INRA, UMR1388, Génétique, Physiologie et Systèmes d'Elevage, 24 Chemin de Borde Rouge BP 52627, 31326 Castanet Tolosan Cedex, France; margot.sabbagh@ifce.fr

At the request of the Fédération des poneys et petits chevaux de France, Inra and Ifce set a new trait for breeding values for sport jumping ponies in order to: (1) improve estimation of young ponies; (2) include results from competitions with small number of ponies; and (3) add subjective measurement of technical difficulty. The trait is the sum of points awarded depending on the place and the technical difficulty (obstacles from 0.40 to 1.35 m) of the test. This trait was combined with the previous trait based on the repetition of ponies rankings in each event in order to provide multiple trait breeding value. Weighting of the two traits was discussed with breeders. In 2014, in jumping, data included 233,685 results from 21,258 ponies from 4 to 30 years old. Rules for the distribution of points according to level of competition were determined according to both expert advice and comparison of ranking of ponies in the different levels of competition. Rules for the distribution of points according to the place were based on normal scores. Model included combination of age and squared height at withers. Results showed a significant interaction between age and height with higher influence of size at middle age and of age at high size. More variability was expressed in young ponies than with the ranking trait. Correlation between the two traits was 0.63. This phenotypic evaluation is now published for breeders and the next objective is to calculate breeding values.

Genetic correlation between fertility and type traits in the Italian Heavy Draught Horse breed

R. Mantovani[1], G. Pigozzi[2] and C. Sartori[1]
[1]Dept. of Agronomy Food Natural Resources Animals and Environment, Viale dell'Universita', 16, 35020 Legnaro (PD), Italy, [2]Italian Heavy Draught Horse Breeders Association, Via Verona, 90, 37068 Vigasio (VR), Italy; roberto.mantovani@unipd.it

This study aimed at assessing the genetic relationship between lifetime fertility (LF) and 15 type traits (TTs) scored in adult Italian Heavy Draught Horse (IHDH). The TTs reflect selection results carried out on meat production and heavy draught, while LF has been recently introduced as an indicator of mares' reproductive efficiency. The LF was expressed as foaling rate at a standard endpoint, i.e. no. of foals produced after 6 reproductive seasons (RS) divided by the no. of foaling opportunities. Fertility data included 3,118 mares with at least 3 registered RS. For mares with less than 6 RS (n=1,413), the foal production at the 6th RS was estimated through a set of predictive equations accounting for age at first foaling, the previous foal(s) production and the number of RS. Estimated foal production was joined with the actual foal production obtained for mares with ≥6 registered RS. The LF records we re merged with data on 6,910 adults scored for 14 linear TTs (9 points, from 1 to 5 with half points) and for an overall score (5 points, from 1 to 5) at about 30 months of age. REML analysis was carried out on a dataset including 7,633 animals, 2,394 of which had both LF and TTs data, and 11,672 animals in pedigree. The estimated genetic correlations (rg) between LF and TTs were generally medium-low but positive. Some exceptions were observed for fleshiness, fore and rear body diameters, and hind legs back view (rg=0.36, 0.41, 0.45, and 0.47, respectively). Low rg were found for head size & expression (rg=0.18) and temperament (rg=0.10), while the rg between LF and the overall score was moderate (rg=0.29). This study indicates that the current selection for meat production and heavy draught in the IHDH has no negative relationship on fertility.

SNP associated with different forms of osteochondrosis

D. Lewczuk[1], A. Bereznowski[2], M. Hecold[2], A. Ruść[3], M. Frąszczak[4], A. Korwin-Kossakowska[1], S. Kamiński[3] and J. Szyda[4]
[1]Institute of Genetics and Animal Breeding, Polish Academy of Science, ul. Postępu 36 A, 05-552 Magdalenka, Poland, [2]Warsaw University of Life Science, ul. Nowoursynowska 166, 02-787 Warszawa, Poland, [3]University of Warmia and Mazury in Olsztyn, ul. Michała Oczapowskiego 2, 10-719 Olsztyn, Poland, [4]Wrocław University of Environmental and Life Science, ul.C.K. Norwida 25/27, 50-375 Wroclaw, Poland; d.lewczuk@ighz.pl

The aim of the study was to find single nucleoide polymorphisms (SNPs) associated with different forms of osteochondrosis without and with presence of joint fragments (OC, OCD). A group of 201 three-year-old Warmblood horses was X-rayed twice (before and after performance test) using digital equipment. For assessment of presence of OC and OCD, ten images of three joints (4 fetlocks, 2 hocks, 2 stifles) were considered. All images were collected and evaluated by the same team of veterinarians. Horses were classified as OC(D) positive and healthy for every investigated joint separately. DNA isolated from blood was used to genotype each horse with llumina Equine SNP genotyping Bead Chip. Of the more than 70,000 SNP markers, 63,946 were retained for analyses. Statistical analysis included Cochran-Armitage test and logistic regression, assuming an additive mode of inheritance and including SNP effect, training center, sex, age, breeder and pedigree information in the model. All SNPs were analysed twice for their association with OC and OCD (before and after training). The number of SNPs associated with OCD was higher than the number of SNPs associated with OC in the first examination for the fetlock and stifle and irregularly for the hock. This research was funded by NCN grant 2011/01/B/NZ2/00893.

Polymorphism of selected candidate genes in relation to osteochondrosis in horses

M. Sokół[1], D. Lewczuk[1], M. Hecold[2], A. Bereznowski[2] and A. Korwin-Kossakowska[1]
[1]*Institute of Genetics and Animal Breeding, Polish Academy of Science, ul. Postępu 36A, 05-552 Magdalenka, Poland,* [2]*Warsaw University of Life Science, ul. Nowoursynowska 166, 02-787 Warszawa, Poland; d.lewczuk@ighz.pl*

The aim of the study was to identify potential mutations affecting the occurrence of osteochondrosis (OCD) in horses. In view of the physiological role of proteins and their specific biological functions, three genes were considered coding: collagen type IX alpha 2 (COL9A2), acyloxyacyl hydrolase (AOAH1) and frizzled related protein (FRZB), respectively. This selection was supported by previous results of QTL analyses revealing associations between these genes and osteochondrosis symptoms. Primer pairs comprising the regions of tested genes were selected using Primer-Blast software based on the equine genome sequence in GenBank. Using appropriate conditions for amplification, few novel polymorphisms in exons of the selected genes were found which were all SNPs. The material consisted of 201 three-years-old Warmblood horses x-rayed twice (before and after performance test) using digital equipment in two training centers. All images and evaluations were done by the same team of veterinarians. Based on the presence of osteochondrosis dissecans in three joints (fetlock, hock, stifle), horses were classified as ill – 1 (with OCD) and healthy (without OCD). DNA isolated from blood obtained was used to genotyping. Analysis of variance was performed to evaluate significant influences of the COL9A2, AOAH1 and FRZB genotypes on the occurrence of osteochondrosis dissecants in the individual joints. The statistical model included fixed effect of year, sex and training center as a well as random effect of the sire. The preliminary results showed a statistically significant effect of genotypes of genes COL9A2 and AOAH1 in the first examination of OCD and COL9A2 and FRZB in the second examination of OCD in individual joints. The obtained results require further investigations. This research was funded by NCN grant 2011/01/B/NZ2/00893.

Analysis of factors influencing coat color intensity in Old Kladruber black horses

B. Hofmanová[1], P. Kohoutová[1], I. Majzlík[1], L. Vostrý[2] and H. Vostrá Vydrová[1]
[1]*Czech University of Life Sciences, Animal Science and Ethology, Kamýcká 129, 165 21 Praha 6 – Suchdol, Czech Republic,* [2]*Czech University of Life Sciences, Genetics and Breeding, Kamýcká 129, 165 21 Praha 6 – Suchdol, Czech Republic; hofmanova@af.czu.cz*

The aim of this study was to evaluate potential factors influencing intensity of coat color in black variety of Old Kladruber horse breed. A total of 236 individuals of both sexes, aged from 1 to 24 years were included in the analysis. The measurement of coat color was performed by Minolta Spectrophotometer 2500d using CIE (L*a*b*) color system. Measurements were taken at 4 body parts (neck, shoulder, belly and back) on every horse, horses were measured twice (during summer and winter season). The GLM procedure (SAS software) was used to examine the influence of effects age, line, sex, season/housing on recorded coat color characteristics (L*a*b*). Age, sex and season/housing were found statistically significant ($P<0.001$) related to all three parameters (L*a*b*). The effect of line reached weak statistical significance ($P=0.05$) for parameters L* and a* only. The results were in a good agreement with empirical evidence of differencies in black coat color in horses according to season and housing system. Especially permanent stay on pasture during summer season may cause changes in coat color intensity due to sunlight. Non-significance of sire line can be explained by the breeding scheme. Lines are not fully closed, each line contains blood (genetic contribution) of other lines. The estimation of heritability in the next step could clarify whether also hereditary factors are involved in observed phenotypic variability in black horses. The prospected genomic analyses could further contribute to clarify the inheritance of fading and non-fading black coat color.

Pedigree analysis of endangered Czech draft horse breeds

L. Vostry[1,2], H. Vostra Vydrova[1,2], B. Hofmanova[2], Z. Vesela[1] and I. Majzlik[2]
[1]Institute of Animal Science, Pratelstvi 815, 10401 Prague – Uhrineves, Czech Republic, [2]Czech University of Life Sciences Prague, Faculty of Agrobiology, Food and Natural Resources, Kamycka 129, 165 21 Prague, Czech Republic; vostry@af.czu.cz

There are two original draft horse breeds in the Czech Republic: Silesian Noriker and Czech-Moravian Belgian horse. Because these two populations are closed, there is a concern about the loss of genetic variation. To identify factors which may affect the genetic variability in breeds studied, genetic diversity and population structure were analyzed based on pedigree information of animals registered in the studbooks. In order to estimate inbreeding coefficients (F_x) trend, effective population sizes and the proportions of inbreed animals in two original Czech draft horse breeds individuals born from 1990 to 2013 were considered. The depth of pedigree was up to 22 and 31 generations for Silesian Noriker and Czech-Moravian Belgian horse respectively, with means of equivalent know n genera tions 11.36 for Silesian Noriker and 11.16 for Czech-Moravian Belgian horse. The effective numbers of founders and ancestors contributing to the current genetic pool were 69.42 and 22.32 for Silesian Noriker and 43.33 and 23.32 for Czech-Moravian Belgian horse. The inbreeding coefficients were 4.6% for Silesian Noriker (with maximum 27%) and 4% for Czech-Moravian Belgian horse (with maximum 17%). Proportions of inbreed animals were high: 98% for Silesian Noriker and 92% for Czech-Moravia Belgian horse. The average rate of inbreeding was 0.5% for both Silesian Noriker and Czech-Moravian Belgian horse, and the corresponding estimates of effective population sizes were 95 for Silesian Noriker and 101 for Czech-Moravian Belgian horse. These results support the concerns about reduction of genetic variability in the Czech draft horse breeds, which could be controlled by managing the use of sires in populations. Supported by the Ministry of Agriculture of the Czech Republic (Project no. QJ1510141).

Influence of the rider-horse interaction in show jumping genetic evaluations

E. Bartolomé[1], I. Cervantes[2], A. Menéndez-Buxadera[3], A. Molina[3] and M. Valera[4]
[1]Faculty of Veterinary Medicine, Lusófona University, 1749-024 Lisbon, Portugal, [2]Department of Animal Production, Complutense University of Madrid, 28040, Spain, [3]Department of Genetics, University of Córdoba, 14071, Spain, [4]Department of Agro-Forestry Sciences, University of Sevilla, 41013, Spain; icervantes@vet.ucm.es

The rider is one of the most relevant environmental factors for sport horses. Horse and rider compete together as a unit and the horse performance can be modified by the rider. The aim of this study was to evaluate the influence of the rider-horse interaction in show jumping genetic evaluation along the age of the horse. Data included 31,129 records from 1101 animals ridden by 606 different riders in 159 competitions. Proportion of riders with 3 or more horses was 63.7%. The heterogeneity of random effects was analyzed across four different age groups previously established (4-6 years old, 7-10, 11-15 and older than 15 years old). The models were grouped in three blocks according to the effects included and the homogeneity/heterogeneity of additive and residual effects (both homogeneous, both heterogeneous or one of each). Within each of the blocks, two models were applied interchanging rider-horse interaction with permanent environmental as random effects. Aside, all models included the event, the sex and the breed as fixed effects and the rider as a random effect. Six different genetic models were finally used. Model fitting was evaluated with different criteria. Results showed that the models of best fit were those including rider-horse interaction as a random effect with heterogeneous variance for both residual and additive effect. Within this model, the percentage of variance explained by the rider and rider-horse interaction showed a decreasing tendency with age, with values that ranged from 10.48 to 6.38% for the rider and from 4.42 to 2.69% for the rider-horse interaction. These results help to find a methodology that allows an appropriate consideration of the rider-horse interaction in routine applications.

Genetic study of stress assessed with infrared thermography during dressage performance

M.J. Sánchez[1], M. Valera[1], A. Molina[2] and E. Bartolomé[3]
[1]University of Seville, Ctra. Utrera km1, 41013 Seville, Spain, [2]University of Cordoba, Ctra. Madrid-Cádiz, km 396a, Cordoba, Spain, [3]Lusófona University, Faculty of Veterinary Medicine, Lisbon, Portugal; v32sagum@gmail.com

Behavioural and physiological parameters in dressage (D) are important traits due to their relation to human-horse accidents, performance and welfare. However, they are only rarely included in horse breeding programs. The assessment of eye temperature (ET) using infrared temperature (IR) has been highlighted as a reliable indicator of stress in horses. In Spain, the Pura Raza Español horse (PRE) Breeding Program focuses mainly on the selection of an adequate D ability. The study of the genetic parameters of ET could highlight the adequacy of this parameter to be included as a selection criterion in the PRE Breeding Program. The aim of this study was to investigate the heritability of ET measured with IR in a real activity related to the Breeding Program. Accordingly, 1,746 ET and D score records (343 PRE) were taken during three final D competitions for young horses in Spain. Genetic parameters were estimated using a Bayesian procedure. The environmental effects that were statistically significant (P<0.05) were used in the genetic models. The pedigree file included 3,350 PRE. The correlations among the ET and D breeding values (BV) were studied. The age, the stud, the transport, the training and the year were significant effects for all the traits. Heritabilities ranged from 0.14 to 0.50. The D BV and the ET BV just after the exercise were correlated with r^2=0.21. The findings indicate that selection for ET in horses is achievable, although it is not clearly related with the D performance in the PRE, and environmental factors have significant impact on ET. So, ET level could be a result of a dynamic interaction between the genetic background of the horse and the environmental factors that affects it. More research including more animals is required, and currently, there is still insufficient proof that ET is suitable for being used as a performance selection criterion.

Genetic analysis of cresty neck in the Pura Raza Español Horse

M.J. Sánchez[1], P.J. Azor[1], A. Molina[2], J.L. López-Rivero[3] and M. Valera[1]
[1]University of Seville, Agro-Forestry Sciences, Ctra. Utrera km1, 41013 Seville, Spain, [2]University of Córdoba, Genetics, Ctra. Madrid-Cádiz, km 396a, 14071 Cordoba, Spain, [3]University of Córdoba, Anatomy and Comparative Pathology, Ctra. Madrid-Cádiz, km 396a, 14071 Cordoba, Spain; v32sagum@gmail.com

The Pura Raza Español Horse (PRE) is a 'Baroque Horses' with a long neck and abundant mane and has a defect known as 'Cresty Neck' (CN) which affects a large number of animals in the population. Horses with this conformational defect lose straightness in the upper neck, and all the soft tissues that support it. CN was analysed as a linear trait on a scale of 9 classes (1-Crest is absent to 9-Crest is so large that it permanently droops to one side). The CN trait was studied alongside other morphological traits such as body condition, height at withers, length of neck, head-neck junction and neck-body junction. The database included 10,929 records from 24 countries. The aims of this study were to determine the prevalence of CN, what effects influence it, its heritability and genetic correlations with other conformational traits. The environmental effects were studied using a General Linear Model and a Tukey's test. A Dunnett's test was carried out in the stud area taking Spain as the control area. The genetic parameters were estimated using a Bayesian procedure. The systematic effects included in the genetic model were those which were significant: age, gender, zone of the stud and coat; and the random effects were animal and random residual. The pedigree information produced a total of 41,055 PRE. Heritability for CN was 0.37, whereas the other conformational heritabilities ranged from 0.25 (Neck-body junction) to 0.80 (height at withers). Genetic correlations between CN and the other conformational traits ranged from -0.06 (with height at withers) to -0.21 (with neck-body junction). So, CN had a moderate heritability which will allow improving this defect by genetic selection.

Designing a selection index for dressage in PRE horse using performance and morphological traits

I. Cervantes[1], M.J. Sánchez[2], M. Valera[2], M.D. Gómez[3], J.P. Gutiérrez[1] and A. Molina[3]
[1]Dpto. Producción Animal, Universidad Complutense de Madrid, 28040, Spain, [2]Dpto. Ciencias Agro-Forestales, Universidad de Sevilla, 41013, Spain, [3]Dpto. Genética, Universidad de Córdoba, 14071, Spain; gutgar@vet.ucm.es

Morphological assessment of Pura Raza Español horse (PRE) for preselecting animals for dressage is a challenge within its breeding program. The morphological linear score system was established in 2008. The objective of this work was to design a selection index to improve the dressage performance using morphological linear traits (LT) as selection criteria. The selection index theory was used to compute genetic indexes including LT traits and performance traits (PT), estimating the relative weights b on genotypes, the equivalent economic values p_{eq} weighting genotypes, and the genetic response using different strategies. Previously, a multivariate animal model was applied for the 6 PT (total score and 5 partial scores: walk, trot, canter, submission and general impression) and 31 LT to estimate the genetic parameters. Also a PLS (partial least square) regression analysis was used over the LT and PT breeding values applying reduced rank regression methodology to select the LT most predictive over PT. According to PLS results the LT variables more related with dressage performance were head length, shoulder length, forearm length, croup length, frontal angle of knee, muscular development, hip-stifle length, buttock-stifle length and lateral hock angle. These 9 traits were included in the selection indexes joint to PT. The genetic response of walk, canter, submission and general impression was lower using the LT as criteria than using PT (50, 3, 11 and 23%, respectively), but higher for trot (28%). Our results suggest that it is possible to preselect the individuals using the conformation traits as animals to be trained in dressage and in a second step select animals using the dressage performance directly. This could optimize the breeding program in terms of time and money.

New variance components for breeding values in the Franches-Montagnes horse breed

C. Flury[1], A. Burren[1], C. Hagger[1], E. Bangerter[1], M. Neuditschko[2], J. Häring[3], S. Klopfenstein[3] and S. Rieder[2]
[1]Bern University of Applied Sciences, School of Agricultural, Forest and Food Sciences HAFL, Länggasse 85, 3052 Zollikofen, Switzerland, [2]Agroscope, Swiss national Stud Farm, Les Longs Prés 191, 1580 Avenches, Switzerland, [3]Schweizerischer Freibergerverband, Les Longs Prés 190, 1580 Avenches, Switzerland; christine.flury@bfh.ch

In 2006 the Franches-Montagnes association introduced estimated breeding values (EBVs) in their breeding program. The EBVs for 28 conformation traits (CT), 12 field test traits (FT) and three traits describing the proportion of white markings (WM) are estimated annually with a BLUP-multiple trait animal model. The amount of phenotypic records has more than doubled since the introduction of routine genetic evaluation. Therefore, variance components (VC) were re-estimated and compared with previous estimates. Furthermore, the effect of the new VC on the ranking of 156 breeding stallions was evaluated. Except for 'height at withers' all other CT-traits showed low to moderate heritabilities. The heritabilities for the FT traits varied from 0.04 ('riding behavior – mounting and dismounting') and 0.32 ('trot under rider'). The heritabilities for WM were high and ranged between 0.56 and 0.73. The changes in the heritabilities for the three trait complexes were generally small. Some of the traits (e.g. 'type' and 'body conformation') showed genetic correlations close to 1. For such traits the phenotypic recording scheme could be reduced without losing relevant genetic information. Depending on the trait the new VC led to changes in the ranking of the active breeding stallions. However, the correlation coefficient of the ranks for the total breeding index was high (0.976). With the implementation of the new VC the improved underlying data is taken into account for routine genetic evaluation.

Identification of unbalanced aberrations in the genome of horse sarcoid cells using CGH technique
M. Bugno-Poniewierska[1], B. Staroń[2], L. Potocki[2], A. Gurgul[1], C. Koch[3], K. Mählmann[3] and M. Wnuk[2]
[1]National Research Institute of Animal Production, Krakowska 1, 32-083, Poland, [2]Centre of Applied Biotechnology and Basic Sciences, Univerity of Rzeszow, Sokolowska 26, 36-100 Kolbuszowa, Poland, [3]Equine Clinic, University of Bern, Länggass-Strasse 124, 3012, Bern, Switzerland; monika.bugno@izoo.krakow.pl

Sarcoid is the most common skin cancer in horses. The etiology of the tumor is associated with BPV infection (BPV-1, -2, -13), which is an inducer of malignant transformation. The comparative genomic hybridization (CGH) technique identifying the unbalanced chromosome aberrations was used to analyze the genome of horse sarcoid cells and to diagnose the chromosome rearrangements involving large deletions or amplification. The results were based on the analysis of 100 metaphases and their kariograms as well as the averaged diagram showing the ratio of green to red fluorescence, using MetaSystems software (Isis). Based on a comparison of of the fluorescence intensity ratios we found duplication in the subtelomeric regions of chromosomes pairs 1, 4, 7, 8 and 23. Duplicated region of chromosome pairs 1 also included the coding region of the rDNA. In the chromosome 23 next to duplication occurring in the centromeric region of q arm (23q11) we also found the presence of deletions involving 23q18-23q19 region. For the pairs of chromosomes from 25 to 31 and the X chromosome software failed to generate CGH diagram, but on the individual kariograms we were able to observe fluorescence signals characteristic for duplication (red), in rDNA regions of chromosome pairs 28 and 31. The study showed that duplications of DNA present in the sarcoid cells are found mainly in the telomeric and rDNA regions. The presence of the duplication of telomeric regions is associated with increased activity of the telomerase enzyme, which is a hallmark of cancer cells, affecting the immortality of these cells. Accordingly, duplications of rDNA coding regions increase activity of nucleolar organizer region which is a tumor marker.

Association analysis of SNPs in MSTN, PDK4 and DMRT3 genes with sporting performance in Trotters
S. Negro[1], M. Valera[2], A. Membrillo[1], M.D. Gómez[2], G. Anaya[1] and A. Molina[1]
[1]University of Córdoba, Edf. Gregor Mendel. Ctra. Madrid-Córdoba, km 396ª, 14072, Córdoba, Spain, [2]University of Seville, Ctra. Utrera, Km 1, 41013, Seville, Spain; z12neras@uco.es

Molecular genetic research in horses has identified several genetic markers to support selection process by allowing to preselect animals most likely to get good results in future competitions. The purpose of the study was to analyze the previously identified polymorphisms of five candidate genes and their possible association with sporting performance in the Spanish Trotter Horse population. Based on their genetic potential for long (endurance) and short (sprinting ability) distance races, 116 horses were selected and genotyped for 5 candidate genes related with sporting performance in horses: MSTN (Myostatine), COX4I2 (Cytochrome C oxidase, subunit IV, isoform 2), PDK4 (Pyruvate Deshydrogenase Kinase, isozyme 4), DMRT3 (Doublesex and mab-3 related transcription factor 3) and CKM (Creatine kinase muscle) genes. An association analysis based on ridge and logistic regression was performed revealing that trotters with genotype TT of g.66493737C>T MSTN SNP are better suited for both, short (P=0.0039) and long (P=0.019) distance races. The AA genotype of DMRT3_Ser301STOP DMRT3 SNP and the GG genotype of g.38973231A>G PDK4 SNP were associated with superior racing ability in short-distance racehorses (DMRT3, P=0.0036; PDK4, P=0.045). Due to the genetic origins of the Spanish Trotter Horse, the results of this study could be relevant for different trotter populations. The screening of the selected SNP markers using a molecular genetic tool like a SNaPshot assay could serve as a very economic tool to preselect trotters for racing competitions. Furthermore, their use may improve selection of future breeding stock by combining information of molecular genetics and sporting results.

Heritabilities of field-recorded fertility traits in Swedish horse breeds

S. Eriksson and Å. Viklund
Swedish University of Agricultural Sciences, Department of Animal Breeding and Genetics, P.O. Box 7023,
75007 Uppsala, Sweden; asa.viklund@slu.se

Horse breeding is seasonal and keeping mares and stallions is costly. It is economically important that mated mares become pregnant and give birth to live foals. Fertility is mentioned in breeding goals, and mating and foaling outcomes are routinely recorded in Sweden. This information has so far not been utilized for genetic analyses or evaluation. The aim of this study was to estimate genetic parameters for field-recorded fertility traits in Swedish Warmblood riding horses (SWB) and Swedish Standardbred and Coldblood trotters. Seasonal data of pregnancy rate and foaling rate were available since 1984 for trotters and since 2000 for SWB. After editing the data contained in total 4,847 stallion-season combinations. Information on all mating dates was recorded since 2000 for trotters and since 2010 for SWB, and in total the edited data comprised 62,656 mare-oestrus observations. Measures of non-return to service after mating, length of insemination period, and foaling outcome were treated in separate analyses as traits of the mare, stallion or expected foal. Pregnancy rate and foaling rate per season were analysed as traits of stallions. Fixed effects of reproductive technique, month of mating, age of mare and stallion, and inbreeding coefficients of foals or stallions were significant. Heritabilities were estimated using linear animal models. Estimates for pregnancy rate and foaling rate per season for stallions were in the range 0.02-0.18. Heritabilities estimated for non-return, foaling outcome and length of insemination period were low, but tended to be higher when analysed as traits of the foal (0.00-0.10) compared with as traits of the mare (0.01-0.04) or stallion (0.00-0.01). The results show that there is a genetic variation in fertility traits in these horse breeds, but that the environmental influence is strong. As most stallions are mated to rather few mares, generally low accuracies of breeding values for fertility traits could be expected.

Welfare, environmental impact and production – conflicting pig breeding goals

L. Rydhmer, T. Mirkena, T. Ahlman and A. Wallenbeck
Sw Univ Agric Sciences, Dept Animal Breeding Genetics, Box 7023, 75007 Uppsala, Sweden;
lotta.rydhmer@slu.se

Economic weights are calculated to achieve high profit for producers and adjusted for societal demands. Improved animal welfare and decreased environmental impact are demands for pig production. Unfavourable genetic correlations create conflicts between breeding goals focused on short term profit and welfare or environmental impact. We used a questionnaire to ask 120 producers which traits they associate with welfare and environmental impact. Based on their answers, genetic parameters and population data we built the alt. breeding programs Welfare and Environment, and compared them with Current. The simulation program ZPLAN was used to calculate genetic progress. In Welfare, disease and parasite resistance, leg health, shoulder ulcers, longevity and piglet survival were allotted 50% of the total economic weight and current weights for growth, feed conversion, meat percentage, fertility and litter size were proportionally reduced. In Environment, growth, feed conversion and disease resistance were allotted 50% of the total economic weight and current weights for piglet growth, meat percentage, leg health, fertility and litter size were proportionally reduced. Compared with Current, yearly genetic progress in Environment was higher for growth (8 vs 23 g/d), feed conversion (0.3 vs 0.4 g/MJ ME) and healthy pigs (0.1 vs 0.4%). This progress was accompanied by negative changes in litter size (0.01 vs -0.01 born alive) and meat percentage (0.1 vs -0.1%). Compared with Current, genetic progress in Welfare was higher for healthy pigs (0.1 vs 0.5%) and pigs free from parasites (0.1 vs 0.5%) and longevity (0.7 vs 2.2 d). The changes in leg health, shoulder ulcers, fertility and piglet survival were unfavourable in Current but favourable in Welfare. This progress in Welfare was accompanied by reduced progress in growth (8 vs 2 g/d) and feed conversion (0.3 vs 0.2 g/MJ ME). Thus societal demands must be related to willingness to pay for animal and environmental friendly production.

The role of welfare-related traits in a breeding goal selecting for production efficiency in pigs

T. Aasmundstad[1], E. Grindflek[1] and O. Vangen[2]
[1]Topigs Norsvin, Research and development, P.O. Box 504, 2304 Hamar, Norway, [2]Norwegian University of Life Sciences, Department of Animal and Aquacultural Sciences, P.O. Box 5003, 1432 Ås, Norway; torunn.aasmundstad@norsvin.no

Modern pig breeds are expected to be efficient in converting feed to meat, have high fecundity and good longevity. If selection of sows for increased productivity is to be performed in a sustainable and ethical way, traits related to welfare should also be included in the breeding goal. In this study, the genetic parameters of three typical production traits (total number piglets born (TNB), lean meat yield (LMY) and growth rate (GR)) and two welfare related traits (shoulder ulcers at weaning (SUW) and osteochondrosis (OC)) were estimated. The dataset analysed consisted of phenotypes from 365,079 purebred Landrace pigs (boars, gilts and sows). GR (n=352,521) was expressed as weight at 150 days old gilts, TNB (n=41,221) and SUW (n=41,191) were expressed in first parity sows, and LMY (n=9,435) and OC (n=5,448) were expressed in 120 kg boars. The results from the five-trait variance component estimation analysis revealed the following heritabilities: h^2_{TNB}=0.08, h^2_{LMY}=0.50, h^2_{GR}=0.31, h^2_{SUW}=0.16, h^2_{OC}=0.34. The genetic correlations in general were low: $r_{g,SUW, TNB}$=0.01, $r_{g,GR, TNB}$=-0.02, $r_{g,LMY, TNB}$=0.02, $r_{g,OC, TNB}$=-0.04, $r_{g,GR, SUW}$=-0.01, $r_{g,LMY, SUW}$=0.13, $r_{g,OC, SUW}$=0.15, $r_{g,LMY, GR}$=-0.39, $r_{g,OC, GR}$=0.40, $r_{g,OC, LMY}$=-0.00. The correlations between lean meat yield and growth rate, and between growth rate and osteochondrosis, were both unfavourable with regard to the desired direction of selection. In this analysis we did not find any significant and unfavourable correlations between shoulder ulcers and production traits. Still, in a breeding goal selecting for increased production, welfare-related traits should be included to observe and counteract unfavourable genetic trends due to genetic relationship between the traits.

Prevalence and risk factors for lesions and lameness in replacement and pregnant gilts

A. Quinn[1], L. Green[2], A. Kilbride[2] and L. Boyle[1]
[1]Teagasc, Pig Development Department, Moorepark, Fermoy, Co. Cork, Ireland, [2]University of Warwick, School of Life Sciences, Coventry, CV4 7AL, United Kingdom; laura.boyle@teagasc.ie

The aim was to determine the prevalence and risk factors for limb lesions and lameness in young sows. Farms (n=68) with >100 sows were visited once in 2012. One pen of replacement gilts and 1 pen or 10 stalls of pregnant gilts were randomly selected. Every gilt was examined if there were ≤10 gilts/pen and 10 gilts were randomly selected and examined if there were >10 gilts/pen. Limbs were examined for: scratches, wounds, swellings, abscesses, calluses and bursitis and scored from 0 to 3 on the basis of size. Locomotory ability was scored from 0 to 5 ('lame'=score ≥2). Materials and dimensions were recorded and a questionnaire was answered by the farmer. Data were analysed using MlwiN 2.27. 518 replacement and 525 pregnant gilts were inspected. The former were in groups (median=10, IQR 8-13) while 335 of the pregnant gilts were in 48 pens and 190 were in stalls. The prevalence of lesions in replacement and pregnant gilts respectively was: scratches (78.1%, 62.7%), wounds (34.7%, 28.8%), swellings (30.3%, 20.2%), abscesses (0.19%, 0.20%), calluses (100%, 99.8%) and bursitis (25.0%, 19.3%). Lameness affected 38.9% of replacement and 41.1% of pregnant gilts. No risk factors associated with flooring were identified for lameness or limb lesions (91.4% of replacement and 99.2% of pregnant gilts were on slatted floors without bedding). The odds of lameness were higher in pregnant gilts in groups (48.1%) compared to stalls (30.4%) (OR 3.66, CI 1.23-3.66). The risk of swellings was higher in replacement gilts separated from finishers at weights of >90 kg compared with those separated at <50 kg (OR 3.1, CI 1.6-7.7). High levels of limb lesions and lameness in young sows is related to the widespread use of slatted flooring. Preferential treatment of replacement gilts could help reduce culling. There is a higher risk of lameness in groups compared to stalls though use of the latter is now prohibited.

Potential genomic biological framework of longevity, osteochondrosis and culling reason in sows

E.M. Van Grevenhof[1], C.A. De Hollander[1], E.F. Knol[2] and H.C.M. Heuven[1]
[1]Wageningen University, Animal Breeding and Genomics Centre, P.O. Box 338, 6700 AH Wageningen, the Netherlands, [2]Topigs-Norsvin, P.O. Box 43, 6640 AA Beuningen, the Netherlands; ilse.vangrevenhof@wur.nl

In the last 10 years, osteochondrosis (OC) potentially influencing longevity has increased in importance in pig breeding. Although, the breeding goal is to improve leg quality and longevity in crossbred (CB) sows, the selection occurs in purebred breeding herds. Genomic selection increases possibilities for selection on CB performance. Sow data, e.g. number of parities, OC and culling reason, are traits of interest and could decrease economic losses and improve welfare. In literature, the correlation of OC with lameness is debated. Culling reason, when recorded in sow management systems, is often unreliable since recording is absent or not highly prioritized. The aim of the study is to distinguish (in)voluntary culling reasons to assess the distribution of the interval from last insemination to culling (IL2C) by estimated genetic parameters and relationships between this interval, OC and longevity. Data, originating from a research farm and commercially culled sows, is collected on culling reason, longevity and dissected joint scores on OC. Additionally, genotypes from 900 CB individuals were collected from culled commercially kept sows. DNA data, dissected joints and sow management data is combined to clarify the genomic biological relationships among traits. Data showed OC is highly prevalent in sows, up to 92%. Preliminary results from statistical analyses showed sows culled for claw problems or trauma had significantly more severe OC compared with sows culled for lameness, production or reproduction. Sows culled for production were culled ~146 d after insemination enabling sows to wean the piglets, while locomotion problems result in culling ~127 d. If OC, which develops during early youth, results in claw problems or trauma and eventually in (early) culling, selection to decrease these high prevalences of OC rises in priority.

Static aggregation of a pig trade network in Northern Germany compared to its temporal counterpart

K. Büttner and J. Krieter
Institute of Animal Breeding and Husbandry, Christian-Albrechts-University, Olshausenstr. 40, 24098 Kiel, Germany; kbuettner@tierzucht.uni-kiel.de

The static aggregation of trade contacts neglects temporal variation in the system. In terms of disease spread, ignoring temporal dynamics of contact structures can lead to over- or underestimation of the speed and extent of an outbreak. This becomes particularly evident if the static aggregation allows for the existence of more paths compared with the time-respecting paths in the temporal networks (i.e. contacts have to be in the right chronological order). Thus, the aim of the study was to reveal differences between temporal and static representation, and assess the quality of the static aggregation compared with the temporal counterpart. From 2006 to 2009 contact data from a pig trade network in Northern Germany were analysed. The data contain information on 4,635 animal movements between 483 farms. A median value of 8.7% (4.6-14.1%) of the nodes and 3.1% (1.6-5.5%) of the edges were active on a weekly resolution. Causal fidelity measures the fraction of paths in a static network which can also be taken in the temporal counterpart. For the whole observation period, 3,005 static and 1,999 time-respecting paths could be obtained, yielding a causal fidelity of 0.67 (i.e. 67% of the time-respecting paths exist in both network representations). For the yearly and monthly aggregation windows, an increase in the causal fidelity to 0.87 (0.82-0.88) and 0.92 (0.8-0.98), respectively, could be observed. The results of the causal fidelity showed that the static aggregation of the pig trade network under investigation captures its temporal characteristics sufficiently well and increased with shorter aggregation windows. Thus, it becomes possible to quantify the simplification made by the static aggregation and to evaluate if the methodology of the static network analysis can be used instead of performing the temporal approach without losing too much information.

Inactivation of porcine epidemic diarrhea virus by heat-alkalinity-time pasteurization

G.V. Quist-Rybachuk[1], H. Nauwynck[1] and I.D. Kalmar[2]
[1]Ghent Univ., Lab. of Virology, Salisburylaan 133, 9820 Merelbeke, Belgium, [2]Veos, Akkerstr 4, 8750 Zwevezele, Belgium; Isabelle.Kalmar@veos.be

Adequate processing of potentially contaminated raw material before use as feedstuff is a crucial aspect in the prevention of feed-borne transmission of porcine epidemic diarrhea virus (PEDV). Spray-drying of plasma and storage at dried conditions were demonstrated to inactivate 4.2 $\log_{10}$ PEDV/ml and up to 2.8 $\log_{10}$ PEDV/g, respectively. We examined the extent of PEDV inactivation by heat-alkalinity-time (HAT) pasteurization, an additional safety step in the production of spray-dried porcine plasma (SDPP). Neutral (pH 7.2) or alkalinized (pH 9.2 or 10.2) cell culture medium (MEM) and plasma samples were mixed with PEDV CV777 to a final titer of 5.5 $\log_{10}$ TCID$_{50}$/ml. Aliquots (n=3) were incubated for up to 120 min at 4, 40, 44 or 48 °C, and the residual virus infectivity was determined via standard endpoint dilution assay on Vero-Ba cells. Decimal reduction time (D-value) of the respective treatments was calculated as the negative inverse of the slope of the semi-logarithmic survival curve. The upper 95% confidence level (UCL95) was calculated as: $\mu + t_{\alpha/2,\,n-1} \sigma/\sqrt{n}$, with n=3 and $t_{\alpha/2,\,n-1} = 4.303$. Data are presented as mean [UCL95]. Irrespective of presence of plasma, PEDV was not sensitive to pH 7.2 to 10.2 at 4 °C. At neutral pH and in absence of plasma, $D_{48\,°C,pH\,7.2}$ was 24.9 [45.5] min. This confirms moderate susceptibility of PEDV to heat treatment at 48 °C. At pH10.2, however, $D_{48\,°C,pH\,10}$ in MEM was reduced to 81 [114] s. Moreover, presence of at least 90% porcine plasma strongly increased sensitivity of PEDV to heat and pH treatment, resulting in $D_{48\,°C,\,pH7.2} = 2.8$ [6.7] min and $D_{48\,°C,\,pH10.2} = 20$ [35] sec. Calculated $H_{48\,°C} A_{pH10.2} T_{10min}$ inactivation of PEDV in plasma and MEM were 17.4 and 5.3 $\log_{10}$ TCID$_{50}$/ml, respectively. Depending on the HAT-conditions, peak loads of PEDV that could potentially be present in abattoir collected porcine blood can be completely inactivated by HAT-pasteurisation.

Effects of milk yield on animal welfare in dairy cattle

D. Gieseke[1], C. Lambertz[2] and M. Gauly[2]
[1]Georg-August-University, Department of Animal Science, Albrecht-Thaer-Weg 3, 37075 Göttingen, Germany, [2]Free University of Bozen-Bolzano, Faculty of Science and Technology, Universitätsplatz 5, 39100 Bolzano, Italy; dgiesek1@gwdg.de

Over the last decades the milk yield of dairy cows increased continuously. Due to antagonistic relationships between production and functional traits, high milk yield is presumed to be a risk factor for health and welfare. However, it is still unknown whether there is a direct correlation between the level of milk yield and these factors in dairy herds. Therefore, the aim of this study was to describe the animal welfare status on 40 conventional dairy cattle farms with different production levels (low: <9,000 kg energy-corrected milk (ECM); medium: 9,000-10,000 kg ECM; high: >10,000 kg ECM) using animal-based measures from the Welfare Quality® Assessment protocol for dairy cattle and official milk recording data. A mixed model was performed to investigate significant differences between the three yield classes. The body condition score (% lean cows) and the mastitis incidence (% SCC>400,000) improved in tendency with increasing milk yield (P>0.05). Diseases like milk fever, vulvar discharge, coughing or diarrhea were found in all three classes in a similar prevalence. In contrast, severe lameness was observed more often in farms with a low (15.9%) and high production level (11.5%) when compared with farms with a medium production level (7.1%) (P=0.04). Severe integument alterations, i.e. lesions and swellings were detected in each category. The numbers were independent of the production level (P>0.05). The overall welfare score was not significantly different between the three yield classes (low: 34.1, medium: 39.0, high: 39.8). In conclusion, milk production level did not directly affect the welfare status of dairy herds. Accordingly, milk yield is not a feasible indicator for the welfare status in dairy cattle.

The use of reticulo-rumen boluses for early fever detection in young male calves

B. Tietgen[1], H.J. Laue[2], M. Hoedemaker[3] and S. Wiedemann[1]
[1]Institute of Animal Breeding and Husbandry, Kiel University, Animal health, Hermann-Rodewald-Str 8, 24118 Kiel, Germany, [2]University of Applied Science Kiel, Germany, Animal Feeding, Grüner Kamp 11, 24783 Osterrönfeld, Germany, [3]University of Veterinary Medicine Hannover, Foundation, Clinic for Cattle, Bischofsholer Damm 15, 30173 Hannover, Germany; swiedemann@tierzucht.uni-kiel.de

In young calves the bovine respiratory disease (BRD) is one of the most important diseases because it largely affects their welfare and performance. An early and reliable detection of the first symptoms of BRD such as fever is of vital importance. The objective of this retrospective study in 150 very young male fattening calves (16.6±3.3 d) was to investigate the feasibility of a bolus which measures the reticulo-ruminal temperature (rerutemp) for the early detection of fever. During the experimental period of 8 weeks an adspection of health was performed at least twice a day and rectal temperature (retemp) was measured in calves showing signs related to BRD. Additionally, retemp was routinely measured in all animals every two weeks. Over the experimental period 99 calves were detected with reticulo-ruminal hyperthermia (reruhyp: rerutemp ≥40 °C) over at least one daytime period. In 31 animals rectal hyperthermia (rehyp: rectemp ≥40 °C) occurred. The average time interval between first reruhyp and first detection of rehyp was 82.9±70.7 h. CUSUM-control charts were further tested for possible application as an alarm system. The sensitivity (Se) and specificity (Sp) of reruhyp to detect rehyp were assessed. The Se of the rerutemp daytime period data was 80% and the Sp was 98%, whereas the CUSUM test revealed a Se of 76% and a Sp of 97%. In conclusion, the reticulo-ruminal temperature measuring bolus is a useful tool for the very early detection of fever in young calves. In this regard, the CUSUM-control chart can be implemented into an alarm system.

Participative monitoring of the welfare of veal calves

V. Pompe and M. Ruis
Van Hall Larenstein University of Applied Sciences, Animal Welfare Group, Post box 1528, 8901 BV Leeuwarden, the Netherlands; vincent.pompe@wur.nl

Welfare monitoring is in development. In the Netherlands several monitoring projects, based on Welfare Quality®, are in the stage of testing the practicality in order to reach broad implementation. 'Welfare Monitoring of Veal Calves' is one of these projects. The project's basic assumption is that Effect (improved welfare) = Quality (science) × Acceptance (farmers). This means that science must meet the social-psychological dimensions of the farmer in the role of entrepreneur, livestock keeper and stockman in order to make welfare monitoring work. Last summer, a number of Dutch veal calves farmers participating in the project were qualitatively interviewed about the economic and managerial aspects of welfare improvement on the farm and the pros and cons of the welfare monitor. The overall positive expectations of the welfare monitor for the veal sector contrast some reservations about the success at farm level. In some aspects science appears to be too remote and business revenues are not in view. A conclusion of the project is a participatory model in which farmers have the possibility to challenge the observation of the welfare assessor. Stimulating this option, famers will get more involved in science and science can be pushed to enrich, refine or reduce welfare indicators. Quality and acceptance become interactive. This project emphasises the need of social (business) science for the success of natural science.

A participatory training approach for farmers to improve pain release during disbudding of calves

A. Brule[1], M. Le Guenic[2], A. De Boyer Des Roches[3], B. Mounaix[1] and L. Mirabito[1]
[1]Institut de l'Elevage, Santé et bien-être des ruminants, 149 rue de Bercy, 75596 Paris cedex, France,
[2]Chambre d'Agriculture Bretagne, Pôle Herbivore, Avenue Général Borgnis Debordes, 56002 Vannes,
France, [3]INRA, UMR1213 Herbivores, Vetagrosup, 63122 Saint-Genès-Champanelle, France;
beatrice.mounaix@idele.fr

A new training program has been elaborated to support farmers in the management of pain during iron-disbudding of calves. This training program was designed in 2013-2014 during a multi-partner project involving researchers in animal science, farmer technicians, teachers, veterinarians. It combines theoretical (e.g. what is pain? how can I reduce it?) and practical training objectives that encourages participative collective expression of farmers to better involve them and to evolve individual attitudes regarding pain management. The originality of this program lies on: (1) It takes into account farmers, veterinarian and technicians representations of animal pain (described in an early step of the project by face to face interviews of farmers and in focus groups, to adapt the training take-home messages according to the different trainee attitudes. (2) It presents experimental results of pain management protocols tested in the project to better explain how farmers can detect pain, how they can reduce it and what are the zootechnical effects of pain management. (3) The practical part involves the trainer and a local veterinarian. This collaboration enhances the change in practice and guarantees long term improvement of pain management on farm. (4) The training also includes a tool for farmers to self-assess their own practices regarding calves disbudding and to monitor individual progress. The performances of this training program have been tested on 40 beef and dairy farms in 2014-2015. The assessment consisted in observation of farmer practices and on interviews pre and post training. Results will be available by June 2015.

Effects of water deprivation on blood components and behaviour in sheep and goats

C. Lambertz[1], M. Hinz[2] and M. Gauly[1]
[1]Free University of Bolzano, Faculty of Science and Technology, Universitätsplatz 5, 39100 Bolzano,
Italy, [2]Georg-August-University, Department of Animal Sciences, Albrecht-Thaer-Weg 3, 37075 Göttingen,
Germany; christian.lambertz@unibz.it

According to EU regulations, sheep and goats can be transported for 14 hours, followed by one hour's rest plus the offer of water, followed by a further 14 hours of travel. However, authorities need tools to confirm that animals had access to water. Therefore, the aim was to study the effects of prolonged water deprivation on blood and behavioural variables of goats and sheep and to evaluate variables for practical use to determine if transport regulations are fulfilled. In total, 15 goats and 15 sheep were assigned to 1 of 3 pens with the water access treatments: AL (*ad libitum*), S (14 h withdrawal – 1 h access – 14 h withdrawal) or L (29 h withdrawal). Each animal received each treatment once. Transport conditions were mimicked (i.e. space allowance). Haematological variables (i.e. packed cell volume (PCV), haemoglobin, red and white blood cells, total protein and serum Na) were analysed at 0, 14, 15, and 29 h. The behaviours standing, lying and eating were recorded, and drinking bouts counted in AL and S (14 to 15 h). L-sheep had lowered PCV compared with AL- and S-sheep at 14 h ($P<0.05$). Na was higher in S- and L- than in AL-sheep at 14 h ($P<0.05$); and higher in L-sheep than in both other groups at 29 h ($P<0.05$). Haemoglobin was higher in S- at 14 and 15 h and in L- at 29 h compared with AL-goats ($P<0.05$). Na increased from 0 to 14 h in S- and L-goats ($P<0.05$). Total eating time was >20 min/h during 0 to 14 h and ~5 min/h during 15 to 29 h ($P<0.05$). The number of drinking bouts in AL-sheep averaged 3.9±1.9 and 3.1±1.1 during 0 to 14 and 15 to 29 h. From 14 to 15 h, S-sheep drank 3.4±1.9 times. The accordant values for AL-goats during 0 to 14 and 15 to 29 h were 8.1±2.6 and 2.3±1.3. S-goats drank 2.9±1.4 from 14 to 15 h. In conclusion, the tested variables showed only limited potential to monitor the fulfilment of transport regulations.

Breed effects and genetic parameters for early live weight and tick count in sheep

K. Thutwa[1,2], J.B. Van Wyk[2], S.W.P. Cloete[3,4], A.J. Scholtz[3], J.J.E. Cloete[4,5] and K. Dzama[4]
[1]Botswana College of Agriculture, P/Bag 0027, Gaborone, Botswana, [2]University of the Free State, Department Animal Wildlife & Grassland Sciences, P.O. Box 339, Bloemfontein 9300, South Africa, [3]Directorate Animal Sciences, P/Bag X1, Elsenburg 7607, South Africa, [4]Stellenbosch University, Department of Animal Sciences, P/Bag X1, Matieland 7602, South Africa, [5]Structured Agricultural Education & Training, P/Bag X1, Elsenburg 7607, South Africa; vanwykjb@ufs.ac.za

Ticks may impair sheep productivity and welfare in a free-range system, transmitting diseases, causing skin damage, lameness, paralysis and udder damage. Resistance in pathogens to chemicals and environmental concerns hamper the long-term sustainability of chemical control measures used at present. Alternative ways of tick control are needed in an integrated control programme. Birth weight (BW), weaning weight (WW) and tick count of Namaqua Afrikaner (NA), Dorper (D) and South African Mutton Merino (SAMM) lambs at the Nortier Research Farm were analysed. Crosses between breeds were also studied: NA rams mated to D ewes, reciprocal cross between the D and SAMM and backcrosses of the latter two genotypes on D or SAMM sires. Animal numbers ranged from 1016 to 1038. NA lambs had lowest BW and WW compared with the other breeds and crosses. NA and NAxD lambs had the lowest tick count of all breed combinations. The heritability (h^2) and dam permanent environmental variance (c^2) for tick count were respectively 0.13±0.06 and 0.07±0.03 in a three-trait analysis with birth and weaning weights. Genetic parameters for weights accorded with the literature. When the effect of breed was excluded to study across-breed genetic variation, values of respectively 0.33±0.07 and 0.06±0.04 were derived for h^2 and c^2 of tick count. There were positive genetic and phenotypic correlations between weights and tick count. Resistance to tick infestation is thus influenced by genetics in sheep. Selecting breeds which are more resistant to ticks will improve tick resistance more in the short term than selecting individuals within breeds.

A *Brachyspira hyodysenteriae* challenge model in pigs: effect of body weight and challenge strain

J.G.M. Wientjes, M.E. Sanders, R. Dijkman, K. Junker, M.M.M. Meijerink and P.J. Van Der Wolf
GD Animal Health, Arnsbergstraat 7, 7418 EZ Deventer, the Netherlands; a.wientjes@gddiergezondheid.nl

Dysentery caused by *Brachyspira hyodysenteriae* (B. hyo) causes economic and welfare problems at pig farms and is associated with high antibiotic use. There is a need for alternative interventions against B. hyo. In order to evaluate effects of interventions, a suitable B. hyo challenge model in pigs needs to be developed. First we studied effects of pig weight (50 vs 10 kg) and B. hyo strain (ATCC B204 vs field strain) on clinical signs and fecal shedding in challenged pigs. Male pigs were assigned to A50 (ATCC B204, 50 kg, n=4), F50 (field, 50 kg, n=4) or F10 (field, 10 kg, n=8) and inoculated for 6 consecutive days with $\geq$2.8×10^9 cfu B. hyo by oral ingestion. Fecal quality was recorded daily and rectal fecal samples were taken before (d-4, 0) and after inoculation (d10, 14, 17, 21) for B. hyo detection by qPCR. Feces quality remained normal until d13 after inoculation, but suddenly changed to bloody and/or slimy diarrhea in 3/4 pigs of A50 (at d14 (euthanized(†) at d15), 18 and 24 resp.), in 1/4 pigs of F50 (at d14; † at d14) and in 3/8 pigs of F10 (at d15 († at d15), d16 and d16 resp.). Fecal samples before inoculation were all negative for B. hyo. For A50, fecal samples were positive for B. hyo in 0/4 pigs on d10, 1/4 pigs on d14, 1/3 pigs on d17 and 2/3 pigs on d21. For F50, positive PCR results were found in 1/4 pigs on d10, 3/4 on d14, 3/3 on d17 and d21. For F10, fecal samples were positive for B. hyo in 3/8 pigs on d10, 6/8 on d14, 5/7 on d17, 7/7 on d21. To conclude, we successfully orally infected pigs with ATCC strain B204 resulting in dysentery and fecal shedding in 75% of the 50 kg pigs. The B. hyo field strain resulted in dysentery in 25% of the 50 kg pigs and 37.5% of the 10 kg pigs, and in fecal shedding in all pigs. Results indicate a similar colonization capacity, but lower pathogenicity of the field strain compared with the ATCC B204 strain, and a comparable response in 10 kg and 50 kg pigs for the field strain.

Genetic parameters for mitral valve disease in the cavalier King Charles spaniel

K. Wijnrocx[1], S. Janssens[1], S. Swift[2] and N. Buys[1]
[1]KU Leuven, Biosystems, Kasteelpark Arenberg 30, 3001 Leuven, Belgium, [2]University of Florida, College of Veterinary Medicine, Small Animal Clinical Sciences, P.O. Box 100126, 2015 SW 16th Ave, Gainesville, FL 32610-0126, USA; katrien.wijnrocx@biw.kuleuven.be

Mitral valve disease (MVD) is the commonest cardiac disease in dogs and is of particular concern in the cavalier King Charles spaniel (CKCS) with high prevalence and early onset. The disease is characterized by thickening of the mitral valve leaflets, which over the years can result in congestive heart failure. Diagnosis can be made by auscultation (presence of a left apical murmur and assessment of murmur grade 1-6) or Doppler echocardiography (assessment of mitral valve leaflet abnormalities and blood flow, score from A-E). There is evidence that MVD is a multifactorial, polygenic disease, with prevalence ranging from 20% to 90%, depending on the population under study. Lewis et al. estimated heritabilities of 0.33-0.67 on a set of 1,250 auscultation records of 4- to 5-year old CKCS in the UK. Because of the exchange of breeding animals, international genetic evaluation is needed. The study objective was to estimate genetic parameters in an international population of the CKCS. Parameters were estimated using both a linear model and a threshold model, allowing comparison between methods. Data were obtained via the Cavaliers for Life project that joins pedigree data and health screening results from CKCS registered in 6 countries. Data comprise 21,000 auscultation scores and 510,000 pedigree files from UK, Denmark, the Netherlands, Belgium, France and Germany. Variance component estimates were obtained using the REML method (VCE6) by means of a linear animal model. Results of the linear model indicate a moderate degree of heritability (0.11). Both age and veterinarian that performed the test were important non-genetic factors. Comparison with the results of the threshold model will be performed.

Reducing the effect of pre-slaughter fasting on the flesh quality of rainbow trout

R. Bermejo-Poza[1], J. De La Fuente[1], C. Pérez[2], S. Lauzurica[1], E. González[1], M.T. Díaz[3] and M. Villarroel[4]
[1]UCM, Avda Puerrta de Hierro s/n, 28040, Spain, [2]UCM, Avda Puerta de Hierro s/n, 28040, Spain, [3]INIA, Ctra Coruña km 7.5, 28040, Spain, [4]UPM, Avda Puerta de Hierro 2, 28040, Spain; rbermejop89@gmail.com

Fasting is a common practice in aquaculture in slaughter previous days to ensure emptying of the digestive system. This study was conducted to evaluate the effect of different intermittent feeding schedules during the last month of fattening and fasting prior to slaughter on flesh quality of rainbow trout (*Oncorhynchus mykiss*). 240 trout were separated into three groups with different final feed (daily, every two days or every four days) and two durations of pre-slaughter fasting (two days of fasting-24.3 degree days, to nine days of fasting-102 degree days). After slaughter a number of biological (slaughter weight, relative growth, coefficient of condition, stomach content, carcass yield, digestive, hepato-somatic indexes) and flesh quality (muscle pH, angle of rigor mortis) parameters were measured. The parameters were analyzed using the SAS software. If the interaction between effects (intermittent feeding, days of fasting and hours post-mortem) was significant, planned comparisons were performed using Bonferroni test ($P<0.05$). Fish fed once every two days and subjected to two days of fasting presented the higher slaughter weight (398 ± 6.29 g) and relative growth ($22.5\pm2.0\%$) (better production). Regarding the flesh quality, the final pH and rigor angle was higher in the flesh of fish subjected to nine days of fasting than two days fasted (6.71 ± 0.01 vs $6.29\pm0.01°$ and 89.81 ± 0.53 vs $85.87\pm0.52°$, respectively) (worst quality). The value of hepato-somatic index was higher in fish fed once every two days after two days of fasting ($18.73+0.92$ and 2.06 ± 0.12, respectively), suggesting that they had lower energy reserves modification and therefore flesh quality could be less affected. Trout fed with every two days feeding frequency subjected to two days of fasting showed better flesh quality without affecting production.

Investigation organic, semi organic and conventional system on blood parameters and oocyst coccidia
A. Salahshour[1], R. Vakili[2] and A. Foroghi[1]
[1]Jihad-e-Agriculture Higher Educational Complex, Organic Animal Husbandry, TV Square, Asian Highway, Mashhad, Iran, [2]Islamic Azad University, Kashmar Branch, Animal Science, Seyed Morteza blvd, Mashhad, Iran; amadsalahshour@yahoo.com

The aim of this study was to investigate the impact of production system (organic, semi organic and Conventional) on Blood parameters and oocyst coccidia (OPG and EPG test) of hen layers. This experiment consisted of 192, eighty week old shaver, over 120 days with three treatments and four replicate. The control group was fed inorganic feed, the organic group was fed organic feed and semi organic group fed organic feed with synthetic amino acid) and vitamins. Evaluated traits were egg and plasma cholesterol, plasma triglyceride and plasma HDL. Production system has no significant effect of cholesterol ($P>0.171$) and triglyceride of plasma ($P>0.838$), but lowest level of them observed in organic and semi organic groups. Treatments has significant effect on HDL in plasma ($P>0.044$). Production system has significant effect on OPG and EPG tests ($P<0.05$). The result of experiment showed organic and semi organic systems reduced cholesterol in yolk and serum and triglycerides in plasma and enhance HDL in serum. Also, organic system enhance oocyst coccidia in laying hens.

Genomic inbreeding coefficients based on the length of runs of homozygosity
L. Gomez-Raya, C. Rodriguez, C. Barragan, W.M. Rauw and L. Silió
Instituto Nacional de Investigaciones Agroalimentarias, Mejora Genética Animal, Carretera de la Coruña km 7, 28040, Spain; gomez.luis@inia.es

Assessment of inbreeding is relevant to estimate the genetic health of a population or individual. Novel genomic inbreeding coefficients are proposed based on the distribution of the length of runs of homozygosity (ROH): (1) Kolmolgorov-Smirnov (based on the test with the same name to compare two distributions); (2) Quantile (quantiles of the individual vs quantiles of the population cumulative distribution function (CDF)); (3) Exponential (fitting of an exponential distribution to the ROH-length distribution); and (4) Normalized exponential (probability of drawing an ROH of length of 1 Mb or more). The new inbreeding coefficients were estimated for 306 Iberian sows, with deep and complete pedigree information. The correlation between pedigree and genomic inbreeding coefficients were 0.40 (Kolmolgorov-Smirnov), 0.57 (Quantile), 0.71 (Exponential), and 0.69 (Normalized exponential). The standard errors ranged between 0.04 and 0.05. The novel genomic inbreeding coefficients make use of the probability density function of the length of ROH and represent a better description of the inbreeding level of a population or individual than pedigree inbreeding coefficients. Genomic inbreeding coefficients can be applied to individual chromosomes or groups under artificial or natural selection. Other advantages and disadvantages of the novel methods to estimate genomic inbreeding coefficients are discussed.

Analysis of resequencing data suggests that the Minipig carries complex disease alleles

C. Reimer[1], C.-J. Rubin[2], S. Weigend[3], K.H. Waldmann[4], O. Distl[4] and H. Simianer[1]
[1]University of Göttingen, Albrecht-Thaer-Weg 3, 37075 Göttingen, Germany, [2]Uppsala University, Husargatan 3, 75123 Uppsala, Sweden, [3]Friedrich-Loeffler-Institut, Höltystraße 10, 31535 Neustadt, Germany, [4]University of Veterinary Medicine, Bünteweg 2, 30559 Hannover, Germany; creimer@gwdg.de

Domestic pig breeds show a remarkable variation in size and the so called 'minipigs' represent breeds with extremely small body size, therefore being particularly suited to study the genetic background of growth. We compared whole genome resequencing (WGS) data of 46 normal sized pigs, either domestic or wild, with data of 11 Göttingen Minipigs, 2 Berlin Minipigs and a pool comprising 10 Berlin Minipigs. Putative signatures of selection were detected by combining expected heterozygosity and the composite likelihood ratio test within a breed, intersected with exaggerated fixation between both groups. Two regions, one on chromosome 5 and one on chromosome 14, with sizes of 4.5 and 9 Mb, respectively, were deemed particularly interesting. They contained in total 119 genes, including BBS10, DGCR2, DGCR14 and DGCR8, which play a role in the Bardet-Biedl-Syndrome (BBS) or the Di-George-Syndrome, respectively, in humans. Seizures, poly- and syndactylia, strabism and reduced growth are symptoms of these disorders in humans and have also been observed in the Göttingen Minipigs. Following functional annotation of SNPs with allele frequency >0.9 in the minipigs inside BBS or DGCR genes showed that DGCR8 carries one, BBS2 carries two missense mutations and in addition an initiator codon variant, all with deleterious effect. BBS10 contains only three exonic SNPs with missense effect and a highly differentiated upstream gene variant. The footprints of selection around the BBS10 and the DGCR genes together with the aforementioned altering mutations and the observation that resembling phenotypes in humans have been linked to these genes make us propose that these loci may play a role in the etiology of low body size in minipigs.

Statistical models to increase disease resilience and uniformity in animal production

H.A. Mulder[1], J.M. Herrero-Medrano[1,2], E. Sell-Kubiak[1], P.K. Mathur[2] and E.F. Knol[2]
[1]Animal Breeding and Genomics Centre, Wageningen University, P.O. Box 338, 6700 AH Wageningen, the Netherlands, [2]TOPIGS Norsvin BV, P.O. Box 43, 6640 AA Beuningen, the Netherlands; han.mulder@wur.nl

Selection for increased disease resistance and tolerance is hampered by the lack of useful data that can be obtained from commercial production systems. A practical approach would be to use in one way or another fluctuations in production of animals as indicators for resilience, i.e. the combination of resistance and tolerance. Here we show two statistical approaches: a reaction norm model (RNM) and a double hierarchical generalized linear model (DHGLM). From a theoretical perspective, RNM is more appropriate for estimating EBV for resilience to disease outbreaks and other challenges. However, RNM needs an environmental covariate, which is not needed in DHGLM. Therefore, the DHGLM could be practically useful for estimating EBV for resilience to endemic diseases. Both models were applied to reproduction traits in pigs such as total number of born piglets and number of piglets born alive in sow lines. For the RNM, we first developed a challenge load indicator to estimate the level of challenge, based on drops in production. Subsequently, we used this challenge load indicator as a covariate in RNM. We found genetic correlations of 0.5-0.85 between healthy and diseased periods indicating reranking of animals, or in other words genetic variation in resilience. We applied a DHGLM to total number of born piglets and found substantial genetic variation in residual variance of litter size with a genetic coefficient of variation in residual variance of 0.17. Using degressed EBV, we found a few highly significant genomic regions affecting the variance of litter size. Both statistical approaches can yield EBV to increase resilience and uniformity of animal production by selection.

Litter weight three weeks after farrowing and genetic correlation to others nursing ability traits

B. Nielsen[1], C. Søgaard[1], B. Ask[1] and T. Mark[2]
[1]Pig Research Centre, Breeding and Genetics, Axeltorv 3, 1609 Copenhaven V, Denmark, [2]University of Copenhagen, Department of Veterinary Clinical and Animal Sciences, Grønnegårdsvej 3, 1870 Frederiksberg C, Denmark; bni@seges.dk

In commercial pig production number of weaned piglets is one of the most important traits to reduce costs and increase productivity. Thus, in many breeding programs litter size has received major selection emphasis during the last decades, and the genetic progress in litter size puts a higher demand on sows to nurse and wean large litters successfully. The objective of this study was to investigate whether litter growth until three weeks of age (LG3W) has added value as an indicator trait of nursing ability. Data was obtained from 22 nucleus herds and consisted of recordings from more than 7,000 purebred litters of 5,532 Danish Landrace sows and 501 boars. Different multi-trait animal models were considered to estimate (co)variances of LG3W in combination with other traits including random, permanent and genetic effects of the sow, and the genetic variances and correlations to other traits associated with nursing ability was identified, i.e. litter size at day 5 (LS5), total number of piglets born (TNB), and litter size at three weeks (LS3W). Preliminary results show that LG3W was heritable showing heritabilities ranging from 0.11 to 0.14. Genetic correlations indicated that LG3W was favorably correlated to LS5 (r_G=0.44) and LS3W (r_G=0.65), but unfavorably correlated to TNB (r_G=-0.2). The preliminary results confirm that selection for LG3W is possible. Similar analyses for the female line of the Danish Yorkshire are ongoing and results of these analyses will also be presented.

Genome wide association study for number of teats and inverted teats in Norsvin Landrace pigs

M. Van Son[1], I.A. Ranberg[1], S. Lien[2], H. Hamland[1] and E. Grindflek[1]
[1]Topigs Norsvin, P.O. Box 504, 2304 Hamar, Norway, [2]Centre for Integrative Genetics (CIGENE), Department of Animal and Aquacultural Sciences, Norwegian University of Life Sciences, P.O. Box 5003, 1432 Ås, Norway; maren.moe@norsvin.no

The number of functional teats in the sow is important, as successful breeding has led to larger litter sizes. In order to get uniform litters with fast-growing piglets the number of teats has become an important fertility trait in pig production. Teat quality is also important and inverted teats are of less value for the suckling piglet due to reduced milk flow. In this experiment, we aimed at detecting loci affecting number of teats and number of inverted teats in the Norsvin Landrace population. For the study, 5,651 pigs were included with their breeding values for these traits. All the pigs were genotyped using the Illumina Porcine 60k BeadChip. A total of 39,733 filtered high quality SNPs were included for genome wide association analysis using the R package GenABEL with a structured association approach. Significant genomic regions associated with number of teats were detected on pig chromosomes SSC7 and SSC14 (P<0.001). For the number of inverted teats in boars, significant QTL were detected on SSC1 and SSC13, whereas SNPs on SSC1 and SSC2 were associated with number of inverted teats in sows. The QTL region on SSC7 for number of teats has previously been found in a Dutch Landrace-based population where VNTR was suggested as a candidate gene whereas the SSC14 QTL is close to one found in a Korean native × Landrace population. This study provides several genomic regions significantly associated with number of teats and inverted teats in the Norsvin Landrace breed. Fine mapping of these regions could provide causal SNPs and putative genetic markers for use in breeding value estimation and selection to ensure sows with a sufficient number of functional teats.

Genome-wide association study for litter size and mortality in Danish Landrace and Yorkshire pigs

X. Guo, O.F. Christensen, M.S. Lund and G. Su
Aarhus University, Department of Molecular Biology and Genetics, Center for Quantitative Genetics and Genomics, Blichers Allé 20, Postboks 50, 8830 Tjele, Denmark; xiangyu.guo@mbg.au.dk

Litter size and piglet mortality are important fertility traits in pig production. The purpose of this study was to identify SNP associated with litter size and piglet mortality in the Danish Landrace and Yorkshire populations. The data from breeding herds and multiplier herds were supplied by the SEGES P/S, Danish Pig Research Centre. The pigs analysed were genotyped using Illumina PorcineSNP60 BeadChip or imputed from lower density chips (8 K). In total, 5,977 Landrace pigs and 6,000 Yorkshire pigs were used for the analysis. The traits in the analysis were total number of piglets born (TNB), litter size at day 5 after birth (LS5) and piglet mortality rate before day 5 (MORT). A single SNP linear mixed model was used to detect association of SNPs with phenotypes of TNB, LS5 and MORT, corrected for the non-genetic effects used in the model for routine genetic evaluation. The association model included a fixed regression of corrected phenotypes on genotypes of a single SNP as a measure of the SNP effect, and in addition, a random polygenic effect accounting for shared genetic effects of related individuals. Significance testing of SNP effects was performed using a two-sided t-test, and a Bonferroni correction was applied to control for false positive associations. The significant level was declared as P-value<0.05/N, where N was the number of SNP loci. Analyses were performed using the DMU package. For Landrace, statistically significant QTL regions were found on chromosome 1, 2, 6, 13 for TNB, on chromosome 1, 7, 9, 10, 13, 17 for LS5 and on chromosome 2, 7, 9, 13, 17 for MORT. For Yorkshire, chromosome 1, 2, 7, 13 for TNB, chromosome 1, 2, 7, 14 for LS5 and chromosome 1, 6, 14, 17 for MORT were observed to include significant QTL regions. Several of these regions were detected in both Landrace and Yorkshire pigs.

Accuracy of genomic breeding values of purebreds for crossbred performance in pigs

A.M. Hidalgo[1,2], J.W.M. Bastiaansen[1], M.S. Lopes[1,3], M.P.L. Calus[4] and D.J. De Koning[2]
[1]Wageningen University, ABGC, Droevendaalsesteeg 1, 6708 PB, Wageningen, the Netherlands, [2]Swedish University of Agricultural Sciences, Dept. of Animal Breeding and Genetics, Ulls väg 26, 750 07 Uppsala, Sweden, [3]Topigs Norsvin, Schoenaker 6, 6641 SZ Beuningen, the Netherlands, [4]Wageningen UR Livestock Research, ABGC, Droevendaalsesteeg 1, 6708 PB, Wageningen, the Netherlands; andre.hidalgo@wur.nl

In pig breeding, the final product is a crossbred (CB) pig. Selection of purebreds (PB) for CB performance, thus, may result in greater genetic progress. The aim of this study was to assess empirically the accuracy of genomic breeding values (GEBV) of PB animals for CB performance. Phenotypes and genotypes were available from sows of three pig populations: 1,668 Dutch Landrace (DL), 2,003 Large White (LW) and 914 CB, an F1 cross between DL and LW. Two traits were analyzed: gestation length (GLE) and total number of piglets born (TNB). Scenarios studied were prediction within population and prediction of PB genetic merit based on CB animals in the training set. Within-population GEBVs were validated by comparison with corrected phenotypes. PB genetic merit for CB performance was validated by comparison to the PB animals' deregressed breeding value (DEBV) estimated from pedigree BLUP with CB phenotypes. GEBVs were estimated with GBLUP using ASReml software. Accuracy was computed as the correlation between the GEBV and the corrected phenotype/DEBV of the validation set animals divided by the square root of the heritability/reliability of the validation population. Accuracy of within-population predictions ranged from 0.36 to 0.66 across the traits and training sets. Accuracy for prediction of PB genetic merit for CB performance for GLE was 0.27 in DL pigs and 0.23 in LW pigs. For TNB accuracy was 0.11 for DL and 0.22 for LW. In sum, an encouraging level of predictive ability was observed for genomic prediction of PB genetic merit for CB performance. Our results show that the prediction of CB performance in PB lines is promising for the traits studied.

Application of single-step genomic evaluation for crossbred performance in pig

T. Xiang[1,2], B. Nielsen[3], G. Su[2], A. Legarra[1] and O.F. Christensen[2]
[1]INRA, UR1388 GenPhyse, 24 Chemin de Borde Rouge, Auzeville CS 52627, 31326 Castanet Tolosan, France, [2]Aarhus University, Center for Quantitative Genetics and Genomics, Department of Molecular Biology and Genetics, Blichers Allé 20, 8830, Tjele, Denmark, [3]Danish Agricultural and Food Council, Pig Research Centre, Axeltorv 3, 1609 Copenhagen V, Denmark; tao.xiang@mbg.au.dk

Crossbreeding is predominant and intensively used in commercial meat production systems, especially in poultry and swine. Genomic evaluation has been successfully applied for breeding within purebreds, but also offers opportunities of selecting purebreds for crossbred performance by combining information from crossbreds. However, it generally requires all relevant animals being genotyped, which is costly and presently does not seem to be a feasible in practice. Recently, a novel single-step BLUP method for genomic evaluation of both purebred and crossbred performance has been developed, which can incorporate marker genotypes into a traditional animal model. This new method has not been validated in real datasets. In this study, we applied this single-step method to analyze data for the maternal trait of total number of piglets born in Danish Landrace, Yorkshire and two-way crossbred pigs. The genetic correlation between purebred and crossbred performance was investigated. The results confirm the existence of a moderate correlation, and it was seen that the standard errors on the estimates were reduced when including genomic information. In addition, validation accuracies of breeding values for crossbred performance were investigated. The results showed an increase in this accuracy when adding genomic information.

Genomic prediction of crossbred performance based on purebred data

H. Esfandyari[1,2], P. Bijma[2], M. Henryon[3], O.F. Christensen[1] and A.C. Sørensen[1]
[1]Center for Quantitative Genetics and Genomics, Department of Molecular Biology and Genetics, Aarhus University, 8830, Denmark, [2]Animal Breeding and Genomics Centre, Wageningen University, 6708, the Netherlands, [3]Danish Agriculture and Food Council, Pig Research Centre, Copenhagen, 1609, Denmark; hadi.esfandyari@mbg.au.dk

The objective of this study was to compare the predictive ability of genomic prediction models with either additive, or both additive and dominance effects, when the validation criterion is crossbred performance (CP). The data used concerned pigs from two pure lines (Landrace and Yorkshire) and their reciprocal crosses, and the trait of interest was total number born in the first parity. Training was carried out on: (1) pure-bred sows of Landrace (2,085) and Yorkshire (2,145); and (2) combined pure lines (4,228) that were genotyped for 38k SNPs. Prediction accuracy was measured as the correlation between genomic-estimated breeding values (GEBV) of boars in pure lines and mean corrected crossbred-progeny performance divided by the average accuracy of mean-progeny performance. Next to a model with additive effects only (MA), we evaluated two different models with both additive and dominance effects (MAD and MADc). GEBV from the MADc model were computed based on SNP allele frequencies of selection candidates in the opposite breed. Compared to the MA model, prediction of crossbred performance based on MAD and MADc models improved prediction accuracy in both breeds. MADc model slightly improved prediction accuracy compared to MAD model. Prediction accuracy for Landrace boars was 0.10, 0.13 and 0.14 for MA, MAD and MADc, respectively. The corresponding values for Yorkshire boars were 0.32, 0.33 and 0.35. Combining animals from both breeds into a single reference population yielded 0.05 to 0.07 higher accuracies than training separately in both pure lines. In conclusion, the use of a dominance models increases the accuracy of genomic predictions of CP that are based on purebred data.

Genome-wide association studies in purebred and crossbred entire male pigs

H. Gilbert[1], Y. Labrune[1], M. Chassier[1], N. Muller[2], M.J. Mercat[3], A. Prunier[4], C. Larzul[1] and J. Riquet[1]
[1]INRA, UMR 1388 GenPhySE, 13326 Castanet-Tolosan, France, [2]INRA, UETP, 35650 Le Rheu, France, [3]IFIP, La Motte au Vicomte, 35651 Le Rheu, France, [4]INRA, UMR1348 PEGASE, 35590 Saint-Gilles, France; helene.gilbert@toulouse.inra.fr

A total of 654 purebred Piétrain entire male pigs and 716 crossbred Piétrain × Large White entire male pigs issued from about 70 Piétrain sires were tested in a French test station for production traits (feed intake, feed efficiency, growth rate, carcass composition and meat quality). All were genotyped with the 60K Porcine SNPchip. Genome wide association studies were run using linear mixed models with a genomic kinship matrix to account for relatedness between individuals, and the fixed effect of each SNP was tested separately. In a first step, separate analyses of the two populations showed suggestive results (P<0.0001) for almost all traits in the two populations. For production traits, eight 1-Mb regions affected multiple correlated traits in the purebred pigs, and only one in the crossbred pigs. Only two regions with P<0.0001 were detected in common in purebred and crossbred individuals after correction for the halothane mutation, on SSC1 and SSC2. Breed differences in linkage disequilibrium between markers and causal variants, or different gene effects due to the purebred vs crossbred polygenic background could explain these discrepancies. Genotypes were phased and chromosome breed origins were identified in all progeny. Analyses were thus run to estimate within breed allelic effects in the crossbred population, and combining the two populations. After accounting for differences in allele frequencies in the two populations, only few SNP estimates showed significantly different allelic effects depending on the genetic background. If confirmed in a larger design, this suggests that genes affecting production traits act similarly in purebred and crossbred commercial pigs, as suggested by high genetic correlations between purebred and crossbred pigs for these traits.

Genetic modelling of feed intake: a case study in growing pigs

I. David, J. Ruesche, L. Drouilhet, H. Garreau and H. Gilbert
UMR INRA / INPT ENSAT / INPT ENVT, Génétique, Physiologie et Systèmes d'élevage, chemin de borde rouge, 31326 Castanet Tolosan, France; ingrid.david@toulouse.inra.fr

With the development of automatic self-feeders and electronic identification, automated repeated measurements of individual feed intake (FI) are becoming available in more species. Consequently, genetic models for longitudinal data can be applied to study FI or related traits. To handle this type of data, several flexible mixed-model approaches exist, such as character process (CP), structured antedependence (SAD) or random regression (RR) models. The objectives of this study were to compare how these different approaches estimate both the covariance structure between successive measurements of FI and genetic parameters and to compare their ability to predict future performances in growing pigs. The dataset consisted in 44,049 weekly averages of daily FI of 3,096 French Large White boars, castrated males and gilts from approximately 67 to 180 days of age. Four approaches were compared: CP, SAD, RR using orthogonal polynomials and RR using natural cubic splines. Based on Bayesian Information Criteria, the SAD and CP models fit the data better than RR models. Estimations of genetic and phenotypic correlation matrices were quite consistent between SAD and CP models, while correlations estimated with the RR models were not. The SAD and CP models provided, as expected and in accordance with previous studies, a decrease of the correlations with the time interval between measurements. The changes in heritability with time (varying from 0.15 to 0.50 depending on the time and the model) showed the same trend for the SAD and RR models but not for the CP model. In comparison with CP model, the SAD and RR models had the advantage of providing stable predictions of future phenotypes one week forward whatever the number of observations used to estimate the parameters. Thus, to study repeated measurements of FI, the SAD approach seems to be very appropriate in terms of genetic selection and real time managements of animals.

Unspecific, maintenance and growth dependent feed efficiency in Duroc pigs

J.P. Sánchez[1], J. Reixach[2], M. Ragab[1,3], R. Quintanilla[1] and M. Piles[1]
[1]IRTA, Torre Marimon, 08140, Caldes de Montbui, Spain, [2]Selección Batallé, S.A, Avd. Segadors, 17421, Riudarenes, Spain, [3]Kafr El Sheikh University, El Gash Street, 33516, Kafr El Sheikh, Egypt; juanpablo.sanchez@irta.es

Estimates of the relative importance of use of feed for different biological functions are reported for a commercial Duroc pig line selected since 1991 for final backfat thickness (BT180) and body weight (BW180). Data consisted on daily feed intake (FI), and weekly body weight (BWG) and back thickness gains (BTG) from 1076 pigs (15-25 weeks of age) controlled since 2003 to 2012 in 4 trials. Weekly average FI was fitted using a random regression (RR) model with animal specific intercepts and linear regressions on $BW^{0.75}$ and BWG. These animal specific RR coefficients comprised both genetic and permanent environmental effects. Batch, pen, and age were included in the model, and also a multiple regression on $BW^{0.75}$, BWG and BTG within trial. Thus, the expected FI is considered throughout this multiple regression and variance component estimates then refer to residual feed intake. Selection bias was accounted for by conducting a 3-variate analysis with BT180 and BW180. Individual variance of FI associated to the intercept (unspecific use of feed) was 0.04(0.003) kg^2, whereas that jointly associated to the intercept and maintenance ranged with age from 0.03(0.003) to 0.14(0.008) kg^2, and that jointly associated to the intercept and growth was constant across ages, 0.04(0.003) kg^2. So, FE individual variation seems to be mainly associated to variations in the use of food for maintenance and unspecific functions, but not to differences in the use of feed for growth. The genetic correlation (rho_g) between those first components of FE was 0.29 (0.09). Improvement of BW180 will slightly reduce maintenance needs (rho_g=-0.29 (0.11)) but will increase needs for unspecific functions (rho_g=0.35(0.08)). Reductions in BT180 will improve unspecific FE (rho_g=0.69(0.09)) and will not modify maintenance FE (rho_g=-0.01(0.12)).

Genetic (co)variance estimation of Gompertz growth curve parameters in Finnish pigs

J.M. Coyne[1], K. Matilainen[2], M.-L. Sevon-Aimonen[2], D.P. Berry[1], E.A. Mantysaari[2], J. Juga[3], T. Serenius[4] and N. Mchugh[1]
[1]Teagasc, Moorepark, Animal & Grassland Research and Innovation Centre, Teagasc, Moorepark, Fermoy, Co. Cork, Ireland, [2]Natural Resources Institute Finland (Luke), Biometrical Genetics, 31600, Jokioinen, Finland, [3]University of Helsinki, Department of Agricultural Science, P.O. Box 28, 00014 University of Helsinki, Finland, [4]Figen Ltd, P.O. Box 319, 01301 Seinäjoki, Finland; jessica.coyne@teagasc.ie

Studies have shown that the Gompertz growth function parameters are heritable and therefore could be incorporated into a breeding program to alter the shape of an animal's growth. The objective of this study was to estimate the genetic (co)variance components of the three Gompertz growth function parameters. A total of 61,715 live weight records from 12,768 pigs, between the years 2006 and 2012, inclusive, were obtained from the Finnish pig breeding company Figen Oy. All animals were on trial in the Längelmäki central test station in Finland. Genetic (co)variance components for the parameters A, B, and k were estimated using a multiplicative model. The fixed effects included in the model were contemporary group (date of entry to test station) and gender; a litter, permanent animal and a sire genetic effect were fitted as random effects. MiX99 was used to obtain the growth curve parameter estimates and EM-REML was undertaken in DMU to obtain variance component estimates. These steps were then iterated until convergence was deemed to have been reached once the variance components changed by less than 0.0001 over successive iterations. The h^2 estimates were moderate to high for all growth curve parameters and ranged from 0.41 for parameter A to 0.83 for parameter B. Weak to moderate genetic correlations existed between the function parameters (-0.38 to 0.08). Exploitable genetic variation exists for the parameters A, B and k in the Gompertz function, and allow for the incorporation of the Gompertz growth function into pig breeding programs.

Horizontal model a novel approach for dissecting genetic background of feed efficiency complex

M. Shirali[1], P.F. Varley[2] and J. Jensen[1]
[1]Center for Quantitative Genetics and Genomics, Aarhus University, 8830, Tjele, Denmark, [2]Hermitage Genetics, Sion Road, Kilkenny, Ireland; mahmoud.shirali@mbg.au.dk

The aim of this study was to develop an approach to allow better modelling of feed efficiency complex by combined analysis of longitudinal feed intake records and single record of production traits of average daily gain (ADG) and lean meat percentage (LMP). Data were available on 3,850 purebred Maxgro pigs from 52 kg body weight (BW) (11 kg, standard deviation (SD)) to 110 kg BW (11 kg, SD) in 7 weeks. In horizontal model, average daily feed intake (FI) per each week of growth in the test period, production trait of ADG in the entire test period and LMP at the end of test period were used in a multi-trait random regression model. FI record of each week was fitted with a random regression model with segmented fixed part of the model and linear animal genetic effect and permanent environmental effect of contemporary group using Legendre polynomials of days on test. Therefore, the genetic and group effects can be used as a function of time with covariance functions. For ADG genetic effects of animal and permanent environment of contemporary group were considered, as for LMP the litter effect was also included. Feed efficiency was estimated using residual feed intake (RFI) by genetically conditioning feed intake for ADG and LMP. The heritability of RFI increased from 0.08 to 0.16 by age. Heritability estimates of weekly FI increased by days on test from 0.15 to 0.27. Heritability estimates for ADG and LMP were found to be 0.20 and 0.50, respectively. Genetic correlations between weekly FI and ADG increased from 0.56 to 0.85 by days on test. The genetic correlations between weekly FI and LMP were moderately negative ranging from -0.55 to -0.42. Genetic selection for feed efficiency should consider different stages of growth. Horizontal model can be a feasible approach for dissecting feed efficiency complex.

REML and Bayesian analysis of association between leg conformation in young pigs and sow longevity

H.T. Le[1,2], P. Madsen[1], N. Lundeheim[2], K. Nilsson[2] and E. Norberg[1]
[1]Aarhus University, Dept. of Molecular Biology and Genetics, Blichers Alle 20, 8830 Tjele, Denmark, [2]Swedish University of Agricultural Sciences, Dept. of Animal Breeding and Genetics, Ulls väg 26, 75007 Uppsala, Sweden; thu.le.hong@slu.se

Longevity is an important trait in pig production regarding economic and ethical aspects. Direct selection for longevity might be ineffective since true longevity can only be recorded when a sow has been culled or has died. Thus, indirect selection for longevity through traits that can be recorded earlier in life and are genetically correlated with longevity might be a promising potential. Leg conformation is now included in many breeding schemes. However, the proof that leg conformation traits to be a good early indicator for longevity remains. Our aim was to study genetic associations between leg conformation traits of young (5 months; 100 kg) SwedishYorkshire pigs in nucleus herds and longevity traits of sows in nucleus and multiplier herds. Data included 97,533 animals with information on conformation (movement and overall leg score) recorded at performance testing; and 26,962 sows with information on longevity. The longevity traits were: stayability to survive upto 2nd parity (STAY12); lifetime number of litters (NoL); lifetime number born alive (LBA). Genetic analyses were performed using REML and Bayesian methods. Heritabilities estimated using the Bayesian method were higher than those estimated using REML, ranging from 0.10 to 0.24 and from 0.07 to 0.20 respectively. All estimated genetic correlations between conformation and longevity traits were significantly favorable (better leg – better longevity). Heritabilities and genetic correlations between conformation and longevity indicate that selection on leg conformation should improve sow longevity. A comparison between the two methods regarding time requirement and parameter estimation suggested that REML method would be advised.

Genetic parameter estimation in purebred and crossbred swine from genomic models

Z.G. Vitezica[1], J.M. Elsen[1], I. Misztal[2], W. Herring[3], L. Varona[4] and A. Legarra[1]
[1]INRA, INPT, UMR GenPhySE, 24 Chemin de Borde Rouge, 31326 Castanet-Tolosan, France, [2]University of Georgia, 425 River Road, Rhodes Center, Athens, GA30602, USA, [3]PIC, 100 Bluegrass Commons Blvd. Ste 2200, Hendersonville, TN 37075, USA, [4]Facultad de Veterinaria, Universidad de Zaragoza, Calle Miguel Servet 177, 50013 Zaragoza, Spain; zulma.vitezica@ensat.fr

Most developments in quantitative genetic theory focus on the study of intra breed/line concepts. With the availability of massive genomic information a need of revisiting the theory for hybrid populations (F1) exists. The objective of this work is to propose an approach to estimate variance components in the F1 population under a genomic model with additive and non-additive (dominance) inheritance. Assuming that marker effects are shared across lines and that the allelic frequencies differ (p, p') between them, we show that the genetic variance in the F1 population involves additive (a) and dominant (d) effects of genes and a term with its covariance. Breeding values of crossbred individuals are generated by substitution effects, where the effects for each parental line depend on the allele frequencies from the other line. In addition, dominance variance is proportional to the product of the heterozygosities of both lines: 4pqp'q'. If a and d effects are considered random with covariance between them equal to zero, genetic variance components of the F1 population can be obtained from these expressions using an equivalent genomic model based on markers. We illustrate these results with variance component estimations in pig data provided by Genus plc (Hendersonville, TN, USA). In the F1 population, estimates of genomic variance due to dominance in litter size ranged from 18 to 25% of the additive genetic variance and were within the range of published pedigree-based estimates for pigs.

Genome-wide association analyses for fat content and fatty acids profile in Iberian pig

N. Ibáñez-Escriche[1], J.F. Tejeda[2], E. Gomez[2] and J.L. Noguera[1]
[1]IRTA, Genètica i Millora Animal, Av. Alcalde Rovira Roure, 191, 25198, lleida, Spain, [2]Universidad de Extremadura, Escuela de Ingenierías Agrarias, Tecnología y Alimentos de Calidad, Avda. Adolfo Suárez s/n, 06007 Badajoz, Spain; noelia.ibanez@irta.es

Iberian pig is characterized by their adipogenic nature that defines the high quality of both, fresh meat and their cured product. Genetics is one of the main factors on the fat muscle deposition and composition in Iberian pig. In fact, there is a higher variability between Iberian pig lines. The aim of this study was performing a genome-wide association (GWAS) analysis using the genomic prediction model Bayes-A for backfat thickness (BFT), percentage of intramuscular fat on longissimus Dorsi (IMF) and fatty acid profile (FA%) composition for IMF and for subcutaneous fat (SCF) in Iberian pigs. The data base used compromised 372 animals genotyped with a 60k single nucleotide polymorphism (SNP) chip. These animals belong from a diallelic experiment with Retinto, Torbiscal and Entrepelado Iberian pig lines growing and slaughtering under commercial environment. The results showed slightly associations of 1 Mb SNP windows with BFT, IMF and FA%. In particular, these associations were higher than 1% in chromosome 1 for BFT in last rib, in chromosome 9 and for IMF and in chromosome 18 for the percentage of Linolenic fatty acid (C18:3 (n-3)) in IMF. Additionally, when the SNP window was expanded up to 4 Mb, QTL region on chromosome 7 was found to explain a higher proportion of the genetic variance (3.4%) for BFT in the 4th rib and for IMF (2%). However, although magnitude of the variance explained was small these regions explained 40 times more percentage of genetic variance that the mean of the windows computed. These results would indicate that the 60k single SNP could not catch up the genetic variability for meat quality in Iberian Pig. However, further studies are needed in order to validate the result s found.

RNA-Seq analysis of porcine reproductive and respiratory syndrome virus infected respiratory cells

M. Pröll[1], C. Neuhoff[1], C. Große-Brinkhaus[1], M.A. Müller[2], C. Drosten[2], J. Uddin[1], D. Tesfaye[1], E. Tholen[1] and K. Schellander[1]
[1]*Institute of Animal Science, University of Bonn, Endenicher Allee 15, 53115 Bonn, Germany,* [2]*Institute of Virology, University of Bonn Medical Centre, Sigmund-Freud-Str. 25, 53127 Bonn, Germany; mpro@itw.uni-bonn.de*

The porcine reproductive and respiratory syndrome (PRRS) causes significant production losses for the global swine industry. The understanding of the responses to porcine reproductive and respiratory syndrome virus (PRRSV) as well as of the genetic elements and functions involved in the immune response to PRRSV is still lacking. Therefore the aims of this study were to investigate the transcriptome profile of respiratory cells of two different pig breeds (Pietrain and Duroc) post PRRSV infection. In order to improve the understanding of genetic components and functions in the responses to PRRSV and to characterize changes in the global transcriptome profile the high-throughput RNA-sequencing technology was used. The RNA-sequencing analysis of PRRSV infected lung dendritic cells (DCs) obtained 20,396 porcine predicted gene transcripts as well as differently expressed gene transcripts for both breeds. The investigations showed that the expression trends proceeded contrarily for both breeds during the first time points post infection. Additionally, the virus sequence alignment exhibited that Lelystad virus (LV) strain infects lung DCs and replicates there. These results represented breed-dependent differences post PRRSV infection for the virus replication, too. The genetic background of Pietrain and Duroc leads to early temporal, different and contrary immune responses post PRRSV infection. Duroc respiratory cells responded more efficiently in the first time points of PRRSV infection than Pietrain respiratory cells. These observed breed differences should be taken into account for breeding strategies and for the improvement of the pigs' health.

Effect of DNA markers on the carcass backfat quality traits in hybrid pigs

O.V. Kostyunina[1], S.M. Raskatova[2], A.A. Traspov[1], V.R. Kharzinova[1], K.M. Shawyrina[1] and N.A. Zinovieva[1]
[1]*L.K. Ernst Institute of Animal Husbandry, Moscow, Podolsk, 142132, Russian Federation,* [2]*'Agrofirma Mtsenskaya', Orel region, 303023, Russian Federation; veronika0784@mail.ru*

Identification of DNA markers affected the carcass traits has a great value in genetic improvements of pigs. The aim of the current study was to evaluate the effect of the polymorphisms of CAST249, CAST638, PRKAG3 (N30T, S52G, I199V, R200Q), MC4R, IGF2 and CCKAR genes on the quality of the carcass backfat in hybrid pigs. DNA was extracted from tissue samples (n=60) using a Nexttech column according to the manufacturer's recommendations. The polymorphisms were determined using PCR-RFLP (CAST249, CAST638, IGF2 and CCKAR genes) and PCR-pyrosequencing (PRKAG3 and MC4R genes) assays. The effect of DNA markers on the quality traits of carcass backfat was carried out by linear regression analysis using Stata 12 Software. No polymorphisms were detected in CAST249 gene and position R200Q of PRKAG3 gene. The significant dominant effect ($P \leq 0,05$) of the MC4R gene variants on the melting point of carcass backfat was detected. Polymorphisms in PRKAG3 (N30T) and MC4R genes were associated with the density of backfat. Additionally N30T substitution in PRKAG3 gene affected the melting temperature of back fat. S52G polymorphism of PRKAG3 was associated with the iodine value of carcass backfat. Our results have important practical applications, in particular the MC4R polymorphism could be applied in a marker assisted selection in the Russian hybrid pigs.

Single marker vs multilocus methods in genome-wide association analysis of pig data

H.P. Kärkkäinen and P. Uimari
University of Helsinki, Department of Agricultural Sciences, P.O. Box 28, 00014 University of Helsinki, Finland; hpkarkka@cc.helsinki.fi

Genome-wide association studies have proven useful in identifying causal loci behind economically important livestock traits. In animal breeding field the most common approach is based on mixed models and REML estimation. However, as the economically important traits commonly have a complex genetic background influenced by multiple genes, such single marker methods, as they ignore the effect of the other causal loci, strictly speaking fit a wrong model. Therefore multilocus methods based on penalized regression models could be preferable. In this work we have considered the performance of two multilocus methods, heteroscedastic ridge regression and Bayesian LASSO, in a genome-wide association study with both real and simulated pig data. As REML estimation is very taxing to computer usage, we have also considered two heavily used less computer intensive single marker methods, GRAMMAR and genomic control. All of the methods recognize the previously known causal loci from the real data; however GRAMMAR and genomic control seem to be less powerful than REML. With the simulated data the result is less clear. The multilocus models are quite powerful but do not provide a method for separating the significant signals from the non-significant ones. The multilocus models are also highly sensitive to prior settings. While under the Bayesian LASSO this is understandable, it impairs the accessibility of the method as none of the objective prior selection methods seem to work properly. As the heteroscedastic ridge regression seemingly does not include any prior information, and hence should not depend upon such, the sensitivity is clearly even more serious failure. Our results indicate that the common restricted maximum likelihood estimation based method seems to hold its place among the single marker methods and that the multilocus models, though are promising, still seem to require further development in order to be readily used for genome-wide association analysis.

Genetic diversity in populations of Czech Large White pigs

T. Urban and B. Černošková
Mendel University in Brno, Department of Animal Morphology, Physiology and Genetics, Zemědělská 1, Brno, 61300, Czech Republic; urban@mendelu.cz

The aim of this thesis was to analyse populations of pigs of breed Czech Large White and to evaluate their genetic variability and diversity. Eleven single-nucleotide polymorphisms (SNPs) were analysed in porcine candidate genes. In total, 139 individuals from two populations of different status in breeding program were evaluated (breed and commercial). The polymorphisms were genotyped by PCR-RFLP in loci: HMGCR, TCF7L2, MC4R, EDG4, HFABP, FTO, LEPR6, LEPR18, PDK4, and two polymorphisms in PLIN. All observed loci were polymorphic. The occurrence of rare allele T of locus EDG4 in breed population was described. All loci in both observed populations were in Hardy-Weinberg equilibrium with the exception of PDK4 locus. Genetic diversity in population of commercial breed was higher (average Het=0.41) than diversity of breeding population (average Het=0.37). The lowest negative values of F_{IS} in breed and commercial populations were in PDK4 locus, -0.276 and -0.507, resp. The highest value of F_{IS} in breeding population was in TCF7L2 loci (0.021) and in commercial population was in HFABP loci (0.029), resp. (Project IGA MENDELU TP 6/2011 and CEITEC no. CZ.1.05/1.1.00/02.0068).

Development and current state of genetics in Swiss fattening pigs

A. Burren, B. Rufer and P. Spring
Bern University of Applied Sciences, School of Agricultural, Forest and Food Sciences HAFL, Länggasse 85, 3052 Zollikofen, Switzerland; beatrice.rufer@anicom.ch

With optimum carcass quality, higher slaughter prices can be realized, that exceed the normal weekly rate and significantly improve the profitability of a pig fattening operation. Lean meat percentage (LMP), carcass weight and fat quality are the three main criteria used in the Swiss payment schedule for slaughter pigs. To assess the situation, the development of LMP and carcass weight were analysed for the period of 2010 to 2013 for 7,487 Swiss operations, providing data from 10 million slaughter pigs. The analysis showed that the mean LMP was 56.7% across farms and was consistent over time. This constant mean LMP value indicated that the continual progress reported in the National breeding programme has not been mirrored at farm level in recent years. In all four years studied, 82-83% of the pigs were classified in the neutral LMP range or the range entitled to a surcharge. In 2013, 90.6% of the carcasses weighed between 74 and 96 kg and were in the weight range desired by purchasers. For other classifications by weight, 2.2% of the animals were too light and 7.2% were too heavy. Underweight animals resulted in deductions of around 1.7 million CHF. The total deductions for overweight animals were significantly higher, estimated at 7.2 million CHF. Many operations could reduce these economic losses significantly by weighing their animals more regularly. It would be worthwhile for some pig producers to increase their average LMP slightly and reduce its variability.

Genetic gain for the reproduction and production traits in Latvian swine populations

L. Paura[1], D. Jonkus[1] and U. Permanickis[2]
[1]Latvia University of Agriculture, Liela 2, Jelgava, 3001, Latvia, [2]'Gensoft' Ltd, Saldus district, 3862, Latvia; liga.paura@llu.lv

The aim of this study was to analyse the genetic gain for the reproduction and production traits of Latvian Landrace and Latvian Yorkshire sows. Since 1999, evaluations for the production and reproduction traits are performed by BLUP animal model. The average daily gain (AGD), test period daily gain (TDG), backfat thickness (BF), number of piglets born alive (NBA1, NBA2+) and 21-day litter weight (LW1, LW2+) at first and later parities, and weaning-conception interval between the first and second parity (WCI) were included in the sows' genetic evaluation. Records of 14,577 first-parity and 8,790 later-parity LL sows and 6,039 first-parity and 12,360 later-parity LY sows born between 2000 and 2012 were included in the analysis. The gilt and sow reproduction traits are evaluated separately: production traits and first parity reproduction traits were analysed by single-trait BLUP animal model; reproduction traits in later parities by repeatability-traits BLUP animal model. During the last decade, production traits were improving in the breeding (nucleus) herds, and genetic response in LL were 51 gr, 109 gr, -0.9 mm in 10 years for AGD, TDG and BF, respectively; and 26 gr, 84 gr, -1.4 mm in 10 years, respectively, for LY sows. Compared to genetic gain for NBA1, genetic gain for NBA2+ were higher in both populations: 1.6 vs1.7 piglets in 10 years for LL and 1.1 vs 1.2 piglets in 10 years for LY. Starting from 2005, the genetic response for LW was positive 0.21 kg/year for LW1 and 0.15 kg/year for LW2+ for LL sows, and 0.18 and 0.19 kg/year, respectively, for LY sows. In the analysed period, a positive response was observed for all traits. Overall, estimates of the genetic trend in LL population were greater than in LY sows population.

Genetic relationship between and within pig nucleus herds

E. Krupa[1], E. Žáková[1], Z. Krupová[1], L. Vostrý[1,2] and R. Kasarda[3]
[1]*Institute of Animal Science, Prague, 10400, Czech Republic,* [2]*Czech University of Life Sciences, Prague, 16521, Czech Republic,* [3]*Slovak University of Agriculture, Nitra, 94976, Slovak Republic; krupa.emil@vuzv.cz*

The monitoring of genetic relationship has to be done first for quantifying of its impact in actual genetic situation of investigated populations The genetic relationship between and within the nucleus herds was computed in two dam (Czech Large White; CLW and Czech Landrase; CLA) pig breeds. Impact of the different field data types (P; production and R; reproduction) and various number of ancestor's generations in reference population (G ranged from 3 to 7) was studied. The average genetic relationship between all herds (GRa), between related herds only (GRr) and within herds (GRw) was computed by indirect methodology of the program CFC 1.0. In total, 81 (CLW) and 33 (CLA) herds were analysed in the period 2008-2013. From these, more than 91% of herds farmed only one breed. For CLW breed, one herd is genetically related to 60.4-68.0 of other herds (varying according to the data type and number of ancestor's generations). There were 6.7% of herds related with less than 50% of other herds and 44.0% of herds with more than 90% of other herds when using of G3. However, appropriate proportion of herds changed to 0.0 and 71.8% when seven generations of ancestors was considered. Lower differences of genetic relationship between and within herds were obtained for reproduction datasets. The GRa and GRr were generally low for both breeds and changed f.e. from 0.94 (CLW P G3) to 1.13 (CLW P G7). Generally, higher relationships were calculated for CLA breed. Considering of upper number of ancestor's generations increased the herd's relationship. The GRw were two times higher than GRa (21.7% CLW R G3 to 25.1%). As positive it is stated that almost all of herds are related to each other, and the testing scheme is well organised from this point of view. Contrary, the high relationship within herds is probably caused by inbreeding. The study was funded by the project NAZV QJ1310109.

Estimation of genetic parameters for performance and body measurement traits in Duroc pigs

C. Ohnishi[1] and M. Satoh[2]
[1]*National Livestock Breeding Center Ibaraki Station, Chikusei, Ibaraki, 3080112, Japan,* [2]*NARO Institute of Livestock and Grassland Science, Tsukuba, Ibaraki, 3050901, Japan; hereford@affrc.go.jp*

Genetic parameters for performance and body measurement traits were estimated from a new line of purebred Duroc pigs selected by restricted best linear unbiased prediction of breeding values for average daily gain, loin muscle area, and backfat thickness by using two multiple trait animal models, which included and excluded litter common random effect. Common litter effects on most of body measurement traits as well as performance traits were important. Heritability estimates for body measurement traits ranged from 0.13 to 0.44. The genetic correlations among body length, withers height, chest depth, and hip height were estimated to be positive (0.17 to 0.86). Those among fore width, chest width, and hind width were also estimated to be positive (0.23 to 0.83). There were negative genetic correlations between the former and the latter traits (-0.82 to -0.30). Many genetic correlations between performance and body measurement traits found to be from mediate to high. Our results indicate that selection using body measurement traits may be more effective to improve performance traits, because body measurement traits are easy to measure.

Genetic variation of beta-lactoglobulin in pig and the effect on the content of bioactive peptides

C. Weimann and G. Erhardt
Justus-Liebig-University, Dept. of Animal Breeding and Genetics, Ludwigstrasse 21B, 35390 Giessen, Germany; christina.weimann@agrar.uni-giessen.de

Beta-lactoglobulin (β-LG) is the most abundant whey protein secreted by porcine mammary gland and belongs to the superfamily of carrier proteins lipocalins. Lipocalins are a group of extracellular transport proteins which are mostly known for their ability to bind small hydrophobic molecules. The porcine β-LG shows about 66% amino acid identity with the bovine β-LG but in contrast to bovine the porcine protein seems not to be able to bind hydrophobic ligands. In the context of the nutritional value of β-LG it is well known that milk proteins in different species are a rich source of bioactive peptides. These peptides have for example antimicrobial, antioxidative or ACE-inhibitory effects. The porcine beta-lactoglobulin (BLG) gene is located on SSC 1 containing 6 exons. The aim of the current study was to estimate the allele frequencies of two non-synonymous SNPs in exon I and exon IV of the porcine BLG gene in different pig breeds and populations. Furthermore it should be demonstrated if the resulting amino acid exchanges have an effect on the occurrence of bioactive peptides. Pigs from different origins (Large White, German Landrace, Wild Pig, Duroc, Pietrain and JSR-Hybrid) were analyzed via PCR-RFLP. Two SNP causing amino acid exchanges were chosen from the ENSEMBL database: rs333403809 (c.81G>C; p.Glu27Asp) and rs55619160 (c.407T>C; p.Val136Ala), respectively. First results showed that glutamin acid at position 27 and valin at position 136 of the protein are most common. Depending on the alleles, occurrence of bioactive peptides was predicted in silico. The glutamic acid instead of aspartic acid on position 27 of the protein leads to one additional bioactive peptide with antioxidative effect. This may induce positive effects on prevention of inflammatory processes. Therefore in further studies the association between porcine BLG polymorphisms and health parameters of piglets during the weaning period will be investigated.

Subjective definition of traits and economic values for selection of organic sows in Denmark

A.C. Sørensen
Aarhus University, Center for Quantitative Genetics and Genomics, Department of Molecular Biology and Genetics, P.O. Box 50, 8830 Tjele, Denmark; achristian.sorensen@mbg.au.dk

The purpose of this study was to assess the difference between organic pig producers' preferences for the genetic make-up of sows and the breeding goal in conventional sow lines (BGconv). Differences can occur for several reasons: (1) it is not necessarily the same traits that are relevant; (2) the traits may have different relative economic importance, e.g. due to price differences of feed and product; and (3) some traits may have non-market value for the organic pig production. Organic farmers assisted in defining a group of eight traits that are potentially economically relevant, heritable, and possible to record. Subsequently, the farmers answered a questionnaire where they were asked to prioritise the eight traits relative to each other in a partial choice design. The answers were analysed using a multinomial logit model (PROC PHREG in SAS9.2), which calculates overall probabilities for prioritisation of each trait. The organic breeding goal (BGorg) included traits that were defined closer to the economically important traits, e.g. number of piglets at weaning, but also a trait like mortality of liveborn, which has no strict economic value when litter size at weaning is included This indicates the presence of non-market values. When compared according to the relative economic value of a phenotypic standard deviation unit of the trait, litter size was less important in BGorg (29 vs 42% in BGconv), while traits related to maternal ability (number of functional teats, early growth, and mortality of liveborn) together were more important in BGorg (42 vs 2% in BGconv). The traits defined for the organic environment requires labour-intensive recording, so there is no straightforward implementation of the organic breeding goal derived in this study. On the other hand, it provides input for the discussion of whether to select sows for organic production from lines selected according to breeding goals for conventional production systems.

Degree of connectedness in Czech Large White and Czech Landrace production and reproduction data

E. Žáková and E. Krupa
Institute of Animal Science, Prague, 10400, Czech Republic; krupa.emil@vuzv.cz

Genetic links between management units (herds or herd-time units) are of great importance for reliable BLUP genetic evaluation. On the degree of connectedness depends the risk of making errors when comparing EBVs of animals across herds. For some traits in genetic evaluation in Czech Republic there is higher level of herd-year-seasons (hys) with low number of observations. The situation is solved by the season prolongation until the required number of observation is reached. The aim of this study is to examine connectedness in datasets used for genetic evaluation and to found the influence of the number of observations within hys on its connectedness. Analyzed dataset of production data consist from 43,500 and 16,398 records for Czech Large White (CLW) and Czech Landrace (CL), respectively. Reproduction data were analyzed separately for the first litters (23,258 and 6,942 records for CLW and CL, respectively) and for second and later litters (73,009 and 20,598 records for CLW and CL, respectively). Seasons were formed in production data monthly, in reproduction data within natural year seasons, for both datasets without prolongation to found the real data connectedness. For each animal were chosen at least 4 generations of ancestors. The connectedness ratio GDV between each combination of hys was computed using Genetic Drift Variance method. For both breeds were found similar results of connectedness, although a slightly better results were for CLW. The best average GDV were found for second and later litters data (0.292±0,178 for CLW and 0.348±0,191 for CL), worse results for production data (0,556±0.241 for CLW and 0.585±0.280 for CL), and the worst for the first litters data (0.598±0.307 for CLW and 0.661±0.326 for CL). For all analyzed data were ascertain the same exponential dependency between number of observation and average GDV for hys. The study was funded by the project NAZV QJ1310109.

Health surveillance in organic dairy farms: comparing indicators of farmers to those of scientists

J. Duval[1,2], N. Bareille[1,2], A. Madouasse[1,2] and C. Fourichon[1,2]
[1]INRA, UMR1300 Biology, Epidemiology and Risk Analysis in animal health, Atlanpôle-La Chantrerie, 44307 Nantes, France, [2]LUNAM Université, Oniris, UMR BioEpAR, Atlanpôle-La Chantrerie, 44307 Nantes, France; julie.duval@oniris-nantes.fr

Today, disease surveillance protocols fail to improve herd health due to a lack of compliance of farmers. An explanatory factor could be that farmers and scientists who design these protocols do not use the same indicators for health surveillance. The aim of this research was to describe how various health indicators differ between farmers and scientists when designing a herd health surveillance tool. Twenty semi-structured research interviews were conducted with French organic dairy farmers and their advisors in animal health. The indicators adopted for herd health surveillance and their characteristics are described and analysed to identify how large the gap is between the indicators used by farmers and those proposed by scientists. Each farmer chose a unique combination of herd health indicators and alert levels. Not one farmer adopted all the indicators as proposed by the scientists. Alternative indicators proposed by farmers can be close to those proposed by scientists or far away, respectively by choosing indicator for earlier detection of disease or indicators measuring very farm specific health management measures for example. Furthermore, analysis of the transcribed interviews allowed identifying a variety of reasons that explain the choice of alternative indicators proposed by farmers. Thus, the results show a great heterogeneity amongst farmers in indicators used for herd health surveillance. Moreover, a gap was identified between what scientists consider as indicators for herd health surveillance and what is adopted in the field. The description of the characteristics of the alternative indicators proposed by farmers and the understanding of some of the reasons for their choices could be used in the design of better herd health surveillance protocols that are more acceptable in field conditions.

Assessment of indirect employment related to livestock farming

A. Lang[1], P. Dupraz[2], Y. Trégaro[3] and P.M. Rosner[1]
[1]CIV, 75012, Paris, France, [2]INRA, 35000, Rennes, France, [3]FranceAgriMer, 93100, Montreuil, France;
agathe.lang@gmail.com

Changes in livestock farming systems that occurred in the past decades have had a strong effect on employment. Many activities that used to take place on the farms have been taken over by external operators: animal breeding, feed processing, products trade, etc. Employment related to livestock farming is therefore larger than direct employment. Many external operators depend on livestock farming and this indirect employment should be considered. Given the current economic situation, employment level is considered as a key indicator to assess the economic weight of a particular sector. However, to date, there is no published study based on reliable data. More yet, there is no accurate and recognized method to assess and describe precisely this indirect employment. The consortium 'Elevages Demain' (Tomorrow's Livestock Farms) which gathers major actors of the French animal sector launched an innovative study to assess employment related to livestock farming in France. In order to identify dependent operators, we have built a quantitative and dynamic method to assess the potential effects on each operator of a change in livestock farming activity. This evaluation is based on three major components: short term effects, potential long term effects (ie asset specificity and relevant market) and geographical constraints. Three classes of dependency were built: strong, moderate and low. This allowed to rank each operator according to the nature and the level of its dependency to livestock farming. Once those dependent operators were identified, we gathered data from professional surveys, official State statistics and professional representatives' knowledge to assess employment. Several data sources were compared to reduce and measure the error margin. We therefore have fully assessed the number of jobs depending on the presence of livestock farming, ranked by their level of dependency. The results are detailed sector by sector, covering all animal productions (dairy, eggs and meat).

Evaluation of the sustainability of the Moroccan sheep meat production systems by the IDEA method

A. Araba and A. Boughalmi
Hassan II Institute of Agronomy and Veterinary Medicine, Madinat Al Irfane-Rabat 6202, Morocco;
a.araba@gmail.com

The characterization of sheep farming system in the Moroccan Middle Atlas has shown mutations in the silvo-pastoral and pastoral ecosystems that may affect their sustainability. To evaluate this hypothesis, 75 farmers from three production systems i.e. silvo-pastoral, pastoral and oasis, were surveyed using the IDEA 'Sustainability Indicators of Agricultural Farms'. 37 sustainability indicators were used, covering the three dimensions of sustainability i.e. agro-ecology, socio-territorial and economic. The preliminary results show that production systems did not affect the economic scale (P>0.5) but affected the agro-ecological and socio-territorial ones (P<0.5). Globally, the overall sustainability of all farms is limited by socio-territorial aspects where intervention could be organized by improving components related to 'Employment and services, quality of life and the contribution to create jobs'. With regards to the agro-ecological parameters, the two extensive systems i.e. silvo-pastoral and pastoral expressed a better sustainability than that do the oasis system because of the high plant and animal diversity and the adequate spatial organization.

Novel dairy systems as an adaptation to diminishing phosphate resources
M.D. March, D.J. Roberts, A.W. Stott and L. Toma
SRUC, Edinburgh, EH9 3JG, United Kingdom; maggie.march@sruc.ac.uk

Increasing demand for protein rich diets and a limited supply of arable land creates a food supply issue which imposes further dependence on phosphorus (P). EU food security depends upon imported P and member states should increase resource use efficiency, reduce losses, and lower total P consumption. We assessed the nutrient efficiency of two contrasting genetic merits of HF dairy cows, within a traditional low concentrate (HF), confined (LF), by-product (BP) and home-grown (HG) dairy feed systems (FS's) by calculating annual farm gate P nutrient budgets. Farm level budgets were combined with system indicators such as land and nitrogen use to fairly represent each regime. Nonparametric Data Envelopment Analysis models were applied to test the eco-efficiency of the dairy FS's, Both models assess P as an undesirable output, and one also considers its additional role as a non-renewable input. Nutrient budgets show efficiencies ranged from 0.29 within a HG FS to 0.49 within a BP FS exporting all manure. A HF system was more P efficient than an LF regime feeding high levels of purchased feeds, at 0.39 vs 0.34. When energy corrected milk production was considered, a HF FS generated lowest average surplus P/litre (0.0021 kg) whilst the HG FS attracted the highest average surplus of 0.003 kg P/litre. Undesirable output orientated DEA model results found a BP system to be most P efficient and average efficiency scores ranged from 0.55 for a superior merit within a HG FS to 0.97 for a superior merit within a BP FS. Whilst a UK traditional low input system was shown to generate the least amount of surplus P per litre, when further resource use indicators are included DEA results show that a BP system exporting manure to be most efficient which could in part be due to no requirement for land or nitrogen. A HG and BP FS could generate a dual production regime in which P is recycled. Manure P exported from the BP system could be utilized within the forages of a HG system, the need for fertilizer is diminished and manure is no longer exported.

Literature review on NH₃ and GHG emitted by pig production – part 1: building emissions
N. Guingand[1], S. Espagnol[1] and M. Hassouna[2]
[1]IFIP, La Motte au Vicomte, 35651 Le Rheu, France, [2]INRA, UMR SAS, 65 rue de Saint Brieuc, 35042 Rennes, France; nadine.guingand@ifip.asso.fr

In an attempt to evaluate emission factors of pig buildings and its possible use for national inventories, an analysis of the published literature was achieved and a database was developed on NH_3, N_2O, CH_4 and CO_2 emissions. Close to 900 references extracted from 120 articles were collected. The main quality of this database is the number of metadata filled: classic criteria like animal category, type of manure, type of floor, feed management are informed but specific criteria concerning the methodology applied (sampling and measurement methods), but also breeding characteristics (ADG, FCR, etc.), room design (density, ventilation rate, etc.) and manure management (pit depth, storage duration, etc.). More than 100 criteria were filled for the characterization of metadata. Collected data were mainly based on slurry system and on fattening pigs but references concerning solid manure systems and others animal category were also integrated. Ammonia is mainly illustrated in the literature with 60% of data collected. The direct comparison of emission factors was not possible because of the diversity of units used. A first work was done by converting data in grams of N or C per pig per day by using metadata collected permitting the comparison of emission factors. Around 20% data was lost after this first step. Average emission factor per gas were calculated (NH_3: 7.9±8.3 g N/d/p – N_2O: 1.9±7.4 g N/d/p – CH_4: 11.7±19.9 g C/d/p) illustrating the great variability of data. This study led us to determine average emission factors per animal category including several breeding management factors (type of floor, manure management, etc.) but also methodological factors. These appeared as one of the main factors explaining differences between references. This state-of-the-art led us also to identify the lack of data for some specific breeding management (scraping system, straw based litter, etc.) but also the importance of metadata in the proper use of references.

Literature review on NH₃ and GHG emitted by pig production – part 2: storage, treatment and spreading

S. Espagnol[1], N. Guingand[1] and M. Hassouna[2]
[1]IFIP, La Motte au Vicomte, 35651 Le Rheu, France, [2]INRA, UMR SAS, 65 rue de Saint Brieuc, 35042 Rennes, France; nadine.guingand@ifip.asso.fr

In order to evaluate the emission factors (EF) for pig production and its possible use in national inventories and environmental assessment, a literature review was conducted on gaseous emissions measurements concerning NH_3, N_2O, CH_4 and CO_2. A database was built and contains 2091 data issued from 229 articles dated from 1990 to 2012. The present article deals only with emissions from storage, treatment and spreading manure with respectively 396, 324 and 483 international references. To explain the variability, 65 metadata were collected for each reference. Some are similar for the different sources: type of manure, season, manure mass balance and gaseous measurement methodology. Others are specific to each source: type of storage, treatment and spreading (application equipment, dose, land occupation). Criteria (scale of experiment, manure composition, slurry management, etc.) were used to select references which could be used to calculate average representative EF for different sources: slurry storage, solid manure storage, composting, biological treatment, slurry spreading, solid manure spreading. Respectively 13, 42 and 10% of the references were kept for the storage, the treatment and the spreading. The EF obtained were analyzed with their standard deviation and the number of references used to their calculation. They were compared to international emissions factors from IPCC (2006) and EMEP EAA (2009). The results show that the number of data available to assess emissions factors for the sources storage, treatment and spreading of the manure is very low and could be considered as insufficient to have representative emissions factors. The average emissions factors obtained have a standard deviation which indicates an important variability of breeding situations and the need to multiply measurements. The methodologies of measurement appeared also to be relevant to explain a part of the variation.

Barriers for improvement of manure management

T. Vellinga[1], K. Andeweg[1], E. Teenstra[2] and C. Opio[2]
[1]Wageningen UR, Livestock Research, P.O. Box 338, Wageningen, the Netherlands, [2]FAO, Rome, 00153, Italy; theun.vellinga@wur.nl

Manure is a valuable source of nutrients, organic matter and renewable energy. It is an important by-product of livestock farming systems. Manure is also a significant contributor to greenhouse gas (GHG) emissions, accounting for 10% of total livestock GHG emissions. Despite its importance, manure management is often poor and as a consequence, nutrients and organic matter are lost, causing environmental and climate problems and threatening public health. Integrated manure management encompasses all activities associated with management of dung and urine; from excretion, collection, housing, and storage, anaerobic digestion, treatment, transport to application. It includes losses and discharge at any stage along the 'manure chain'. Technologies for and knowledge of integrated manure management are available; however implementation is often a challenge. To-date, very little is known about manure management at a global scale. A global assessment has been performed with responses from 34 countries to improve insight on manure management at farm level and barriers to adoption of integrated manure management. Four main barriers have been identified: (1) limited awareness of the importance of integrated manure management in contributing towards food security and reducing climate emissions, (2) the level of knowledge, (3) limited access to financial credits and other financial incentives, (4) ineffective (and even contradictory) manure policies and legislation. The assessment provided an insight in potential strategies to improve manure management in practice which focuses on the added value of manure as a fertiliser to enhance efficiency in agricultural production, close nutrient cycles and reduce overall environmental impact of livestock. 6 pilots are identified in Asia, Africa and Latin America, which will work on uptake of knowledge and technologies. The presentation will discuss the first outcomes of the pilots, and how to improve uptake of innovation in livestock farming systems.

Water footprint of extensive sheep systems in the semi-arid region of Chile

P. Toro-Mujica, C. Aguilar and R. Vera
Pontificia Universidad Católica de Chile, Departamento de Ciencias Animales, Vicuña Mackenna 4860,
Santiago, Chile; pmtoro@uc.cl

Sheep production in Chile is largely based on native rangelands whose productivity depends on rainfall and stocking rate. A simulation model was developed to assess the influence of the above variables on system's water footprint. The model integrates two existing models that simulate forage growth and lamb production respectively. The forage model was modified to incorporate the Penman-Monteith equation that calculates the reference evapotranspiration and the K_s coefficient that represents the effect of water stress on crop evapotranspiration. The lamb model was modified to incorporate ewe's requirements for pregnancy (fertility and birth weights) and lactation (milk production and its energy value). Random effects with a normal distribution were included for dry matter and water intake, lamb birth weight and monthly rainfall. A typical 100 ha farm was simulated. A 3^2 factorial experiment including three sheep stocking rates (0.5, 1, and 1.5 sheep/ha, SR) and three annual rainfalls (dry=380 mm; average=702 mm; rainy=1,081 mm) representatives of the observed range for the years 2007, 1993 and 2005 respectively. Lambs were weaned and sold at 4 months of age. Fifty reps were run for each scenario. Response variables included, among others, the water footprint/kg meat (WM) and kg of lamb/ha (KL). The average KL was 7,103±635 l/kg (97±3% green water), and that of KL was 23.8±8.9 kg/ha. The SR-rainfall interaction was significant only for KL. The lowest value of WM (6,414±211 l/kg) corresponded to an average year at the high SR, and the highest value for KL (35.4±0.9 kg/ha) was obtained for a rainy year with the highest SR. It was concluded that the model is useful for estimating technological outputs and the effect of climate (rainfall) change on water use efficiency.

Lambs performance with limited access to grazing: stocking rates and supplementation effects

L. Piaggio[1], M.D.E.J. Marichal[2], M.L. Del Pino[1], H. Deschenaux[1] and O. Bentancur[2]
[1]Secretariado Uruguayo de la Lana, Investigación y Desarrollo, Rambla B. Brum 3764, 11800,
Uruguay, [2]UdelaR, Facultad de Agronomía, Avda. Garzón 780, 12900 Montevideo, Uruguay;
mariadejesus.marichal@gmail.com

An experiment was conducted to evaluate the performance of lambs when grazing was limited to short periods and lambs were supplemented with an energy concentrate. Thirty six female Poll Dorset × Merino lambs (30.4±3.6 kg BW; 3.4±0.2 condition score (CS)) and 36 male Merino lambs (27.9±2.7 kg BW; 3.1±0.2 CS) were allotted to 12 groups (6 lambs/group, 2 groups/treatment). The experiment lasted 133 days, divided into 89 winter period days (WP) and 44 spring period days (SP). Pasture was a native pasture improved with *Lolium multiflorum*, *Trifolium repens* and *Lotus corniculatus*. In the WP different time of access to grazing (TAG: 2, 4, 6 and 8 h/day) and stocking rates (SR) were applied. In SP lambs continuously grazed at the same SR than in WP. Energetic supplementation (S), 300 g/lamb/day was evaluated shorts grazing times. Treatments were: 2H:TAG 2 h/d, 30 lambs/ha; 2HS:TAG 2 h/d + S, 36 lambs/ha; 4H:TAG 4 h/d, 24 lambs/ha; 4HS: TAG 4 h/d +S, 29 lambs/ha; 6H: TAG 6 h/d, 16 lambs/ha; 8H: TAG 8 h/d, 12 lambs/ ha. The highest ADG was observed in 8H (146 g/lamb/day) and the lowest in 2H, 2HS and 4H (88.3, 97.5 and 100.2 g/lamb/day, respectively), 4HS and 6H presented intermediate values (107.9 and 132.7 g/lamb/ day, respectively).The largest ($P<0.05$) animal productivity (kg of BW/ha) was registered in 2HS and 4HS and the lowest ($P<0.05$) in 8H (428.5, 382.5 and 215.4 kg BW/ha, respectively), intermediate values were observed in 2H, 4H and 6H (320.4, 293.9 and 291.0 kg BW/ha). Short times of grazing in winter at high SR with supplementation, followed by continuous grazing in spring at the same SR and supplementation, appears to be effective management and feeding strategies to duplicate lamb productivity.

Evaluation of different covers to reduce gaseous emissions from pig liquid manure

R. Matulaitis, V. Juskiene and R. Juska
Lithuanian University of Health Sciences, Institute of Animal Science, R. Zebenkos 12, 82317, Baisogala, Lithuania; violeta@lgi.lt

Livestock manure is a significant source of greenhouse gas emissions, other pollutants and can cause environmental damage such as soil acidification, eutrophication or has a negative impact on the health. The covers can control gaseous emissions from manure, but the impact is still under discussion. The aim of this study was to evaluate the effects of covering pig liquid manure on methane (CH_4), nitric oxide (NO), hydrogen sulfide (H_2S), ammonia (NH_3), carbon monoxide (CO), and carbon dioxide (CO_2) emissions. Six types of floating covers were tested: light expanded clay aggregate (leca), peat, sunflower oil, sawdust, straw, and plastic film. The manure was stored at 5, 15, and 25 °C for 37 d. Gaseous emissions were measured from headspaces of dynamic chambers. The results of our study showed that both the covering and temperature have a noticeable impact on gas emission from pig liquid manure. The plastic film cover was the most efficient at all temperatures tested, because it reduced the emissions of all measured gases. In this instance, average emission reductions were: CH_4 91.5% (P<0.01), NO 92.0% (P<0.05), H_2S 78.1% (P<0.05), NH_3 54.7% (P<0.01), CO 98.4% (P<0.01), and CO_2 67.1% (P<0.01). Other covers had inconsistent impact on separate gas emissions. However, covers generally helped to decrease the emissions of NH_3, H_2S, and CO_2.

Preliminary study on Alpine ibex and livestock distribution in Gran Paradiso National Park

M. Zurlo[1], E. Avanzinelli[2], B. Bassano[1] and N. Miraglia[3]
[1]Gran Paradiso National Park, Via della Rocca 47, 10123 Torino, Italy, [2]CNR, Institute of Atmospheric Sciences and Climate, Corso Fiume 4, 10133 Torino, Italy, [3]Università degli studi del Molise, Department of Agriculture, Environment and Feeds, Via Francesco de Sanctis 1, 86100 Campobasso, Italy; miraglia@unimol.it

In the last decades there was a drastic decrease in Gran Paradiso National Park ibex population linked with a reduction of stable occupied territories. Causes are still not completely clear but drastic decline is in partly related to recent climate changes. The objectives of this work are: (1) understand ibex distribution in GPNP in 1985-2009 period and describe livestock distribution in the same area in 2000-2009; (2) assess relation between distribution pattern and ibex population trend in 2000-2009. To understand distribution patterns 5 landscape ecology metrics are been selected to assess the composition and spatial configuration of occupied areas. Spearman's rank correlation coefficient was used to test composition and configuration metric trends and their relation with ibex population size. Results showed a reduction of ibex occupied territories from 4,587.50 hectares in 1985 to 2,331.25 ha in 2009 (r_s=-0,818; P<0,001). Number of patches increased from 130 to 224 units (r_s=0,784; P<0,001). Livestock distribution didn't show a particular trend (r_s≈0 or P>0,05). The relation between changes in ibex population trend and distribution patterns was not proven (all P>0,005). These results suggest that probably ibex distribution was influenced by different combined factors (landscape changes, climate change, anthropic activities) and they show how landscape ecology approach may become an useful tool to understand the degree of fragmentation and connectivity of landscape defined on species distribution. In conclusion, the understanding of processes behind Alpine ungulates distribution have to consider the influence of landscape patterns on environmental processes to improve the conservation efforts at management level.

Treatment efficiency of the plug-flow anaerobic digester for swine farm wastewater in Thailand

C. Nuengjamnong[1] and P. Rachdawong[2]
[1]Chulalongkorn University, Animal Husbandry, Henri Dunant Rd., Patumwan, 10330 Bangkok, Thailand,
[2]Chulalongkorn University, Environmental Engineering, Phayathai Road, Patumwan, 10330 Bangkok,
Thailand; chackrit.n@chula.ac.th

The objective of this research focused on the removal efficiency of the plug-flow anaerobic digester for a swine farm in Thailand. Eighty four wastewater samples from the digester were collected for nine months. Influent and effluent regions were selected for wastewater collection. The removal efficiency of system was evaluated by key parameters such as chemical oxygen demand (COD), soluble chemical oxygen demand (SCOD), total solids (TS) and total suspended solids (TSS). These parameters were analyzed using the Standard Method for Examination of Water and Wastewater. The average removal efficiencies of COD, SCOD, TS and TSS were 80.6, 85.6, 58.7 and 82.7% respectively. For the COD removal, the solid part was removed by sedimentation process while the soluble part was done by anaerobic digestion process. The TSS removal efficiency was quite high, indicating that the system could separate suspended solid from wastewater. These results indicated that the system could eliminate solid and soluble portions of swine farm wastewater.

A comparison of egg production from hens reared under organic, semi organic and conventional systems

A. Salahshour[1], R. Vakili[2] and A. Foroghi[1]
[1]Jihad-e-Agriculture Higher Educational Complex, Organic Animal Husbandry, TV Square, Asian Highway,
Mashhad, Iran, [2]Islamic Azad University,Kashmar Branch, Animal Science, Seyed Morteza blvd, Mashhad,
Iran; afroghi@yahoo.com

The aim of this study was to investigate the impact of production system (organic, semi organic and conventional) on egg yolk and production performance of laying hens. This experiment consisted of 192, eighty week old shaver, over 120 days with three treatments and four replicate. The control group was fed inorganic feed, the organic group was fed organic feed and semi organic group fed organic feed with synthetic amino acid and vitamins. Evaluated traits were egg production, egg mass, feed intake, feed conversion rate and body change weight. Production system has significant effect of egg production.egg mass, feed intake, feed conversion rate and body change weight (P<0.05). The best of performance observed in control group that has significant difference with organic group and semi organic group (P<0.05). The lowest level of cholesterol in yolk egg observed in semi organic and organic group (P<0.0006). The result of experiment showed organic and semi organic systems reduced production performance and cholesterol in egg yolk.

Innovative grazing system with sheep for sustainable management of the alpine cultural landscape
F. Ringdorfer, R. Huber, T. Guggenberger and A. Blaschka
Research and education centre, Raumberg 38, 8952, Austria; ferdinand.ringdorfer@raumberg-gumpenstein.at

The innovative approach of this project was that a large flock of sheep was kept by a shepherd with the assistance of his dogs specifically to the slopes and alpine meadows of the Hauser-Kaibling. In Austria it is traditionary that sheep move free on the alpine pastures. Sheep are foodies and as such they always seek out the best grasses and herbs and move on to seek new. The less palatable plants remain. By herding the flock to the point all plants should be eaten up by the sheep. Flock size was between 650 to 950 sheep and lambs in the different years. The aim of the project: effects of grazing on the plant population and forage yield; study of animal performance; impact on animal health. Results: In the 6 years of the duration of the project total 4914 sheep and lambs have been forced onto the mountain in total. The average grazing period was 104 days. This herd has consumed about 4 million kg of biomass in total. The grazing has meant that, on the one hand the dwarf shrubs were pushed back, on the other hand have grown grasses and herbs. On the intensively grazed areas, the dwarf shrubs proportion of initially around 30% has been reduced to 5%, the grasses and herbs increased to 65% in return from the original 40%. On a zero variant surface, that is, without grazing the dwarf shrubs have increased from 30% to 70%. The forage yield has increased after 4 years of grazing of 880 kg DM/ha to 1150 kg DM/ha. The performance of the lambs, measured in daily gains was in a range 90 to 210 grams. If the sheep guarded, then the increases are lower than if the sheep can move freely. Alpine Lambs have compared to lambs from livestock housing not as pronounced meaty carcasses, but the inner quality of the Alpine Lambs is better. By this is meant that the proportion of valuable fatty acids (omega-3 fatty acids) is higher in lambs from alpine pasture than fattening lambs from stable. Many sheep from many farm kept together in one flock results in higher problems with parasites and claw diseases.

Genetic heteroscedastic model to study genetic variability of sheep litter size
S. Fathallah, I. David and L. Bodin
INRA, GenPhyse, Chemin de Bordé Rouge Auzeville-Tolosane, 31320, France;
samira.fathallah@toulouse.inra.fr

In order to provide tools to maintain a stable production in various environments; several authors have developed heteroscedastic models to jointly estimate the influence of genetic and environmental factors on both the mean and the variability of a trait. These models mainly target continuous traits and are not theoretically appropriate for the analysis of discrete traits. To overcome this drawback, we have proposed recently a heteroscedastic threshold model suitable for the genetic analysis of ordinal data. This model is an extension of the classical threshold model in which non-genetic and genetic factors of heterogeneity of the variance of its underlying variable are included. It permits to break down the link between mean and variance that holds with the classical homoscedastic threshold model and to estimate genetic values for the mean and the variance of the underlying variable. In the present study, we applied this model to analyze litter size variability in three breeds with very different mean litter size in order to covers the full range of prolificacy observed in the French breeds. We analyzed the litter size of 18,888 Romane, 36,076 Rouge de l'Ouest and 6,404 Suffolk ewes. The genetic variance calculated for the mean and the residual variance on the underlying scale were respectively 0.070 and 0.016 for the Romane, 0.10 and 0.022 for the Rouge de l'Ouest and 0.079 and 0.008 for the Suffolk.

Milk composition of dairy goat fed with de-stoned olive cake silages
G. Keles and V. Kocaman
Adnan Menderes University, Faculty of Agriculture, Animal Science, 09100, Aydin, Turkey;
gurhan.keles@adu.edu.tr

As an agro-industrial by-product, olive cake (OC) presents high potential as ruminant feedstuff, considering both widespread feed shortages and its increasing production. Effective technologies used for reducing kernel level allow OC to use it more efficiently in ruminant nutrition. Together with lower kernel level, the remaining oil after extraction process makes the OC a valuable feedstuff for improving ruminant performance and product quality. Therefore, the present study was carried out to test the feeding value olive cake silage (OCS) after effective and applicable de-stoning process for lactating dairy goat. The OC de-stoned in fresh form by using 3.5 mm sieve and was ensiled directly in 120 l drums equipped with a lid that enables the release of the fermentation gases. A total of 18 Saanen crossbred goats (late lactation) divided into three groups and were individually fed total mixed ration (TMR) that contains OCS at DM proportions of 0.0 (C), 0.10 (OC10) and 0.20 (OC20). The TMRs contained 70% forage and were isonitrogenous. The experiment continued for three weeks after 12 d adaptation period. Performance and milk composition were determined in the first and the last week of feeding trial. The data were analyzed in a completely randomized design using a model that accounted for the main effects of sampling time and the level OCS used in factorial arrangement. Apparent digestibility of ration's crude protein, ether extract and NDF were increased in each increment of the OCS. Goats fed OC10 had lower ($P<0.05$) dry matter intake than the C group. The ECM yield of Goat offered C, OC10 and OC20 were 1.02, 1.07 and 1.10 kg/d ($P>0.05$), respectively. Inclusion of OCS in the diets at higher ratio increased ($P<0.05$) milk fat by approximately 16% (3.7 vs 4.4%). The results indicated that using two-phase de-stoned OCS at a level that could be considered high for lactating ruminant does not cause any detrimental effect on intake of Goats, but considerably increases the milk fat.

Diet supplementation with 18:0 does not alleviate fish oil-induced milk fat depression in dairy ewes
P.G. Toral[1], G. Hervás[1], D. Carreño[1], J.S. González[1], J. Amor[2] and P. Frutos[1]
[1]Instituto de Ganadería de Montaña, CSIC-ULE, Finca Marzanas, 24346 Grulleros, León, Spain, [2]INATEGA S.L., Ctra. Valdefresno 2, 24228 Corbillos de la Sobarriba, León, Spain; pablo.toral@csic.es

The supplementation of ewe diet with fish oil (FO) may improve the fatty acid (FA) profile of milk but induces milk fat depression (MFD). This latter effect has been associated with a shortage of 18:0 for mammary cis-9 18:1 endogenous synthesis and its possible impact on the maintenance of milk fat fluidity. On this basis, this study was conducted to test the hypothesis that supplemental 18:0 would alleviate FO-induced MFD in sheep. The assay followed a 3×3 Latin square design (4 animals/group) with 3 periods of 4 weeks each and 3 experimental diets: non-supplemented (control), and supplemented with 2% FO (FO) or with 2% FO plus 2% 18:0 (FOSA). Diets were offered *ad libitum*, and milk production and composition (including a complete FA profile) were analyzed on the last 3 days of each period. At the end of the trial, the digestibility of additional 18:0 was estimated using 6 lactating sheep. Milk yield was not affected by the inclusion of fish oil in the diet ($P>0.10$), despite supplementation with FO alone tended to decrease feed intake ($P<0.10$). However, both FO and FOSA, compared with the control, reduced milk fat content in a similar proportion (20%; $P<0.05$), which suggests that the addition of stearic acid does not alleviate FO-induced MFD. Additional 18:0 showed a relatively low digestibility coefficient but this data does not seem enough to fully explain the negative results, because the decrease in milk cis-9 18:1 concentration linked to dietary marine lipids was partially compensated by this supplement. It is therefore hypothesized that variations in other metabolites (such as odd- and branched-chain and biohydrogenation-derived FA) might contribute to explain the unsuccessful response to the addition of 18:0.

The cholesterol content in meat, fat and giblets of lambs depending on the breed and feeding

B. Borys[1], R. Niznikowski[2], M. Swiatek[2] and B. Kuczynska[2]
[1]National Research Institute of Animal Production, Krakow, Experimental Station, Koluda Wielka, 88-160 Janikowo, Poland, [2]Warsaw University of Life Sciences, SGGW, Ciszewskiego str. 8, 02-786 Warsaw, Poland; bronislaw.borys@onet.eu

The study was conducted on muscle longissimus lumborum (LL), external fat, and giblets (liver, lung, heart and kidney) collected from 30 ram lambs of Koluda Sheep (50%) and its crossbreeds with Ile de France (50%) fattened at body weight 32-37 kg. The lambs were feeding intensively (IN) *ad libitum* using dry mash (15% of rape cake RC and 15% DDGS + 5% of linseed) or semi-intensively (SIN) using dry mash – 3% of body weight with hay *ad libitum* or with pasturing (5 h/day). The dry mash in SIN contained 55% oil components – 50% RC or DDGS + 5% of linseed. The fat extraction was performed according to the method of Röse-Gottlieb and cholesterol content was performed using the Agilent Technologies gas chromatograph, type 7890A, column HP-5 (30 m × 320 μm × 0,25 μm) – injector temp. 250 °C, split sample injector 1:25, detector temp. 300 °C. The tested tissues and organs differed in cholesterol content – the highest in the lung (633.2 mg/100 g), then in the liver and kidney (242.8 and 201.2 mg/100 g), external fat (71.5 mg/100 g) and the lowest in LL and heart; 46.4 and 41.3 mg/100 g. Breed origin did not affected on content of cholesterol. The method of fattening influenced on cholesterol content in external fat (IN>SIN; 58.8 vs 36.5 mg/100 g; P≤0.01). Pasturing of lambs compared with the use of hay increased the cholesterol content in the liver (290.7 vs 209.6 mg/100 g; P≤0.01), with tendency to increase the concentration of this component in the LL and lung (about 20%). The use of DDGS compared with RC, increased the cholesterol content in the lungs (721.0 vs 513.0 mg/100 g, P≤0.01), with a tendency to a lower concentration of this component in the external fat and heart (average about 15%).

Mature size of Saanen goats

A.K. Almeida, M.H.R.M. Fernandes, K.T. Resende and I.A.M.A. Teixeira
Unesp/Univ Estadual Paulista, Animal Science, Via de Acesso Prof. Paulo Donato Castellane, 14884-900, Brazil; almeida.amelia@gmail.com

Mature size has been recalled in many areas of animal nutrition such as intake predictions and nutritional recommendations. Conceptually, chemical maturity of the animals is reached when their crude protein concentration in fat-free dry matter becomes constant or, according to Australian feeding system, when skeletal maturity is reached. However, the evaluation of the growth of goats aiming to determine their maturity has not been yet defined. For this purpose, a database with body composition of 76 animals was used, gathering information of five studies with female Saanen goats from 4.6 to 59.4 kg of body mass, from which the percentage of ash or protein:fat ratio during growth, were adjusted to a non-linear logistic function: $Y=\beta_0 \times exp^{(-\beta_1 \times EBM)}+\beta_2$; here β_2 is the asymptotic percentage of ash or protein:fat ratio. Logistic functions were fitted using the SAS macro %NLINMIX. The between-study variation was considered by introducing a random parameter for each parameter estimates. Mature empty body mass (EBM) was assumed to be the asymptote which was estimated iteratively by the upper limit of the confidence interval of β_2. On the other hand, the logistic function fitted with the percentage of ash estimated the mature EBM as 16.9 kg. As animals grows, the body favors fat deposition at the expense of lean-tissue growth, the fitted logistic equation of protein:fat ratio showed the asymptote at 15.9 kg EBM. These results are consistent and correspond to the puberty period of these females. The shift in the hormonal profile, triggered during puberty may play a role in the asymptote estimates. It is worth to point out that this representation is not as a standard value of a breed, gender or nutrition planes, because such factors are likely to affect the mature size (FAPESP, Grant # 2014/14734-9 and 2014/14939-0).

Ash and protein deposition patterns in the body of Saanen goats
A.K. Almeida, M.H.R.M. Fernandes, I.A.M.A. Teixeira and K.T. Resende
Unesp/Univ Estadual Paulista, Animal Science, Jaboticabal/SP, Brazil, Via de Acesso Prof. Paulo Donato
Castellane, 14884-900, Brazil; almeida.amelia@gmail.com

Animal growth has long been characterized by the change in body mass (BM) per unit of time (age). Growth affects most management decisions in a herd, mainly regarding intake projections and nutritional requirements of a given group of animals. As the animal grows, a shift of priority of tissue deposition occurs and will influence the efficiency for gain of the animal, thus the body composition should be a better predictor of the phenomena of growth. Thus, the aim of this meta-analysis was to fit a growth curve that describes the increase of protein and ash contents in function of the empty body mass (EBM). For this purpose, a database with body composition of 76 female Saanen goats weighing from 4.6 to 59.4 kg BM were used, gathering information of five studies. Gompertz growth functions were fitted using a nonlinear mixed model methodology by the SAS macro %NLINMIX. This function was used because showed best fit among growth functions tested. The between-study variability was modeled by introducing the parameters u_1, u_2 and u_3 to the β_0, β_1 and β_2 parameters, respectively. The equations for protein and ash content were Protein, $g = 300.1 \pm 24.5 \times \exp^{3.15 \pm 0.0801 \times (1-\exp{-0.0772 \pm 0.00480 \times EBM, kg})}$, $\sigma^2_e = 0.00158$; and Ash, $g = 53.7 \pm 7.10 \times \exp^{3.232 \pm 0.119 \times (1-\exp{-0.0922 \pm 0.00793 \times EBM, kg})}$, $\sigma^2_e = 0.00544$. The estimated accumulation rate was higher for ash than for protein (3.23 vs 3.15), indicating that the ash content reach its asymptote at lower EBM than protein. It is important to consider it when defining which nutrient will dictate the mature size; some authors consider protein content and other ash content. Moreover, the estimated inflection point occurred at 13 kg of EBM or 10.8 kg EBM, according to protein or ash equations, respectively. It is expected that from this point onward, there is a decrease in growth rate, demanding changes in feeding strategy. FAPESP, Grant # 2014/14734-9 and 2014/14939-0.

The vitamin A and E content in meat, fat and giblets of lambs depending on the breed and feeding
M. Swiatek[1], B. Borys[2], R. Niznikowski[1] and K. Puppel[1]
[1]Warsaw University of Life Sciences – SGGW, Ciszewskiego str. 8, 02-786 Warsaw, Poland, [2]National Research Institute of Animal Production, Experimental Station, Koluda Wielka, 88-160 Janikowo, Poland; roman_niznikowski@sggw.pl

The study was conducted on meat (longissimus lumborum; LL), cover fat and giblets (heart, liver, kidney, lung) collected from ram lambs – Koluda Sheep and its crossbreeds with Ile de France fattened at body weight 32-37 kg. The lambs were feeding intensively (IN) *ad libitum* using dry mash – 15% of rape cake (RC) and DDGS + 5% of linseed) or semi-intensively (SIN) using dry mash – 3% of body weight with hay *ad libitum* or with pasturing (5 h/day). The dry mash in SIN contained 55% oil components – 50% RC or DDGS + 5% of linseed. Vitamins were determined using reversed-phase high-performance liquid chromatography method (RP-HPLC). Differences in the level of vitamin A and E in tested organs were observed. The highest level of vitamin A (4.39 mg/100 g) as well as vitamin E (8.71 mg/100 g) were observed in liver. The lowest level of both vitamins were find in lung (vit. A: 0.007; vit. E: 0.28 mg/100 g). In case of LL were the following results (vit. A: 0.91; vit. E: 1.52 mg/100 g). Genotype did not affected on content of vitamin A and E in tested organs. The applied methods of fattening influenced on level of vit. E in kidney (IN: 9.3 mg/100 g; SIN: 1.83 mg/100 g; P≤0.01). Content of vit. A comparing IN and SIN were leveled. Pasturing of lambs compared with use of hay increased level of; vit. A in LL (1.23 vs 0.64 mg/100 g; P≤0.01), lungs (0.010 vs 0.004 mg/100 g; P≤0.01) heart (0.26 vs 0.09 mg/100 g; P≤0.01) and vit. E in lung 1.69 vs 0.70 mg/100 g; P≤0.01). Conversely use of hay increased level of vit. A in liver (5.11 vs 3.48 mg/100 g; P≤0.01), kidney (2.53 vs 1.83 mg/100 g; P≤0.01) and vit. E in kidney (3.50 vs 0.17 mg/100 g; P≤0.01). Addition of RC and DDGS affected only on level of vit. E in kidney (RC>DDGS; 3.25 vs 0.41 mg/100 g; P≤0.01) and in lung (RC<DDGS; 0.84 vs 1.56 mg/100 g; P≤0.05).

Fast and sensitive identification of species-specific milk using PCR and PCR-RFLP techniques

S. Abdel-Rahman
Genetic Engineering and Biotechnology Research Institute, Department of Nucleic Acid Research, City of Scientific Research and Technological Applications, Post Code 21934, Alexandria, Egypt; salahmaa@yahoo.com

For the fast, specific and sensitive identification of buffalo's, camel's, cattle's, goat's and sheep's milk, species-specific PCR and PCR-RFLP techniques were developed. DNA from small amount of fresh milk (100 µl) was extracted to amplify the gene encoding species-specific regions (SSR) and the mitochondrial DNA segment (cytochrome-b gene). PCR amplification size of the gene encoding SSR regions was 603 bp in both buffalo's and cattle's milk, 300 bp in camel's milk, 855 bp in goat's milk and 374 bp in sheep's milk. On the other hand, PCR amplification size of the gene encoding cytochrome-b gene was 359 bp in both buffalo's and cattle's milk. Polymerase chain reaction-restriction fragment length polymorphism (PCR-RFLP) technique was used to differentiate between buffalo's and cattle's milk. Restriction analysis of PCR-RFLP of the mitochondrial cytochrome-b segment (359 bp) analysis showed difference between buffalo's and cattle's milk. Where, the fragment length generated by TaqI restriction enzyme was 191 and 168 in buffalo's milk, whereas no fragments were obtained in cattle's milk. The proposed PCR and PCR-RFLP assays represent a fast and sensitive method applicable to the detection and authentication of milk species-specific.

Growth Curve in male and female lambs of Baluchi Sheep by Nonlinear Growth Models

M. Hosseinpour Mashhadi
Department of Animal Science, Mashhad Branch, Islamic Azad University, Mashhad, 91735-413, Iran; mojtaba_h_m@yahoo.com

Baluchi is the most common native breed of sheep in Iran, comprising 29% of the sheep population. This breed is native to the eastern part of the country, which has a dry and hot climate. The animals have had to adapt to the harsh environment. The purpose of this study is to describe growth curve in Baluchi Sheep by application of nonlinear growth models. The weight records of 1,228 and 676 Baluchi male and female lambs from birth weight, one-month, two-month, three-month, four-month, six-month, nine-month and one year of age were studied. These data had been collected by Baluchi Sheep Breeding Center in Mashhad (Abbas Abad) during 2004 to 2009. Brody, Von Bertalanffy, Gompertz and Logistic nonlinear models were fitted to describe the growth curve. The most suitable model was determined by R2, Root MSE, as well as mean absolute error (MAE). For data analysis, NLIN procedures were used. The results of this research for male lambs show that Brody model with R2 (90.4276%), Root MSE (8.35) and mean difference between expected and observed value (MAE) (5.6578) is the best model compared with others, this value for best model for female lamb were 96.85% (R2), 4.54 (Root MSE) and 3.16 (mean difference between expected and observed value). The correlation between A and K parameters for Von Bertalanffy, Brody, Gompertz and Logistic were -0.36, -0.54, -0.3 and -0.21 respectively.

Linear type traits in dairy goats
P. Herold[1], D. Birnbaum[2], J. Bennewitz[2] and H. Hamann[1]
[1]Landesamt fuer Geoinformation und Landentwicklung Baden-Wuerttemberg, Abteilung 3, Referat 35, Stuttgarter Str. 161, 70806 Kornwestheim, Germany, [2]Universitaet Hohenheim, Institut fuer Nutztierwissenschaften, Garbenstr. 17, 70593 Stuttgart, Germany; pera.herold@lgl.bwl.de

Dairy goat farming is a growing alternative branch in German livestock production. Therefore, a breeding value estimation based on selection goals of dairy goat farmers is to be set up. Overarching goal is to establish a breeding value estimation for lifetime performance. Breeding values for milk traits were published in 2014 for the first time. In a next step, the precise recording of type traits is essential as underlying information on conformation: Strong correlations between conformation / type traits and productive lifetime are known from dairy cattle and are also assumed for goats. A rating sheet with 10 traits on stature and form and 8 traits to describe the udder was developed. In 2013 and 2014 in total 194 German Fawn and 48 Thueringer Wald goats in 1st lactation were characterized on basis of the rating sheet. In 2013, goats were recorded at two farms by 2 judges; in 2014 goats were recorded on 10 farms by 9 judges. Herd sizes ranged from 5 to 270 dairy goats with 3 to 63 in 1st lactation. First results suggest that there are correlations between udder traits, especially udder depth and fore attachment, and milk performance. Heritabilities ranged from 0.03 for teat position to 0.83 for fore attachment in udder traits, from 0.03 for abdomen width to 0.69 for withers height in stature traits and from 0.08 for leg posture to 0.40 for regularity of teeth. Because of the small data set herd had a major effect on most type traits. Herd and judge were also compound effects. Overall, the first results seem promising. In a next step, data recording should be broadened to more farms and animals.

Investigation of ensiling possibilities of high moisture oak acorn with molasses
A. Kamalak[1], O. Canbolat[2], C.O. Ozkan[1], M. Sahin[1] and Y. Akagunduz[1]
[1]Kahramanmaras Sutcu Imam University, Animal Science, Avsar Campus, 46100 Kahramanmaras, Turkey, [2]Uludag University, Animal Science, Gorukle Campus, Bursa, Turkey; akamalak@ksu.edu.tr

The fruit of oak trees is a nut called an acorn which has similar content to that of cereal grains. The success in ensiling of these grains with high moisture content encouraged to test the idea of ensiling of oak acorn with high moisture. So far there is no attempt to ensile the acorn of oak trees in the literature. Therefore the aim of the current study was to investigate the possibilities of high moisture oak acorns with molasses. The acorn was ensiled with molasses (0, 1.5, 3.0, 4.5 and 6.0 on fresh basis) quadruplicate in plastic silo with 3 kg capacity. The molasses had a significant (P>0.001) effect on the chemical composition and fermentation parameters of acorn silage. Dry matter (DM), crude protein (CP) and crude ash (CA) contents of the acorn silages increased with increasing level of supplementation of molasses whereas ether extract(EE), neutral detergent fiber (NDF) acid detergent fiber (ADF), acid detergent lignin (ADL) and condensed tannin (CT) contents were decreased. The decrease in EE, NDF, ADF, ADL and CT of the resultant acorn silage is due to dilution effect of molasses which does not contain any NDF, ADF, ADL fractions or lower CT contents than acorn. Supplementation of molasses increased the lactic acid and propionic acids contents whereas the supplementation of molasses decreased ascetic, butyric acids contents and pH of the resultant acorn silage. The increase in lactic acids and decrease in pH of the resultant acorn silage are associated with the high content of water soluble sugar content of molasses. As a conclusion, the supplementation of molasses improved acorn silage quality. Ensiling of high moisture acorn with molasses offers an excellent alternative to dry acorn for small ruminant animals. This study was supported by The Scientific and Technological Research Council of Turkey (TUBITAK), Project No: 213 O 220).

Ultrasound measurements of longissimus dorsi and subcutaneous fat in two Romanian local breeds
E. Ghita, R. Pelmus, C. Lazar, M. Gras, C. Rotaru, T. Mihalcea and M. Ropota
National Research Development Institute for Animal Biology and Nutrition, Animal Biology, Calea Bucuresti nr.1, Balotesti, Ilfov, 077015, Romania; elena.ghita@ibna.ro

The local Romanian breeds are dual purpose breeds; until some 2 decades ago they have been bred mainly for wool production, meat and milk being regarded as secondary. Since the demand for wool decreased, it is necessary to improve these breeds for milk and meat production. The purpose of our experiment was to study the aptitudes for meat production of two local breeds (Teleorman Blackhead; TBH and Tsigai; T) belonging to three farms (F). We evaluated the properties of Longissimus Dorsi muscle (depth, area and perimeter) and the thickness of the subcutaneous fat on live animals using ultrasound. We surveyed 45 TBH lambs (F1) aged 150 days and weighing 28.19 kg; 37 TBH (F2) lambs aged 270 days and weighing 41.0 kg and 60 T lambs (F3) aged 260 days and weighing 29.84 kg). The ultrasound measurements were performed with an Echo blaster 64 using LV 7.5 65/64 probe, supplied by TELEMED ultrasound medical systems. The ultrasound images were recorded using Echo Wave II software version 1.32/2009. The first measurement point was 5 cm from the spine, at the 12th rib; the second measuring point was between 3rd and 4th lumbar vertebrae. The average values of the subcutaneous fat layer thickness, LD muscle depth, area and perimeter were 1.74 mm; 19.91 mm; 9.08 cm^2; 130.75 mm for F1 lambs; 2.10 mm; 22.98 mm; 9.96 cm^2; 131.55 mm for F2 lambs and 1.95 mm; 23.47 mm; 9.88 cm^2; 129.69 mm for F3 lambs. These results show that there are very significant differences (P<0.001) between the lambs of the two breeds and between the lambs of the same breed depending on age and body weight. Although T lambs had lower gains, their LD muscle had better properties than TBH lambs. There were significant phenotypic correlations between the body weight and the ultrasound measurements, the correlation coefficients ranging between 0.240 and 0.705.

Genetic parameters of ewe reproduction, body weight and scrotal traits in the Elsenburg Merino flock
P.A.M. Matebesi-Ranthimo[1], S.W.P. Cloete[2,3], J.B. Van Wyk[4] and J.J. Olivier[2]
[1]National University of Lesotho, P.O. Roma 180, Roma, Lesotho, [2]Directorate Animal Science, P/Bag X1, Elsenburg, 7607, South Africa, [3]University of Stellenbosch, P/Bag X1, Matieland, 7602, South Africa, [4]University of the Free State, P.O. Box 339, Bloemfontein 9300, South Africa; vanwykjb@ufs.ac.za

Lamb output depends on the fertility of males and efficient reproduction of females. Ewe productivity (number or weight of lambs weaned per lambing opportunity) contributes significantly to lamb production. This study estimated the heritability of, and genetic correlations of, ewe reproduction with body weight and scrotal traits of South African Merinos. Data consisted of 4905 animals, the progeny of 241 sires and 1,502 dams (born 1986-2012). Ewe reproduction traits were (heritability in brackets): first parity number of lambs born (NLB1) (0.10), number of lambs weaned (NLW1) (0.07), and total weight weaned (TWW1) (0.10); totals over three parities for number of lambs born (NLB3) (0.25), number of lambs weaned (NLW3) (0.12) and total weight weaned (TWW3) (0.18). Body weight traits were: birth weight (BW) (0.17) within 24 hours of birth, weaning weight (WW) (0.16) at roughly 100 days and hogget live weight (LW) (0.36) at yearling age. Scrotal traits were: scrotal circumference (SC) (0.32) and testicular diameter (TD) (0.23). Genetic correlations (r_g ±SE) of WW and LW with ewe reproduction were moderate to high and favourable, ranging from 0.41±0.15 between LW and NLW3 to 0.86±0.17 between WW and TWW1. Positive and significant r_g were estimated for NLB1 and NLB3 with both measures of scrotal size, suggesting that rams with higher measurements of SC and TD are likely to father daughters with an improved lambing rate. Rams with an increased SC and TD are also likely to sire daughters that wean more lambs, as suggested by high favourable rg with NLW1 and NLW3. TWW1 and TWW3 were also favourably correlated with SC on the genetic level. Selection for ewe reproduction will result in favourable correlated responses in body weight and testis size.

Effects of flavonoids dietary supplementation on yoghurt antioxidant capacity

P. Simitzis[1], M. Goliomytis[1], M. Charismiadou[1], T. Massouras[2], K. Moschou[1], C. Ikonomou[1], V. Papadedes[1], S. Lepesioti[1] and S. Deligeorgis[1]
[1]Agricultural University of Athens, Department of Animal Breeding and Husbandry, 75 Iera Odos, 11855 Athens, Greece, [2]Agricultural University of Athens, Department of Food Science and Technology, 75 Iera Odos, 11855 Athens, Greece; pansimitzis@aua.gr

An experiment was conducted to examine the effects of dietary hesperidin or naringin supplementation on greek traditional yoghurt antioxidant capacity. Hesperidin and naringin are bioflavonoids that are abundant in inexpensive by-products of citrus cultivation such as citrus pulp. Thirty-six multiparous ewes (in their second lactation period) were assigned into 4 experimental groups of 9 ewes each. One of the groups served as control (C) and was given a commercial basal diet, without bioflavonoid supplementation, whereas the other three groups were given the same diet further supplemented with hesperidin at 6 g/kg (H) or naringin at 6 g/kg (N) or α-tocopheryl acetate at 0.2 g/kg (E) of concentrated feed. Yoghurt was manufactured by milk collected from ewes after 0, 7 and 21 days of dietary supplementation. Apart from the determination of colour parameters, pH values, syneresis and rheological characteristics (texture properties), measurements of antioxidant capacity were performed in yoghurt samples after refrigerated storage at 4 °C for 10 and 20 days. In general, oxidative stability of yoghurt, expressed as ng MDA/ml milk, was not influenced by the bioflavonoids' dietary supplementation. According to the findings of the present study, flavonoids do not seem to improve the quality characteristics of yoghurt. This research project was implemented within the framework of the project 'Thalis – The effects of antioxidant's dietary supplementation on animal product quality', MIS 380231, funding body: Hellenic State and European Union.

Relationship between β-LG variants and fatty acids composition in milk of Polish Lowland sheep

A. Radzik-Rant[1], A. Rozbicka-Wieczorek[2] and W. Rant[1]
[1]Warsaw University of Life Sciences, Department of Animal Breeding, Ciszewskiego 8, 02-786 Warsaw, Poland, [2]Polish Academy of Sciences, The Kielanowski Institute of Animal Physiology and Nutrition, Instytucka 3, 05-110 Jabłonna, Poland; witold_rant@sggw.pl

Beta-lactoglobulin (β-LG) is the major whey protein in ruminants milk. Its biological functions are still unclear. It is known that B-LG has ability to bind hydrophobic ligands such fatty acid or vitamins. Therefore the aim of this study was to evaluate the relationship between B-LG polymorphism and fatty acid composition as well as basic milk components in Polish Lowland ewes. The study was carried out on 30 ewes aged 3-4 years. Milk samples were collected during the 4th week of lactation and analyzed for basic components (fat, protein, casein, lactose, total solids), whey protein content and fatty acid composition. Whey proteins were examined using high-performance liquid chromatography and fatty acid by gas chromatography. Detailed analysis of milk proteins enabled β-LG variants to be determined. Three β-LG genotypes (AA, AB, BB) were identified in the studied ewes. The frequency of variant A was twice as high as that of variant B. There was no effect of the β-LG variant on basic components of milk. Milk with genotypes BB and AA was characterized by higher (P≤0.05, P≤0.01) content of MUFA compared to AB genotypes. In this group of fatty acid significant differences were related to the content of C17:1, C18:1C9 and C20:1. The highest (P≤0.01) content of polyunsaturated fatty acid n-3, including C18:3 n-3 and the smallest (P≤0.05) PUFA n-6, including C18:2 n-6 in the milk was associated with the AB genotype, although this genotype differed only with the AA. The highest content of LCFA, including C18:1C9 and C18:2 c9, t11 has been found in the milk of ewes with the genotype BB. The results, although ambiguously, showed relationship between β-lactoglobulin variant and some fatty acids content. In order to establish the role of B-LG in lipid metabolism in mammary gland the further study are required.

Health beneficial omega-3 fatty acids levels in lambs from artificial and traditional rearing

M. Margetín[1,2], M. Oravcová[1], O. Bučko[2], L. Luptáková[2] and J. Tomka[1]
[1]NAFC, Research Institute for Animal Production Nitra, Hlohovecká 2, 951 41 Lužianky, Slovak Republic,
[2]Slovak University of Agriculture, Tr. A. Hlinku 2, 949 76 Nitra, Slovak Republic; ondrej.bucko@uniag.sk

The meat quality of 120 light lambs of both genders from artificial (AR, n=40) and traditional rearing (TR, n=80) were assessed on the basis on essential and health beneficial fatty acids of intramuscular fat (IMF). The average weight of lambs before slaughter was 17.30 kg and age of lambs was 57.3 days. Milk powders with 24% of refined vegetable oil (coconut and palm) were given to AR lambs; ewe's milk was given to TR lambs. After 7 days from birth, both groups were offered commercial concentrates and hay. Fatty acid (FA) profiles were determined in M. longissimus dorsi by gas chromatography. Differences between AR and TR lambs were calculated using GLM model with effects of way of rearing (R), breed nested within R, sex (S) and interaction R×S. No statistical differences in IMF were found between AR and TR (3.90 vs 3.51 g/100 g meat; P>0.05). Highly significant differences in essential FA (linoleic, α-linolenic) were between AR and TR lambs (10.28 vs 6.82 and 9.88 vs 35.98 mg/100 g meat). The level of omega-6 FA was significantly higher (P<0.001) and the level of omega-3 FA was significantly lower in AR lambs than in TR lambs. The level of health beneficial long chain omega-3 FA (EPA, DPA, DHA) was significantly higher in TR lambs (P<0.001). The level of EPA+DHA was 5.32±2.02 mg/100 g meat in AR lambs; however, the level of EPA+DHA was 36.07±1.43 mg/100 g in TR lambs. The level of rumenic acid (C18:2 cis9, trans11) was significantly higher in TR lambs than in AR lambs (4.41 vs 23.71 mg/100 g meat). Thus, meat of TR lambs appears to be a rich source of omega-3 FA. In contrast, meat of AR lambs fed milk powders is of a lower level of omega-3 FA. These also have the unfavorable ratio omega-6/omega-3 FA (15.15 vs 3.28). The study was supported by the projects APVV 0458-10 and VEGA 1/0364/15.

Genetic variability in the closed nucleus of Iranian Moghani sheep estimated by pedigree analysis

M. Sheikhlou[1] and F. Bahri Binabaj[2]
[1]University of Tabriz, Ahar faculty of agriculture and natural recourses, Department of Animal Science, Ahar, Iran, [2]University of Gonbad Kavoos, Department of Animal Science, Gonbad Kavoos, 5177645659, Iran; m.sheikhlou@yahoo.com

A study was conducted to characterize the genetic structure and variability of the closed nucleus of Iranian Moghani sheep. Herdbook information collected from 1986 to 2014 was used to estimate the generation interval, inbreeding coefficient and average relationship coefficients between animals. Ewes that was born at last 4 years of the pedigree file with both parents known was considered as a reference population for estimating realized effective population size and parameters derived from probability of gene origins, including founders and effective number of founders, effective number of ancestors and effective number of founder genomes. The mean generation interval was 4.18 years in the studied period. The average complete generation equivalents of the reference population as an indicator of the pedigree completeness level was 4.18. Number of founders of the reference population were 460 animals, while the effective number of founders, effective number of ancestors and effective number of founder genomes of the reference population were 110, 83 and 42 respectively, indicating an unbalanced contribution of the parents to the next generations due to the selection of animals, creation of bottlenecks and random genetic drift. Average inbreeding coefficient of the reference population was 0.82 and realized effective population size estimated from individual increase in inbreeding was 179 animals. The rams with lowest average relationship coefficients to the current animals were identified using the scatter plot of the average relationship coefficient of the ram to other rams and of the ram to ewes, which could be used as donors of genetic material for development of germplasm cryoreserves to reintroduce genetic diversity at a later point in time. The results of this study indicated that the population under study has fairly good genetic variability.

Substituting supplemental sunflower oilcake meal with urea in sheep fed low-quality roughage
H. Mynhardt, W.A. Van Niekerk, L.J. Erasmus and R.J. Coertze
University of Pretoria, Department of Animal and Wildlife Science, Private Box X20, Hatfield, 0028, South Africa; vanniekerk@up.ac.za

The aim of this study was to determine whether a true rumen degradable protein source (sunflower oilcake meal; SFOC) can be substituted with a non-protein nitrogen source (NPN; urea) without impacting negatively on intake, digestibility, rumen fermentation and microbial protein supply (MPS) in sheep consuming a low-quality *Eragrostis curvula* hay. Five rumen cannulated wethers were used in a 5×5 Latin square design format. The animals were fed twice-daily *ad libitum* a low-quality *E. curvula* hay and infused in equal proportions into the rumen, one of five iso-nitrogenous and iso-energetic supplements. The supplements differed in the ratio of RDP supplied by SFOC and urea and were as follows: T0 (100% SFOC, 0% urea); T15 (85% SFOC, 15% urea); T30 (70% SFOC, 30% urea); T45 (55% SFOC, 45% urea) and T60 (40% SFOC, 60% urea). Supplementation increased the CP-balance of the sheep from -10.3 g CP/day to 40.8 g CP/day. Roughage intake and total DM digestibility did not differ between treatments ($P>0.05$); however, forage NDF digestibility was higher in the wethers supplemented with the higher urea-substituted treatments (T45 and T60) compared to T15 ($P<0.05$). Neither rumen pH (mean 6.48) nor total rumen VFA concentrations (mean 75.12 mmol/dl) differed between treatments ($P>0.05$). The mean rumen ammonia nitrogen (RAN) concentration of T60 was higher than T30 (9.35 vs 7.41 mg/dl; $P<0.05$); however, no differences ($P>0.05$) were observed in the MPS (mean 59.6 g CP/day) or efficiency of MPS (mean 93.6 g MPS/kg digestible organic fermented in the rumen) between treatments. Results suggest that up to 60% of the RDP supplied by SFOC can be substituted with urea, without affecting intake, digestibility or rumen efficiency in wethers receiving low-quality *E. curvula* hay.

Selected trace element concentrations in goat milk
F. Fantuz[1], S. Ferraro[2], L. Todini[1], R. Piloni[2], P. Mariani[1], A. De Cosmo[1] and E. Salimei[3]
[1]Università di Camerino, Scuola di Bioscienze e Medicina Veterinaria, Camerino, 62032, Italy, [2]Università di Camerino, Scuola di Scienze e Tecnologie (Sezione Chimica), via Gentile III da Varano, Camerino, 62032, Italy, [3]E. Salimei, via De Sanctis, Campobasso, 86100, Italy; luca.todini@unicam.it

Goat milk is used for human consumption in both developed and developing country and is suggested as a dietary supplement for consumers with inflammatory conditions and for elderly people. As milk from other species goat milk contains, besides known essential trace elements, also potentially toxic trace elements and other less known minor trace elements whose role, if any, is not known. The aim of this study was to measure the concentrations of Zn, Fe, Cu, Mn, Co, Se, Mo, Li, Ti, V, Co, As, Rb, Sr, Cd, Cs, and Pb in goat milk. Individual milk samples were obtained by machine milking (8:30 am) from 18 Saanen goats at approximately 30 days from parturition. The milk concentrations of the aforementioned elements were analyzed by Inductively Coupled Plasma-Mass Spectrometry. The average (± SD) milk concentrations (μg/l) of the investigated elements were as follow: Essential Zn 2138±356; Fe 326±65.6; Cu 48.7±14.2; Mn 19.9±6.1; Co 1.4±0.2; Se 9.3±1.4; Mo 111±36.7; Li 23.7±6.3; Ti 299±24.0; V 0.67±0.09; As 0.57±0.09; Rb 1144±102; Sr 1001±136; Cs 0.9±0.18; Pb 3.6±0.6. The milk concentration of Cd was below the limit of detection, calculated at 0.3 μg/l of milk. Significant ($P<0.05$) correlation coefficients were observed between some investigated elements: the correlations were positive between Fe and Co (r=0.81), Ti (r=0.65), V (r=0.58), and Sr (r=0.55), between Cu and Mn (r=0.53), between Co and Ti (r=0.71), V (r=0.66), and Sr (r=0.49), and between Cs and Sr (r=0.66), whereas Li was negatively correlated with Mn (r=-0.53) and Ti (r=-0.52). The current study added new data and provided novel information on the concentration of essential, potentially toxic, and minor trace elements in goat milk.

Genomic evaluation of snow sheep (*Ovis nivicola*) using OvineSNP50 BeadChip

T.E. Deniskova[1], E.A. Gladyr[1], I.M. Okhlopkov[2], V.A. Bagirov[1], G. Brem[1,3] and N.A. Zinovieva[1]
[1]L.K. Ernst Institute of Animal Husbandry, Russia, 142132, Russian Federation, [2]Institute of Biological Problems Cryolithozone, Russia, 677890, Russian Federation, [3]Institute of Animal breeding and Genetics, Austria, VMU, Austria; horarka@yandex.ru

Snow sheep (*Ovis nivicola*) is one of the rare species of mountain ungulates in North-East Asia. As the object of agricultural interest snow sheep is the most obvious potential source of biodiversity for introduction in domestic sheep breeds. However, the capabilities for genome study of snow sheep as both non-model and wild organism are limited. In this case, cross-species utilization of commercial SNP chip created for domestic sheep would be useful tool for understanding the genetic of O. nivicola. The aim of our work was to study the potential utility of using OvineSNP50 BeadChip in genome assessment of snow sheep (O. nivicola). Tissue samples were collected from 20 snow sheep from the different isolated areas of Yakutia including Moma Range (n=2), Sietindensky Range (n=3), Orulgan Ridge (n=11), Verkhoyansky Range (n=3) and Oimyakon area (n=1). Animals with genotype call rates less than 90% were excluded from evaluation (2 samples of 20). SNPs were discarded if more than 10% of SNP calls were missing. The total number of SNPs used for the study was 49478 (91.2%). The number of polymorphic SNPs was 1024, which represented 2.1% of the total number of SNPs. The observed heterozygosity for polymorphic SNPs was 0.281 ± 0.007 while the expected heterozygosity was 0.277 ± 0.005. Minor allele frequency (MAF) was 0.42% of the total number of scored SNPs and 20.4% of polymorphic SNPs. ROH values ranged from 10.1 to 16.5 Mb. The F values varied from -0.096 to 0.147. A negative F value might be associated with the small number of samples. Our study is the first attempt to apply SNP chip for genome evaluation of O. nivicola and the results will be used in comparative studies with domestic sheep in the future. Study was performed under financial support of Russian scientific foundation, project no. 14-36-00039.

The effect of body weight at slaughter and breed on histochemical profile of two sheep muscles

W. Rant and A. Radzik-Rant
Warsaw University of Life Sciences, Department of Animal Breeding, Ciszewskiego 8, 02-786 Warsaw, Poland; witold_rant@sggw.pl

The development and differentiation of muscle fibres types during growth affects muscle mass in slaughter animals and by their diversity in terms of function and morphology in the living or after slaughter muscle may also have an impact on meat quality. Therefore, the aim of this study was the analysis of muscle fibres profile in two muscles in sheep of diverse genotypes. The experiment was carried out on ram lambs of Polish Lowland (PL) and Wrzosówka (W) breeds in a frame of research project no. N N311081240. The animals were maintained at the same conditions and slaughtered at two different weight standards: 23-25 kg (10 for each breed) and 35-40 kg (10 for each breed). Within 1 hour after slaughter the meat samples were taken from longissimus lumborum (LL) and musculus semitendinosus (SM) and frozen in liquid nitrogen. Than samples were cat on 10 μm slices and subjected to combined histochemical staining to classify types of fibres: slow twitch oxidative (SO), fast twitch oxidative-glycolytic (FOG) and fast twitch glycolytic (FG). The fibre diameters as well as percentage distribution of each fibres type were measured from 1 mm^2 surface. The animals slaughtered at 23-25 kg showed the higher diameter of SO ($P\leq0.05$), FOG ($P\leq0.05$), and FG ($P\leq0.01$) fibres in SM muscle in comparison to LL, while the distribution of fibre types was similar. At higher body weight at slaughter (35-40 kg) the larger diameter ($P\leq0.01$) was observed for FG fibres in SM muscle. At this group the higher proportion of FG fibres in SM muscle ($P\leq0.05$) compared to LL have been found. The animals slaughtered at higher body weight showed higher fibre diameter of all fibre types in both muscles. The rams of W breed compared to PL regardless of the body weight at slaughter, were characterized by smaller diameter of SO, FOG and FG fibres as well as higher content of SO fibres in LL and SM muscles. In order to establish the influence of muscle fibre types on meat quality the further study are required.

Sheep and goat production systems and local breeds dynamics in a mountainous area of Greece

C. Ligda[1], D. Tsiokos[2], E.N. Sossidou[1] and I. Tzouramani[3]
[1]Veterinary Research Institute, P.O. Box 60272, 57001 Thessaloniki, Greece, [2]Alexander Technological Educational Institute, Sindos, 57400 Thessaloniki, Greece, [3]Agricultural Economics Research Institute, Ilissia, 115 28 Athens, Greece; chligda@otenet.gr

The aim of this research work was to examine the current situation of sheep and goat production in a mountainous area of Greece, with particular focus on transhumant farming systems. The data was collected in the frame of the DoMEsTIc project (ARIMNet) with the use of a structured questionnaire and personal interviews of sheep and goat farmers. Information about the production system, farmer's profile, farm management system, production aspects and strategies for marketing and promotion of the products was collected from 32 farmers. The objective was to assess the links between the structure of the farming systems, farmer practices and the characteristics of the sheep and goat breeds with the sector's resilience and sustainability. In the studied area, the sector of sheep and goat production is of high territorial importance, not only for its economic contribution but also from the point of social cohesion and the maintenance of the environment. In terms of the livestock diversity, the main reason of farmers' preferences to the specific local breeds, according to the survey, is their adaptation to the production system and various stresses and constraints of the environment, while especially in the cases of breeds with small population, heritage and tradition is considered as the most important factor for the choice of breeds. In terms of population dynamics, the trends of the population size, the management of genetic diversity were studied, in relation also to the existing valorization processes and the organisation of the sector. In conclusion, possible interventions and alternative policies are outlined in respect to the production system, the breeding strategies, the products and value chain, and organisation of the stakeholders to ensure the sustainable future of the sector.

Genotype association of the MNTR1A gene with fertility in a Greek indigenous sheep breed

O. Stoupa[1], I.A. Giantsis[1], G.P. Laliotis[2] and M. Avdi[1]
[1]Aristotle University of Thessaloniki, Department of Animal Production, University Campus, 54124 Thessaloniki, Greece, [2]Agricultural University of Athens, Department of Animal Breeding and Husbandry, Iera Odos 75, 11855 Athens, Greece; igiants@agro.auth.gr

Sheep's reproductive physiology in temperate latitudes such is Greece, is characterized by seasonality and regulated by photoperiodic exposure. Melatonin is the key hormone involved in this phenomenon, influencing positively the reproduction. However, the melatonin secretion and therefore the ewes reproductive activity underlies variation, proposed to be linked with the melatonin receptor subtype 1A (MNTR1A) gene structure. This study was designed to investigate the polymorphism of the MNTR1A gene in a Greek indigenous sheep breed and to determine its potential association with fertility. Two groups of ewes from a farm located in Agrinio (western Greece), were chosen. The first one was consisted by 30 ewes that showed reproductive activity in late spring, while the second by 30 ewes showing reproductive activity in middle summer. All the animals were in good health condition and were fed with the identical diet. Males were introduced in both groups at late spring. Blood samples were taken from the jugular vein and total DNA was extracted using the Nucleospin blood kit (Macherey-Nagel, Germany). A 824-bp DNA fragment of the MTNR1A exon 2 was amplified using PCR and was further digested using the RsaI restriction endonuclease. The elecrtophoretic procedure revealed three genotypes, C/C, C/T and T/T. Specifically, 44 animals showed the C/C genotype (28 from the first group and 16 from the second), 14 the C/T genotype (2 from the first and 12 from the second) and 2 animals had the T/T genotype (both from the second group). Statistical analysis using the SPSS package indicated a positive correlation between genotype and fertility, with CC genotype playing a crucial role. In conclusion, particular genotypes seem to have an additive effect in sheep's fertility in regard to the occurrence reproductive activity.

DNA divergence between two mitochondrial haplogroups of Polish Wrzosowka sheep

A. Koseniuk, T. Rychlik and E. Słota

National Institute of Animal Production, Animal Cytogenetics and Molecular Genetics, Krakowska street 1, 32-083, Poland; anna.koseniuk@izoo.krakow.pl

Until today, mitochondrial genomes of modern *Ovis aries* breeds have been clustered into 5 distinct clades A, B, C, D and E which are distributed through all geographical latitudes. In Europe the most frequent is haplotype B, although haplogroup A has been identified there, too. The goal of the study was to determine the genetic diversity between two haplogroups of Polish Wrzosowka breed based on mitochondrial Control Region polymorphism. Wrzosowka is the old native breed included in the sheep genetic conservation programme in Poland. DNA was extracted from whole blood of 33 Wrzosowka sheep from 2 flocks. The previously described PCR and sequencing primers (1CR_F 5'AGAAGCTATAGCCCCA and 1CR_R 5' ACTTAAAGCAAGGCAC) were used in the survey. The sequences obtained were aligned with GenBank assembled sequences of Wrzosowka haplogroup B (DQ242454, DQ242450, DQ242451) and A (DQ242454) as well as reference sequences of haplogroups B (AF010406) and A (AF010407). The DnaSp 5.0 was used to generate haplotypes, number of polymorphic sites, average number of nucleotide differences between populations (K) and average number of nucleotide substitutions per site between populations (D). Genetic distance was calculated based on p-distance algorithm implemented in Mega5. Thirty-nine sequences of 327 bp were obtained. Based on 34 polymorphic sites there were identified 20 haplotypes which have been divided into two groups according to genetic distance results. The HA group included three haplotypes which comprised the reference sequence for haplogroup A (AF010407), the Wrzosowka sequence classified as haplogroup A (DQ242454) and two Wrzosowka sequences obtained in this work. The HB group comprised the remaining 17 haplotypes. The average number of nucleotide differences and average number of nucleotide substitutions per site between two haplogroups was 17.078 and 0.50 respectively. Supported by Ministry of Science and Higher Education grant no. N N311 541840.

Effect of native pasture allowances on productive performance of ewes

R. Pérez-Clariget[1], M.J. Abud[1], A. Freitas-De-Melo[2], S. Ramírez[3] and A. Álvarez-Oxiley[1]

[1]Facultad de Agronomía, Garzon 780, 12900, Uruguay, [2]Facultad de Veterinaria, Las Places 1620, 11600, Uruguay, [3]Universidad Autónoma Agraria Antonio Narro, Periférico y Carretera a Santa Fe, A.P. 940, Mexico; raquelperezclariget@gmail.com

In order to investigate the effect of two native pasture allowances (PA) from 23 days before conception until 120 days of gestation on the productive performance, 124 multiparous, pregnant Corriedale ewes [live weight (LW): 46.5±0.5 kg; body condition score (BCS): 2.74±0.03 units] were used. The ewes grazed on: high (HPA; single: n=43; twin: n=23): 2.9-3.8 kg forage/kg of LW or low (LPA; single: 42; twin: 16): 1.4-2.6 kg forage/kg of LW. Three pens divided in two were used (each treatment had 3 replications). The ewes were shorn at 122 days of gestation and LW, BCS and fleece weight (FW) were recorded. Five days before shearing until lambing the ewes received daily 200 g/animal of rice bran and 50 ml of glycerine. After shearing, all animals grazed together on Festuca arundinacea (2.8 kg forage/kg of LW). On day 145 of gestation LW, BCS were registered and udder volume was estimated. Birth weight and sex of the lambs were recorded. Data was analyzed using ANOVA. The interaction treatment × litter size did not affect any of the variables studied. On days 122 and 145 of gestation, HPA ewes were heavier (P<0.05) and had greater BCS (P<0.05) than LPA. Ewes carrying twins were heavier (P<0.05) than those carrying single, but no differences in BCS were observed. Ewes of HPA and carrying twins had heavier (P=0.03) FW than LPA and ewes carrying single. Treatments did not affect udder volume. Birth weight of lambs were influenced by treatments (HPA: 3.71±0.13 vs LPA: 3.11±0.24 kg; P=0.05), type of birth (Single: 4.05±0.15 vs Twins: 2.78±0.25 kg; P=0.0003) but not by sex. Lambs born from gestation <146 days were lighter (P<0.0001) than lambs born from longer gestations. Pasture allowances until 23 days of lambing influenced LW, BCS, FW and BW of lambs, but did not affect udder volume.

Analysis of lamb losses in Suffolk and Romney sheep flocks
M. Milerski, J. Schmidová and G. Malá
Institute of Animal Science, Přátelství 815, 104 00 Prague 10 – Uhříněves, Czech Republic;
milerski.michal@vuzv.cz

Profitability in sheep farming is largely dependent on the number of lambs weaned per ewe each year, which is determined by litter size and lamb mortality. The aim of this study was to analyze the systematic factors affecting lamb mortality in selected Romney and Suffolk flocks under extensive conditions of the whole-year pasture production system. The significant effect of dam age on their lambs losses was confirmed especially regarding lamb mortality during rearing period compared to perinatal lamb losses. The age of dams affected also lamb birth-weight, which is connected with the lamb losses as well. In Suffok flock the strong relationship between ewe behavior during lambs tagging and losses of lambs was found, while in Romney this was not a case. This aspect, however, requires further investigations. Lamb losses caused by predators are highly site-specific. In observed flock of Suffolk losses was caused by foxes (6.3% of total lamb losses), while in the Romney flock the problems with ravens prevailed (9.2% of total lamb losses). In both breeds also lamb losses caused by improper udder shape of the ewe were noted.

L-carnitine content in longissimus dorsi and heart muscle in lambs
J. Pekala[1], R. Bodkowski[2], B. Patkowska-Sokola[2], K. Czyz[2], P. Nowakowski[2] and M. Janczak[2]
[1]Wrocław University of Technology, Department of Bioorganic Chemistry, Wybrzeże Wyspiańskiego 27, 50-370 Wrocław, Poland, [2]Wroclaw University of Environmental and Life Sciences, Institute of Animal Breeding, Chelmonskiego 38c, 51-630 Wroclaw, Poland; piotr.nowakowski@up.wroc.pl

L-carnitine is amino acid essential for proper muscles performance, and is especially important for heart muscle proper functioning. The aim of the study was to evaluate the content of total and free L-carnitine in longissimus dorsi and heart muscle of lambs. The study was conducted on the lambs of Polish Merino sheep, fed with pasture green, complete fodder and good quality hay. The samples of longissimus dorsi muscle and heart muscle were collected from 10 animals kept on Agrominor farm. Analyses of L-carnitine content were performed using spectroscopic method, after prior samples preparation, in the Laboratory of Organic Chemistry, Wroclaw University of Technology. The results obtained point that heart muscle was characterized by significantly higher content of total and free L-carnitine compared to longissimus dorsi muscle. The content of total form of L-carnitine in heart muscle was ca. 2-fold higher compared to longissimus dorsi muscle (223.71 vs 107.96 mg/100 ml), while the content of free form of L-carnitine was higher ca. 1.6-fold (151.19 vs 92.50 mg/100 ml). In turn, the ratio of free to total L-carnitine was ca. 30% higher in longissimus dorsi (0.86) compared to heart muscle (0.67). The results of the study point that heart muscle contains higher concentration of L-carnitine, both in total and free form, which is consistent with the fact that this compounds plays especially important role in this muscle functioning. In general, lamb meat may be considered as a valuable source of L-carnitine in human diet.

Nitrogen balance and coefficient of residue in sheep fed babassu meal

J.M.B. Ezequiel[1], O.R. Serra[1,2], J.R.S.T. Souza[2], A.L. Lima[2] and E.H.C.B. Van Cleef[1]
[1]São Paulo State University, Jaboticabal, 14884-900, Brazil, [2]Maranhão State University, São Luiz, 65055-000, Brazil; janembe_fcav@yahoo.com.br

The objective of this study was to evaluate increasing concentrations of babassu (Orbignya phalerata) meal partially replacing Tifton-85 bermudagrass hay on nitrogen balance and coefficient of residue of feedlot lambs. Twenty-seven crossbred lambs (90 d of age, 19.57±0.41 kg BW) were randomly assigned to one of three experimental isoenergetic and isonitrogenous diets. Treatments were composed of ground corn, soybean meal, mineral premix, Tifton-85 bermudagrass hay, and 0 (T0), 15 (T15), or 30% (T30) babassu meal. Animals were fed once daily (08:00 h) for 70 d, and between d 37 and d 41 this trial was conducted. Feed and ort samples were collected each morning and composited by animal. Total daily production of feces and urine were measured, and aliquots (10%) were composited and used for determination of N contents. Coefficient of residue (CR) was calculated as total daily fecal production (g of DM) as function of average daily gain. Regarding to N balance, N intake as well as fecal and urinary N were used to calculate N retained by animals. Data were analyzed as a completely randomized design by using mixed models. The partial replacement of Tifton-85 bermudagrass hay by babassu meal at 0, 15, or 30% did not affect N intake (31.9 g/d), fecal N (5.4 g/d), urinary N (10.4 g/d), or retained N (16.0 g/d). However, the CR was linearly decreased as the concentration of babassu meal was increased (3.0, 2.6, 2.0, respectively for T0, T15, and T30; P<0.05). The data indicate that babassu meal is a suitable replacement for Tifton-85 bermudagrass hay in diets for crossbred finishing lambs and it can replace up to 30% the roughage without adverse effects on nitrogen balance, and decreasing the amount of manure output per unit of meat produced.

Genetic resources of goat in Algeria

N. Adili[1], M. Melizi[1] and H. Belabbas[2]
[1]Institute of Veterinary Sciences and Agricultural Sciences, University of El-Hadj Lakhdar, Veterinary Medicine Department, Batna, 05000, Algeria, [2]University of Mohamed Boudiaf, Microbiology and Biochimistry Department, M'sila, 28000, Algeria; nezar.adili@yahoo.fr

This work is a synthesis of several reports and preliminary studies, it aims to make an identification of goat breeds present in Algeria; and to know the different mechanisms and livestock management. The study focuces mainly on the recognition of the state of genetic diversity and livestock management techniques; it describes also the production systems among Algerians goats. The analysis of the results obtained is performed using Microsoft Office Excel 2010. Goat population is estimated at 3 million heads in which 50% are goats; it is conducted within the extensive and semi-intensive systems, mainly for the production of meat and milk. The results recorded allow classifying goats into three types: the local population, imported breeds and crossbred product (local × exotic). The local population is represented by three breeds: the Arabic (most widespread in terms of number estimated to 60%), the Kabyle (14,23%) and the M'zabit (20,25%). Both Mediterranean breeds Alpine and Saanen, are the most imported populations; they have been the subject of breeding attempts in purebred specialized in dairy production. There is in some regions frequent crossing with the exotic materials; and this to enhance the performance of the local dairy goat. Goat breeding in Algeria is not well exploited, and appropriate measures for the preservation of livestock are needed to ensure the durable use of production systems, which represent vital resources and essential components of food security and rural development for the country.

Geographical differiantation of gestational practices of transhumant sheep and goat farms in Greece

V. Lagka[1], A. Siasiou[2], A. Ragkos[1], I. Mitsopoulos[1], A. Lymperopoulos[1], S. Kiritsi[1], V. Bampidis[1] and V. Skapetas[1]
[1]Alexander Technological Educational Institute of Thessaloniki, Department of Agricultural Technology, P.O. Box 141, 57400, Greece, [2]Democritus University of Thrace, Agricultural Development, Pantazidou 193, 68200, Orestiada, Greece; lagka@ap.teithe.gr

Transhumance constitutes a traditional agropastoral activity, which is, nonetheless, still alive in Greece, as in other Mediterranean countries. The structure of transhumant farms and their main production characteristics are heavily affected by their geographical position. The purpose of this study is to explore the herd management practices of transhumant sheep and goat farms in Greece by emphasizing on their geographical variation. We seek to examine whether the level of introduction of new technologies in milking and reproduction varies among the Greek regions. Furthermore, the degree to which the nutritional management of transhumant farms changes in terms of geographical criteria (summer domicile, winter domicile, distance of movements) is investigated. The analysis is based on primary data from a large-scale questionnaire survey of Greek transhumant sheep and goat farms throughout the country. The results of the analysis indicate considerable variation in the aforementioned practices. Farmers in the Region of Thessaly tend to adopt more innovative technologies and to use concentrates in animal nutrition, while in southern parts of the country (Peloponnese) animals graze more and new technologies such as machine milking and oestrus synchronization are rather rare; in addition, the types of feedstuff vary, as in some places farmers use by-products of agricultural production (e.g. orange skins in Peloponnese). The assumption that these variations affect the productivity of transhumant farms is also tested. Acknowledgement: This paper is part of the project 'The dynamics of the transhumant sheep and goat farming system in Greece. Influences on biodiversity' which is co-funded by the European Union (European Social Fund) through the Action 'THALIS'.

Mechanical properties of alpaca, merino and goat fibers and yarns

D. Jankowska, B. Patkowska-Sokola, P. Nowakowski, K. Czyz, A. Wyrostek and R. Bodkowski
Wroclaw University of Environmental and Life Sciences, Institute of Animal Breeding, Chelmonskiego 38c, 51-630 Wroclaw, Poland; anna.wyrostek@up.wroc.pl

Animal wool fibers constitute a valuable material for the textile industry as a high quality natural protein structures. Knowledge of mechanical and structural properties of fibers is crucial in yarns properties understanding. Strength and elasticity are the most important parameters determining fibers usefulness, while spatial arrangement of fibers in yarns affect manufacturing profitability and development. In this work we investigated single fibers and yarns from 3 animal species: alpaca, merino and goat. The samples were examined using light microscope Zeiss Axio Imager M1m, and average diameters were 30 ± 7 µm for fibers, and 796 ± 121 µm for yarns. Uniaxial tensile tests of yarns and fibers were made on a MTS Synergie 100 testing machine. Initial length of each specimen between clamps was approximately 20 mm. Samples were conducted at constant speed of 20 mm/min until rupture. Stress-strain curves were determined from non-linear characteristics: breaking stress, breaking strain and Young's modulus. All date were tested by student's t-test with significance level $P<0.05$. The results demonstrated that with similar breaking strain value we obtained differences in mechanical properties between the species. Breaking stress in goat fibers was significantly higher compared to merino. Goat fibers Young's modulus was significantly higher than that of merino and alpaca. Breaking strain in goat and alpaca yarns were lower than for merino yarn. Breaking stress of yarns did no differ significantly between animal groups, but the yarns parameters were 3-fold lower than fibers parameters. However, the values for alpaca was smaller compared to other groups. Young's modulus in yarns was 6-fold lower than in fibers. For goat yarns it was significantly higher than for alpaca. Mechanical characteristics test indicated that natural animal fibers and yarns are the best suitable to practical application in textile industry.

Variability in fibre diameter distribution and genetic gain on fineness in French Angora goats
D. Allain[1] and P. Martin[2]
[1]INRA, UMR1388 GenPhySE, CS52627, 31326 Castanet Tolosan, France, [2]Capgènes, Agropole, 2135 Route de Chauvigny, 86550 Mignaloux Beauvoir, France; daniel.allain@toulouse.inra.fr

A national selection scheme of Angora goat is running in France from 1990. Selection criteria are mean fibre diameter (MFD), kemp (coarse fibre) content and fleece weight. OFDA methodology is widely used for measuring MFD and fibre diameter distribution (FDD) with a good accuracy (4,000 fibres measured per fleece sample). Considering hair follicle development in angora goats, fibres are produced by three kinds of hair follicles: primaries, secondaries and derived formed by branching from secondaries. It is assumed that primaries and derived produce the coarsest and finest fibres respectively while secondaries produce intermediate fibres, but globally a near normal FDD is usually observed. A total of 5,429 fleece samples from 5,020 animals were measured from 1999 within the French selection scheme. OFDA MFD and FDD parameters: CV of fibre diameter (CVFD), skewness (S) and kurtosis (K) were determined on each sample. FDD shape showed large variability between animals from a normal symmetric distribution to non-symmetric skewed or excess kurtosis up to bimodal distributions suggesting a mixture of Gaussian subpopulations. FDD data were analysed with Mclust R package in order to determine the presence of a mixture distribution and if applicable the different cluster parameters: number (CN), proportions, means and variances. Within the selection scheme, MFD decrease from 30 to 26μ with a genetic gain of 1.9 σg without any variation of CVFD but with a decrease of K and increases of S and CN within FDD shape. Genetic parameters of FDD criteria were estimated using VCE software. High heritability estimates for MFD, CVFD, S and K (0.36 to 0.47) and a low heritability estimate of CN (0.05) with moderate unfavorable genetic correlations (0.23 to 0.39) between MFD and FDD parameters were observed. Genetic relations between MFD and FDD dispersion criteria and impact for genetic improvement of Angora goat were discussed.

Effect of breed Hamra and Ouled Djellal on physicochemical parameters and protein profiles in sheep
A. Ameur Ameur
University of Bouira, Agronomic sciences, Rue DRISSI Yahia. Bouira, BP 10000, Algeria; ameurabdelkader@gmail.com

In order to characterize the sheep's milk, we analyzed and compared, in a first stage of our work, the physical and chemical characteristics in two Algerian sheep breeds: Hamra race and race Ouled Djellal breeding at the station the experimental ITELV Ain Hadjar (Saïda Province). Analyses are performed by Ekomilk Ultra-analyzer, they focused on the pH, density, freezing, fat, total protein, solids-the total dry extract. The results obtained for these parameters showed no significant differences between the two breeds studied. The second stage of this work was the isolation and characterization of milk proteins. For this, we used the precipitation of caseins phi (pH 4.6).For this, we used the precipitation of caseins Phi (pH 4.6). After extraction, purification and assay, both casein and serum protein fractions were then assayed by the Bradford method and controlled by polyacrylamide gel electrophoresis (PAGE) in the different conditions (native, in the presence of urea and in the presence of SDS). The electrophoretic pattern of milk samples showed the presence similarities of four major caseins variants (αs1-, αs2-β-and k-casein) and two whey proteins (β-lactoglobulin, α-lactalbumin) of two races Hamra and Ouled Djellal. But compared to bovine milk, they have helped to highlight some peculiarities as related to serum proteins (α La β Lg) as caseins, including αs1-Cn.

Evaluation of nutritional systems for maximizing milk production

M.S.A. Khattab[1], H.M. El-Sayed[2], S.A.H. Abo El-Nor[1], A.A. Abo Amer[1], A.M. Kholif[1], H.M. Saleh[3], I.M. Khattab[4] and M.M. Khorsheed[2]
[1]National Research Center, Dairy Science Department, El-Behouth Street, 12622 Dokki, Giza, Egypt, [2]Animal Production Department, Faculty of Agriculture, Ain shams University, 12622 Giza, Egypt, [3]Biological Applications Department, Nuclear Research Center, Atomic Energy Authority, 12622 Cairo, Egypt, [4]Department of Animal Nutrition, Desert Research Center, 12622 Cairo, Egypt; msakhattab@yahoo.com

The aim of this work was to assess the impact of incorporating the nutrient synchrony concept into the traditional least cost ration. Fifteen multiparous lactating Barki ewes (35.53±1.75 kg), were randomly assigned to three groups (5 ewes each). The first group fed the least-cost ration (R1); The second group fed synchronous least-cost ration (R2); The third group fed synchronous least-cost ration with minimal amount of berseem (R3). Results revealed that there were no significant effect of dietary treatment on in OM, CP, CF, NFE digestion coefficients and nutritive values. R1 and R2 showed a considerable decrease ($P<0.05$) in EE digestibility. Blood serum total protein, globulin, AST and triglycerides concentration were not significantly affected and were in normal range. However, R3 had a significantly higher serum albumin than R1. Feeding on synchronous rations resulted in an increase in serum ALT, R1 had the lowest serum glucose concentration (32.79 mg/dl). In contrast, R3 has the highest serum glucose concentration (58.58 mg/dl). R3 serum cholesterol concentration was significantly increased, Daily milk yield and its composition and characteristics did not significantly affected among dietary treatments; however, synchronous least cost ration (R2) had the highest daily milk yield, 4% FCM and milk fat%. Feeding of synchronous diets resulted in an insignificant increase in milk urea-N and significant increase in serum urea-N. The highest milk production efficiency and net income were for R2 followed by R3 then R2, being 154.08, 140.43 and 80.38, respectively.

The brain-gut axis in nutrient sensing

R. Zabielski[1] and A. Kuwahara[2]
[1]Warsaw University of Life Sciences, Faculty of Veterinary Medicine, Department of Physiological Sciences, ul. Nowoursynowska 159c, 02-776 Warsaw, Poland, [2]University of Shizuoka, Institute og Enivronmental Sciences, to be filled out, to be filled out, Japan; rzabielski@plusnet.pl

The bidirectional communication between the gut and the brain, the gut-brain/brain-gut axis (BGA), is crucial in controlling food intake and digestion. The role of BGA is to integrate all digestive functions aiming at ingesting and preforming the food into absorbable nutrient essential for growth and keeping body functions, and nutrient sensing at the gut level is one of the newly suggested BGA ativities. The BGA is a complex mechanism involving the vagal nerves, enteric nerves and a number of gut regulatory peptides released locally by the diffuse neuroendocrine system (DNES) cells. A rise in nutrients occurs in the small intestine after food ingestion, and nutrients are released into the circulation after their absorption. While still in the preabsorptive state, nutrients activate sensing mechanisms by affecting taste cells, DNES cell and/ or vagal afferents in the duodenum to trigger gut-brain-driven negative mechanisms to inhibit exogenous nutient intake and endogenous nutrient production. Thus the small intestine can sense nutrients such as glucose and short chain fatty acid influx and trigger a number of feedback loops involving cholecystokinin, leptin, obestatin as well as a vast number of other regulatory peptides in the gut mucosa to synchronize food intake with metabolic homeostasis. Finally, a role of gut microbiota affecting intestinal nutrient sensing mechanisms will be discussed.

The effects of niacin on FoxO1 and genes involved in hepatic glucose production in dairy cows

A. Kinoshita[1], K. Hansen[2], L. Locher[1], U. Meyer[3], S. Dänicke[3], J. Rehage[1] and K. Huber[2]
[1]University of Veterinary Medicine Hannover, Clinic for Cattle, Bischofsholer Damm 15, 30173 Hannover, Germany, [2]University of Veterinary Medicine Hannover, Department of Physiology, Bischofsholer Damm 15, 30173 Hannover, Germany, [3]Friedrich-Loeffler-Institute, Institute of Animal Nutrition, Bundesallee 50, 38116 Braunschweig, Germany; Asako.Kinoshita@tiho-hannover.de

Forkhead box protein O1 (FoxO1) is an important transcription factor involved in hepatic glucose production (HGP) which is deactivated by insulin-induced phosphorylation. Niacin could affect HGP, e.g. by modifying the expression of genes. The objective of this study was to investigate the effects niacin on expression of hepatic FoxO1 and genes involved in HGP in dairy cows during the transition period. Methods: 21 pluriparous German Holstein cows were assigned in 4 groups and fed with diets containing 0 g (C) or 24 g (N) niacin and 30% (L) or 60% (H) of concentrate on dry matter basis from -42 days related to calving (d-42) to calving. The concentrate proportion was then increased from 30% up to 50% within 16 days for CL and NL and within 24 days for CH and NH. Liver biopsies were taken at d-42, d3, and d21. Protein expression of FoxO1 and phosphorylated FoxO1 at serine 256 (pFoxO1) was measured semi-quantitatively by western blotting. The mRNA of FoxO1, glucose-6-phosphatase (G6P), pyruvate carboxylase, phosphoenolpyruvate carboxykinase (PCK1), and propionyl CoA carboxylase (PCCA) was measured by Real-time RT-PCR. Data was evaluated by mixed model for repeated measures. Results: The protein and mRNA expression of FoxO1 and the protein expression of pFoxO1 were affected neither by time nor by diet. In cows fed with niacin, the relative quantities of mRNA of G6P and PCCA were higher at d21. Conclusion: Dietary niacin altered the gene expression, potentially modulating hepatic gluconeogenesis in the transition period of cows. However, the regulation of HGP by FoxO1 seemed to be of less importance on the levels of mRNA, protein and phosphorylation.

Nutrient signalling receptors for free fatty acids and hydroxycarboxylic acids in farm animals

M. Mielenz
Leibniz Institute for Farm Animal Biology (FBN), Institute of Nutritional Physiology 'Oskar Kellner`, Wilhelm-Stahl-Allee 2, 18196 Dummerstorf, Germany; mielenz@fbn-dummerstorf.de

During the last decade two receptor families were characterised which bind the energy sources free fatty acids and hydroxycarboxylic acids as ligands, namely the family of free fatty acid receptors (FFAR) and the family of hydroxycarboxylic acid receptors (HCA). In different models these receptors affect insulin release or sensitivity, lipolysis, immune response or the release of gut hormones. The short chain fatty acid butyrate is a ligand for FFAR1 and FFAR2 but also for HCA2. Beta-hydroxybutyrate but also niacin activates HCA2. Data on the importance and the pharmacology of these receptors in farm animal species are scarce. However, effects of butyrate e.g. on growth performance and the activity of the immune system were shown. In the porcine intestine the mRNA of FFAR2 and FFAR3 is differentially expressed, highest in the small intestine. Comparing the protein abundance of both receptors in pig, highest FFAR3 abundance was detected in the ileum but for FFAR2 in spleen. Bovine FFAR2 was firstly described in the rumen. In addition, propionate affects gene expression in ovine and bovine adipose tissue (AT) *in vivo* and *in vitro*, respectively and stimulates bovine neutrophil granule release. In ovine subcutaneous AT propionate induces the mRNA expression of the adipokine leptin. Furthermore, both receptors but also HCA2 and FFAR1 mRNA, the latter binding long-chain fatty acids, are differentially expressed in bovine AT and liver throughout lactation. FFAR2 and HCA2 act anti-lipolytic. The HCA2 ligand niacin reduces the phosphorylation and abundance of the hormone-sensitive lipase ex vivo and niacin increases the secretion of adiponectin in differentiated bovine adipocytes *in vitro*, which may improve insulin sensitivity. The interest in the regulation of metabolism by members of both receptor families is gaining in importance. However, in different farm animal species more functional data should be provided by future studies.

The effect of feeding level on genes related to lipid metabolism in the mammary tissue of sheep

E. Tsiplakou[1], E. Flemetakis[2] and G. Zervas[1]

[1]Agricultural University of Athens, Nutritional Physiology and Feeding, Iera Odos 75, 11855, Greece, [2]Agricultural University of Athens, Agricultural Biotechnology, Iera Odos 75, 11855, Greece; eltsiplakou@aua.gr

Fat synthesis is under the control of a large number of genes whose nutritional regulation is still poorly documented. In this study, we examined the effect of long term under- and over-feeding on the expression of genes (acetyl-CoA carboxylase: ACC, fatty acid synthetase: FAS, lipoprotein lipase: LPL, stearoyl-Co A desaturase 1: SCD, peroxisome proliferator activated receptor γ2: PPARγ2, sterol regulatory element binding protein-1c: SREBP-1c and hormone sensitive lipase: HSL) related to milk FA metabolism in sheep mammary tissue (MT). Twenty four lactating sheep were divided into three homogenous sub-groups and fed the same ratio in quantities covering 70% (underfeeding), 100% (control) and 130% (overfeeding) of their energy and crude protein requirements. MT samples were collected on the 30th and 60th experimental day of each animal. The data were analysed using a general linear model (GLM) for repeated measures analysis of variance (ANOVA) with dietary treatments and sampling time as fixed effects. The results showed a significant reduction on mRNA of ACC, FAS, LPL and SCD gene in the MT of the underfed sheep compared with the respective control and overfed. Additionally, significant was also the down regulation on the mRNA transcript accumulation of the PPARγ2 and SREBP-1c gene in the MT of underfed sheep compared with the overfed ones. Further to that, the HSL transcript level in sheep MT did not differ between the three dietary treatments. The SCD was the most abundant transcript relative to the other classical lipogenic genes. Finally, the ACC, FAS, LPL, SCD, PPARγ2 and SREBP-1c mRNA levels in MT of sheep were reduced significantly between the two sampling times. In conclusion, a lower feeding level and consequently lower nutrients availability, leads to lower rate of lipid synthesis in sheep MT by down regulation the lipogenic gene expression.

Effects of sedation and general anesthesia on plasma metabolites and hormones in pigs

G. Das, A. Vernunft, S. Görs, E. Kanitz, J. Weitzel, K.-P. Brüssow and C.C. Metges

Leibniz Institute for Farm Animal Biology (FBN), Wilhelm-Stahl-Allee 2, 18196, Dummerstorf, Germany; gdas@fbn-dummerstorf.de

Plasma metabolites are often determined during or after general anesthesia (GA) or sedation (S). We investigated whether the metabolites and hormones are influenced by GA induced with ketamine (K) and two different sedatives, namely azaperone (A) and xylazine (X). Female pigs (n=6) fitted with jugular vein catheters were rotationally sedated/anaesthetized with each of 4 different treatments following a basal-day control measurements (CON). Each pig received a single administration of A or X or AK or XK on different days, separated with a washout day. Plasma samples collected during 6 h were analysed to determine concentrations of glucose, lactate, non-esterified fatty acids (NEFA), triglycerides (TG), glucagon, insulin and cortisol by spectrophotometry, RIA, and ELISA, respectively. Data were analysed with repeated measures ANOVA using a mixed model. Except for glucagon, all metabolites and hormones were affected ($P<0.05$) by GA/S. Glucose was increased ($P<0.05$) by A and X with the latter exerting a stronger effect ($P<0.05$). It was more rapidly increased due to X than A, and the increase lasted longer. Lactate was elevated by A, and ketamine (AK) intensified this effect ($P<0.05$). Administration of X alone and as XK did not influence ($P>0.05$) lactate level, and prevented the ketamine-associated increase when administered as XK. NEFA concentrations were increased by A and AK. TG concentrations were elevated by A and AK, however remained unaffected by X in either form (X, XK). Cholesterol was increased ($P<0.05$) only by the combination treatment AK. Insulin concentration was reduced only by X ($P<0.05$) administered either singly or as XK. Cortisol was elevated by A, and even more so by AK ($P<0.05$), while it showed no change ($P>0.05$) with X. In conclusion, with the exception of glucagon, A and AK strongly increased plasma levels of all the metabolites and hormones, likely mediated through the GA/S effects on cortisol.

Organic pollutant release from adipose to blood in response to lipomobilisation in the ewe

S. Lerch[1], C. Guidou[1], G. Rychen[1], A. Fournier[1], J.P. Thomé[2] and S. Jurjanz[1]
[1]Université de Lorraine, INRA USC 340, UR Animal et Fonctionnalités des Produits Animaux, 2 avenue de la Forêt de Haye, 54518 Vandoeuvre-lès-Nancy, France, [2]Université de Liège, Laboratoire d'Ecologie Animale et d'Ecotoxicologie, 11 allée du 6 Août, 4000 Sart-Tilman Liège, Belgium; sylvain.lerch@univ-lorraine.fr

Animal production has faced sanitary crises involving contamination of products by persistent organic pollutants (POP). In order to avoid the disposal of products originating from contaminated herd, innovative strategies aiming to hasten POP removal should be identified. The first step is to release POP from their storage site, the adipose tissue (AT), to the blood, in order to be metabolized or excreted. We hypothesized that POP would be released from AT to blood along with lipids, during body-fat mobilization over short- (1 h) or medium-term (1 wk) periods due to β-agonist challenge or underfeeding, respectively. In order to test these hypotheses, 3 adult non-lactating Romane ewes were contaminated over 6 wk with a diet spiked with indicator polychlorobiphenyls (0.21 µg/kg BW/d of each PCB 28, 52, 101, 138, 153 and 180) and chlordecone (11.6 µg CLD/kg BW/d). Indicator PCB and CLD were chosen because of their high lipophilicity and contrasted distribution: mainly in AT for PCB, and in liver for CLD. Over the first 5 wk, diet met energy requirements, whereas during the 6th wk ewes were underfed (35% of energy requirements). β-agonist challenges (4 nmol/kg BW of isoproterenol i.v.) performed at the end of wk 5 and 6 of the trial, induced fast (5-10 min) increase in plasma non-esterified fatty acid (NEFA) concentration (P=0.05), but did not affect serum POP concentrations. Conversely, short-term (2-7 d) underfeeding increased NEFA, PCB 138, 153 and 180, and CLD (P<0.01) in serum (×1.3-1.5 for POP, 0.11 to 0.17 ng/g after 7 d for PCB 180), without affecting PCB concentrations in AT and body fatness. These results suggest that POP are released from AT to blood through lipomobilisation in response to underfeeding, but not by acute lipolysis due to β-agonist challenges.

Insulin as a suppressive factor of GH release in the ruminant

K. Katoh[1], R. Kobayashi[2], K. Nishihara[3], Y. Suzuki[4], K. Suzuki[5] and S.G. Roh[6]
[1]GSAS, Animal Science, Tohoku University, Aoba-ku, Sendai, 981-8555, Japan, [2]GSAS, Animal Science, Tohoku University, Aoba-ku, Sendai, 981-8555, Japan, [3]GSAS, Animal Science, Tohoku University, Aoba-ku, Sendai, 981-8555, Japan, [4]GSAS, Animal Science, Tohoku University, Aoba-ku, Sendai, 981-8555, Japan, [5]GSAS, Animal Science, Tohoku University, Aoba-ku, Sendai, 981-8555, Japan, [6]GSAS, Animal Science, Tohoku University, Aoba-ku, Sendai, 981-8555, Japan; kato@bios.tohoku.ac.jp

An explanation for the mechanism of the postprandial suppression of blood GH concentrations in the ruminant was that increased concentrations of short-chain fatty acids (SCFAs) in blood exert direct suppressing actions on somatotrophs in the anterior pituitary gland (Rumeno-Pituitary axis), as suggested by *in vivo* and *in vitro* studies. However, these studies have also suggested that practical and postprandial increases in SCFA concentrations could not be sufficient to suppress GH release. The aim of the present study, therefore, was to investigate whether or not insulin has the suppressive action on GH release in the ruminant. Intravenous injection of insulin or glucose caused a transient and significant increase in plasma GH concentrations in goats. In addition, either injection caused a significant increase in plasma glucose or NEFA concentrations, respectively, which was also accompanied with increased cortisol concentrations. However, high insulin condition without changing plasma glucose and NEFA concentrations induced by the glucose clamp technique caused a significant decrease in plasma GH concentrations. In addition, insulin caused a concentration-dependent decrease in GH release from pituitary slices. From these results, we conclude that: (1) insulin has a direct inhibitory action on GH release from pituitary somatotrophs; and (2) changes in plasma insulin concentrations, which are concomitant with changes in glucose or NEFA concentrations, may cause an increase in GH release because of activation of the hypothalamus-pituitary-adrenal gland axis.

Responses of weaned pigs to organic acids on growth performance and intestinal mucosa morphology
S. Khempaka, P. Pasri, S. Okrathok and W. Molee
School of Animal Production Technology, Institute of Agricultural Technology, Suranaree University of Technology, 111 University Avenue, Suranaree Subdistrict, Mueng, Nakhon Ratchasima, 30000, Thailand; khampaka@sut.ac.th

The weaning period is a critical stage of life for piglets, it is often associated with reduced feed intake, little or no weight gain, insufficient enzyme production, change in intestinal morphology, diarrhea, morbidity and death. The addition of organic acids to diets for pigs has been reported to solve these problems. Therefore, this study aimed to investigate the effect of organic acids in the form of formic acid and butyric acid in weaned pig diets on growth performance and intestinal mucosa morphology. A total of ninety six 21-day, weaned crossbreds (Large White × Landrace × Duroc) pigs (48 castrated males and 48 females) with average initial body weight 6.0±1.92 kg was subjected in a Completely Randomized Block Design (CRBD). All pigs were randomly divided over 4 treatments with 4 replicate pen per treatments (6 pigs per pen). The experimental diets composed of control, 0.3 formic acid, 0.1% butyric acid and 0.3% formic acid + 0.1% butyric acids. All pigs were provided feed and water *ad libitum* throughout the experimental period (28 days after weaning). Feed intake and body weight were recorded. At the end of the experimental period, one pig from each replicate was sampled and slaughtered for intestinal mucosa morphology mesurement. The results showed that the growth performance of weaned pigs was not significantly affected by treatments. However, pigs fed organic acids tended to have greater ADG and feed efficiency than those fed the control diet. In addition, feeding butyric acid and formic acid + butyric acid can cause a significant enhancement of jejunal villus height.

MicroRNA expression changes after diet treatment in the liver of pigs
M. Oczkowicz, K. Pawlina, M. Świątkiewicz and M. Bugno-Poniewierska
National Research Institute of Animal Production, Krakowska 1, 32-083 Balice, Poland; klaudia.pawlina@izoo.krakow.pl

The aim of our study was to evaluate the expression changes of selected miRNAs in the liver of pigs fed with different source of fat. The animals were divided into four feeding groups (6 animals per group). All groups obtained isonitrogenous and isoenergetic diet, covering the nutritional requirements of pigs. The diet differed among groups with the presence of corn dried distilled grains with solubles (DDGS) (groups II, III, IV) and the type of fat used (rapeseed oil – group I and II, beef tallow – group III, coconut oil – group IV). Samples of the liver of all animals were collected during the slaughter. After RNA isolation, we performed reverse transcription using TaqMan MicroRNA RT kit, followed by qPCR with commercially available TaqMan MicroRNA Assays specific for ssc: let-7a, miR-103, miR-122a, miR-148a, miR-26a, miR-16, miR-92a, miR-335. GeNorm analysis revealed that the most stable are miR-16, miR-26a and miR-148a, thus they were used as endogenous control. The statistical analysis was performed using SAS Enterprise Guide 5.1 employing GLM Procedure. The analysis with diet and sex as fixed factors showed that expression of none of the miRNAs was affected by the sex. Among the feeding groups only the expression of miR-122a was different (P<0.03). The abundance of miR-122a was significantly lower in the group obtaining DDGS and coconut oil than in pigs obtaining DDGS and beef tallow (P<0.027), the group fed the diet without DDGS but with the addition of rapeseed oil (P<0.0052) and the group obtaining DDGS and rapeseed oil (P<0.035). We did not find significant changes in the expression of the other analysed miRNAs, however, some trends were observed for miR-335 (P<0.096) – the mean expression was 2-fold lower in the -DDGS+rapeseed oil group. These results for the first time show that nutrition may affect miRNA expression in the liver of pigs.

Effects of colostrum vs formula feeding on mRNA expression of haptoglobin in neonatal calves

B. Getachew[1], H. Sadri[1], J. Steinhoff-Wagner[1,2], H.M. Hammon[2] and H. Sauerwein[1]
[1]University of Bonn, Institute of Animal Science, 53115 Bonn, Germany, [2]Leibniz Institute for Farm Animal Biology (FBN), Institute of Nutritional Physiology, 18196 Dummerstorf, Germany; hsadri@uni-bonn.de

The transition from intra- to extra-uterine life is associated with dramatic changes in the environmental conditions including exposure to various antigens. Haptoglobin (Hp) is one of the major acute phase proteins (APP) in cattle; it binds free hemoglobin released from erythrocytes and thereby inhibits its oxidative activity. Hp release is stimulated by proinflammatory cytokines, however, in newborns the innate immune system (IIS) is immature and the endogenous production of Hp is assumingly low. Colostrum intake might be able to stimulate the maturation of the IIS including Hp gene expression and we thus hypothesized that the mRNA expression of Hp in hepatic and extrahepatic tissues of newborn calves will be greater if receiving colostrum as compared to formula. German Holstein calves were fed either pooled colostrum (COL) obtained during the first 3 days (d) of lactation or milk-based formula (FOR) with comparable nutrient composition as colostrum until slaughter at d4 of life (n=7/group). As determined by ELISA, FOR contained less Hp (0.5 µg/ml) than COL colostrum (69, 94, and 20 µg/ml from d1 to d3, respectively). Expression of Hp mRNA in liver, adipose tissue, duodenum and ileum was quantified by qPCR. Formula contained much less Hp (0.5 µg/ml) than COL colostrum (69, 94, and 20 µg/ml from d 1 to d 3, respectively). The mRNA abundance of Hp in liver and adipose tissue was 3- and 2.2-fold greater (P≤0.02) in FOR than in COL calves, respectively. The ileal expression of Hp mRNA was numerically greater (P=0.13) in FOR vs COL calves, whereas duodenal Hp mRNA was not different between groups. Contrasting our hypothesis, FOR but not COL feeding seems to stimulate Hp mRNA expression in liver and adipose tissue indicating that maturation of the IIS might be accelerated by FOR feeding due to lack of immunoglobulin intake.

Effect of dietary fish oil on the mRNA abundance of mammary lipogenic genes in dairy ewes

D. Carreño, G. Hervás, P.G. Toral, T. Castro-Carrera, M. Fernández and P. Frutos
Instituto de Ganadería de Montaña, CSIC-ULE, Finca Marzanas, 24346 Grulleros, León, Spain; p.frutos@csic.es

In dairy cows, diet-induced milk fat depression (MFD) has been related to the downregulation of certain lipogenic genes in the mammary tissue. However, information in this regard is very scant in dairy ewes, and results are inconclusive. Therefore, this assay was conducted in sheep with the aim of studying the relationship between changes in the milk fatty acid (FA) composition and the mRNA abundance of genes involved in mammary lipogenesis, in response to a diet known to induce MFD. A total of 12 ewes were divided into two treatments (n=6) and received a TMR supplemented with 0 (control) or 17 (FO) g of fish-oil/kg DM, for 31 days. On days 0, 7 and 30 on treatments, milk samples were collected to analyze the FA profile by gas chromatography. On days -13, 8 and 31, samples of mammary secretory tissue were removed by biopsy and used for a candidate gene expression approach by quantitative reverse transcription-PCR. Data were subjected to ANOVA for repeated measures using a statistical model that included the fixed effects of diet, time, and their interaction, and the initial record measured at the beginning of the trial as covariate. The FO diet did not affect feed intake and milk yield (P>0.10) but, as expected, induced MFD and modified milk FA composition. Compared with the control, reductions in milk fat concentration and yield were not detected on day 7, but reached up to 25 and 22%, respectively, on day 30 (P<0.05). Increases in some putative antilipogenic FA (e.g. trans-10 18:1, trans-10 cis-12 18:2, and trans-9 cis-11 18:2; P<0.05) were accompanied by reductions in the mRNA abundance of genes such as ACACA, FASN, SCD1 and SREBF1 (P<0.05), which would support that the nutritional regulation of milk fat content and composition is mediated by transcriptional control mechanisms. Most changes in gene expression were identified on day 8 (i.e. on early stages on the treatments) and stayed relatively stable afterwards.

Chemical composition of Olkuska sheep milk depending on fertility rate

B. Patkowska-Sokola, P. Nowakowski, M. Iwaszkiewicz, K. Czyz, R. Bodkowski and K. Roman
Wroclaw University of Environmental and Life Sciences, Institute of Animal Breeding, Chelmonskiego 38c,
51-630 Wroclaw, Poland; katarzyna.roman@up.wroc.pl

Milk composition was investigated in prolific Olkuska Steep depending on lambing type. Milk composition was examined at 2, 12, 22 and 56 day after lambing. In was noted that in All samplings milk from mothers of twins was characterized by lower content of dry matter (16.1-16.5%), non-fat dry matter (9.4-9.8%), fat (6.4-6.7%), protein (3.8-5.4%) compared to mothers of single lambs (17.7-18.7, 9.9-10.9, 7.8-8.3 and 3.4-6.6%, respectively). Only lactose content in particular samplings was on a slightly higher level in milk from mothers of twins. Attention should be paid to similar content of unsaturated fatty acids in all samplings in milk from mothers of twins and single lambs. Relatively high values were noted in the range of very valuable, biologically active fatty acids, i.e. CLA, EPA, DHA and VA in milk of the examined sheep. Milk composition did not depend on fertility rate, but only on the stage of lactation. Lower content of basic milk components (protein, fat, dry matter) in milk from mothers of twins points the necessity of additional feeding of lambs from multiple lambing, and special care of them.

Dimensions of milk fat globules depending on sheep breed and lactation stage

B. Patkowska-Sokola, P. Nowakowski, K. Czyz, R. Bodkowski, M. Iwaszkiewicz, K. Roman and A. Wyrostek
Wroclaw University of Environmental and Life Sciences, Institute of Animal Breeding, Chelmonskiego 38c,
51-630 Wroclaw, Poland; katarzyna.roman@up.wroc.pl

The aim of the study was to examine the differences in milk fat globules dimensions and share depending on sheep breed and lactation stage. The study was conducted during pasture grazing on the ewes of Polish Mountain Sheep, Friesian and Karagouniko breeds. All the animals were fed entirely with pasture grass, with free access to salt-licks and water. Milk samples were collected at the beginning of lactation (about 50 days after lambs weaning) and in final phase of lactation from 60 sheep from each breed. Fat globules were examined under scanning microscope in the Laboratory of Electron Microscopy, Wroclaw University of Environmental and Life Sciences. After suitable samples preparation, SEM images were made using Zeiss 435 VP scanning electron microscope in two magnifications (3,000× and 5,000×). This allowed to measure the diameters of fat globules using AutoCad software. Fat globules diameters ranged from 3.47 to 4.32 μm. The smallest globules were observed in Friesian sheep, while the largest in case of Polish mountain sheep, and concurrently, an increase in fat globules dimensions was noted in all the breeds at the end of lactation period. The highest percentage was noted for fat globules of diameter within the range 3-6 μm (from ca. 65 to 78%), then 1-3 μm (from ca. 13 to 28%), and the least numerous were globules of the highest diameter, i.e. 6-9 μm (from ca. 2 to 14%). The results obtained point that the most profitable, when regards technological properties reflected in fat globules size and share.

Plasma insulin, glucose and amino acids responses to duodenal infusion of leucine in dairy cows
H. Sadri[1], D. Von Soosten[2], U. Meyer[2], J. Kluess[2], S. Dänicke[2], B. Saremi[3] and H. Sauerwein[1]
[1]Institute of Animal Science, Physiology & Hygiene Unit, University of Bonn, Germany, [2]Institute of Animal Nutrition, Friedrich-Loeffler-Institute, Braunschweig, Germany, [3]Evonik Industries AG, Animal Nutrition Research Group, Hanau, Germany; hsadri@uni-bonn.de

Carbohydrates are not the only macronutrients triggering insulin release following food intake. Besides other components, leucine (Leu) has been demonstrated to induce insulin secretion in humans and in laboratory rodents. We hypothesized that Leu stimulates the release of insulin in dairy cows, and thus our objective was to test the effects of a single-dose of Leu infused intraduodenally on the concentrations of insulin, glucose and free amino acids (AA) in blood plasma as compared to infusions with either glucose or saline. Six duodenum-fistulated Holstein cows were studied in a replicated 3×3 Latin square design with 3 periods of 7 d, in which the treatments were applied at the end of each period. The treatments were duodenal bolus infusions of Leu (DIL; 0.15 g/kg BW), glucose (DIG; at Leu equimolar dosage) or saline (SAL). Blood samples were taken at -15, 0, 10, 20, 30, 40, 50, 60, 75, 90, 120, 180, 210, 240 and 300 min relative to the infusion and the concentrations of insulin, glucose and AA were determined in plasma. In DIG, insulin and glucose concentrations peaked at 30-40 and 40-50 min after the infusion, respectively. Insulin concentrations were greater ($P<0.05$) from 30-50 min in DIG than DIL and SAL. In DIG, glucose concentrations were greater ($P\leq0.01$) from 30-75 and 40-50 min than in DIL and SAL, respectively. In DIL, Leu concentrations peaked 50-60 min after infusion, reaching 20 and 15-fold greater values than that in DIG and SAL, respectively. The plasma concentrations of total AA minus Leu were affected by treatment ($P<0.0001$), resulting in lowest mean concentrations of total AA minus Leu in DIL, followed by DIG and SAL. The data suggest that Leu infusion did not elicit an apparent insulin response, but may stimulate the tissue uptake of AA by mechanisms yet to be elucidated.

Determination of endotoxin activity in rumen fluid with the chromogenic LAL assay
C. Emsenhuber, N. Reisinger, N. Schauerhuber, S. Schaumberger and G. Schatzmayr
BIOMIN Research Center, Technopark 1, 3430 Tulln, Austria; caroline.emsenhuber@biomin.net

Endotoxins are known to have a negative impact on the health of ruminants. High-grain diets, used to increase energy intake and productivity, can lead to pH-shift and bacterial imbalance in the rumen. As a consequence, proliferation and death of Gram-negative bacteria can increase, resulting in an increased endotoxin load in the rumen. High levels of endotoxins can penetrate into the blood stream through an impaired rumen wall, inducing inflammatory processes. Possible consequences are rumenitis, metabolic acidosis, laminitis, reduced feed intake, and lower productivity. The objective of our study was to evaluate, if it is possible to reproducibly determine the endotoxin activity in rumen fluid with the limulus amoebocyte lysate (LAL) assay. Furthermore, it was tested, if sample preparation has an influence on the measured endotoxin activities. Rumen fluids from two animals were sampled at slaughter house and centrifuged. Different endotoxin concentrations (0-500,000 EU/ml) were added to the supernatants and stored at -20 °C. For testing, supernatants were thawed, heat inactivated (100 °C), filtrated (0.45 μm) and diluted (50,000-fold). Endotoxin concentration was measured with the LAL assay according to the manufacture's guidelines. The LAL reagent was reconstituted with Glucan Inhibiting Buffer. Specificity, linearity, precision and accuracy were determined to evaluate the suitability of the LAL assay in rumen fluid. Endotoxin concentrations added to supernatants were recovered with 70 to 140% (specificity), a linearity of $R^2=0.9$ and a precision of <16%. The accuracy was below 25%. The limit of detection and the limit of quantification were defined as 0.05 and 2,500 EU/ml. The average of endotoxin activities in rumen fluids were 132,379±32,334 and 21,434±2,975 EU/ml. Results show that it is possible to measure reproducible endotoxin activities in rumen fluid with the LAL assay and sample preparation has no influence on endotoxin activity.

Selection for feed efficiency: direct and correlated responses in two rabbit lines

H. Garreau[1], L. Drouilhet[1], J. Ruesche[1], A. Tircazes[1], M. Theau-Clément[1], T. Joly[2], E. Balmisse[3] and H. Gilbert[1]

[1]INRA, Animal Genetics, UMR1388 Génétique, Physiologie et Systèmes d'Elevage, 31326 Castanet-Tolosan, France, [2]ISARA, AGroécologie, Environnement, Marcy l'Etoile, 69280 Lyon, France, [3]INRA, Animal Genetics, UE 1322 PECTOUL, 31326 Castanet-Tolosan, France; garreau@toulouse.inra.fr

Two alternative traits to feed to gain ratio (FCR) were studied in growing rabbits to compare strategies for the genetic improvement of feed efficiency: Residual Feed Intake (RFI) and Average Daily Gain (ADG) under fixed restricted feeding (ADG_R) (80% of *ad libitum*). One line has been selected for each trait (generations 0 to 9) from the same population G0. Records comprised about 2,450 rabbits per line, for body weights at weaning (30 days of age) (BW30) and slaughter (63 days of age) (BW63) and feed consumption, and ADG, FCR and RFI were computed. Selection criteria showed similar heritabilities (0.21±0.04 for RFI, 0.29±0.05 for ADG_R) and responses to selection in the two lines (-0.34 genetic standard deviations (σ^2_a) per generation for RFI, +0.29 σ^2_a per generation for ADG_R). Responses to selection were -0.30 σ^2_a per generation on FCR in both lines, and genetic correlations with selection criteria were close to unity. Genetic correlations between selection criteria and BW30 and BW63 differed between lines. To further examine responses to selection, in generation 9 the two selected lines have been compared with the G0 control population (using frozen embryos) under *ad libitum* and restricted feeding by testing 30 individuals per line and feeding level combination. When fed *ad libitum* RFI animals grew at the same rate as G0 rabbits and ADG_R animals grew faster than these groups (P<0.001), whereas both selected lines grew faster than G0 animals when restricted (P<0.001). As a result, despite different feed intakes and BW63, independently from the feeding level both selected lines had a similar improved FCR (2.62±0.02) compared to the G0 line (2.82±0.02).

The effect of tannins on the production parameters and quality of pork in Large white breed

O. Bučko[1], A. Lehotayová[1], P. Juhás[1], J. Petrák[1], M. Margetín[1], K. Vavrišinová[1], I. Bahelka[2] and O. Debrecéni[1]

[1]Slovak University of Agriculture, Faculty of Agrobiology and Food Resources, Department of animal husbandry, Trieda A. Hlinku 2, 94976 Nitra, Slovak Republic, [2]National agricultural and food centre, Research institute for animal production Nitra, Institute of Animal Husbandry Systems, Breeding and product quality, Hlohovecká 2, 95141 Lužianky, Slovak Republic; ondrej.bucko@uniag.sk

The aim of the experiment was to analyse the effect of the tannins (Farmatan plus) to the growing rate, feed consumption, carcass composition and the quality of meat in Large white breed. The experiment included 19 pigs. The pigs were divided into a control group of 9 pigs (4 barrows and 5 gilts) and 10 pigs in the experimental group, which consisted of 5 barrows and 5 gilts. The experimental group was fed by feed mixture enriched with Farmatan plus (5 g/kg feed mixture). The results showed that there was no statistically significant difference in the growth rate between the control and the experimental group, but there was a lower feed consumption/kg daily gain in the experimental group (P≤0.01). The parameters of the carcass value showed that the weight of the thigh in % from the weight of the half carcass was higher in the experimental group, the difference was statistically significant at P≤0.01. The weight of the fat and backfat thickness was significantly lower (P≤0.05) in the experimental group than in the control group. It was found out that the weight of the kidney fat was significantly higher in the control group of the pigs at P≤0.01. The parameters of physical and technological quality showed a statistically significant difference at P≤0.05 in the parameters pH1, colour CIE a * 24 hours post mortem and colour CIE b * 7 days post mortem. There was a significantly lower values in the experimental group of the pigs in the parameter shear force (P≤0.01). The results of our experiment showed that tannins have a positive effect on the feed conversion and on the some parameters of the carcass value. This work was supported by projects VEGA 1/0364/15.

Analysis of factors affecting length of productive life of Latvian brown cows breed

L. Cielava, D. Jonkus and L. Paura
Latvia University of Agriculture, Liela 2, 3001 Jelgava, Latvia; daina.jonkus@llu.lv

The aim of this study was to analyse the factors affecting productive life of Latvian Brown (LB) breed cows using Proportional hazard model of Survival analysis. The following factors were evaluated: cow's birth year, farm, age at first calving, energy-corrected milk yield and somatic cells score in the first lactation. Records of 9,544 LB cows born between 2002 and 2006 were included in the analysis. 2,040 cows (21.4%) were alive at the end of the study and were considered as censored. The length of productive life of the cows excluded from herd was in average 1,467.7 days or 4.05 standard lactations; and was in range from 334 to 3,200 days. All factors had a significant affect to the length of productive life of cows. The risk decreased with age at first calving. In the study was found that significantly less length of productive life expected for cows that calved before 24 months of age. Cows with age at first calving more than 30 month characterised by longer productive life, but at the same time by lower life productivity. Only 40% of cows with energy-corrected milk yield less than 3,999 kg reached 3,000 days of productive life. Significantly higher risk of being excluded from the herd was for cows with somatic cell score in the first lactation more than 2.5.

Effect of aluminosilicate adsorbents on milk aflatoxin M_1 concentration of Holstein dairy cows

M. Mojtahedi[1,2], M. Danesh Mesgaran[1] and A.R. Vakili[1]
[1]Department of Animal Science, Ferdowsi University of Mashhad, 91775-1163, Mashhad, Iran, [2]Department of Animal Science, University of Birjand, 97175-331, Birjand, Iran; mojtahedi@birjand.ac.ir

Aflatoxin M_1 (AFM_1) residues in milk are regulated in many parts of the world and can cost dairy farmers significantly due to lost milk sales. Additionally, due to the carcinogenicity of this compound contaminated milk can be a major public health concern. An experiment was conducted to determine the efficacy of 3 aluminosilicate adsorbents, named S1, S2 and S3, in reducing AFM_1 concentrations in milk of dairy cows fed diet contaminated with aflatoxin B1 (AFB_1). Twenty mid-lactation dairy cows averaging 154 d in milk and 30.5 kg/d milk production, were used in a complete randomize design with repeated measurements. Cows were randomly assigned to the dietary treatments for 23 days. Dietary treatments included AFB_1 [136 µg of AFB_1/kg of diet dry matter (DM)]; AFB_1+ 120 g S1 adsorbent; AFB_1+ 120 g S2 adsorbent; and AFB_1+ 120 g S3 adsorbent per day per cow. Milk samples were collected on d 15 and 23 of the experimental period to evaluate changes in milk production, composition and AFM_1 concentrations. Results indicated that feed intake, milk production, milk fat percentage, milk protein percentage and milk lactose percentage were not affected by dietary treatments (P>0.05) and averaged 22.33 kg/d of DM, 27.46 kg/d, 3.43%, 3.01%, and 4.80%, respectively, across all treatments. The addition of S3 and S2 aluminosilicate adsorbents to the AFB_1 diet resulted in a significant (P<0.05) reduction in milk AFM_1 concentrations (S2, 16%; S3, 53%). In contrast, S1 adsorbent was not effective in reducing milk AFM_1 concentrations (P>0.05). Results of present study indicated that different aluminisillicate adsorbents have variable binding and detoxification capacity for AFB_1 and efficiency of an adsorbent must be approved with *in vivo* experiments.

Metabolic response to diet energy restriction in ewes selected for high and low mastitis resistance

J. Bouvier-Muller[1], C. Allain[1], F. Enjalbert[1], P. Hassoun[2], D. Portes[3], C. Caubet[4], C. Tasca[4], G. Tabouret[4], G. Foucras[4] and R. Rupp[1]
[1]INRA, UMR1388, 31326 Toulouse, Castanet-Tolosan, France, [2]INRA, UMR868, 34060, Montpellier, France, [3]INRA, UE0321, Domaine de la Fage, Roquefort sur Soulzon, France, [4]INRA-ENVT, UMR1225, 31326 Toulouse, Castanet-Tolosan, France; juliette.bouvier-muller@toulouse.inra.fr

Global climate change and extension of dairy systems using grazing and forage may lead to increased competition for nutrient and energy resources in the future. Negative energy balance (NEB) in early lactation is considered to increase susceptibility to mastitis. The aim of this study was to determine the effect of NEB on metabolism and immune response in dairy ewes with a different genetic background of susceptibility to mastitis. Accordingly, 48 early lactation dairy ewes from high and low somatic cell score (SCS) genetic lines were allocated in two homogeneous subgroups: a NEB group which was energy restricted to 60% of their energy requirements during 15 days and a control-fed group. Ewes were monitored for milk production, SCS, body condition and blood metabolites. Energy restriction resulted in a decrease of milk yield, weight and body condition score and an increase in milk fat-to-protein ratio. Accordingly, plasma non-esterified fatty acids and blood β-hydroxybutyrate (BHBA) concentrations were increased during restriction, reflecting the body reserve mobilization. Noteworthy, we evidenced an interaction between genetic line and feed restriction on metabolic parameters and body condition. Indeed high-SCS ewes showed a higher loss of weight and increase in BHBA concentration than low-SCS ewes, when facing NEB. NEB in early lactation led therefore to extensive mobilization of body reserve and ketosis in mastitis susceptible sheep. These preliminary results reinforce the hypothesis of a genetic association between mastitis susceptibility and energy metabolism and open the way to further studies on the biological basis of this association.

Response to selection for intramuscular fat content and correlated responses in other muscles

M. Martínez-Álvaro, V. Juste, A. Blasco and P. Hernández
Universidad Politécnica de Valenica, Instituto de Ciencia y Tecnología Animal, Camino de Vera s/n, 46071 Valencia, Spain; mamaral9@upv.es

A divergent selection experiment for intramuscular fat content (IMF) in Longissimus dorsi muscle (LD) in rabbits was carried out during 6 generations. Direct and correlated responses in carcass weight, scapular and perirenal fat weight and IMF of LD, Bíceps femoris, Supraspinatus and Semimambranosus propius muscles at 9 and 13 weeks of age were estimated as the differences between lines. Perirenal and scapular fat weights and IMF of all muscles increased with age. Response to selection for IMF was successful. The difference between high and low lines in IMF of LD at 9 weeks was 0.32 g/100 g, representing a 30.4% of the mean. A positive correlated response was shown in IMF of all studied muscles. Selection for IMF showed a positive correlated response with fat deposits.

Fitting nutrition to genetic selection: dietary energy source and fitness response in rabbit does

A. Arnau-Bonachera, C. Cervera, E. Blas, E. Martínez-Paredes, L. Ródenas and J.J. Pascual
Institute for Animal Science and Technology, Universitat Politècnica de Valencia, Camino de Vera, s/n.
46022 Valencia, Spain; alarbo@upv.es

Genetics and nutrition are usually studied as separated topics. Assuming that better animals express their whole genetic potential when fed with the best diet, is expected that both evolve together. However, different genetic types (GT) with different breeding goals could also have different main priorities. Addressed nutrition could improve their biological efficiency. 203 rabbit females belonging to 3 different GT were fed with 2 different diets from the 1st to the 6th parturition. GT were, line H, founded for hyper-selection in litter size at birth; line LP, founded for hyper-selection in reproductive longevity; and line R, founded and selected for growth rate. Diets only differed in their main energy source, animal fat (AF) to promote milk yield or cereal starch (CS) to promote body reserves gain. Litter size was standardized at parturition. Survival rate, fertility and productivity were determined. LP animals showed the highest survival rate (on av. +106%; P<0.05) and productivity (on av. +39.3%; P<0.05) after 6 parturitions, and their fertility was higher than H animals (+24%; P<0.05). The improvement of body condition of LP females when fed with CS diet increased productivity respect AF diet (+5.6 kits weaned per year; P<0.05). R animals had higher fertility than H animals (+23.5%; P<0.05), but no significant differences were found on survival and productivity between them. H females showed the lowest fertility, specially, between the 1st and 2nd parturition (on av. -41%; P<0.05), those fed with CS diet got fat and decreased their survival rate at 6th parturition (-44.5%; P<0.05) and fertility from the 2nd to the 6th parturition (-34.6%; P<0.05) respect to females fed with AF diet. In conclusion, GT could show different fitness response in function of the dietary energy source. To fit nutrition to genetics, understanding how animals manage resources is needed.

Use of genomic recursions in single-step genomic BLUP with a large number of genotypes

I. Misztal[1], B. Fragomeni[1], D.A.L. Lourenco[1], Y. Masuda[1], I. Aguilar[2], A. Legarra[3] and T.J. Lawlor[4]
[1]University of Georgia, Animal and Dairy Science, Athens, GA 30602, USA, [2]INIA, Canelones, 90200, Uruguay, [3]INRA, UMR1388, 31326 Castanet Tolosan, France, [4]Holstein Association, Brattleboro, VT 05302, USA; ignacy@uga.edu

The purpose of this study was to evaluate accuracy of genomic selection in single-step genomic BLUP (ssGBLUP) when the inverse of the genomic relationship matrix (G) is derived by the Algorithm for Proven and Young (APY). This algorithm implements genomic recursions on a subset of 'proven' animals. Only a matrix due to the subset needs to be inverted and extra costs of adding 'young' animals are linear. Analyses involved 10,102,702 Holsteins with final scores on 6,930,618 cows. A total of 100k animals with genotypes included 23k sires (16k with more than 5 progenies), 27k cows, and 50k young animals. Genomic EBV (GEBV) were calculated with a regular inverse of G, and with the G inverse approximated by APY. Initially, animals in the 'proven' subset included only sires or cows. Later, animals in the 'proven' subset were randomly sampled from all genotyped animals in sets of 5k, 10k, and 20k; each sample was replicated four times. Genomic EBV with APY were accurate when the number of animals in the subset was ≥10k, with little difference between the ways of creating the subset. Numerical properties as shown by the number of rounds to convergence were best with random subsets. The properties of the APY algorithm can be explained using the concept of a finite number of independent chromosome segments. Assume that each segment has a fixed value, that a fraction of each segment has a value proportional to the length of that segment (infinitesimal model), and that a genome of an individual is composed of a fraction of each segment. Subsequently, in the absence of confounding, n animals allow for identification of a population with n segments. The ssGBLUP with APY can accommodate a large number of genotypes at low cost and with high accuracy.

Genomic predictions with approximated G-inverse from large-scale genotyping data

Y. Masuda[1], I. Misztal[1], S. Tsuruta[1], D.A.L. Lourenco[1], B. Fragomeni[1], A. Legarra[2], I. Aguilar[3] and T.J. Lawlor[4]
[1]University of Georgia, 425 River Road, Athens, GA 30602, USA, [2]INRA, UR631 SASA, BP 52627, 31326 Castanet-Tolosan Cedex, France, [3]INIA, Canelones, 90200, Uruguay, [4]Holstein Association USA Inc., Brattleboro, VT 05301, USA; yutaka@uga.edu

The objective of this study was to compare the accuracy of genomic predictions in final score for young Holstein bulls calculated from single-step GBLUP models with the regular G^{-1} and approximated G^{-1} (G^{-1}_{ap}) matrices. The G^{-1}_{ap} was calculated with recursions on a small subset of animals. The predictor data set consisted of 77,066 genotyped animals, 9,009,998 pedigree animals, and 6,384,859 classified cows born in 2009 or earlier. For calculation of G^{-1}_{ap}, 9,406 high accuracy bulls or 16,828 high accuracy bulls and cows were used as the small subset. Genomic predictions (GEBV2009) were calculated for predicted bulls who had no classified daughters in 2009 but did in 2014. The validation data set contained phenotypes and pedigree recorded up to March 2014. Daughter yield deviations (DYD2014) were calculated for the predicted bulls with at least 30 daughters in 2014 (n=2,948). Coefficient of determination (R^2), calculated from a linear regression of DYD2014 on GEBV2009, was 0.44 with the regular G^{-1} and 0.45 with G^{-1}_{ap} for either subset. Genomic predictions using all available 569,404 (570k) genotypes were also calculated with G^{-1}_{ap} on 9,406 bulls. The computation was performed using 16 CPU cores and 61 G bytes of working memory. Setting up G^{-1}_{ap} took 1.8 h, setting up matrices associated with A_{22}^{-1} took 7 min, and iterations took 4.6 h, resulting in 6.4 total hours. In contrast, BLUP computations without genotypes took 1.6 h in total. The R^2 value from 570k genotypes was similar to the result from GEBV2009. Genetic predictions can be obtained with substantially less computational cost but without loss of reliability using G^{-1}_{ap}. The single-step GBLUP with G^{-1}_{ap} is applicable to very large genotyped populations.

Large-scale single-step genomic BLUP evaluation for American Angus

D.A.L. Lourenco[1], S. Tsuruta[1], B.O. Fragomeni[1], Y. Masuda[1], I. Aguilar[2], A. Legarra[3], J.K. Bertrand[1], D.W. Moser[4] and I. Misztal[1]
[1]University of Georgia, Animal and Dairy Science, 425 River Rd, 30602 Athens, GA, USA, [2]INIA, Km 10 Rincon del Colorado, 90200 Canelones, Uruguay, [3]INRA, 24 Chemin de Borde Rouge, 31326 Castanet Tolosan, France, [4]Angus Genetics Inc., 3201 Frederick Avenue, 64506 St. Joseph, MO, USA; danilino@uga.edu

This study investigated the feasibility of single-step genomic BLUP (ssGBLUP) for American Angus genetic evaluation. Over 6 million records were available on birth weight (BW) and weaning weight (WW), and 3.4 million on post-weaning gain (PWG). Genomic information was available on 51,883 animals. Realized accuracies were based on a validation population of 18,721 animals born in 2013. Traditional and genomic EBV were computed by BLUP and ssGBLUP using a multiple-trait model. Two methods for handling large numbers of genotyped animals were tested: indirect prediction based on SNP effects derived from ssGBLUP, and algorithm for proven and young (APY) that uses genomic recursions on a small subset of reference animals to invert the genomic relationship matrix (G). All genomic methods were based on reference populations of different sizes. With BLUP, realized accuracies were 0.48, 0.67, and 0.52 for BW, WW, PWG, respectively. With ssGBLUP and the 2k (33k animals) reference population, the accuracies were 0.55, 0.71, and 0.60 (0.62, 0.78, 0.65), respectively. With 8k reference population, index of indirect prediction with parent average was as accurate as prediction from regular ssGBLUP. With 33k reference population, indirect prediction alone was as accurate as prediction from regular ssGBLUP. Using APY with recursions on 4k (8k) animals gave 97% (99%) gains of regular ssGBLUP; the cost of APY inverse of G is 1% (4%) of the regular inverse. The genomic evaluation in beef cattle with ssGBLUP is feasible while keeping the same models already used in regular BLUP. Indirect predictions allow for low cost interim evaluations. The use of APY allows for inclusion of unlimited number of genotyped animals in the main evaluation.

Single-step approaches: from conceptual simplicity to implementation complexity

V. Ducrocq

INRA, UMR 1313 GABI, 78352 Jouy-en-Josas, France; vincent.ducrocq@jouy.inra.fr

Single step (SS) approaches are sometimes presented as an obvious choice which solves many technical issues that current genomic evaluations are facing: SS combines all sources of information (genotypes, phenotypes, pedigree) in a natural and balanced way. It propagates genomic information to related ungenotyped animals. By including all data on which selection is based, it should correct for pre-selection biases. Unfortunately, implementation is not trivial. New software have been made available but they are restricted to simple models when in fact, one major issue of the past 30 years has been to develop sophisticated genetic evaluation models (e.g. threshold/survival/random regression models with heterogeneous genetic and/ or residual variances) to better account for data characteristics. Simplifying models of analysis to avoid revisiting the existing software should be avoided. A strict application of SS also imposes constraints such as the automatic inclusion of records of animals potentially preferentially treated or the exclusion of information from foreign animals when their individual records are not available. These constraints can be removed with a flexible approach iterating between a sophisticated (national) genetic evaluation after proper adjustment to account for genomic evaluation and any genomic evaluation (without restriction to a unique approach: GBLUP) which may exclude animals (e.g. bull dams) and include foreign information. A lot of progress has been made recently to solve some issues such as the calibration of the classical and genomic relationship matrices or the inclusion of 'meta-founders' to mimic unknown groups. Among the challenges still ahead, the major one is to cope with hundreds of thousands of genotyped animals in the near future. This requires at least to move to a marker model and to avoid the use of sparse computation when this is not justified. As an example, we found that iteration on genomic data stored in memory leads to very fast solution of large genomic evaluations.

Exclusion or inclusion of information of culled bulls on the single-step genomic evaluation

M. Koivula[1], I. Strandén[1], G.P. Aamand[2] and E.A. Mäntysaari[1]

[1]Natural Resources Institute Finland (Luke), Myllytie 1, 31600 Jokioinen, Finland, [2]NAV Nordic Cattle Genetic Evaluation, Agro Food Park 15, 8200 Aarhus N, Denmark; minna.koivula@luke.fi

Number of genotyped animals has increased rapidly creating computational challenges genomic evaluation. In animal model BLUP, a candidate animal without progeny and phenotype do not contribute information to the evaluation and can be discarded. In theory, genotyped candidate animal without progeny can bring information into single-step BLUP (ssGBLUP), and can affect estimation of other breeding values. To quantify this, we studied the effect of inclusion or exclusion of genomic information of young bulls with lowest GEBVs. A ssGBLUP was computed using animal model deregressed genetic evaluations (DRP). EBVs for all Red dairy cattle (RDC) animals were obtained from the May 2014 Nordic NAV evaluations for milk, protein, and fat. We calculated DRPs for 3.2 million RDC cows with records. DRPs were thereafter used in the ssGBLUP. Full genotype data included 15,148 RDC animals. Candidate bulls were chosen to be 869 genotyped bulls born 2006-2010. The reduced DRP data was created from the full DRP data by excluding records of daughters of candidate bulls. ssGBLUP was computed using reduced DRP data and full genomic data to obtain genomic breeding values, named reduced GEBVs. Next, we removed all data from 30% of candidate bulls that had the lowest reduced GEBVs. These were called culled validation bulls. From the full DRP data we exluded the DRPs of daughters of culled validation bulls. Then we computed ssGBLUP both using the full genomic information and reduced genomic information. The effect of data exlusion was quantified by comparing GEBVs of selected genotyped validation bulls, and GEBVs of dams of culled validation bulls in different setups. The correlations between GEBVs of validation bulls from all genomic evaluations were >0.999. Also the bull dam GEBVs did not change notably as the correlations were 0.99. Thus, it seems that culled bulls can be safely excluded from the ssGBLUP.

Effect of SNP chip density on model reliability of genomic evaluations

I. Strandén and E.A. Mäntysaari
Natural Resources Institute Finland (Luke), Green Technology, Myllytie 1, 31600 Jokioinen, Finland;
ismo.stranden@luke.fi

Computational work in calculation of model reliabilities for GBLUP increases cubically with increase in number of genotyped animals. In SNP-BLUP, size of the mixed model equations (MME) is limited by the number of included SNP markers. In GBLUP, when the number of animals is more than the number of markers, reliability calculations may be affected by numerical constraints when generalized inverse of the genomic relationship matrix needs to be calculated. SNP-BLUP does not have such computational issues. In practice, SNP-BLUP will offer computational advantages over GBLUP if model reliabilities can be approximated with low number of SNP markers. We studied the effect of chip density on model reliability of young candidate bulls in GBLUP. Genotype data included 2,180 Red dairy cattle animals of which 710 were candidates. Candidate bulls were born 2000-2006. Number of SNP markers was 410,559, at highest. Heritabilities were 0.1 and 0.5. In general, average model reliability increased with decrease in the number of used markers. Correlation was 0.999 or higher using at least 20,000 markers irrespective of heritability. Correlation decreased more the less markers were used with higher heritability although was above 0.990 with at least 5,000 markers for both heritabilities. Average difference of model reliability between full and lower marker densities increased rapidly and, even more so for high heritability. Model reliability was 1.8% (2.7%), 4.2% (6.2%) and 9% (13.7%) higher than with full chip density by using 20,000, 10,000 and 5,000 markers with 0.1 (0.5) heritability. A simple linear regression can be used to adjust model reliabilities by 10,000 markers or more to approximate full SNP marker density.

Generalizing the Gaussian infinitesimal model when using information from genomic markers

R.J.C. Cantet
Universidad de Buenos Aires, CONICET, Departamento de Producción Animal, Facultad de Agronomía,
Av. San Martín 4453, 1417 C.A.B.A. (Buenos Aires), Argentina; rcantet@agro.uba.ar

Prediction of breeding value (BV) from the classic infinitesimal animal model has a well-developed quantitative genetics framework. A generalization of the Gaussian infinitesimal model using identity by descent (IBD) information on genomic markers is presented based on causal theory, a recent and generalized version of path coefficients. Normality of BV is assumed after the work of K. Lange or M. Abney. Direct inheritance from parents is viewed as 'causal' regression and IBD sharing to collateral relatives (non-ascendants and non-descendants) as correlations. Detected correlations above their expected value are later avoided to obtain a 'directed acyclic graph', by including direct paths to common ancestors into the vector of BV. Zero off-diagonal elements of the covariance matrix of BV with genomic information (G) indicate marginal independence, whereas zero elements of the inverse covariance matrix (G^{-1}) indicate conditionally independence on all remaining animals. Taking it as positive definite, G has a modified Choleski decomposition as the additive relationship matrix A. Therefore, G^{-1} can be written as $(I - B')$ $D^{-1}(I - B)$ where B is a matrix of regression coefficients, and the diagonal matrix D includes the Mendelian residual variances. Elements of B result from regressing the BV of any individual (X, say) on half the parental BV, plus the BV of other related and, possibly, non-related individuals that are conditionally dependent with X due to sharing IBD information from distant recombinations. The BV of grandparents enter into the regression if the pairwise genomic relationships with X are above or below their expectation. Other individuals previously born to X, and marginally independent, may enter into the equation when there have progeny in common with ancestors of X and become conditionally dependent. An algorithm is presented to compute both G and G^{-1}, animal by animal, ordered by date of birth.

Prediction with major genes partially genotyped using gene content multiple trait BLUP

A. Legarra and Z.G. Vitezica
INRA, UMR 1388 GenPhySE, 31526 Castanet Tolosan, France; andres.legarra@toulouse.inra.fr

In pedigreed populations with a segregating major gene influencing a quantitative trait, it is unclear how to use pedigree, genotypes and phenotypes when not all individuals are genotyped. We propose to consider the genotype at the major gene, coded numerically as gene content, as a second trait correlated to the quantitative trait, in a gene content multiple trait BLUP (GCMTBLUP). The genetic covariance between the trait and the gene content (genotype at the gene) is a function of the effect of the gene on the trait. The genetic covariance of genotypes, scored as gene content, and phenotypes can be written in a multiple-trait form, which can accommodate any pattern of missingness. Major gene alleles' effects and the covariance can be estimated using standard EM-REML or Gibbs sampling. Prediction can be undertaken for very large data sets using multiple trait BLUP software. We simulated two scenarios: a selected trait (heritability=0.05), and an unselected trait with heritability=0.5. In both cases the major gene explains half the genetic variation. Competing methods are use of imputed gene contents, either by Gengler et al. method or by iterative peeling; those methods do not consider information on the quantitative trait for genotype prediction, but GCMTBLUP does. GCMTBLUP gives unbiased estimation of the gene effect (Gengler et al. and peeling are biased by selection), and less bias and equal accuracy in prediction than competing methods. GCMTBLUP improves the estimation of genotype in non-genotyped individuals, in particular if they have own phenotype and for highly heritable traits. Ignoring the major gene in genetic evaluation leads to serious biases and decreased accuracy. GCMTBLUP is an extension and improvement on Gengler et al. method for gene content prediction, and very similar to Single Step GBLUP used in genomic evaluation. The method can cope with biallelic or multiallelic genes.

Split-and-merge Bayesian variable selection enables efficient genomic prediction using sequence data

M.P.L. Calus[1], C. Schrooten[2] and R.F. Veerkamp[1]
[1]Wageningen UR Livestock Research, Animal Breeding and Genomics Centre, P.O. Box 338, 6700 AH Wageningen, the Netherlands, [2]CRV BV, P.O. Box 454, 6800 AL Arnhem, the Netherlands; mario.calus@wur.nl

Simultaneous use of more than 10,000,000 SNP imputed from sequence data hardly improved the accuracy of genomic prediction, even with commonly used Bayesian variable selection models. One reason may be that the overparametrization problem is more severe than with e.g. 50k SNPs. We hypothesize that split-and-merge Bayesian variable selection may provide a solution to overcome this issue. Our application of split-and-merge, also known as divide and conquer, combined with Bayesian variable selection involves two steps. The first step divides the SNPs in ~300 subsets of 40k SNPs. Subsets are formed by going through the list of SNP ordered by their position on the genome and assigning each next SNP to the next subset in line. In each subset, BayesC is used for genomic prediction, and SNPs are ranked based on their posterior probabilities indicating their likelihood to be strongly associated with the investigated trait. In the second step, from each subset a few hundred SNPs with the largest posterior probability will be selected into a final set of SNPs that are used to build the final genomic prediction model using BayesC. Next to attempting to alleviate the overparametrization problem, an additional practical benefit of this modelling procedure is that the first step comprises ~300 analyses with ~40,000 SNPs each, rather than one analysis with >10,000,000 SNPs. Since all these analyses can be run in parallel, computation time will be a matter of hours instead of more than a month. Results will be presented at the conference.

Comparing genomic prediction and GWAS with sequence information vs HD or 50k SNP chips

R.F. Veerkamp[1], R. Van Binsbergen[1,2], M.P.L. Calus[1], C. Schrooten[3] and A.C. Bouwman[1]
[1]Animal Breeding and Genomics Centre, Wageningen UR Livestock Research, P.O. Box 338, 6700 AH Wageningen, the Netherlands, [2]Wageningen UR Biometris, P.O. Box 16, 6700 AA Wageningen, the Netherlands, [3]CRV BV, P.O. Box 454, 6800 AL Arnhem, the Netherlands; roel.veerkamp@wur.nl

Earlier work showed that using whole genome sequence information did not improve the accuracy for genomic prediction, because there are too many SNPs in close LD to pinpoint the functional SNP accurately. In this study we therefore compared the single SNP GWAS results using imputed whole genome information (from run4 of the 1000 bull genomes project) to the GWAS results obtained using either the 50k or777k HD SNP chips. For the analysis, (imputed) HD genotypes were available on 5,549 Dutch bulls, of which 3,416 were used for GWAS and subsequent training. Single SNP GWAS was performed using the genomic relationship matrix (based on HD) to account for population structure. In the sequence information there were 28,076,109 SNP imputed, but 10,258,688 where monomorphic in our training population. For protein yield 2,241 SNPs were significant (-log10(p)>5), and 28 (160) of those were present on the 50k (HD) SNP chips. For somatic cell score the equivalent number of SNPs were 1,545, 90 and 7 using sequence, HD or 50k, and for 'interval first to last insemination' the equivalent number was 952, 27 and 4 SNPs, respectively. Fitting all SNPs together (using Bayesian Variable Selection) the HD and 50k SNPs gave clear evidence for QTL, but using sequence information the signal was spread across many SNP in high LD. In conclusion, although more significant SNP were found using sequence information, relatively few new regions were identified, and every significant SNP was accompanied by several others in high LD. Therefore, to benefit from sequence data in genomic selection, more sophisticated methodology is required than currently used for genomic prediction.

A comparison of short and long range phasing methods

C.F. Baes[1,2], B. Gredler[1], B. Bapst[1], F.R. Seefried[1], H. Signer-Hasler[2], C. Flury[2], C. Stricker[3] and D. Garrick[4]
[1]Qualitas AG, Chamerstrasse 56, Zug, Switzerland, [2]Bern University of Applied Sciences, School of Agriculture, Forestry and Food Sciences, Länggasse 85, Zollikofen, Switzerland, [3]agn Genetics, Börtjistrasse 8b, Davos, Switzerland, [4]Iowa State University, Department of Animal Science, Kildee 225, Ames, USA; christine_baes@gmx.de

The affordability of low (19k, 19,000 SNP), medium (50k; 54,609 SNP) and high-density (HD; 777,962 SNP) SNP arrays, and more recently whole genome sequence data (WGS), is revolutionising animal breeding. In dairy cattle populations, the population structure comprising large half-sib families facilitates imputation of animal genotypes from array-based densities up to sequence level using WGS information of key ancestors. Imputation exploits haplotype information. Haplotypes are sets of associated SNP alleles on a single chromatid of a chromosome pair. Genotypes can be phased to reconstruct the haplotypes that represent the original allelic combinations that an individual received from its parents. Although methods have been developed for phasing population-based samples, most exploit correlations of SNP alleles within linkage disequilibrium (LD) blocks (local or short-range phasing). In addition to phasing being computationally intensive and slow, distantly separated SNPs will be in different LD blocks and cannot be reliably phased using local phasing. Long range phasing may provide a viable method of phasing alleles over multiple LD blocks and has been shown to be much faster than local phasing. It also allows the detection of possible sequencing errors that are evident when two haplotypes are identical except for a single variant call. In this study we compared haplotypes determined using short and long range phasing methods. Haplotype concordance was evaluated by comparing short and long range LD, 50k, HD and WGS haplotypes with eachother. Furthermore, the comparison of actual sequence with sequence-derived haplotypes provides a measure of sequencing error.

GATK and Beagle 4.0 as the two different software for haplotype phasing for NGS data

M. Frąszczak[1], M. Mielczarek[1], J. Szyda[1], G. Minozzi[2], E.L. Nicolazzi[2], C. Diaz[3], A. Rossoni[4], C. Egger-Danner[5], J. Woolliams[6], L. Varona[3], J. Williams[2], H. Schwarzenbacher[5], C. Ferrandi[2], T. Solberg[7], F. Seefried[8], D. Vicario[2] and R. Giannico[2]
[1]Department of Genetics, Wroclaw University of Environmental and Life Sciences, Kożuchowska 7, 51-631 Wroclaw, Poland, [2]Fondazione Parco Tecnologico Padano, Cascina Codazza 12, Lodi, Italy, [3]Universidad de Zaragoza, Pedro Cerbuna 12, Zagaroza, Spain, [4]Associazione Nazionale Allevatori Bovini della Razza Bruna, Bussoleng, Verona, Italy, [5]ZuchtData EDV-Dienstleistungen GmbH, Dresdner Straße 89/19, Vienna, Austria, [6]Roslin BioCentre, EH25 9RG, Roslin, United Kingdom, [7]GENO Breeding and A.I. Association, Holsetgata 22, Hamar, Norway, [8]Qualitas AG, Chamerstrasse 56, Zug, Switzerland; magdalena.fraszczak@up.wroc.pl

Determination of haplotype phase has become increasingly important especially in the area of the large-scale sequencing. The methods for haplotype phasing are still developing, what can be seen in case of Beagle software, which has been adapted to VCF format of data. The aim of this study is to look at the different tools to phasing NGS data and try to compare them. DNA sequences of 160 bulls from 9 breeds were obtained within the framework of Gene2Farm project by Illumina HiSeq 2000 Next Generation Sequencing platform. Filtered reads were aligned to the UMD3.1 reference genome using BWA. SNPs were detected using several software, including GATK. For haplotype phasing two different software packages were used: Beagle 4.0 version and GATK tool called Read Backed Phasing. Both applied packages differ substantially in terms of the applied haplotyping methodology as well as in the exploiting the haplotype information contained in the data. The comparison of those programs will include the regions and percentage of identically phasing SNPs for each chromosome separately. In based on results of haplotype phasing, the similarity of two different animals will be presented. Some of the technical aspects of using of those software packages will be compared either.

Genomic selection in sheep in Ireland – plans for the future based on experiences from the past

D.P. Berry[1], K. Moore[2], E. Wall[3], J. Conington[2], T. Pabiou[3], M. Coffey[2], K. Mcdermott[3] and N. Mchugh[1]
[1]Teagasc, Animal & Grassland Research and Innovation Centre, Fermoy, Co. Cork, Ireland, [2]SRUC, Roslin Institute Building, Easter Bush, Midlothian EH25 9RG, United Kingdom, [3]Sheep Ireland, Bandon, Co Cork, Ireland; donagh.berry@teagasc.ie

Genomic selection was launched for Holstein dairy cattle in Ireland in 2009 and retrospective analysis shows that it is up to 56% more accurate than traditional evaluations. The genomic selection reference population consisted of 985 high reliability Holstein-Friesian sires; in 2015 the training population now consists of over 5,000 high reliability bulls and also includes pure Friesian bulls. Multi-breed national genomic evaluations for beef cattle are expected to be delivered in 2015 with a reference population consisting of over 100,000 pure and crossbred, pedigree and commercial animals. International collaboration (i.e. sharing of genotypes and phenotypes) was, and continues to be fundamental to the success of both genomic selection schemes. Farmers and breeders have also directly contributed financially. Ireland has its own genotyping platform for all cattle which covers parentage (in)validation, major genes, genetic disorders and sufficient genomic coverage to generate accurate genomic predictions. Ireland is currently genotyping a multi-breed purebred population of over 10,000 rams and ewes with phenotypic records on a combination of high density (±600,000 genetic markers), medium density (54,241 genetic markers) and low density (±15,000 markers) genotyping platforms. Genotype sharing with the UK is underway. Joint analysis of Irish and UK sheep breeding values indicate moderate to strong correlations between some traits suggesting that, similar to in dairy (INTERBULL) and beef (INTERBEEF), exploiting such information in a genomic prediction algorithm could be beneficial. Projects are underway in Ireland to improve connectedness internationally to facilitate such initiatives.

QTL detection and heritability estimations for two undesired coat color in French Saanen goats
P. Martin, G. Tosser-Klopp, I. Palhière and R. Rupp
INRA, INPT ENSAT, INPT ENVT, UMR1388 Génétique, Physiologie et Systèmes d'Elevage, 31326 Toulouse, Castanet-Tolosan, France; pauline.martin@toulouse.inra.fr

In the framework of the national 'Phénofinlait' (www.phenofinlait.fr) and the EU '3SR' (www.3srbreeding.eu) projects, a large daughter design has been carried out in commercial farms for mapping traits of interest in French dairy goats. Using these data, we mapped QTL for two types of undesired coat color phenotypes. One fifth of the Saanen females is indeed phenotyped 'pink' (8.0%) or 'pink neck' (11.5%). Consequently these colored goats are not included in the breeding scheme as elite animals. A total of 810 goats and their 9 Animal Insemination sires were genotyped with the Illumina GoatSNP50 BeadChip. QC included SNP call rate (>99%), animal call rate (>98%), minimum allele frequency (>1%), Hardy Weinberg equilibrium and pedigree consistency. After edits, a total of 49,647 out of 53,347 SNPs were validated. We conducted a Genome-Wide Association Study based on a polygenic model including single SNP as fixed effects. Heritability was estimated both with a binomial logistic and a linear animal mixed model on 103,443 females born between 2004 and 2010 and scored by the breeding organization. A highly significant signal (-log10pvalue=16.0) was found as associated with the 'pink neck' phenotype on CHI11 which supported the presence of a major gene. For the 'pink' phenotype, two highly significant signals were found on chromosomes 5 and 13 (-log10pvalue equal to 8.2 and 11.6, respectively). The peak on chromosome 13 was is in the region of the ASIP gene, which is well known for its association with coat color phenotypes. The heritabilities estimated were 0.25 for 'pink' and 0.15 for 'pink neck' on the binary scale and 0.59 and 0.50 on the linear scale, respectively. Fine mapping is ongoing. Altogether results showed strong genetic control for coat color in French Saanen goats and pave the way for eliminating undesired colored animals from the breeding using SNP information.

Genetic parameters for conformation traits in dairy goats
S. Mucha, A. Mclaren, R. Mrode, M. Coffey and J. Conington
Scotland's Rural College, Roslin Institute Building, SRUC, EH25 9RG, United Kingdom; sebastian.mucha@sruc.ac.uk

Improving the functional fitness of dairy goats for traits such as animal mobility and structural correctness is important for ensuring continued animal health if genetic selection for increased productivity is actively pursued. The aim of this work is to quantify genetic and phenotypic properties of udder, teat, legs and feet scores together with milk yield in UK dairy goats. The data set comprised 2,429 conformation records (udder, legs, and feet) and 126,262 milk yield records on 2,429 first lactation mixed breed dairy goats, scored for conformation in 2013. The pedigree file contained 30,139 individuals. In total 10 conformation traits were scored on a linear scale (1-9) by one scorer from captured photographic images. Covariance components were estimated with the AI-REML algorithm in the DMU package with an animal model containing farm, lactation stage, and birth year as fixed effects. Heritability estimates for conformation traits, ranged from 0.02 to 0.45. Significant genetic correlations were detected between udder furrow and udder depth (0.74), udder attachment (0.66), teat shape (0.96), and back legs (0.57). Teat angle was correlated with teat placement (0.60), back legs (0.48), and back feet (0.67). The phenotypic correlations ranged from 0 to 0.39. Genetic correlations estimated between milk yield and udder and teat conformation traits were negative, ranging from -0.60 to -0.20, and from -0.50 to -0.20, respectively. Genetic correlations with feet and leg conformation were between -0.30 and 0.30. Udder and teat conformation traits were low to moderately heritable and feet and legs traits less heritable. The genetic correlations between the conformation traits and milk yield indicate that breeding programmes for dairy goats should include conformation scores, so that selection for productivity is not accompanied by deterioration in functional fitness especially in udder, teat and leg conformation.

Corriedale PRO: a genetic and organizational innovation case study

G. Ciappesoni, G. Banchero, A. Vázquez, E.A. Navajas and A. Ganzábal
INIA Uruguay, Animal Breeding, Ruta 48 km 10, Rincón del Colorado, 90200, Uruguay;
gciappesoni@inia.org.uy

The technological change was illustrated originally with the linear model of innovation, including three phases: invention, innovation and diffusion. Nowadays, the model of technological change involves innovation at all stages of research, development, diffusion and utilization. The development of the Corriedale PRO, a maternal line of the Corriedale breed of Uruguay, is an example of this process. Innovation in research: A joint research project between INIA and Central Lanera Uruguay (wool farmer's cooperative) and their farmers, was executed between 2006 and 2012. Previously in the year 2004, prolific sheep breeds were imported by INIA: Finnsheep (Finn) and new lines of East Friesian (EF). After five years of evaluations of six biotypes (Corriedale, Finn, EF and their crosses) based on 2.771 reproductive a productive records of 967 ewes, the more promissory breed combinations were proposed to the Corriedale Breeders Association of Uruguay (SCCU) for their evaluation in semi-intensive production systems. Innovation in development and use: The SCCU accepted the challenge and implemented with INIA and the Wool Uruguayan Secretariat (SUL) the crossbreeding system to develop the new genetic line. Six stud-flocks, two commercial flocks and SUL research station are involved in the current stages, with the incorporation of INIA research stations and more commercial flocks in the near future. The research at this stage includes reproductive, phenotypic and molecular characterization of the new line. Between 2012 and 2015, 4,000 Corriedale ewes were mated with Finn-EF rams to produce Corriedale PRO ewe-lambs. From 2013 to 2015, about 1,400 Corriedale PRO ewes were joined with rams of the same line. Weaning percentages of the new line were between 100-130% for ewe-lambs and 130-160% for 2-years old ewes. These results are significantly higher than the national average (65-75%) and better than those obtained in the best commercial flocks (100-120% in similar production systems).

Is annual genetic gain influenced by the rate of registered paternity in sheep and goat breeding programs?

J. Raoul[1,2], I. Palhière[1], J.M. Astruc[2] and J.M. Elsen[1]
[1]INRA, Animal Genetic, 31326, Castanet-Tolosan, France, [2]idele, 31321, Castanet-Tolosan, France;
jerome.raoul@idele.fr

In sheep and goat French commercial farms, female sires are generally unknown In breeding programs, rate of paternal filiation is highly variable. This rate ranged from situations where pedigrees are well known (mostly in terminal sire breeds) to situations where they are scarcely known (hardy breeds). In this context, paternity assignation using SNP markers, became an attractive alternative for breeders. The annual genetic gain (AGG) expected from a breeding program depends on the precision of genetic evaluations: in populations with a low rate of paternal filiation, number of known relatives of a given individual is small and candidate EBVs, based on this information, are imprecise. However, the quantitative consequences of partial paternal filiations, in terms of expected genetic progress, have never been evaluated. This is our objective. We developed a deterministic model in Fortran language, to assess, for various existing breeding schemes, the influence of known paternity rate of females on AGG. First a demographic model describes the accumulation of known performances for individuals and their closed relatives depending on different categories. Female categories were determined by their parity and sire status (unknown, proven sire, etc.). Male categories were defined according to their age and their own selection process. More than 50 female and 10 male categories were defined. Then, outputs of this demographic model were used to compute an EBV average accuracy per category. Gene flow between categories after one reproductive cycle was described using standard matrix methodology. Contributions of each sire category to the new born males, as well as the differential selection along the sire-male path depends on the breeding program assessed. Contributions of each dam category to the new born males was computed assuming a single truncation threshold. Following an iterative process we finally obtained asymptotic AGG.

Development of a SNP parentage assignment panel for French sheep breeds

F. Tortereau[1], C. Moreno[1], G. Tosser-Klopp[1], L. Barbotte[2], L. Genestout[2] and J. Raoul[1,3]
[1]INRA, Genphyse, campus INRA, 31326 Castanet-Tolosan, France, [2]Labogena-DNA, campus INRA, 78350 Jouy-en-Josas, France, [3]Institut de l'Elevage, Génétique&Phénotypes, campus INRA, 31321 Castanet-Tolosan, France; jerome.raoul@idele.fr

Parentage assignment is a priority for the French sheep breeding programs. Around 750 individuals from 30 French breeds were genotyped with the Illumina OvineSNP50 and OvineHD BeadChips (54,000 or 600,000). Genotypes were filtered on several criteria (call-freq, Minor Allele Frequency (MAF), Hardy-Weinberg equilibrium, Mendelian compatibilities, genomic distribution) to produce a single SNP panel for parentage assignment useful in a large variety of French breeds. Each SNP had the following MAF characteristics: higher than 0.10 in each of the 30 populations, higher than 0.30 in at least 20 populations out of the 30 analyzed, between 0.20 and 0.30 in a maximum of 10 populations, between 0.10 and 0.20 in a maximum of one population. A final set of 249 SNPs was retained. These SNPs were first confirmed on a simulated population. They have then been validated on a real population (809 lambs, 105 sires). The genotyping was performed with the Sequenom technology. A total of 192 SNPs for parentage assignment and 19 causal mutations segregating in French breeds could be integrated into 4 plexes. In this population, we could assigned 90 to 95% of the lambs. Assignation was performed using maximum likelihood. In addition, the MAF of the 249 SNPs were checked on different international populations. For example, in the Spanish breeds, around 80% of the SNPs have a MAF higher than 0.30 whereas this percentage falls down to 55% in the Scottish breeds. The final French panel will be integrated into the international LD chip, in construction within the International Sheep Genomics Consortium. The challenge is now to propose to the French breeding programs a low-cost genomic tool enabling parentage assignment and major genes genotyping.

An update on goat genomics

E.L. Nicolazzi[1], P. Ajmone Marsan[2], M. Amills[3], H.J. Huson[4], P. Riggs[5], M.F. Rothschild[6], R. Rupp[7], B. Sayre[8], T.S. Sonstegard[9], A. Stella[1], G. Tosser-Klopp[7], C.P. Vantassel[9], W. Zhang[10] and The International Goat Genome Consortium[7]
[1]Parco Tecnologico Padano, Via Einstein, 26900, Italy, [2]Universita Cattolica S. Cuore, Piacenza, 29100, Italy, [3]Center for Research in Agricultural Genomics, Barcelona, 08193, Spain, [4]Cornell University, Ithaca, NY 14850, USA, [5]Texas A&M University, Texas, TX 77843, USA, [6]Iowa State University, Ames, IA 50011, USA, [7]INRA, INPT ENSAT, ENVT, UMR1388 GenPhySE, Toulouse, 31326, France, [8]Virginia State University, Petersburg, VA 23806, USA, [9]USDA-ARS, Beltsville, MD 20705, USA, [10]Kumming Institute of Zoology, Kunming, 650000, China, P.R.; ezequiel.nicolazzi@tecnoparco.org

Goats are specialized in dairy, meat and fiber production, being adapted to a wide range of environmental conditions and having a large economic impact in developing countries. In the last years, there have been dramatic advances in the knowledge of the structure and diversity of the goat genome/trascriptome and in the development of genomic tools, rapidly narrowing the gap between goat and related species such as cattle and sheep. Major advances are: (1) publication of a de novo goat genome reference sequence; (2) development of whole genome high density RH maps; and (3) design of a commercial 50k SNP array. Moreover, there are currently several projects aiming at improving current genomic tools and resources. An improved assembly of the goat genome using PacBio reads is being produced, and the design of new SNP arrays is being studied to accommodate the specific needs of this species in the context of very large scale genotyping projects (i.e. breed characterization at an international scale and genomic selection) and parentage analysis. As in other species, the focus has now turned to the identification of causative mutations underlying the phenotypic variation of traits. In addition, since 2014, the ADAPTmap project (www.goatadaptmap.org) is gathering data to explore the diversity of caprine populations at a worldwide scale by using a wide variety of approaches and data.

Genetic diversity of Swiss goat breeds

M. Neuditschko[1], A. Burren[2], H. Signer-Hasler[2], I. Reber[3], C. Drögemüller[3] and C. Flury[2]
[1]Agroscope, Swiss National Stud Farm, Les-Longs-Prés 191, 1580 Avenches, Switzerland, [2]Bern University of Applied Sciences, School of Agricultural, Forest and Food Sciences HAFL, Länggasse 85, 3052 Zollikofen, Switzerland, [3]University of Bern, Institute of Genetics, Bremgartenstrasse 109a, 3001 Bern, Switzerland; markus.neuditschko@agroscope.admin.ch

Indigenous livestock resources are important in the socio-economics of rural agricultural systems to ensure farming in remote mountain areas. To conserve unique genetic resources for future generations, a better understanding of genetic variation within livestock breeds is required. We report the first analysis of genetic diversity of Swiss goats breeds based on genome-wide SNP genotypes. We collected genome-wide SNP genotypes for 241 goats using the GoatSNP50 Genotyping BeadChip®. The samples originate from 10 different breeds namely Appenzell goat, Grisons striped goat (GSG), Capra grigia, Coloured Chamois (CHG), Valais goat including the strains copper necked (VCN) and black necked (VBN), Nera Verzasca, Peacock goat, Saanen goat (SAN), Booted goat, and Toggenburger goat (TOG). After quality control filtering, 241 samples and 48,750 SNPs were included in the genetic diversity analysis. The average genetic distance between goats was highest in GSG (0.312 ± 0.019) and lowest in TOG (0.273 ± 0.016). Gene diversity varied from 0.275 in SAN to 0.262 in VBN. CHG showed highest proportion of polymorphic SNPs (98%), while the least proportion was found in VBN (89%). Investigating the population structure conducting a PCA revealed that the breeds originate from different genetic resources, while breeds/strains from the same geographical region remain more closely related (e.g. VCN and VBN). Therefore, we suggest the implications of managing improvement and conservation strategies to conserve the genetic resources of Swiss indigenous goats.

Diversity and population structure in three Angora goat populations using SNP data

C. Visser[1], S. Lashmar[1], E. Van Marle-Koster[1], M. Poli[2] and D. Allain[3]
[1]University of Pretoria, Animal and Wildlife Sciences, Private Bag x 28, Hatfield, 0028, South Africa, [2]INTA, Instituto de genetica, CICVyA, CC25, CP 1712-Castelar, Argentina, [3]INRA, UMR1388 GenPhySE, CS52627, 31326 Castanet Tolosan, France; carina.visser@up.ac.za

The Angora goat populations in Argentina (AR), France (FR) and South Africa (SA) have been kept geographically and genetically distinct. Due to country-specific selection and breeding strategies there is a need to characterize the populations on a genetic level. In this study we analysed genetic variability of Angora goats using the standardized 50k Goat SNP Chip to assess the effect of genetic and geographical isolation. A total of 104 goats (AR:30; FR:26; SA:48) were genotyped, 3 AR goats were removed with a sample rate of <95%. Between 11.8% (AR) and 15.7% (SA) of SNP were removed based on standard QC (call rate<95%, MAF<5%, HWE, P<0.001). PLINK was used to calculate heterozygosity values as well as inbreeding coefficients across all autosomes per population. Diversity, as measured by expected heterozygosity ranged from 0.371 in the SA population to 0.397 in the AR population. The SA goats were the only population with a positive average inbreeding coefficient value of 0.009. After merging the three datasets, standard QC and LD-pruning were performed and 15,105 SNPs remained for analyses. Principal component analysis was used to visualize individual relationships within and between populations. All SA Angora goats were separated from the others and form a tight, unique cluster, while outliers were identified in the FR and AR breeds. Clustering analyses was performed with Admixture. At K=3 geographic grouping was observed with the three populations' individuals allocated into three distinct clusters. Apparent admixture between the AR and FR populations was observed, while both these populations showed signs of having some common ancestry with the SA goats. Results confirmed that geographic isolation and differing selection strategies caused genetic distinctiveness between the populations.

Comparison of across countries sheep breeding objectives in New Zealand and Ireland

N. Mc Hugh[1], P.R. Amer[2], T.J. Byrne[2], D.P. Berry[1] and B.F.S. Santos[2,3]
[1]Animal & Grassland Research and Innovation Centre, Teagasc, Moorepark, Fermoy, Co. Cork, Ireland,
[2]AbacusBio Limited, P.O. Box 5585, Dunedin, New Zealand, [3]School of Environmental & Rural Science,
University of New England, Armidale, Australia; noirin.mchugh@teagasc.ie

Breeding objectives underpin the direction, extent, and economic implications of genetic selection in livestock populations. The definition of breeding objectives may differ between countries and production systems. However objective comparisons of alternative indexes may inform producers how livestock selected through specific breeding objectives are likely to perform in alternative systems. The objective of this study was to calculate correlations between national selection indexes in Ireland and New Zealand and gain a deeper understanding of the factors influencing responses in economically important traits. Two national sheep breeding objectives are published for both countries: a maternal and terminal index, with a similar range of traits recorded in the both national breeding objectives. To calculate index correlations and the response to selection for individual traits, selection index methodology was used. Moderate to strong correlations (0.48 to 0.86) were calculated among indexes within and between countries, with the strongest correlation observed between the New Zealand and Irish maternal indexes. The expected annual economic responses to selection between comparable traits in New Zealand and Ireland were of a similar direction for most traits, but in general greater economic responses were recorded in the New Zealand breeding objectives. Responses to selection in both maternal indexes are largely driven by growth traits; ewe mature weight also accounted for an important proportion of overall response and has a significant trait relative emphasis in both maternal indexes. The majority of emphasis in terminal indexes of both countries is on growth and meat traits. Results indicate that differences between both national breeding objectives are unlikely to be a barrier to the exchange of gene stocks among countries.

Lamb survival is lowly heritable in Norwegian White Sheep

J.H. Jakobsen[1] and R.M. Lewis[2]
[1]Norwegian Association of Sheep and Goat Breeders, Box 104, 1431 Ås, Norway, [2]University of Nebraska,
Animal Science Department, Lincoln, NE 68583, USA; jj@nsg.no

The economic efficiency of sheep enterprises can be accentuated by increasing litter size in order to market more lambs per ewe. Such has been the case in Norway, where average litter size in Norwegian White Sheep (NWS) has increased over the last decades. However, with the increase in numbers born, early lamb deaths have risen with welfare concerns. The aim of the study was to estimate the amount of genetic variation in lamb survival in NWS, and thereby opportunities for its incorporation in the breeding program. Data on 497,085 lambs born in 1996 to 2013 were extracted from the Norwegian Sheep Recording database. Lamb survival was defined as a binomial trait (0-1), with lambs stillborn or dead within 24 hours defined as failed to survive. The frequency of lamb loss was 5.8%. Analyses were done fitting either a linear model, assuming 0-1 data were normally distributed, or a logistic regression, assuming data were binomially distributed and using a logit link function. Either animal or sire-maternal grandsire (MGS) pedigree relationships were used. Fixed effects were flock-year, lambing difficulty, the interaction of litter size and age of dam, and the covariate day of birth deviated from the flock-year median. Based on log-likelihood ratio tests (when assuming data normally distributed), the random terms were direct (or sire) and maternal (or MGS) additive effects, permanent and temporary environmental effects, and residual. Sire and MGS variances were converted to direct and maternal additive variances; variance ratios from logistic regression were transformed to the 0-1 scale. Regardless of analytical approach, direct heritabilities were extremely low (0.001 to 0.003), with maternal heritabilities only slightly higher (0.006 to 0.007). At least within NWS, additive variance for lamb survival is low. Coupled with the relatively low frequency of lamb loss, justification for integrating lamb survival into the breeding goal is tenuous.

Contrary to the Booroola (BMPR1B) the Lacaune gene (B4GALNT2) shortens the onset of sheep oestrus

N. Debus[1], C. Maton[1], F. Bocquier[1] and L. Bodin[2]
[1]INRA-SUPAGRO UMR868 Selmet, 2 place Viala, 34060 Montpellier, France, [2]INRA-INPT UMR 1388 GenPhySE, CS 52627, 31326 Castanet-Tolosan, France; loys.bodin@toulouse.inra.fr

The effects of the Booroola gene (BMPR1B) on the ovulation control of the heterozygous (F+) and homozygous (FF) ewes are well known while the effects on other traits are less documented but likely weaker. The effects of the Lacaune gene (B4GALNT2) are rather similar to those of the Booroola in terms of phenotypic expression of ovulation: the ovulation rate of heterozygous (L+) ewes is suitable but that of homozygous (LL) is excessive, although the involved physiological mechanisms are different. However it has been shown that the Lacaune gene might have other effects on reproductive traits. Features of oestrus onset after a classical hormonal treatment (FGA vaginal sponge during 14 days) were studied on 3 groups of ewes bearing different combinations of these genes: (++/++ (MA); n=15), (FF/++ (FF); n=13), (FF/L+ & FF/LL (FFLL); n=9). At sponge withdrawal (D0), ewes were joined with teaser rams equipped with Alpha-R® electronic harness. All mounts occurring until D6 were registered and ovulation were controlled by laparoscopy at D8. Data were analysed by chi square test and ANOVA. The 3 teasers (A, B, C) have detected a similar number of ewes in heat (A=35; B=32; C=34) but did not perform the same number of mounts (A=311; B=371; C=233; P<0.001) since they had preferred females. There was a large variability between females for the number of mounts (μ=24.4; σ=19.4) and the total duration of mounts (μ=39.1 sec; σ=35.3) but no difference between genotypes. Age of ewe and genotype group had significant effects on the onset of oestrus. This onset began on average 31.4 hours after sponge withdrawal for the FFLL; that was 9.8 h (P<0.01) before the FF (μ=41.2 h) and 12.1 h (P<0.01) before the MA (μ=43.5). However there was no group difference for the duration of oestrus (μ=13.1 h σ=6.4 h) and within genotype group the ovulation rate had no effect on the oestrus features (onset, duration).

Using genomic technology to reduce mastitis in meat sheep

A. Mclaren[1], W. Sawday[2], S. Mucha[1], J. Yates[2], M. Coffey[1] and J. Conington[1]
[1]SRUC, Scotland's Rural College, Easter Bush, Midlothian, EH25 9RG, Edinburgh, United Kingdom, [2]Texel Sheep Society, National Agricultural Centre, Stonleigh Park, Kenilworth, CV8 2LG, Warwickshire, United Kingdom; joanne.conington@sruc.ac.uk

Mastitis is a problem encountered by many livestock species throughout the world. Most information currently available refers to how the disease affects dairy breeds of cattle and small ruminants, with few acknowledging its impact in meat sheep breeds. However, the disease can have a considerable effect on meat sheep, costing the industry millions of pounds each year due to poorer animal performance, increased medical costs and the premature culling of animals. The aim of this project, which commenced in 2015 in collaboration with the British Texel Sheep Society, is to investigate the genetic aspects of this disease in more detail using genomic technologies to reduce the prevalence of mastitis and allow the genetic selection of mastitis-resistant animals. Approximately 3,500 pure- and crossbred Texel sheep, selected based on their influence in the population, will be genotyped. The high density (700k) and low density (50k and 1k) single nucleotide polymorphism (SNP) beadchips will be used for genotyping. The mastitis resistance traits analysed will include: (1) farmer records of mastitis; (2) udder and teat conformation; (3) somatic cell counts; and (4) California Mastitis Test. Genomic estimated breeding values (GEBVs) for mastitis resistance, in addition to current breeding goal traits, will be generated with the BLUPf90 package using the single-step method. Phenotyping strategies deemed the most appropriate for predicting mastitis resistance, and which are suitable for breeders to carry out on-farm, will be also identified. Overall, it is anticipated that the ability to deliver the first GEBVs for mastitis resistance in meat sheep, in addition to identifying the most appropriate phenotypes for breeders to record, will provide the tools to achieve improved animal welfare and production and increase overall sustainability.

Polymorphism of the PRNP gene in Romanov sheep breed in Poland

A. Piestrzynska-Kajtoch, G. Smołucha and B. Rejduch[†]
National Research Institute of Animal Production, Animal Cytogenetics and Molecular Genetics, Krakowska
1, 30-083 Balice, Poland; agata.kajtoch@izoo.krakow.pl

Sheep is natural host for scrapie, which is fatal neurodegenerative prion disease. There are two types of scrapie: classical and atypical. After exposure to the classical scrapie-causing agent, susceptibility to the disease is linked to polymorphisms in three codons of PRNP gene (amino acid change: A136V, R154H, R171Q/H). The $V_{136}R_{154}Q_{171}$ allele is associated with susceptibility to classical scrapie and ARR allele is associated with resistance. The aetiology of atypical scrapie differs from that of classical scrapie. The disease is often observed in sheep with the AHQ or $AF_{141}RQ$ alleles (L141F). The aim of the study was to analyze the PRNP polymorphism in the Romanov sheep-breed under the Genetic Resources Conservation Program in Polnad in which one atypical scrapie case was diagnosed. The combined allelic discrimination assay and RFLP method were used to analyze the PRNP polymorphism in 481 Romanov sheep. 263 animals had only 3 codon analyzed (136, 154, 171) and for 218 animals analysis of codon 141 has been additionally performed. 11 different PRNP genotypes were found in the whole population studied. The most frequent genotype was ARR/ARQ (41.37%) and the least frequent was VRQ/VRQ (0.21%). The frequency of VRQ allele was 4.57% and 41.48% for ARR allele. The most frequent allele was ARQ (47.09%). 31 Romanov sheep were in G5 susceptibility class (the most susceptible to classical scrapie) and 86 individuals were in G1 susceptibility class (the least susceptible to classical scrapie). The F allele was present only in one animal. One herd of Romanov sheep breed have been under regular scrapie genotypes monitoring since 2009. The results show that selection based on scrapie genotypes has effectively decreased the frequency of VRQ allele (from 4.88% in 2009 to 0% in 2014) and increased the frequency of ARR allele (from 42% in 2009 to 50% in 2014) among annually studied lambs. The study was financed by statutory project no. 4-007.1.

Sheep biodiversity in transhumant farming in Greece

D. Loukovitis[1], D. Chatziplis[1], A. Siasiou[2,3], I. Mitsopoulos[3], A. Lymberopoulos[3], Z. Abas[2] and V. Laga[3]
[1]Alexander Technological Educational Institute of Thessaloniki, Laboratory of Agrobiotechnology and Inspection of Agric. Products, Dept. of Agricultural Technology, P.O. Box 141, Sindos, Thessaloniki, 57400 Sindos, Thessaloniki, Greece, [2]Democritus University of Thrace, Department of Agricultural Development, Pantazidou 193, Orestiada, 68200 Orestiada, Greece, [3]Alexander Technological Educational Institute of Thessaloniki, Department of Agricultural Technology, School of Agricultural Technology, Food Techn. and Nutrition, P.O. Box 141, Sindos, Thessaloniki, 57400 Sindos, Thessaloniki, Greece; chatz@ap.teithe.gr

Preliminary results of biodiversity in transhumant sheep populations from three Greek mainland geographical districts (Epirus, Thessaly and Sterea Ellada) are presented. Genetic variation regarding the above populations was calculated using various statistical methods and was compared to that of nine rare pure sheep Greek breeds. In total, 368 samples were genotyped using a set of ten microsatellite markers. The polymorphism of the ten microsatellites was medium to high, ranging from 8 to 32 alleles. The total observed and expected heterozygosity was 0.69 and 0.73, respectively. According to pairwise F_{st} estimates, most genetic differentiation was distributed among pure breeds (F_{st} range 0.0231-0.1525). Differentiation between pure breeds and transhumant sheep populations was lower (F_{st} range 0.0089-0.0695), whereas differentiation between the three transhumant sheep populations was the lowest (F_{st} range 0.0014-0.0036). Results of microsatellite DNA analysis herein revealed an obvious isolation of the pure breeds between each other as well as from the transhumant sheep. On the other hand, the three transhumant sheep populations seem to form a unique group. Further analysis is under way as more populations from various geographical sites are added to the sample panel.

Effect of introgression of Argali blood on growth and metabolic profiles of Romanov breed lambs

V.A. Bagirov, B.S. Iolchiev, V.G. Dvalishvili, I.V. Gusev, T.E. Deniskova, E.A. Gladyr and N.A. Zinovieva
L.K. Ernst Institute of Animal Husbandry, 142132, Russian Federation; horarka@yandex.ru

Hybridization of domestic and wild animals is an attractive way to improve the growth and to enhance the adaptation capacities of farm animal species. The aim of our work was to study the growth rate and to evaluate the blood metabolic and clinical profiles of the purebred Romanov lambs (ROM) and their contemporary carrying 12.5% blood of Argali (*Ovis Ammon*) of Pamir origin (ROM_A). The lambs were housed at the same environmental conditions within one facility and were fed by the same diets. The lambs were weighted at the age of 4, 6 and 8 months. The blood samples for clinical and metabolic analysis were collected at the age of 9 months. We found significant differences in growth rates of lambs: the daily gains from 4 to 6 months and from 6 to 8 months were significantly higher in ROM_A lambs: +56.7 g (P<0.01) and +75.2 g (P<0.01), respectively. No significant differences were observed for blood erythrocytes' and leukocytes' numbers, but the hemoglobin content, hematocrit and erythrocyte mean cellular volume were significantly higher in ROM_A lambs comparing to ROM lambs: +7.72 g/l (P<0.001), +5.55% (P<0.01) and +1.91 (P<0.01). Measurement of metabolites in blood of lambs revealed the significantly lower concentrations of the total protein (-5.33 g/l, P<0.001), globulin (-3.83, P<0.001), urea (-0.73 mM/l, P<0.05) and alkaline phosphatase (-244.5 IE/l, P<0.001) but the higher concentrations of creatinine (+7.75 mcM/l, P<0.05) and glucose (+0.32 mM/l) in ROM_A comparing to ROM lambs. No significant differences were observed in blood concentrations of ureic acid, triglycerides, cholesterol, ALT and AST enzymes. Thus, the introgression of Argali blood resulted in increase of growth rate and led to alterations of some clinical and metabolic parameters of blood of Romanov' lambs. Study was performed under financial support of Russian scientific foundation, project no. 14-36-00039.

Association of calpain, small subunit 1 gene polymorphism with drip loss and cooking loss in sheep

E. Grochowska[1], B. Borys[2], D. Lisiak[3] and S. Mroczkowski[1]
[1]UTP University of Science and Technology, Department of Genetics and General Animal Breeding, Mazowiecka 28, 85-084 Bydgoszcz, Poland, [2]National Research Institute of Animal Production, Experimental Station Kołuda Wielka, Parkowa 1, 88-160 Janikowo, Poland, [3]Institute of Agricultural And Food Biotechnology, Rakowiecka 36, 02-532 Warsaw, Poland; grochowska@utp.edu.pl

Recently, number of studies have shown effects of polymorphisms of calpains genes not only on post mortem tenderization of meat but also on other carcass and meat quality traits in livestock. The aim of the study was to detect the polymorphism in calpain, small subunit 1 (CAPNS1) gene (also known as CAPN4) and determine the occurrence of associations between these polymorphisms and natural drip loss and cooking loss. In total 102 Coloured Polish Merino sheep were analysed. Animals were maintained in Experimental Station of NRIAP in Kołuda Wielka. Male lambs were slaughtered at 105±10 days of live. Meat quality traits: natural drip loss (%) and cooking loss (%) were measured on samples of longissimus dorsi. After DNA extraction a 190-bp fragment of CAPNS1 (exon 5 and 6 including intron) was amplified in PCR. Amplicons were subjected to MSSCP method using 10% polyakrylamide gels (5% glycerol) at 15, 10 and 5 °C for 800 Vh for each thermal phase. Samples representing certain band profiles were sequenced in Genomed, Poland. Association analysis were performed between the observed genotypes and traits of interest using GLM procedure of SAS with fixed effects of genotype and the year of observation. Five genotypes named G1, G2, G3, G4 and G5 were found in the investigated group of animals. The frequencies of genotypes were: 17.65, 41.18, 13.73, 14.71, 12.75%, respectively. In our analysis we detected new genotypes of ovine CAPNS1 gene. A high significant (P<0.01) and significant (P<0.05) associations were observed between natural drip loss and cooking loss and CAPNS1 genotypes, respectively. Project no. NN311521440 was foundrd by National Science Centre.

Effects of linseed and hemp seed supplementation on gene expression and milk quality in Alpine goats

S. Chessa[1], P. Cremonesi[1], F. Turri[1], L. Rapetti[2] and B. Castiglioni[1]
[1]National Research Council, Institute of Agricultural Biology and Biotechnology, Via Einstein, 26900 Lodi, Italy, [2]University of Milan, Department of Agricultural and Environmental Sciences, Production, Landscape, Agroenergy, via Celoria, 20133 Milano, Italy; chessa@ibba.cnr.it

In the last years great interest was given to the effect of nutrition on ruminant products, in particular on fatty acid (FA) composition, because of the effects of FA on human health. The inclusion of less than 4% oil in cattle and sheep diets increases the polyunsaturated fatty acid content of milk. Linseed fat supplementation improved goat milk performance, and enhanced the nutritional profile of milk FAs, by increasing the vaccenic acid, CLA, omega-3 concentrations, and the indexes used as markers of dietary fat quality. Less is known on the effects of hemp seed supplementation on lactating goats. We report the results of linseed (L) and hemp (H) seed supplementation on milk traits and milk somatic cells expression profile of 18 pluriparous goats. The goats were fed the same diet from parturition to 22 days in milk (DIM) and then equally divided into 3 groups: control diet group (C), and L and H seeds supplementation groups. A total of 150 ml of milk was collected for both milk fatty acid composition analysis and RNA extraction from somatic cells at 21 (T0) and at 150 (T1) DIM 2 hours after morning milking. Real-Time -PCR was performed on the cDNA retro-transcribed from 3 goats of each group both at T0 and T1 the RNA. In order to verify if the two lipid supplementations had different effects on mammary gland cells' activity the expression level of the following genes was analysed: ACACA, AGPAT6, CD36, DGAT1, FABP3, FASN, LPL, NR1H3, PLIN2, PLIN3, PNPLA2, PPARG, SCAP, SCD, and SREBF1. Significant differences were found both in milk FA composition and in the expression level of some of the analysed genes between T0 and T1 and among groups, confirming the important role of the different nutrients on animal metabolism and milk composition.

Associations between microsatellite markers and footrot in German Merinoland sheep

C. Weimann, S. Huebner, G. Luehken and G. Erhardt
Justus-Liebig-University, Dept. of Animal Breeding and Genetics, Ludwigstrasse 21B, 35390 Giessen, Germany; christina.weimann@agrar.uni-giessen.de

Footrot is one of the most costly and impacting animal welfare disease in small ruminants. Resistance to footrot varies between populations and phenotypes. Estimated heritability values ranging from 0.1 to 0.3 depending on the parameters used indicate a genetic background of the disease and the possibilities for selection. It is known that alleles of the DQA2 gene within the major histocompatibility complex (MHC) are associated with footrot risk in sheep. In addition associations between the microsatellite BMC5221 and footrot are described. The aim of the present study was to test the associations between microsatellites mainly located within the MHC and footrot in German Merinoland sheep. DNA from a total of 348 sheep from two farms was analyzed for 4 microsatellites (BMC5221, OMHC, DYMS1 and OLADRB1). The footrot phenotypes of these sheep were available as field data (clinically footrot positive or negative). There was a significant (chi-square test, $P \leq 0.05$) difference between allele frequency of footrot positive and negative animals for microsatellite OMHC1. A variance analysis with footrot as binomial distributed trait and the fixed effects farm and microsatellite alleles showed significant effects of farm for foot rot prevalence. No significant allele substitution effect was observed for any of the microsatellites included in the study. For further studies it would be useful to have a more detailed phenotypic assessment of footrot which will be more time consuming. Furthermore the association between DQA2 alleles and the foot rot prevalence in German sheep breeds will be investigated.

Analysis of microsatellite DNA markers in parentage testing of sheep in Poland

A. Szumiec, A. Radko and D. Rubiś
National Research Institute of Animal Production, Department of Animal Cytogenetics and Molecular
Genetics, Krakowska 1, 32-083 Balice, Poland; anna.radko@izoo.krakow.pl

The parentage testing and identification of sheep in Poland has been carried out since 1973 by the National Research Institute of Animal Production based on blood group system. From 2016, sheep pedigree data in Poland will only be validated based on DNA technique. The aim of the study was to evaluate the usefulness of 12 markers recommended by ISAG (AMEL, CSRD247, ETH152, INRA005, INRA006, INRA063, INRA172, MAF065, MAF214, McM042, McM527, OarFCB20) for individual identification and pedigree testing of sheep in Poland. The study used genomic DNA isolated from whole blood of 281 sheep of the following breeds: Old Type Polish Merino (93), Olkuska (88), and Wielkopolska (100). The PCR products were electrophoresed on an ABI 3130xl sequencer and the results were analysed using GeneMapper Software 4.0. The largest number of alleles (12) was obtained for INRA005 in Wielkopolska and the lowest for ETH 152 (4 alleles for each breed). H_O and H_E values for all the breeds were high (>0.5161) except INRA172 in Merino (H_O=0.2903, H_E=0.3307) and INRA006 in Olkuska (H_O=0.3750). The highest values were observed for INRA63 in Merino (H_O=0.8710, H_E=0.8370). The highest differences between H_O and H_E (Fis) were observed in Olkuska (from -0.2235 for ETH152 to 0.3028 for INRA006). PIC ranged from 0.3171 for INRA172 to 0.8181 for INRA63. The high exclusion probability values showed that the animals' pedigree data can be assigned with more than 97.89% probability when the genotype of one parent is known (CPE_1) and with more than 99.79% probability when the genotypes of both parents are known (CPE_2). The high combined power of discrimination (PD_C), close to unity, is indicative of the high usefulness of the tested set of markers for individual identification. The analysed panel of 12 microsatellite markers showed a high degree of polymorphism in the loci of the analysed sheep breeds, which indicates that it can be used for evaluating the genetic structure of sheep.

Identification of two mitochondrial haplogroups in native sheep breeds from Poland

A. Koseniuk, T. Rychlik and E. Słota
National Institute of Animal Production, Animal Cytogenetics and Molecular Genetics, Krakowska street
1, 32-083, Poland; anna.koseniuk@izoo.krakow.pl

The study of mitochondrial DNA (mtDNA) control region has been commonly used in phylogenetics. Mitochondrial Haplogroup B predominates in Europe. However, in several surveys haplogroup A, which is common in Asia, was also determined in the European sheep population. The aim of the study was to determine the number of female lineages in the Polish sheep breeds. The material of the survey was constituted of DNA extracted from blood of 143 sheep of the following breeds: Wrzosowka (wrzos, n=33, from 2 flocks), Swiniarka (swin, n=35, from 2 flocks), Pomorska (pom, n=40, from 3 flocks) and Wielkopolska (wlk, n=35, from 1 flock). For both PCR and sequencing reaction two sets of primers were used. The first one was used for probes of all but Wielkopolska breed whereas the second set of primers was used for samples of Wielkopolska breed only. The sequences obtained were aligned with Type B reference sequence (GeneBank accession no. AF010406). Genetic distance was calculated with Kimura's 2-parameter and the Neighbor Joining (NJ) phylogenetic tree was constructed for haplotypes. There were obtained 561 bp mitochondrial CR sequences spanning *Ovis aries* reference sequence from 15,959 to 16,519 bp. Sixty-five haplotypes were identified. The calculation of genetic distance enabled identification of two major sequence types in the analyzed sheep group. The NJ phylogenetic tree revealed that 3 haplotypes clustered into clade A and the remaining 62 with clade B. These 3 haplotypes, which pooled together with cluster A, included 2 sequences of Wrzosowka, 5 sequences of Pomorska and 1 sequence of Wielkopolska breed. All haplotypes of Swiniarka clustered into haplogroup B. Supported by Ministry of Science and Higher Education grant no. N N311 541840.

Evaluation of SNPs using DNA arrays developed for cattle and sheep in reindeer

V.R. Kharzinova[1], A.A. Sermyagin[1], E.A. Gladyr[1], G. Brem[2] and N.A. Zinovieva[1]
[1]L.K. Ernst Institute of Animal Husbandry, Moscow, Podolsk, 142132, Dubrovitcy 60, Russian Federation, [2]Institute for Animal Breeding and Genetics, Vienna, Veterinaerplatz 1, A 1210, Austria; veronika0784@mail.ru

Semi-domesticated reindeers (*Rangifer tarandus*) are one of the most important elements of the ecosystem of Russia and the main object of livestock in Siberia. Recent studies have demonstrated that commercially developed SNP chips are possible means of SNP discovery for non-model organisms. We aimed to apply two sets of commercial chips developed for cattle (*Bos taurus*) and sheep (*Ovis Aries*) to whole-genome analysis of reindeer *R. tarandus* inhabiting Russia. Tissue samples were collected from four female animals bred in south-east region of Siberia and used for DNA extraction. Genotyping was performed using Bovine SNP50 v2 BeadChip and Ovine SNP50 BeadChip. All subsequent analysis was conducted with PLINK. Among 54,609 SNPs on the Illumina BovineSNP50 BeadChip, 23,481 (43.0%) were fully scored on all animals and 6,771 SNPs (12.6%) were partially genotyped on various numbers of animals. Only 1,257 SNPs (2.3% of the total; 5.3% of fully scored loci) were polymorphic. The observed heterozygosity of polymorphic SNPs was 0.91 ± 0.01 and the expected heterozygosity was 0.48 ± 0.143. Among 5,4241 SNPs on the Illumina Ovine SNP50 BeadChip, 25,512 SNPs (47.0%) were fully scored on all four animals and 4,242 SNPs (7.9%) had calls on one to three animal samples. The number of polymorphic SNPs was 519 (1.0% of the total and 2.0% of fully scored loci). The observed heterozygosity of polymorphic SNPs was 0.79 ± 0.01 and the expected heterozygosity was 0.44 ± 0.004. Our study confirms that SNP chips developed for domestic animal species can be used as a tool for SNP discovery even in widely diverged non-model species including such unique species as *R. tarandus*. Supported by the Russian Scientific Foundation, project no.14-36-00039.

Whole genome scan to identify loci associated with carpet wool quality

M. Rastifar[1], A. Nejati-Javaremi[2] and M.H. Moradi[3]
[1]University of Tehran, Karaj, Iran, [2]University of Tehran, Karaj, Iran, [3]Arak University, Arak, Iran; Hoseinmoradi@ut.ac.ir

Iran is one of the most important countries in wool production while almost all of its producing is classified as coarse and low quality wool and used in production of many high volume commercially available carpets. The aim of the present study was to whole genome scan for identifying the loci associated with wool quality traits. The blood samples were collected from 94 Iranian sheep samples and the DNA was extracted and genotyping was performed using Ovine 50k BeadChip arrays. Quality control filters were applied to initial genotyping data and finally, 48,056 SNPs belongs to 90 animals were used in the final analysis. The wool samples of these animals were collected and were analyzed for mean fiber diameter, the proportion of fiber that are equal or more than 30 μm and curvature traits using OFDA technology. The fixed effects of herd, age and sex were examined and then the genome wide association study was performed using PLINK software and FDR correction was used to adjust the error rate. The results of this study revealed that the sex and herd had significant effect on the curvature while only the sex of animals had significant effect on the fiber diameter and the proportion of fiber that are equal or more than 30 μm ($P<0.05$). Considering these significant effects in the genomic association analysis, three SNPs on chromosomes 1 and 6 (two loci) were identified that significantly affect the trait of proportion of fiber that are equal or more than 30 μm, which is closely related with coarse wool, and also two SNPs on chromosomes 6 and 8 was identified that significantly affect the curvature. Genes identified in this study also showed the presence of major genes affecting the quality of wool as CAP in these regions. In general, the results of the present study can play an important role in identifying causal mutations affecting the quality of the wool in Iranian sheep breeds.

Genetic diversity within casein genes in European and African sheep breeds

S. Chessa[1], I.J. Giambra[2], E.M. Ibeagha-Awemu[3], H. Brandt[2], A.M. Caroli[4] and G. Erhardt[2]
[1]National Research Council, UOS di Lodi, Institute of Agricultural Biology and Biotechnology, via Einstein, Località Cascina Codazza, 26900 Lodi, Italy, [2]Justus-Liebig-University, Ludwigstrasse 21B, 35390 Giessen, Germany, [3]Dairy & Swine Research & Development Centre, 2000 College Street, Sherbrooke, QC J1M 0C8, Canada, [4]University of Brescia, Viale Europa 11, 25123 Brescia, Italy; chessa@ibba.cnr.it

The aim of this study was to analyze variability within casein genes in dairy and non-dairy sheep breeds. In total, 30 single nucleotide polymorphisms (SNPs), one indel, and one microsatellite, all located in milk protein genes (CSN1S1, CSN1S2, CSN2, CSN3) were genotyped. The analysis included 478 samples derived from 14 different European (Italy and Germany) and African (Nigeria and Cameroon) sheep breeds, and 32 European Mouflon (EM) samples. Allele and genotype frequencies were calculated using PopGene V 1.31. Intragenic haplotypes were calculated for each casein gene using the expectation and maximization (EM) algorithm of the HAPLOTYPE procedures of SAS® 9.3. Frequencies of casein haplotypes between dairy and non-dairy breeds were compared using a principal component analysis (PCA) with the PRINCOMP procedure of SAS® 9.3. Out of 30 SNP analysed 3 were found monomorphic in all breeds. Moreover in a further SNP T allele could be considered rare as it occurred only in Sopravissana breed at a very low frequency (0.0161). Considering the casein genes it was not possible to find clear differences in haplotype distribution between dairy and non-dairy breeds. Some haplotypes of one of the four genes were fixed or rare in some breeds, whereas in the same breeds different haplotype distribution was found within the other genes. Only 2 haplotypes were found in EM and at least 4 breeds showed more than 7 haplotypes. The most frequent haplotype occurred in all breeds including EM. The lowest H_o values were observed for the three African breeds. In the future, association studies should be made including milk performance traits by phenotyping the animal material.

Effect of melatonin treatment on reproduction of noire de Thibar rams

R. Taboubi[1] and M. Rekik[2]
[1]Institut Supérieur Agronomique de Chott Meriem, Sousse Tunisia, Animal Production, 12 street Marsa Borj El Baccouche Ariana, 2080 Ariana Tunisia, Tunisia, [2]International Centre for Agricultural Research in the Dry Areas (ICARDA), Amman, Jordan, Animal production, Amman Jordan, 1000 Jordan, Tunisia; rahmaca@hotmail.com

The aim of this study was to investigate the effect of melatonin treatment on sperm production and sexual behavior of Noire de Thibar rams. A total 16 Noire de Thibar rams aged between 16 months and 4 years were allocated to two balanced groups according to live weight. Rams of melatonin group were treated with 3 melatonin subcutaneous implant (Melovine®; Laboratoires CEVA). There was a significant increase ($P<0.05$) in scrotal circumference of rams; scrotal circumference was clearly higher for melatonin-treated rams than Control rams. For Noire de Thibar rams, the circumference reached average values of 36.2±0.96 and 31.9±1.92 cm for respectively melatonin and control animals ($P<0.001$). Measured semen characteristics (volume of ejaculate, sperm concentration and individual activity score of spermatozoid) were not significantly different between two treatments. But the mass activity score was significantly different between two groups at the end of the trial ($P<0.001$) when average values reached 4.75±0.42 and 4.25±0.25 for respectively melatonin and control group. As to sexual behavior, melatonin-treated Noire de Thibar rams revealed an ardor and sexual aggressiveness more pronounced than the Control counterparts. This was clear through the young Noire de Thibar rams. It is concluded that treatment with melatonin improves the reproductive potential of rams in spring.

Herd factors affecting longevity in Swedish dairy cattle

E. Strandberg[1], A. Roth[2] and U. Emanuelson[3]
[1]Swedish University of Agricultural Sciences, Department of Animal Breeding and Genetics, P.O. Box 7023, 75007 Uppsala, Sweden, [2]Växa Sverige, Box 288, 75105 Uppsala, Sweden, [3]Swedish University of Agricultural Sciences, Department of Clinical Sciences, P.O. Box 7054, 75007 Uppsala, Sweden; erling.strandberg@slu.se

The length of life of a dairy cow has a large impact on herd profitability, mainly by spreading the heifer raising cost over a longer productive life. Currently, the productive life in Sweden is about 2.6 lactations, which corresponds to a replacement rate of about 40%. However, there is large variation among herds, and there are herds with replacement rates of only 15%. The aim of this study was to find what herd-related factors influence the average longevity in Swedish dairy herds. The analysis was based on cows calving from September 2004 through August 2011. The average productive life (PL) of cows culled in the last 3 years of data was calculated and was used to define herds with short or long PL (lowest or highest quartile). To be defined as a herd with short (long) PL, the herd must have at least 5 culled cows per year in each of the last 3 years and belong to the lowest (highest) quartile for at least 2 of these 3 years. This resulted in 765 short PL and 638 long PL herds. A logistic regression model evaluated the probability of being a long PL herd, where several herd-related predictor variables were tested using a forward stepwise selection procedure. Large herds and herds increasing in size by more than 50% were more likely to have short PL. High average milk yield was not a risk factor for short PL but herds with the lowest yield had higher odds of being long PL herds. Herds with long PL had lower EBV Milk, most likely due to a slower turnaround. A high proportion of cows culled early in first lactation had a large effect on the odds of being a short PL herd. Overemphasis on keeping short calving intervals or low SCC can result in short PL. Having many cows with long interval from calving to first insemination, high cow mortality and stillbirth rate increased the risk of short PL.

Genetic analysis of infectious diseases: estimating gene effects for susceptibility and infectivity

M.T. Anche, P. Bijma and M.C.M. De Jong
Wageningen University, Drovendaalsesteeg 1, 6708 PB Wageningen, the Netherlands; mahlet.teka@gmail.com

Genetic selection of livestock against infectious diseases can be considered as complementary to existing infectious disease control interventions. Current genetic approaches aiming to reduce disease prevalence assume that an individual's disease status (infected/not-infected) is solely a function of its susceptibility to a particular pathogen. However, also an individual's infectivity affects the risk and prevalence of an infectious disease in a population. Among-host variation in susceptibility and infectivity affects transmission of an infection in the population, which is usually measured by the value of the basic reproduction ratio R0. R0 is important epidemiological parameter that determines the risk and prevalence of infectious diseases. An individual's breeding value for R0 is a function of its genes for both susceptibility and infectivity. Thus, to estimate effects of genes on R0, we need to estimate the effect of genes on individual susceptibility and infectivity. Here we developed a generalized linear model to estimate the relative effects of genes for susceptibility and infectivity. A simulation study was performed to investigate the bias and precision of the estimates, and effect of R0, the size of the susceptibility and infectivity effects of genes, and relatedness among group mates on the bias and precision of the estimates. The model was developed from an equation that describes the probability of an individual to become infected as a function of its own susceptibility genotype and of the infectivity genotypes of its infected group mates. Simulation results showed that estimated gene effects were always conservative and fairly close to simulated values. Smaller bias was obtained when R0 is between approximately 4 and 7, and with higher relatedness among group mates. With larger effects both absolute and relative standard deviations becomes clearly smaller, but relative bias stayed the same.

Production, health and fertility performance of Norwegian Red and Norwegian Red crosses in Ireland

E. Rinell[1], R. Evans[2], F. Buckley[3] and B. Heringstad[1]
[1]Norwegian University of Life Sciences, P.O. Box 5003, 1430 Aas, Norway, [2]Irish Cattle Breeding Federation, Highfield House, Shinagh, Bandon, Cork, Ireland, [3]Teagasc, Moorepark, Fernmoy, Cork, Ireland; ellen.rinell@nmbu.no

The objective of this study was to evaluate the performance of Norwegian Red dairy cattle (NR) and NR crosses (NRX) on commercial farms in Ireland in comparison with their Holstein (HO) counterparts. Data were provided by the Irish Cattle Breeding Federation. Only herds with at least 30 NR or NRX records were kept. NRX cows were at least 25% NR. Only records from years in which NR or NRX were in the herd were retained. There were 108 herds with 1,252 NRF, 5,980 NRX and 37,521 HO cows in the production dataset. In the health dataset, there were 95 herds with 618 NRF, 3,187 NRX and 15,200 HO cows. In the dataset with calving interval and somatic cell count (SCC), there were 107 herds with 2,259 NRF, 8,252 NRX and 42,426 HO cows. Some animals had records from multiple lactations, some from only one lactation. Significant differences (P<0.01) between the least squares means (LSM) of HO vs NR and NRX were found for all traits mentioned below. HO had a higher 305-day milk, fat and protein yield than NR and NRX. The respective LSM were 5,120, 5,057 and 5,047 kg milk, 220, 218 and 218 kg fat, and 184, 182 and 182 kg protein. HO had a higher LSM for lactation mean SCC than both NR and NRX, 167, 151, and 157 (×1000 cells) respectively. HO also had longer calving intervals than NR and NRX, with LSM of 382, 378 and 379 days, respectively. The LSM of average mastitis score were significantly higher for HO (0.26) than NR and NRX (0.20 and 0.24). HO also had a higher average for lameness score (0.16) than NR and NRF (0.14). Consistent with previous studies, NR and NRX were slightly less favorable than HO in terms of milk, fat and protein production. However, this was compensated for with a lower somatic cell score, shorter calving intervals and a lower incidence of lameness and mastitis in contrast to their HO herd mates.

Factors affecting the serum protein pattern in multi-breed dairy herds

T. Bobbo[1], G. Stocco[1], C. Cipolat-Gotet[1], M. Gianesella[2], E. Fiore[2], M. Morgante[2], G. Bittante[1] and A. Cecchinato[1]
[1]University of Padova, Department of Agronomy, Food, Natural Resources, Animals and Environment (DAFNAE), Viale dell'Università 16, 35020 Legnaro (PD), Italy, [2]University of Padova, Department of Animal Medicine, Productions and Health (MAPS), Viale dell'Università 16, 35020 Legnaro (PD), Italy; tania.bobbo@studenti.unipd.it

This study investigated the effect of breed, production level, somatic cell count, days in milk and parity on serum protein pattern in multi-breed herds. Milk and blood samples of 1,527 cows from 41 herds were collected. Six breeds were considered: Brown Swiss (BS), Holstein Friesian (HF), Jersey (Jer), Simmental (Si), Grey Alpine (GA) and Rendena (Ren). For all cows, somatic cell count (SCC) was measured and blood samples were analyzed for serum protein pattern: Total Protein (TP), Albumin and Globulin content. Analysis of variance was performed using the MIXED procedure of SAS, with a linear model including the effects of production level (2 levels), breed (6 levels), SCC (7 levels), DIM (6 levels) and parity (4 levels) and the random effect of the herd. All effects were important sources of variation for the investigated traits, except for the production level that only affected Albumin content (P<0.001). As expected, higher levels of Globulin have been reported for cows with higher parity order (41.0 vs 43.7 g/l for the first and fourth parity order, respectively). The same pattern has been observed for the SCC, with TP (and Globulin) increasing from 72.4 (40.9) to 75.1 (44.3) g/l for group of cows belonging to class 1 (SCC lower than 18,000 cells/ml) and class 7 (SCC higher than 450,000 cells/ml) respectively. On the other side, Albumin content decreased in cows with high SCC. The higher producing breeds (BS, HF, Si) exhibited higher levels of Globulin (greater than 43 g/l) respect to GA, Jer and Ren. In conclusion, such serum protein pattern might be used for monitoring the welfare status of dairy cows.

Early postnatal plane of nutrition affects subsequent milk production of Holstein calves

S. Wiedemann[1], H.-J. Kunz[2] and M. Kaske[3]
[1]Kiel University, Animal Breeding and Husbandry, Olshausenstr. 40, 24098 Kiel, Germany, [2]Chamber of Agriculture, Futterkamp, 24327 Blekendorf, Germany, [3]University of Zurich, Department for Farm Animals, Winterthurerstr. 260, 8057 Zurich, Switzerland; swiedemann@tierzucht.uni-kiel.de

Two different feeding strategies during early postnatal life of Holstein calves were assessed to study the impact on first lactation milk production. During the first 4 weeks (wk) of life, calves were fed either *ad libitum* (AdL; *ad libitum* feeding of whole milk during wk 1 and milk replacer (MR; 160 g/l) from d 8-28; n=38) or according to a restrictive feeding protocol (RES; 4 l milk/d during wk 1, 6 l MR (120 g/l) from d 8-28; n=30). Feeding was similar in both groups after the 4th wk of life. All animals were kept individually during the first wk of life and in groups therafter. The feed intakes were analyzed during the first 10 wk of life in all animals and during first lactation in 37 animals. Daily milk yield and monthly milk composition were recorded. Total energy intake during the first 4 wk was higher in AdL-calves compared to RES-calves (16.6 vs 10.2 MJ ME/d, respectively; P<0.01). Thereafter, no difference in energy intake was observed until the 10th wk of life. In AdL-calves the average daily gain was higher compared to RES-calves during the first 4 wk of life (0.72 vs 0.45 kg/d; P<0.001), while age at first calving did not differ (765 vs 777 d; P=0.30). In the AdL- and RES-group, 21 and 26 animals remained on the farm for a full 305-d first lactation, respectively. The FCM yield was higher in AdL-animals compared to RES-animals (29.2±0.4 vs 28.0±0.4 kg FCM/d; P<0.05); but milk composition did not differ. The higher FCM yield was accompanied by a higher feed intake in AdL-animals (19.3 vs 18.8 kg DM/d; P<0.01). The results indicate that an increased feeding intensity during early life has positive long-term effects on the milk production potential in the first lactation.

Milk production of the mother is associated with the birth weight rather than the sex of the calf

M.H.P.W. Visker[1], Y. Wang[1], M.L. Van Pelt[2] and H. Bovenhuis[1]
[1]Wageningen University, Animal Breeding and Genomics Centre, P.O. Box 338, 6700 AH Wageningen, the Netherlands, [2]CRV BV, Animal Evaluation Unit, P.O. Box 454, 6800 AL Arnhem, the Netherlands; marleen.visker@wur.nl

The amount of milk that mammals produce can be different after giving birth to offspring of different sexes. Usually, if there is any sex-bias, giving birth to male offspring leads to a higher milk volume and/or milk with a higher energy content. However, a study on US dairy cows showed higher 305-day milk yield after giving birth to female calves. The aim of our study was to investigate the effect of the sex of the calf on the subsequent 305-day milk production in Dutch dairy cows. Milk production records of 1,615,765 lactations of 861,273 Holstein Friesian cows from 7,303 herds throughout the Netherlands were used for analysis. Plain averages suggest that giving birth to male calves results in higher 305-day milk production: differences up to 90 kg are found in lactation 2 and 4. This changes dramatically when including gestation length, lactation length, calving ease and birth weight as fixed effects, and sire of cow and herd-year-season as random effects in the analysis. The effect of the sex of the calf on the subsequent 305-day milk yield is significant only for lactation 1 and 3, with a small positive effect after giving birth to a female calf in the first lactation and a small positive effect after giving birth to a male calf in the third lactation. The effect of the birth weight of the calf is significant in all lactations and seems more substantial: a calf of 55 kg instead of 25 kg at birth is associated with a 305-day milk production that is about 800 kg higher. What at first glance appears to be an effect of the sex of the calf is actually a confounding effect of the birth weight of the calf, because male calves tend to be heavier than female calves. It is not clear whether previous reports of sex-biased milk production can also be attributed to differences in birth weight of the offspring of different sexes.

Differences between cloned and non-cloned cows during first lactation

F. Montazer-Torbati[1], M. Boutinaud[2], N. Brun[1], C. Richard[1], V. Hallé[1], A. Neveu[1], F. Jaffrezic[1], D. Laloë[1], M. Nguyen[1], D. Lebourhis[1], H. Kiefer[1], H. Jammes[1], J.P. Renard[1], S. Chat[1], A. Boukadiri[1] and E. Devinoy[1]
[1]INRA, UMR1313 GABI, UMR1198 BDR, UE1298 UCEA, INRA, 78350 Jouy-en-Josas, France, [2]INRA UMR1348 PEGASE, Agrocampus Ouest, 35590 Saint-Gilles, France; eve.devinoy@jouy.inra.fr

High variability in bovine milk production depends both on genetic and environmental factors. Whereas genomic selection has contributed to greatly improve milk production, mechanisms underlying the influence of the environment are mostly unknow. They can be studied using cloned cows (obtained through nuclear transfer), presumed to have identical genomes. Nine Prim'Holstein cloned cows and nine non-cloned cows were produced at similar periods of the year, raised in similar conditions, and inseminated at the age of 18 months. Eight cloned and 7 non-cloned cows delivered. Body condition score at calving was lower ($\Delta BSC=-0.6$) for cloned cows (P=0.03). During the first 200 days in milk (DIM), milk yield varied greatly among animals but was similar between the two groups, whereas milk fat and protein contents varied less in cloned cows (P=0.004 and 0.09). Around 67 DIM, fat, protein and lactose contents (P<0.05) were lower in cloned cows. Mammary biopsies were then collected from a rear quarter of the udder, but only five biopsies in each group contained more than 60% of mammary epithelial tissue. These were from a sub-selection of cows with a milk production that was representative of that observed among the 9 or 8 cows in each group. Relative levels of transcripts involved in casein, fat or lactose syntheses did not differ, when a cow with a high somatic cell count in milk collected from one of the fore udders at 65 DIM was omitted. It is known that renewal of mammary cells is slow during lactation and mammary specific DNA methylation profiles are likely to be maintained by DNMT1. Here, DNMT1 RNA levels were higher (P=0.05) in cloned cows. This variation may be related to modifications of global DNA methylation profiles and may explain the modified milk production in cloned cows.

Productive and reproductive performance of dairy cows under once-a-day milking in a university herd

N. Lopez-Villalobos, F. Lembeye, J.L. Burke, P.J. Back and D.J. Donaghy
Massey University, Private Bag 11222, 4442 Palmerston North, New Zealand; n.lopez-villalobos@massey.ac.nz

Milking cows once-a-day (OAD) for the entire lactation has been practiced by approximately 3% of dairy farmers in New Zealand since 2000, to reduce labour costs and improve farmer life style. This is despite reductions in milk yield per cow. The economic viability of OAD grazing systems, however, remains uncertain as OAD increases reproductive performance and improves animal welfare, resulting in higher survival rates and potentially higher within-herd genetic gains. To evaluate long term effects of OAD on overall farm productivity and profitability, and also the environmental impact, the milking regime on a Massey University dairy farm was changed from twice-a-day (TAD) to OAD milking in July 2013, with the feeding system based mainly on pasture. This study reports the productive and reproductive performance during the first year of the experiment of 39 Friesian (F), 44 Jersey (J) and 89 FxJ crossbred cows and compares this herd with the national herd under TAD milking. Compared to the national herd, the OAD milking herd had a significantly shorter lactation length (241 vs 266 days), lower lactation yields of milk (3,549 vs 4480 kg, P<0.001), fat (184 vs 212 kg, P<0.001) and protein (140 vs 170 kg, P<0.001). The OAD milking herd had significantly better reproductive performance: a higher submission rate at 21 days after the start of the mating period (SM) (96 vs 79%, P<0.001), a higher conception rate at 42 days after SM (85 vs 65%, P<0.001) and a lower empty rate at the end of the mating period (5 vs 13%, P<0.001). F cows had higher milk yields than FxJ and J cows (P<0.001), while the three breed groups had similar fat yields, and F and FxJ had higher protein yields than J cows (P<0.001). The three breed groups had similar reproductive performance. These results can be used for simulation studies to evaluate the profitability of OAD milking systems in New Zealand that are using crossbreeding.

Biosecurity in French cattle farms: attitudes and expectations of farmers to improve

B. Mounaix, M. Thirion and V. David
Institut de l'Elevage, Santé et bien-être des ruminants, 149 rue de Bercy, 75596 Paris cedex, France;
beatrice.mounaix@idele.fr

Biosecurity appears as one main challenge for farmers, a cornerstone of the 2013 proposal for EU Animal Health Law, and a key point for antibiotic reduction within French Ecoantibio 2017 plan. To help cattle farmers to better manage biosecurity, understanding on farm implementation with regards to farmers' attitudes is necessary, as well as promoting efficient practical solutions. During 45 face-to-face surveys conducted in 4 French regions in 2014, farmers were indirectly asked about their knowledge on biosecurity and what would help in improving, i.e. what could be better biosecurity practices, the necessary conditions for the implementation of these better practices, and obstacles to implementation. All interviewed farmers expressed a low level of understanding of the word 'Biosecurity' and frequent misunderstandings were observed, due to the other meanings of the word in French and its novelty in cattle farming. Although they had good theoretical skills regarding external biosecurity, implementation was low, excepted for regulated diseases. Main condition for improving would be to promote the homogeneity of practices among pairs, most interviewed farmers considering that the benefit of good practices could be jeopardized by low biosecurity in other farms, neighbours or traders. Economical costs was considered as the lowest obstacle to implementation. This was observed both in dairy and beef cattle production. Internal biosecurity was not considered at risk by farmers, cattle herd being frequently considered as a global entity, with the frequent housing of young and adult animals in the same premises. This study concluded on the need for practical solutions adapted to cattle farming to manage the biosecurity. This study has been completed by a large scale internet survey to search for on farm innovative solutions to manage biosecurity. Results will be available by June 2015 and disseminated.

Integrated data management as a tool to prevent diseases

C. Egger-Danner[1], R. Weissensteiner[1], K. Fuchs[2], F. Gstoettinger[3], M. Hoermann[4], R. Janacek[5], M. Koblmueller[6], M. Mayerhofer[1], J. Perner[7], M. Schaginger[2], G. Schoder[8], T. Wittek[9], K. Zottl[6], W. Obritzhauser[9] and B. Fuerst-Waltl[5]
[1]ZuchtData, Dresdner Str. 89/19, 1200 Vienna, Austria, [2]AGES, Spargelfeldgasse 191, 1220 Vienna, Austria, [3]RZV, Sportplatz 7, 4840 Vöcklabruck, Austria, [4]Chamber of Agriculture, SChauflergasse 6, 1010 Vienna, Austria, [5]Animal Health Organisation, Landhausplatz 1, 3109 St. Pölten, Austria, [6]LKV, Dresdner Str 89/19, 1200 Vienna, Austria, [7]Chamber of Veterinarians, Hietzingerkai 83, 1130 Vienna, Austria, [8]University of Natural Resources, Gregor-Mendel-Str. 33, 1180 Vienna, Austria, [9]University of Veterinar Medicine, Veterinärplatz 1, 4021 Vienna, Austria; egger-danner@zuchtdata.at

Dairy farmers and veterinarians are faced with many relevant data which are scattered across different databases on both an organizational and farm level. Increasing automation at the farm level would offer an opportunity for easier herd management. The problem is that the different systems have different standards and often don't communicate with each other. Therefore relevant data are not directly linked at present, and farmers are forced to document or insert data more than once into separate systems. The disjointedness of different data sources is a hindrance to efficient ways of working on the prevention of health disorders. For timely reaction and prevention, reliable data and benchmarks for triggering alerting systems would be needed. Within the Comet K-Project ADDA – Advancement of Dairying in Austria the basis for an integrated data management tool shall be established. A first step is the assessment of stakeholder needs. For this purpose, internet surveys were developed to gather feedback from farmers and veterinarians about the use of the herd management tools currently available and present status of documentation and recording of relevant data. Based on the needs of the stakeholders an overall concept for an integrated data management tool is elaborated subsequenly.

Effects of culling rates on measures of ecological sustainability in dairy cattle

S. Ebschke and E. Von Borell

Institute of Agricultural and Nutritional Sciences; Martin-Luther-University Halle-Wittenberg, Theodor-Lieser-Strasse 11, 06120 Halle, Germany; stephan.ebschke@landw.uni-halle.de

Herd data are routinely used to evaluate dairy cattle efficiency, health and welfare. The aim of this study was to estimate on how changing culling rates affect carbon footprints in dairy farms. Intra-farm ecological sustainability was based on outputs of methane (CH_4), ammonia (NH_3) and nitrous oxide (N_2O) in relation to housing system, pasture and type of manure storage as well as on energy intensity. TIER 2 and 3 methods of IPCC (2006) were used to analyse individual farm data. Emission factors were chosen from literature or calculated from available data. We focused on culling rates and rearing progeny to estimate the reduction potential of global warming (GWP as CO_2-equivalents) associated with milk production. A total of 22 dairy cattle farms located at different German regions were surveyed. In a first step each farm was analysed as a whole, then was determined for its required progeny (considering 3% of rearing losses) and finally was standardised at a culling rate of 23% as a target for dairy production. A mean of 0.956 (SD±0.186) CO_2-eq. per kg ECM (Energy Corrected Milk) was estimated ranging from 0.739 to 1.337. These footprints were further transformed into CO_2-eq. per kg eatable protein (eP). Farms ranged from 20.3 to 31.3 CO_2-eq. kg eP-1. GWP decreased up to 10.3% when considering the targeted progeny. In cases of outsourced rearing, we obtained reverse results with an increase of up to 15.9%. Standardisation of culling rates to 23% resulted in decreased GWP from 22.0 to 21.9 and 25.4 to 22.78 CO_2-eq/kg eP (0.7-10.3%) respectively. The overall potential for standardised progeny and culling rate ranged from 0.09 to 4.8%. It was shown that culling rates directly interact with ecological sustainability. The potential to mitigate GWP is highest in dairy farms with poor culling rates. Those with low culling rates can further improve their carbon footprints by selling surplus progeny at an early state.

Manipulating diet nitrogen content to reduce urine nitrogen excretion in lactating dairy cows

T. Yan

Agri-Food and Biosciences Institute, Hillsborough, Co. Down, BT26 6DR, United Kingdom; tianhai.yan@afbini.gov.uk

The objective was to evaluate the effect of diet nitrogen (N) content on urine N excretion in lactating dairy cows. The reduction of urine N output not only increases N utilisation efficiency, but also decreases environment footprint (e.g. nitrate pollution and ammonia and nitrous oxide emissions). The dataset used was obtained from 470 Holstein-Friesian lactating cows in 26 total diet digestibility studies undertaken at this Institute. Animals were of various genetic merits, parity (1 to 9), live weight (385 to 781 kg) and milk yield (6.1 to 49.1 kg/d). Thirty eight cows were offered grass silage as the sole diet and remaining cattle (n=432) given grass silages with various proportions of concentrates (294 to 876 g/kg DM). All animals were offered either silage or the complete diet *ad libitum* for at least 22 days before a 6-day total collection of faeces and urine. The linear regression technique was used to develop relationships between diet N content and urine N excretion with experimental effects removed. There was a large range in N intake (155 to 874 g/d) and N output in faeces (48 to 241 g/d), urine (70 to 452 g/d) and milk (24 to 231 g/d). A number of equations were developed by relating diet N content (Nc) to urine N excretion rates, including: urine N/N intake = 0.0057 Nc + 0.269 (r^2=0.37); faecal N/N intake = -0.0068 Nc + 0.491 (r^2=0.45); urine N/manure N = 0.0086 Nc + 0.345 (r^2=0.47); urine N/faecal N = 0.0571 Nc – 0.131 (r^2=0.48). These results indicate that increasing diet N content can reduce faecal N/N intake, but increase urine N/N intake, urine N/faecal N and urine N/manure N. Feeding dairy cows with high N diets can change the N output pattern towards excreting more N into urine. Manipulating diet N content is an effective approach to increase N utilisation efficiency and reduce environment footprint in dairy production, especially for grazing cattle with which urine N output is a considerable source for nitrate pollution and ammonia and nitrous oxide emissions.

Are reticular temperatures correlated to body temperature in dairy cows?

S. Ammer[1], C. Lambertz[2] and M. Gauly[2]
[1]*Georg-August-University Göttingen, Livestock Production Systems, Department of Animal Science, Albrecht-Thaer-Weg 3, 37075 Göttingen, Germany,* [2]*Free University of Bolzano, Faculty of Science and Technology, Universitätsplatz 5, 39100 Bolzano, Italy; sammer1@gwdg.de*

The objectives of this study were to compare methods of measuring body temperature in dairy cows and to evaluate certain effects on the measured values. Therefore, rectal (REC), vaginal (VAG) and reticular (RET) temperature of 12 Holstein-Friesian cows were measured three times per day (7:30 am, 12:00 pm and 17:30 pm) during a warm (June) and cold (October) 5-day period. The animals were kept in a non-insulated loose-housing barn. Ambient temperature (AT, °C) was measured inside the stable at 15-minute intervals. For RET, the mean (RET-MEAN) and median (RET-MED) values of 2 hours preceding the respective measurement were compared with the other methods in addition to the single values recorded when REC and VAG was determined (RET-SIN). During the study periods, AT averaged 23.0±4.3 °C in June and 14.6±1.7 °C in October. Furthermore, the mean body temperatures amounted in June compared to October: REC 38.3±0.6 and 37.9±0.4 °C, VAG 38.3±0.5 and 38.1±0.3 °C, RET-SIN 38.9±1.0 and 38.5±0.6 °C, RET-MEAN 38.6±0.8 and 38.4±0.5 °C, RET-MED 38.9±0.6 and 38.6±0.4 °C, respectively. Overall, RET-SIN, RET-MEAN and RET-MED values were significantly (P<0.001) higher than REC and VAG. Within reticular temperatures, RET-SIN and RET-MED were significantly higher than RET-MEAN values (P<0.001). VAG and REC were higher correlated in the warm (r=0.81) than in the cold (r=0.52) period. Compared with RET-SIN and RET-MEAN, RET-MED showed higher correlations with REC (r=0.53) and VAG (r=0.60) in June (P<0.001). In conclusion, the RET were significantly higher (P<0.001) in contrast to VAG and REC measurements, while the use of RET-MED was most related to them. Body temperatures were influenced by AT and showed significantly higher values in June (P<0.001).

Measures related to energy balance in Dairy Gyr cows

G.H.V. Barbosa[1], A.E. Vercesi Filho[1], L.G. Albuquerque[2], M.E.Z. Mercadante[1], A.R. Fernandes[3] and L. El Faro[1]
[1]*Animal Science Institute, Beef Cattle Center, Carlos Tonanni, KM 94 Sertãozinho, SP, 14174-000, Brazil,* [2]*São Paulo State University, Animal Science, Jaboticabal, SP, 14884-900, Brazil,* [3]*Gyr Cattle Association, Uberaba, MG, 38022-330, Brazil; lenira@iz.sp.gov.br*

This experiment measured body and milk production traits in 26 primiparous zebu cows, contemporary at calving and managed on pasture during lactation, in cooperation with the Brazilian Association of Dairy Gyr Breeders (ABCGIL). Milk yield (MY), body weight (BW), body condition scores (BCS, 1 to 9) and ultrasound subcutaneous fat thickness (FT) measured at 5 different body locations, were evaluated during lactation (10 monthly MY records). The FT measures were: FT1 (longitudinal at three different sites between the 11th and 13th rib); FT2 (longitudinal in the rump region); FT3 (transverse in the flank); FT4 (transverse from the hip bone to the tip of the pin bone – rump fat thickness); FT5 (transverse between the 12th and 13th rib – backfat thickness). Stepwise multiple regression analysis was performed to determine the variables responsible for MY, BW and BCS during lactation using Mallow's C(p) criterion. The average curve for MY increased from calving to 60 days of lactation when a discrete peak lactation was observed. Cows calved at a BW of 470 kg and BCS of 7, decreasing to 400 kg and 4.5, respectively, in the fourth month of lactation. The same declining trend was observed for FT after calving, with FT4 showing the lowest values throughout lactation and the most pronounced declines. The variables that most influenced BW after calving were BCS and FT1, with regression coefficients of 13.9 and 9.7 kg, respectively. For BCS, the most influential variables were BW and two FT measures (4 and 5), with coefficients of 0.006, 0.19 and 0.14, respectively. For MY, the variable selected were BCS, FT4 and lactation month, with coefficients of 0.87, -0.86 and -0.74 kg, respectively. A positive association exists between BW, BCS and some FT measures. The analysis for MY suggests loss of fat reserves during lactation.

Silage intake and milk production in early lactation dairy cows fed cereal grains

V. Endo[1], T. Eriksson[2], E. Spörndly[2], K. Holtenius[2] and R. Spörndly[2]
[1]Sao Paulo State University, Animal Science, Via de Acesso Prof. Paulo Donato Castellane, s/n, 14884-900, Brazil, [2]Swedish University of Agricultural Sciences, Animal Nutrition and Management, Box 7024, 750 07, Sweden; viviane.endo@slu.se

The cost of organically produced protein feeds is high and the supply is limited. Forage and cereal grains produced and used on the farm provide a stable production economy since they are not influenced by fluctuations in market prices. This communication reports results from the first three lactation months of a major project with the aim to evaluate the potential to produce milk with protein supplied only from forage and cereals. Twenty cows (Holstein and Swedish Red Breed) were distributed in a completely randomized design in two groups and fed grass silage (OMD 81%; 18.5% CP of DM) *ad libitum*. Concentrate allotment was restricted to 50% of ration DM to comply with regulations for organic farming. Group 1 (G1) was supplied with 9.9 kg cereal grains/d (rolled barley, wheat and oats; 10.3% CP of DM). Group 2 (G2) was supplied with 7.6 kg cereal grains/d and protein feeds/d (mainly soy beans and rapeseed meal; 32.3% CP of DM) according to Swedish feeding standards. Intake, milk production and BW were recorded daily; milk composition was recorded monthly. Data were subjected to analysis of variance by PROC GLM of SAS. In the first month of lactation, silage intake was unaffected by treatment (12.5 kg DM/d). In the second and third months of lactation, however, G1 cows consumed more (p <0.05) silage (15.5 and 17.4 kg DM/d) than G2 cows (13.3 and 15.0 kg DM/d). Diets did not differ (P>0.05) regarding production of raw milk (36.7 and 38.8 kg/d for G1 and G2, respectively) or energy corrected milk (36.9 and 37.5 kg/d for G1 and G2, respectively). In early lactation, feeding cereal grains and high quality silage to cows was sufficient to maintain milk production.

Modelling targeted supplementation to dairy cows under different New Zealand farm systems

P. Gregorini, B. Delarue and C. Eastwood
DairyNZ, Cnr. Ruakura and Morrisnville Rds, 3240 Hamilton, New Zealand; pablo.gregorini@dairynz.co.nz

This modelling study aimed to explore criteria to identify cows and opportunities for targeted individual supplementation, determining milk solids (MS) response to feeding criteria. The study compared 3 typical pastoral New Zealand dairy systems, using in-shed supplementary feed in late (System 3), early and late (System 4) and throughout (System 5) the lactation. Cows in each system were supplemented at a herd level (CTL) or individually according to body condition score (BCS), MS, genetic merit (GM) or Age. Concentrate was allocated within each criteria according to 4 treatments (TRT), where the CTL cows received the same (3 kg) and TRT cows received varied amounts of concentrate (0, 2, 3, 4 and 6 kg DM/cow per day) depending on their rank in the herd (percentile) within the criteria. TRT × percentile combination were: cows <25%, 0, 6, 4 and 2 for TRT1-4 respectively; cows 25-75%, 3 for all TRT; and cows >75%, 6, 0, 2 and 4 for TRT 1-4 respectively. The total amount of concentrate remained the same for all TRT. The simulations were conducted using the DairyNZ Whole Farm Model. This model includes a metabolic cow and a climate driven pasture growth models. Each combination of System and TRT was run for 5 years and replicated in 10 herds, giving 170 model simulations runs per 'System'. The outputs of the model were MS (kg/cow/year) and operating farming profit (OP, NZ$/ha/year). The means of the best outcome of MS for each TRT × criteria × systems were the following, System 3: 367, 363 for CTL and BCS and 365 for MS, GM and Age; System 4: 382 for CTL and BCS and 381, 38 for MS, GM and Age; and System 5: 454, 456, 453, 455 and 453 for CTL, BCS, MS, GM and Age respectively. OP under each System was virtually the same too, independently of the criteria and TRT, with 2,235±30, 2,624±18 and 3,281±32 NZ$/ha/year for systems 3, 4 and 5 respectively. This exercise suggests that typical pastoral dairies of New Zealand might not benefit from targeted supplementation.

A post-crisis assessment of soil and milk contamination by hexachlorocycloesane in a dairy district

B. Ronchi, R. Bernini, I. Carastro, P. Carai and P.P. Danieli
University of Tuscia, Department of Agriculture, Forests, Nature and Energy (DAFNE), Via S. C. de Lellis snc, 01100, Italy; ronchi@unitus.it

In 2005, a dairy district in Central Italy was threatened due to the environmental contamination of the Sacco River by isomers of hexaclorocyclohexane (HCH), an obsolete pesticide. In 2013, a post-crisis study program aimed at investigating the residual contamination by HCH in the area. In this report, some preliminary results on soil and bovine milk contaminations are presented. Ten dairy farms were involved in the study. In each of them, from one to four soil sampling sites were selected within officially banned and unbanned areas along the river path. Soils were sampled from July to September 2013; cow milk samples were collected from June to July 2013. The samples were analysed thrice by GC-ECD and confirmed by GC-MS. Statistical analysis was done by ANOVA. The α, β and γ isomers of HCH were detected in 16 out 26 soil samples with mean concentrations up to 0.299, 0.437 and 0.385 mg/kg, respectively. Overall, the highest levels were detected in banned areas and about 30% of soil samples were above the National law limit of 0.02 mg/kg (total HCH). An uneven occurrence of contaminations among the farms (P<0.01) was seen and four farms (namely A, B, C and M) had rural soils contaminated above the law limit. Among them, in farms A and C the contamination by HCH was found to be lower than those observed in 2005, during the emergency monitoring program. In milk samples, the β-HCH was detected only in three farms (B, C and L) at concentrations ranging from the EU limit of 0.003 mg/kg (farms B and L) to 0.005 mg/kg (Farm C). It is also interesting to notice that in two dairy farms that were high in soil contamination by HCH (farms B and C), cow milk was found contaminated by β-HCH. In conclusion, these preliminary results suggest that soils are recovering their potential for a safe forage production. Episodic detection of β-HCH in cow milk implies that the emergency is not completely resolved in that area.

Fatty acid profile of organic and conventional halloumi cheese

O. Tzamaloukas, M. Orford, D. Miltiadou and C. Papachristoforou
Cyprus University of Technology, Department of Agricultural Sciences, Biotechnology and Food Science, P.O. Box 50329, 3603 Lemesos, Cyprus; ouranios.tzamaloukas@cut.ac.cy

Lipids of milk fat, have distinctive effects on nutritional, textural and organoleptic properties of milk and dairy products, while unsaturated and, especially, polyunsaturated fatty acids (FA) have been reported to possibly confer some beneficial biological effects on human health. The objective of the present work was to study the FA profile of halloumi cheese produced using organic or conventional milk from Cyprus farms. The study was conducted over a period of two years. A total of 118 samples of organic and conventional halloumi cheese were collected at two-month intervals from products available in supermarkets and other retail outlets. The samples were analysed by gas chromatography-mass spectrometry for fatty acid profile. The effect of the farming system on halloumi cheese constituents was investigated by ANOVA analysis using the SPSS statistical package. The results showed that organic halloumi cheese had higher concentrations of the beneficial for human health FA compared with the conventional product. Thus, organic halloumi had 36% higher concentrations of total polyunsaturated fatty acids than the conventional one (7.1 vs 5.2% wt/wt of total FA, respectively; P<0.01). Further analysis on specific health-related FA showed an increase of linoleic acid (4.7 vs 3.5% wt/wt of total FA; P<0.01), α-linolenic acid (0.81 vs 0.66% wt/wt of total FA; P<0.01) and conjugated linoleic acid (cis-9, trans-11 CLA, 1.14 vs 0.81 g/kg of total FA; P<0.001) in organic as compared to conventional halloumi cheese. These differences are more likely attributed to the different feeding practices followed by organic and conventional farms in Cyprus.

Alternative method for birth weight data consistency in Nellore cattle

J.A.I.V. Silva[1], L.E.C.S. Correia[1], E.G. Moraes[2], L.F.N. Souza[2], L.G. Albuquerque[1], R.A. Curi[1] and L.A.L. Chardulo[1]
[1]*Universidade Estadual Paulista, CP 560, 16618-970, Brazil,* [2]*Qualitas Consultoria, CP, 74934-605, Brazil; jaugusto@fmvz.unesp.br*

In the tropics, beef cattle production is based on pastures and, in some large herds, about 15% of calves in a contemporary group (CG) can present the same birth weight (BW). It is possible that these calves weights were attributed based on their body size at birth. The objective of this work was to verify the influence of this large percentage of animals with equal BW on genetic parameter estimates and breeding value predictions for BW in Nellore cattle. Definition of CG was herd, year, management group and sex at birth. Animals with BW above or below 3.5 standard deviations from CG average were deleted. Two datasets, keeping (BW1) or not (BW2) animals with the same BW record were used. For BW2, animals showing the same BW, and belonging to CG with more than 15% of these equal records, were deleted. The number of records were 67,394 (BW1) and 55,233 (BW2), respectively, both with the same BW mean (31.1 kg). An animal model including additive direct and maternal genetic plus maternal permanent environmental effects as random and fixed effects of CG and dam age at calving was implemented by Bayesian inference. A total of 1,100,000 samples were generated,with a burn-in period of 100,000 samples and time interval of 20 cycles. Direct heritability estimate was slightly higher excluding animals with equal BW from the data set (0.28) than not (0.25). Spearman correlations between sires' breeding values obtained in both data sets, considering all sires, top 10 sires and sires with more than 1000 progenies, were below 0.35, indicating large changes in sires' rank. Genetic trends for BW were positive and larger when using BW2 (9.0 g/y) compared to that obtained with BW1 (3.0 g/y). It is concluded that the removal of animals with equal birth weights records from the analyses affects genetic parameter estimates and can increase sires' BW breeding value accuracies.

The role of growth factors in regulating cellular events during ovarian follicular development

L.J. Spicer
Oklahoma State University, Department of Animal Science, 114 Animal Science, 74078, USA; leon.spicer@okstate.edu

Development of the ovarian follicle(s) destined for ovulation is a process in which antral follicles undergo a recruitment, selection and subsequent dominance phase that involves a balance of cell proliferation and differentiation. In addition to endocrine signals, several complex intraovarian or autocrine/paracrine regulatory mechanisms have been evoked to explain these processes. Several families of growth factors participate in these mechanisms, and include the insulin-like growth factors (IGFs), hedgehog proteins, fibroblast growth factors (FGFs), and the transforming growth factor-beta (TGFB) superfamily all of which regulate the response of the two major follicular cell types, granulosa (GC) and theca cells (TC), to gonadotropins. For example, Indian hedgehog proteins (IHH) induce TC proliferation and LH-induced androstenedione production, and increased free IGF1 in dominant follicles suppresses expression of IHH mRNA in GC and their receptors in TC; these effects are regulated in a paracrine way by estradiol and other intra- and extra-gonadal factors. Another example is hormonally regulated GC FGF9 production that acts in an autocrine/paracrine way to inhibit follicular differentiation via decreasing gonadotropin receptors, the cAMP signaling cascade, and steroid synthesis while stimulating proliferation. The TGFB members, growth differentiation factor-9 (GDF9) and bone morphogenetic protein-4 simultaneously promote GC and TC proliferation while preventing premature differentiation of GC and TC during growth of follicles. Microarray studies have identified a plethora of possible new intracellular regulatory factors. In conclusion, several families of growth factors work in concert with gonadotropins at the cellular level having significant effects on follicle selection and development. It is the integration of these intra-follicular factors and endocrine signals that determine whether a follicle will continue to develop, become cystic, or undergo atresia.

Pre and post evolution of ram sperm parameters and fatty acids in dietary containing of fish oil

A.R. Alizadeh[1], V. Esmaeili[2] and A.H. Shahverdi[2]
[1]Department of Animal Science, College of Agriculture, Saveh Branch, Islamic Azad University, Saveh, Iran,
[2]Department of Embryology at Reproductive Biomedicine Research Center, Royan Institute for Reproductive
Biomedicine, ACECR, Tehran, Iran; masouleh7@yahoo.com

The effects of dietary fish oil (FO) on semen quality and sperm fatty acid (FA) profiles during consumption of n-3 FA as well as the persistency of FA in ram's sperm after removing dietary oil from the diet were investigated. Nine Zandi's rams were randomly assigned to two groups (constant level of vitamin E): Control (CTR; n=5) and Fish oil (FO; n=4) for 70 days during breeding season. Semen was collected once per 2 weeks. After treatment period, all rams were fed a conventional diet and semen samples were collected one and two months. The sperm parameters and FA profiles were measured by computer assisted semen analyzer (CASA) and gas chromatography (GC), respectively. Data were analyzed using the SPSS 16. Dietary FO had significant positive effects on all sperm quality and quantity parameters compared with CTR during feeding period (P<0.05). Interestingly, the positive effects of FO on sperm concentration and total sperm output were observed one and two months after removing FO (P<0.05). At the commencement of the study, C14, C16, C18, C18:1 and C22:6 were the major ram's sperm FA. Inclusion FO in the diet increased sperm DHA (30 vs 9.5% for FO and CTR, respectively), and C18:0 (14 vs 6% for FO and CTR, respectively) during feeding period (P<0.05). Sperm saturated FA was significantly affected by last feeding strategy; whereas DHA concentrations remained similar one month after removing FO with a negligible change two month after removing FO. In conclusion, manipulation of sperm FA profiles by dietary FA has a biological rhythm. Dietary FO had a significant direct or indirect impact on FA profiles of the sperm during feeding period and after removing FO from the diet, which might lead to physical and chemical changes in mammalian sperm characteristics.

Introduction of bucks during the late luteal phase of female goats modifies progesterone pattern

R. Ungerfeld[1] and A. Orihuela[2]
[1]Facultad de Veterinaria, Universidad de la República, Lasplaces 1620, Montevideo 11600, Uruguay,
[2]Facultad de Ciencias Agropecuarias de la Universidad Autónoma del Estado de Morelos, Avenida
Universidad 1001, Cuernavaca, Morelos 62210, Mexico; rungerfeld@gmail.com

The sudden introduction of bucks to cyclic does induces an increase in LH pulse secretion, and thus of oestradiol secretion. As an increase in oestradiol secretion triggers luteolysis in cyclic ruminants, the aim was to determine the changes in progesterone profile after the introduction of bucks during the advanced luteal phase of does. Fourteen does received intra-vaginal sponges containing impregnated with fluorogestone for 12 days, and a prostaglandin injection 2 days before sponge removal. Fifteen days after sponge withdrawal, one buck was introduced in one of the pens (BE group; n=6), while the female goats in the other pen remained as controls (CON group; n=8). Each buck was replaced every 24 h, alternating their presence until the end of the experiment. Progesterone concentration was daily measured, and compared with ANOVA for repeated measurements. From Day 16 to 22 after sponge withdrawal there was an interaction between treatment and day (P=0.02). While progesterone concentration increased from day 15 to day 16 in BE does (8.2±0.7 to 11.1±1.5 ng/ml; P=0.01), there were no changes in CON does on those days (8.6±1.1 vs 7.5±1.5 ng/ml; P=0.2). On the other hand, progesterone concentrations decreased in BE does from day 18 to day 19 (9.1±0.8 to 6.4±1.2 ng/ml; P=0.02), without changes in CON does (8.0±1.6 vs 7.5±1.9 ng/ml; P=0.6). Finally, there was a sharp decrease from day 19 to day 20 in both, BE (P=0.0009) and CON (P<0.0001) does. Our results demonstrated that the introduction of bucks during the late luteal phase of isolated does can induce changes in the progesterone pattern, showing an early increase followed by a pronounced withdrawn. These results open the possibility of studying the inclusion of the male effect in estrous synchronization treatments.

Effect of selenium with vitamin E on reproductive hormones and performance of Ghezel ewe
H. Daghigh Kia, S. Saedi and A. Hossein Khani
University of Tabriz, 29 Bahman Blvd, 5166614766, Tabriz, Iran; hdk6955@gmail.com

Forty-four Ghezel ewes were allocated to four groups (n=11) to study the supplementation effect of selenoprotein and sodium selenite in combination with vitamin E in flushing period on estrogen, progesterone, insulin hormones, lambing and twining rate. The ewes received one of the following treatments: Groups A- control (basal diet); Groups B – barley grain; Groups C- barley grain, vitamin E + selenoprotein; groups D- barley grain, vitamin E+ sodium selenite. Flushing period started 4 weeks before and 2 weeks after artificial insemination. Estrus was synchronized using CIDR. The ewes were inseminated 48 h after CIDR removal. The number of lambs born, lambing rate, and daily gain were recorded. Blood samples collected 24, before and after CIDR removal and 21 days after insemination. Estrogen, progesterone and insulin levels were determined. Data were analyzed using Mixed model of SAS. Blood hormones levels changed significantly during reproductive cycle (P<0.05). Comparing to the other groups, flushing treatments had considerable fluctuation in estrogen. Treatment D and C with 37.43±1.45 and 37.66±1.45 pg/ml estrogen at estrus time and with 5.2±0.13 and 5.12±0.13 ng/ml, progesterone at 21 days of gestation, had the highest estrogen and progesterone level compared to groups A and B (P<0.01). Serum insulin concentration showed significant differences in treatments at all times, except for the start of the experiment (P<0.01). During estrus, the highest ratio of estrogen to progesterone was observed in group C (0.059) which had the highest lambing and twinning rate. Treatments D and C had the highest birth weight and daily gain compared to the control group (P<0.01). The ratio of male to female lambs born in treatments C and D were 1.5 and 2.5 respectively, which has a significant difference to the control group (P<0.01). Supplementation of selenoprotein and sodium selenite in combination with vitamin E in the ewe' diet, before and after mating and during pregnancy improves reproductive performance.

The effect of seasonal daily live weight gain in females calves on reproductive performance
G. Quintans[1], J. Echeverría[1], G. Banchero[1] and F. Baldi[2]
[1]National Institute for Agricultural Research, beef and wool, Ruta 8 km 281, 33000, Uruguay, [2]FCAV Unesp, Via de Acesso Prof. Paulo Donato Castellane s/n, 14884-900, Jaboticabal, São Paulo, Brazil; gquintans@inia.org.uy

The aim of this experiment was to evaluate two different winter daily live weight gain (DLWG) in females calves and its effect on their reproductive performance. Forty nine female calves (AA×HH; 252±1.73 days of age=Day 0) with 195.7±1.65 kg of life weight (LW) were assigned in winter to two treatments during 98 days: (1) grazing native pastures offered at 14% of LW (90 g CP and 2.05 MCal of ME/kg DM, CON, n=25); and (2) grazing native pastures and supplemented individually with a concentrate at 1.5% of LW (131 g CP and 2.1 MCal of ME/kg DM, SUP, n=24). Calves grazed together in the same paddock during the period treatments were applied and also thereafter until the end of the experiment (mating period). Pregnancy diagnosis was performed 2 months after the end of mating which lasted 45 days. The heifer pregnancy was analyzed using a no liner model, assuming a binomial distribution and a probit link function. The model included the effect of winter feeding management. A multivariate discriminate analysis was performed to evaluate which period of DLWG during the whole period was best to separate pregnant from non-pregnant heifers. The DLWG for winter were 0.074±0.02 and 0.757±0.02 for CON and SUP, P<0.05; in spring were 0.758±0.02 and 0.601±0.02 for CON and SUP, P<0.05; in summer were 0.331±0.02 and 0.247±0.02, P<0.05 and in autumn were 0.217±0.02 and 0.216±0.02 kg/a/d, P>0.05. Pregnancy rate was higher in SUP than in CON cows (88 vs 36%; P<0.01). The multivariate analysis demonstrated that winter DLWG was the main variable (P=0.000002) that discriminated between pregnant and non pregnant heifers, respect to spring, summer and autumn DLWG (P=0.023; P=0,046 and P=0.12, respectively). In summary, in range conditions first winter feeding management after weaning is the main responsible for reproductive success in beef heifers.

Environmental impacts on the developing conceptus
S.E. Ulbrich
ETH Zurich, Animal Physiology, Inst. Agric Sci, Universitaetstr. 2, 8092 Zurich, Switzerland;
susanne.ulbrich@usys.ethz.ch

The maternal uterine environment is modulated in a well-orchestrated qualitative, quantitative, temporal and spatial manner. The differential gene and protein expression of the endometrium that has been thoroughly characterized throughout the estrous cycle and the early preimplantation phase enable the timed secretion of the intrauterine histotroph. The histotroph plays a crucial role in providing a milieu permissive to the needs of the developing embryo. It is here that embryonic and maternal secretions compile and countercurrent signaling for inevitable embryo-maternal communication takes place. The endocrine network regulating reproduction events reacts very sensitive to subtle external perturbations. Disturbances may arise from nutritional impacts, metabolic imbalances or endocrine disruptors increasingly found in the environment, and lead to changes of the maternal histotroph. Maternal exposure to estrogens is known to induce long-term epigenetic effects in the offspring. We recently demonstrated changes in body composition and bone density after gestational low-dose estradiol-17β (E2) treatment in pigs. To elucidate the possible direct impact of E2 during the preimplantation period, we analyzed maternally exposed day 10 single blastocysts and the respective endometrium. Overall, by applying RNA Seq, both endometrium and conceptus displayed differential gene expression in the treatment groups. Developing blastocysts showed sex-specific low-dose estrogen effects. Due to the specific mRNA affected, an indirect effect by E2 causing a perturbed histotroph seems likely. Thus, the environment in which the embryo develops has a profound influence on development and differentiation processes. Disturbances may lead to developmental impairment. The results highlight the vulnerability of the conceptus and the possibility of lasting epigenetic effects originating during early preimplantation development and causing adverse effects later in life.

***In vitro* maturation of camel oocytes using different media and sera**
A.E.B. Zeidan[1], M. El Hammad[2], M.H. Farouk[3], S.M. Shamiah[1], E.A.A. Ahmadi[1], W.M.A. Nagy[1] and S.A. Gabr[2]
[1]Animal Production Research Institute, Camel Research Deptartment, Dokki, Giza, Egypt, [2]Faculty of Agriculture, Tanta University, Department of Animal Production, Tanta, Al Gharbiyah, Egypt, [3]Faculty of Agriculture, Al-Azhar University, Department of Animal Production, Nasr City, 1184 Cairo, Egypt; mhfarouk@azhar.edu.eg

A total number of 350 pairs of ovaries collected from 175 non-pregnant dromedary camels at 5 to 10 years of age and 500-600 kg body weight were used in the present study. The experimental work aimed to investigate the effect of different media (tissue culture medium: TCM 199 and minimum essential medium: MEM) and sera (fetal dromedary camel serum: FDCS; bovine serum albumin: BSA and estrus sheep serum: ESS) addition to maturation medium on *in vitro* maturation efficiency of follicular camel oocytes, during breeding and non-breading season. The obtained results showed that, maturation rate percentage of the dromedary oocyte reaching metaphase II stage was significantly ($P<0.05$) higher, while the degenerated oocytes revealed significantly ($P<0.05$) lower during breeding than non-breeding season. The maturation rate percentage of camel oocyte reaching germinal vesicle, germinal vesicle breakdown and metaphase I stage was insignificant either breeding or non-breeding season. MEM medium resulted higher distribution of germinal vesicle, germinal vesicle breakdown and lower percentage of degenerated camel oocytes than that TCM 199 medium. Maturation rate percentage (oocyte at metaphase II) was significantly ($P<0.05$) higher with FDCS addition to *in vitro* maturation medium than that BSA or ESS.

Effect of tailectomy, testicular ventilation and fat-tail elevation on Awassi ram spermatogenesis

S. Abi Saab[1], W. Tarabay[1], E. Tabet[1] and P.Y. Aad[2]
[1]USEK, Ag. Scie, Kaslik, 0, Lebanon, [2]NDU, Scie, Louaize, 0, Lebanon; paad@ndu.edu.lb

Awassi sheep have high tolerance for heat stress (HS) when reproduction starts. Common HS alleviation include shearing the Awassi rams, but the presence of the fat tail (FT) covering the testicles plays a major role against this alleviation. Therefore 2 experiments evaluated the effect of tailectomy (exp 1), ventilation (Vent), or lifting (Lift) the FT concomitant with shearing (exp 2) on testicular microclimate and semen quality. In exp 1, 6 adult Awassi rams were divided into control or tailectomy (TE), semen quality evaluated before and after treatment. In exp 2, 12 awassi rams were divided in shorn, ventilated testis (control) or lifted fat tail with or without sheering, and semen quality evaluated once per week for 49 days before, during and after treatments. Testicular microclimate was measured at the middle of the testicles. Semen was collected via an electroejaculator and evaluated for volume, appearance, motility, concentration and abnormalities. Data were analyzed using SPSS. Ambient temperature ranged from 27 to 29 °C with relative humidity of 80 to 85%. No difference (P>0.10) in ram body weight and temperature, respiration and heart rates or testicular circumference and volume in both experiments, attesting to the adaptability of Awassi rams. However, testicular length decreased in all treatment groups as an indication of HS alleviation of Vent, Lift or amputation of FT. The microclimate of the testicles improved (P<0.05) with testicular temperature decreasing by 1.5 to 2 °C across all treatments vs pre-experimental counterparts. Semen V, conc and motility decreased following TE, but then recovered 29 d later; these improved following Vent, Lift as compared to pre-exp, whereas abnormalities decreased (P<0.05). We conclude that increased heat stress conditions highly affects spermatogenesis and any alleviation technique will increase sperm counts, mobility and volume, while decreasing sperm abnormalities.

Influence of boar's age and season on parameters of semen quality assessed by advanced methods

B. Szczesniak-Fabianczyk, M. Bochenek, P. Gogol, M. Trzcinska, M. Bryla and Z. Smorag
National Research Institute of Animal Production, Department of Biotechnology of Animal Reproduction, ul. Krakowska 1, 32-083 Balice/Kraków, Poland; piotr.gogol@izoo.krakow.pl

The aim of the study was to examine the influence of boars' age and a year's seasonality on semen quality. The semen examination included sperm motility and morphology, DNA fragmentation (SCSA), membrane integrity (PI/SYBR-14), oxidative stress (luminescence method), mitochondrial transmembrane potential (JC-1) and apoptosis (YO-PRO-1/PI). In the study 171 ejaculates of fresh and stored in BTS extender collected from 16 boars were used. Boars age varied from 7 months to 7 years. The ANOVA was used for statistical calculations. Higher motility (approx. by 5%), higher ratio of sperm morphologically normal (approx. by 2%), more spermatozoa with an intact membrane (approx. by 4.5%), less apoptotic spermatozoa (approx. by 5%) as well as a lower level of sperm oxidative damages were observed during an October-March period compared to an April – September period. Longer lifetime (over 1 day) was observed in semen stored at +17 °C during fall-winter season. However, statistically significant differences between fall-winter and spring-summer seasons were observed in sperm DNA fragmentation (higher by 0.5% in spring-summer, P≤0.05) and membrane integrity (more moribund spermatozoa by 3% in spring-summer, P≤0.01) only. The highest level of motility, membrane integrity, mitochondrial potential, morphologically normal spermatozoa and the lowest level of DNA fragmentation and oxidative damages were observed in semen collected from boars at age up to12 months. The level of these parameters for adult boars (in age of 12-48 months) was similar but semen lifetime (down to 30% motility) was statistically significant longer by 2.5 days (P≤0.05). The worst results was observed in boars aged more than 48 months. A practical conclusion that may be drawn from the above results is that both an age of a boar and a season of a year are factors influencing quality of boar's semen.

Correlation room temperature – reproductive response of sows inseminated with semen post-cervically
J.M. Romo, J.A. Romo, R. Barajas, H.R. Guemez and J.M. Uriarte
Universidad Autónoma de Sinaloa, Facultad de Medicina Veterinaria y Zootecnia, GRAL. Angel Flores
s/n, 80080, Culiacán, Sinaloa, Mexico; jumanul@uas.edu.mx

Two hundred and twenty three multi parturient sows were utilized to determinate the effect of the applying of semen doses spiked with oxytocin, using the technique intrauterine artificial insemination, on reproductive performance of sow serviced during summer-autumn season in the North-West of Mexico. (months of May to October with maximum registered ambient temperature of 44 °C and 28 °C minimum, and the weekly average temperature was correlated with fertility). Sows were assigned randomly to be serviced twice with semen from same boar(s) in one of two treatments: (1) Served with a reduced semen dosage equivalent to 1.5×10^9 viable spermatozoa cell diluted in 40 ml of semen dose (CTRL, n=111); and (2) CTRL plus addition of 4 IU oxytocin to semen at service time (OX, n=112). Sows were serviced from Mayo to October, using an intrauterine semen delivery device. Total number born (including mummified fetuses) and total number born alive were counted at farrowing. Results were compared by ANOVA for a completely randomized design; farrowing date was compared using χ^2 analyses. The fine no significant correlation between ambient temperature was found to fertility (P=0.71). Treatments have no effect on litter size with 11.0 vs 10.8 (P=0.577), neither number born livepiglet with 9.8 vs 9.6 for CTRL and OX respectively. OX enhances (P=0.03) 9.6% the farrowing rate, with means values of 84.68 vs 93.75 for CONT and OXY, respectively. These results suggest that the addition of oxytocin to semen at service time improves farrowing rate of multi parturient sows serviced during summer-autumn in the North-West of Mexico, without effect on other reproductive performance variables.

Exposure to high altitude impairs ram semen quality
E. Cofré[1], L.A. Raggi[1], O. Peralta[1], B. Urquieta[1], J. Palomino[1], I. Ceppi[1], A. González-Bulnes[2] and V.H. Parraguez[1]
[1]Faculty of Veterinary Sciences, University of Chile, Santa Rosa 11735, 8820808 Santiago, Chile, [2]Comparative Physiology Lab, SGIT-INIA, Avda. Puerta de Hierro s/n, 28040 Madrid, Spain; vparragu@uchile.cl

Sheep breeding is an important farm activity in the highlands around the world. Reproductive efficiency, however, is low due to effects of high altitude (HA), mainly hypoxia, on different sheep reproductive physiological processes. This study report some effects of HA exposure on ram semen quality, with the purpose of knowing the contribution of males to reproductive deficiency in ovine flocks kept for different periods in the highlands. Twenty four mixed breed, 1-2 years old rams were used: HA (≥3600 m) natives (group 'HH', n=8), low altitude natives exposed to HA (3600 m) for a period of 30 days (group 'LH', n=8) and low altitude natives maintained at sea level (control group 'LL', n=8). The experiment was done during the reproductive season, but the rams were maintained in sexual abstinence. Semen samples were obtained by electro-stimulation and analyzed by a computer assisted semen analysis system. Results were compared among groups by means Kruskal-Wallis followed by Dunn`s Multiple Comparison test. The results showed that exposure to HA did not affect the ejaculate volume (LL=1.7±0.1, LH=2.2±0.3, HH=1.9±0.6; ns) or total sperm motility (LL=72.0±3.7, LH=63.0±6.3, HH=64.4±3.0%; ns). However, HA exposure decreased sperm concentration (LL=3.8±0.5, LH=2.0±0.3, HH=1.7±0.2×10⁹ sperms/ml; P<0.001), progressive motility (LL=39.9±7.0, LH=16.6±3.3, HH=16.0±1.6%; P=0.013), curvilinear sperm velocity (LL=129.6±9.7, LH=79.7±3.0, HH=78.6±4.1 µm/seg; P<0.001), rectilinear sperm velocity (LL=81.6±9.7, LH=37.7±2.9, HH=37.3±2.9 µm/seg; P<0.001) and beat frequency of sperm flagellum (LL=10.7±0.7, LH=8.9±0.6, HH=9.1±0.6 Hz; P=0.031). It is concluded that long- or short-term exposure of rams to HA induce negative effects on seminal characteristics, which may contribute to the low fertility of the species bred in the highlands. Support: Grant FONDECYT 1130181, Chile.

Effect of enzymatic and hormonal supplement on some reproductive parameters of Ossimi sheep

A.Q. Al-Momani, E.B. Abdalla, F.A. Khalil, H.M. Gado and F.S. Al-Baraheh
Jerash Private University, Animal Production, P. O. Box 426002, 11140 Amman-Jabal-Alnaser, Jordan;
ahmadmomani2000@yahoo.com

Fifty-seven, multiparous, adult Ossimi ewes were randomly assigned to four groups,each group fed ration supplemented with one of the following components: melatonin (G1; n=14), ZADO (G2; n=14), melatonin and ZADO (G3; n=15) and ration with no additives (G4; n=14). All ewes were fed maintenance ration for one month (during June) followed by flushing ration for five weeks (two weeks before and three weeks after ram introduction). Melatonin (3 mg/h/d) and ZADO (15 g/h/d) mixed with total ration started to be added with maintenance and flushing rations, respectively. Fertile rams were introduced for two successive estrous cycles. Before flushing serum leptin concentration was slightly higher in melatonin treated groups (G1 and G3) than non-supplemented groups (G2 and G4). It wasn't expected to find variation between the four groups before flushing especially that ewes were originated from the same flock, they were at the same physiological status (non pregnant), and consume the same rations despite that ration for G1 and G3 were supplemented with melatonin. Overall mean concentration of leptin in different groups. Leptin concentration was higher (P<0.01) in ZADO supplemented groups (G2 and G3). It was the highest in G3 and the lowest in G4, while G1 was intermediated between G2 and G4. Percentage of ewes expressed estrus was significantly different (P<0.05) among the treatment groups. All ewes expressed estrus in G1, while the lowest percentage (71%) was detected in G4. The interval from ram introduction to the onset of estrus differed (P<0.01) among treatment groups. Ewes received melatonin in G1 and G3 expressed estrus earlier than ewes didn't receive melatonin in G2 and G4. Neither body weight nor body condition score were different at the first estrus following ram introduction, while average daily gain was promoted by treatments particularly melatonin, but the difference was not significant.

The effect of *Origanum vulgare* extract on kinetics and viability of cryopreserved Holstein semen

H. Daghigh Kia, R. Farhadi, A. Hossein Khani and I. Ashrafi
University of Tabriz, Faculty of Agriculture, Department of Animal Science, 29 Bahman Boulevard,
5166614766, Tabriz, Iran; hdk6955@gmail.com

Reactive oxygen species generated during the freeze-thawing process may reduce sperm quality. This study evaluated the effects of *Origanum vulgare* extract supplementation as an antioxidant in the semen extender on post-thaw parameters of kinetics and viability of bull spermatozoa. After extraction, *O. vulgare* extract was added to the citrate-egg yolk extender to yield six different final concentrations: 0, 2, 4, 8, 12, 16 and 20 ml in dl diluents solution. Ejaculates were collected from six Holstein bulls via artificial vagina. The ejaculates were then mixed in a pool to remove individual differences in bulls. Then, the semen was cooled slowly to 5 °C. The sample was packaged in 0.5 ml straws. The straws were sealed and the semen was frozen with liquid nitrogen vapor. After thawing, kinematic parameters were evaluated by means of a computer-assisted semen analysis (CASA) and the viability of the sperm was assessed using eosin-nigrosin stain. The results showed that semen extender supplemented with various doses of *O. vulgare* extract increased (P<0.05) total motility, progressive motility, linearity, lateral head displacement, straight linear velocity, curvilinear velocity, average path velocity after the freeze-thawing process. The most effective concentration of *O. vulgare* extract in kinetical parameters of the bull sperm freezing extender was 4 ml/ dl. The viability was increased by inclusion of 2 and 4 ml/dl *O. vulgare* extract in the semen extender. In conclusion, supplementation of 2 and 4 ml/dl *O. vulgare* extract in the semen extender improved the quality of post-thawed semen, which may associate with a reduction in oxidative stress.

Early embryonic survival and development in two strains of rabbits

R. Belabbas[1], H. Ainbaziz[2], M.L. García[3], A. Berbar[1], G.H. Zitouni[4], M. Lafri[1], Z. Boumahdi[1], N. Benali[2] and M.J. Argente[3]
[1]LBRA, University Saad Dahleb, 09000 Blida, Algeria, [2]Laboratory of research: H.A.P, ENSV, 16200, Algeria, [3]Departamento de Tecnología Agroalimentaria, Universidad Miguel Hernández, 03312, Spain, [4]ITELV, Algeria, 16200, Algeria; r_belabbas@yahoo.fr

The aim of this study was to analyse ovulation and fertilization rates, embryo development and early embryonic survival in rabbits of synthetic line (S) and local population (L). Rabbit does from L (n=23) and S (n=24) were used in this study. The females were first mated at 18 weeks of age and at 10 days after parturition thereafter. On their fourth gestation, nonlactating females were mated and sacrificed at 72 h postcoitum (p.c). The oviduct and the first one-third of the uterine horn were flushed for recovering embryos. All statistical analyses were performed using Bayesian methodology. The S showed a higher value of litter size than L (+1.32) and guaranteed values of difference with a probability of 80% (k80%) were very high (1 kits). In the present experiment, we considered an amount of 0.7 kits as a relevant difference between S and L (R). The probability of the difference being greater than R was high and relevant (PR=0.93). The females of S released 50% more ova than those of L (+4.42 ova, P(DS-L>0) =1). The probability of the difference being higher than R was very high (PR =1). S showed 50% more embryos at 72 h of gestation than L (+3.92 embryos, PR=1). Otherwise, differences for fertilization rate and early embryo survival were not relevant between S and L. The S females presented less advanced embryonic stage of development. The S showed higher number of early morulae (EM) (+23.14%, P(DS-L>0=0.98) and lower compact morulae (CM) (-27.36%, P(DS-L>0=0.99) than L. Considering a relevant value of 8% for EM and CM, the probability of this difference being greater than R was high for both traits (PR=0.97). In conclusion, this crossbreeding has improved litter size, ovulation rate but modify embryo development at 72 h p.c.

Components of litter size in rabbit of local Algerian population

R. Belabbas[1], H. Ainbaziz[2], M.L. García[3], A. Berbar[1], G.H. Zitouni[4], M. Lafri[1], Z. Boumahdi[1] and M.J. Agente[3]
[1]Laboratory of Biotechnologies related to Animal Reproduction, Institute of Veterinary Sciences, University Saad Dahled, Bliba, Algeria, 09000, Algeria, [2]Laboratory of research Health and Animal Production, ENSV, 16200, Algeria, [3]Departamento de Tecnología Agroalimentaria, Universidad Miguel Hernández de Elchè, 03312 Alicante, Spain, [4]Institute Technical of breeding, Bab Ali, 16200, Algeria; r_belabbas@yahoo.fr

The objective of this work was to characterize the components of litter size in rabbit does of local Algerian population (ovulation rate and prenatal mortality). 36 females were used in this experiment. All does were first mated at 18 weeks of age and thereafter 10 days after parturition. On their third gestation, laparoscopy was performed at d12 postcoitum and the number of corpora lutea, implanted embryos, live and dead fetuses was noted. The data were analyzed using the GLM procedure. The litter size at birth was 6.46 born alive with high variability between females. The number of implanted embryos at 12 d of pregnancy was 7.30 (6.70 alive and 0.6 resorbed). Embryonic and fetal survival were high (86 and 93% respectively) in the females of local population as far as the prenatal survival (82%). The females of local population have shown interesting performances and which could be improved more by selection or crossbreeding with other races and strains.

Effect of monensin sodium and *Saccharomyces cerevisiae* on reproduction performance of Ghezel ewes

A. Hosseinkhani, L. Ahmadzadeh and H. Daghigh Kia
University of Tabriz, Animal science, 29 Bahman Bolvard, Faculty of Agriculture, 5166614766, Iran;
a.hosseinkhani@tabrizu.ac.ir

Monensin sodium (MS) and *Saccharomyces cerevisiae* (SC) are feed additives which can manipulate rumen fermentation toward propionate production. Due to low twining rate of Ghezel ewes (22%), usage of these additives along flushing ration may be appropriate to increase reproduction performance. Forty four purebred Ghezel ewes were allotted to 4 dietary treatments representing: (A) grazing on pasture + flushing diet + MS (30 mg/ewe/d); (B) grazing on pasture + flushing diet + SC (4×10^9 cfu/ewe/d); (C) grazing on pasture + flushing diet; (D) control (only grazing on pasture). Flushing diet contained 78.1% TDN and 13% CP which fed to the ewes during 2 weeks before and 3 weeks after mating. The amount of flushing diet was 450 g/ewe/d which was offered to ewes after grazing on pasture every evening. All of the ewes were synchronized using CIDR and then were mated with purebred Ghezel rams 48 hours after CIDR removing. Experimental treatments resulted in 150, 136, 138 and 109% lambing, respectively. Results from blood plasma analysis showed that estrogen level was higher in treatments A and B (43.7 ± 2.6 and 41.2 ± 2.6 pg/ml, respectively) compared with the other treatments in the estrus phase ($P<0.05$). Treatment A had the maximum level of estrogen in follicular phase and thereby maximum lambing rate (150%). Also results showed that treatments A and D had the highest and lowest levels of blood progesterone at 21st day of pregnancy (6.3 vs 3.3 ng/ml, respectively) ($P<0.05$). Treatments A and B showed maximum levels of insulin (0.81 and 0.73 ng/ml. respectively) at 24 h after CIDR removing ($P<0.01$). Moreover lambs were born from ewes consumed treatments A and B had maximum birth weight compared with other treatments ($P<0.01$). Overall results from this study revealed that using MS and SC in breeding season may have positive effects on endocrine status and thereby ewes reproduction performance.

Session 55   Theatre 1

Effects of perinatal environment on later life resilience in farm animals

B. Kemp, H. Van Den Brand, A. Lammers and J.E. Bolhuis
Wageningen University, Dept. of Animal Sciences, Adaptation Physiology Group, P.O. Box 338, 6700 AH Wageningen, the Netherlands; bas.kemp@wur.nl

Evidence is accumulating that perinatal experiences have long-term effects on the ability to cope with environmental challenges later in life. Recent studies suggest that also the resilience of farm animals is shaped by their prenatal and early postnatal conditions. Environmental conditions in the prenatal period to which the mother is exposed appear to effect the offspring. For instance, severe feather damage and hig corticosterone levels in mother hens in parent stock flocks were related to increased levels of fearfulness and feather pecking in offspring laying flocks. Also more direct effects of prenatal conditions have been shown. High incubation temperatures in broilers were e.g. demonstrated to hamper heart development with lasting effects on susceptibility to develop ascites in later life. In early postnatal life of chickens and pigs, a timely development of nutritional independence seems important. Poultry systems that provide chicks with feed and water immediately after hatch have beneficial short and long term effects on growth and immune development and likely also on microbiota composition. Also in piglets, ealry feed sampling, already ealry in lactation, may have important developmental effects. Enrichment of the environment with foraging substrates and facilitating transfer of information from the sow about what and were to eat seem important to facilitate early experience of piglets with feed. These strategies subsequently result in better performance, less diarrhoea and lower stress responses around weaning, but may also have longer term effects on development of injutious behaviours like ear and tail biting. These examples illustrate the importance of perinatal conditions for robustness and resilience and emphasize the need for a whole chain approach to optimze life time functioning of farm animals, in which husbandry conditions in early stages of life need to be reconsidered.

Long-lasting effects of thermal manipulations during embryogenesis in broiler chickens

T. Loyau[1], S. Métayer-Coustard[1], C. Berri[1], S. Mignon-Grasteau[1], C. Hennequet-Antier[1], M.J. Duclos[1], S. Tesseraud[1], C. Praud[1], N. Everaert[2], M. Moroldo[3], J. Lecarbonnel[3], P. Martin[4], V. Coustham[1], S. Lagarrigue[5], S. Yahav[6] and A. Collin[1]
[1]INRA, UR83 Recherches Avicoles, 37380 Nouzilly, France, [2]University of Liège, Gembloux Agro-Bio Tech, Animal Science Unit, Gembloux, Belgium, [3]INRA, CRB GADIE, Domaine de Vilvert, 78350 Jouy-en-Josas, France, [4]INRA, UMR1313, GABI, 78350 Jouy-en-Josas, France, [5]INRA, Agrocampus Ouest, UMR1348 Pegase, 35590 Saint Gilles, France, [6]ARO, The Volcani Center P.O. Box 6, 50250 Bet Dagan, Israel; thomas.loyau@tours.inra.fr

Climate change may affect animal physiology, having deleterious consequences for poultry farms. Thermal manipulation (TM) during embryogenesis has been shown to improve heat acclimation in the long term in broiler chickens. TM consists of increasing cyclically incubation temperature from 37.8 °C to 39.5 °C from day (d) 7 to d16 of embryogenesis. Our study aimed to determine the effects of TM and of a subsequent heat challenge (HC) on chickens performance, physiology and metabolism, in order to better understand heat adaptation acclimation mechanisms. In our conditions, TM did not modify hatchability and performance until d28 post hatch, but slightly decreased body weight and composition at d34. TM decreased body temperature (bT) until d28 as well as plasma triiodothyronine concentrations. Our results suggest that TM changes respiratory physiology and decreases stress response during HC at d34. TM associated with low bT affected the expression of genes involved in muscle energy metabolism, or mitigated the effects of acute heat stress. The transcriptomic study performed on breast muscle also pointed out new pathways modified by TM in the long term regulating vascularization and epigenetic mechanisms. TM animals activated more genes in response to HC than controls, possibly explaining their better heat adaptation. This integrated study demonstrated that TM induced long term physiological and metabolic adaptations to changes in the thermal environment of chickens.

***In ovo* stimulation of the chicken embryo with synbiotics influences adult's molecular responses**

M. Siwek, A. Slawinska, A. Plowiec, A. Dunislawska and M. Bednarczyk
University of Technology and Sciences, Animal Biochemistry and Biotechnology, Mazowiecka 28, 84-085 Bydgoszcz, Poland; siwek@utp.edu.pl

Chicken embryo development takes place inside the egg and is influenced by the genetic background, composition of the egg yolk (e.g. nutrients and IgY antibodies deposited by the hen) and conditions during incubation (e.g. temperature and humidity). However, in ovo technology allows the supply of additional stimuli to the growing embryo to optimise conditions for embryonic and post-hatch growth and development. In commercial settings, the newly hatched chicks are kept in a sterile environment, separate from the adults. Therefore, the neonates are deprived of the stimuli needed for the development of a healthy intestinal microflora. With the properly selected combinations of pre- and probiotics (synbiotics) delivered in ovo, it is possible to modulate the optimal composition of the chicken intestinal microflora, fully developed at hatching. In this study we analyzed the immune-related gene expression regulation in adult broiler chickens (Cobb) that had been treated in ovo at 12th day of embryo development with different synbiotic combinations. Intestinal (jejunum, cecal tonsils) and splenic tissues were harvested at several time points post hatching (day: 1, 7, 14, 21, 35). Gene expression regulation was analyzed using RT-qPCR to determine the expression of the selected gene panel, including IL-4, IL-6, IL-8, IL-18, IL-12p40, cluster of differentiation (CD) 80, interferon-beta (IFN-β) and interferon-gamma (IFN-γ). Preliminary results indicate that in ovo modulation of the microbiome results in the response of the host at the molecular level. Results of this study will provide an insight into development of the host-microbiome relations throughout the chickens' lifespan in terms of molecular responses of the periphery of the immune system (spleen), intestine (jejunum) and the gut-associated lymphoid tissue (cecal tonsils) to in ovo delivered synbiotics. Acknowledgements: FP7-KBBE-2012-6-singlestage project from European Commission.

Effects of DSS on intestinal morphology, immune development and immune response in chickens

K. Simon, B. Kemp and A. Lammers
Wageningen University, De Elst 1, 6708 WD Wageningen, the Netherlands; kristina.simon@wur.nl

Disturbance of intestinal homeostasis early in life may affect immune development as well as immune responses later in life. The present study investigated dextran sodium sulphate (DSS) as a model for early life disturbances of intestinal homeostasis and its effects on intestinal immune development and immune response in chickens. Broiler and layer chicks received 2.5% DSS in drinking water during d 11-18 post hatch (DSS) or plain drinking water (control, C). As immunological challenge birds received a combination of *Escherichia coli* LPS and HuSA i.m. at d 35. Effects of DSS on intestinal morphology and intestinal cytokine and Ig mRNA expression levels were examined. Antibody titers against LPS, HuSA, and KLH were measured to investigate effects of intestinal inflammation on the immune response later in life. 3 d after the start of DSS treatment, DSS birds of both breeds showed significantly lower bodyweight (BW) than C birds. DSS broilers had recovered regarding BW by d 28, while DSS layers continued to show lower BW through d 42. Broilers also showed faster recovery regarding colon and caecum length. Histological examination of intestinal samples of DSS birds showed shortening and damage of villi in ileum and colon, and complete absence of villi in caeca. Effects of DSS on intestinal morphology were again less severe in broilers, which also showed a lower mortality in response to DSS than layers. Effects of DSS on the immunological challenge later in life were limited and were mainly observed for antibody titers against LPS. Interestingly DSS broilers showed lower anti-LPS titers than C broilers, while layers showed opposite results. In conclusion DSS may serve as an experimental model to simulate disturbances of intestinal homeostasis in early life, although more research on the appropriate dose is necessary and is likely to differ between breeds. Results indicate that broilers are either less susceptible to DSS or have a better capacity to recover from disturbances in intestinal morphology.

Effect of prenatal overfeeding on post-weaning behavior of rabbits

P. Simitzis[1], A. Kiriakopoulos[1], G. Symeon[2], A. Kominakis[1], I. Bizelis[1], S. Chadio[1], O. Pagonopoulou[2] and S. Deligeorgis[1]
[1]Agricultural University of Athens, Department of Animal Breeding and Husbandry, 75 Iera Odos, 11855, Athens, Greece, [2]Democritus University of Thrace, Department of Physiology, 5th km Alexandroupoli, Makri, 68100 Alexandroupoli, Greece; pansimitzis@aua.gr

Prenatal overfeeding in rabbit is a frequent phenomenon, particular when does are fed at libitum, and can be a factor of embryonic programming, causing permanent changes in the metabolism, the physiology and the behavior of the offspring. An experiment was therefore conducted to estimate the effect of does' overnutrition during two different periods of pregnancy on the behavior of kits after weaning. Thirty does of the hybrid Hyla NG were randomly allocated into three experimental groups: control group (diet at the level of 100% of maintenance needs during the whole period of pregnancy), group O1 (diet at the level of 150% of maintenance needs between the 7th and 19th day) and group O2 (diet at the level of 150% of maintenance needs between the 20th and 27th day). At the age of 50 and 65 days, the behavior of 24 randomly selected rabbits into cages was recorded, while an open field test was also implemented in other 24 rabbits; a test that constitutes a good indicator of fear and stress. As it was demonstrated, maternal overnutrition did not have a significant effect on the display of the recorded behavioral elements. No significant effects of prenatal overfeeding on the exhibition of behavior were also found during the open field test. Some differences were found between sexes, different ages and different hours of the day. As it is concluded, no differences were found in the exhibition of offsprings' behavior as a result of maternal overnutrition during pregnancy. This research project was implemented within the framework of the Action 'Supporting Postdoctoral Researchers' of the Operational Program 'Education and Lifelong Learning' (Action's Beneficiary: General Secretariat for Research and Technology), and is co-financed by the European Social Fund (ESF) and the Greek State (LS9 (1678)).

Post-weaning performance of piglets raised in a multi-suckling system vs farrowing crates
S.E. Van Nieuwamerongen[1], J.E. Bolhuis[1], C.M.C. Van Der Peet-Schwering[2] and N.M. Soede[1]
[1]Wageningen University, Animal Sciences, De Elst 1, 6708 WD Wageningen, the Netherlands, [2]Wageningen UR, Livestock Research, De Elst 1, 6708 WD Wageningen, the Netherlands; sofie.vannieuwamerongen@wur.nl

Multi-suckling (MS) systems, in which lactating sows and their litters are group housed, may provide a better transition from group housing during gestation than individual housing. We developed an MS system for five sows with a communal feeding area where piglets can learn to eat with the floor-fed sows. We hypothesized that contact among multiple litters and more environmental complexity in the MS system facilitates piglets' adaptation to weaning. In 5 batches with 80 piglets each, post-weaning performance was investigated from 4-9 weeks of age. Post-weaning, MS piglets were housed in groups of 40 and received some straw, hessian sacks, ropes and chew toys. Performance was compared with piglets raised with a crated sow and housed post-weaning with 10 littermates and a chew toy. Post-weaning floor area and feeder area were equal per piglet. Mixed models (SAS 9.2) were used for analyses, with 40 piglets within system and batch as experimental unit. MS piglets grew faster from weaning (weaning weight 8.4±0.2 vs 8.3±0.2 kg, ns) until 9 weeks of age (25.4±1.4 vs 22.0±0.6 kg, P<0.01), with a similar feed conversion (1.45±0.03 vs 1.48±0.03, ns). MS piglets also played more (freq/h 4.0±0.3 vs 2.8±0.3, P<0.05) and showed less damaging oral behaviors (freq/h 1.8±0.3 vs 3.5±0.4, P<0.01) like tail biting. On day 6 post-weaning, a lower proportion of MS raised piglets had diarrhea (0.26±0.10 vs 0.46±0.07, P<0.05). In short, MS piglets had a better post-weaning performance than conventional piglets. Although there was a higher early postnatal piglet mortality in the MS system which may have had an influence, we propose that the MS system and the post-weaning pen (both with more enrichment and a larger group to facilitate eating behavior) contributed to improved post-weaning performance.

The effect of ketoprofen administered post farrowing on pre weaning mortality
S.H. Ison and K.M.D. Rutherford
SRUC, Scotland's Rural College, Animal Behaviour & Welfare, The Roslin Institute Building, Easter Bush, EH25 9RG, Roslin, Midlothian, United Kingdom; sarah.ison@sruc.ac.uk

Recent research into the use of non-steroidal anti-inflammatory drugs (NSAIDs) post farrowing demonstrates some benefits to sow health, welfare and piglet performance. These benefits were more obvious in certain parity groups and for farms with a high incidence of post-partum dysgalactia syndrome. This study investigated the use of the NSAID ketoprofen using a randomised, blinded, placebo controlled trial with 24 gilts and 32 sows, including 17 sows of parity two to four, 11 five to seven and 4 eight plus. Gilts and sows were randomly allocated to receive Ketoprofen (3 mg/kg bodyweight; KET) or a saline placebo (CON) of equivalent volume by intra-muscular injection 1.5 hours after the birth of the last piglet. Piglet mortality was recorded and piglets were weighed six hours after the injection was given (6 hr BW, kg), on day three post-farrowing and at weaning. Live born mortality (LBM, %) was calculated with cross fosters included with the foster mother. Data were analysed using linear mixed models in Genstat. LBM was greater in the control group for all but the eight plus parity group (Gilts: KET=7.62±2.23, CON=14.21±3.10; two to four: KET=15.68±4.80, CON=21.36±7.19; five to seven: KET=6.63±3.04, CON=11.88±4.85; eight plus: KET=13.25±2.14, CON=0.00±0.00). However, the 6 hr BW was greater in the KET group for all but the eight plus parity group (overall: KET=1.38±0.02, CON=1.24±0.02; P<0.1), therefore with 6 hr BW as a random variable, there was no significant differences in LBM by treatment or treatment × parity group (P>0.05). Despite individuals being randomly allocated to treatment groups pre-farrowing, piglets in the KET were heavier than those in the CON group, possibly accounting for the difference in LBM between groups. A larger sample size, balanced for all other factors that influence pre weaning mortality would confirm this result.

Cross-comparison of lamb mortality in two Mediterranean countries
J.M. Gautier[1], S. Ocak[2], S. Ogun[3], O. Yilmaz[4], E. Emsen[5] and F. Corbière[6]
[1]*Institut de l'Elevage, BP 42118, 31321 Castanet Tolosan Cedex, France, [2]Nigde University, Department of Animal Production and Technologies, Faculty of Agricultural Sciences and Technologies, 5100 Nigde, Turkey, [3]Redrock Agricultural Pastoral Company, Kibris Cad. No:8 Zeugma Center, 1/107, 34365 Istanbul, Turkey, [4]Adnan Menderes University, Department of Animal Science, Faculty of Agriculture, 09100 Aydın, Turkey, [5]Ataturk University, Department of Animal Science, Faculty of Agriculture, 2540 Erzurum, Turkey, [6]National Veterinary School of Toulouse, 23 chemin des Capelles, BP 87614, 31076 Toulouse Cedex, France; jean-marc.gautier@idele.fr*

The aim of this study was to compare the lamb mortality rate (until weaning) and the management factors associated with lamb mortality in two different Mediterranean regions (Turkey and France). The lamb mortality rate was respectively 10 and 13.6% in Turkey and in France. Production system type (intensive vs pastoral) had significant effect on lamb mortality in both France and Turkey even when similar management practices were applied. Distribution of lamb mortality by age was not uniform in the two regions nor in their varying production systems. In Turkey higher early mortalities (22% in 0-7 days) were witnessed in intensive systems as opposed to extensive production systems (12% in 0-7 days). However in France 57% of lamb mortality was witnessed within 10 days post-partum irrespective of the rearing system. Although production systems differed between the two regions, sufficient similarities were seen in their management procedures. Farmer intervention for artificial suckling in both cultures was dependent on the observation of the mother, offspring interaction and the colostrum intake. When the lamb exhibited signs of weakness or lethargy additional postnatal nutrition was provided which in turn had a positive impact on the weaning survivability. These observations suggest the need to have a global approach in understanding the resilience of lambs (soon after birth) and improving intrinsic maternal behaviour of the dam which may impact on lamb survivability.

Effect of colostrum alternative and milk replacer on lamb performance, health and rumen development
A. Belanche[1], C.L. Faulkner[1], E. Jones[1], H.J. Worgan[1], J. Cooke[2] and C.J. Newbold[1]
[1]*IBERS, Aberystwyth University, SY23 3DA, United Kingdom, [2]Volac International Ltd, Orwell, Cambridgeshire, SG8 5QX, United Kingdom; aib@aber.ac.uk*

The use of colostrum alternatives (CA) and milk replacer (MR) has been suggested over recent years to maximize the number of lambs weaned, but its effects on the animals beyond weaning are still unknown. This experiment evaluated the long term effects of CA and MR compared to ewe-rearing. A total of 24 pregnant ewes carrying triplets were used and lambs were randomly allocated to the experimental treatments: (1) EE: ewe's colostrum and ewe's milk; (2) EA: ewe's colostrum and MR (Lamlac Instant, Volac); and (3) AA: CA (Lamb Volostrum) and MR. Each treatment was placed in independent pens with free access to ryegrass hay and creep feed. At 45 d of age all lambs were sampled, weaned, reunited in the same ryegrass pasture and resampled at 5 months of age. During lactation MR lambs had greater incidence of mild diarrhoea than EE lambs. At weaning, MR lambs had no rumen protozoa but a greater concentration of rumen bacteria than EE lambs. This was accompanied by greater plasma levels of calcium, HDL, LDL and alkaline phosphatase, and lower blood cells levels as a result of the high milk intake in late lactation (2.9 l/d). Contrary, EE lambs showed a greater ruminal development characterized by presence of protozoa and higher concentration of total VFA and butyrate molar proportion. EE lambs also had greater plasma levels of amylase and β-hydroxybutyrate due to their greater creep intake. After weaning all these differences vanished and all treatments showed similar rumen function. In terms of animal performance, in early lactation EE lambs had greater growth rates than those fed MR; these differences disappeared in late lactation due to their lower milk intake resulting in similar weaning weights across treatments (18.6 kg), but differences reappeared again after weaning highlighting the importance of maximizing the creep intake and rumen development before weaning in order to facilitate their adaptation to fibrous diets.

Fetal programming of high salt water hematological and intraruminal mechanisms
W.G. Mehdi, H. Ben Salem and S. Adibi
National Institute of Agronomic Research of Tunisia, Forage and Animal Productions Laboratory, Street
hedi Karray, 2049 Ariana, Tunisia; wiem_mehdi2000@yahoo.com

In order to study the programmation fetal of high sodium water (10% NaCl), we used two groups of ewes (C and S-ewes) from the conception day until lactation period and their offspring (C-lambs and S-lambs) from the day the parturition day until 8-months-old and we investigated to determine the hematological and ruminal fermentation changes. There were no significant changes in several blood hematological parameters associated with salt load. Animals can subsist drinking high salt water for a relatively long period without exhibiting adverse effects on health. Adult lambs were programmed to adapt to salt stress. Moreover, results showed no treatment effect on rumen pH, rumen sodium and potassium concentration, protozoa and ammoniac-N concentration ($P>0.05$) in Barbarine sheep. Drinking saline water increased the proportion of the Endiplodium and Polyplastium in rumen S-ewes compared with C-ewes ($P<0.05$). On the other hand, in adult Barbarine lambs, there was no high salt water effect on rumen condition except ($P>0.05$) that S-lambs had high significantly lower rumen pH at the time of rumen samples (0 h) ($P<0.05$). Drinking high salt water reduced ruminal fluid pH on 0 h post-feeding ($P<0.05$) but had no effect on its at 3 h post-feeding ($P>0.05$). Protozoa quantification and generic composition were not affected by treatment and hour post-feeding ($P>0.05$). Protozoa composed similary for C-lambs and S-lambs on 60.5% of Endiplodium, 38.5% of Epidimium and 0.93% of Polyplastium. This study indicated that Barbarine male offspring born to ewes that drinking high salt water from pregnancy to lactation period had beneficial adaptation that enable them to cope better with salty stress without adverse effects on health or performance illustrating that fetal programming didn't change the temporal pattern of how the offspring adapt to a load of drinking salt.

Effect of dietary supplementation of milk replacer on growth and muscle metabolism of light piglets
J.G. Madsen[1,2], E. Seoni[2], M. Kreuzer[1] and G. Bee[2]
[1]ETH Zurich, Institute of Agricultural Sciences, Universitätstrasse 2, 8092, Zurich, Switzerland,
[2]Agroscope, Institute for Livestock Sciences, Route de la Tioleyre 4, 1725, Posieux, Switzerland;
johannes.madsen@agroscope.admin.ch

Decreased average birth weight, as a consequence of increased litter size in high prolific sows, has been shown to negatively affect offspring's viability and growth performance in the suckling period. The study objective was to compare the impact of dietary supplements involved in muscle metabolism to a milk replacer offered *ad libitum* to growth retarded piglets born from large litters (>15 total born) on growth performance and glycolytic (lactate dehydrogenase; LDH) and overall oxidative (citrate synthase; CS) capacity of the semitendinosus muscle. A total of 36 male and female piglets with the lowest d 7 body weight of the litters (BW: 2.08±0.30 kg) were selected and assigned for 21 d to three milk replacer diets containing either no supplements (C), carnitine (CAR, 0.480 g/d) or arginine (ARG, 21.8 g/kg DM). Feed was offered *ad libitum*, and BW and feed intake were determined weekly and daily, respectively. At d 28 of age, piglets were euthanized and complete Semitendinosus muscles were excised. Data were analyzed using the MIXED procedure of SAS. Only one out of the 36 piglets died during the experiment. Compared to C and CAR, ARG numerically had a 7.5% greater BW at d 28 (6.4 kg), 2.6% greater BW gains (196 g/d), 7.8% greater feed intake (236 g/d), and similar feed efficiency. The relative importance of glycolytic with respect to overall oxidative capacity was numerically greater in the semitendinosus muscle of CAR compared to C pigs and was greater in females than castrated pigs. In conclusion, although both supplements failed to display significant effects, the study provides enough indications to further test the potential to improve growth and muscle maturation. This study is part of ECO-FCE and has received funding from the European Union's Seventh Framework Programme for research, technological development and demonstration.

Factors affecting the survival and live weights of Cyprus Damascus kids from birth to weaning

G. Hadjipavlou[1], L. Sofokleous[2] and C. Papachristoforou[2]
[1]Agricultural Research Institute, P.O. Box 22016, 1516 Lefkosia, Cyprus, [2]Cyprus University of Technology, P.O. Box 50329, 3603 Lemesos, Cyprus; c.papachristoforou@cut.ac.cy

The present study examined several possible effects on the viability of Cyprus Damascus kids at birth and until weaning, and factors influencing live weight at birth and at weaning. The dataset comprised of 1,826 Damascus kid phenotypes, born from November 2009 to March 2013 at the Agricultural Research Institute Farm. Information on live weights at birth and at weaning was analyzed using linear regression models, whereas a logistic regression model and a binomial analysis were used for survival phenotypes. The factors that significantly affected viability at birth and at weaning, were date of birth within kidding season and year (P<2.0×10^{-2} at birth, P<2.0×10^{-4} at weaning), gestation length (P<3.5×10^{-14} at birth, P<3.2×10^{-3} at weaning) and birth weight (P<1.1×10^{-11} at birth and P<2.7×10^{-9} at weaning). Additionally, dam age had a significant effect, mainly on survival at birth (P<0.03). Kids born to multiparous (three years or older) dams were more likely to survive at birth than kids from primiparous dams (odds ratio 0.18-0.43). With respect to the live weight analysis outcomes, kid sex and age of dam, significantly affected both live weight at birth (P<2.0×10^{-16}) and at weaning (P<1.5×10^{-7}). Gestation length and litter size had significant effects on birth weight (P<2.0×10^{-16} and P<1.3×10^{-6}, respectively), whereas rearing system (kids fed either by natural suckling or with milk substitute until weaning) significantly affected weaning weight (P<2.1×10^{-5}). The main cause of death after birth and until weaning was enteritis, as a result of a variety of infections. The overall findings of this study indicate the need for further research work on some of the causes of kid losses, and for improved management practices during gestation, at birth and until weaning, especially for primiparous ewes, in order to reduce mortality rates of Cyprus Damascus kids.

Effect of protein content in milk replacer and weaning method on performance of dairy calves

P. Górka[1], T. Radecki[1], J. Kański[1], J. Flaga[1], W. Budziński[2] and Z.M. Kowalski[1]
[1]University of Agriculture in Krakow, Department of Animal Nutrition and Dietetics, al. Mickiewicza 24/28, 30-059 Krakow, Poland, [2]Polmass SA, ul. Fordońska 40, 85-719 Bydgoszcz, Poland; p.gorka@ur.krakow.pl

The aim of the study was to determine the effect of protein content in milk replacer (MR) and weaning method (WM) on performance of calves. Eighty Holstein calves (8.6±0.1 day old) were allocated to four treatments (20 calves/treatment) according to 2×2 factorial study and fed MR containing 21 or 25% of crude protein (CP) in MR powder (MR21 and MR25, respectively) and weaned gradually over a period of two weeks (by decreasing volume of MR offered in wk 6 of the study from 6 to 5 l/day and in wk 7 of the study from 5 to 4 l/day; WMA) or one week (by decreasing volume of MR offered by half from 6 to 3 l/day in wk 7 of the study; WMB). Calves were completely weaned in wk 8 of the study. MR intake over a study period was equal for all treatments. Pelleted starter mixture was offered free of choice from the first day of the study. Feed intake and health status of calves were recorded daily and body weight was recorded weekly. On day 0, 21 and 63 of the study blood samples were collected and serum total protein, glucose, IGF-1, insulin, β-hydroxybutyric acid and amylase concentrations were determined. Each calf was in a study over a period of 9 wk. MR25 tended to increase average daily gain (ADG) and feed efficiency between wk 1 to 5 of the study. WMA calves had lower ADG in wk 6 and higher in wk 9 of the study, as compared to WMB calves (WM × time interaction). WMB calves consumed more starter mixture in wk 7 of the study, as compared to WMA calves, but no difference in whole study period was shown (WM × time interaction). There was no effect of CP content in MR neither WM on investigated blood parameters. This study indicates that CP content in MR affects performance of calves during pre-weaning period whereas WM affects performance of calves after weaning. The latter one may have long term effect on performance of rearing heifers.

Season of birth affects growth rate and body condition in Holstein Friesian heifers
M. Van Eetvelde, M.M. Kamal and G. Opsomer
Faculty of Veterinary Medicine, Ghent University, Salisburylaan 133, 9820 Merelbeke, Belgium;
mieke.vaneetvelde@ugent.be

Growth and body condition were followed in 51 female Holstein Friesian calves. Calves were weighed at birth and their wither height (WH) and diagonal length (DL) were recorded. The body condition index (BCI) was calculated as birth weight/(WH×DL) ands used as an index for body mass. All calves were kept indoors during the first year and were fed according to their requirements for maintenance and growth. Every 3 months until first calving body weight, body measurements and body condition score (BCS) were recorded. Season of birth did not affect weight (39.1±4.26 kg) nor BCI (73.5±4.43 kg/m^2) at birth. During the first year of life, calves born in fall grew 934 g/day; which was significantly more (P<0.05) than calves born in spring. This resulted in fall calves being respectively 42.0 and 61.0 kg heavier at first insemination (14.9 months) than spring and summer calves (P<0.01). Spring calves showed the lowest growth rate during the first year and grew 120.2 g/day less than fall calves and 82.4 g/day less than winter calves. After 12 months, growth slowed down in all calves and an opposite seasonal effect was noticed: heifers born in spring and summer grew 175.8 and 170.5 g/day more than those born in fall (P<0.01). The great early growth in the heifers born in fall resulted in a lower BCI at calving compared to heifers born in spring (256.3 vs 279.5, P<0.05), caused by a numerically higher WH (+1 cm) and DL (+4.8 cm) at a similar weight. Besides, heifers born in fall had an average BCS at first calving of 3.3, which was 0.7 less than heifers born in spring (P<0.01). Dams of calves born in fall completed their gestation during the hottest months of the year, which is known to have a negative effect on fetal growth. Although this didn't result in lower birth weights, an early catch-up growth during first year of life is suggested in these calves; resulting in heifers with a larger body frame and less body fat at calving.

Effects of dietary conjugated linoleic acid supplementation on performance of calves
J. Flaga, M. Belica, P. Górka and Z.M. Kowalski
University of Agriculture, Department of Animal Nutrition and Dietetics, al. Mickiewicza 24/28, 30-059
Krakow, Poland; j.flaga@ur.krakow.pl

The beneficial effect of conjugated linoleic acid (CLA) on animal health is well established. Rumen is one of the main place of CLA production and after entering the blood circulation it is transferred to milk, which contains its considerable amount. However, in milk replacers (MR), popularly used in calves rearing instead of whole milk, CLA is lacking or its content is negligible. Thus we hypothesized that MR supplementation with CLA could have a positive influence on performance of calves. Thirty female calves (9.9±2.0 day-old; mean ± SD) were randomly allocated into two experimental groups (15 animals per group): (1) control; (2) MR supplemented with 3 g of CLA rich oil/day (~1 g of CLA/day; Lutalin®, BASF). CLA was supplemented by direct addition of CLA oil into MR once per day. Liquid and solid feed intake, health status and occurrence of diarrheas were recorded daily and body weight was recorded weekly during 9 weeks of experiment (8 weeks of feeding with MR and one week after weaning). MR supplementation with CLA had no effect on liquid and solid feed intake, daily body weight gains, health status and frequency of diarrhea. However, there was a tendency (P=0.08) to better feed efficiency in the group receiving CLA between week 4 and 6 of the study. Thus, in our study CLA dietary supplementation had no distinct effect on performance of calves. It is possible that the results of CLA administration may occur later in calves life but confirmation of this hypothesis needs further studies.

Monitoring of correctness of dairy cow feeding in the transition period
Z.M. Kowalski
University of Agriculture in Krakow, Al. Mickiewicza 24/28, 30-052 Krakow, Poland; rzkowals@cyf-kr.edu.pl

Transition period, defined as a period of last 3 weeks prepartum and first 3-4 weeks postpartum, is a time of tremendous biological changes for dairy cow organism including calving and start of lactation. These changes together with welfare challenges (e.g. changes of groups, boxes, etc.) create a very high risk of health problems. Moreover, dynamic changes in the requirements for energy and nutrients within this period cannot be fully followed by diet changes since the rumen environment needs 2-3 weeks for adaptation to the new diet composition. Although there has been a dramatic increase in understanding of the transition period challenges over few last decades, leading to introducing new feeding and welfare strategies as well as herd management systems, still the prevalence of metabolic diseases, typical for transition period, such as ketosis, fatty liver syndrome, milk fever, and gastrointestinal disorders, such as displaced abomasum, is too high. The consequences of above health problems increase the prevalence of secondary health disorders, reproduction failures, rate of culling and finally decrease the effectiveness of milk production. The importance of transition period challenges has provoked many research groups all over the world to study the monitoring methods, including monitoring of correctness of dairy cow feeding. Some of them are still on the level of 'laboratory stage', but many have been introduced into the practice. This review will concentrate on those methods which are related to feeding and are already in use, also on Polish dairy farms. Special attention will be paid to the models used in the prediction of negative energy balance, including blood or milk analyses for selected chemical parameters. The models used in body condition scoring will also be discussed (e.g. models based on the use of digital camera). Finally, a short description of the Polish unique method of subclinical ketosis monitoring will be presented. The method was successfully introduced into the practice on 1 April, 2013, for all cows being in the monthly recording system. Today we monitor about 750,000 cows and about 20,000 herds. What we have learnt about ketosis in Poland from such a monitoring of cows and herds, will also be discussed, with special attention on risk factors.

DairyCare: monitoring of the welfare and health of dairy cows
C.H. Knight
University of Copenhagen, SUND IKVH, Dyrlægevej 100, 1870 Frederiksberg C, Denmark; chkn@sund.ku.dk

Animal husbandry involves monitoring animal health, wellbeing and productivity and then responding in an appropriate way when problems are noticed. It is still largely done by humans using skills that have not changed significantly in many years, but this approach is increasingly difficult to sustain, for two related reasons: in many parts of Europe animal production units are getting larger and/or the cost of skilled labour is increasing. Both of these factors reduce the opportunity for husbandry staff to monitor animals effectively, so that husbandry becomes more difficult. Technology is used in many different ways for monitoring and improving our own, human, wellbeing, and the benefits that this provides to society in, for example, improved disease diagnosis, improved security, improved mobility and generally in improved quality of life are well recognized and accepted. The principle of using appropriate technologies within animal production to improve animal wellbeing by focusing the valuable time of animal husbandry staff onto those animals that most require attention is undeniable and compelling. This was recognized in 2014 by the award of networking funding to DairyCare, COST Action FA1308 (www.dairycareaction.org). DairyCare is focused on three types of monitoring technology: Biomarkers of welfare and health, Activity-based measures of welfare and health, Systems-level technologies that identify and integrate key data to obtain a measure of current welfare and health status. This review will concentrate on the first two groups. Biomarkers include traditional indices of stress (cortisol, for example) measured in novel ways (multi-matrix analysis, for example) as well as protcomic, peptidomic and metabolomic discovery of new biomarkers. Activity measures include video analysis, positioning analysis by GPS and tri-axial accelerometer analysis, where the emphasis is to develop multi-use outputs from a single monitor (feeding activity and lameness as well as estrus detection, for instance).

African Swine Fever in Europe – monitoring and control of the spreading
K. Śmietanka[1], Ł. Bocian[1], G. Woźniakowski[2] and Z. Pejsak[2]
[1]*Department of Epidemiology and Risk Assessment, National Veterinary Research Institute, Al. Partyzantów 57, 24-100 Puławy, Poland,* [2]*Department of Swine Diseases, National Veterinary Research Institute, Al. Partyzantów 57, 24-100 Puławy, Poland; zpejsak@piwet.pulawy.pl*

The emergence of African Swine Fever (ASF) in Georgia in 2007 with subsequent unprecedented spread to other countries of the Caucasus regions, Russia, Ukraine, Belarus, Baltic states and Poland, has marked a new era in ASF history. During 15 months after the first detection of ASF in Poland, 48,440 of samples were tested virologically and serologically. So far (as of 5th of May 2015), 57 ASF cases involving more than 100 wild boar as well as three outbreaks in backyard pigs have been found. The total affected area covers approximately 1000 km^2. A strong positive correlation between wild-boar density and the number of ASF cases was demonstrated. The mortality in wild boar is thought to exceed 90% while lethality is <5%. As opposed to other infectious diseases (e.g. classical swine fever, foot-and-mouth disease), ASF does not seem to be highly contagious and the spread is very slow. Factors that contribute to the long-term occurrence of ASF in the region appear to be a high wild-boar population density in the affected area, potential re-introductions of the virus from neighboring countries (especially with unknown status regarding ASF) and long environmental persistence of the virus. On the contrary, the determinants that hamper the fast virus spreading beyond the infected territory involve a very specific social behavior of wild boar (small spatial dispersal ranges), low wild-boar population density in most of the areas surrounding the affected zone and a low number of domestic pig holdings in the region. However, poor biosecurity measures in backyard pig holdings pose the major risk for the further escalation of the epidemic. Therefore, a new legislation introducing strict biosecurity standards in all pig holdings has been passed recently.

Monitoring of endemic diseases in pig herds
D. Maes
Unit of Porcine Health Management, Department of Reproduction, Obstetrics and Herd Health, Faculty of Veterinary Medicine, Ghent University, Salisburylaan 133, 9820 Merelbeke, Belgium; dominiek.maes@ugent.be

Endemic diseases such as porcine respiratory disease complex and intestinal disease cause major production losses in intensive pig production systems. To implement proper control measures, it is necessary to monitor endemic diseases in pig herds. Clinical symptoms can be monitored by means of indexes e.g. coughing index or by automated systems. Treatment incidences of antimicrobials are also relevant and can be used for benchmarking. Changes in feed and water intake may be early signs of disease onset. Production data such as weight gain, feed conversion and mortality are important economic parameters. Necropsies of dead pigs may be very helpful. However, lesions may not appear typical for one pathogen as mixed infections are common. Prevalence and severity of respiratory tract lesions can be monitored by performing slaughter checks. This is cheap and simple, and also allows to detect subclinical infections. Taking blood samples of different age groups for serology (cross-sectional and/or serial sampling) allows to investigate the infection pattern in groups of pigs. Different age groups may also be sampled to detect pathogens (virus, bacterium) or the genetic material (PCR) in blood (e.g. viremia for PRRSV), tonsil, nasal, pharyngeal, tracheal samples, BAL fluid, feces, etc. This allows to further characterize the pathogen, and in case of bacteria, to test antimicrobial sensitivity. Blood samples or other samples may be pooled to save costs, and for some pathogens (e.g. PRRS, PCV2), oral fluids have become more popular. In conclusion, there are several ways to monitor endemic diseases in pig herds, each of them providing slightly different information. The pathogens to be included and the intensity of monitoring depend on the herd situation and the specific aims e.g. to determine whether a herd is infected with a pathogen, and if so, which age groups are affected, or to monitor the progress of disease control and eradication programs.

Introduction to genomic selection: current knowledge and troublespots for commercial implementations

I. Misztal
University of Georgia, Animal and Dairy Science, Athens, GA 30602-2771, USA; ignacy@uga.edu

Genomic selection (GS) was applied to all major farm species including dairy, beef, pigs and chicken. Initial methodology was a multi-step procedure consisting of BLUP evaluation, extraction of pseudo-observations for genotyped animals, running genomic predictions via estimation of SNP effects or genomic relationships, and combining genomic and conventional predictions in an index. A newer single-step (SS) procedure works as BLUP with combined pedigree and genomic relationships. Experiences with GS indicate that: (1) a few thousand genotyped animals are required for a reasonable increase in accuracy over BLUP; (2) SNP effects are not QTLs; (3) differences between methods based on estimation of SNP effects and genomic relationships are small; (4) gains with SNP chip >50k are minimal; (5) gains from genotyping low-accuracy animals are small; (6) across breed/line prediction does not work; (7) genomic predictions decay over time requiring constant phenotyping; (8) single-step procedure is much simpler to use than multi-step and usually more accurate. Experiences are in agreement with a theory that breeding values are defined by 10,000 or fewer independent chromosome segments. Gains in accuracy due to genomics for different types of animals can be explained by decomposition of GEBV due to phenotypes, pedigree and genomic information. Practical applications of GS indicate many potential problems and potentially genomic prediction that is less accurate than the conventional prediction. Problems include genotyping errors, pedigree errors, incorrect imputation, issues with deregression, inaccurate index, etc. Any application of GS requires a proper validation; validation technique used in dairy may be unsuitable for pigs. For commercial use, the most flexible yet least trouble-prone methodology is a single-step procedure that, with recently developed algorithm for inversion of genomic relationship matrix, can accommodate any number of genotypes at low cost.

Genomic selection in practice

T.J. Lawlor
Holstein Association USA, 1 Holstein Place, Brattleboro, VT 05302, USA; tlawlor@holstein.com

Genomic selection of dairy cattle has proved to be highly successful. The use of DNA markers to track the inheritance of both desirable and undesirable segments of a chromosome has a great intuitive appeal to farmers. Current uptake of this technology exceeds more than 27,000 newly genotyped Holsteins per month with over 800,000 animals genotyped as of June, 2015. Genomic predictions are used for the selection of elite animals to be parents of the next generation; and for making management decisions within a commercial herd on which animals to keep as a replacement, as well as the quality and characteristics of the mate to be used. Parent discovery or pedigree correction is made for about 10% of the commercial cattle. Results are provided weekly to a herd or breeding organization for preliminary decisions on animals that should be kept or culled. Sourcing high genetics is assisted by the monthly ranking of all newly tested animals. All genetic information is refreshed at the full genetic update conducted very 4 months. Mating decisions are aided by easy access to qualitative genetic information and the relationship amongst potential mates. Reporting of haplotypes and individual gene tests is freely available via an online 3-generation Family Tree. An inbreeding calculator is one of the most-used resources on our web site, with more than 7,000 matings investigated in March 2015. All farmers are benefiting from the shortening of the generation interval and more rapid genetic gain. A decrease in the number of farmers contributing elite genetics is occurring. Producing top genetics requires access to early semen of the top young bulls; the financial resources to pay for advanced reproductive technology, such as, IVF and ET; and the availability of a large number of recipients.

Lessons from practical implementation of genomic evaluation across species

D.A.L. Lourenco
University of Georgia, Department of Animal and Dairy Science, Athens, GA, 30602, USA; danilino@uga.edu

Practical results from implementing genomic selection (GS) in dairy cattle, beef cattle, chicken, and pigs are presented; the single-step GBLUP is the method of choice. In dairy genomic evaluations, bias is observed when genotyped animals have different length of pedigree. We tested the impact of removing old generations of pedigree and phenotypes on genomic evaluations of young animals for US and Israeli Holsteins, and commercial pigs. Retaining only 3 generations of phenotypes and 2 extra generations of pedigree did not decrease accuracy. Removing old data revealed issues due to population structure. One common issue in GS is whether populations with a small number of genotyped animals can benefit. For a small Israeli Holstein population with 1,300 genotyped bulls, accuracy of genomic evaluation compared to traditional evaluation increased up to 10 points, depending on the trait; adding genotypes for 300 cows had a positive but small effect on accuracy. In another species, genotyping females can be beneficial if they have high contribution to the population. For broiler chickens, prediction accuracy on young birds increased up to 20 points if both sexes were genotyped. If only one sex was genotyped, the benefit was higher for that sex. Another issue in GS is how to do crossbred evaluations, since breeds may have different allele frequencies. We tested breed-specific genomic relationship matrices for GS in a pig population; however, the standard genomic relationship matrix that uses an average allele frequency was still the best option. Genotyping strategy is not well defined in all species. For American Angus, we tested the effect of genotyping more and less influential animals. We observed that the increase in prediction accuracy given by 2,000 top accuracy bulls was the same as given by 30,000 lower accuracy animals. As the number of genotyped animals increases fast for American Angus, we tested 2 methods: the algorithm for proven and young and indirect predictions for young animals. Both methods gave the same accuracies as the regular ssGBLUP at lower computing cost.

Current status of genomic selection in Polish dairy cattle breeding

M. Pszczola
Poznan University of Life Sciences, Department of Genetics and Animal Breeding, Wolynska 33, 60-637 Poznan, Poland; mbee@up.poznan.pl

Genomic selection (GS) revolutionized cattle breeding industry. Research on implementing GS in Polish dairy cattle breeding started in 2008. In 2012, Poland joined EuroGenomics. Due to law restrictions, using genomic breeding values as official proofs was not possible until 2014. First official genomic evaluation in Poland was published in in summer'2014. The genomic breeding values are estimated in three steps: (1) direct genomic values (DGV) are estimated using deregressed national proofs; (2) DGV are combined with conventional national breeding values (EBV) to national genomically enhanced breeding values (GEBV); (3) national GEBV are provided to Interbull for estimating international GEBV using Genomic Multiple Across Country Evaluation (GMACE). Bulls in Polish genetic evaluation are ranked based on international GEBV, national GEBV or Interbull's international conventional EBV. International GEBV are official genomic proofs in Poland and only when these are not available, national GEBV can be used instead. In the December'2014 Interbull's evaluation, about 8% (815) of all evaluated bulls were Polish. National Polish selection index (PF) includes production, functional and health traits. Top ten bulls evaluated in Poland (April'2015 evaluation) ranked on PF had genomic proofs and no daughters. Their average PF value was 147.3 (min. 144, max. 153). Top ten bulls ranked based on international EBV had an average PF of 137.5 (min. 135, max. 142). The average international GEBV for PF for top ten bulls evaluated outside of Poland was 154.3 (min. 153, max. 158). The average reliability for national GEBV for milk production was 0.65 and for international GEBV was 0.68. Young genomically proven bulls are better than daughter-proven ones, yet, their reliability is lower. High breeding values promise the uptake of the genomic technology by the Polish farmers. However, lower reliability and high semen price can slow down this process. Thus special efforts to communicate benefits of GS to farmers are needed.

Herd health challenges in high yielding dairy cow systems

R.F. Smith
University of Liverpool, School of Veterinary Science, Leahurst Campus, Neston Cheshire, United Kingdom;
robsmith@liv.ac.uk

In high yielding cows the delay in peak dry matter intake compared to the nutritional demands of peak milk production coupled with the digestive requirements for adequate fibre to maintain rumination and rumen pH results in a period of negative energy balance. This in turn may result in body condition score loss and an increased risk of lameness and reduced fertility. There is a growing appreciation of the interactions between dry period and lactation nutrition, body condition score and the key health issues of reproduction, lameness and mastitis. There are both genetic and environmental risk factors for these diseases and the success of management of these risks determine the welfare of cows and commercial success of farming systems. Herd management has a large influence over the possibility of an animal and herd meeting their genetic potential for production and fertility. The term environment ranges from local geography, determining crops, growing and grazing season; to building design, determining cow comfort and risk of many diseases. Food security will drive the development of farming systems but, whilst cost and availability are key issues, a proportion of consumers are influenced by perceptions of animal welfare in different systems. Retailers are starting to require objective welfare assessments as evidence of successful management of animal genotype and environment interactions to meet these consumer expectations. Contrasting data regarding yield and fertility will be used as an example of the complex interactions that need to be appreciated to meet the GplusE objectives to produce a road map to achieve high welfare, sustainable high yielding dairy systems.

Phenotypic interrelationships between parameters predominately in milk – the GplusE project

J.K. Höglund, M.T. Sørensen and K.L. Ingvartsen
Aarhus University – Foulum, Dept. of Animal Science, Blichers Allé 20, Box 50, 8830 Tjele, Denmark;
kli@anis.au.dk

A cow's ability to reproduce and stay healthy is essential for efficient milk production and is a key biological feature on which the dairy industry is based. Risk profiles of particular diseases in particular periods have been described and the period after calving represents a higher risk of disease. In fact, the risk for all production diseases is highest during the first weeks after calving. The phenotyping of cows so far has primarily been focused on performance data such as milk yield and composition, reproductive data, veterinary records etc. These phenotypes are highly influenced by management strategies and are often suboptimal and have not been able to address challenges to combat subclinical states causing increased risk of disease, suboptimal performance and reproduction failure. This calls for physiologically based measures from easily accessible samples, i.e. milk, which can be collected and analysed automatically and used in real-time. Challenges for future disease prevention and management of individual dairy cows are believed to include monitoring of physiological imbalance and understanding how e.g. nutrition and management of the individual cow should be changed to bring the cow in balance. To address these challenges WP3 of the GplusE project will focus on milk biomarkers that can be used as predictors of subclinical diseases, create milk MIR equations as predictors of key phenotypes of interest, provide phenotypes for genetic studies and develop management protocols and also provide tissue samples in order to provide a deeper biological understanding of the underlying physiology. A detailed study that includes measures of milk biomarkers and cow phenotypes is carried out on 200 cows placed in 6 countries. Based on data analysed from these 200 cows it will be decided which milk biomarkers and cow phenotypes shall be measured on 10000 cows that will be enrolled at a later stage to substantiate the data foundation.

On the use of novel milk phenotypes as predictors of difficult-to-record traits in breeding programs

C. Bastin[1], F. Colinet[1], F. Dehareng[2], C. Grelet[2], H. Hammami[1], H. Soyeurt[1], A. Vanlierde[2], M.L. Vanrobays[1] and N. Gengler[1]
[1]University of Liège, Gembloux Agro-Bio Tech, 5030 Gembloux, Belgium, [2]Walloon Agricultural Research Centre, 5030, Gembloux, Belgium; catherine.bastin@ulg.ac.be

In the genomic era, the routine collection of accurate phenotypes remains a major challenge, especially for difficult- or expensive-to-record traits such as health, fertility and environmental footprint. Previous research demonstrated the opportunity of using milk biomarkers as predictors of these traits. Hence, milk biomarkers can be useful in the prediction of genetic merit for direct fertility, health, and environmental footprint traits as long as they are easier to measure, heritable, and genetically correlated. Recently, mid-infrared (MIR) spectrometry was recognized as a rapid and cost-effective tool to collect routinely a wide range of novel milk phenotypes including fine milk composition, milk technological properties and cow's physiological status. Investigations on the genetic relationship between fertility and milk phenotypes indicated that milk phenotypes related to the negative postpartum energy balance and the body fat mobilization (e.g. fat to protein ratio, fatty acids profile) could be used to improve fertility. Moreover, contents in milk of ketone bodies could be useful for breeding cows less susceptible to ketosis. Although the genetic association of milk phenotypes other than somatic cell count and mastitis should be further investigated, a wide range of traits (e.g. lactoferrin, minerals, citrate) are worth to be considered as potential indicators of udder health. Finally, fatty acids can be used as predictors of methane production and MIR-prediction of methane production has been demonstrated as genetically variable and heritable. Further studies will allow to (1) grasp underlying associations among novel milk phenotypes and health, fertility, and environmental footprint traits, (2) conduct genetic and genomic studies for these traits, and (3) include these traits in broader breeding strategies.

Author index

Apper, E.	123	Bapst, B.	296, 506
Appuhamy, J.A.D.R.N.	364	Barajas, R.	183, 534
Araba, A.	467	Baratte, C.	301
Arango, J.	259, 305	Barba, C.	209
Argente, M.J.	536	Barbari, M.	145
Arias, G.	421	Barbat, A.	153, 214
Arias, R.	209	Barbat, M.	251
Arizala, J. Angulo	360	Barbe, F.	119
Arnau-Bonachera, A.	501	Barbosa, G.H.V.	526
Arnesson, A.	418	Barbotte, L.	510
Arney, D.	373	Bareille, N.	466
Arnott, G.	384	Barget, E.	195
Asa, R.	239, 240	Barnard, L.P.	409
Ashfaq, U.	415	Baro, J.A.	245, 310
Ashrafi, I.	535	Barragan, C.	452
Ask, B.	214, 454	Barrantes, O.	208
Astruc, J.M.	509	Barras, E.	223
Atalay, A.I.	413	Bartolomé, E.	188, 440, 441
Atxaerandio, R.	184, 272, 273, 422, 423	Bartoň, L.	211
Aubron, C.	136	Bartyzel, B.J.	183
Auer, W.	298, 304, 323	Bassano, B.	471
Aufrère, J.	270	Basso, B.	217
Avanzinelli, E.	471	Basson, R.	182
Avdi, M.	484	Bastiaansen, J.W.M.	246, 259, 310, 311, 352, 432,
Awadalla, I.	429		455
Ayrle, H.	266	Bastien, D.	269
Azor, P.	98	Bastin, C.	353, 354, 550
Azor, P.J.	441	Battacone, G.	148, 268
		Battaglini, L.M.	379, 388
B		Bauer, E.A.	324
Baban, M.	187	Bauer, J.	407
Babenko, O.	185	Baumont, R.	410
Babicz, M.	171	Baumung, R.	96, 339
Back, P.J.	523	Baur, A.	248
Back, W.	432	Bayat, A.	133
Baes, C.F.	224, 296, 506	Beaujean, N.	157
Baeten, V.	147, 400	Bebe, B.O.	342
Bagirov, V.A.	367, 483, 515	Bébin, D.	203
Bagnall, A.	309	Beck, P.	316
Bagnicka, E.	125	Bedere, N.	276
Bahelka, I.	112, 498	Bednarczyk, M.	538
Bahri Binabaj, F.	481	Bee, G.	117, 169, 375, 381, 542
Bailoni, L.	126, 379	Beerg, P.	214
Bajorinaitė, A.	166	Behmaram, R.	396
Bąk, A.	171	Beiko, R.G.	155
Baker, P.E.	178	Belabbas, H.	105, 426, 487
Bakke, K.A.	287	Belabbas, R.	536
Balcioglu, M.S.	283, 318	Belanche, A.	374, 382, 541
Baldi, A.	119	Belica, M.	544
Baldi, F.	156, 238, 252, 254, 259, 531	Bellabes, R.	424
Ball, A.J.	228	Bellagamba, F.	119
Balmisse, E.	498	Bem, N.T.	238
Bampidis, V.A.	343, 344, 345, 488	Benali, N.	420, 424, 536
Banchero, G.	509, 531	Benis, N.	263
Bangerter, E.	442	Benkhaled, A.	426
Bani, P.	133	Bennewitz, J.	97, 178, 316, 352, 478
Banos, G.	227, 309	Benoit, M.	206, 249

Ben Salem, H.	542	Blasco, A.	219, 500
Bentancur, O.	470	Blasco, I.	208
Bérard, J.	231	Blas, E.	501
Berbar, A.	536	Błaszczyk, B.	370
Berchielli, T.T.	419	Bloemhof, J.M.	202
Berentsen, P.B.M.	333	Blouin, C.	437
Beretta, V.	233, 388, 424, 425	Blunk, I.	309
Bereznowski, A.	438, 439	Bobbo, T.	521
Bergamaschi, M.	358	Bochenek, M.	533
Berger, B.	101	Bocian, Ł.	546
Bergfelder-Drüing, S.	397	Bocquier, F.	513
Bergfelder, S.	351	Bodin, L.	473, 513
Berglund, B.	215, 276, 293	Bodkowski, R.	139, 176, 235, 426, 486, 488, 496
Berg, P.	100, 403	Boegheim, I.J.M.	432
Bergsma, R.	249, 310	Boettcher, P.	96, 339
Berg, W.	89, 322	Bogaert, H.	142
Bermejo-Poza, R.	451	Bogas, C.	344
Berndt, A.	135, 239	Bogoliubova, N.V.	367
Bernini, R.	528	Bogucki, M.	323
Bernués, A.	136, 338	Boichard, D.	153, 214, 248, 346, 347, 355, 359
Berri, C.	538	Boison, S.A.	348, 351, 397
Berry, D.P.	146, 213, 227, 236, 255, 279, 458, 507, 512	Bojkovski, D.	99
		Bokkers, E.A.M.	121, 141
Bertolini, F.	152, 349	Bolado-Carrancio, A.	315
Bertol, T.M.	414, 420	Bolhuis, J.E.	109, 537, 540
Berton, M.	202	Boligon, A.A.	252
Bertrand, J.K.	502	Bonaudo, T.	204
Besbes, B.	96	Bonetti, L.	225
Bessa, R.J.B.	190, 416	Bonilha, S.F.M.	243, 293
Bessei, W.	178	Bonin, A.	134
Bestman, M.	177	Bonin, M.N.	156
Bewley, J.M.	278, 318, 320, 322	Bonnet, M.	227
Bezabih, M.	382	Bonny, S.P.F.	229
Biagini, D.	93, 413	Bonow, S.	434
Biasioli, F.	358	Bontempo, L.	364
Bidanel, J.P.	217, 307	Boonanuntanasarn, S.	372
Bielak, A.	92	Borba, H.	135, 239
Bieniek, J.	316	Borlido-Santos, J.	145
Biffani, S.	212, 358	Borrisser-Pairó, F.	162
Bigler, A.	223	Borthwick, F.	195
Bigot, G.	195	Borys, B.	475, 476, 515
Bijma, P.	179, 217, 306, 350, 392, 456, 520	Bosi, P.	152
Bikker, P.	260	Boudina, H.	427
Bilić-Šobot, D.	383	Bouffioux, A.	103
Billon, Y.	307, 312	Boughalmi, A.	467
Birnbaum, D.	478	Boukadiri, A.	523
Biscarini, F.	132, 212	Boumahdi, Z.	536
Bittante, G.	138, 358, 521	Bourgon, S.	235, 279, 288
Bizelis, I.	539	Boussaha, M.	153, 214, 346
Bjerring, M.	241, 332	Boussarhane, R.	180
Black, R.A.	318	Bouthors, D.	274
Blanc, G.	437	Boutinaud, M.	282, 360, 523
Blanchet, B.	164	Bouvier-Muller, J.	500
Blanco, M.	313, 315	Bouwman, A.C.	276, 346, 506
Blanken, K.	319	Bouyeh, M.	174, 175
Blanquet, V.	227	Bouyer, C.	227
Blaschka, A.	197, 473	Bouzidi, S.	424

Bovenhuis, H.	91, 131, 226, 352, 392, 522	Butty, A.	224
Bovolenta, S.	207	Buys, N.	99, 100, 103, 110, 113, 166, 451
Bovo, S.	152	Byrne, T.J.	512
Bowman, P.J.	151		
Boyle, L.	173, 445	**C**	
Boysen, T.J.	291	Cabello, A.	102
Bozkurt, Y.	231, 337, 345	Cabrera, V.E.	207
Braga, M.T.	199	Cabrita, A.R.J.B.	90, 416, 417
Branco, R.H.	243	Cadet, A.	401
Brand, B.	361	Caldeira, R.M.	190
Brändle, J.	143	Calnan, H.B.	199, 216
Brandt, H.	519	Calò, D.G.	152, 349
Brard, S.	432	Calus, M.P.L.	98, 246, 251, 310, 311, 350, 455,
Braz, C.U.	238, 254		505, 506
Brem, G.	483, 518	Calvo, J.H.	313, 315
Bremm, C.	199	Camargo, G.M.F.	252, 254
Bresolin, T.	254, 259	Cambra-López, M.	417
Brigden, C.V.	192	Camerlink, I.	384, 400
Brinker, T.	179	Camin, F.	364
Brinkmann, J.	435	Campidonico, L.	380
Brocard, V.	209, 329, 330	Campos, N.M.F.	199
Brochard, M.	153, 214, 355	Cañas-Álvarez, J.J.	245, 310
Brøndum, R.F.	346	Canbolut, O.	478
Bruckmaier, R.M.	151	Čandek-Potokar, M.	383
Brügemann, K.	295, 434	Candrák, J.	394
Bruininx, E.M.A.M.	410	Canesin, R.C.	293
Brule, A.	449	Cantet, R.J.C.	504
Brumby, O.	237	Cant, J.	123
Bruni, M.	274	Caorsi, C.J.	424
Brun, N.	523	Capelletti, M.	364
Bruno, J.	339	Capitan, A.	151
Brunsch, R.	89, 322	Carabaño, R.	122
Brunschwig, G.	206	Carai, P.	528
Brusselman, E.	321	Carastro, I.	528
Brüssow, K.-P.	492	Cardoso, D.F.	348
Bryla, M.	533	Cardoso, E.O.	148
Brzozowska, A.M.	425	Cardoso, E.S.	148
Buckley, F.	521	Carillier-Jacquin, C.	308
Bučko, O.	386, 389, 481, 498	Carlsson, G.	205
Budziński, W.	543	Caroli, A.M.	519
Bueno, R.S.	156	Carreño, D.	474, 495
Buergisser, M.	236	Carriquiry, M.	244, 274, 371
Bugno-Poniewierska, M.	221, 222, 285, 286, 404,	Carthy, T.R.	255
	405, 443, 494	Carvalheiro, R.	252, 254, 259, 351
Bulumulla, P.B.A.I.K.	313	Carvalho, M.E.	156
Bunnakit, K.	243	Casabianca, F.	138
Buntinx, S.E.	409	Casasús, I.	313, 315, 338
Bunz, A.	255	Cassandro, M.	142, 359
Burdych, J.	207	Cassar-Malek, I.	227
Burek, J.	142	Caste, G.	227
Burgos, A.	223	Castellano, R.	164
Burgstaller, J.	290	Castellanos, M.	102
Burke, J.L.	523	Castiglioni, B.	516
Burren, A.	223, 236, 242, 296, 442, 463, 511	Castilhos, Z.M.S.	199
Burton, I.	288	Castrillo, C.	127
Buttazzoni, L.	349	Castro-Carrera, T.	495
Büttner, K.	385, 446	Castro Filho, E.S.	389, 418

Castro, J.M.	199	Coffey, M.P.	227, 309, 356, 507, 508, 513
Caubet, C.	500	Cofré, E.	534
Cavali, J.	419	Cohoe, D.	111
Cavero, D.	246	Colaço, J.	188
Ceacero, T.M.	293	Colinet, F.G.	99, 103, 147, 353, 400, 550
Cecchinato, A.	358, 521	Collier, R.J.	362
Celi, P.	362	Collin, A.	538
Ceppi, I.	534	Colliver, K.	296
Cerisuelo, A.	417	Cone, J.W.	143, 147
Černošková, B.	462	Conington, J.	400, 507, 508, 513
Cervantes, I.	218, 223, 431, 433, 440, 442	Connor, M.L.	115
Cervera, C.	501	Cooke, J.	541
Cesaro, G.	202	Cools, A.	422
Chadio, S.	539	Copani, G.	380, 381
Chagunda, M.G.G.	264	Corazzin, M.	207
Challois, S.	164	Corbière, F.	541
Chapoutot, P.	261, 265, 269, 270, 274	Cornale, P.	379, 388
Chardulo, L.A.L.	238, 254, 259, 529	Cornelissen, J.M.R.	334
Charfeddine, N.	290	Cornu, A.	358
Charismiadou, M.	480	Correddu, F.	268
Charneca, R.	115	Correia, L.E.C.S.	529
Charton, C.	141	Corte, R.R.S.	244
Chassier, M.	457	Costa, C.N.	406
Chat, S.	523	Costa, R.B.	252, 254
Chatziplis, D.	343, 344, 514	Costa, R.L.D.	159
Chaudhry, A.S.	264	Cottrell, J.J.	120, 362, 365
Chenier, T.	279	Cournut, S.	341
Chen, X.J.	336	Coustham, V.	538
Cherrane, M.	424	Coutinho, J.	94
Chessa, S.	516, 519	Coyne, J.M.	458
Chevaux, E.	167	Craig, A.	116
Chikuni, K.	167	Craigon, J.	133
Chilibroste, P.	274	Crane, E.	235
Chirane, M.	420	Cremonesi, P.	516
Choi, C.	241	Crespi, R.	421
Cho, I.C.	157, 159	Croiseau-Leclerc, H.	141
Choi, H.B.	118	Croiseau, P.	153, 211, 214, 222, 248
Choisis, J.P.	206	Crosson, P.	94
Christen, A.-M.	290	Croué, I.	399
Christensen, O.F.	247, 455, 456	Cruywagen, C.W.	182
Christiansen, K.	433	Cruz, A.	223
Christodoulou, V.	345	Cuevas, F.J.	102
Ciappesoni, G.	258, 509	Cuitiño, M.J.	421
Cielava, L.	499	Cunha, J.A.	415
Cipolat-Gotet, C.	358, 521	Curik, I.	311
Cividini, A.	99	Curi, R.A.	529
Claramunt, M.	244, 371	Cyrillo, J.N.S.G.	243, 252, 293, 348
Clarke, I.J.	362	Czopowicz, M.	125
Clément, F.	284	Czyż, K.	139, 176, 235, 426, 486, 488, 496
Cloete, J.J.E.	450		
Cloete, S.W.P.	450, 479	**D**	
Closter, A.M.	228	Dadousis, C.	358
Cobuci, J.A.	406	Daetwyler, H.D.	151
Cocks, B.G.	225	Dagnachew, B.S.	392
Coelho Filho, R.C.	208	Dahl, G.E.	361
Coertze, R.J.	482	Dahmani, Y.	420, 424
Coffey, E.L.	279	Dalla Riva, A.	142

Dall'olio, S.	152	Depuydt, L.	163
Dalmau, A.	232	Derewicka, O.	316
Dal Prà, A.	364	Derno, M.	92, 376
Dalt, L. Da	126	De Roest, K.	230
Damasceno, F.A.	318	De Roos, A.P.W.	251
Damborg, V.K.	262	Deschenaux, H.	470
Danesh Mesgaran, M.	499	Desire, S.	400
Dänicke, S.	354, 491, 497	De Smet, S.	161, 166
Danielak-Czech, B.	171	Desrues, O.	375
Danieli, P.P.	528	Dessauge, F.	360
Danvy, S.	437	Dessie, T.	339
Dardenne, P.	132, 147, 353, 354, 400	Devinoy, E.	186, 523
Darimont, C.	354	Devries, A.	297
Das, G.	387, 492	Devyatkin, V.A.	367
Dauguet, S.	411	Diana, A.	173
Daumas, G.	165	Dias, B.G.	184
Davenel, A.	164	Diaz, C.	152, 507
David, I.	457, 473	Díaz, C.	245, 310
David, J.	305	Diaz, I.D.P.S.	252
David, V.	524	Diaz, M.M.	252
Davière, J.B.	354	Díaz, M.T.	451
Day, G.B.	318	Dibbits, B.W.	432
D'eath, R.B.	400	Diel Dc Amorim, M.	279
De Boer, H.C.	319, 320	Diercles, D.F.	252
De Boer, I.J.M.	88, 140, 141, 202, 260, 333	Difford, G.F.	92, 398
De Boever, J.	271, 411	Digiacomo, K.	120, 362, 365
Debournoux, P.	282	Dijkman, R.	450
De Boyer Des Roches, A.	449	Dijkstra, J.	143, 147
Debrecéni, O.	386, 389, 498	Diniz, F.T.	225
De Bruijn, C.	121	Dirx, N.	109
Debus, N.	513	Disenhaus, C.	276
De Campeneere, S.	112, 271, 411	Distl, O.	453
De Cara, M.A.R.	257	Doaré, S.	284
De Cosmo, A.	194, 387, 482	Dodenhoff, J.	111
Deeb, N.	349	Do, D.N.	153
Degezelle, I.	163	Dogan, C.	231
De Haas, Y.	90, 289	Dohme-Meier, F.	375, 381
Dehareng, F.	132, 147, 353, 354, 400, 550	Domig, K.J.	143
De Hollander, C.A.	446	Domingues, J.P.	204
Dekkers, J.C.M.	259, 305	Domínguez-Clavería, L.	162
Delaby, L.	276	Dominiak, K.N.	297
De La Fuente, G.	127, 374	Donaghy, D.J.	523
De La Fuente, J.	451	Döpfer, D.	290
Delarue, B.	527	Doumandji, W.	427
Delgado, R.	122	Dourmad, J.-Y.	414, 420
Deligeorgis, S.	480, 539	Driehuis, F.	319
Dell'orto, V.	119	Drögemüller, C.	511
Del Pino, M.L.	470	Drosten, C.	461
Del Prado, A.	89	Drouilhet, L.	457, 498
Demeyer, P.	321	D'souza, D.N.	162
De Moor, C.	422	Duchemin, S.I.	357
Denholm, S.J.	309, 356	Duclos, M.J.	261, 538
Deniskova, T.E.	483, 515	Ducro, B.J.	432, 433
De Oliveira, P.A.V.	414, 420	Ducrocq, V.	211, 248, 276, 399, 401, 406, 503
Depellegrin, T.	114	Dufrasne, I.	329, 330, 331
De Planell, G.	232	Duijvesteijn, N.	109, 306
Depreester, E.	142	Dunbar, K.A.	290

Dunislawska, A.	538	Estelles, F.	89
Dunshea, F.R.	120, 162, 362, 365	Estima, I.M.	271
Dupraz, P.	467	Ettema, J.F.	403
Dūrītis, I.	378	Evans, R.D.	227, 255, 279, 325, 521
Duroy, S.	196	Everaert, N.	538
Duteurtre, G.	341	Eynard, S.E.	98
Duthie, C.A.	197, 302	Ezequiel, J.M.B.	268, 389, 418, 487
Duval, J.	466		
Dvalishvili, V.G.	515	**F**	
Dwyer, C.M.	400	Fabien, C.	195
Dzama, K.	450	Fablet, J.	305
		Fahri, F.T.	162
E		Fair, M.D.	95
Eaglen, S.	101	Faivre, L.	123
Eaglen, S.A.E.	251	Fanelli, F.	152
Eastwood, C.	527	Fangmann, A.	351
Ebschke, S.	525	Fantin, R.L.	225
Echeverría, J.	531	Fantuz, F.	193, 194, 482
Eckelkamp, E.A.	318, 322	Farahat, E.S.A.	269
Edel, C.	352, 394	Fargetton, M.	282
Eding, H.	331	Farhadi, R.	535
Egger-Danner, C.	152, 290, 402, 507, 524	Farid, A.H.	176
Ehsani, A.	181	Farinós, B.	417
Eisersiö, M.	191	Farish, M.	384
Elahi Torshizi, M.	295	Farmer, L.J.	229
El-Hage Scialabba, N.	204	Faro, L. El	159, 526
El-Hofy, M.	265	Farouk, M.H.	532
Ellen, E.D.	179	Fathallah, S.	473
El-Nomeary, Y.A.	377	Faulkner, C.L.	541
El-Nor, S.A.H. Abo	414, 490	Faure, J.	164
El-Rahman, H.H.	377	Faverdin, P.	201, 301
El-Sayed, H.M.	414, 490	Fawcett, R.H.	122, 144
El-Sayed, R.R.	269	Fels, M.	385
Elsen, J.M.	460, 509	Fenelon, M.A.	146
El-Tanboly, E.	265	Férard, A.	410
Emamgholi Begli, H.	181	Ferenčaković, M.	311
Emanuelson, U.	520	Ferguson, N.	203
Emarloo, A.	173	Fernandes, A.R.	526
Emmerling, R.	352, 394	Fernandes, M.H.R.M.	475, 476
Emsen, E.	541	Fernandes, R.	190
Emsenhuber, C.	497	Fernández, J.	98, 371, 393
Endo, V.	271, 527	Fernández, M.	495
Englishby, T.M.	227	Fernando, R.L.	259, 305
Enjalbert, F.	500	Ferrand-Calmels, M.	355
Erasmus, L.J.	482	Ferrandi, C.	152, 507
Erbe, M.	155, 212, 246, 348, 351	Ferrari, F.B.	135, 239
Erhardt, G.	465, 516, 519	Ferraro, S.	193, 194, 482
Eriksson, S.	434, 444	Ferraz, J.B.S.	156
Eriksson, T.	527	Ferreira-Dias, G.	190
Ermilov, A.N.	253	Ferrer, P.	417
Ertl, P.	144, 205	Fiedler, A.	290
Ertugrul, M.	194	Fiedler, M.	322
Esfandyari, H.	456	Fiems, L.O.	271
Esmaeili, V.	530	Figueiredo, M.P.	148
Espagnol, S.	411, 468, 469	Fikse, W.F.	224, 293, 357
Espigolan, R.	238, 252, 254, 259	Fiore, E.	521
Espinoza, J.A.	149	Fiorentine, G.	419

Fischer, J.	322	**G**	
Flaga, J.	543, 544	Gabai, G.	126
Flament, J.	209	Gabrielle, B.	204
Flemetakis, E.	492	Gabr, S.A.	532
Flockhart, J.	264	Gado, H.M.	535
Flury, C.	296, 442, 506, 511	Gaillard, C.	301
Folborski, J.	182	Galama, P.J.	319, 320
Foley, C.	333, 334, 335	Galic, A.	292
Fondevila, M.	127, 428	Galimberti, G.	152, 349
Fonseca, A.J.M.	90, 416, 417	Gallo, L.	202
Fontanesi, L.	152, 349	Gallo, M.	349
Font-I-Furnols, M.	162	Gamboa, S.	181
Fontoura, A.	235, 288, 296	Gandini, G.	137
Formoso-Rafferty, N.	218	Gangnat, I.D.M.	231
Fornal, A.	285, 314, 315, 317, 436	Ganzábal, A.	509
Foroghi, A.	452, 472	García, A.	149, 209
Foroudi, F.	376	García-Cortés, L.A.	257, 371, 393
Fortina, R.	379, 388	Garcia, D.A.	254
Fortin, F.	111	García, J.	122
Foster, R.	279	Garcia, J.F.	351
Foucras, G.	500	García, M.L.	536
Fouilloux, M.N.	251, 399	García-Robledo, A.	409
Founta, A.	344	García-Rodríguez, A.	184, 266, 272, 273, 422, 423
Fourichon, C.	466		
Fournier, A.	493	Garcia, S.C.	335
Fowkes, R.C.	215	Gardner, G.E.	198, 199, 216, 229, 237
Fradinho, M.J.	190	Garnsworthy, P.C.	133
Fragomeni, B.O.	247, 252, 501, 502	Garreau, H.	457, 498
Francisco, P.	190	Garrick, D.J.	259, 296, 305, 506
Francois, J.	330	Gaspa, G.	148
François, L.	99, 100, 103	Gasperi, F.	358
Franco, J.	424	Gasselin, M.	282
Frąszczak, M.	350, 438, 507	Gasteiner, J.	126, 299, 321
Fredeen, A.	235	Gatti, F.C.	271
Freitas, A.	115	Gaudré, D.	172
Freitas-De-Melo, A.	367, 369, 485	Gaughan, J.B.	362
Fremaut, D.	163	Gauly, M.	384, 387, 447, 449, 526
French, P.	242	Gautier, J.M.	196, 541
Frevert, H.	434	Gawełczyk, A.T.	280
Friedrich, J.	361	Gelé, M.	214, 354, 355
Friggens, N.C.	141, 301, 412	Gelencsér, T.	266
Frischknecht, M.	433	Genestout, L.	510
Fritz, S.	153, 214, 248, 359	Gengler, N.	99, 103, 132, 147, 353, 354, 400, 550
Frobose, H.	162	Georgoudis, A.	343, 344
Froidmont, E.	132	Gerl, C.	342
From, T.	137	Gervais, F.	209
Früh, B.	266	Getachew, B.	495
Frutos, P.	380, 474, 495	Getya, A.	168
Frys-Żurek, M.	405	Gheorghe, A.	267
Fuchs, K.	524	Ghita, E.	267, 479
Fuerst, C.	140, 397, 402	Gholamhosseini, B.	124
Fuerst-Waltl, B.	140, 402, 524	Giambra, I.J.	519
Fulton, J.E.	259, 305	Gianesella, M.	521
Furger, M.	133	Giannico, R.	152, 350, 507
Fürst-Waltl, B.	101	Gianola, D.	358
Furtado, S.	421	Giantsis, I.A.	484
		Gieseke, D.	447

Gilbert, H.	248, 312, 457, 498	Grimberg, C.G.E.	385
Gillet, P.	269, 270, 274	Grindflek, E.	164, 445, 454
Gil, M.	162, 232	Grochowska, E.	515
Gimeno, A.	127, 428	Groenen, M.A.M.	246
Ginane, C.	380, 381	Groeneveld, E.	100, 104
Giorgi, M.	312	Groeneveld, L.F.	100
Girard, M.	375, 421	Grosse Brinkhaus, A.	381
Gittins, J.	178	Große-Brinkhaus, C.	111, 154, 397, 461
Gjerlaug-Enger, E.	164	Gross, J.J.	151
Gladyr, E.A.	253, 483, 515, 518	Gstoettinger, F.	524
Glantz, M.	357	Guan, L.	123
Gleeson, D.	281	Gubbiotti, M.	387
Glendenning, R.	237	Guemez, H.R.	183, 534
Goachet, A.G.	123	Guenic, M. Le	449
Godber, O.F.	337	Guggenberger, T.	197, 321, 473
Goddard, M.E.	346	Gugolek, A.	182
Gogol, P.	533	Guidou, C.	493
Gogué, J.	307	Guillomot, M.	157
Goldberg, V.	258	Guillou, D.	167
Golian, G.	383	Guimarães, A.L.	293
Goliomytis, M.	480	Guinard-Flament, J.	141
Gomez, A.	290	Guingand, N.	468, 469
Gómez Cabrera, A.	303	Guldbrandtsen, B.	104, 150, 278, 346, 347, 357
Gomez, E.	460	Guldimann, K.	321
Gómez, E.A.	417	Gul, M.	210, 337, 345
Gómez, M.D.	442, 443	Günther, R.	180
Gomez-Raya, L.	452	Guo, X.	455
Gómez Raya, L.	371	Gurel, F.	226
Gómez Rodríguez, J.	303	Gurgul, A.	221, 222, 285, 286, 404, 405, 443
Gondeková, M.	112	Gusev, I.V.	367, 515
González-Bulnes, A.	534	Gutiérrez, J.P.	218, 223, 442
González, E.	451	Gutzwiller, A.	421
González, J.S.	474	Guzzo, N.	126
González-Recio, O.	225		
Gonzalez-Rivas, P.A.	120	**H**	
González-Rodríguez, A.	245, 310	Haas, J.H.	300
Goossens, K.	271, 411	Habeanu, M.	267
Gordo, D.G.M.	238, 252, 254, 259	Hăbeanu, M.	270
Górka, P.	543, 544	Haddadi Bahram, A.	174, 175
Görs, S.	492	Hadjipavlou, G.	543
Gottardo, P.	146, 355	Hadlich, F.	361
Götz, K.U.	111, 288, 352, 394	Hagger, C.	442
Gourdine, J.L.	312	Hajati, H.	383
Gouw, A.	332	Halachmi, I.	89
Govignon-Gion, A.	153, 214, 346, 355	Hallé, V.	523
Graczyk, M.	102	Hamann, H.	309, 478
Grandl, F.	132, 133	Hamland, H.	454
Gras, M.	479	Hammad, M. El	532
Graunke, K.L.	361	Hammami, H.	353, 550
Gredler, B.	132, 296, 506	Hammon, H.M.	92, 495
Greeff, A. De	121	Hamouda, M. Ben	317
Green, L.	445	Hanafy, M.A.	269
Gregg, D.	189	Hansen, K.	491
Gregorini, P.	527	Hansen, M.B.	150
Gregson, E.	133	Han, S.H.	157, 159
Grelet, C.	354, 550	Ha, N.-T.	151
Grignani, C.	94	Hanzal, V.	183

Harichaux, G.	261	Hoffmann, I.	96, 137, 339
Häring, J.	442	Hofmanová, B.	439, 440
Harmon, R.J.	322	Hogeveen, H.	140
Harstad, O.M.	94	Höglund, J.K.	158, 357, 549
Hart, K.J.	374	Holand, Ø.	196
Hashemi, S.M.	181	Holgersson, A.L.	189
Hashemi, S.R.	380	Hol, J.M.G.	319
Haskell, M.J.	302, 360	Holtenius, K.	527
Haslgrübler, P.	197	Holthausen, A.	377
Hassanabadi, A.	383	Holtz, J.	196
Hassenfratz, C.	110	Homer, E.	133
Hassouna, M.	468, 469	Hong, J.K.	395
Hassoun, P.	500	Hong, J.S.	118, 120, 267, 372, 416
Haugaard, K.	356	Hoofs, A.	109
Hawken, R.	247, 395	Horan, B.	279
Hayes, B.J.	151, 225	Horecká, E.	437
Hecold, M.	438, 439	Horecký, Č.	437
Hedayat, N.	312	Hořín, P.	437
Hedlund, S.	418	Hornick, J.L.	331
Heidaritabar, M.	246, 259	Horn, M.	126, 299
Hellebuyck, S.	166	Hosseinkhani, A.	537
Hellinga, I.	432	Hosseinpour Mashhadi, M.	292, 477
Hellwing, A.L.F.	93	Hostens, M.	142
Hempel, S.	89, 322	Hostiou, N.	341
Henderson, A.	233, 388	Howard, D.M.	349
Hendriks, W.H.	143, 147, 375, 382	Hozé, C.	248, 346
Hennequet-Antier, C.	538	Huba, J.	210, 404
Henryon, M.	214, 289, 393, 456	Huber, G.	321
Heo, P.S.	118, 267	Huber, K.	354, 491
Heringstad, B.	287, 290, 291, 356, 521	Huber, R.	197, 473
Herlin, A.	327	Huebner, S.	516
Hermansen, J.E.	229	Huhtanen, P.	133
Hermans, K.	142	Hulsegge, I.	99, 100, 104
Hernández, P.	219, 500	Hulst, M.M.	412
Herold, P.	478	Hulst, S.	236
Herrero-Medrano, J.M.	453	Hurtado, A.	184, 422, 423
Herring, W.	460	Hurtaud, C.	355
Hervás, G.	380, 474, 495	Huson, H.J.	510
Heuß, E.	111	Huyen, L.T.T.	340, 341
Heuvelink, A.E.	128	Huyen, N.T.	375
Heuven, H.C.M.	446		
Heuzé, V.	261	**I**	
Heyrman, E.	110, 112, 113	Ibáñez-Escriche, N.	218, 460
Hicks, V.	356	Ibba, I.	148
Hidalgo, A.M.	455	Ibeagha-Awemu, E.M.	519
Hiemstra, S.J.	98, 100, 103, 104, 137	Iguácel, L.P.	313, 315
Hietala, P.	398	Iheshiulor, O.O.M.	353
Hillestad, B.	348	Ikinger, C.	326, 327
Hinrichs, D.	255, 291	Ikonomou, C.	480
Hinz, M.	449	Ilvonen, J.	193
Hirooka, H.	91	Imsland, F.	436
Hjortø, L.	403	Inger Lise, I.L.	117
Hocquette, J.F.	229	Ingvartsen, K.L.	549
Hoedemaker, M.	448	Invernizzi, G.	119
Hoermann, M.	524	Iolchiev, B.S.	515
Hoff, J.L.	346	Irano, N.	252
Hoffmann, G.	89	Iribarne, R.	233, 388

Ismael, A.	215	Jong, M.C.M. De	520
Ison, S.H.	540	Jonkus, D.	405, 463, 499
Iten, A.	236	Joosten, T.	332
Ito, D.	219	Joshi, J.	155
Ivanković, A.	187, 233	Jouneau, L.	157, 359
Iwaisaki, H.	314	Joy, M.	338
Iwaszkiewicz, M.	496	Judge, M.M.	236
Izumi, K.	365	Juga, J.	398, 458
		Juhas, P.	386
J		Juhás, P.	389, 498
Jacob, R.H.	216	Julliand, V.	123
Jäderkvist, K.	431	Junge, W.	300
Jaeger, M.	295	Jung, S.W.	118, 120
Jafariahangari, Y.	380	Jung, W.L.	118
Jafarikia, M.	111	Junior, G.A. Fernandes	252
Jaffrezic, F.	523	Júnior, G.A. Fernandes	254, 259
Jagannathan, V.	433	Junior, G.A. Oliveira	156
Jagusiak, W.	294	Junior, H.A. Santana	148
Jain, N.	155	Junior, L.C. Roma	423
Jakobsen, J.H.	512	Junker, K.	450
Jammes, H.	157, 186, 282, 359, 523	Jurjanz, S.	493
Janacek, R.	524	Juska, R.	471
Janczak, M.	176, 486	Juskiene, V.	471
Janczarek, I.	191, 192, 286, 366	Juste, V.	219, 500
Jang, J.C.	118, 120, 372		
Janiszewski, P.	182, 183	**K**	
Janke, D.	322	Kaba, J.	125
Jankowska, D.	488	Kadarmideen, H.N.	153, 158
Jankowska, M.	323	Kaddour, R.	427
Janns, L.	352	Kadlečík, O.	221, 394, 404
Jánová, E.	437	Kadowaki, H.	156
Jansman, A.J.M.	263	Kaić, A.	187
Janssens, G.P.J.	422	Kaiser, M.G.	154
Janssens, S.	99, 103, 110, 113, 166, 433, 451	Kalmar, I.D.	447
Jarczak, J.	125	Kamada, H.	282
Jasielczuk, I.	221, 222, 404, 405	Kamalak, A.	478
Javandel, F.	181	Kamal, M.M.	544
Jaworski, N.W.	262	Kåmark, I.	191
Jensen, D.B.	297, 303	Kamiński, S.	438
Jensen, J.	181, 307, 395, 459	Kämmerling, J.D.	180
Jensen, S.K.	262	Kamp, A.J. Van Der	332
Jeong, J.H.	267	Kang, Y.J.	157, 159
Jessop, N.	122	Kanitz, E.	492
Jessop, N.S.	144	Kański, J.	543
Jianlin, H.	313	Kapkowska, E.	102
Jiménez Sobrino, L.	302	Karaca, S.	201
Jin, S.S.	118, 120, 372	Karakach, T.	288
Jin, X.H.	267	Kara, N. Karslioglu	292
Ji, Y.J.	416	Karatzia, M.A.	366
Joerg, H.	223, 242	Kargo, M.	228, 241, 277, 403
Johanson, S.	328	Kärkkäinen, H.P.	462
Joly, T.	498	Karli, B.	337, 345
Jonas, D.	248	Karrow, N.	235
Jónás, D.	211	Kar, S.K.	263
Jones, E.	541	Karsli, T.	283, 318
Jong, G. De	331	Karsten, S.	323
Jong, G.L. De	332	Kasarda, R.	221, 311, 394, 464

Kaske, M.	522	Klopčič, M.	324, 406
Kasper, G.	319	Klopfenstein, S.	442
Kataria, K.	189	Kluess, J.	497
Katoh, K.	493	Kluivers-Poodt, M.	109
Kaufmann, F.	180	Klukowska-Rötzler, J.	286
Kaya, E.	413	Kmiecik, M.	316
Kearney, J.F.	213	Knap, P.W.	349
Kebreab, E.	364	Knaust, J.	361
Kedachi, N.	424	Knaus, W.	144, 205
Kędzierski, W.	191, 192, 286, 366	Knegsel, A.T.M. Van	140
Keles, G.	474	Kneifel, W.	143
Kelly, A.K.	230	Kneubuehler, J.	223
Kelman, K.R.	199	Knight, C.H.	363, 545
Kemp, B.	140, 537, 539	Knol, E.F.	109, 249, 306, 352, 446, 453
Kemper, K.E.	255	Knoll, A.	437
Kemper, N.	180, 385	Kobayashi, R.	493
Kenyon, F.	195	Koblmueller, M.	524
Kern, E.L.	406	Kocaman, V.	474
Kerrisk, K.L.	335	Koch, C.	286, 443
Kessler, A.M.	420	Koenig, J.E.	155
Khalil, F.A.	535	Koerkamp, P. Groot	131
Khani, A. Hossein	531, 535	Koerkamp, P.W.G. Groot	121
Kharitonov, S.N.	253	Kofler, J.	290
Kharzinova, V.R.	461, 518	Kohoutová, P.	439
Khattab, I.M.	414, 490	Koivula, M.	503
Khattab, M.S.	414	Kojima-Shibata, C.	156
Khattab, M.S.A.	490	Kok, A.	140
Khaw, H.L.	217	Kominakis, A.	539
Kheiri, F.	378	Komninou, E.	343
Khempaka, S.	177, 343, 494	Kompan, D.	99, 101
Khodayi-Motlagh, M.	312	Kongsro, J.	164
Kholif, A.M.	269, 414, 490	König, S.	295, 401, 434
Kholif, S.M.	269	König Von Borstel, U.	295, 434
Khorshed, M.M.	414	Koning, D.J. De	114, 347, 357, 392, 455
Khorsheed, M.M.	490	Koning, K. De	332
Kia, H. Daghigh	531, 535, 537	Konjačić, M.	233
Kiarie, E.	370	Kor, A.	201
Kiefer, H.	157, 186, 282, 359, 523	Korpa, V.	285
Kilbride, A.	445	Korwin-Kossakowska, A.	438, 439
Kim, D.	142	Koseniuk, A.	485, 517
Kim, D.W.	395	Koster, J. De	142
Kim, S.	241	Kostyunina, O.V.	461
Kim, Y.K.	157, 159	Kovalchuk, I.	185
Kim, Y.M.	395	Kowalik, S.	286
Kim, Y.Y.	118, 120, 267, 372, 416	Kowalski, Z.M.	543, 544, 545
Kinal, S.	176	Kozioł, K.	316
Kinoshita, A.	491	Kozubska-Sobocińska, A.	160, 171
Kirchner, R.	221	Krattenmacher, N.	399, 435
Kiriakopoulos, A.	539	Kravchenko, O.	168
Kiritsi, S.	488	Kreis, A.	223
Kišidayová, S.	128, 129	Kreuzer, M.	117, 132, 133, 231, 375, 542
Kjaer, J.	178	Kreżel-Czopek, S.	323
Kjellberg, L.	193	Krieter, J.	107, 298, 304, 323, 385, 446
Klambeck, L.	180	Kristensen, A.R.	298, 303
Kleinman-Ruiz, D.	393	Kristensen, L.M.	92
Klemetsdal, G.	348	Kristensen, T.	88
Klootwijk, C.W.	333	Krogmeier, D.	288

Kron, R.	189
Krpálková, L.	207, 211
Kruczek, K.	171
Kruijt, L.	263
Krupa, E.	280, 464, 466
Krupová, Z.	280, 464
Krzyścin, P.	232, 240
Kuba, J.	370
Kubale, V.	383
Kuchida, K.	239, 240
Kuczynska, B.	475
Kuhla, B.	92
Kühn, C.	361
Kumagai, H.	91
Kunz, H.-J.	522
Kunz, P.	236
Kurt, O.	413
Kuwahara, A.	490
Kvapilík, J.	207, 211
Kwiatkowski, S.	369
Kwon, H.	241
Kyriazakis, I.	203, 408

L

Labrune, Y.	457
Lacasse, P.	360
Lafri, M.	536
Lagarrigue, S.	538
Laga, V.	514
Lagka, V.	344, 488
Lagriffoul, G.	196
Laliotis, G.P.	484
Lalöe, D.	523
Lamadon, A.	206
Lambe, N.	195
Lambertz, C.	384, 387, 447, 449, 526
Lambton, S.L.	178
Lammers, A.	537, 539
Lamont, S.J.	154
Lam, S.	123, 279, 288, 362
Lang, A.	467
Langbein, J.	361
Lanzén, A.	184
Larrán, A.M.	371
Larroque, H.	141, 308
Larzul, C.	110, 248, 457
Lashmar, S.	511
Lassen, J.	90, 92, 130, 289, 293, 398
Lasseur, J.	136
Latacz, A.	363, 369
Laterza, A.	396
Laue, H.J.	448
Laustriat, M.	411
Lauwaerts, A.	422
Lauzurica, S.	451
Lawlor, T.J.	245, 501, 502, 547
Lawson, C.	215

Lazar, C.	479
Lazzaroni, C.	93, 413
Lebourhis, D.	157, 523
Lebret, B.	164
Lecarbonnel, J.	538
Lecchi, C.	119
Leclerc, H.	234
Leconte, Y.	217
Le Dividich, J.	115
Leeb, C.	108
Leeb, T.	433
Lee, C.H.	416
Leegwater, P.A.J.	432
Lee, J.	241
Leen, F.	163
Lefebvre, R.	234
Le Floch, N.	172
Lefter, N.A.	267
Legarra, A.	247, 248, 456, 460, 501, 502, 505
Legrand, I.	229
Lehotayová, A.	498
Le, H.T.	459
Leiber, F.	175, 266
Leinonen, I.	203, 408
Leisen, M.	300
Lekeux, P.	328
Lembeye, F.	523
Leme, P.R.	238
Le Morvan, A.	410
Leonard, N.	173
Lepesioti, S.	480
Lerch, S.	493
Leroy, G.	96, 97, 224, 339
Leso, L.	145
Lessire, F.	331
Lessire, I.	330
Lessire, M.	261
Letaief, R.	222
Leurent-Colette, S.	276
Leury, B.J.	120, 162, 362, 365
Lewczuk, D.	438, 439
Lewis, R.M.	512
Lewis, S.	224
Lherm, M.	203
Li, B.	293
Lidauer, M.H.	293
Lien, S.	454
Liepa, L.	378
Ligda, C.	343, 484
Lillehammer, M.	390
Lima, A.L.	268, 487
Lind, A.	108
Lindahl, C.	108
Lindgren, G.	431
Link, Y.	323
Lisiak, D.	515
Liu, H.	393

Matebesi-Ranthimo, P.A.M.	479	Meuwissen, T.H.E.	96, 97, 116, 186, 250, 253, 348,
Mateus, L.	190		353, 390, 391, 392, 431
Mathur, P.K.	453	Meyer, C.	385
Matilainen, K.	458	Meyer-Hamme, S.	384
Maton, C.	513	Meyer, U.	491, 497
Matsuda, H.	314	Mézec, P. Le	249
Matsui, Y.	282	Mezzullo, M.	152
Matsumoto, M.	156	Micek, P.	425
Mattalia, S.	234	Michaličková, M.	112, 280
Matthews, L.	308	Michie, C.	197
Matulaitis, R.	471	Mielczarek, M.	350, 507
Maurer, V.	266	Mielenz, M.	491
Maxa, J.	299	Migdał, A.	316
Maximo, B.	271	Migdał, Ł.	316
Mayeres, P.	103	Migdał, W.	316
Mayerhofer, M.	524	Migliorini, E.	225
Mayer, M.	309	Mignon-Grasteau, S.	305, 538
Mazanek, K.	256	Migose, S.A.	342
Mazaraki, K.	344	Mihalcea, T.	479
Mazon, M.R.	238	Mihina, S.	210
Mazzoni, G.	152	Mikkelsen, A.	241
Mcbean, D.	195	Milerski, M.	139, 486
Mccracken, D.	195	Miller, S.	123, 279, 288, 296, 362
Mcdermott, A.	146	Millet, S.	110, 112, 113, 163, 166
Mcdermott, K.	507	Miltiadou, D.	528
Mcgilchrist, P.	198, 228	Mimosi, A.	379, 388
Mchugh, M.	196	Minero, M.	325
Mchugh, N.	458, 507, 512	Minery, S.	355
Mclaren, A.	195, 508, 513	Minet, J.	331
Mcneilly, T.N.	309	Minozzi, G.	152, 350, 507
Mcparland, S.	132, 146, 356	Mirabito, L.	449
Medugorac, I.	101	Miraglia, N.	283, 471
Meerburg, B.G.	260	Miranda, G.	214
Megens, H.J.	246	Mirczuk, S.	215
Mehdi, W.G.	542	Mirkena, T.	402, 444
Meier, K.	390, 391	Mirzai, M.	175
Meijerink, M.M.M.	450	Miszczak, M.	171
Meikle, A.	244, 371	Misztal, I.I.	245, 247, 256, 460, 501, 502, 547
Melizi, M.	105, 487	Mitchell, M.C.	309
Membrillo, A.	443	Mitsopoulos, I.	344, 345, 488, 514
Mendes, E.D.M.	135, 239	Miyauchi, K.	219
Mendes, F.B.E.	148	Miyazaki, Y.	219
Mendes, L.B.	321	Moakes, S.	208
Menéndez-Buxadera, A.	440	Mobæk, R.	196
Mengistu, G.	382	Modesto, V.C.	419
Menoyo, D.	122	Moerman, S.	142
Menz, C.	89	Mogensen, L.	88, 229
Mercadante, M.E.Z.	243, 252, 293, 348, 526	Mohamad Salleh, S.	158
Mercan, L.	105	Mohamed, A.G.	150
Mercat, M.J.	110, 164, 248, 457	Mohamed, M.	429
Merckx, W.	422	Mohamed, M.I.	377
Meslier, E.	410	Mojtahedi, M.	430, 499
Messad, S.	341	Molee, A.	161, 343
Mészáros, G.	213, 351, 397	Molee, W.	177, 343, 494
Métayer-Coustard, S.	538	Molina, A.	245, 310, 436, 440, 441, 442, 443
Metges, C.C.	376, 492	Molina, M.	281
Metz, G.A.	185	Molina, M.G.	396

Mollenhorst, H.	260	**N**	
Moll, J.	224	Nádaský, R.	237
Monroe, G.R.	432	Nadeau, E.	418
Monsignati, I.	389, 418	Nagata, R.	124
Montanholi, Y.	123, 235, 279, 288, 296, 362	Nagy, W.M.A.	532
Montazer-Torbati, F.	523	Nakajima, I.	167
Monteiro, A.N.T.R.	414, 420	Na-Lampang, P.	169
Monteiro, F.M.	243	Nam, S.O.	372, 416
Monziols, M.	165	Napoléone, M.	136
Moore, K.L.	227, 507	Narcy, A.	261
Moradi, M.H.	312, 518	Nasr, J.	378
Moraes, E.G.	529	Nassiri-Moghaddam, H.	383
Morante, R.	223	Nassiry, M.R.	383
Morantes, M.	209	Natonek-Wiśniewska, M.	232, 240
Moravčíková, N.	168, 221, 237, 311, 394	Nauwynck, H.	447
Moravej, H.	428	Navajas, E.A.	258, 509
Morek-Kopec, M.	407	Navarro, A.N.O.	271
Moreno, C.	310, 510	Navrátilová, A.	168
Moreno, S.	102	Nawaz, H.	415
Morgan-Davies, C.	195	Negro, S.	436, 443
Morgan, K.	191, 193	Negussie, E.	132
Morgante, M.	521	Nejati-Javaremi, A.	518
Moroldo, M.	538	Neju, W.	323
Morshed, W.	145	Neser, F.W.C.	95, 401
Morsy, T.A.	150	Netshipale, A.J.	346
Moschou, K.	480	Neuditschko, M.	296, 433, 442, 511
Moser, B.	340	Neuenschwander, S.	224
Moser, D.W.	502	Neuhoff, C.	154, 461
Mostert, P.F.	141	Neves, H.H.R.	351
Mottet, A.	260	Neveu, A.	523
Mottram, T.T.F.	300	Neveux, A.	282
Moulin, C.H.	136	Newbold, C.J.	374, 379, 382, 541
Mounaix, B.	449, 524	Nguyen, M.	523
Moureaux, S.	249	Nguyen, T.L.T.	229
Mouresan, E.F.	245, 310	Nicholas, P.K.	208
Mroczkowski, S.	515	Nicodemus, N.	122
Mrode, R.	508	Nicolazzi, E.L.	152, 350, 358, 507, 510
Mucha, S.	508, 513	Nicol, C.J.	178
Muche, S.	418	Nicoloso, C.S.	208
Mueller, A.	242	Niderkorn, V.	380, 381
Mulder, H.A.	217, 306, 308, 453	Nielsen, B.	454, 456
Muller, A.	204	Nielsen, P.	290
Müller, K.	107, 290, 298	Nielsen, U.S.	250
Müller, M.A.	461	Nielsen, V.H.	104
Muller, N.	110, 209, 457	Niero, G.	355, 359
Müller, U.	100	Ni, G.	212
Munilla, S.	245, 310	Nijman, I.J.	432
Munk, A.	403	Nilsson, K.	216, 459
Munro, J.	123, 235, 279, 362	Nirea, K.G.	391
Murani, E.	185	Nishihara, K.	493
Murawska, D.	182, 183	Nishio, M.	156
Muroya, S.	167	Nishiura, A.	294
Murphy, B.	230	Nistor, E.	345
Muth, P.C.	340	Nitas, D.	345
Mu, W.	202	Niżnikowski, R.	200, 475, 476
Mynhardt, H.	482	Noguera, J.L.	460
		Norberg, E.	228, 459

Romanov, V.N.	367	Sahraoui, L.	427
Romanzin, A.	207	Saidj, D.	420, 424
Romo, J.A.	183, 534	Saintilan, R.	251, 399
Romo, J.M.	183, 534	Sakai, E.	219
Ronchi, B.	198, 528	Sakamoto, L.S.	135, 239
Rooke, J.A.	144	Sakoda, K.	239
Röös, E.	205	Salaet, I.	417
Ropota, M.	267, 270, 479	Salahshour, A.	452, 472
Rosa, G.J.M.	254, 358	Salau, J.	300
Roschinsky, R.	340, 342	Salazar-Villanea, S.	410
Rosé, R.	312	Saleh, H.M.	414, 490
Rosner, P.M.	467	Salimei, E.	193, 194, 482
Ross, D.W.	197	Salles, F.A.	423
Rossi, C.M.	198	Salles, M.S.V.	423
Rossoni, A.	152, 507	Salman, F.M.	377
Rotaru, C.	479	Salvador, S.	207
Roth, A.	520	Salvaing, J.	157
Rothschild, M.F.	510	Samorè, A.B.	152, 349
Roubos-Van Den Hil, P.J.	121, 128	Sánchez, J.P.	220, 458
Rouillé, B.	269, 270, 274, 355	Sánchez, M.J.	441, 442
Roulenc, M.	203	Sanchez, M.P.	153, 214, 222, 346
Royer, E.	167, 172	Sandberg, S.	191
Rozbicka-Wieczorek, A.	480	Sanders, M.E.	450
Rubin, C.-J.	453	Santana, M.C.A.	148, 419
Rubiś, D.	106, 517	Santana, M.H.A.	156
Ruesche, J.	457, 498	Santana, R.S.S.	415
Rufer, B.	463	Santina, N.V.	225
Ruis, M.	448	Santis, L.	281
Ruiz, I.	117	Santos, A.S.	94, 188
Ruíz, R.	184, 272, 273, 422, 423	Santos, B.F.S.	512
Rupp, R.	500, 508, 510	Santos, D.J.A.	351
Ruść, A.	438	Saran Netto, A.	423
Russell, G.C.	309	Saremi, B.	497
Russo, V.	349	Sargolzaei, M.	346
Russo, V.M.	120	Sarto, P.	313, 315
Rust, M.	223	Sartori, C.	438
Rutherford, K.M.D.	360, 540	Sasaki, K.	167
Ruuls, L.	121	Sasaki, N.	129
Rychen, G.	493	Sasaki, O.	294
Rychlik, T.	485, 517	Satoh, M.	156, 294, 464
Rydhmer, L.	391, 402, 444	Sato, S.	124
Rykov, R.A.	367	Sauerwein, H.	354, 495, 497
		Sauvant, D.	261, 265
S		Savoini, G.	119
Saastamoinen, M.	285	Sawa, A.	323
Sabbagh, M.	437	Sawday, W.	513
Sabboh-Jourdan, H.	261	Sayin, C.	210
Sadi, F.	428	Sayre, B.	510
Sadri, H.	495, 497	Scarpino-Van Cleef, F.O.	389
Saedi, S.	531	Schader, C.	204
Sáenz, L.	281	Schaginger, M.	524
Sæther, N.H.	100	Schatzmayr, G.	497
Saetnan, E.R.	374	Schauerhuber, N.	497
Saha, C.	322	Schauf, S.	127
Sahana, G.	130, 150, 278, 346, 347, 357	Schaumberger, S.	497
Sahin, E.	283	Scheel, C.	304
Sahin, M.	413, 478	Schellander, K.	111, 154, 397, 461

Takada, L.	238, 254
Takahagi, Y.	156
Takasuga, A.	314
Takeda, H.	294
Takemoto, M.	219
Takeo, M.	240
Tamou, C.	338
Tanghe, S.	166
Taniguchi, Y.	314
Tapio, I.	134
Taraba, J.L.	318, 320, 322
Tarabay, W.	533
Tasca, C.	500
Tascioglu, Y.	210, 337, 345
Taylor, J.F.	346
Tazkrit, N.	420
Teenstra, E.	469
Teinturier, C.	274
Teixeira, I.A.M.A.	475, 476
Tejeda, J.F.	460
Tejerina, F.	102
Tello, E.	338
Temim, S.	420, 424, 427
Tempelman, R.J.	289
Tenghe, A.M.M.	276
Tesfaye, D.	461
Tesseraud, S.	538
Tessier, M.	222
Tetens, J.	151, 399, 433, 435
Teuscher, F.	158
Thaller, G.	151, 255, 291, 300, 399, 433, 435
Theau-Clément, M.	498
Theis, S.	112
Thénard, V.	206
Theodoridou, K.	427
Thirion, M.	524
Thirstrup, J.	390
Tholen, E.	111, 154, 351, 397, 461
Thoma, G.	142
Thomas, C.	89
Thomasen, J.R.	403
Thomas, G.	290
Thomas, S.	301
Thomé, J.P.	493
Thompson, J.M.	228
Thorp, Z.	215
Thumanu, K.	161
Thurman, K.	112
Thurner, S.	299
Thutwa, K.	450
Tichit, M.	204
Tietgen, B.	448
Tircazes, A.	498
Tixier-Boichard, M.	95, 305
Todini, L.	193, 194, 482
Toft, N.	303
Toma, L.	468

Tomás, C.	371
Tomaszewska, K.	183
Tomka, J.	481
Tonhati, H.	252, 348
Tonon, A.	364
Tontini, J.F.	199
Tonussi, R.L.	254, 259
Toral, P.G.	380, 474, 495
Tormo, E.	270, 274
Toro, M.A.	371, 393
Toro-Mujica, P.	470
Torres, Y.	149
Tortereau, F.	510
Tosser-Klopp, G.	508, 510
Tost, J.	157
Trakooljul, N.	185
Trakovická, A.	168, 221, 237, 311, 394
Tran, G.	261, 269
Traspov, A.A.	461
Traulsen, I.	107, 298, 304
Trégaro, Y.	467
Trevisi, E.	133
Trevisi, P.	152
Tribout, T.	251, 307
Trindade, H.	94
Troch, T.	147, 400
Troquier, C.	203
Trou, G.	209
Troy, S.	197
Trzcinska, M.	533
Tsiokos, D.	484
Tsiouris, V.	125
Tsiplakou, E.	492
Tsuruta, S.	245, 247, 256, 502
Tullio, R.R.	135, 239
Tůmová, E.	419
Turner, S.P.	360, 384, 400
Turri, F.	516
Tusell, L.	248
Tuyttens, F.	110, 113
Tuzun, C.G.	231
Tzamaloukas, O.	528
Tzouramani, I.	484

U

Uberti, M.	145
Udała, J.	370
Uddin, J.	461
Uenishi, H.	156
Uhlířová, L.	419
Uimari, P.	114, 462
Ulbrich, S.E.	532
Umstatter, C.	302
Ungerfeld, R.	367, 369, 530
Urban, T.	462
Urgeghe, P.	148
Uriarte, J.M.	183, 534

Printed in the United States
by Baker & Taylor Publisher Services